T0405837

Global Tropical Cyclogenesis
(Second Edition)

Eugene A. Sharkov

Global Tropical Cyclogenesis

(Second Edition)

Professor Eugene A. Sharkov
Head of "Earth Exploration from Space" Department
Space Research Institute
Profsouznayastr., 84/32
Russian Academy of Sciences
Moscow
Russia, 117997

SPRINGER–PRAXIS BOOKS IN ENVIRONMENTAL SCIENCES

ISBN 978-3-642-13295-7 e-ISBN 978-3-642-13296-4
DOI 10.1007/978-3-642-13296-4
Springer Heidelberg Dordrecht London New York

Library of Congress Control Number: 2011936539

Cover design: Jim Wilkie
Project management: OPS Ltd., Gt. Yarmouth, Norfolk, U.K.

Printed on acid-free paper

Springer is part of Springer Science+Business Media (www.springer.com)

Contents

Preface to the Second Edition

This book represents the revised and enlarged Second Edition of E. A. Sharkov's *Global Tropical Cyclogenesis*, published by Springer/Praxis in 2000. The principal feature of the First Edition was the development of a methodological approach, first proposed by the author, to studying tropical cyclones (TCs), based on the concept of tropical cyclogenesis evolving as a stochastic flow of homogeneous events in the lifetime of each TC. A series of studies following on from the First Edition has shown the fruitfulness of such an approach. A number of new scientific results, based on this approach, are presented in this edition.

The necessity for a revised and enlarged Second Edition is the result of ongoing growing interest—not only from the science community but also from the industrial and administrative bodies of many countries—in the problems that occur both as a result of the appearance and in forecasting catastrophic atmospheric phenomena; these principally include global climate change, irreversible damage to ecological processes, and transformation of energy transfer and mass transfer in the ocean–atmosphere system all of which could result in changing the current favorable conditions for life.

The reason for studying atmospheric catastrophes is explained by a number of things. Principally, these atmospheric processes represent a direct physical threat to humankind and to infrastructure, in addition to the many administrative–economical problems that arise as a consequence. In some cases, they even bring about political and state initiatives in attempts to control them. Humankind has long considered tropical cyclones as the most destructive elements of the ocean–atmosphere system causing considerable loss of life and material losses. Serious efforts have been undertaken (mainly in the U.S.A. where, it has to be said, they have been completely ineffective) to limit the amount of damage caused by TCs by any means. These attempts are still ongoing even now.

The process that gives rise to these vortical systems in the Earth's atmosphere, long considered on the basis of standard meteorological approaches, was supposed to be a purely meteorological phenomenon. It was only after 1983, when a series of studies was carried out under the leadership of Academician R. Z. Sagdeev and Professor S. S Moiseev of the Russian Academy of Sciences (RAS), hat it became clear that serious progress in studying such complicated systems could only be achieved using new physical approaches, both theoretical and experimental. So, from the viewpoint of forecasting such complicated systems, it is first necessary to clearly understand the spatiotemporal picture of this global phenomenon as a multi-faceted process—this especially relates to ascertaining the determinate components both globally and regionally. Moreover, recent investigations using space remote-sensing data and the latest achievements in complex systems theory indicate a principally different view of tropical cyclogenesis: we can state, with a high degree of probability, the determining role played by tropical cyclones in carrying out global mass transfer and energy transfer in the global ocean–atmosphere system and in regulating the greenhouse effect, which is so important for biological (including human beings) life on the Earth. Thus, global tropical cyclogenesis most likely represents the necessary and, probably, determining factor in the ecological balance (understood in a broad sense) that exists both in the geophysical ocean–atmosphere system and in the ecological systems of the Earth.

Catastrophic atmospheric vortices represent a unique mechanism that effectively removes excessive heat from the atmosphere under conditions in which the usual mechanisms, principally turbulent convection and global circulation, clearly become insufficient. So, catastrophic phenomena play an important role (as paradoxically as it may seem) that is vital for humankind in regulating the climatic temperature regime of the Earth (the greenhouse effect), removing excessive heat and preventing the planet from overheating in the tropical zone (Chapter 4). The stable integral regime of plural cyclogenesis generation, both in the cyclone-generating basins of the World Ocean and in the basins of northern and southern hemispheres, is revealed.

A key discovery in the initial cyclogenesis and intensification of various forms of tropical cyclones is the quick-response energy source that provides extremely fast intensification and formation of mature forms of tropical cyclones. The point of view that only those basins of oceans that experience high surface temperatures can be such a source has a long history and many adherents. However, contradictory views toward such a viewpoint have recently appeared, mainly as a consequence of the catastrophically fast intensification of TC "Katrina" which subsequently struck the coastal areas of the U.S.A. Chapter 4 presents results that indicate the main power sources that fuel tropical cyclones are water vapor regions with heightened integral concentration, not only captured by a cyclone from the tropical zone by means of monsoonal circulation of the atmosphere but also retained by the cyclone throughout the whole stage of its evolution (the so-called "capture" effect and the "camel" model).

In contrast, ever since the end of the 19th century, researchers have undertaken numerous attempts to establish a relationship between tropical cyclone appearance and solar activity, which seemed obvious at that time. However, this did not lead to unequivocally and obviously interpretive results from the physical point of view. The reasons relate (as have now become clear) to the multiscale and nonlinear character of the process of the interaction between solar activity and cyclogenesis, as well as with the flawed mathematical procedures then used. By using modern wavelet analysis the response (with a correlation of nearly 100%) of 27-day solar activity variations in the time series of global cyclogenesis was found for the first time (Chapter 5).

The idea that research into how TCs interact with the ocean–atmosphere system should not be limited to the troposphere but, instead, should be based on considering the large-scale crisis state as a global phenomenon influencing various geophysical media, starting with the ocean surface and troposphere and ending with the ozonosphere and ionosphere, was first proposed in 1996 by the author in collaboration with researchers at the Space Research Institute of the Russian Academy of Sciences (SRI RAS). Research into the kinematic, thermodynamic, and electrodynamic relationships between elements of the ocean–troposphere–upper atmosphere–ionosphere system at crisis states should undoubtedly be a major component of space research; attempts are now being undertaken to organize complex investigations using rocket-based and radar ionospheric sounding. The analysis of ionospheric disturbances carried out in Chapter 6 confirms, to a high degree of probability, the direct and rapid effect that powerful vortical systems in the troposphere have on the overlying lower and middle ionosphere.

Appendix A contains unique quantitative detailed information on the spatiotemporal features of global and regional tropical cyclogenesis between 1983 and 2010.

Based on the response to the First Edition, the present edition will be useful not only for professional experts on tropical meteorology, but also for researchers and engineers of government organizations and research institutes engaged in forecasting emergency situations and monitoring natural environments.

The author wishes to thank his many colleagues for useful discussions and remarks which have improved the text of the book. The author is especially grateful to Dr. I. V. Pokrovskaya for fruitful cooperation over many years.

The author is also grateful to Mr. Yu. V. Preobrazhenskii for translating the manuscript into English and to Mrs. Nataly Komarova for typing services.

Keeping in mind the complexity of the present publishing project, the author welcomes comments from interested readers (email: *easharkov@iki.rssi.ru*).

Preface to the First Edition

It is common knowledge that the tropical zone of the Earth plays a crucial role in the evolution of the global thermodynamic surface–atmosphere system and in the socio-ecological processes on our planet. In this case, it should be primarily pointed out that the ocean–atmosphere system of Earth's tropical zone possesses the unique characteristic of generation of organized mesoscale vortical structures—tropical cyclones (TCs)—from atmospheric turbulent chaos. TCs are among the most destructive natural phenomena on Earth and are serious ecological hazards for humanity. Material damage is often accompanied by a considerable loss of life; so TCs rank next (after earthquakes) among natural disasters in the number of human lives lost. Vulnerability to tropical cyclones is becoming more pronounced because the fastest population growth is in tropical coastal regions.

However, there exists another important aspect of this problem. Recent remote-sensing satellite investigations have clearly demonstrated, on the one hand, the significance of the tropics for the planet's life and, on the other hand, the importance to humankind of understanding the fine features of teleconnections between the tropics and the rest of the globe. In this case, there is good reason to believe (and recent results strongly support it) that global tropical cyclogenesis plays a decisive role in the maintenance of global mass-and-energy exchange. Moreover, it becomes apparent that the TC is not an exotic but a necessary (appropriate) element in "working" the tropical atmosphere–ocean system.

If we remember the cliché that "the tropics is the fire box of the atmosphere", so there can be no doubt that "the tropical cyclone is a harsh flame".

Understanding TC genesis, development, and characteristic features has been a challenging subject in meteorology over the last several decades. However, the pool of experience led us to conclude with certainty that it is hardly probable that the problem can be resolved strictly in the framework of meteorological notions. There is no question that for an understanding of the behavior of such complicated physical phenomenon as tropical cyclogenesis it is necessary to invoke the notions of modern non-linear physics and chaotic dynamics developed over many years.

The principal purpose of this book is to demonstrate the new capabilities and results of the proposed methodological approach based on the concept of tropical cyclogenesis evolution as a stochastic flow of homogeneous events.

The aim of Chapter 1 of the book is to substantiate the scientific and applied rationales for global cyclogenesis study; in so doing we point to both the natural scientific tasks and socioeconomic problems.

The basic idea of investigation of statistical characteristic of global tropical cyclogenesis formed as a signal of a stochastic intermittent flow by the mathematical apparatus of chaotic dynamics are introduced in Chapter 2. Emphasis has been laid on investigations into the hierarchical structure of global tropical cyclogenesis by retention of all interaction timescales. The type of probability model revealed varies radically with the choice of interaction timescale. We first examine the time evolution of initial forms of TC as a stochastic process, and then features of the rate of hurricane generation.

The spatiotemporal features of regional tropical cyclogenesis are pursued further in Chapter 3 [of the First Edition]. The emphasis is on the evolution of a statistical model of Pacific tropical cyclogenesis and its interactions with the environment of the atmosphere–ocean system. In this chapter [of the First Edition] we are concerned with the problem of the hierarchy and clusterization of tropical convective systems in the context of the concept of self-organization in open systems.

Chapter 4 [of the First Edition] deals with existing physical models and the results of simulations of global tropical cyclogenesis. In addition to the well-known (at least in Western literature) approaches—such as the statistical synoptical forecasting and modeling of disturbances in global climate models—here we crucially exmine new physical approaches in tropical cyclogenesis problems, in particular the concept of self-organization of large-scale structures in helical turbulence and the kinetic diffusion approach, presenting the birth–death processes of ordered elements in an active fluctuating medium.

In Chapter 5 [of the First Edition] we look at the basic principles of data gathering and acquisition for the development of global tropical cyclogenesis databases including descriptions of the design principles, structures of datasets, cataloguing procedures, and computer architectures. Much attention is given to state-of-the-art systematic and chronological catalogues and hydrometeorological archives.

The purpose of Chapter 6 [of the First Edition] consists of considerations of passive and active remote sensing for tropical monitoring. Emphasis has been laid on satellite activity for tropical studies. The chapter covers the current status of operational satellite systems and gives a synopsis of projected developments for special missions for tropical study and of international efforts for tropical cyclogenesis monitoring.

Appendix A [of the First Edition] has much potential for yielding information about the spatiotemporal features (quantitative data) of global and regional tropical cyclogenesis for 1983–1999.

I wish to thank my many colleagues whose efforts and discussions have contributed to the present work. I specifically wish to thank my colleague Dr. Irina Pokrovskaya for her close and fruitful collaboration for many years. I am thankful to Dr. Nataly Astafyeva for her contribution to the intriguing problem of stochastic behavior for global tropical cyclogenesis. I have also benefited from useful discussions with Prof. S. S. Moiseev.

I would like to express my sincere thanks to Dr John Mason for his helpful comments and suggestions in editing the manuscript. I also wish to express my thanks for the support and encouragement received from Clive Horwood, Chairman of Praxis. Thanks also to Prof. Robert Houze, Dr. Guosheng Liu, and Dr. Kevin Walsh for permission to reproduce their figures. The typing of the intricate draft by Nataly Komarova is also appreciated.

Figures

Tables

Abbreviations and acronyms

AAA	Aircraft Accessible Array
AATSR	Advanced Along-Track Scanning Radiometer
AC	Anticyclone Circulation
ADEOS	ADvanced Earth Observation Satellite (Japan)
AFWA	Air Force Weather Agency
AGCM	Atmospheric General Circulation Model
AGW	Acousto-Gravitational Wave
AIREPS	Aircraft (Weather) Reports
AIRS	Advanced InfraRed Sounder
ALADIN	Atmospheric LAser Dopper INstrument
ALPEX	ALPine EXperiment
ALT	ALTimeter
AMI	Active Microwave Instrument
AMOS	Automated Meteorological Observing Station
AMSU	Advanced MSU
AMTS	Advanced Moisture and Temperature Sounder
AO	Announcement of Opportunity
AOI	Announcement of Opportunity Instrument
AOR	Area Of Responsibility
APT	Automatic Picture Transmission
ARGOS	International Service for Drifting Buoys
ARISTOTELES	Application and Research Involving Space Techniques Observing The Earth fields from Low Earth Spacecraft
ASAR	Advanced Synthetic Aperture Radar
ATLID	ATmospheric LIDar
ATN	Advanced TIROS-N
ATS	Application Technology Satellite
ATSR	Along-Track Scanning Radiometer

AVE	Atmospheric Variability Experiment
AVHRR	Advanced Very High Resolution Radiometer
AVNIR	Advanced Visible and Near Infrared Radiometer
AW	Acoustic Wave
AWN	Automated Weather Network
BE	Baroclinically Enhanced
BM	Ballistic Missile
BOMEX	Barbados Oceanographic and Meteorological EXperiment
BU	Booster Unit
CAO	Central Aerologic Observatory
CAPE	Convective Available Potential Energy
CCM	Community Climate Model
CDA	Command and Data Acquisition
CERES	Cloud and Earth's Radiant Energy System
CF	Cumulative Function
CGCM	Coupled ocean–atmospheric General Circulation Model
CISK	Conventional Instability of a Second Kind
CLIVAR	CLImate VARiability and Predictability Program, WCRP
CLOUDS	European satellite mission
CMM	Commission for Marine Meteorology (WMO)
CMOD	Compact Meteorological and Oceanographic Drifter
CNES	*Centre National d'Etudes Spatiales*
CO_2	Carbon dioxide
COARE	Coupled Ocean–Atmosphere Response Experiment
COMNAVMETOCCOM	COMmander, NAVal METeorology and OCeanology COMmand
COSPAR	COmmittee on SPAce Research
CPHC	Central Pacific Hurricane Center
CSP	Central Surface Pressure
CTH	Cloud-Top Height
CTI	Cyclone Threshold Intensity
CTT	Cloud-Top Temperature
CVW	Cloud Vector Wind
CZCS	Coast Zone Color Scanner
DAIS	Diagnostic Airglow Imaging System
DAS	Doppler Atmospheric Scatterometer
DCP	Data Collection Platform
DIAL	DIfferential Absorption Lidar
DMSP	Defense Meteorological Satellite Program
DoD	Department of Defense
DORIS	Doppler Orbitography and Radio-positioning Integrated by Satellite

DS	Dwell Sounding
DSS	Disturbed Sea Surface
ECHAM	General circulation model (tropospheric version)
ECMWF	European Centre for Medium-range Weather Forecasts
EEM	Earth Explorer Mission
ENSO	El Niño–Southern Oscillation
ENVISAT	ENVIronmental SATellite (ESA)
EOS	Earth Observing System
ERBE	Earth Radiation Budget Experiment
ERS	European Remote Sensing
ESA	European Space Agency
ESCAP	Economic and Social Commission for Asia and the Pacific (U.N.)
ETM +	Enhanced Thematic Mapper Plus
EUMETSAT	EUropean Organization for the Exploitation of METeorological SATellites
EWM	Earth Watch Mission
FFNMOC	Fleet Numerical Meteorology and Oceanography Center (USN)
FGGE	First GARP Global Experiment
FLENUMETOCCEN	FLEet Numerical METeorology and OCeanography CENter (USN)
FNMOC	= FLENUMETOCCEN
FOV	Field Of View
GARP	Global Atmospheric Research Program
GATE	GARP Atlantic Tropical Experiment
GAW	Global Atmosphere Warning
GCC	Global Climate Change
GCM	General Circulation Model *or* Global Climate Model
GDOS	Global Disaster Observation System
GEO	GEOsynchronous Orbiting
GEOWARN	Global Emergency Observations and WARNing System
GEWEX	Global Energy and Water Cycle EXperiment
GFDL	Geophysical Fluid Dynamics Laboratory
GHRC	Global Hydrological Research Center
GIS	Geographical Information System
GMS	Geostationary Meteorological Satellite (Japan)
GMT	Greenwich Mean Time
GOES	Geostationary Operational Environmental Satellite
GOME	Global Ozone Monitoring Experiment
GOMOS	Global Ozone Monitoring by Occultation of Stars
GOMS	Geostationary Operational Meteorological Satellite
GPCP	Global Precipitation Climatology Project

GREBEN	Russian precision radiometer
GSM	Global Spectral Model
GTC	Global Tropical Cyclogenesis
H	Hurricane
HD	Hurricane Day
HDF	Hierarchical Data Format
HDP	Hurricane Destruction Potential
HF	High Frequency
HIRS	High-resolution Infrared Radiation Sounder
HRV	High Resolution Visible sensor
HSA	Hawaii Solar Astronomy
ICE	ISO Cloud Ensemble
ICSU	International Council of Scientific Unions
IDBP	Integrated Drifting Buoy Plan
IDNDR	International Decade for Natural Disaster Reduction
IFA	Intensive Flux Array
IFOV	Instantaneous FOV
IGBP	International Geosphere–Biosphere Program
IGW	Inner Gravitational Wave
IH	Intense Hurricane
IHD	Intense Hurricane Day
IKAR	Russian microwave radiometric instrument
IKI RAN	Space Research Institute of the Russian Academy of Sciences
ILAS	Improved Limb Atmospheric Spectrometer
IMG	Interferometric Monitor for Greenhouse Gases *or* International Monitor for Greenhouse Gases
IOP	Intensive Observing Period
IPCC	Intergovernmental Panel on Climate Change
IR	InfraRed radiation
IRIS	InfraRed Interferometer Spectrometer
ISCCP	International Satellite Cloud Climatology Project
ISO	Intra-Seasonal Oscillation
ISTOK	Russian IR spectrometric instrument
ITCZ	Inter-Tropical Convergence Zone
ITD	Initial Tropical Disturbance
ITOS	Improved TIROS Operational System
ITPR	Infrared Temperature Profile Radiometer
IUGG-CAS	International Union of Geophysics and Geodesy–Committee on Atmospheric Sciences
JAROS	JApan Resources Observation System Organization
JGOFS	Joint Global Ocean Flux Study
JMA	Japanese Meteorological Agency
JTWC	Joint Typhoon Warning Center
Landsat	NASA Earth Observing satellite mission

LEO	Low Earth Orbiting
LIDAR	Laser Infared Detecting And Ranging
LIS	Lighting Imaging Sensor
LOCSS	LOCal Space Service
LSA	Large Scale Array
LST	Local Space Time
LT	Local Time
LTE	Local Thermodynamic Equilibrium
LV	Launch Vehicle
MCS	Mesoscale Convective System
MEGHA-TROPIQUES	European satellite mission
MERIS	MEdium-Resolution Imaging Spectrometer
Meteosat	European geostationary meteorological satellite
METOP	European METeorological OPerational Satellite
MHAT	Mexican HAT wavelet
MID	Moving Ionospheric Disturbance
MIMS	Meteosat Infrared and Microwave Sounder
MIPAS	Michelson Interferometer for Passive Atmospheric Sounding
MIT	Massachusetts Institute of Technology (U.S.A.)
MITI	Ministry of International Trade and Industry
MJO	Madden–Julian Oscillation
MLIM	Multiple Linear Interdependent Model
MM	Mesoscale Model
MOF	Maximum Observed Frequency
MOM	Modular Ocean Model (GFDL)
MOP	Meteosat Operational Program
MOS	Modular Optoelectronic Scanner
MOS-OBZOR	Russian visible and IR spectrometric instrument
MP	Meteorological Probe
MPI	Maximum Potential Intensity
MSG	Meteosat Second Generation
MSI	MultiSpectral Imaging
MSLP	Minimum Sea Level Pressure
MSS	Multi-Spectral Scanner
MSU	Microwave Sounding Unit
MWR	MicroWave Radiometer
NASA	National Aeronautics and Space Administration
NASDA	NAtional Space Development Agency (Japan)
NAVOCEANO	NAVal OCEANographic Office
NCDC	National Climate Data Center
NCEP	National Centers for Environmental Prediction (U.S.)
NCSA	National Center for Supercomputing Applications
NDVI	Normalized Difference Vegetation Index
NEMS	Nimbus E Microwave Spectrometer

NEP	North-East Pacific
NESDIS	National Environmental Satellite Data and Information Service
NESR/NEdN	Noise Equivalent Spectral Radiance
NESS	National Environmetal Satellite Service
NH	Northern Hemisphere
NHC	National Hurricane Center
NIMBUS	NASA satellite mission
NIO	Northern Indian Ocean
NIPRNET	Non-secure Internet Protocol Router NETwork
NOAA	National Oceanic and Atmospheric Administration
NOGAPS	Navy Operational Global Atmospheric Prediction System
NRL	Navy Research Laboratory
NS	Named Storm
NSCAT	NASA SCATterometer
NSD	Named Storm Day
NWP	Numerical Weather Prediction
NWS	National Weather Service (U.S.A.)
OAGCM	Ocean–Atmosphere General Circulation Model
OBZOR	Russian visible and IR spectrometric instrument
OCTS	Ocean Color and Temperature Sensor
OGCM	Ocean General Circulation Model
OLS	Operational Linescan System
OPS	Optical Sensor
OSS	Outwards Spiraling (Stratospheric) Surge *or* Operations Support Squadron
OZON	Russian spectrometric instrument
P/PO	Poleward/Poleward Oriented
PACAF	PACific Air Force
PACOM	PAcific COMmand (U.S.A.)
PAGASA	Philippine Atmospheric Geophysical Astronomical Meteorological Service
PBL	Planetary Boundary Layer
PC	Parachute Capsule
PCN	Position Code Number
PHC	Percent High Cloudiness
PHTF	Past Hurricane Track File
PMR	Pressure Modulated Radiometer
PMT	Photo Multiplier Tube
POLDER	POLarization and Directionality of the Earth's Reflectance
PP	Power Plant
PR	Precipitation Radar
QBO	Quasi-Biennial Oscillation

r.m.s	root mean square
RA	Radar Altimeter *or* Regional Association
RHF	Rate of Hurricane Formation
RIS	Retroreflector In Space
RR	Reference Region
RS	Research Ship
RSA	Russian Space Agency
RSC	Rocket and Space Complex
RSMC	Regional and Specialized Meteorological Center
RSSI	Russian Space Science Internet
RU	Repeater Unit
S&R	Search and Rescue package
S/DR	Standard pattern/Dominant Ridge region
SA	Solar Activity
SAR	Synthetic Aperture Radar
SAR TRAVERS	Russian SAR mission
SARSAT	Search And Rescue SATellite
SATOPS	SATellite OPerationS
SCAMS	SCAnning Microwave Spectrometer
ScaRaB	Scanning Radiation Budget
SCIAMACHY	SCanning Imaging Absorption spectroMeter for Atmospheric CartograpHY
SCS	SuperConvective System
SEM	Space Environment Monitor
SF	Structure Function
SH	Southern Hemisphere
SHIPS	Statistical Hurricane Intensity Prediction Scheme
SIO	Southern Indian Ocean
SIRS	Satellite InfraRed Spectrometer
SLP	Sea Level Pressure
SLR	Side-Looking Radar
SMS	Synchronous Meteorological Satellite
SNC	Space Nose Cone
SNR	Signal-to-Noise Ratio
SOC	Self-Organized Criticality
SOI	Southern Oscillation Index
SPCZ	South Pacific Convergence Zone
SPOT	Système Pour l'Observation de la Terre (France)
SPRM	Solid Propellant Rocket Motor
SRI RAS	Space Research Institute of the Russian Academy of Sciences
SSC	Scientific Steering Committee
SSE	Special Sensor E
SSH	Special Sensor H
SSM/I	Special Sensor Microwave/Imager

SSM/T	Special Sensor Microwave/Temperature
SSP	SubSatellite Point
SST	Sea Surface Temperature
SSU	Stratospheric Sounding Unit
STC	Simulated Tropical Cyclone
STR	SpatioTemporal Rate
SWP	South-West Pacific
TB	Tropospheric Body
TC	Tropical Cyclone
TCFA	Tropical Cyclone Formation Alert
TCP	Tropical Cyclone Project
TD	Tropical Disturbance
TDE	Tropical DEpression
TDI	Tropical DIsturbance
TIPS	Typhoon Intensity Prediction Scheme
TIR	Thermal IR
TIROS	Television and InfraRed Observation Satellite
TM	Thematic Mapper
TMI	TRMM Microwave Imager
TO	Tropical Only
TOGA	Tropical Oceans and Global Atmosphere
TOMS	Total Ozone Mapping Spectrometer
TOVS	TIROS Operational Vertical Sounder
TPC	Tropical Prediction Center
TRAVERS	Russian synthetic aperture radar
TRMM	Tropical Rainfall Measuring Mission
TROPIQUES	French satellite mission
TS	Tropical System
UA	Undisturbed Atmospheric state
UARS	Upper Atmosphere Research Satellite
UKMO	UK Meteorological Office
U.N.	United Nations
UNEP	United Nations Environment Program
UNESCO	United Nations Educational, Scientific and Cultural Organization
USCINCPACFLT	U.S. Commander-IN-Chief PACific (FLT—Fleet)
USGCRP	U.S. Global Change Research Program
USPACOM	U.S. PAcific COMmand
UTC	Universal Coordinate Time (= GMT)
VAS	VISSR Atmospheric Sounder
VHRR	Very High Resolution Radiometer
VIRS	Visible InfraRed Scanner
VIS-IR	VISible/InfraRed imaging radiometer
VISS	Visible Infrared Spin Scan radiometer
VISSR	VISible/infrared imaging SpectroRadiometer

VTPR	Vertical Temperature Profile Radiometer
WCP	World Climate Program
WCRP	World Climate Research Program (ICSU/WMO)
WEDOS	World Environment and Disaster Observation System
WEFAX	Weather facsimile
WMO	World Meteorological Organization
WO	World Ocean
WSD	Wind Speed and Direction
WSR	Weather Surveillence Radar
WT	Wavelet Transfer
WV	Water Vapor
Z	Zulu time (GMT/UTC)
ZWA	Zonal Wind Anomaly

1

Introduction: Scientific and applied rationales for the study of tropical cyclogenesis

Tropical regions are the primary source of the Earth's weather and atmospheric circulation patterns, a fact first recognized by the English scientist G. Hadley in the 18th century. Inquiring into the causes of the general trade winds, Hadley explained these by the ascent of heated equatorial air, which moves toward the poles at high altitudes and is replaced by cooler air flowing toward the equator at lower levels. Later studies confirmed the existence of "Hadley cells" as the dominant features of north–south atmospheric circulation.

Modern research has underscored the importance of equatorial regions to global climate and the planet's environments. We now know that water vapor evaporated over half the Earth's surface is carried by low-level Hadley circulation into a narrow equatorial belt called the Inter-Tropical Convergence Zone (ITCZ). There, the solar energy stored in water vapor is released through condensation into tropical rainfall. Moreover, because of the role of latent heat release in driving general atmospheric circulation, the variability in tropical rainfall is an important determinant of circulation variability and short-term climate changes (GCOS, 1995; JSTC, 1995; Huffman *et al.*, 1997; IPCC, 1996; Arkin and Xie, 1994; Alcamo *et al.*, 1998; Chahine *et al.*, 1997; Sharkov *et al.*, 2008a, b, 2009; Sharkov, 2010).

A large body of observational and modeling evidence has accumulated during the past decade, which shows significant relationships, particularly in the Pacific basin, between changes in tropical sea surface temperature (SST), tropical convective activity and global circulation, and short-term climate variability.

The multiscale convective activity in the Earth's tropical zone is the centerpiece of the hydrological cycle in the tropics. Tropical convective activity exhibits expanding spatiotemporal characteristics and structural modes.

First of all it should be pointed out that the ocean–atmosphere system of Earth's tropical zone possesses the unique characteristic of generation of ordered mesoscale vortical structures with a complicated topological system of streamlines—tropical cyclones (TCs)—from atmospheric turbulent chaos. TCs are among the most

---

E. A. Sharkov, *Global Tropical Cyclogenesis* (Second Edition).

destructive natural phenomena on Earth and are serious ecological hazards for humanity.

There is good reason to believe (and recent results strongly support it) that global tropical cyclogenesis plays a decisive role in the maintenance of global water, mass exchange, and energy exchange. Moreover, it becomes apparent that the tropical cyclone is not an exotic but a necessary (appropriate) element in "working" the tropical ocean–atmosphere system.

Because of the enormous impacts that tropical cyclones have (e.g., in 1995 total mainland U.S. hurricane damage averaged in the order of $5 billion annually—Pielke and Landsea, 1998), it is essential that detailed studies be made of observed TC activity. An understanding of such activity plays an important role in both public and private policy decisions (Diaz and Pulwarty, 1997; Pielke and Pielke, 1997; Willoughby and Black, 1996; Fernandez-Partages and Diaz, 1996; Simpson, 1998; Pielke and Landsea, 1999). Additionally, TC activity has rather large interannual and interdecadal variations, which are extremely important for their own sake and which could turn out to have a greater impact relative to changes forced by greenhouse warming, as suggested by Lighthill *et al.* (1994). A reliable assessment of what the future holds for tropical cyclone activity would also have significant policy utility.

1.1 INTERNATIONAL EFFORTS ON TROPICAL CYCLOGENESIS STUDY

It is an indubitable fact that the elaboration of contemporary methods of observations, research, and forecasting of TCs and also the forming of an optimal strategy of safety under conditions of atmospheric catastrophes are actual and international problems (Bryant, 1993; Tarakanov, 1980; Elsberry, 1987; WMO, 1995a; Choudhury, 1994).

The study of global tropical cyclogenesis was primarily supported by such important international organizations as the United Nations (U.N.) and the World Meteorological Organization (WMO).

The General Assembly of the U.N., in Resolution 2733 D(XXV) of 1970, called upon the WMO to take appropriate action with a view to obtaining basic meteorological data and discovering ways and means of mitigating the harmful effects of tropical cyclones (U.N., 1992; Alves, 1992). In response to this resolution, the Sixth WMO Congress (April 1971) established a tropical cyclone project which was subsequently upgraded in 1980 to become the WMO Tropical Cyclone Program (TCP). The WMO long-term plan for the TCP (1992–2001) approved by the 11th WMO Congress (1991) serves as a basis for the development and implementation of the program, in particular in association with the U.N. International Decade for Natural Disaster Reduction (IDNDR).

The planned activities of the TCP are (U.N., 1992):

(a) identification of the requirements of national meteorological/hydrological services during TC threats for specialized products and advisory information as needed from regional and subregional meteorological centers (RSMCs);
(b) promotion of the effectiveness of the functioning of RSMCs with activity specialization in TCs to provide the products identified above, including the strengthening of communication links between those services and RSMCs;
(c) development of the services through training of personnel and provision of special facilities required for TC and flood-warning services;
(d) enhancement of international and regional cooperation and coordination in meteorology and hydrology;
(e) acceleration of transfer of technology and knowledge related to TCs and flood warnings as well as the community response thereto;
(f) promotion of public information, awareness, and education on TCs;
(g) surveys, studies, project formulation, and implementation related to improvements in the existing arrangements and systems within the region concerned for provision and dissemination of warnings and related advisory information.

The long-term plan is being implemented partly through the transfer of technology. An example of this process is the preparation of reports by small groups of experts on specific subjects such as meteorological satellites, cyclone preparedness, and partly by means of regionally organized programs. Regional activities are organized through five regional cyclone bodies. Of these, two bodies—the ESCAP/WMO Typhoon Committee and the WMO/ESCAP Panel on Tropical Cyclones for the Bay of Bengal and the Arabian Sea—are intergovernmental bodies, and three—the Regional Association (RA) I Tropical Cyclone Committee for the South-West Indian Ocean, the RA IV Hurricane Committee- and the RA V Tropical Cyclone Committee for the South Pacific and South-East Indian Ocean—are bodies established by regional associations of the WMO.

Significant progress has been made in the formulation and implementation of the programs of each of these regional bodies, with substantial attention being given to meteorological aspects, including the acquisition and use of data from polar-orbiting and geostationary meteorological satellites, hydrological aspects, disaster prevention and preparedness, research, and training. Particular attention has been given to group-training arrangements and provision of fellowships. For example, regional training seminars and workshops, including the use of satellite data for TC detection and forecasting, were held in Nadi (Fiji, 1987), Calcutta (India, 1987), Miami (U.S., 1989), Colombo (Sri Lanka, 1990), Vanuatu (1990), Miami (1991), Suva (Fiji, 1991), Maputo (Mozambique, 1991), and St-Denis (Réunion, 1991).

The study of global tropical cyclogenesis was supported by such governmental and non-governmental organizations as UNESCO UNEP (United Nations Educational, Scientific and Cultural Organization; United Nations Environment Program) and the International Council of Scientific Unions (ICSU). In the joint ICSU/WMO project on TC disasters it has been well appreciated that ICSU's contribution can most appropriately be concentrated on oceanographic and the air–sea-interaction aspects of TC disasters, in ways that will complement the in-depth expertise of the WMO in meteorological and hydrographic aspects and in the effective delivery of forecasts and warnings. Accordingly, during the 1997 preparations for the WMO/ICSU Fourth International Workshop on Tropical Cyclones (Haiku, China, April 21–30, 1998), ICSU representatives readily consented to provide the preliminary discussion document related to air–sea interactions. This document (Lighthill, 1998), circulated in advance to all workshop participants, was aimed at stimulating active discussion of the following suggestion: that present day forecasts of the track to be followed by the eye of a TC will need in the future to be supplemented with forecasts of intensity changes resulting from air–sea interaction.

This suggestion is given detailed support as follows (Lighthill, 1998):

(i) by drawing attention to cases of grave intensification of TCs resulting from their passage over warm ocean areas shortly before landfall;
(ii) by emphasizing the current availability of relevant ocean data in real time for use as initial data in forecasting models; and
(iii) by showing how a good atmospheric model for numerical weather prediction can improve forecasts of intensity changes if it is coupled with a fine mesh multilayer ocean model making use of just such initial data.

Although, in addition, some contributions are made to continuing the investigation of likely influences of global climatic changes on TC frequency and intensity, the discussion document is above all directed at the above suggestion; namely, the oceanographic and air–sea interaction information will need to be used to enhance the precision of forecasts so that coastal populations threatened by tropical cyclogenesis may become increasingly likely to trust the resulting warnings and take the measures recommended for their protection.

The U.N.'s Intergovernmental Panel on Climate Change (IPCC) has speculated that climate change due to increasing amounts of anthropogenic "greenhouse" gases may result in increased tropical SSTs and increased tropical rainfall associated with a slightly stronger ITCZ (IPCC, 1996). Because TCs extract latent and sensible heat from warm tropical oceans and release the heat in its upper-tropospheric outflow to fuel the storm's spin-up, early work of the IPCC expressed concern that warmer SSTs will lead to more frequent and intense hurricanes, typhoons, and severe TCs (Holton and Alexander, 1999; Philander, 1999; Taylor and Edwards, 1991). Any changes in TC activity are intrinsically also tied to large-scale changes in the tropical atmosphere. As a result, SSTs by themselves cannot be considered without corresponding information regarding moisture and stability in the tropical troposphere.

In addition to thermodynamic variables, changes in tropical dynamics also play a large role in determining changes in TC activity. For example, if the vertical wind shear over the tropical North Atlantic moderately increased during the hurricane season in a world affected by increased CO_2 (carbon dioxide)—as typically seen during El Niño–Southern Oscillation (ENSO) warm phases (El Niño events)—then we would most likely see a significant decrease in TC activity. This is due to the Atlantic basin having a marginal climatology for TC activity because of its sensitivity to changes in the vertical wind shear and lack of an oceanic monsoon trough (Gray, 1993). In other less marginal TC basins, changes in the vertical shear profile typically result in alterations in the preferred location of development (Gray *et al.*, 1997; Landsea *et al.*, 1999).

These complications, along with conflicting global circulation modeling (GCM) runs, compelled the IPCC in 1995 (IPCC, 1996) to express greater uncertainty about the nature of TCs in an enhanced CO_2 environment.

The formation of TCs depends not only on SST, but also on a number of atmospheric factors. Although some models now represent tropical storms with some realism for the present day climate, the state of the science does not allow assessment of future changes.

Henderson-Sellers *et al.* (1998) address a few of the TC/greenhouse-warming problems. Their first conclusion was that "there is no evidence to suggest any major changes in the area or global location of tropical cyclone genesis in greenhouse conditions." Another conclusion suggested by Henderson-Sellers *et al.* (1998) was an increase in maximum potential intensity (MPI) of 10% to 20% (in central pressure or 5% to 10% in maximum sustained winds) for a doubled CO_2 climate but the known omissions (ocean spray, momentum restriction, and possibly also surface to 300 hPa lapse rate changes) all act to reduce these increases.

Since the production of the 1996 IPCC reports, our knowledge has advanced to permit the following summary (Henderson-Sellers *et al.*, 1998):

- there are no discernible global trends in TC number, intensity, or location from historical data analyses;
- regional variability, which is very large, is being quantified slowly by a variety of methods;
- empirical methods do not have skill when applied to TCs under greenhouse conditions;
- global and mesoscale model-based predictions for TCs under greenhouse conditions have not yet demonstrated prediction skill.

1.2 PRESENT STATE OF THE ART OF TROPICAL CYCLOGENESIS STUDY

Understanding TC genesis, development, and associated characteristic features has been a challenging subject in meteorology over the last several decades.

In the framework of present meteorological insights, the problem of tropical cyclogenesis is stated in the following way (Henderson-Sellers *et al.*, 1998).

Tropical cyclone is the generic term for a non-frontal synoptic-scale low-pressure system originating over tropical or subtropical waters with organized convection and definite cyclonic surface wind circulation. TCs with maximum sustained surface winds of less than 17 m s^{-1} are generally called "tropical depressions". Once a TC achieves surface wind strengths of at least 17 m s^{-1} it is typically called a "tropical storm" or "tropical cyclone" and assigned a name. If the surface winds reach 33 m s^{-1} the storm is called a "typhoon" (in the northwest Pacific Ocean), a "hurricane" (in the North Atlantic Ocean and the northeast Pacific Ocean), or a "severe tropical cyclone" (the southwest Pacific Ocean and southeast Indian Ocean) (see Appendix B).

TCs derive energy primarily from evaporation from the ocean and the associated condensation in convective clouds concentrated near their center as compared with midlatitude storms that primarily obtain energy from horizontal temperature gradients in the atmosphere. Additionally, TCs are characterized by a "warm core" (relatively warmer than the environment at the same pressure level) in the troposphere. The greatest temperature anomaly generally occurs in the upper troposphere around 250 hPa. It is this unique warm core structure within a TC that produces very strong winds near the surface and causes damage to coastal regions and islands through extreme wind, storm surges, and wave action.

Research efforts focused on assessing the potential for changes in TC activity under a greenhouse-warmed climate have progressed since the earlier efforts that were the basis of this IPCC assessment (which were undertaken in 1994 and early 1995). Henderson-Sellers *et al.* (1998) gather input from the members of the Steering Committee of the WMO Commission for Atmospheric Sciences and reflect recent experimental results in a summary of the new findings in this field.

This review should be read in the context of what is currently known about TC predictions in a greenhouse-warmed world.

1.2.1 What do we know?

(1) TCs are devastatingly severe weather events.
(2) Humankind's vulnerability to TCs is increasing because of increasing populations living along tropical coasts.
(3) TC formation and intensity change are currently very difficult to predict.
(4) The costs associated with TC impacts are increasing because of the increasing costs of infrastructure and increasing insurance claims on private and public funds.
(5) The balance of evidence indicates that greenhouse gas emissions are producing climate change.
(6) Concern about possible future changes in TC activity relates to changes in (i) the frequency of occurrence, (ii) the area of occurrence, (iii) mean intensity, (iv) maximum intensity, and (v) the rain and wind structure.

1.2.2 What do we not know?

(1) How to predict TCs today: genesis, maximum intensity.
(2) How those environmental parameters that appear to be important for TC genesis will change.

(3) How the large-scale circulation features that appear to be linked to TC climatology—especially the Quasi-Biennial Oscillation (QBO) and ENSO—will change.
(4) How the upper-ocean thermal structure, which acts as the energy source for TC development, will change.

1.2.3 What tools are available to us now?

(1) Coupled ocean–atmosphere general circulation models (OAGCMs). These provide useful information on the general characteristics of climate change, but they currently suffer from coarse resolution (about 500 km) and unproven skill for present day TCs.
(2) Atmospheric general circulation models (AGCMs)—linked to mixed layer ocean submodels or employing SST predictions from OAGCMs—have better resolution (about 100 km) but are still too coarse for mesoscale dynamics and share the latter two drawbacks of OAGCMs—as in (1).
(3) Mesoscale models driven offline from the output of OAGCMs or AGCMs have better resolution (about 20 km) but still share the other drawbacks—as for (1) and (2).
(4) Empirical relationships—such as Gray's genesis parameters or simply SSTs on their own—suffer from the drawbacks associated with empiricism.
(5) "Upscaling" thermodynamic models—such as those of Emanuel (1991) and Holland (1997)—are known not to capture all processes of importance.

This review is phrased in terms of doubled CO_2 climate conditions for simplicity and because the evaluations assessed pre-date any attempt to consider the additional impacts of sulfate or other aerosols on tropical cyclones. However, all the assessments are equally applicable to greenhouse conditions modified by either or both other greenhouse gases and atmospheric aerosols.

Gray (1997) presents his personal (very interesting and emotional) view of tropical meteorology—in particular, on TC problems—covering the last 40 years. In his view in the previous few decades the most notable findings were:

A. The discovery of the stratospheric QBO and its physical explanation as a result of upward-propagating Kelvin and mixed Rossby–gravity waves.
B. The Madden–Julian Oscillation (MJO) whose exact physics is still being researched.
C. The fundamental nature of the large compensating up-and-down mass recycling of tropical weather systems. Total low-level up-and-down vertical motion is an order of magnitude larger than mean (net) low-level vertical motion. Hence, closing of the mass, moisture, and energy budgets in the tropics requires great amounts of mass-compensating vertical motion.
D. The large diurnal variation of cumulonimbus convection and heavy rainfall and clear region subsidence (late morning maximum–early evening minimum) over tropical ocean regions where no significant diurnal lapse rate variation occurs.

This must be driven by the influence of day versus night differences in net tropospheric radiative cooling.

E. That condensation warming from rainfall occurs indirectly from return flow subsidence in remote locations rather than locally. This seems simple now but it was not obvious 30–40 years ago.
F. Advances in our understanding of the physics and structure of TCs and the significant improvements in numerical track predictions at and beyond 2-day lead times.
G. The surprisingly persistent seasonal to year-in-advance teleconnection and predictive signals for much tropical weather (other than the Walker circulation associated with ENSO).
H. The influences that tropical circulation and weather systems have on the evolution of midlatitude weather and climate.

There have also been some disappointments as regards certain anticipated advances which have not materialized. These include:

A. Rainfall enhancement from cloud-seeding activities and attempts to reduce hurricane intensity using this method within the hurricane's inner-core area.
B. Significant improvements in 1 to 2-day prediction of tropical rainfall and easterly wave development has not occurred. In general, the formation and dissipation of individual tropical weather systems still cannot be well predicted. The great expectations of 50 years ago along these lines have not been realized and likely will not be realized in the future; factors influencing the timing and location of convective weather breakouts in the tropics are just too complicated.
C. Most theoretical ideas on how the tropical atmosphere functions (including CISK and easterly wave theories) appear not to have prevailed against the test of time. The tropical atmosphere is now recognized to be more complex (or perverse!) than most armchair meteorologists of the past (or present) were or are able to envisage.
D. A unified theory for the primary factors governing TC genesis and intensity change has not yet been developed.

An important step forward in the direction of the "local" approach was made by Moiseev *et al.* (1983a). The authors crucially propose a new physical mechanism called the vortical dynamo in hydrodynamics. Under conditions of unstable thermal stratifications this mechanism provides the possibility of principal reconstruction of the thermoconvection structure, with the appearance of a large-scale ordered vortical system including the non-trivial topological structure of streamlines. In other words, a case in point is the generation of a large-scale structure with helical turbulence via the inverse cascade process (non-Kolmogorov type). The concept has gained wide acceptance among hydrodynamic specialists (Levich and Tzvetkov, 1985; Branover *et al.*, 1999; Moiseev *et al.*, 2000).

Note also that during all periods of observation the TC and its characteristic features were considered strictly as meteorological phenomena. However, experience led us to conclude with certainty that it is hardly likely that the problem can be resolved strictly within the framework of meteorological notions. There is no doubt that for an understanding of the behavior of such a complicated physical phenomenon as a tropical cyclogenesis it is necessary to invoke the notions of modern non-linear physics and chaotic dynamics developed over many years (Ghil *et al.*, 1985, 1991; Glantz *et al.*, 1991; Gaspard, 1997; Dubois, 1998; Chang *et al.*, 1996; Duane, 1997).

As a consequence, a number of Russian scientific groups have proposed and developed physical directions—namely, self-organized approaches in hydrodynamics and fluctuation kinetics. These directions differ in a crucial respect from the known and above-mentioned approaches (i.e., GCM simulations and the synoptical statistical method).

Problems relating to the construction of a physical model of the formation of stable vortex systems, against the background of the turbulent chaos of the atmosphere, can be addressed in at least two ways:

- a "local" approach can be used in which a single (individual) vortex structure is formed out of the turbulent chaos under conditions in which a local ocean–atmosphere system is highly out of equilibrium;
- a "global" approach can be used to consider the formation of eddy systems in the World Ocean as a set of centers of relaxation oscillations in the active medium of the ocean–atmosphere system (where the latter is viewed on a global scale) under weakly non-equilibrium conditions.

As for the "global" approach, note that it is possible to use another notion—namely, describing the occurrence and functioning of TCs as ordered elements in the active medium of the ocean–atmosphere system. The success that the kinetic and diffusion approach has had at describing out-of-equilibrium systems in hydrodynamics and chemical kinetics (see Nicolis and Prigogine, 1977; Polak and Michailov, 1983) is well known. Here, the kinetic characteristics of a system are defined as the birth and death processes of the elements of the system (reaction mechanism) and diffusion is defined as random walks between an element of the volume and adjacent elements.

Using these analogies, it was expedient to propose and justify (using experimental data) the kinetic and diffusion approach to describe global tropical cyclogenesis as a discrete Markov process. It is clear that, in this case, reaction and diffusion mechanisms may be interpreted as follows: the birth and death processes of elements of the system correspond to the creation (birth) of a TC and its disappearance (dissipation, death), while diffusion may be interpreted as random walks on the functional lifetime of individual TCs and as a random time series of the instants at which they occur.

This approach was first proposed and developed by the present author (Sharkov, 1995, 1996d, 1997, 1999a).

Determination of the extent to which the active medium of the ocean–atmosphere system is out of equilibrium in terms of the generation of coherent structures is of great ecological importance, since this is related to the possible reconstruction of the regime of mesoscale generation of a sequence of individual TCs into the global synchronous catastrophic regime of the generation of supertyphoons, as happens in the atmosphere of Venus (Arnett, 1999). While this is apparently a hypothetical notion, it is nevertheless fully supported by certain experimental evidence.

The above-mentioned new physical insights have allowed investigations to be made of global tropical cyclogenesis by new multiscale processing approaches. In subsequent chapters we will demonstrate the new capabilities and results of the proposed methodological approach based on the concept of tropical cyclogenesis evolution as a stochastic flow of homogeneous events.

The basic idea of investigation of the statistical characteristics of global tropical cyclogenesis, formed as a signal of stochastic intermittent flow by the mathematical apparatus of chaotic dynamics, gives an insight into the hierarchical structure of global tropical cyclogenesis by retention of all temporal scales of interactions.

2

Global tropical cyclogenesis as a stochastic process

The methodological approach described in this chapter is based on the concept of tropical cyclogenesis evolution as stochastic flows of homogeneous events. The basic idea of the investigation and the formation of probability characteristics for global tropical cyclogenesis, as a signal of interdependent structure by the mathematical apparatus of random flow theory, was first proposed and developed by the present author. At first glance it would seem that there is no correlation between the proposed approach and the approaches commonly used with the aim of determining the quantitative characteristics of global cyclogenesis (e.g., year-averaged or month-averaged numbers of tropical cyclones or TCs). What actually happens is that this is not the case. The proposed method is a more generalized approach as the outlined approximations fit naturally into this scheme.

2.1 INFORMATION SIGNAL MODEL: SIMULATION, CUMULATION

The formation and accumulation of the analyzed process are produced as follows. By a uniform (indistinguishable) event in the given study we mean a functioning TC within its lifetime—from generation till the phase of dissipation—without detailed account of its energy and thermodynamic features. In such an event random flow is such a stochastic process, in which the temporal set consists of unit differences (impulses), taking non-negative integer values. The unit positive "jump" corresponds to a moment of onset of the unit event (generation of the TC) and the negative one is the disappearance of the event (dissipation of the TC), but the number of unit differences indicate the number of appearances (or disappearances) of events at these moments. The number of pulses occurring (events) in a time interval (24 hours in our case) is therefore a natural physical parameter, the "instantaneous"

E. A. Sharkov, *Global Tropical Cyclogenesis* (Second Edition).

intensity of cyclogenesis, which determines the energetics of the ocean–atmosphere interrelationship (Pokrovskaya and Sharkov, 1993a).

However, on the strength of this, the time domain of operation of a concrete TC (lifetime) is the stochastic process (possibly with its complex statistics), the probabilistic features of global cyclogenesis will to a certain degree be modified depending on accepted values of the temporal interval (temporal window of observations). In the language of statistical procedures, this approach is a method for counting the number of events and taking into account event lifetimes (Cox and Lewis, 1966; Apanasovich *et al.*, 1988).

It is important to note here that completion of the operation of a TC as a geophysical system—or, in other words, "refusal' of a system to function (work) in the terminology of popular service theory or the theory of queues—usually has external reasons (presence of land in the path of a TC or its involvement in more large-scale circulation), and such a flow of events is generally identified as a "censorial" process (Khinchin, 1963; Cox and Lewis, 1966; Cox and Oakes, 1984; Gnedenko and Kovalenko, 1987).

Since we are not interested in the detailed structure and dynamics of each individual tropical structure, on the time axis we shall represent each tropical disturbance as a pulse with unit amplitude and a random length (corresponding to the time over which the TC is active) with a random time of occurrence (generation of an individual TC).

Mathematically, this procedure for forming a signal may be written as follows:

$$N(t) = \sum \Theta(t - t_i; \tau_i) \tag{2.1}$$

where $N(t)$ is the instantaneous intensity of cyclogenesis (the number of TCs active in a 24-hour period); and $\Theta(t)$ is the bounded Heaviside function:

$$\Theta(t - t_i; \tau_i) = \begin{cases} 1 & t_i < t \leq t_i + \tau_i \\ 0 & t_i + \tau_i < t < t_i \end{cases} \tag{2.2}$$

where τ_i is the lifetime of an individual TC; t_i is the time of its formation (generation); and $t_i + \tau_i$ is the time of its dissipation.

From the physical standpoint, this procedure structures the temporal evolution of global tropical cyclogenesis as a sequence of impulses that are equal in absolute value and are shifted in time. At first glance it seems that this temporal construction of events is rather strange. However, as we will show below, it is possible to learn much about the internal (inherent) correlation features of global tropical cyclogenesis by the proposed mathematical procedure in forming such a temporal construction.

The sequence of pulses formed in this way is just an integer-valued random temporal flow of indistinguishable events. Thus, we shall proceed using a representation of the temporal sequence of the intensity of tropical cyclogenesis as a statistical signal with a complex structure. Of course, this approach is highly simplified especially when referred to actual conditions occurring in nature. Nevertheless, as noted later, it does enable us to identify the important statistical mechanisms governing global cyclogenesis and its intra-annual variability and to detect the structural characteristics of this process on timescales from several days to 5 years (see Sections 2.2–2.4, 2.7).

To determine a concrete form of the probabilistic model of cyclogenesis (amplitude characteristics) for each studied basin for each year (in a 3-month interval), experimental histograms of the number of events $N(t)$ were constructed and their statistical distributions approximated, with subsequent analysis of their degree of fit.

To study the possible systematic changes in time variation of the intensity of the flow of events, we shall use a graph of the function of the cumulative number of events—cumulative function (CF) (Cox and Lewis, 1966) of active TCs in a 24-hour interval)—over the period of observation (1 year):

$$F(t) = \sum_{k=1}^{N(t)} \Theta_0(t - t_k) \tag{2.3}$$

where $N(t)$ corresponds to (2.1); t_k is the time of event generation; and $\Theta_0(t - t_k)$ is the Heaviside function, given by:

$$\Theta_0(t - t_k) = \begin{cases} 1 & t \geq t_k \\ 0 & t < t_k \end{cases} \tag{2.4}$$

where event generation is defined as just a positive increment of $N(t)$ $(\Delta N(t_k) > 0)$.

As we will indicate below, the application of the CF approach (Eqs. 2.3–2.4) in processing time series of events is highly advantageous for detecting and revealing the fine structure in the time evolution of a system's dynamics. However, there exist many such cases (e.g., economic and administrative problems during natural disasters; Pielke and Pielke, 1997) when one way of looking at a system's dynamics is more restricted—namely, under temporal averaging of the order of 1 month (or 1 year). Under these conditions, the fine dynamical features of temporal evolution (and thus the dynamical interactions that are responsible for it) can be eliminated using a proper averaging procedure.

As we have already indicated (in Chapter 1), to ascertain the number of problems of climatic interactions, it is just as necessary to obtain the quantitative values of intensities of cyclogenesis as for primary and mature forms under temporal averaging of the order of 1 month (or 1 year) or, in other words, performing the following procedure:

$$N_0(t_i; \Delta t) = \frac{1}{\Delta t} \int_{t_i}^{t_i + \Delta t} N(t)\, dt \tag{2.5}$$

where t_i is the beginning of the current month (year); and Δt is the duration of the corresponding month (year). When the life of each individual structure is less than the month, the value of the parameter for "instant" intensities of flows that occurred within the month studied will go into the simple individual tropical event amount. Thereby, information on dynamic interactions in the global system on scales from 1 day up to 1–3 months will be lost. As we will show, with the help of wavelet analysis (Section 2.7), it is precisely these scales of interactions that are of great concern in the dynamics of global systems. Nevertheless, the monthly averaged (or yearly averaged) approach allows us to reveal a number of interesting moments that are commonly used (e.g., Elsner and Kara, 1999; Henderson-Sellers *et al.*, 1998; Gray *et al.*, 1997).

2.2 ANNUAL SINGLE-COMPONENT STOCHASTIC MODEL OF GLOBAL TROPICAL CYCLOGENESIS

On the basis of processing temporal sets of intensities of global cyclogenesis for annual cycles 1988–1992, the purpose of this section is to present experimental results pointing to the existence of firm Poisson laws of distribution in the annual cycle of tropical cyclogenesis and making possible determination of their parameters (Pokrovskaya and Sharkov, 1993a, 1994a).

2.2.1 Time series and cumulative functions of global tropical cyclogenesis

Figure 2.1 presents the graphs of temporal evolutions of flows of global cyclogenesis intensity (daily averaging) for a 5-year term from 1988 to 1992. The raw data for 1988–1992 for global tropical cyclogenesis were taken from the systematized dataset "Global-TC" (see Chapter 5 for details and Pokrovskaya *et al.*, 1993; Pokrovskaya and Sharkov, 1999d).

From Figure 2.1 it is not difficult to see that, despite external distinctions of concrete temporal sets, flows of intensity represent typical telegraphic processes from the standpoint of casual processes. Moreover, in the annual temporal cycle of tropical cyclogenesis we can recognize some general properties: between August and October of each year (for the explored 5-year cycle) there comes a period of obvious increased activity when in the World Ocean Basin there are six to ten TCs acting simultaneously (this has occurred every year since September 26, 1992). Such a high concentration of TCs in the process of tropical cyclogenesis suggests the possibility of space and time correlations as a result of the excitement and operation of vortical systems. In Section 2.2.2 we will show experimental findings of this. At the same time, it is not difficult to see that global cyclogenesis fades considerably from December on. But, despite specific intra-annual variability, it is possible to try and formulate a united probabilistic model of global cyclogenesis. The question of intra-annual variability in the flow of cyclogenesis intensity requires separate consideration (see Section 2.3).

All the mentioned temporal features of global tropical cyclogenesis have effectively emerged as a result of observations of cumulative functions (Figure 2.2a–e). To better understand the temporal behavior of tropical cyclogenesis, its time series have been on a reduced scale than those represented in Figure 2.1. We should note that the shape of the cumulative function has effectively revealed the large-scale structure of the temporal evolution of random flows.

2.2.2 Probability model and its parameters

To get concrete-type probabilistic models of the physical process considered, histograms of integer parameter $N(t)$ were built. Histograms are statistical analogues of the selective density of probability in accordance with known rules. For each group of arrays (1988–1992) sample averages and variances were calculated. The results of processing are presented in Table 2.1. External collation of sample histograms that possibly approximate the Poisson law (in the linear scale in Figure 2.3) already favor the hypothesis that the Poisson nature of the time sequence of tropical cyclogenesis can be calculated. So, in accordance with the Pearson criterion of agreement, the measure of divergence between the theoretical distribution and the experimental histogram (for 1991) is $X^2 = 2.9$ (four degrees of freedom[1]) and satisfies the inequality $X^2 < \chi^2(0.05; 4)$. The last-mentioned fact indicates that the experimental sample set is compatible with the general set of Poisson distributions with $\lambda = 1.7$. In the same way, the theoretical Poisson distribution with $\lambda = 2.0$ is satisfactorily compatible with experimental data for 1989. Here $X^2 = 10.3$ and $\chi^2(0.05; 6) = 12.6$.

However, for other annual cycles (1988, 1990, and 1992 for which X^2 are 34, 80, and 56, respectively), theoretical distributions are in poor agreement with observations (at the 5% confidence level). It is easy to find the reasons for divergences if we analyze the experimental histograms themselves (Figure 2.3a, c, e). Detailed consideration of histograms allows us to understand that divergences in theoretical and experimental distribution are attributable to histograms that show the number of days that are free from functioning TCs (i.e., days with $N = 0$) and the lack of days when $N = 1$ and 2. So, in 1988 there were 99 days free from TCs and in 1992, 84 days. If we rearrange the excess zeroes in the histograms to $N = 1$, it is possible to show that the theoretical Poisson distribution will satisfactorily be in agreement with experimental data. The physical explanation specified in the above divergences is connected with the strong inhomogeneity (understood, of course, in the statistical sense) in distributing the events (functioning TC). In other words, it is possible that the effects of clustering in the probabilistic structure of the event sequence may come into existence.

Study of the tails of the experimental distribution is of prime interest for the study of clustering effects in flows of Poisson type. Under the linear scale (Figure 2.3) these effects cannot practically be revealed. To resolve these questions we will use another presentation of experimental histograms—one on a semilogarithmic scale, as follows.

Comparison of experimental histograms and theoretical approximations executed on the semilogarithmic scale (Figure 2.4) shows that the Poisson character of the distribution of a value N is valid when the number of events (in the day) equal 6. At values $N > 6$ the experimental histogram is greatly distinguished from Poisson distribution. This circumstance testifies to the fact that originally

[1] In accordance with statistical rules (Cox and Lewis, 1966), the number of degrees of freedom is defined as the number of categories in the experimental histogram minus the number of independent conditions (relationships), assessed on frequencies (for the Poisson process the number of relationships is two).

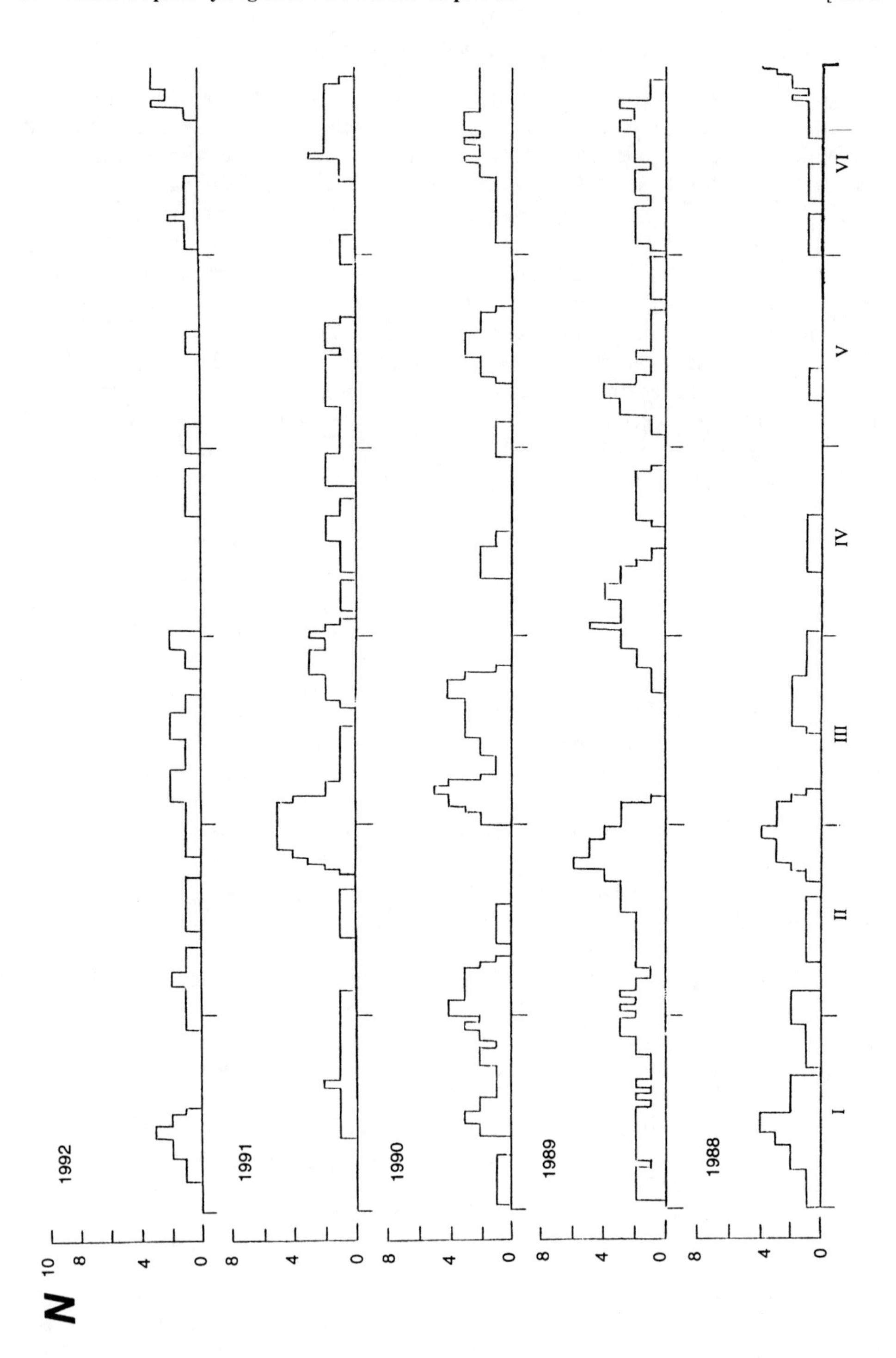
N
10
8
4
0
1992
1991
1990
1989
1988
I
II
III
IV
V
VI

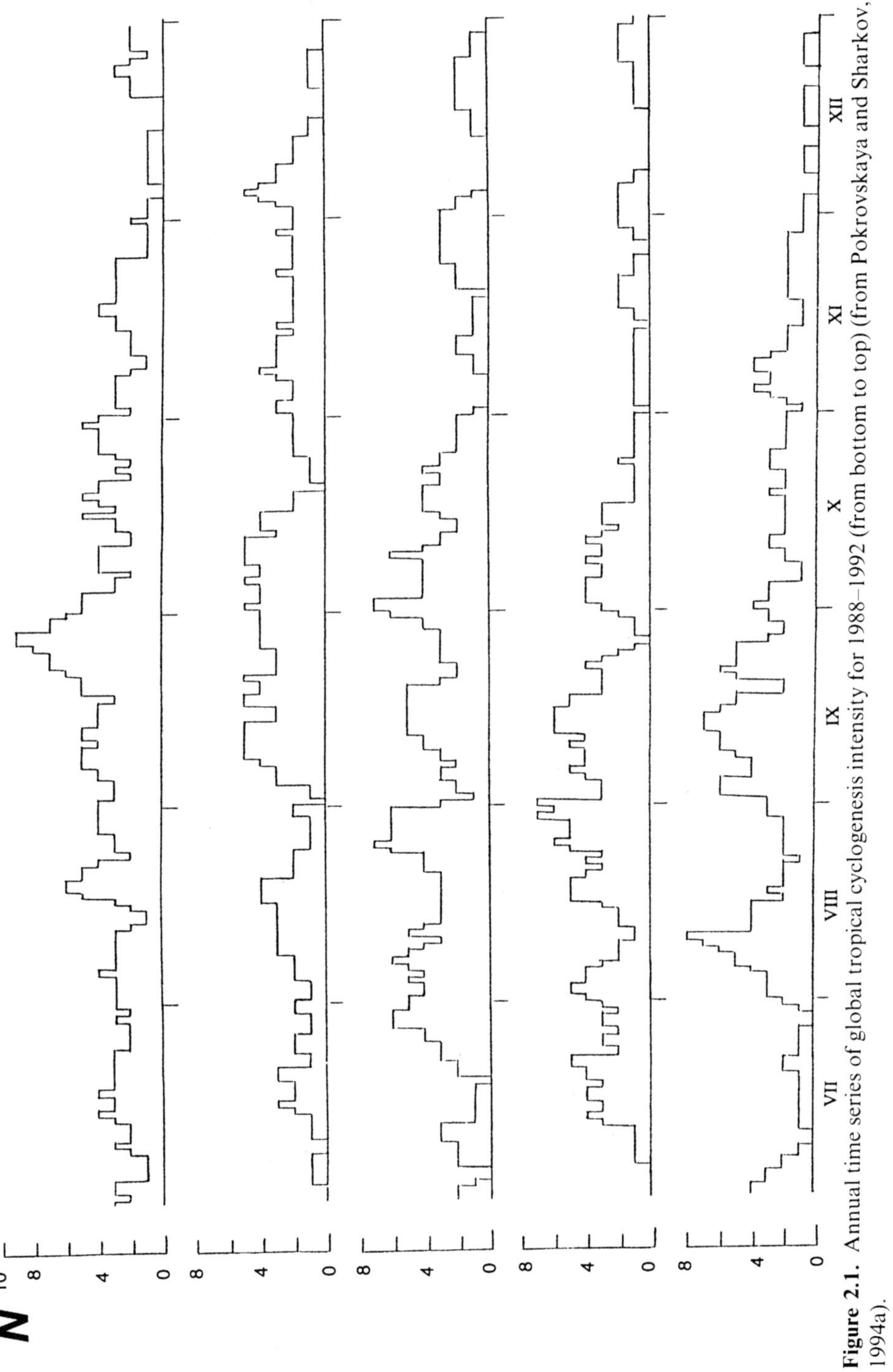

Figure 2.1. Annual time series of global tropical cyclogenesis intensity for 1988–1992 (from bottom to top) (from Pokrovskaya and Sharkov, 1994a).

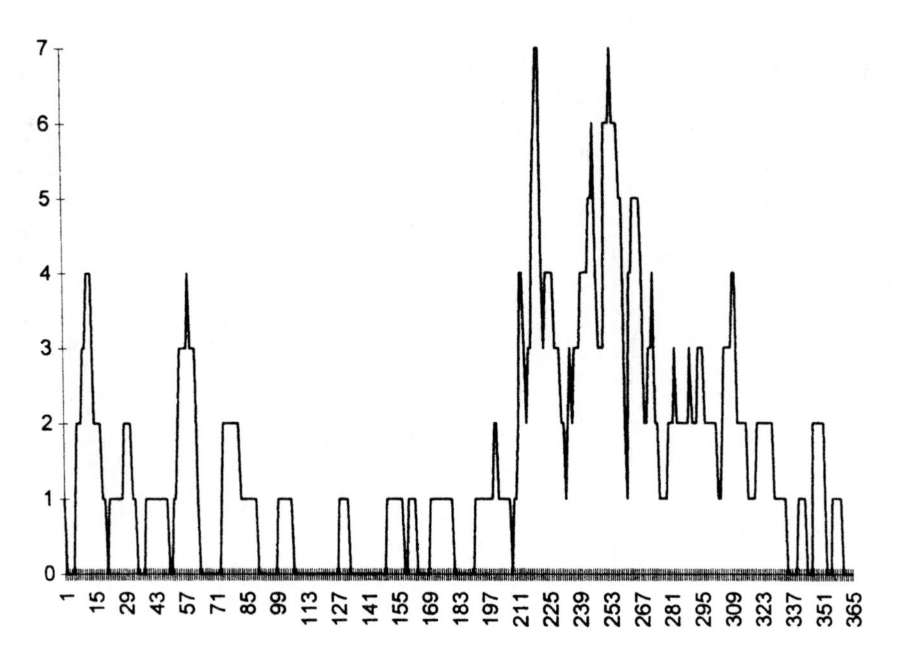

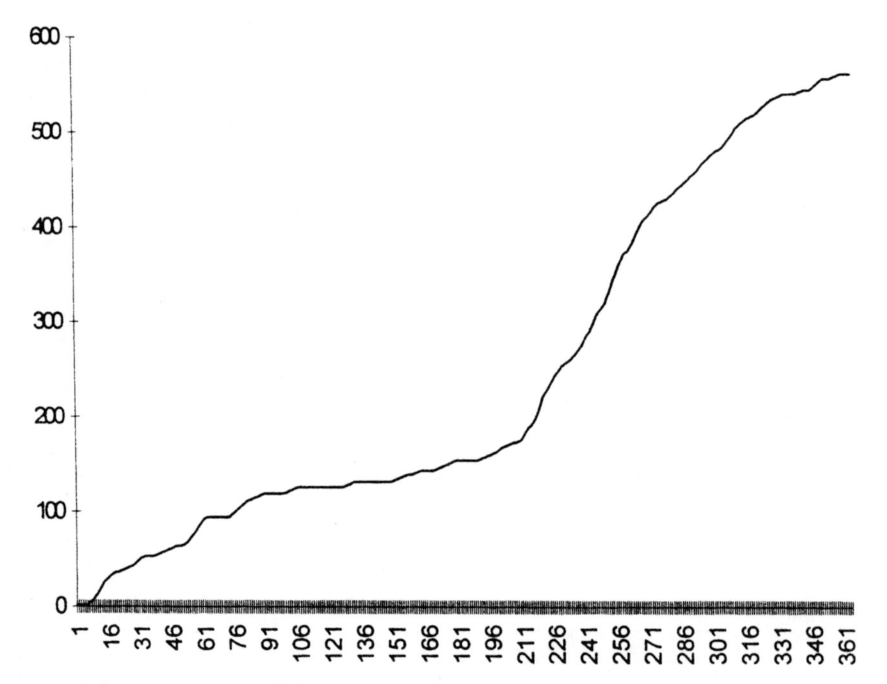

Figure 2.2a. Annual time series of the intensity (top) and cumulative function (bottom) of global tropical cyclogenesis for 1988.

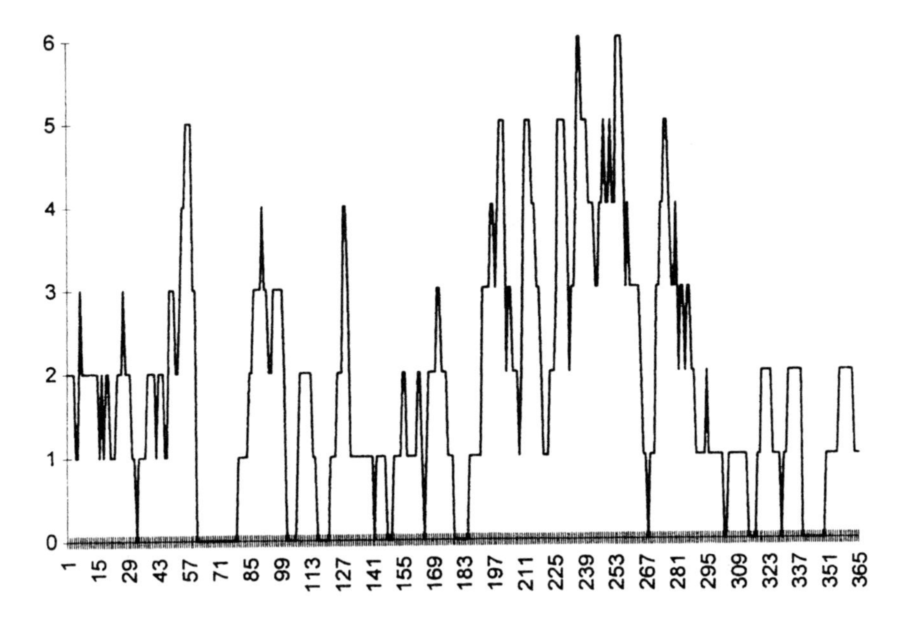

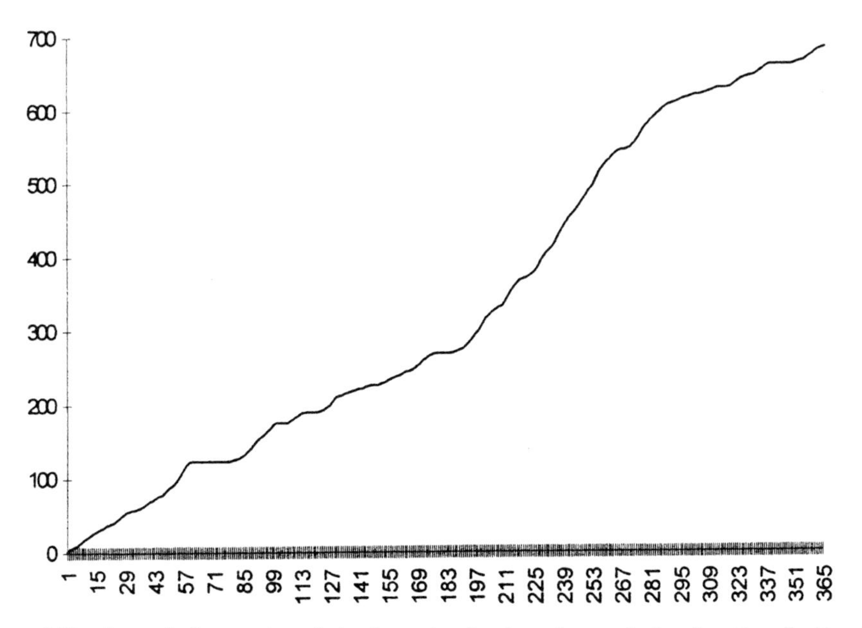

Figure 2.2b. Annual time series of the intensity (top) and cumulative function (bottom) of global tropical cyclogenesis for 1989.

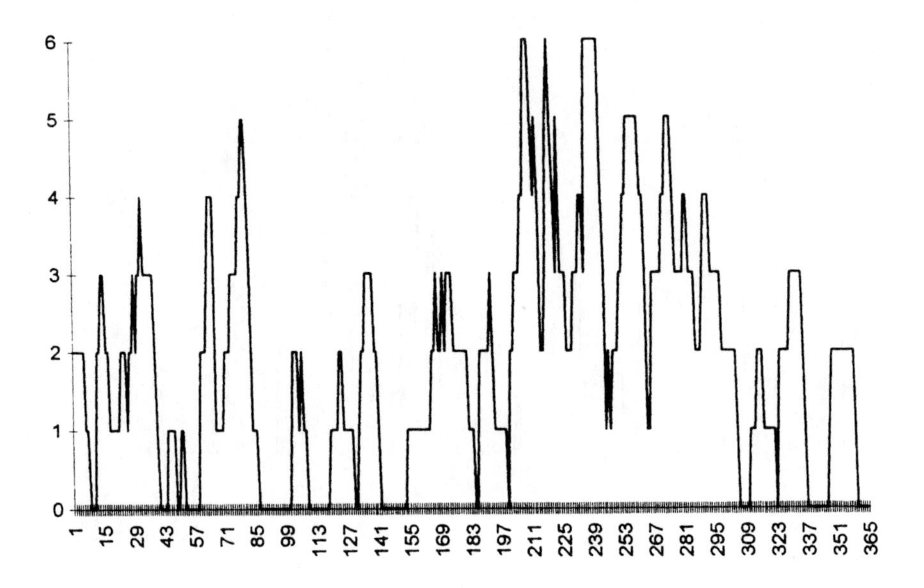

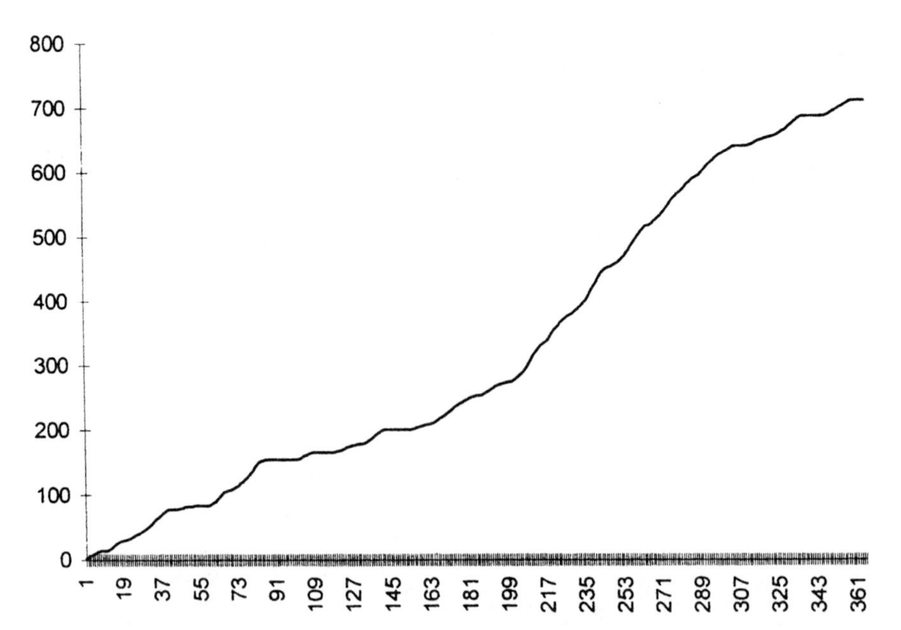

Figure 2.2c. Annual time series of the intensity (top) and cumulative function (bottom) of global tropical cyclogenesis for 1990.

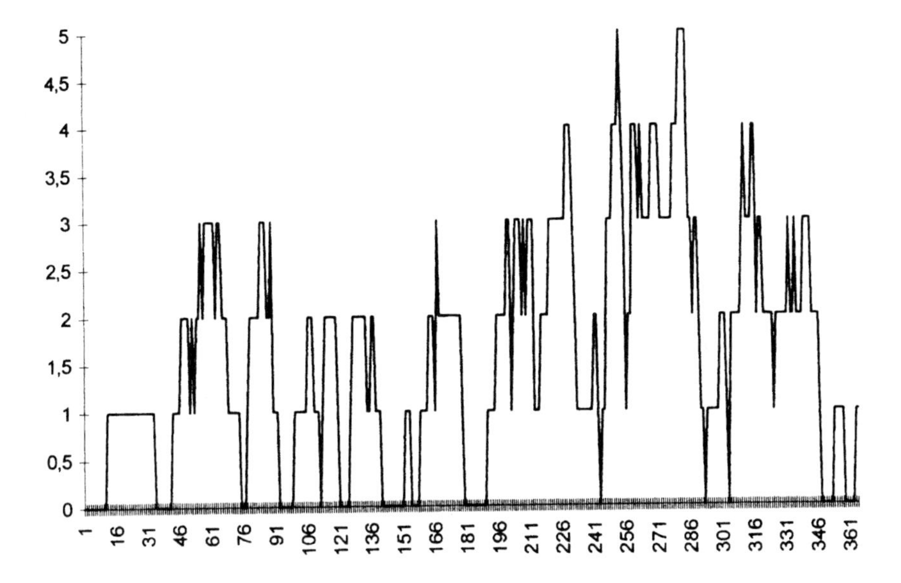

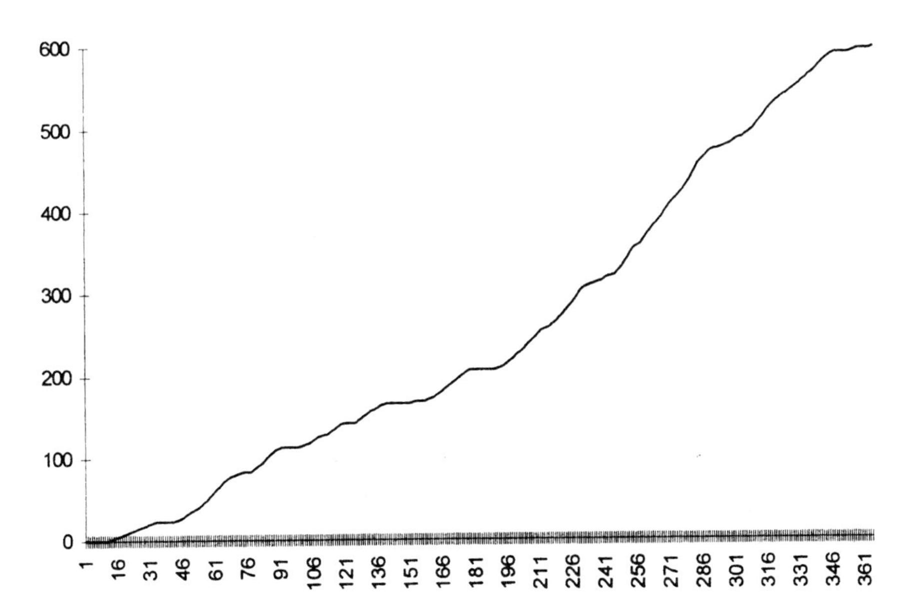

Figure 2.2d. Annual time series of the intensity (top) and cumulative function (bottom) of global tropical cyclogenesis for 1991.

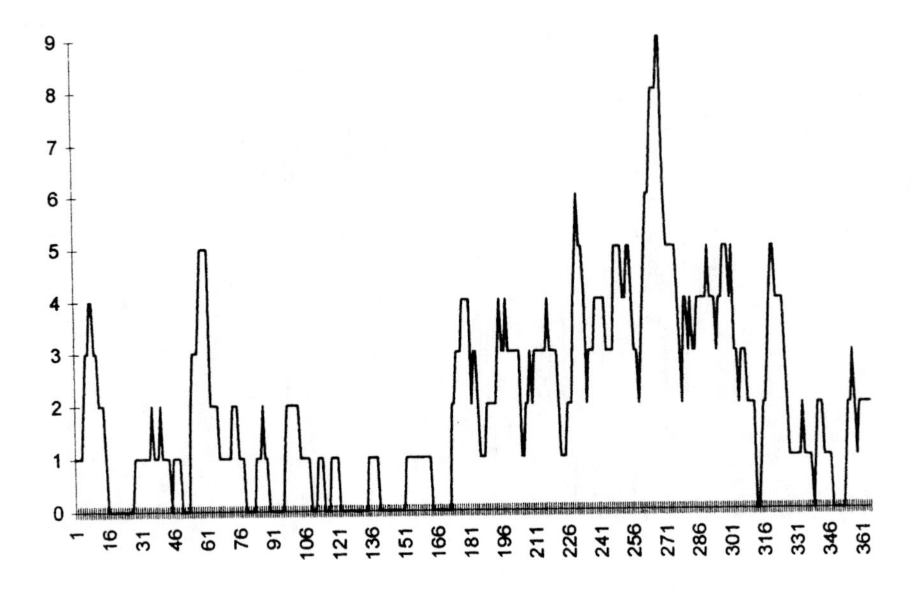

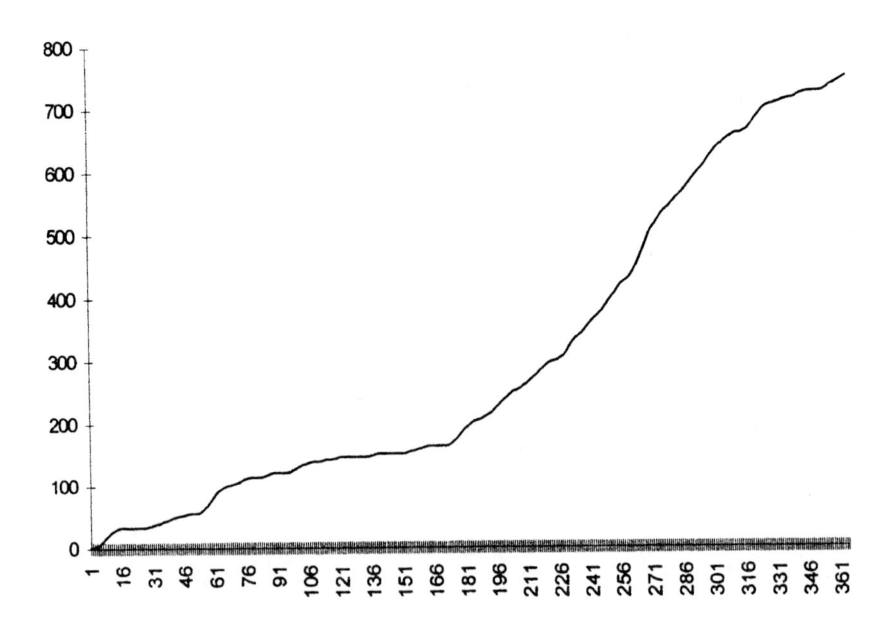

Figure 2.2e. Annual time series of the intensity (top) and cumulative function (bottom) of global tropical cyclogenesis for 1992.

Table 2.1. Parameters of distributions of global cyclogenesis intensity.

Year	*Total number of occurrences*	*Main numerical characteristics of histograms*		*Parameter of approximating Poisson law*
		Mean	*Variance*	
1988	571	1.57	2.30	1.6
1989	744	2.04	2.22	2.0
1990	741	2.03	2.75	2.0
1991	626	1.66	1.60	1.7
1992	722	1.97	3.10	2.0

Note: Total number of runs is 365 days. Parameter λ is the flow intensity. Total number of occurrences is the total amount of registered events (TCs per day) for the full period (365 days).

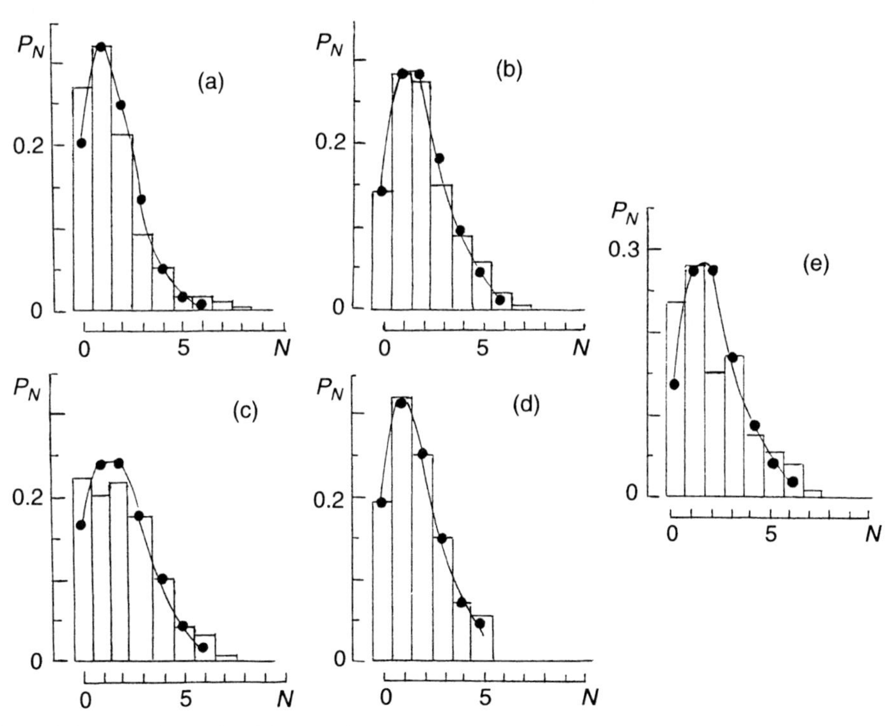

Figure 2.3. Sample probability densities (histograms) of cyclogenesis intensity and theoretic Poisson distributions on the linear scale. The stepped curves represent the experimental histograms. The full circles and the solid lines represent the theoretic Poisson law with the value given in Table 2.1. (a) Data for 1988; (b) for 1989; (c) for 1990; (d) for 1991; (e) for 1992 (from Pokrovskaya and Sharkov, 1994a).

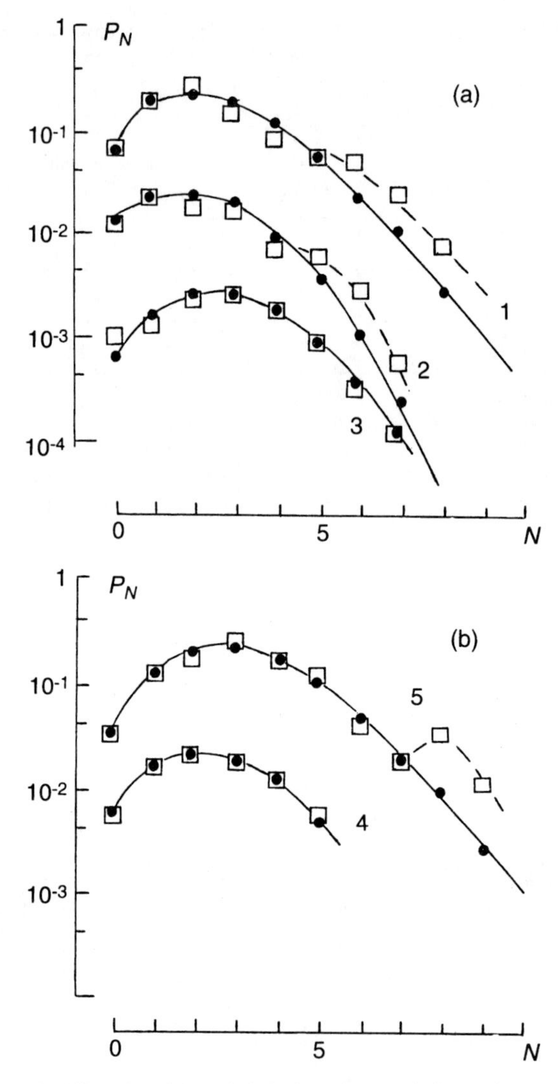

Figure 2.4. Sample probability densities of global cyclogenesis intensity and theoretic Poisson laws on semilogarithmic scales. The open squares and dotted lines represent the experimental histograms. The full circles and solid lines represent the theoretic Poisson law with the value given in Table 2.1. (a) Data for 1988–1990; (b) for 1991–1992; (1) data for 1988; (2) data multiplied by 10^{-1} for 1989; (3) data multiplied by 10^{-2} for 1990; (4) data multiplied by 10^{-1} for 1991; (5) for 1992 (from Pokrovskaya and Sharkov, 1994a).

independent events (recall the functioning TC) gain correlation relationship (either in space or time). Under these circumstances the Poisson characteristic of the ordinariness of flow breaks down (see Section 2.2.4). A similar situation (deviation from the Poisson distribution under $N = 6$) is observed for the experimental histograms of 1988 and 1989 (for 1991 $N > 6$ was not observed). In both the process of development and in the operation of tropical vortical systems clustering effects were discovered earlier by experiment (Minina, 1983; Pokhil, 1990, 1996; Carlowicz, 1995).

The particularities discovered require more detailed analysis of the structure of Poisson-type flows regarding not only amplitude but also temporal probabilistic features.

In view of the significant number of recorded events in experimental data ($n > 500$, see Table 2.1), point evaluation of the intensity of Poisson distribution flow λ and its confidence levels can be executed by means of a method based on normal approximation for the Poisson distribution with the equation:

$$\lambda t_0 = n + 1/2C_{\alpha/2}^2 \pm C_{\alpha/2}n^{1/2} \tag{2.6}$$

where t_0 is the total number of runs; and C_α is the upper α-quantile of unit normal distribution. Using this correlation under confidence levels of probability of 95%, we can evaluate the confidence level intervals for λ. They are equal to $\pm$(0.12–0.14). Mean values of λ are provided in Table 2.1; the root-mean square (r.m.s.) of λ is 0.6–0.7.

2.2.3 Interannual variabilities

In this section we will consider the question of interannual variations in the intensity of Poisson flow, which is important from the standpoint of global energy exchange in the ocean–atmosphere system. Essentially, it is necessary to take or reject a hypothesis on the accessories of all revealed Poisson flows in the united general set. To resolve this problem we use a method that requires a fixed number of events when using dispersion relations and the F-distribution, respectively.

We know (Cox and Lewis, 1966) that if the total number of event occurrences of matching flows is great ($n \gg 1$), it is possible to use a logarithmic transformation for the dispersion relation:

$$Z = \log R = \log(\lambda_2/\lambda_1) + \log(n_0' t_0''/t_0' n_0'') \tag{2.7}$$

where λ_1 and λ_2 are the parameters of two independent Poisson processes; and t_0' and t_0'' are observed time runs ($t_0' = t_0'' = 366$), for which n_0' and n_0'' events have occurred accordingly. In this case the value Z will have a normal distribution with the average value and standard variance, respectively $\frac{1}{2}(1/n_0' - 1/n_0'')$ and $(1/n_0' + 1/n_0'')^{1/2}$. Using this procedure to analyze most of the distinguishing intensities for 1988 and 1989 (see Table 2.1), we get 95% confidence levels for the correlation λ_2/λ_1 as (1.62; 1.006). It is not difficult to see that the selective equation λ_2/λ_1 is 1.29 and that it falls in the specified confidence interval. This means that the significant difference (within the framework of the specified approach) in correlations of the intensities of flows is not observed. A similar conclusion holds for correlations of the other annual intervals under investigation.

2.2.4 Poisson random model

We emphasize from the outset that the case in point is a measured parameter—the intensity of tropical cyclogenesis—possessing a sufficiently complex procedure of accumulation (shaping). The value of this parameter characterizes not only the intensity of processes of the birth and death of individual TCs but also contains the statistics of TC timescales. By the structured element of the time sequence under investigation is meant a variation in intensity of the unit ($\Delta N = 1$). This change is not, generally speaking, associated directly with the generation of individual TCs.

In spite of the rather complicated structure of signal accumulation, as a whole the amplitude features of tropical cyclogenesis can be described within the framework of sufficiently simple physical Poisson-type models (e.g., fluctuation density in an ideal gas and fluctuation noise in electrical circuits). Of course, the process can be described within the framework of the univariate controlling probabilistic equation of birth–death (Bharucha-Reid, 1960). From equilibrium resolution it appears that the average value of the intensity of flow is the ratio of probabilities of both birth and death.

The Poisson approximation is useful for estimating probabilities associated with rare or extreme events (Bharucha-Reid, 1960; Feller, 1968; Zolotarev, 1983). The Poisson distribution describes the number of times that some event occurs as a function of time, where the events occur at random times.

The simplest whole number Poisson point process $\{M(t), 0 \leq t < +\infty\}$ is specified by three conditions:

1. The process is ordinary (i.e., events must occur only once in any increasingly small time interval):

$$P\{M(\Delta t) = 1\} = \lambda \cdot \Delta t + o(\Delta t) \tag{2.8}$$

$$P\{M(\Delta t) > 1\} = o(\Delta t) \tag{2.9}$$

   where λ is some positive quantity. The symbol $o(\Delta t)$ represents the term that has a higher order value than Δt. In other words, this condition of Poisson flows means that the events in Poisson processes are very scarce and, hence, these events do not interact with each other.
2. The process is steady state (i.e., probability characteristics are independent of time).
3. The process possesses the following property: the number of events occurring in any time interval is independent of the number occurring in any other non-overlapping interval ("a lack of consequences").

Based on these conditions it can be shown (Bharucha-Reid, 1960) that the probability $P_k(t)$ of exactly k occurrences (points) over a time semi-interval $(0, t]$ is given by the cumulative probability distribution function (Poisson law):

$$P_k(t) = P[M(t) = k] = \frac{(\lambda t)^k \cdot e^{-\lambda t}}{k!}, \qquad k = 0, 1, 2, \ldots, \infty \tag{2.10}$$

where the factorial k is denoted by $k!$

From this equation it follows that the probability of the lack ($k = 0$) of an event occurring in the semi-interval τ is:

$$P_0(\tau) = \exp(-\lambda\tau) \tag{2.11}$$

The expected value m and the variance D of the Poisson distribution are equal to one another:

$$m = D = \lambda t \tag{2.12}$$

As m is the average number of occurrences in the time semi-interval $(0, t]$, so the parameter:

$$\lambda = \frac{m}{t} \tag{2.13}$$

may be considered as the average number of events occurring in the unit interval. Hence, λ is often denoted as the Poisson process intensity.

A number of various extensions of the Poisson distribution have been studied (Bharucha-Reid, 1960; Feller, 1968; Korn and Korn, 1961; Cox and Lewis, 1966). We consider two such extensions for the purpose of detailed processing of the global tropical cyclogenesis time series:

- an inhomogeneous Poisson process; and
- a filtered Poisson process.

The Poisson distribution is spoken of as an inhomogeneous process if its intensity is dependent on time $\lambda(t)$. For such an inhomogeneous process the probability $P_k(t)$ of exactly k occurrences over a time semi-interval $(0, t]$ is given by the following equation:

$$P_k(t) = \frac{1}{k!}\left(\int_0^t \lambda(\tau)\,d\tau\right)^k \cdot \exp\left(-\int_0^t \lambda(\tau)\,d\tau\right), \qquad k \geq 1 \tag{2.14}$$

The expected value $m(t)$ and the variance $D(t)$ of this Poisson distribution are equal:

$$m(t) = D(t) = \int_0^t \lambda(\tau)\,d\tau \tag{2.15}$$

The filtered Poisson process $\{\xi(t); t \geq 0\}$ can be expressed by the following equation:

$$\xi(t) = \sum_{i=1}^{M(t)} h(t, t_i, a_i) \tag{2.16}$$

where $\{M(t), t \geq 0\}$ is, in the general case, an inhomogeneous Poisson process with inhomogeneous intensity $\lambda(t)$; $\{a_i\}$ is a sequence of random functions that are independent of $\{N(t); t > 0\}$; and $h(t, t_i, a_i)$ is a deterministic function of the three variables.

For practical implementation, the parameters of the equation $\xi(t)$ allow the following interpretation:

- t_i is the time when the random event occurs;
- a_i is the amplitude of the elemental signal associated with this event;
- $h(t, t_i, a_i)$ is the value of the elemental signal (at time t) governed by this event;

- $\xi(t)$ is the value (at t) of the sum of elemental signals governed by the events that occurred over the temporal semi-interval $(0, t]$.

These processes are often termed "fluctuation noise" or "Schottky noise" (Bharucha-Reid, 1960; Korn and Korn, 1961).

This was the very same approach that was used to establish the information signal for global tropical cyclogenesis (see Section 2.1). For the instantaneous intensity of cyclogenesis (Eq. 2.1), $h(t, t_i, a)$ was the bounded Heaviside function. The amplitude a_i of the elemental signal associated with the event is equal to 1 and the event's time t_i is the time of TC onset.

For the cumulative function (Eq. 2.3), $h(t, t_i, a)$ was the Heaviside function. The event time t_i is the time of positive increment of $N(t)$ ($\Delta N(t) > 0$) as distinct from the case of instantaneous intensity.

As mentioned in Section 2.1, the qualitative and quantitative analysis of cumulative functions has been applied with much success to the study of possible systematic variability in the temporal variation of time series intensity (Cox and Lewis, 1966). The efficiency of this method depends upon the fact that the shape of the cumulative function is acutely sensitive to temporal variation. Using time-dependent plots of cumulative functions, estimations of the integral and differential intensities of random processes may be determined from graphical displays of cumulative experimental data.

Striking examples of what can be achieved using analyses of cumulative functions are presented in Sections 2.2.1 and 2.6.1.

As a direct consequence of the Poisson approximation, tropical cyclogenesis possesses features known from the theory of Poisson flows: variation in intensities does not depend on the number of existing structured elements, on their previous history, or on the system's condition. Also, the process satisfies ordinariness characteristics.

Variation in experimental histograms from the purely Poisson models explored above is highly symptomatic. There can be, as stated above, two types: an excess of zeroes and an excess of experimental tails of histograms above the Poisson branch of distribution. The first type of variation is connected with the time unsteady state of the process; the second can be interpreted as a break in ordinariness characteristics. In other words, the probability of the simultaneous origin of two or more events is not small and is finite. This is easily observable by analyzing the temporal sets of global cyclogenesis (Figures 2.1 and 2.2) for the July–September period of any annual cycle.

Note that the obtained results refer to daily interval partitioning of the time axis. By averaging for grid intervals, the statistics of the amplitude features of global cyclogenesis intensity can be greatly transformed so as to account for the change in the "contribution" statistics of lifetime (it can, in general, drop out of consideration) and to account for the statistics of time of TC genesis as well.

Besides the technique used in this approach for the accumulation of data employing a method that accumulates the number of events, there exist many other areas that need resolution. So, for example, the formation of a flow by recording the moments of occurrence of separate events (TC generation)—the method of recording the births of events—is of concern in the solution of prognostic tasks. The flow generated by recording the births of events will contain more "pure" information on the statistics of TC generation and its properties as a whole will differ from the properties of the flows of intensity mentioned above.

2.3 ANNUAL TWO-COMPONENT STOCHASTIC MODEL OF GLOBAL TROPICAL CYCLOGENESIS

As we might expect, the attempt to generate unified (within the limits of a single year) probabilistic models of global tropical cyclogenesis meets with particular difficulties that can be bypassed, having considered two timeframes in the annual cycle that clearly have different intensities in the process of cyclogenesis: from January till July (interval I) and from August till December (interval II) (Pokrovskaya and Sharkov, 1994b).

2.3.1 Two-component probability model and its parameters

For detection of a particular kind of probabilistic model of the physical process under consideration, histograms of the integer parameter $N(t)$, being statistical images of the sample density of probabilities, were constructed according to known rules. For each group of sets for intervals I and II for each year (1988–1992) a sample mean and variances were calculated. Processing outcomes are shown in Table 2.2. External comparison of sample histograms with the possible (probable) approximating Poisson law (in a linear scale, see Figures 2.5 and 2.6) already favors the hypothesis that the Poisson character of the temporal sequence of tropical cyclogenesis can be considered. So, in line with a goodness-of-fit test of the Pearson criterion, the deviation of a theoretical distribution and experimental histogram for the data for interval II (1990–1991) satisfies the inequality $X^2 < \chi^2(0.05)$. This proves that experimental sampling is compatible with the general set of Poisson distributions having the values of λ indicated in Table 2.2. The satisfactorily theoretical Poisson distribution is in agreement with the experimental dataset for interval II (1988–1989), but at a broader window of reliability ($\alpha = 0.01$).

As for the timeframe of interval I, the situation here is worse than with interval II. Here, the parent distributions of the Poisson distribution satisfactorily agreed with the experimental histograms for 1989 ($\alpha = 0.05$) and 1990 ($\alpha = 0.01$). But it poorly agreed with the outcomes of observations for 1988, 1991, and 1992.

It is an easy matter to find the reasons for divergences if we analyze experimental histograms (Figure 2.6). Detailed consideration of histograms allows us to understand that the divergences between theoretical distributions and the experimental dataset are explained by some kind of "non-uniformity" (understood in the statistical sense, of course) of the temporal distribution of events (operating TCs).

Table 2.2. Parameter of distribution of global tropical cyclogenesis (two-component model).

Year	*Intra-annual interval*	*Total number of occurrences*	*Main numerical characteristics of histograms*		*Parameter approximating Poisson law*	*Measure of deviation of Pearson's criterion*
			Mean	*Variance*		
1988	I	232	1.09	0.96	1.10	12.06
	II	376	2.46	3.17	2.50	17.23
1989	I	387	1.83	1.68	1.80	5.14
	II	337	2.20	2.55	2.20	12.21
1990	I	312	1.48	1.82	1.50	15.26
	II	387	2.53	3.01	2.50	10.28
1991	I	265	1.26	0.90	1.26	25.83
	II	373	2.45	1.80	2.50	4.13
	I	246	1.16	1.06	1.16	26.31
	II	495	3.21	3.09	3.20	9.86

Note: Total number of runs for interval I (January–July) is 213 days and for interval II (August–December) is 153 days.

In other words, the effect of some clusterization in a probabilistic temporal sequence of events is possible. So, in the data of 1988, the excess of the "four-in-hand" ($N = 4$) structural elements of sampling is observed. In 1992, on the contrary, there is an excess of triple units ($N = 3$) and disadvantage twin ($N = 2$) units. In 1991 there was a grouping (clusterization) (Figure 2.5) of five units ($N = 5$) within 1 week (end of February–beginning of March) or, in other words, an eruption in the activity of cyclogenesis against the background of a rather languid process. These effects are also responsible for the divergence of theoretical models and observational data (for interval I).

Studying the tails of experimental distributions, which at the linear scale (Figures 2.5 and 2.6) practically cannot be detected, is of most interest from the point of view of research into clusterization effects in Poisson-type flows. For the solution of these problems we shall take advantage of another representation of experimental histograms: namely, a semilogarithmic scale.

The study of comparability of experimental histograms and theoretical approximation models for the active period of the year, made on a semilogarithmic scale (Figure 2.7), has shown that the Poisson character of a density function of value $N(t)$ is retained if the values of the number of events (per day) equal 6–7. At values $N > 7$ experimental histograms differ essentially from the pure Poisson model. This proves that originally independent events (recall operating TCs) gain correlation connections (either in space or time) and, thus, the Poisson property of ordinariness of flow (see Section 2.2.4) is upset. Such a clusterization of generation and operation of tropical vortical systems was earlier detected from observational

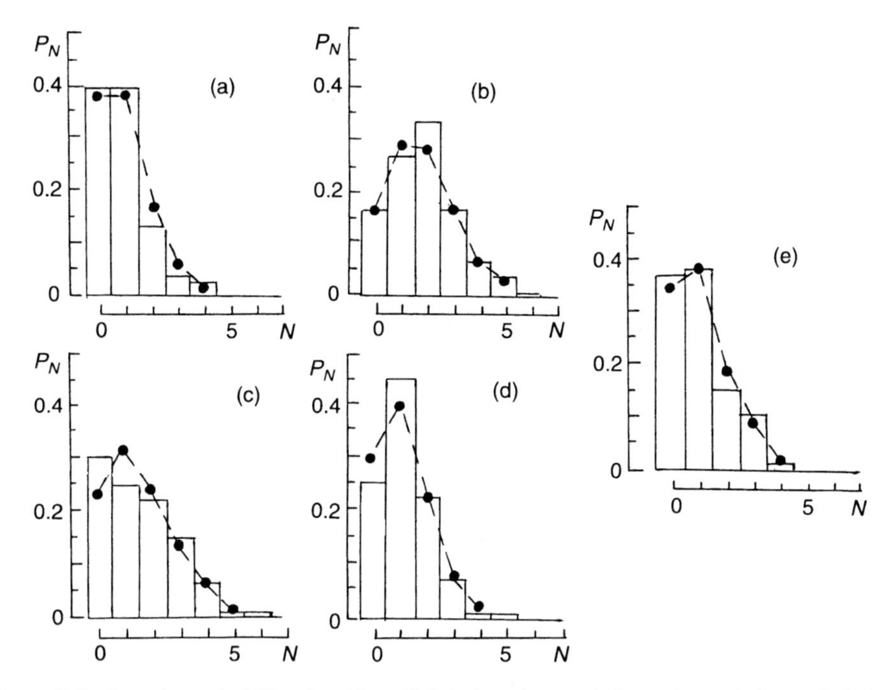

Figure 2.5. Sample probability densities of global cyclogenesis intensity and theoretic Poisson laws on linear scales for the intra-annual interval I. Stepped curves represent experimental histograms. Full circles and dotted lines represent theoretical Poisson laws with the values given in Table 2.2. (a) Data for 1988; (b) for 1989; (c) for 1990; (d) for 1991; (e) 1992 (from Pokrovskaya and Sharkov, 1994b).

data of the conditions for cyclogenesis in the Pacific Ocean (Minina, 1970; Pokhil, 1990, 1996).

These features require more detailed study not only of the amplitude but also the temporal probabilistic characteristics of Poisson-type flow patterns.

In view of the considerable sample size of the number of events in the experimental data ($n > 200$, see Table 2.2) point estimation of the intensity of a flow of Poisson distribution λ and its confidence level boundaries can be executed (made) by using a method grounded on the normal approximation of Poisson distributions (Cox and Lewis, 1966), by asking the question:

$$\lambda t_0 = n + \tfrac{1}{2} C_{\alpha/2}^2 \pm C_{\alpha/2} n^{1/2} \tag{2.17}$$

where t_0 is the total number of runs (see Table 2.2); and C_α is the upper α-quantile of a unit normal distribution. Using this ratio at a confidence level probability of 95%, we get estimations of confidence intervals for λ. They were $\pm(0.14\text{–}16)$ for interval I and $\pm(0.22\text{–}0.28)$ for interval II. The average values of λ are listed in Table 2.2; the r.m.s of λ is equal to 0.06–0.07 for interval I and 0.11–0.14 for interval II.

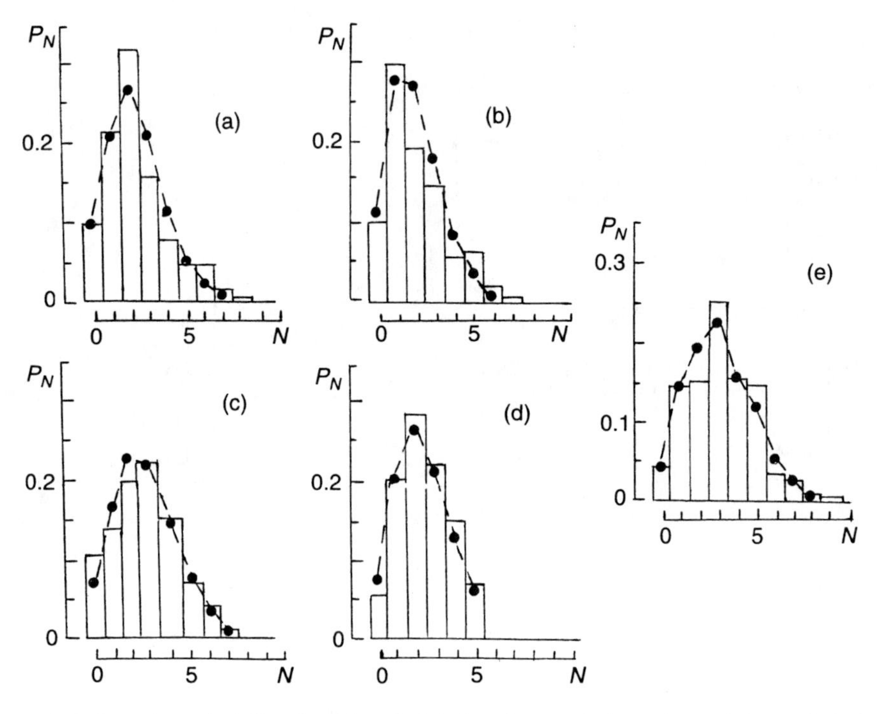

Figure 2.6. Sample probability densities of global cyclogenesis intensity and theoretic Poisson laws on linear scales for intra-annual interval II. See symbols and notations in Figure 2.5 (from Pokrovskaya and Sharkov, 1994b).

2.3.2 Interannual variabilities for the two-component model

Let us consider the important problem of interannual variations in the intensity of a Poisson flow simulating tropical cyclogenesis from the point of view of global exchange in the ocean–atmosphere system. Essentially, it is necessary to accept (or reject) a hypothesis about fitting all Poisson models to a united (for 5 years) general set. To solve this we use a method based on the time of occurrence of a fixed number of events by using a variance ratio and F-distributions, respectively.

We know (Cox and Lewis, 1966) that if the sampling volume of the number of events for related flows is great ($n \gg 1$), it is possible to take advantage of the logarithmic transformation for a variance ratio:

$$Z = \log R = \log(\lambda_2/\lambda_1) + \log(n_0' t_0''/t_0' n_0'') \tag{2.18}$$

where λ_1 and λ_2 are the parameters of two Poisson independent processes; and t_0' and t_0'' are the observable timeframes ($t_0' = t_0'' = 213$ for interval I and 153 for interval II), for which n_0' and n_0'' events have also taken place, respectively. Thus, Z will have a normal distribution with average value and reference deviation of $\frac{1}{2}(1/n_0' - 1/n_0'')$ and $(1/n_0' + 1/n_0'')^{1/2}$, respectively. Using this procedure for the

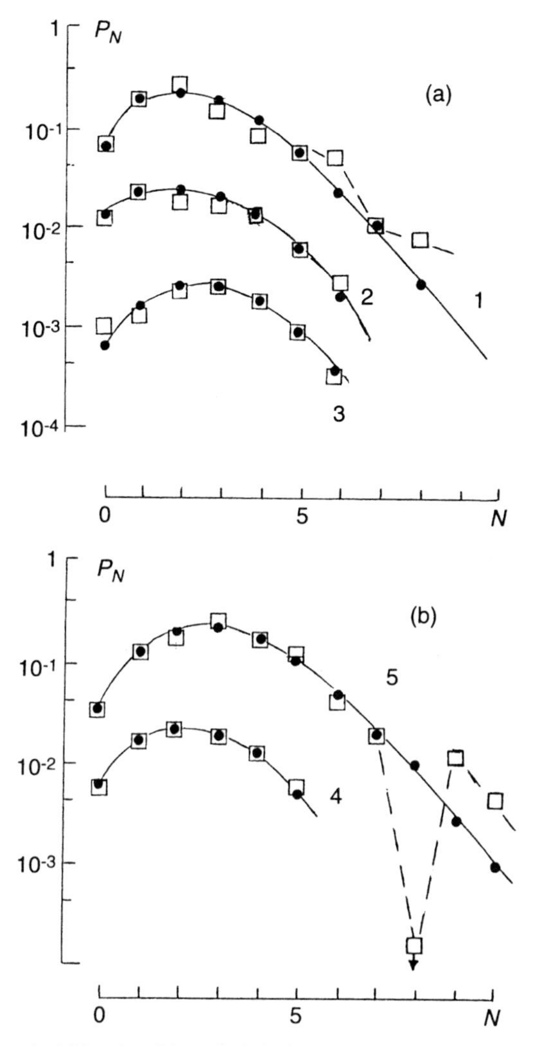

Figure 2.7. Sample probability densities of global cyclogenesis intensity and theoretic Poisson laws on semilogarithmic scales for intra-annual interval II. Open squares and dotted lines represent experimental histograms. Full circles and solid lines represent the theoretic Poisson law with the value given in Table 2.2. (a) Data for 1988–1990; (b) data for 1991–1992; (1) data for 1988; (2) data multiplied by 10^{-1} for 1989; (2) data multiplied by 10^{-2} for 1990; (4) data multiplied by 10^{-1} for 1991; (5) data for 1992 (from Pokrovskaya and Sharkov, 1994b).

most different intensities for 1988 and 1989 (interval I) we find that the 95% confidence levels for ratio λ_2/λ_1 are equal to (2.6; 1.2). It is not difficult to see that the sample relation λ_2/λ_1 is equal to 1.63 and falls in the indicated confidence interval. By this we mean that no significant difference (within the framework of the indicated approach) in the ratio of the intensity of flow is observed.

For interval II the greatest differences in values of the intensity of flow are observed between 1989 and 1992. By following a similar procedure to Table 2.2, we find that the sample value λ_2/λ_1 is equal to 1.43, whereas the 95% confidence interval is put in the range 2.18 and 1.14.

However, we emphasize that the criterion used to distinguish flow intensity is too "soft" and it makes sense to take advantage of other procedures. It is not difficult to see that application of the double t-criterion for the above-mentioned sampling would specify significant differences between the second intervals for 1988 and 1989. Thus, the problem of interannual variability requires further and more detailed analysis.

It has been found experimentally that the probabilistic characteristics of the amplitude of global cyclogenesis exhibit steady intra-annual variability and can be described within the framework of two series of Poisson approximations with steady parameters for flow intensity.

2.4 TROPICAL CYCLOGENESIS OF THE NORTHERN AND SOUTHERN HEMISPHERES

The first purposeful studies of cyclogenesis as a stochastic flow have shown a complex hierarchical structure of global tropical cyclogenesis and have allowed the formation of preliminary statistical–quantitative models (see Sections 2.2 and 2.3).

As for tropical cyclogenesis, considering each hemisphere separately, qualitative considerations can be applied over time to the non-uniform contribution to global cyclogenesis areas of northern hemisphere (NH) and southern hemisphere (SH) basins. At the present time, the problem of showing distinctions in the intensity of cyclogenesis of both hemispheres is currently connected with determination of the role of global circulations and ENSO (El Niño–Southern Oscillation) phenomena in forming tropical cyclogenesis (Henderson-Sellers *et al.*, 1998; Gray *et al.*, 1997).

On the basis of processing temporal sets of the intensity of tropical cyclogenesis, determined in the basins of the two hemispheres over a 15-year term (1983–1997), Pokrovskaya and Sharkov (1999b) have presented experimental findings on particularities in the structure of signals which reflect the operation of hemisphere cyclogenesis, as well as on intra-annual and interannual particularities of the contribution by both hemispheres to global tropical cyclogenesis.

For the procedure of accumulation and formation of a stochastic signal we use the approach developed in Section 2.1 and in Chapter 5.

Raw data for 1983–1997 on global cyclogenesis were selected from the systematized database Global-TC of remotely observed data on global tropical cyclogenesis (Pokrovskaya and Sharkov, 1999d).

2.4.1 Time series and cumulative functions for hemisphere cyclogenesis

Figure 2.8 presents graphs showing temporal evolutions of flows of intensity $N(t)$ and of functions of an accumulating number of events $F(t)$ for tropical cyclogenesis

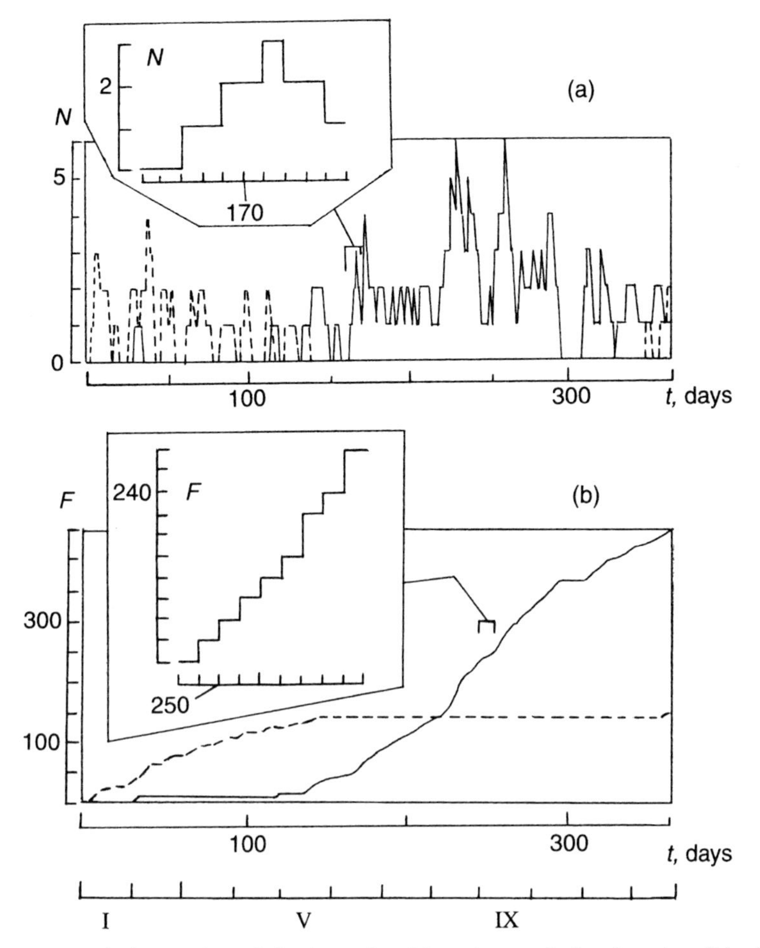

Figure 2.8. Annual time series of the intensity (a) and cumulative function (b) of tropical cyclogenesis for northern and southern hemispheres for 1986. Solid and dotted lines represent tnorthern and southern hemisphere data accordingly (from Pokrovskaya and Sharkov, 1999b).

in the NH and SH (daily averaging) for 1986. This year is chosen as a distinctive example. Analysis of all 15-year (from 1983 to 1997) temporal cycles conducted by the present author shows that all explored annual samples of flows of intensity present themselves as typically telegraphic processes (Figure 2.8a). The detailed structure of the intensity and functions of a number of accumulating events are presented in increasing (daily) scale on insets to the corresponding graphs (Figure 2.8a–b).

In the annual cycle of hemispherical tropical cyclogenesis some general lines are tracked. So, in August–December of each year in NH basins a period of increased activity clearly exists, but for SH basins the activity is clearly reduced. For the first half of a given year the reverse is true for NH basins where cyclogenesis is weak in comparison with its maximum in the second half of the year, but for SH basins the season of increased activity sets in. These particularities manifest themselves particularly graphically when considering the time evolutions of accumulation functions of a number of events (Figure 2.8a–b), where the time derivative of $F(t)$ points to the differential intensity of the process.

Another important particularity of tropical cyclogenesis is its intermittent nature, which particularly graphically manifests itself in the analysis of temporal sets of accumulation functions of the number of events. Such an effect was first discovered during analysis of the time dynamics of tropical cyclogenesis in the Pacific for 1990–1991 (Pokrovskaya and Sharkov, 1996a).

Analysis of a 15-year period of the evolution of global cyclogenesis shows that intermittent effects in cyclogenesis on both a daily and monthly scale are universal phenomena that can be considered in the same way as global (or hemispherical) cyclogenesis (the same applies for their regional component). To show the quantitative features of this process, Pokrovskaya and Sharkov (1996a) proposed using two approaches: a differential one and an integral one (for corresponding time intervals). We will use these approaches in Section 2.4.3.

2.4.2 Probability model and its parameters

To show a concrete type of probabilistic model for cyclogenesis (amplitude features) in the NH and SH and for each temporal interval (interval I from January to July and interval II from August to December during 1983–1997), experimental histograms showing integer numbers of events $N(t)$ and approximating their statistical distribution—by subsequent analysis and degrees of agreement—were built. Moreover, the average and variances for each group of arrays were calculated. Processing results are presented in Tables 2.3–2.6.

Collation of selective histograms with the possible approximating Poisson law on linear and semilogarithmic scales (Figure 2.9) favors the hypothesis of the Poisson nature of amplitude fluctuations of intensity $N(t)$ for tropical cyclogenesis. So, in accord with the Pearson criterion of agreement, analyzing the measure of divergence in the theoretical distribution and experimental histograms for a given cyclogenesis for NH and SH for 1983–1997 points to the fact that the experimental samples are compatible with the general set of Poisson distributions with values λ, as specified in Figure 2.9a–d. However, detailed analysis of the histograms has shown that in a number of cases unsatisfactory compatibility with theoretical approximations is observed.

Here we note that significant divergences occur exactly when the values of parameters reach their greater values (the tails of the experimental distribution). However, these areas of distribution are very interesting from the standpoint of the study of clusterization effects in Poisson-type flows. In the linear building of histograms, the tail branches of the distribution cannot practically be detected. To resolve these problems we will use another presentation of experimental histograms of a semilogarithmic type, as follows.

Table 2.3. Parameters of the distribution of cyclogenesis intensity over basins of the northern hemisphere for 1983–1997 (January to July of given year).

Year	*Total number of occurrences*	*Main numerical characteristics of histograms*		*Parameter approximating Poisson law* λ (day^{-1})	*Measure of deviation of Pearson's criterion*
		Mean	*Variance*		
1983	73	0.34	0.46	0.3	28 (7.8)
1984	112	0.52	1.12	0.5	141 (11.1)
1985	132	0.61	1.0	0.6	46 (9.5)
1986	125	0.59	0.59	0.6	16 (7.8)
1987	121	0.57	0.89	0.6	63 (9.5)
1988	63	0.29	0.28	0.3	22 (9.5)
1989	142	0.67	1.19	0.7	59 (11.1)
1990	166	0.78	1.21	0.8	284 (12.6)
1991	132	0.62	0.59	0.6	19 (7.8)
1992	146	0.69	0.78	0.7	67 (9.5)
1993	87	0.41	0.62	0.4	42 (9.5)
1994	135	0.64	1.34	0.6	553 (12.6)
1995	92	0.43	0.85	0.4	140 (9.5)
1996	121	0.57	0.82	0.6	25 (9.5)
1997	220	1.03	1.67	1.0	63 (12.6)

Note: Total number of runs is 212 days. The figures in parens give the tabulated values of the function $\chi^2(\alpha, f)$ for Pearson's criterion at the 95% confidence level for a corresponding number of degrees of freedom.

Study of the comparability between experimental histograms and theoretical approximations for an active period, executed on a semilogarithmic scale (Figure 2.9e–g), shows that the Poisson nature of density in sharing a value $N(t)$ is retaines if the values of a number of events (in the day) equal 4–6. At values $N < 5$ the experimental histogram is distinguished from the Poisson distribution; moreover, both an "excess" and "deficit" of values of experimental histograms with respect to pure Poisson models are observed. Here the evident and significant excess of the experimental values of histograms about the theoretical distribution under $N > 5$ are practically the same as the universal phenomena for cyclogenesis in both the NH and SH.

Table 2.4. Parameters of the distribution of cyclogenesis intensity over basins of the northern hemisphere for 1983–1997 (August to December of given year).

Year	*Total number of occurrences*	*Main numerical characteristics of histograms*		*Parameter approximating Poisson law* λ (day^{-1})	*Measure of deviation of Pearson's criterion*
		Mean	*Variance*		
1983	279	1.80	1.68	1.8	8.4 (11.1)
1984	300	1.96	2.08	2.0	12.6 (12.6)
1985	339	2.21	2.90	2.2	45.0 (16.9)
1986	318	2.06	1.90	2.1	4.1 (12.6)
1987	304	1.99	2.31	2.0	24.0 (12.6)
1988	350	2.29	2.67	2.3	22.0 (14.1)
1989	325	2.13	2.89	2.1	28.0 (12.6)
1990	373	2.44	2.87	2.4	14.0 (14.1)
1991	290	1.88	2.03	1.9	15.0 (11.1)
1992	449	2.93	3.50	3.0	43.0 (16.9)
1993	308	2.01	2.99	2.0	75.0 (14.0)
1994	381	2.50	3.04	2.5	30.0 (12.6)
1995	339	2.21	3.69	2.2	70.0 (14.0)
1996	326	2.13	1.95	2.1	4.6 (11.0)
1997	426	2.78	3.16	2.8	30.0 (12.6)

Note: Total number of runs is 153 days. The figures in parens give the tabulated values of the function $\chi^2(\alpha, f)$ for Pearson's criterion at the 95% confidence level for a corresponding number of degrees of freedom.

So, for instance, for cyclogenesis in the NH (for interval II in 1985) for values $N = 6, 7, 9$ an excess of experimental values over theoretical ones exists, but for $N = 8$ the reverse situation is true (the number of events with $N = 8$ is zero and, accordingly, the experimental value of the density of distribution is also zero). For cyclogenesis in the SH a distinctive excess is also inherent in the experimental values of histograms over theoretical ones: so, for $N = 5$ (SH, 1983, interval I and SH, 1992, interval I) experimental values are more than theoretical ones by a factor of 100 (Figure 2.9f, g). It is precisely this circumstance that may be responsible for significant divergences in the Pierson criterion.

Table 2.5. Parameters of the distribution of cyclogenesis intensity over basins of the southern hemisphere for 1983–1997 (January to July of given year).

Year	*Total number of occurrences*	*Main numerical characteristics of histograms*		*Parameter approximating Poisson law* λ (day^{-1})	*Measure of deviation of Pearson's criterion*
		Mean	*Variance*		
1983	124	0.60	1.03	0.6	380.0 (11.1)
1984	175	0.82	1.30	0.8	107.0 (11.1)
1985	141	0.67	1.16	0.7	84.0 (9.5)
1986	126	0.60	0.75	0.6	13.8 (9.5)
1987	121	0.57	0.81	0.6	26.0 (9.5)
1988	119	0.55	0.78	0.5	207.0 (9.5)
1989	200	0.83	1.38	0.8	57.0 (11.1)
1990	156	0.74	1.22	0.7	64.0 (9.5)
1991	122	0.58	0.59	0.6	1.7 (7.8)
1992	121	0.57	1.10	0.6	281.0 (11.1)
1993	125	0.60	0.66	0.6	4.2 (7.8)
1994	156	0.73	0.22	0.7	42.0 (7.8)
1995	110	0.52	0.12	0.5	21.0 (9.5)
1996	137	0.64	0.82	0.6	25.0 (7.8)
1997	260	1.22	2.06	1.2	108.0 (12.6)

Note: Total number of runs is 212 days. The figures in parens give the tabulated values of the function $\chi^2(\alpha, f)$ for Pearson's criterion at the 95% confidence level for a corresponding number of degrees of freedom.

We can show that exactly such a situation is also true for the remaining samples (e.g., for SH, 1984, interval I; SH, 1988, interval I; NH, 1994, interval I; and so on).

In view of the significant volume of the number of events in experimental data ($n > 150$), point evaluations of the intensity of flow of Poisson distribution λ and its confidence level can be executed by means of a method based on normal approximation for the Poisson distribution (Cox and Lewis, 1966). With a confidence level probability of 95%, we can evaluate the confidence level intervals for λ. They reach ± 0.14 for interval I and ± 0.16 for interval II. Point evaluations and confidence intervals of the average value of λ for tropical cyclogenesis hemispheres for 1983–1997 are shown in Figure 2.11 (p. 46).

Table 2.6. Parameters of the distribution of cyclogenesis intensity over basins of the southern hemisphere for 1983–1997 (August to December of given year).

Year	*Total number of occurrences*	*Main numerical characteristics of histograms*		*Parameter approximating Poisson law* λ (day^{-1})	*Measure of deviation of Pearson's criterion*
		Mean	*Variance*		
1983	50	0.32	0.33	0.30	5.20 (6.0)
1984	26	0.17	0.16	0.17	0.34 (6.0)
1985	13	0.08	0.08	0.10	0.18 (3.8)
1986	15	0.10	0.13	0.10	2.50 (6.0)
1987	31	0.20	0.23	0.20	1.90 (6.0)
1988	30	0.20	0.19	0.20	1.50 (6.0)
1989	16	0.10	0.09	0.10	0.18 (3.8)
1990	14	0.09	0.08	0.10	0.02 (3.8)
1991	52	0.34	0.46	0.30	37.00 (7.8)
1992	32	0.21	0.23	0.20	25.00 (6.0)
1993	24	0.16	0.13	0.16	0.05 (3.8)
1994	22	0.14	0.12	0.14	0.03 (3.8)
1995	34	0.22	0.25	0.20	8.80 (6.0)
1996	82	0.53	0.74	0.50	1.60 (7.8)
1997	48	0.31	0.28	0.30	0.68 (6.0)

Note: Total number of runs is 153 days. The figures in parens give the tabulated values of the function $\chi^2(\alpha, f)$ for Pearson's criterion at the 95% confidence level for a corresponding number of degrees of freedom.

2.4.3 Intermittency coefficient and "true" intensity

As already noted, to show the physical reasons for significant differences between average values of the intensity of the process, it is necessary to undertake a more detailed analysis of the temporal set of the sequence of events and, primarily, analyze the functions of the accumulating number of events (see Eq. 2.3).

Figure 2.10a–c presents (on a specially increased scale) the functions of a number of accumulating events for tropical cyclogenesis of the NH and SH for 1995–1997. Analysis of the data shown in Figure 2.10, as well as functions $F(t)$ for all other explored periods (not shown in the drawing because of insufficient

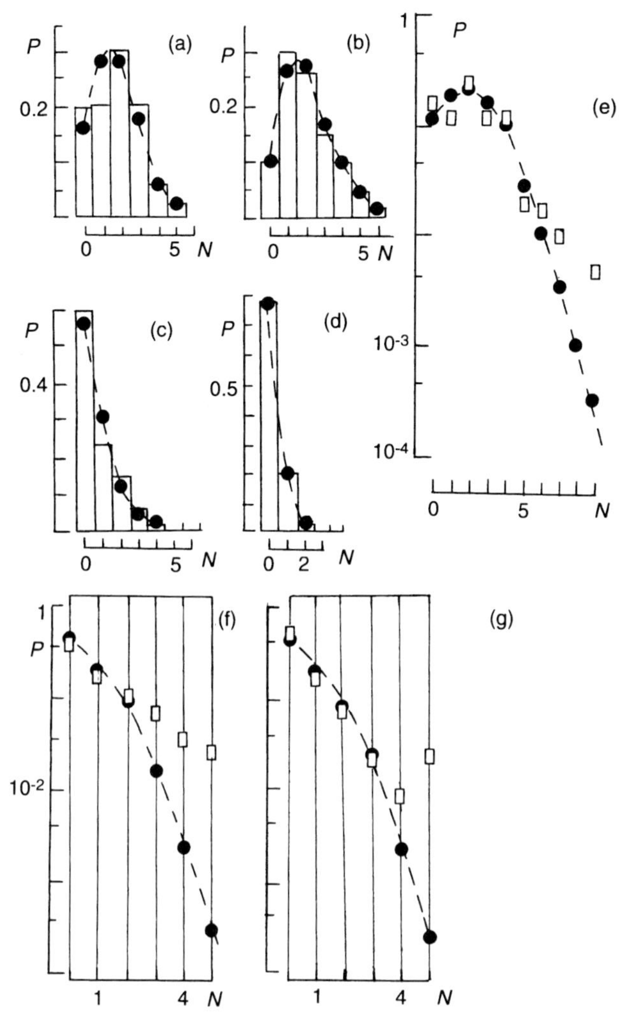

Figure 2.9. Sample probability densities of cyclogenesis in the northern hemisphere (NH) and southern hemisphere (SH) and theoretic Poisson laws on linear scales (a–d) and on semilogarithmic scales (e–g). Stepped curves and open squares represent the experimental histograms. Dotted lines and full circles represent the Poisson distributions with values given in Tables 2.3–2.6. (a) Data of NH for 1983, II interval; (b) data of NH for 1986, II interval; (c) data of SH for 1986, I interval; (d) data of SH for 1988, II interval; (e) data of NH for 1985, II interval; (f) data of SH for 1983, I interval; (g) data of SH for 1992, I interval (from Pokrovskaya and Sharkov, 1999b).

room) shows that the behavior of $F(t)$ for 1995–1997 is highly characteristic of the whole explored 15 years (1983–1997). The main particularity of the temporal set of the functions of the accumulation of events is its intermittent nature. In other words, intervals of generation and operation of cyclogenesis—where the derivative on a time

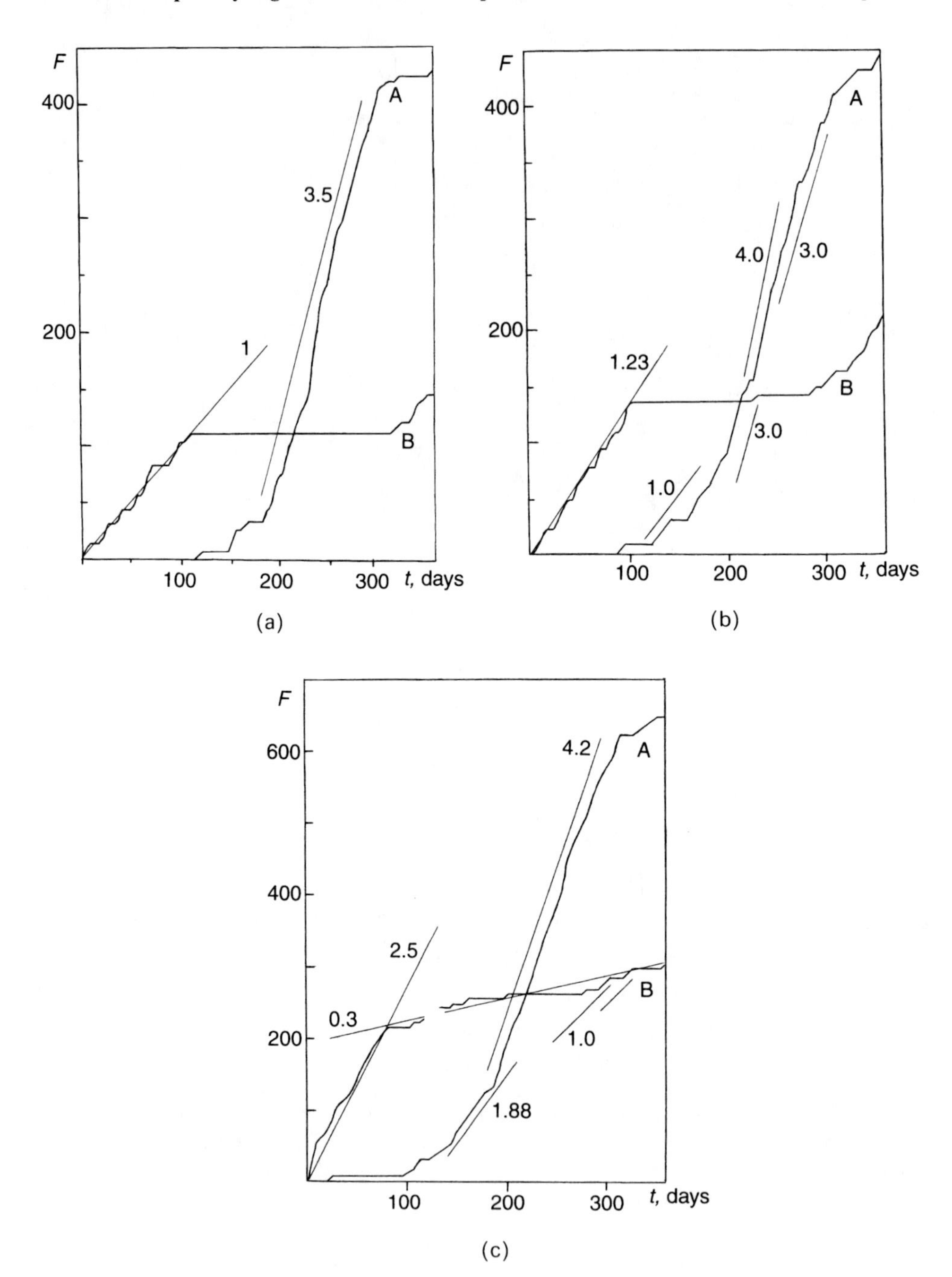

Figure 2.10. Annual time series of cumulative functions for northern (A) and southern (B) hemisphere cyclogenesis. (a) Data for 1995; (b) data for 1996; (c) data for 1997. Figures next to the straight lines represent the values of the intensity of Poisson processes (from Pokrovskaya and Sharkov, 1999b).

from $F(t)$ is different to zero—alternate in a random manner with the intervals of "silence"—where $dF(t)/dt = 0$. Moreover, these intervals can range greatly in size—from days to months (the so-called "languid" season). So, analysis of the temporal set of accumulation functions for 1986 (Figure 2.8b) and for 1995–1996 (Figure 2.10a–c) shows that the active season for the NH is from May to November of the current year, while for the SH it is from December to March, and in the following winter (for the SH) it shows the silence area from March to November (the languid season). As a whole, such a structure for cyclogenesis is sufficiently well known. However, this methodology of building accumulation functions allows us to get new information on the modes of generation and operation of cyclogenesis.

Hence, we know (Cox and Lewis, 1966) that the ratio of the increment of accumulation functions ΔF for the Poisson process to the time of observing $(\Delta F/\Delta t)$ is more or less the average value of the intensity of a Poisson flow of events (for the given time of observing Δt). Of course, the time of observing (averaging) will correspond to the scale of events over which the studies are conducted. So, the values of intensity shown in Table 2.3 correspond to periods of accumulations in interval I (212 days) and interval II (153 days). But, if the time of accumulation is chosen sufficiently small, it is possible to speak of a kind of differential mode of values having an intensity of flows of λ_0 and their temporal evolutions during $\lambda_0(t)$ up to the integral value if for Δt we take the whole term of observations. So, analysis of Figure 2.10a–b shows that cyclogenesis in the NH during its active period (May–October) is highly irregular in its rate of generation: periods with a high rate of generation (for 1995 $\lambda_d = 3.5\,\text{day}^{-1}$) alternate with relatively small rates (over 5–7 days) of sharply lowered intensity (for 1995 $\lambda_d = 1.0\,\text{day}^{-1}$). The reverse holds for cyclogenesis in 1996 (Figure 2.10b), when there was an interval (230–274 days of the current year) with a sky-high rate of generation ($\lambda_d = 4.0\,\text{day}^{-1}$) in between other intervals where activity dropped from $3\,\text{day}^{-1}$ down to $1\,\text{day}^{-1}$ inclusive of the silence season.

A qualitative change was observed for cyclogenesis in 1997 (Figure 2.10c). First, it is necessary to note the almost complete absence of winter silence areas, which are traditional for cyclogenesis in the SH. This distinctive branch is divided into small areas that include active periods with differential intensity of $\lambda_d = 1.0\,\text{day}^{-1}$. SH cyclogenesis is divided into two large periods (from January to March with integral intensity of $\lambda_d = 2.5\,\text{day}^{-1}$ and from April to December with integral intensity of $\lambda_d = 0.3\,\text{day}^{-1}$). As for NH cyclogenesis, it has its own particularities: cyclogenesis began much earlier (from the middle of January) and in its most active period (196–300 days of the year) it reached a sky-high (for the 15 years under discussion) differential rate of generation (it measured $\lambda_d = 4.2\,\text{day}^{-1}$). Such a mode of cyclogenesis generation was called the "pipeline" of cyclones by Carlowicz (1995). Such high differential rates of generation are responsible for the greater values of accumulation functions—for the NH $F(365) = 646$ and for the SH $F(365) = 308$—whereas in the "normal" mode these values are 1.5–2 times smaller (see Figures 2.8b and 2.10a, b). The physical reason for such serious global transformations in hemisphere cyclogenesis was probably the ENSO phenomena (the 1997–1998 episode), which actively developed during the whole of 1997 and at the beginning of 1998. As is well known, during active ENSO episodes ocean temperature in the Pacific Ocean and the large-scale Walker circulation significantly increase and, in turn, this is greatly reflected in global cyclogenesis.

For a quantitative description of chaotic intermittent processes there are complex mathematical methods in common usage (Kadanov, 1993; Dubois, 1998). For our purpose we will use simpler approaches: we will define an intermittency coefficient and the true intensity. The first parameter γ will be defined as the ratio between the "clean" time of generation (t_g) and the time of the full observation period (t_0) or the volume of samples taken. The second parameter will be defined as the ratio of the number of events N to the time of clean generation of the process (i.e., where $\Delta F/\Delta t$ is not zero). Hence, $\lambda_t = N/t_g\,\text{day}^{-1}$ is the value of the true intensity.

Calculated from experimental data for 1983–1997, the values of the intensity of the intermittency coefficient and the true intensity of flows of cyclogenesis of the NH and SH are presented in Table 2.7 and Figures 2.11–2.13.

Special calculations have shown that the values of the intensity of Poisson flows found earlier (Tables 2.3–2.6) are connected to the true intensity with a good degree of accuracy (better than 5%) by the equation $\lambda_t = \lambda_0/\gamma$. In other words, the physical sense of the true intensity is reached by recovering the average value of intensity of the condition of uniform generation for the whole period observed (or, in other words, without considering the contribution from the silence period).

These results, which might at first sight appear to be striking and uncommon, actually represent the usual and natural regime of the operation of tropical cyclogenesis under the assumption that this natural process can be conceived of as a radiophysical oscillator. For example, it can be a multivibrator that operates in pulse mode with equal amplitudes of pulses but with durations of pulses and a quenching oscillator (regime of intermittency) that are chaotic. In other words, for the case in point the compounded regime incorporates both a free-running and a triggered regime.

Thus, the intermittency coefficient is the ratio between the "pure" time of oscillations (a free-running regime) and the total interval of the operation. The term "true intensity" is the number of pulses in a unit time when the oscillator is operating (a free-running regime).

We will repeatedly use this demonstration and instructive analogue in other places in this book.

2.4.4 Interannual variability between northern hemisphere and southern hemisphere cyclogenesis

The general picture of this temporal (15 years) set of values of cyclogenesis intensity (averaged on the temporal terms under investigation) in both hemispheres is presented in the Figure 2.11. Close inspection of these data shows that tropical cyclogenesis in both hemispheres possesses a clear time regularity and is determined by stability (of its intensities). In the NH maximum cyclogenesis is reached in the second half of the year, while in the SH the maximum is formed in the first half of the

Table 2.7. Intermittency coefficient (γ) and the "true" intensity (λ_t) of cyclogenesis over the basins of the northern and southern hemispheres for 1983–1997.

	Northern hemisphere				*Southern hemisphere*			
	Interval I		*Interval II*		*Interval I*		*Interval II*	
Year	λ_t	γ	λ_t	γ	λ_t	γ	λ_t	γ
1983	1.43	0.24	2.27	0.80	1.38	0.42	1.28	0.250
1984	2.07	0.25	2.34	0.84	1.96	0.42	1.00	0.170
1985	1.94	0.32	2.78	0.77	1.88	0.35	1.00	0.085
1986	1.56	0.38	2.36	0.91	1.56	0.38	1.25	0.078
1987	1.59	0.36	2.43	0.82	1.61	0.35	1.15	0.170
1988	1.10	0.27	2.63	0.87	1.48	0.37	1.07	0.180
1989	1.77	0.38	2.60	0.82	2.04	0.46	1.00	0.100
1990	1.76	0.44	2.87	0.85	1.92	0.38	1.00	0.092
1991	1.55	0.40	2.42	0.78	1.37	0.42	1.33	0.250
1992	1.78	0.39	3.51	0.83	1.66	0.35	1.45	0.140
1993	1.52	0.27	2.42	0.83	1.44	0.41	1.00	0.150
1994	1.98	0.32	3.05	0.83	1.79	0.41	1.00	0.140
1995	1.88	0.23	2.92	0.75	1.51	0.34	1.26	0.170
1996	1.57	0.36	2.45	0.87	1.61	0.40	1.34	0.400
1997	1.96	0.53	3.22	0.86	2.04	0.60	1.17	0.270

year. Based on the qualitative aspect of the problem, this statement is sufficiently well known. However, the quantitative evaluation of cyclogenesis intensity that we get is highly non-trivial: the intensities of NH and SH cyclogenesis for interval I practically coincide with annual values on average for 15 years, whereas for interval II they are different by a factor of 10 (on average). Moreover, attention is drawn to the fact that there is significant stability in the values of cyclogenesis intensity, with the probable exclusion of interval II of 1992 (NH), when cyclogenesis intensity sharply increased by more than 30% (on average).

Having expected a symmetrical distribution (normal type) for the values of intensity for these 15 years, it is possible to form a preliminary statistical model of a

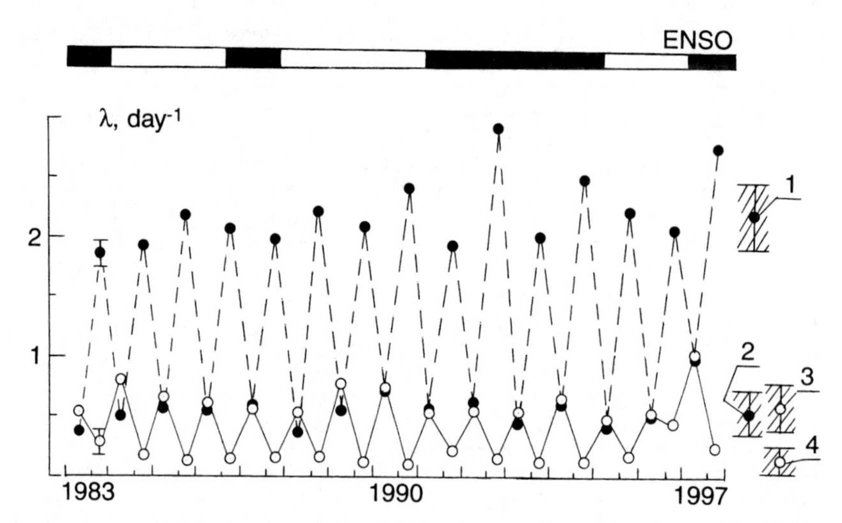

Figure 2.11. Intra-annual and interannual variability in experimental values of hemisphere cyclogenesis intensities for corresponding intervals (I and II) of the given year during 1983–1997. Full circles and dotted lines represent NH data. Open circles and solid lines represent SH data. Vertical line segments represent estimates of year-averaged intensity at the 95% confidence levels. (1) Multiyear (15 years) statistical model of NH cyclogenesis intensity (I interval); (3) model of SH cyclogenesis intensity (I interval); (4) SH, II interval. Solid and open bars indicate the active and suppressed periods of ENSO, respectively (from Pokrovskaya and Sharkov, 1999b).

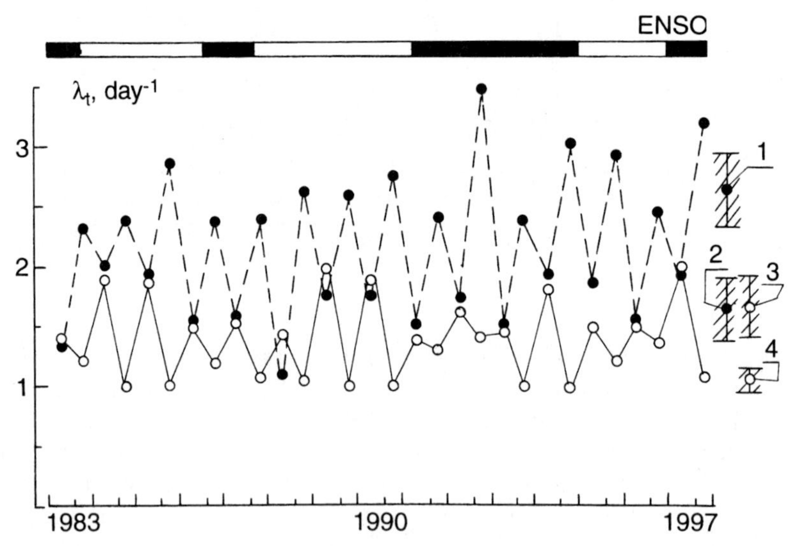

Figure 2.12. Intra-annual and interannual variability in experimental values of hemisphere cyclogenesis "true" intensities for corresponding intervals (I and II) of the given year during 1983–1997. See symbols and notations in Figure 2.11 (from Pokrovskaya and Sharkov, 1999b).

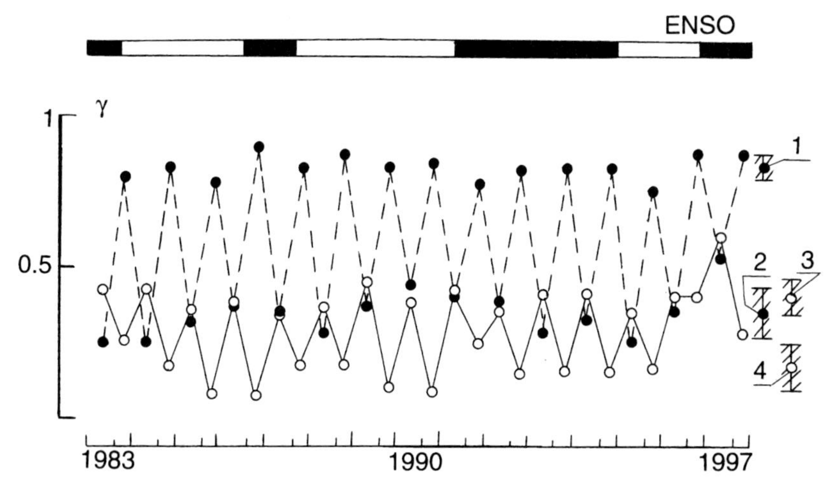

Figure 2.13. Intra-annual variability in the experimental value of the hemisphere cyclogenesis intermittency coefficient. See symbols and notations in Figure 2.11 (from Pokrovsakya and Sharkov, 1999b).

long-standing sample set for the NH and SH:

$$\text{I interval}\begin{cases} \text{NH} & \lambda_0 = 0.58 \quad \sigma(\lambda) = 0.17 \\ \text{SH} & \lambda_0 = 0.68 \quad \sigma(\lambda) = 0.17 \end{cases} \tag{2.19}$$

$$\text{II interval}\begin{cases} \text{NH} & \lambda_0 = 2.22 \quad \sigma(\lambda) = 0.32 \\ \text{SH} & \lambda_0 = 0.21 \quad \sigma(\lambda) = 0.12 \end{cases} \tag{2.20}$$

where σ is the non-biased evaluation of r.m.s. These models are displayed in graphic form in Figure 2.11.

Let us consider the important question of interannual variations in the intensity of Poisson flow in making a prototype for tropical cyclogenesis hemispheres. To do this, it is necessary to accept or reject the hypothesis that all annual Poisson models belong to the united (15-year) general collection set. Generally speaking, analysis can be done by using different approaches on the assumption that long-standing normal models of NH and SH tropical cyclogenesis exist (see Eqs. 2.19–2.20).

To check the hypothesis that the parameters of samples (taken on corresponding temporal terms) are comparable with the parameters of the general (15-year) set (Eqs. 2.19–2.20) $H_0 : \lambda_i = \lambda_0$, we will consider (following Feller, 1968) the sample function Z:

$$Z = \frac{\lambda_i - \lambda_0}{\sigma_0}\sqrt{t_i} \tag{2.21}$$

where t_i and λ_i are the total number and sample average set under study; and σ^2 and λ_0 are values of the variance and average of the general set. The formation and analysis of sample functions for the values of cyclogenesis parameters, over the considered period, for the NH and SH have shown that, under the given value

$\alpha = 0.05$, hypothesis H_0 has a 95% probability of being accepted for samples both for the NH and SH, with a few exceptions. So, samples from interval I for 1997 and 1990 (NH) and for 1984 and 1989 (SH), as well as from interval II for 1996 and 1983 (SH), cannot be referred to the general set (on the specified confidence level). This is very clear from analysis of Figure 2.11. The question of comparability of these samples with the general set requires more detailed consideration.

To resolve this problem, other statistical procedures—in particular, the double *t*-criterion (Feller, 1968) and the comparability of intensities of two Poisson processes method (Cox and Lewis, 1966)—were used. These approaches produced results similar to the ones we have just discussed.

Figures 2.12 and 2.13 show the temporal sets of values of the true intensity and the intermittency coefficient of cyclogenesis in the NH and SH for 1983–1997. As would be expected, just like the values of the true intensity, the temporal picture as a whole is similar to the one that corresponds to the average values of the intensity (Figure 2.11) with the difference that the values of the true intensity are greater than the values of the average intensity. The general regular picture of the time evolutions of the true intensity breaks down in the period 1991–1993, when a strong ENSO episode was observed. It is interesting to note that values of the true intensity in period I of the year under study for the NH and SH practically coincide, whereas in period II they are different by as much as 2.4–2.5 times. As for the time evolution of the intermittency coefficient, an amazing regularity and stability in the values of the intermittency is observed here. Any dependence on ENSO episodes were not observed.

Having expected a symmetrical distribution (normal type) for values of the true intensity and the intermittency coefficient for the 15-year period it is possible to form preliminary statistical models for samples from the NH and SH:

$$\text{I interval}\begin{cases} \text{NH} & \lambda_t = 1.68 \quad \sigma(\lambda_t) = 0.25 \\ \text{SH} & \lambda_t = 1.68 \quad \sigma(\lambda_t) = 0.24 \\ \text{NH} & \gamma = 0.34 \quad \sigma(\gamma) = 0.083 \\ \text{SH} & \gamma = 0.40 \quad \sigma(\gamma) = 0.064 \end{cases} \tag{2.22}$$

$$\text{II interval}\begin{cases} \text{NH} & \lambda_t = 2.69 \quad \sigma(\lambda t) = 0.37 \\ \text{SH} & \lambda_t = 1.15 \quad \sigma(\lambda_t) = 0.15 \\ \text{NH} & \gamma = 0.83 \quad \sigma(\gamma) = 0.042 \\ \text{SH} & \gamma = 0.17 \quad \sigma(\gamma) = 0.086 \end{cases} \tag{2.23}$$

All these models are shown in graphic form in Figures 2.12 and 2.13.

In spite of the complex structure of signal accumulation, including the full lifetime of a single event (TC), it turns out that the amplitude features of the intensity of hemisphere tropical cyclogenesis can by and large be described within the framework of simple physical Poisson-type models (Brownian motion) with very clear parameters and weak interannual variability. As is clear from the Poisson model, tropical cyclogenesis of both hemispheres possesses characteristics known from the theory of Poisson flows: changes in intensity do not depend on the number of existing structured elements, on their previous histories, or on the system's condition; moreover, the process satisfies the ordinariness characteristic (see Section 2.2.4).

Variations of experimental histograms from these purely Poisson models are highly symptomatic. This has to do with the fact that, from the standpoint of building physical models of cyclogenesis, these variations from the purely Poisson mode embody information on the structure of the object, because different-scale internal correlations in the system produce a break in Poisson characteristics. Note that these deflections reveal themselves as significant excesses by experimental tails in the histograms above the Poisson branch. In other words, these deflections manifest themselves by frequent and strong drops in cyclogenesis intensity. Such a type of clusterization of TC generation and operation (here separated in space by significant distances) was discovered earlier after experimental study of cyclogenesis in the Pacific and Atlantic Oceans (see Chapter 3).

As is the case with features of time series of cyclogenesis intensity, seasonal variability in cyclogenesis may possibly be interpreted as large-scale intermittency, while variability (with distinctive scales of the order of several day) within active seasons can be conceived of as small-scale intermittency. Such behavior of the time evolution of system parameters is characteristic of chaotic trigger systems. To demonstrate the comparability of cyclogenesis with a corresponding type of chaotic generator it is first necessary to make a detailed analysis of the particularities and the type of intermittencies.

2.5 EVOLUTION OF TROPICAL CYCLONE INITIAL FORMS AS A STOCHASTIC PROCESS

Remote-sensing study of the primary forms of TCs occupies a special place in programs that remotely monitor tropical disturbances (TDs) for a variety of reasons. First, there is a need to examine how medium-range (72 hours) forecasting transforms an individual primary TD into its developed stage (TC) and a need for detailed remote study of the structured, dynamic, and thermodynamic particularities of TDs at the moment of reaching maturity. Despite the obvious difficulty of making observations (in view of the multiscale process of dynamic interaction), the first successful attempts at complex study of the process of TC formation were carried out in the 1990s (Ritchie and Holland, 1997; Simpson *et al.*, 1988). Second, it is important to present another aspect of the problem regarding the structured particularities and correlation of temporal flows of the primary forms of TDs and the developed forms of TC that arise from them. Consideration of these questions is an important aspect of the study of the ocean–atmosphere system when studying the contribution that intensive vortical disturbances make to the thermodynamics and kinematics of the tropical atmosphere on different timescales. This problem is closely allied to the study of climate change and the study of the influence of large-scale atmospheric circulation and ENSO phenomena on cyclogenesis (Henderson-Sellers *et al.*, 1998; Lighthill *et al.*, 1994).

However, attempts at remote-sensing investigation of primary forms of tropical disturbances have run into a variety of difficulties. In spite of significant efforts by researchers to observe and register separate (individual) tropical vortices (e.g., see Sharkov, 1998), final remote-sensing criteria about just how close an individual tropical disturbance is to its time of transition to the mature form are lacking. Difficulties emerge when we consider TDs as stochastic flows of events. They are associated with the absence of remote, meteorological, and chronological databases, created by the study of the primary forms (reasons for the cause of the last problem are discussed in Chapter 5). These databases must be formed, just as space–time stochastic crisis-situation models must be formed. Such models, in turn, can form the basis of resolving corresponding problems of ballistics provision for the remote experiments (Avanesov *et al.*, 1992; Anfimov *et al.*, 1995), and to resolve problems in predicting crisis situations.

The first goal-directed studies of mature forms of cyclogenesis (considered on the global scale as a stochastic flow of events) made possible the formation of preliminary statistical–quantitative models (see Sections 2.2 and 2.3).

As for cyclogenesis with respect to primary forms of TDs, semiquantitative considerations on the rates of generation of developing forms (TCs) from primary stages in different basins of the World Ocean have long been proposed (Khromov, 1948, 1966; Minina, 1970). However, it should be noted that no serious experimentally motivated proof of this has been cited. No absolute quantitative data on the intensity of the primary forms of cyclogenesis have been obtained.

On the basis of processing a time series of the intensity of the tropical cyclogenesis of both primary and mature forms, determined from the basins of both hemispheres for 1997, the purpose of this section is to present experimental results, pointing to features in the structure of the signal that reflect the temporal operation of hemisphere cyclogenesis, as well as to intra-annual particularities of the contribution to primary and mature forms in both hemispheres of tropical cyclogenesis, which in our view is important for understanding temporal variability in the correlation of primary and mature forms.

2.5.1 Simulation of an information signal

In this work we develop a methodical approach, based on presenting the temporal flow of the intensity of primary and mature forms of tropical cyclogenesis as a random flow of uniform events (see Section 2.1). The uninteresting detailed structure and track record of each individual tropical formation (in equal degrees either in its primary or mature form) will be placed on the time axis of each individual TD as a pulse of single amplitude with random duration (corresponding to the time of operation of TDs) and with random moments of TD appearances (generation of individual TDs). The number of received pulses (events) in a single temporal interval (here 1 day) is in such an event a natural physical parameter—an instant intensity in cyclogenesis, defining the energy of the interaction in the ocean–

atmosphere system. In the language of statistical procedures, this approach is identified as a means of calculating the number of events with due regard to the lifetime of an event (Apanasovich *et al.*, 1988).

The mathematically offered procedure of forming a signal may possibly be written in the same way as for mature forms (see Section 2.1).

Raw data for 1997 on the global cyclogenesis of developed forms (TCs) were not obtained from the systematized database Global-TC of remote observations for global tropical cyclogenesis (Pokrovskaya and Sharkov, 1999a). The database of primary forms of TDs in the World Ocean Basin for 1997 was put together on the computer platform Global-TC with analysis and systematization of source information on conditions in the tropical area, regularly (daily) updated during the whole of 1997 on the Internet (see Chapter 5).

2.5.2 Time series of the intensity of initial and mature forms

Figure 2.14 presents graphs of time series of the flows of intensity $N(t)$ of tropical cyclogenesis in the NH and SH (daily averaging) for 1997 for primary forms of disturbances (Figure 2.14a) and for developed forms of TCs (Figure 2.14b). Note that on the graph of the intensity of primary forms (Figure 2.14a) the primary forms of mature forms of cyclones are not entered. They are, however, included as a natural period of evolutions when building up the time series of the intensity of mature forms. Analysis of Figure 2.14 shows that annual samples of the flows of intensity are presented as typical telegraphic processes both for primary and mature forms of cyclone. The detailed structure of the intensity of flows of primary forms is presented on an increased (daily) scale on the inset to the corresponding graph (Figure 2.14a). The main difference in the flow of intensity of primary forms from that of mature forms of cyclones is exemplified by the sharply fluctuating type of temporal set of signal. This is connected with the short (compared with mature forms of cyclones) lifetime of primary forms (1–3 days) and, accordingly, the reduced time of correlations of the process.

In the annual cycle of tropical cyclogenesis in both hemispheres some general lines are tracked. From August to November (day 170–day 320 of the year) in 1997 in NH basins, there is of course a period of increased activity both for primary forms and mature forms, but for SH basins there is correspondingly lowered activity. For the first half of the year (day 1–day 130) and end of the year (day 350–day 365), the inverse situation is observed: for NH basins cyclogenesis fades from its maximum in the second half of the year, but for SH basins there is a season of increased activity. In particular, graphically specified particularities reveal this when considering the time evolutions of the accumulation functions of a number of events (Figure 2.15a, b), where the derivative points to the differential intensity of the process. In essence, the question arises about how to collect two Poisson-type processes with greatly different intensities. For this reason we will select the annual temporal movement of these processes over two intervals (interval I is January–May and December and interval II is June–November) and use them to conduct some statistical procedures.

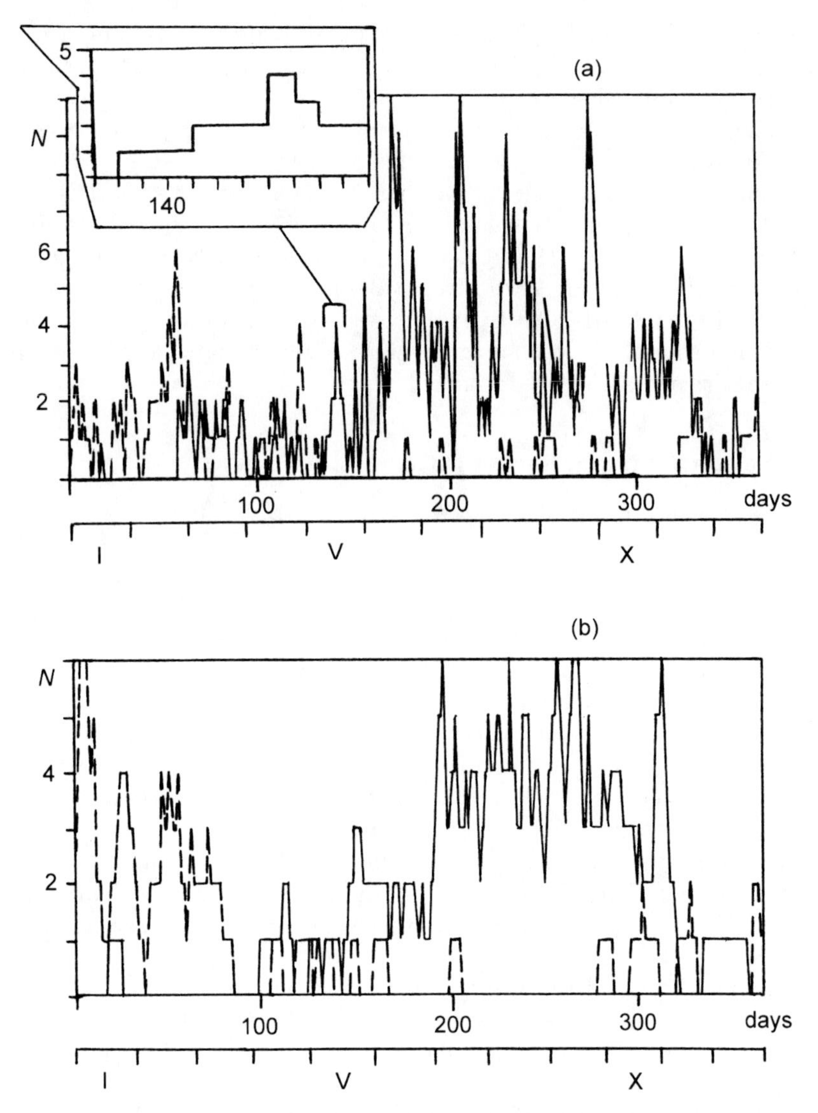

Figure 2.14. Annual time series of cyclogenesis intensity for initial forms (a) and for mature forms (b) for 1997. The solid line represents data from the northern hemisphere. The dotted line represents data from the southern hemisphere (from Pokrovskaya and Sharkov, 1999c).

Another important particularity of tropical cyclogenesis is the intermittent nature of generation that particularly graphically reveals itself when analyzing the temporal set of accumulated functions of a number of events. Pokrovskaya and Sharkov (1996a) first discovered such an effect when analyzing the time dynamics of tropical cyclogenesis in the Pacific for 1990–1991. To make the quantitative

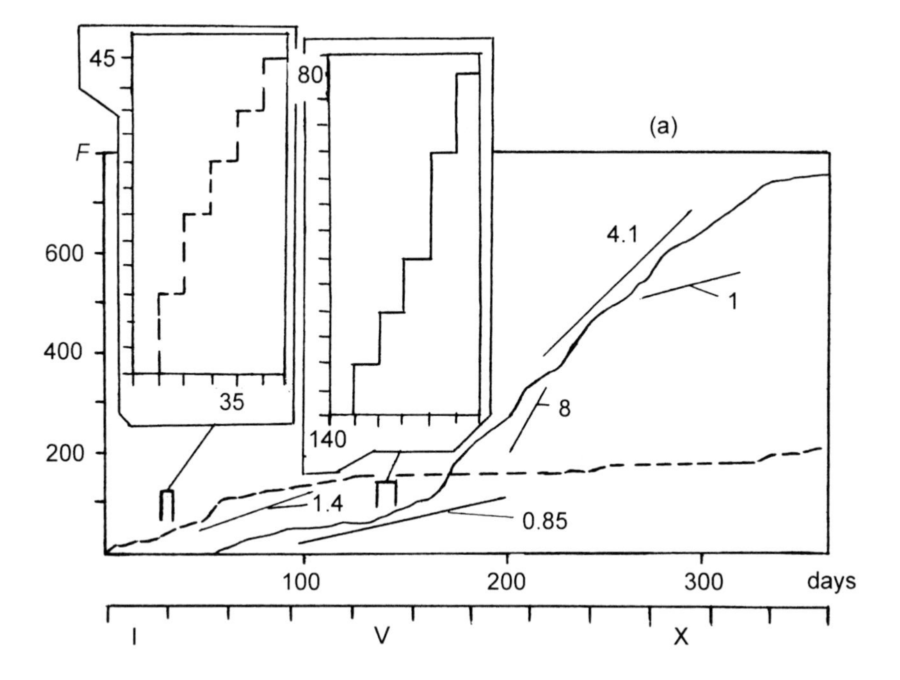

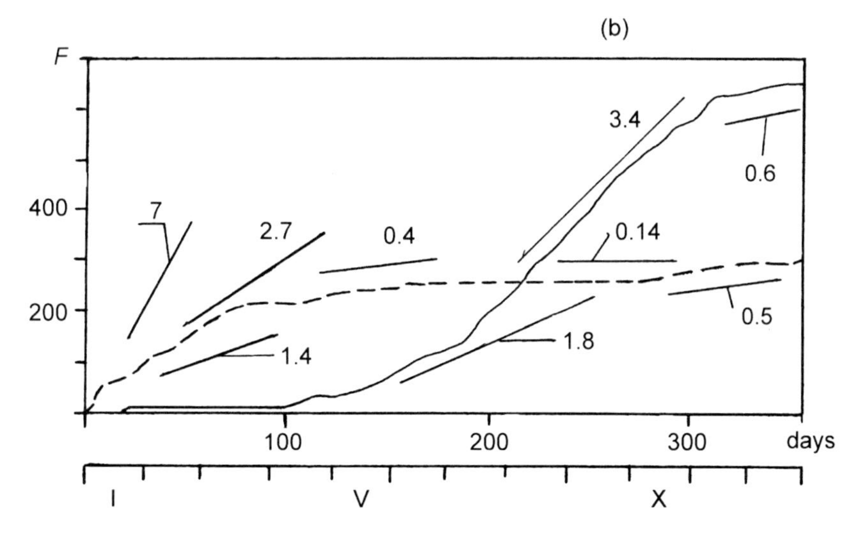

Figure 2.15. Annual time series of cumulative functions for initial forms (a) and for mature forms (b) for 1997. Solid and dotted lines represent data from the northern and southern hemispheres. The figures next to the straight lines represent values of the intensity of Poisson processes (from Pokrovskaya and Sharkov, 1999c).

features of this process known they proposed consideration of both a differential mode of cyclogenesis generation and an integral (for the considered time lag). We will use these approaches below.

Detailed analysis (at the pixel level) of the structures of accumulation functions (refer to insets on Figure 2.15) shows that the microstructure of accumulation functions possesses a non-uniformly step-like character that, strictly speaking, should be expected from microstructure intensities of cyclogenesis having both primary and mature forms of cyclones. Such a nature of temporary processes is well known in chaotic dynamics and is called the "devil's staircase" process (see Section 2.7.1).

2.5.3 Probability models of the intensity of initial and mature forms

To demonstrate a concrete type of probabilistic model (amplitude features) for cyclogenesis of developed and primary forms for the NH and SH over interval I (January to July and December of the current year, 1997) and interval II (August to November of the current year) experimental histograms of the integer number of events $N(t)$—which approximate their statistical distribution by analyzing the degree to which they agree—were built. Moreover, sample averages and variances for each group of arrays were calculated. The numerical results of this are presented in Table 2.8.

The collation of sample histograms using a Poisson-type approximating law and semilogarithmic scales (Figure 2.16) shows the possibility of using the hypothesis on

Table 2.8. Parameters of the distribution of the intensity of initial and mature forms of cyclogenesis over basins of the northern and southern hemispheres.

		Interval	*Total number of occurrences*	*Main characteristics of histograms*		*Parameter of approximating Poisson law* λ (day^{-1})	*Measure of deviation of Pearson's criterion*
				Mean	*Variance*		
TCs	NH	I	86	0.47 (0.10)	0.42	0.5	8.03 (6.0)
		II	560	3.11 (0.25)	2.41	3.1	12.69 (11.1)
	SH	I	258	1.39 (0.17)	2.18	1.4	23.30 (11.1)
		II	50	0.27 (0.07)	0.21	0.3	3.10 (3.8)
TDs	NH	I	96	0.53 (0.10)	0.63	0.5	5.50 (7.8)
		II	650	3.58 (0.28)	4.83	3.6	11.80 (16.9)
	SH	I	171	0.93 (0.14)	1.23	0.9	4.00 (9.5)
		II	38	0.20 (0.06)	0.20	0.2	0.21 (3.8)

Note: Total number of runs is 182 days for interval I (January to May and December) and 183 days for interval II (June to November). Figures in parens give the tabulated values of the function $\chi^2(\alpha, f)$ for Pearson's criterion at the 95% confidence level for a corresponding number of degrees of freedom. In the "Mean" column the figures in parens give absolute values of confidence limits at the 95% confidence level.

the Poisson nature of the size of the fluctuation intensity $N(t)$ of tropical cyclogenesis both for primary and developed forms. So, in accordance with the Pierson criterion of agreement, analysis of the measure of divergence between the theoretical distribution and experimental histograms of data on cyclogenesis for the NH and SH for 1997, proves (Table 2.8) that the experimental samples are by and large compatible with the general set of Poisson distributions with the parameter values shown in Table 2.8. This can be graphically demonstrated by collating the theoretical distribution and experimental histograms (Figure 2.16). It is important to note here that the statistical amplitude of the intensity of signal for tropical cyclogenesis for (strictly speaking) totally different physical systems (TDs and mature cyclones) practically coincide: in Figure 2.16 the left column of histograms presents the signal formed during the evolution of developing forms (TCs), and the remaining histograms present data on primary forms. It is easy to see that the statistical structure of signals and (interestingly) the numerical values when evaluating their parameters are very similar (see Table 2.8).

Here, detailed analysis of the histograms has shown that, in a number of primary forms, divergence between experimental histograms and theoretical distribution is observed; moreover, it will be necessary to note that such divergence (either for signals or developing mature forms) is defined at the greater values of the parameters (i.e., tails) of experimental distribution (Figure 2.16e, j).

Comparison between experimental histograms (primary forms) and theoretical approximations for intervals I and II of observation, executed on a semilogarithmic scale (Figure 2.16e, j), shows that the Poisson nature of the density of distributions of value $N(t)$ is retained if the value of the number of events in a day equal 5. At values $N > 5$ the experimental histogram is distinguished from the Poisson distribution; moreover, the significant excess of values of experimental histograms with respect to clean Poisson models for cyclogenesis is observed in both the NH and SH.

For instance, primary forms of cyclogenesis in the NH (for interval II of 1997) for values $N = 8$–9 have an excess of experimental values over theoretical ones of 7–8 times. Cyclogenesis in the SH also has a clear excess of experimental values of histograms over theoretical ones: so for $N = 6$ (SH, 1997, interval I) experimental values of the order of 1.5 above theoretical ones.

As a result of that, the effect of amplitude clusterization is inherent in both developed forms of TDs (this is the reason we repeatedly use experimental data, see Sections 2.2 and 2.3) and primary forms.

In view of the significant size of the sample number of events in experimental data ($n > 180$, see Table 2.8), point evaluation of the intensity of flow of the Poisson distribution λ and its confidence levels can be executed by means of a method based on the normal distribution drawing near a Poisson distribution (Cox and Lewis, 1966). With a confidence level probability of 95%, the evaluations of confidence intervals for λ, executed on the specified strategy, are presented in Table 2.8. It is interesting to note that in absolute values of the intensity of the time process both mature and primary forms practically coincide (with provision for confidence intervals).

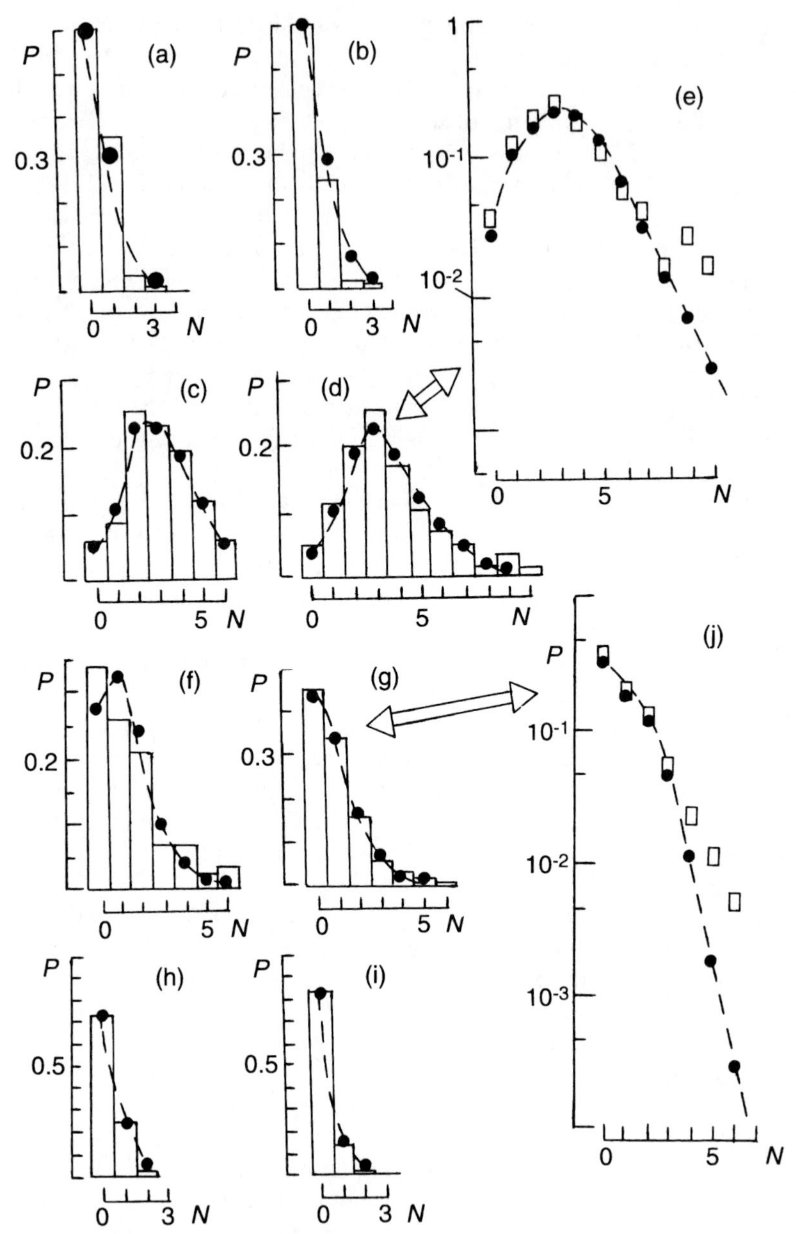

Figure 2.16. Sample probability densities of the cyclogenesis of initial and mature forms and the theoretical Poisson law on linear (a–d, f–i) and on semilogarithmic (e, j) scales. Stepped curves and open squares represent the experimental histograms. Full circles and dotted lines represent the Poisson distributions with the values given in Table 2.8. (a–b) Data of NH, I interval; (c–e) data of NH, II interval; (f–g, j) data of SH, I interval; (h–i) data of SH, II interval; (a), (c), (f), (h) represent the data of mature forms; (b), (d), (e), (g), (i), (j) represent the data of initial forms (from Pokrovskaya and Sharkov, 1999c).

2.5.4 Integral intensity of processes

As already noted, to demonstrate the physical reasons for the significant difference in average values of intensities of process, it is necessary to subject the time series of the flow of events to more detailed analysis and at the same time to analyze the functions of the accumulating number of events.

The ratio between increases in the accumulation function ΔF for the Poisson process and observation time $(\Delta F/\Delta t)$ is quite simply the average value of the intensity of the Poisson flow of events (for the given time of observation Δt). Of course, the time of observation (averaging) will correspond to scaling of these events for which studies have been conducted. But if the time of accumulation is chosen sufficiently small, it is possible to speak of a kind of differential mode of values for the intensity of flows λ_0 and their evolutions during $\lambda_0(t)$, up to integral value, if we take the whole term of observations for Δt.

Figure 2.15 presents the functions of the number of accumulating events of tropical cyclogenesis of primary and mature forms in the NH and SH for 1997. The main particularity of the temporal set of accumulation functions of events is the intermittent and non-uniform nature on scales of the order of several days and more. In other words, the intervals of generation and operation of cyclogenesis—where the derivative of $F(t)$ to time is different from zero—alternate with the intervals of silence—when $dF(t)/dt = 0$. Moreover, these intervals can be highly significant—from days to months (the so-called "faded" season). Within the active season, we can observe the highly non-uniform structure of the signal, which is characterized by the scattering of values of differential intensity across rather broad intervals. As a whole, the large-scale behavior of accumulation functions both for primary and mature forms are very similar, though differences do exist in the details. So, analysis of the temporal set of accumulation functions for 1997 shows that the active season for primary forms (NH) began at the beginning of March and took place in two stages: the first stage includes March to the middle of June at an intensity of $0.85\,\text{day}^{-1}$; the second stage begins in the middle of June and ends in November (day 339 of the current year). The differential intensity of the process in this period varies in a wide range from $1\,\text{day}^{-1}$ to $8\,\text{day}^{-1}$ with an average value of $4.1\,\text{day}^{-1}$ (Figure 2.15a). The SH active stage occupies the period from January to April (day 1–day 120 of the year) with significant differential intensity fluctuations from $0.5\,\text{day}^{-1}$ to $6\,\text{day}^{-1}$ and later in the year (for the SH) by silence areas from June to November (faded season) with the values of differential intensities within $0.1\,\text{day}^{-1}$ and $0.3\,\text{day}^{-1}$ (Figure 2.15a).

The cyclogenesis of mature forms in the NH began in 1997 relatively early (at the end of January), then between April and June generation faded (the intensity over this term was $1.8\,\text{day}^{-1}$), and then in July to November (day 196–day 326 of the year) experienced a high rate of generation (for this term the intensity was $3.4\,\text{day}^{-1}$). In December 1997 generation dropped dramatically to the observed level of $0.6\,\text{day}^{-1}$. As for the cyclogenesis of developed forms for 1997 (SH), it is necessary to note the almost complete absence of winter silence areas, which are traditional for SH cyclogenesis. This winter silence area is divided into small areas that have active periods with differential intensity of $\lambda_d = 1.0\,\text{day}^{-1}$. The cyclogenesis of developed forms in the SH is divided into two large areas: the first from January to March with integral intensity $\lambda = 2.7\,\text{day}^{-1}$ and a broad range of variation in differential intensity from $1.4\,\text{day}^{-1}$ to $7\,\text{day}^{-1}$ and the second from April to the beginning of June with integral intensity $\lambda = 0.4\,\text{day}^{-1}$ and from June to October with intensity $0.14\,\text{day}^{-1}$. In November and December cyclogenesis became more active with intensity reaching $0.5\,\text{day}^{-1}$ (Figure 2.15b). Such an annual structure of variability in cyclogenesis of qualitative mature forms is by and large well known. However, for primary forms such experimental information is received first. Likewise, the large-scale structures of the annual cycle of cyclogenesis of developed and primary forms turn out to be similar (at least from 1997 data).

Analysis of the proposed structure of the accumulation signal, including the full lifetime of a single event (TC and primary form), has shown that the amplitude features of the intensity of tropical cyclogenesis (primary and developed forms) for both hemispheres can be described by simple physical Poisson-type models (Brownian motion) with very firm parameters. As a direct effect of the Poisson distribution drawing near, tropical cyclogenesis of both forms possesses characteristics known from the theory of Poisson flows: changing intensity does not depend on the number of existing structured elements, nor on their previous histories, nor on the system's condition; moreover, the process satisfies the ordinariness characteristic.

These deviations of experimental histograms from purely Poisson models are highly symptomatic. This is connected with the fact that—from the standpoint of building physical models of cyclogenesis—it is exactly these deviations from purely Poisson mode that carry the main load, since internal correlation relationships of different scale in the system cause a break in Poisson characteristics. Note that these deviations reveal themselves basically as significant excesses (i.e., experimental tails) in histograms on the Poisson branch. In other words, these deviations reveal themselves as fairly frequent and strong short outbreaks of cyclogenesis intensity for developed and primary forms. For developed forms (TCs) such a kind of clusterization (space and time) is a well-known experimental fact ("pipeline" of cyclones; Carlowicz, 1995), but for primary forms such an experimental fact is observed for the first time.

As for the particularities of temporal sets of cyclogenesis intensity, seasonal variability in cyclogenesis can certainly be interpreted as large-scale intermittency (two temporal intervals), while intermittency (with distinctive scales of the order of several days) within active seasons can possibly be interpreted as small scale. Such behavior in time evolution by the parameters of this system is highly characteristic of the chaotic trigger-type systems. To make the parameters of cyclogenesis correspond to a type of chaotic generator it is necessary to make a detailed analysis, primarily, of particularities and the type of intermittency.

It is also important to present a consideration of the physical reasons underlying the sharply different (1.5 times) efficiency in generating developing forms for SH and NH basins.

On the basis of the proposed and developed strategy of shaping and accumulating statistical signals on remote data of the tropical cyclogenesis of primary and developed forms, considered for each hemisphere separately, it has been experimentally shown that the amplitude features of cyclogenesis intensity can be satisfactorily described by Poisson processes of alternating type. The cyclogenesis of primary and developed forms posesses a sharp intra-annual variability (revealed as two terms) and can be described by two Poisson models with firm parameters. We have created numerical models for the intensities of primary and developed forms for 1997. As a result, we have discovered the experimental effect of the amplitude clusterization of the flow intensity of primary forms.

2.5.5 Rate of hurricane formation

For a number of problems relating to climate interaction, it is necessary to obtain quantitative values of the cyclogenesis intensities of primary and developed forms by time averaging of the order of months (or even years) $N_0(t_i; \Delta t)$ or, in other words, by carrying out the following procedure:

$$N_0(t_i; \Delta t) = \frac{1}{\Delta t} \int_{t_i}^{t_i+\Delta t} N(t)\, dt \tag{2.24}$$

where t_i is the beginning of current month; and Δt is the duration of the corresponding month.

As already indicated (Section 2.1), the correct averaging procedure, as shown in Eq. (2.24), eliminates the fine small-scale features of the temporal evolution of the tropical cyclogenesis dynamical system from consideration and yet can reveal the presence of large-scale features. This can be illustrated by the ratio between developing and mature stages of TCs (Pokrovskaya and Sharkov, 2000).

Numerical expression of the rate of hurricane formation (RHF) of developed forms will be defined as follows:

$$G(t_i) = N_{\mathrm{TC}}(t_i)/N_{\Sigma}(t_i) \tag{2.25}$$

where N_{TC} is the monthly average number of developing forms; and $N_{\Sigma}(t_i)$ is the average monthly number of TDs at any stage of life.

The database of primary and developed forms of TDs in the World Ocean Basin for 1997–1998 was created on the Global-TC computing platform with the help of pre-processing, systematization, and archiving of raw information about conditions in the tropical area, regularly (daily) obtained during all of 1997 and 1998 from the Internet (Pokrovskaya and Sharkov, 1999a, d).

The main results of monthly averaging of the instant intensity of flows of both mature and primary forms observed for 1997 and considered for the basins of both hemispheres are shown in Figure 2.17. Figure 2.17a, b also shows the time series of monthly cyclogenesis intensity in the NH and SH for both mature and primary forms. An allowance is made in both tropical systems for primary forms that transform into mature forms and for systems that remain unchanged. It is

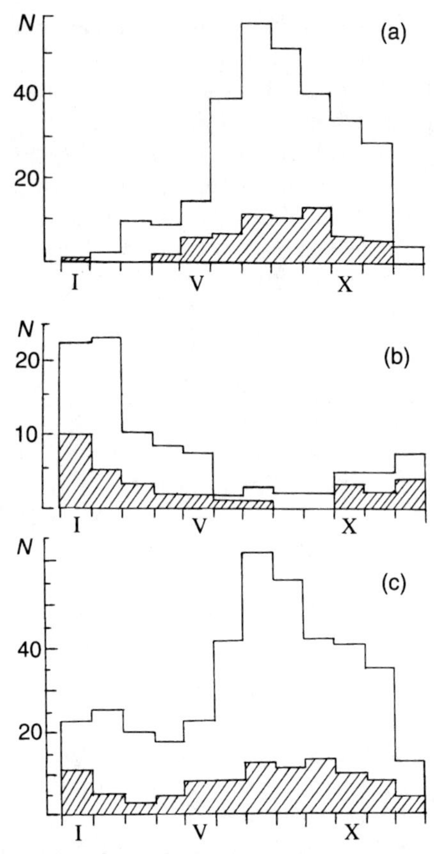

Figure 2.17. Annual time series of month-averaged intensities for mature and initial forms of tropical disturbances for 1997. Open stepped curves represent the data of initial forms. Stepped cross-hatched curves represent the data of mature forms. (a) Data of NH cyclogenesis; (b) data of SH cyclogenesis; (c) data of global tropical cyclogenesis (from Pokrovskaya and Sharkov, 1999c).

important to note that the time series of the intensity of developing forms almost exactly corresponds to the time series of primary forms, though there are determined particularities. There is also a determined number of primary forms (5–10 disturbances) in time intervals when TCs did not generally form: worthy of note are February, March, and December 1997 (for the NH) and August–September 1997 (for the SH). Moreover, we note that primary forms in both the NH and SH are generated every month, despite being statistically very lumpy. So, for the NH the maximum number of primary forms are observed between June and September and for the SH they are observed in January–February and October–December (Figure 17a, b). These are the times that maximum cyclogenesis of mature forms occur.

The picture throughout the World Ocean of the correlation between primary forms (where primary forms are born and subsequently when they become mature

Table 2.9. Monthly average rates of initial and mature form generation for 1997–1998.

	Years	*Basins of NH*	*Basins of SH*	*World Ocean basins*
Total number of TDs	1997 1998	24.6 21.4	8.0 11.5	32.7 32.9
Initial forms of TDs	1997 1998	19.7 17.1	5.2 8.5	25.0 25.6
TCs	1997 1998	4.9 4.3	2.7 3.0	7.7 7.3

forms) and mature forms is shown in Figure 2.17 and Table 2.9. As would be expected, intra-annual variability in the cyclogenesis of mature forms has greatly smoothed out, whereas for primary forms it has remained observable. The total number of TCs for 1997 was 92 and, hence, the average number of cyclones per month was 7.7, the average number of primary forms that did not transform into mature forms was 25 month^{-1} (and so exceeds the rate of generation of mature forms by more than three times), and the average monthly number of all registered forms of tropical outbreaks was 32.7 month^{-1} (total number of TDs for 1997 was 392). Collation of these parameters into data for 1998 (see Table 2.9) shows a striking interannual (1997–1998) stability in monthly average rates of the generation of TDs: variations in the values of parameters, describing the rates of generation of disturbances at different stages, were less than 13%. However, in this case the asymmetry in the values of monthly average rates of TC generation in the basins of both the NH and SH (by 1.4–1.5 times) is noteworthy.

The total number of disturbances, however, do not yet fully represent the equilibrium of the global ocean–atmosphere system. In our opinion a more appropriate parameter of the stability study can serve as the correlation between mature and initial forms. Analysis of the data shown in Figure 2.17 shows that seasonal variability determined in the general number of disturbances that appear in the World Ocean Basin per month is greatly flattened when the rate of births (Eq. 2.25) of mature forms from primary ones is considered. The results of calculating the generation factor are shown in Figure 2.18 (for 1997 data). Analysis of the results leads to a number of interesting conclusions. The annual birth rate of mature forms in both the NH and SH has denominated periods: active and break (faded) periods (Figures 2.18a, b). In this case it is important to note that during active periods the generation factor has a stable value. In the NH its value is 0.2 and in the SH it is 0.3 with a very small monthly variation.

The total pattern throughout the World Ocean is shown in Figure 2.18c. From analysis of the data it follows that the annual value of the generation factor—which is considered the process of generation of developing forms in all active basins of the World Ocean (in other words, cyclogenesis is presented as a united global process)—is amazingly stable. The average value for 1997 is 0.23 with very small monthly

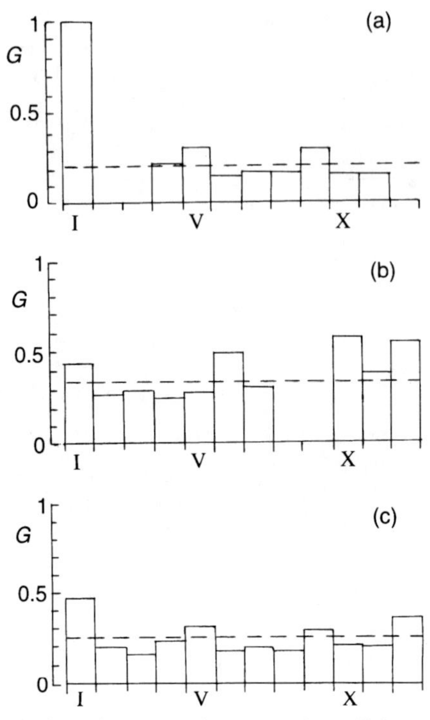

Figure 2.18. Annual time series of the month-averaged coefficient of hurricane formation for 1997. (a) Data of NH cyclogenesis; (b) data of SH cyclogenesis; (c) data of global tropical cyclogenesis. Dotted lines represent values of the year-averaged coefficient of hurricane formation (from Pokrovskaya and Sharkov, 1999c).

variations and for 1998 it is 0.22. If cyclogenesis is considered separately for each hemisphere, the average (for 1997) value of the generation factor for cyclogenesis in the NH is 0.20, but in the SH it is 0.33. A very close situation is observed for 1998: in the NH the generation factor is 0.20 and in the SH it is 0.26. The significant asymmetry in the year-averaged values of the generation factor in basins of the NH and SH has caught our attention.

These models and numerical features of the intensity of cyclogenesis (not only regional but also hemispherical) can be used in the study of global interactions in the ocean–atmosphere system; they can also be used for the operative planning of space experiments alongside existing systems for monitoring crisis situations.

2.5.6 Regional features of Pacific cyclogenesis

Figure 2.19 presents graphs of the time evolutions of the flows of intensity $N(t)$ of tropical cyclogenesis (monthly averaging) for 1997 for primary forms of disturbances and for mature forms (TCs) for three active areas of the Pacific Ocean: northwest Pacific (NWP), northeast Pacific (NEP), and southwest Pacific (SWP).

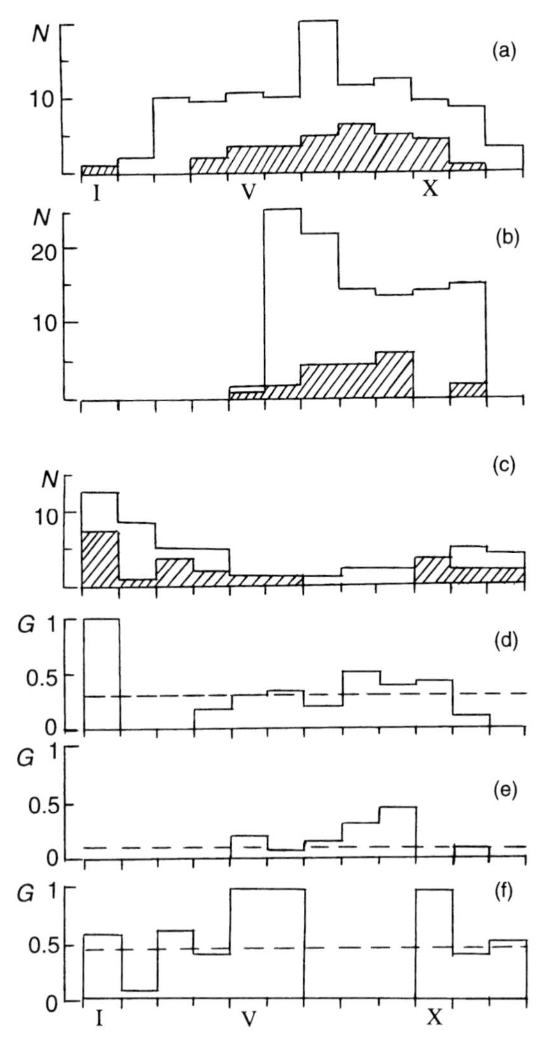

Figure 2.19. Annual time series of month-averaged intensities for mature and initial forms (a–c) and of the coefficient of hurricane formation (d–f) for cyclogenesis in the Pacific for 1997. (a, d) Data of northwest Pacific; (b, e) data of northeast Pacific; (c, f) data of southwest Pacific. Open stepped curves represent the data of initial forms. Stepped cross-hatched curves represent the data of mature forms. Dotted lines represent values of the year-averaged coefficient of hurricane formation.

Analysis of Figure 2.19a–c shows that annual samples of intensity flows present themselves as typically telegraphic processes (under corresponding time averaging) both for primary and mature forms of cyclones. The distinctive particularity of the temporal flow of the intensity of primary and mature forms is concluded by and large to have similar annual cycles of both forms of cyclogenesis. However, it is easy to verify that the detailed structure of temporal flows has a variety of particularities.

So, there are both general lines and particularities found in the annual cycle of tropical cyclogenesis in the active area of the NWP. The active season of mature forms in the SWP was in April and continued until November 1997 (Figure 2.19a); moreover, it reached its maximum in August. By January 1997 the cyclogenesis of mature forms resulted in TC "Hannah" (9701) in the NWP. The average monthly rate of mature form generation in the active period was 3.2. At the same time, the cyclogenesis of primary forms in 1997 more or less existed throughout the whole calendar year and were intensive at 10 events (tropical depressions per month) reaching a maximum in July (20 events). The total number of TDs of both types in the SWP for 1997 was 108 and, as a result, the average monthly rate of generation of disturbances was 9.

In basins of the NEP the situation was very different: the cyclogenesis of primary and mature forms functioned synchronously and over a limited period (May–November). In this timegap the cyclogenesis of primary forms was very intensive with 15 to 25 events per month (Figure 2.19b) and the cyclogenesis of mature forms dropped to an average 3 month^{-1}. The total number of events in the NEP was 105, which means the average monthly rate of generation of disturbances was 8.75 (a numerical value close to the rate of generation in the NWP).

As is well known, in SH basins of the Pacific Ocean one season of increased activity occurs in summer and maximum activity is in January. These particularities are clearly shown in Figure 2.19c. However, it should be noted that the cyclogenesis of mature forms for 1997 was sufficiently extended throughout the year and lasted for large parts of the year (with the exclusion of May to September) with a sufficiently faded rate of generation of 2.4 month^{-1}. The cyclogenesis of primary forms in the SWP occurred throughout the year (without a break) with the evident maximum in January (12 events). The total number of outbreaks in the SWP in 1997 was 49 corresponding to an average rate of generation of 4.08 month^{-1}. Collation of the rates of generation in NH basins of the Pacific Ocean reveals that activity in southern parts of the Pacific with respect to the generation of TDs is twice as small.

Let us analyze annual variation in the generation factor of mature forms (Figure 2.19d–f). For basins in the NWP the nature of change is rendered sufficiently firm by the average value 0.3 (within an active period) and by small variation. The exception was January ($G = 1$) when one TC occurred. The average annual generation factor of mature forms for the NWP reached 0.28. For basins in the NEP, temporal evolution of the rate of generation of mature forms was different and increased from May to September. The average value for an active period was 0.22; the average annual value was considerably less at 0.11. Unlike basins in the North Pacific Ocean, the time series of the generation factor for basins in the SWP exhibits a totally different nature. In spite of the faded cyclogenesis of mature forms, practically every second depression transforms into the mature form and, hence, a very intensive generation process is realized: in active periods the average value of the generation factor is 0.62 which is 2.5–3 times greater than for NH basins. The average annual value of the generation factor for basins is 0.46, which is also significantly greater than similar features for northern basins of the Pacific Ocean.

2.5.7 Regional features of Indian Ocean cyclonic activity

The time series of the cyclogenesis intensity of primary and mature forms for 1997 that formed in the basins of the Indian Ocean are shown in Figure 2.20a, b. The active season of mature forms in basins in the southern part of the Indian Ocean (SIO) is from December to February which Figure 2.20a illustrates very well. For the

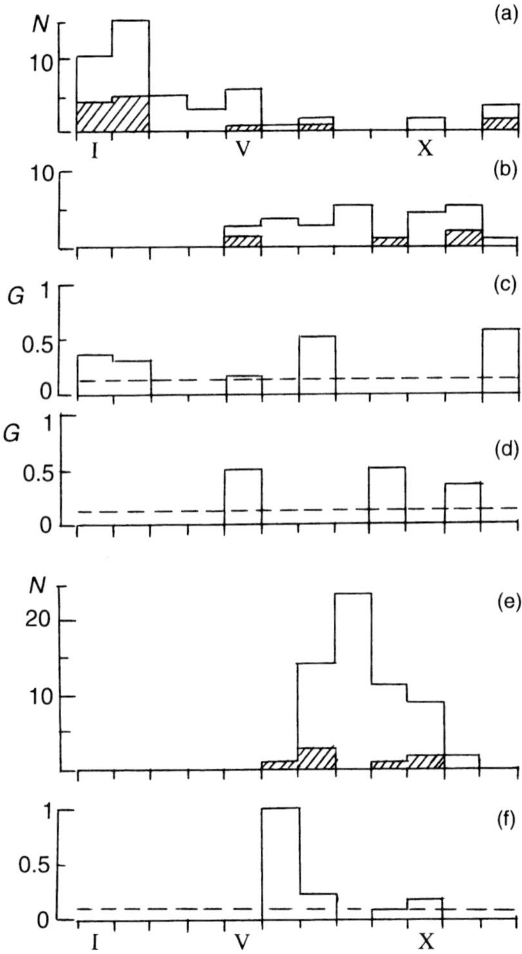

Figure 2.20. Annual time series of month-averaged intensities of mature and initial forms (a, b, e) and of the coefficient of hurricane formation (c, d, f) for cyclogenesis of the Atlantic (e, f), the North Indian Ocean (a, c), and South Indian Ocean (b, d). See symbols in Figure 2.19.

specific period, 9 events (TC) occurred causing the average monthly generation rate of mature forms to be 3. This value is close to the generation rate in the basins of the Pacific Ocean. The cyclogenesis of primary forms operated for large parts of the year (with the exclusion of August–September and November), though it was highly non-uniform: the maximum came in January–February (10 and 15 events, respectively). In the whole SIO for 1997, 47 events (TDs) occurred giving an average annual generation rate of disturbances of all forms of 3.9 $month^{-1}$ and an average annual generation rate of mature forms of 0.92 $month^{-1}$.

Cyclogenesis of the northern part of the Indian Ocean (NIO) is significantly determined by the monsoon particularities of the regional climate. In 1997 in this part of the ocean four full-blown TCs occurred within 3 months (May, September, and November), giving a generation rate of developing forms of 1.3 $month^{-1}$. At the same time, the cyclogenesis of primary forms in the active season, corresponding to the period from December on gave a generation rate of 3 $month^{-1}$ and "worked" continuously. In all the basins of the NIO for 1997, 24 events occurred. The average annual generation rate for disturbances was 2 $month^{-1}$ and that of mature forms was 0.3 $month^{-1}$.

Let us analyze the generation rate of mature forms in the basins of the Indian Ocean (Figure 2.20c, d). The intermittent generation of mature forms in the basins of the Indian Ocean means the annual cycle of the generation factor breaks down into a number of separate (monthly averaged) areas with sufficiently high values of the generation factor: from 0.3 to 0.67 for the basins of the SIO and 0.4–0.5 for the basins of the NIO. The average annual values of the generation factor turn out to be moderate: 0.16 for basins of the SIO and 0.11 for basins of the NIO.

2.5.8 Regional features of Atlantic Ocean cyclogenesis

Figure 2.20e presents graphs of the time evolutions of the intensity flows $N(t)$ of tropical cyclogenesis (monthly averaged) for 1997 for primary forms of TDs and for mature forms (TCs) for basins in the Atlantic Ocean.

The cyclogenesis of mature forms in the Atlantic in 1997 is worthy of note in that it had an unusually "faded" nature: cyclogenesis "worked" for as little as 4 months (June–July, September–October); moreover, the total number of TCs was 7 and, accordingly, the average monthly generation rate was 1.75. This value is twice as small as that of the basins of the Pacific Ocean. Unlike the cyclogenesis of mature forms, the cyclogenesis of primary forms in 1997 was highly "energetic" and compact from the temporal viewpoint: for 4 months (July–October) there were 57 events with most occurring in August (22 events). In other words, the average monthly generation rate of primary disturbances was 14.25 (in an active period) and the average annual rate was 4.9 $month^{-1}$.

Analysis of annual variation in the generation factor of mature forms (Figure 2.20f) shows that its annual cycle is divided into two areas: June–July and September–October with a moderate value of the generation factor of 0.1–0.2, as an exception in June, when one TC occurred ($G = 1.0$). The average monthly value of the generation factor is 0.17 (in an active period) which, in general, should be expected because of the intensive cyclogenesis of primary forms for the interval under investigation. The average annual (in 1997) value of the generation factor in the Atlantic was 0.12.

2.6 LARGE-SCALE STRUCTURE OF GLOBAL TROPICAL CYCLOGENESIS

In this section the term "large-scale structure" is used to refer to the variability of global tropical cyclogenesis on the understanding that data are monthly (and yearly) averaged. As noted above, the different time-averaged sizes of readings for complicated fractal systems may, broadly speaking, give quite different results. We will demonstrate these intriguing features in the time series of global tropical cyclogenesis time below.

2.6.1 Spatiotemporal variability in global cyclogenesis

Vortical structures in the tropical atmosphere may occur at any period of the year in tropical areas of oceans (with the exception of the southeast Pacific and the South Atlantic) given a certain combination of hydrometeorological settings in the atmosphere and on the Earth's surface (see Chapter 4). This combination of settings substantially varies with each region and from year to year. Nonetheless, there exists a certain regularity, which is considered below.

The origination centers, frequency, and recurrence of TCs have all been determined—and basic pathways along which they advance have been predicted—as a result of climatological processing of satellite data.

The World Ocean (WO) Basin is rather conventionally subdivided into six regions, each with its own specific spatiotemporal features of tropical cyclogenesis (Minina, 1970; Neumann, 1993; Pielke and Pielke, 1997).

Four regions are in the NH:

- the NWP bounded by 0–40°N and 100–180°E;
- the NEP bounded by 0–40°N and 100–180°W;
- the northern part of the Atlantic Ocean (NA) bounded by 0–40°N and 20–100°W;
- the NIO bounded by 5–25° and 50–100°E.

Two regions are in the SH:

- the SWP bounded by 5–30°S and 30–130°E;
- the SIO bounded by 5–30°S and 30–130°E.

There are authors who hold the view that there should be a more detailed spatial separation of global cyclogenesis.

In this section we consider the distribution and contribution of each of the six basins of the World Ocean within a 10-year period of cyclogenesis based on the information provided by the Global-TC dataset (Pokrovskaya and Sharkov, 1994c, 1997a, 1999a, d). The major information parameter is the total number of TCs occurring monthly or yearly (the cyclogenesis rate) during a 10-year cycle (1983–1992) both over the entire basin of the World Ocean and certain active basins (expressed as a percentage of the total number of TCs on a global scale) (Pokrovskaya and Sharkov, 1995a).

Global cyclogenesis overall (annual average indices) and its regional components within the 27-year period of 1983–2010 is shown in Table A.1 (Appendix A). Analysis of the data presented shows that between 75 and 90 TCs per year (an average 86 TCs per year) occur throughout the World Ocean Basin, and two-thirds of these occur in the southern and eastern hemispheres. Cyclogenesis was abnormally intensive in 1992 (99) and weak in both 1988 (75) and 1993 (75). The number of TCs distributed (on average) throughout the six cyclogenesis regions is (per annum): NWP above 30 TCs; NEP 18; NA 9; NIO 4; SWP 15; and SIO 10 TCs (Figure 2.21).

Despite mass-media reports about the destructive consequences of TCs in the NIO basin, from the standpoint of cyclogenesis intensity the contribution of this basin is the lowest (4.6% over the 10 years). Cyclogenesis in this region is stable with slight annual variation. Cyclogenesis is predominantly controlled by the monsoon features of the regional climate.

The NA and SIO basins likewise provide a low contribution (NA 10.6% and SIO 11.6% on average) to global cyclogenesis. Annual variation is also similar and rather stable. The SWP basin provides a substantially higher contribution to cyclogenesis (17.9%) and is free of any considerable annual variation.

During the 10 years under consideration, it should be noted that we observed considerable stability in the cyclogenesis processes taking place in the NWP basin, which provides 32.9% of the total number of TCs to global cyclogenesis. The last basin (NEP), in our opinion, is the most interesting for two reasons: first, it provides a considerable contribution to cyclogenesis (22.1%) and, second, because of its peculiar destabilizing effect on global cyclogenesis variation. For instance, the above-mentioned abnormally weak (1988) and intensive (1992) cyclogeneses were mostly the consequence of sharp variation that occurred in this basin. Elucidation of the physical reasons for such a material variation in NEP cyclogenesis is a subject worthy of further investigation. One probable cause is the rather sharp change in the character of the atmospheric circulation in the SH which took place in 1987–1988 and 1991–1992 (ENSO phenomena) (Henderson-Sellers *et al.* 1998; Revell and Goulter, 1986).

Let us now consider the same 10-year time series of global tropical cyclogenesis in a different way using the cumulative function (Section 2.1). In the case being considered, all the timescales of interaction will be taken into account to depict the structure of the cumulative process. Figure 2.22 illustrates the 10-year temporal evolution of the cumulative functions of global tropical cyclogenesis and of NH and SH cyclogeneses.

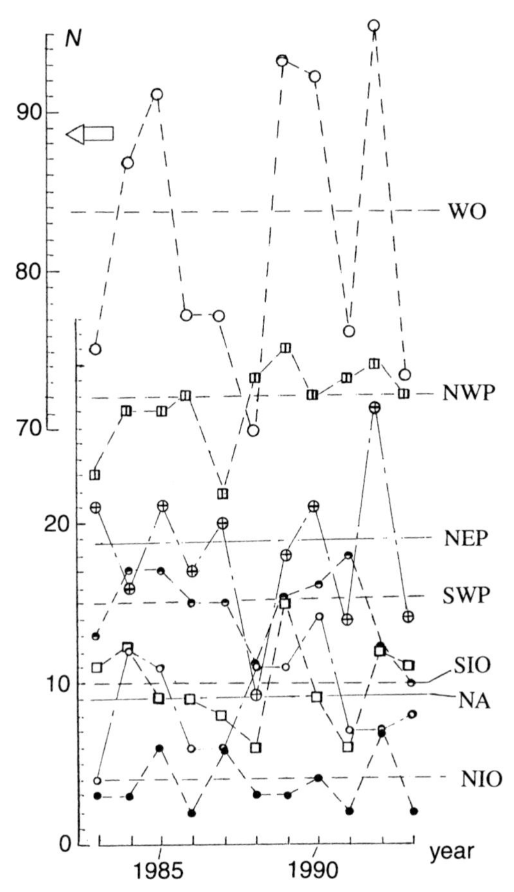

Figure 2.21. Temporal evolution of the year-averaged intensity of global and regional cyclogenesis for the 10-year period 1983–1992. See symbols in the text of Section 2.6.1 (from Pokrovskaya and Sharkov, 1994c).

As expected, the temporal evolution of cumulative functions differs fundamentally from year-averaged temporal patterns. The cumulative functions of both global cyclogenesis and hemisphere cyclogenesis are linear functions with small variation and are the consequence of intra-annual features.

These functions are best demonstrated by the NH pattern (Figure 2.22). Global and SH cumulative functions have less pronounced intra-annual features.

It is significant that the cumulative patterns exhibit the following characteristics:

- high degree of regularity;
- total absence of interannual variation of any values;
- stable values of averaged slopes of approximated straight lines.

This characteristic (i.e., averaged slope) is of great importance in cyclogenesis dynamics since it may be physically interpreted as the global rate of cyclogenesis

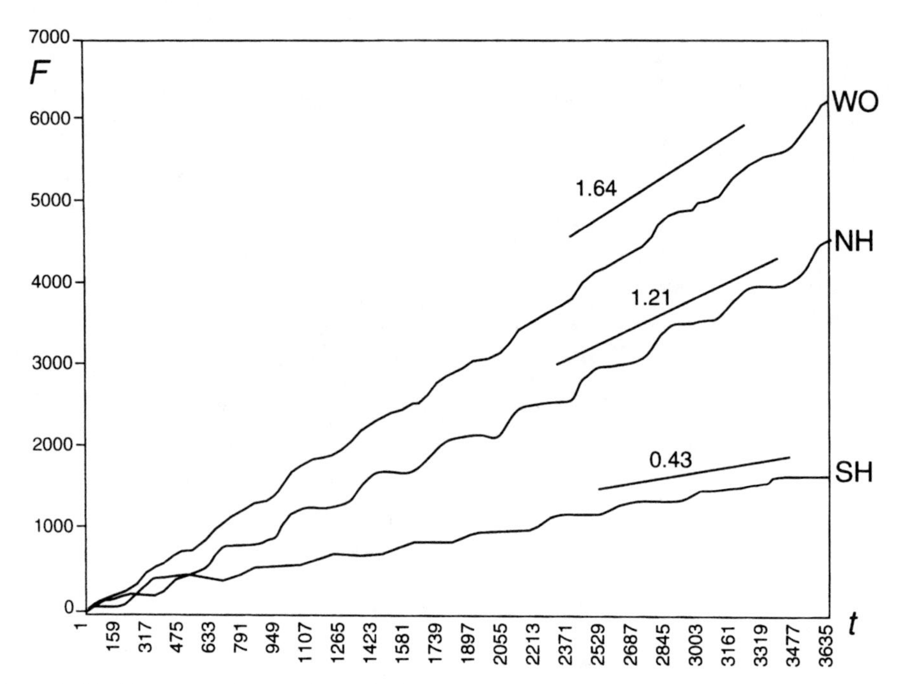

Figure 2.22. Ten-year time series of the cumulative functions (F) of global tropical cyclogenesis (WO) and of NH and SH cyclogeneses. The timescale is in days. Figures next to the straight lines represent values of the intensity of the Poisson process.

generation. As for global tropical cyclogenesis for the 10-year period (1983–1992), the global rate of cyclogenesis generation is equal to 1.64 events/day, while the rates of NH and SH cyclogenesis equal 1.21 and 0.43 events/day, respectively. Thus, the global rate of NH cyclogenesis far exceeds (2.82 times as great) these characteristics for the SH.

Furthermore, it must be emphasized that the intra-annual variation for NH and SH cyclogenesis has the property of clearly defined harmonicity. The first harmonics for hemisphere cyclogenesis are phase-shifted by about half a year. This feature of global cyclogenesis is well known (Minina, 1970; Neumann, 1993; Pokrovskaya and Sharkov, 1994c, 1995a).

2.6.2 Intra-annual variation of global cyclogenesis

Let us now turn to considering (monthly averaged) annual variation of global cyclogenesis (Pokrovskaya and Sharkov, 1994c, 1995a). Analysis of the intensity of global cyclogenesis (monthly averaged) presented in Figure 2.23 provides evidence of some of its peculiarities. We can see that two maxima occurred in the 10-year period (in February and September). The first is due to activity in basins of the SWP and SIO and is as much as 5 TCs per month. We can see a sharp increase in cyclogenesis in August–September (up to 12 TCs per month) which is mainly due to

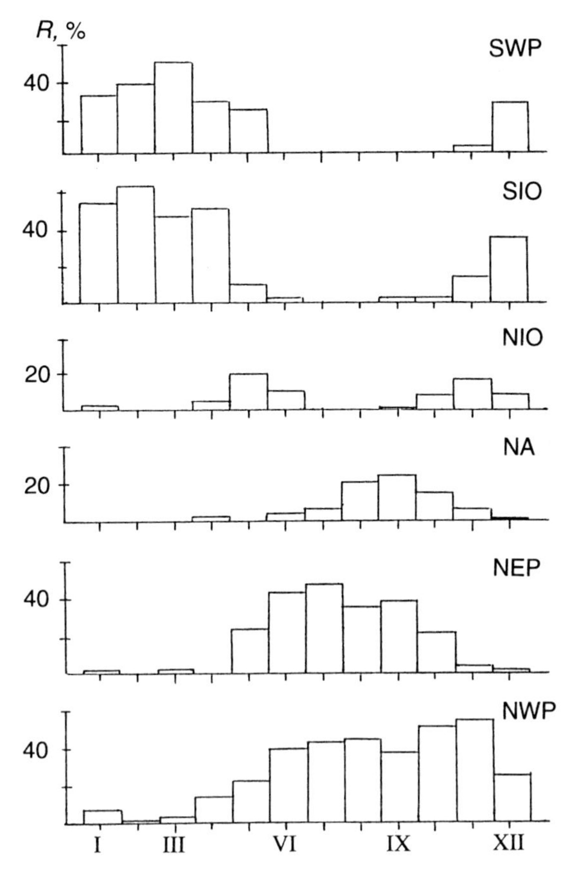

Figure 2.23. Relative contribution (R) of the active basins of the World Ocean to global cyclogenesis (monthly averaged). See symbols in Section 2.6.1 (from Pokrovskaya and Sharkov, 1995a).

activity in the NEP and NWP. When considering the relative contribution to tropical cyclogenesis from oceans located in the NH and SH (Figure 2.24), we can see an interesting regularity in that the contribution from SH oceans dominates (more than 80%) in January to April, whereas in June to October the situation radically differs where contribution from the NH ocean predominates (more than 90%). May and December can be viewed as transition periods where the contribution of each hemisphere is about 50%.

2.6.3 Spatial structure of generation centers

In order to assess the TC generation capability of basins, let us introduce a quantitative parameter: the spatiotemporal rate (STR) of cyclogenesis K over a $1{,}000 \times 1{,}000$ km basin per month in an active cyclogenesis season in a given

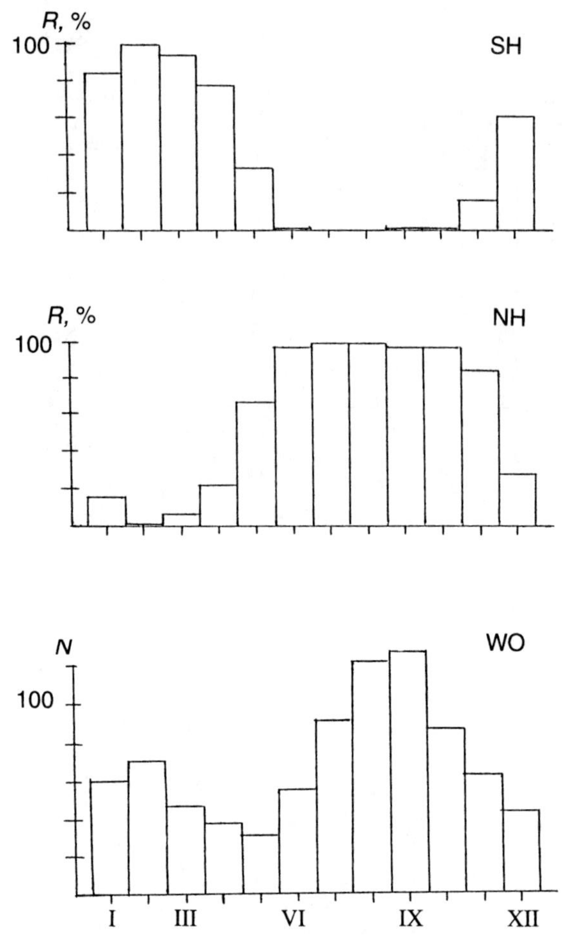

Figure 2.24. Global cyclogenesis intensity in 1983–1992 (monthly averaged) and the relative contributions (R) of basins in the NH and SH. The values of N are given in absolute units (number of TCs per month from 1983 to 1992); the values of R are given as percentages (from Pokrovskaya and Sharkov, 1995a).

region (Pokrovskaya and Sharkov, 1995a). Based on the spatiotemporal characteristics of this parameter, we can make certain conclusions about space-monitoring strategy and tactics.

Let us start the analysis of spatial features of metastable zones with the NWP basin, which makes a large contribution to global cyclogenesis. Figures 2.25 and 2.26 present spatial maps of tropical cyclogenesis origination centers in the eastern and western hemispheres. Blackened points designate the geographical locations of TC initial phases, which subsequently convert to tropical storms (TSs). Analysis of the geographical spatial distribution of origination centers shows the sharply irregular concentration of origination centers. We can clearly trace zones where there is an

Table 2.10. Values of the spatiotemporal rate parameter for cyclogenesis in the northwest Pacific Ocean basin.

Block coordinates (°N, °E)	*Parameter values*
10–20, 110–120	0.38
15–25, 130–140	0.36
15–25, 140–150	0.42
15–25, 150–160	0.22
5–15, 130–140	0.70
5–15, 140–150	0.67
5–15, 150–160	0.47

increased concentration of these centers: intensive cyclogenesis zones about 300×300 km in size. For instance, we can distinguish at least three active zones in the NWP with the following centers: 10°N, 145°E; 13°N, 132°E; and 17°N, 133°E.

Also noteworthy is the great number of TDs (27 of 274 within a 10-year period) that convert to advanced phases in the immediate vicinity of a coast, at a range of 20 km to 100 km. Table 2.10 presents estimates of the spatiotemporal rate for the most heavily filled blocks in the NWP basin. As expected, southern blocks in this basin turn out to be most informative. The cyclogenesis rate sharply decreases to the north and east of these blocks (by an order of magnitude). Note that a closed basin in the South China Sea also experiences intensive cyclogenesis ($K = 0.38$).

A far higher concentration of origination centers is exhibited in regional cyclogenesis in the NEP (Figure 2.25), where a vast (500×500 km) active zone is found confined by the coordinates of 15°N and 100°W. It is as close as 250 km from the Mexican coastline. Moreover, the occurrence of TD effects in the immediate vicinity of North America has been likened to those in the NWP. So, in the offshore zone (200 km) of the NW basin of the Pacific Ocean nearly 30 TCs have originated (i.e., 17% of the total number of regional TCs).

When assessing the STR for the most heavily filled blocks—those with coordinates of 10–20°N, 100–110°W (polygon A), 10–20°N, 110–120°W (polygon *B*), and 5–15°N, 90–100°W (polygon *C*)—we get the following values: $K = 1.7$ for polygon *A*; $K = 0.7$ for polygon *B*; and $K = 0.76$ for polygon *C*. In other parts of this basin the STR sharply decreases by an order of magnitude or more (e.g., $K = 0.18$ for polygon *D*, with coordinates of 10–20°N and 145–155°W). This suggests that blocks *A*, *B*, and *C* are of most interest to space monitoring. This is so because the frequency of spacecraft intersecting the possible origination center of a vortex structure is greatest here.

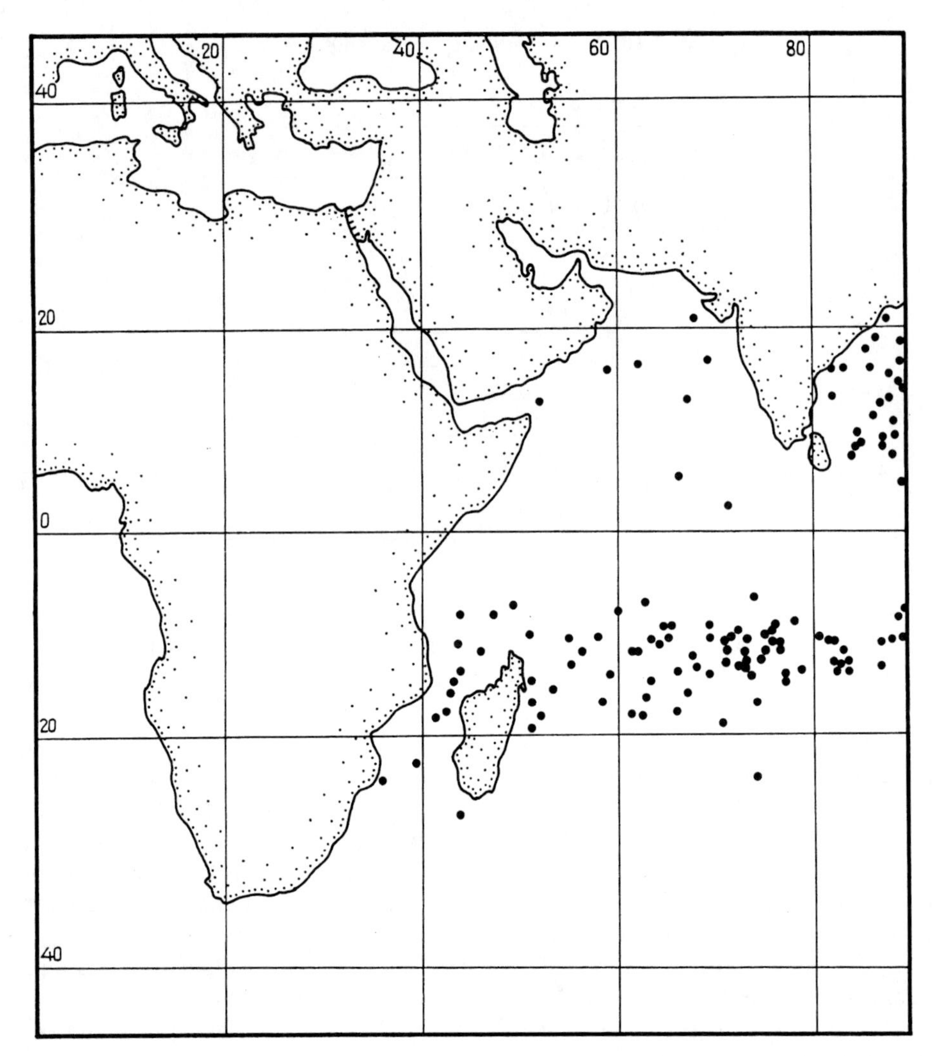

Figure 2.25. Spatial maps of tropical cyclone generation centers in the eastern hemisphere for 1983–1992. Points designate geographical locations of the generation stage of TCs (from Pokrovskaya and Sharkov, 1995a).

From the standpoint of tropical cyclogenesis, the Atlantic basin of the World Ocean (Figure 2.26) has the following features: advanced forms of TC occur only in its northern part (i.e., the North Atlantic); cyclogenesis is "scattered" rather uniformly throughout the vast basin; and no activity centers have revealed themselves (within a 10-year cycle). STR values are in the range $K = 0.13$–0.3 both in the open sea and in closed basins (e.g., $K = 0.26$ in the Gulf of Mexico). Much as in other active basins, in this basin a great number of TCs (up to 10% of the total) have originated in the immediate vicinity of a coast. This effect is most pronounced for TC

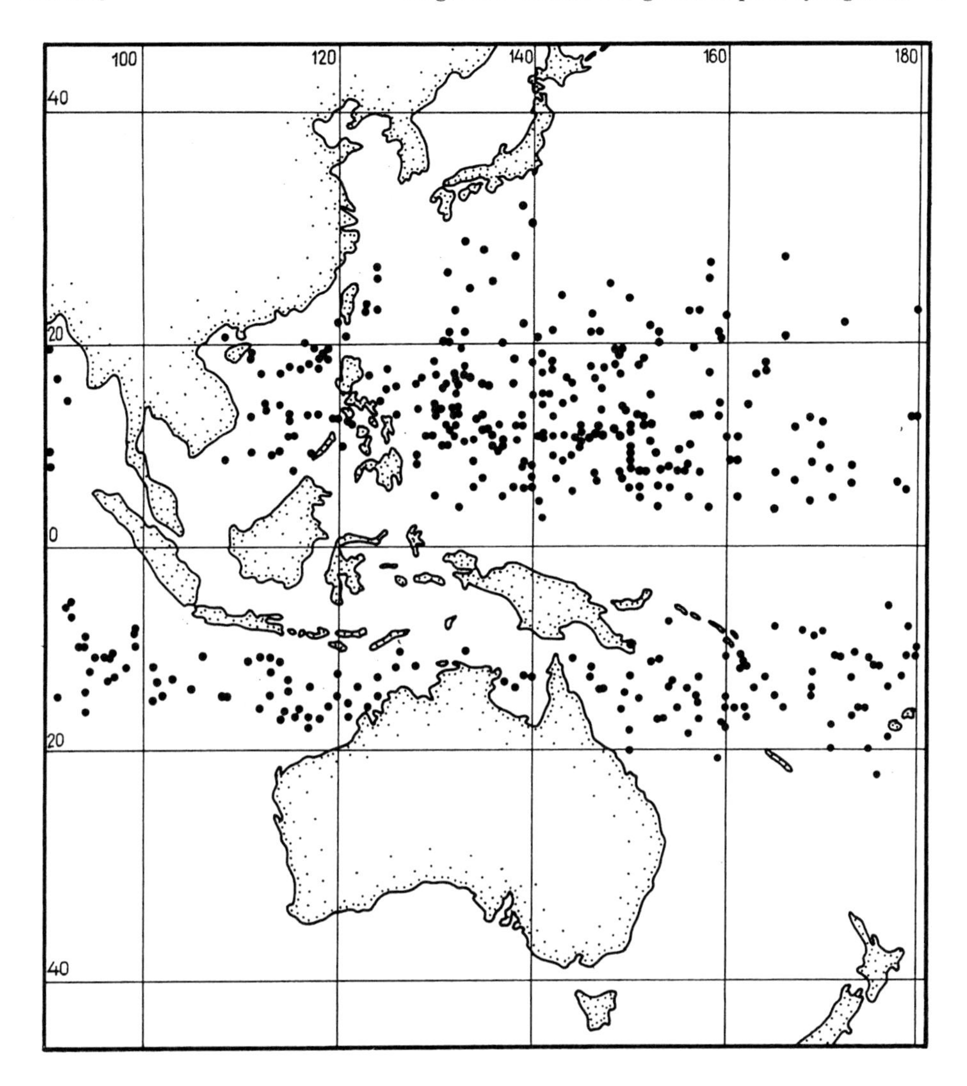

origination in closed basins (i.e., the Gulf of Mexico and the Caribbean Sea). The origination and formation of TCs at upper latitudes right up to 37–38°N are important and distinctive features of Atlantic cyclogenesis; they are most likely due to the Gulf Stream.

Though the NIO (Figure 2.26) is characterized by weakly defined cyclogenesis (4.5% of total number of TCs in the World Ocean over 10 years), we can clearly see an active zone located in the area with coordinates of 9°N and 85°E (Bay of Bengal). What is more, estimation of the STR in this basin provides a rather high value of

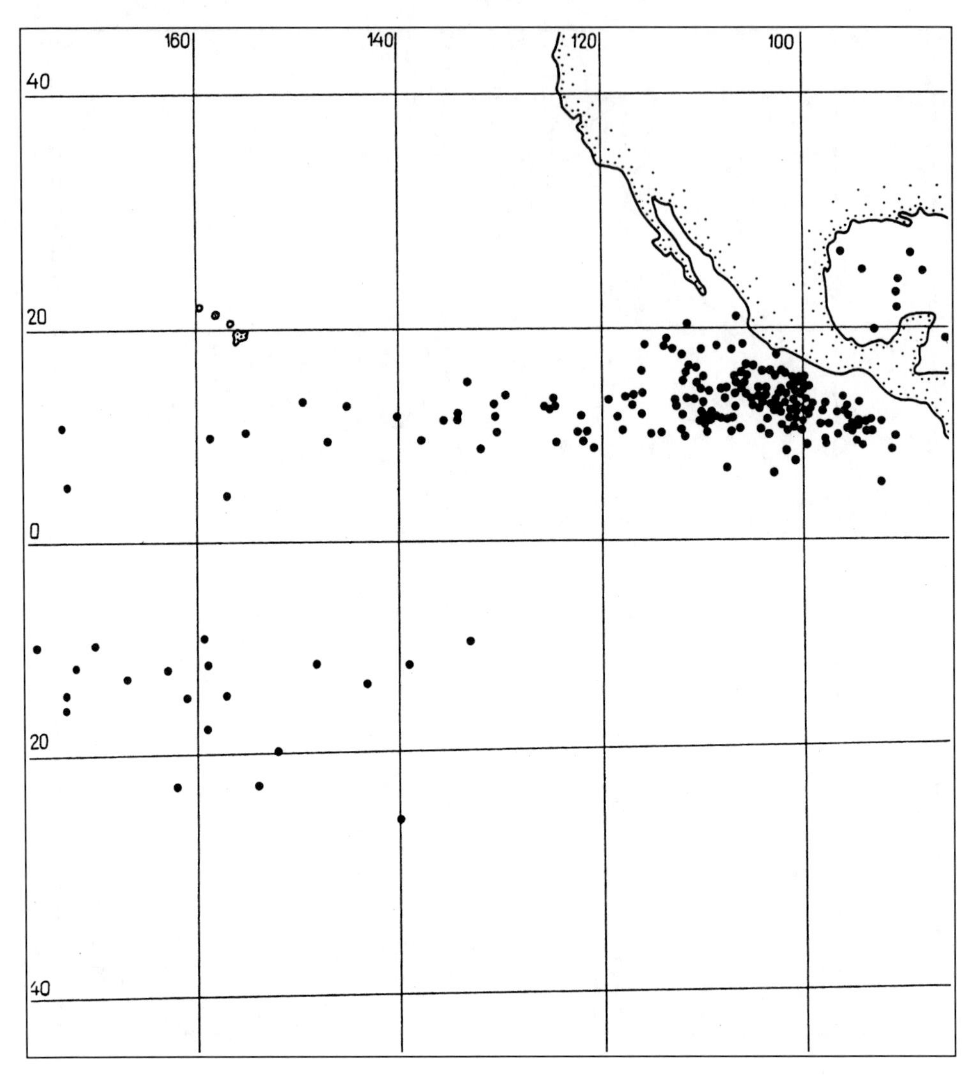

Figure 2.26. Spatial maps of tropical cyclone generation centers in the western hemisphere for 1983–1992. Symbols are the same as in Figure 2.25 (from Pokrovskaya and Sharkov, 1995a).

parameter K ($K = 0.65$) because of the short duration of cyclogenesis activity during monsoon periods. About 30% of TCs occur in the immediate vicinity of the coast (200 km). Practically all TCs that originate in the Bay of Bengal are serious hazards to the heavily populated areas of Bangladesh and India. Another noteworthy feature of NIO cyclogenesis is that, in particular cases, TCs have even originated near desert areas such as the Arabian Desert (i.e., in regions with distinctly reduced humidity).

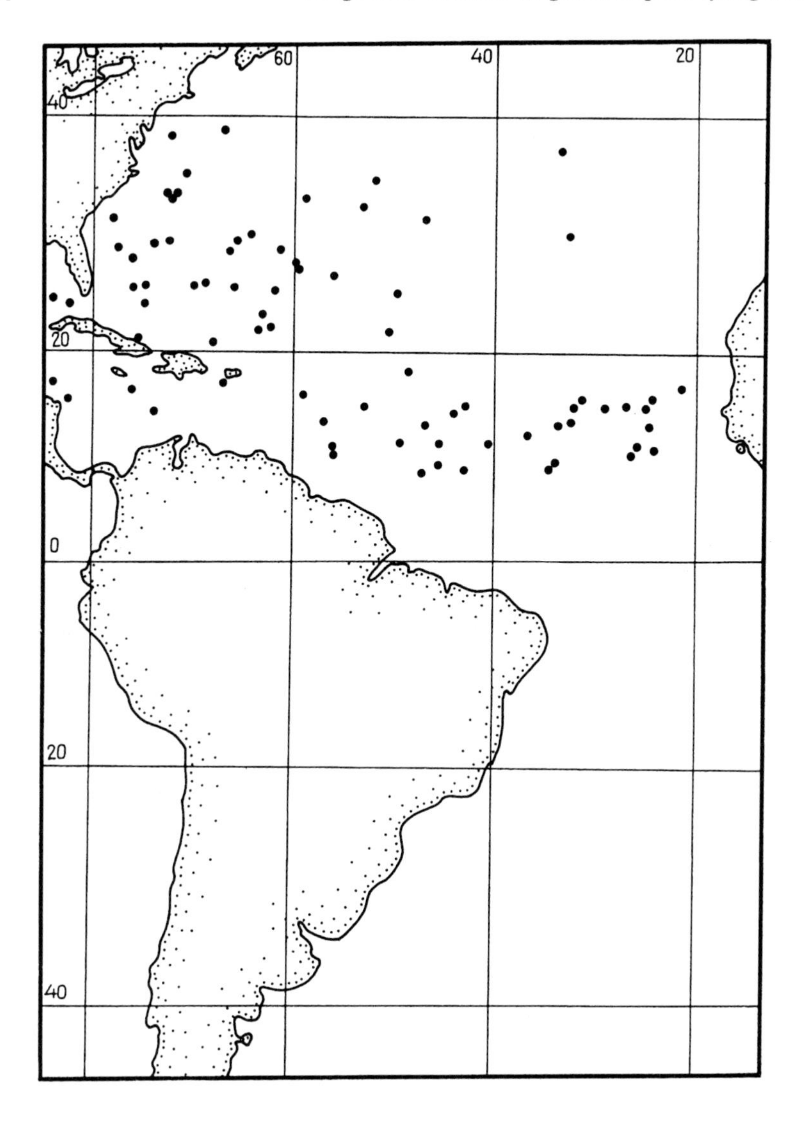

In southwestern parts of the Pacific and Indian Oceans, TC origination centers are distributed more or less uniformly, without sharply defined active zones. Estimates of the STR are $K = 0.1$–0.4 for the SWP and $K = 0.2$–0.4 for the SIO. From the observational standpoint, of most interest are studies of intensive cyclogenesis ($K = 0.6$) in the offshore zone of the Indian Ocean near the western coast of Australia (Great Sand Desert), in closed and shallows seas (Timor and Arafura Seas), and the closed shallow Gulf of Carpentaria.

2.7 HIERARCHICAL STRUCTURE OF GLOBAL TROPICAL CYCLOGENESIS

Global tropical cyclogenesis is a good example of a chaotic process, in which the resulting picture is formed as a consequence of a whole series of thermohydrodynamic interactions of different intensities, occurring in a very broad range of temporal and spatial scales and resulting in a rather complex hierarchical structure.

Despite this obvious physical evidence, the approach involving the intentional apportionment and separation of the unitary global tropical cyclogenesis process into two (or several) timescales (daily, monthly averaged, or yearly averaged) is used widely (e.g., see Gray *et al.*, 1997; Wilson, 1997; Landsea *et al.*, 1999; Smith, 1999; Elsner and Kara, 1999). We have also used this approach in the study of the time series of global tropical cyclogenesis (see Sections 2.2–2.3 and 2.6). However, it is obvious that a number of timescales of interactions are entirely lost under these conditions. To ease comprehensive study of these difficulties, a mathematical method for processing time series can be used to perform the analysis across wide timescales.

Astafyeva *et al.* (1994a, b, c) proposed and developed a modern approach called the wavelet transform method for the purpose of processing time series of global tropical cyclogenesis and, thus, revealing the effective scales of interactions. The wavelet transform method consists of expanding the signal under study in terms of the hierarchical basis that is attractive for direct analysis of complicated signals.

As this approach for processing observational data has not gained widespread acceptance in remote-sensing and geophysical investigations, we consider it opportune to carry out a concise review of its main properties.

2.7.1 Main properties and examples of wavelet transform

The notion of a wavelet (literally, a small wave) began in the 1980s when Grossmann and Morlet applied it to analysis of the properties of seismic and acoustic signals (Chui, 1992; Astafyeva, 1996; Bouman and Newell, 1998). The family of analytical functions dubbed wavelets is being increasingly used to address problems in pattern recognition, in processing and synthesizing of various signals (e.g., speech); in the analysis of images of any kind (satellite images of clouds, a planet's surface, mineral-bearing rocks, etc.), in the study of turbulent fields, in the contraction (compression) of large volumes of information, and in many other spheres of interest.

The wavelet transform of a one-dimensional signal involves its decomposition over a basis obtained from a solution-like function (wavelet), possessing some specific properties, by means of dilation and translation. Each of the functions of this basis emphasizes both a specific spatial (temporal) frequency and its location in physical space (time).

Thus, unlike the Fourier transform traditionally used in signal analysis, the wavelet transform offers a two-dimensional expansion of a given one-dimensional signal, with the frequency and the coordinate treated as independent variables. As a result, the possibility emerges of analyzing the signal simultaneously in physical (time, coordinate) and frequency spaces. All of this can readily be generalized to multidimensional signals or functions.

The area in which wavelets can be used is not confined to analysis of the properties of signals and fields of arbitrary nature obtained numerically, experimentally, or observationally. Wavelets are beginning to be used in direct numerical simulations as hierarchical bases that are well suited to describe the dynamics of complex non-linear processes characterized by the interaction of perturbations across wide ranges of spatiotemporal scales.

Many researchers refer to wavelet analysis as the "mathematical microscope", as this term accurately conveys the remarkable capability of the method to offer good resolution at different scales (Astafyeva, 1996). The capability of this microscope to reveal the internal structure of an essentially inhomogeneous process (or field) and expose its local scaling behavior has been demonstrated through many classical examples such as fractal Weierstrass functions and the probability measures of the Cantor series, to mention a couple. Application of wavelet analysis to a turbulent velocity field in a windtunnel under large Reynolds numbers offered for the first time a vivid confirmation of the Richardson cascade. The analogy between the energy cascade and the structure of the multifractal inhomogeneous Cantor series was explicitly settled. Even more efficient was the application of wavelet analysis to the multifractal invariant measures of several well-known dynamical systems that model the transitions to chaos observed in dissipative systems (Nicolis and Prigogine, 1977; Kadanov, 1993; Dubois, 1998).

So, wavelet analysis can be successfully applied to solve various problems. It is, however, not widely known to researchers dealing with the analysis of experimental and observational data. In this section, an attempt is made to outline the main properties needed in practical applications of the wavelet transform to processing signals of various types.

For practical applications it is important to know the criteria that a function should satisfy in order to be a wavelet; we (following Astafyeva, 1996) present them here and consider, by way of examples, certain well-known functions by ascertaining whether they observe these criteria:

- *Localization.* Unlike the Fourier transform, the wavelet transform is based on a localized basis function $\psi(t)$. A wavelet must be localized both in time and frequency.
- *Zero mean*

$$\int_{-\infty}^{\infty} \psi(t)\, dt = 0 \tag{2.26}$$

In applications, it is frequently necessary to additionally have the first m-zero moments:

$$\int_{-\infty}^{\infty} t^m \psi(t)\, dt = 0 \tag{2.27}$$

Such a wavelet is referred to as an m-order wavelet. Wavelets possessing many zero moments enable us to ignore most regular polynomial components of signals and analyze their small-scale fluctuations and high-order features.

- *Boundedness*

$$\int |\psi(t)|^2 \, dt < \infty \tag{2.28}$$

Possessing these properties, the soliton-like function (wavelet) forms the basis for designing the necessary base for the family of functions, obtained from a single function $\psi(x)$:

$$\psi_{a,b}(x) = |a|^{-\eta} \psi\left(\frac{x-b}{a}\right), \qquad a, b \in R, \quad a < G \tag{2.29}$$

A characteristic feature of a wavelet transform basis is its self-similarity. All wavelets of a given family $\psi_{a,b}(t)$ have the same number of oscillations as the basis wavelet $\psi(t)$; they are derived from it by scale transforms and translations. It is for this reason that the wavelet transform is successfully used in fractal analysis.

The continuous wavelet transformation of a one-dimensional function $f(x)$ is of the form:

$$W(a,b) = C^{-1/2} a^{-1/2} \int \psi\left(\frac{x-b}{a}\right) f(x)\, dx \tag{2.30}$$

where $C = \int |k|^{-1} |\psi(k)|^2 \, dk$ is a normalizing constant; and $\psi(k)$ is the Fourier transform of $\Psi(x)$.

Decomposition of the analyzing signal on scales is realized by means of extending or compressing the analyzing wavelet before its combination with the signal. This brings about the fact that wavelet transformation has good agreement both on large and small scales.

If the analogy of a "mathematical microscope" is pursued further, then the translation parameter b fixes the focusing point of the microscope, the scale factor a the magnification, and finally the choice of the basis wavelet ψ determines the optical quality of the microscope (Astafyeva, 1996).

When performing calculations under these conditions the wavelet transform is reversible in so far as it can recover a function on its wavelet transform. The formula for reconversion is of the form:

$$f(x) = C^{-1/2} \int \psi\left(\frac{x-b}{a}\right) W(a,b)\, da\, db/a^2 \tag{2.31}$$

where $da\, db/a^2$ is a surface element that does not change at either shift or zoom.

Let us write down the basic elementary properties of the wavelet transform of function $f(t)$—the designation $W[f] = W(a,b)$ will be adopted for brevity:

- *Linearity*

$$W[\alpha f_1(t) + \beta f_2(t)] = \alpha W[f_1] + \beta W[f_2] = \alpha W_1(a,b) + \beta W_2(a,b) \tag{2.32}$$

It therefore follows that the wavelet transform of a vector function is a vector with components that are wavelet transforms of the respective components of the vector analyzed.

- *Translational invariance*

$$W[f(t-b_0)] = W(a, b-b_0) \tag{2.33}$$

This property leads to the commutativity of differentiation, in particular, $\partial_t W[f] = W[\partial_t f]$ (here $\partial_t = \partial/\partial t$). Together with the first property it implies permutability for vector analysis derivatives.

- *Invariance to dilations (contractions)*

$$W\left[f\left(\frac{t}{a_0}\right)\right] = \frac{1}{a_0} W\left(\frac{a}{a_0}, \frac{b}{a_0}\right) \tag{2.34}$$

This property makes it possible to determine whether the function analyzed has singularities and to investigate their character.

Apart from these three elementary properties, which are independent of the choice of analyzing wavelet, the wavelet transform has a few others. In our opinion, the most important and helpful among them is the following:

- *Differentiation*

$$W[\partial_t^m f] = (-1)^m \int_{-\infty}^{\infty} f(t)\, \partial_t^m [(\psi_{a,b}^*(t))]\, dt \tag{2.35}$$

So, if we ignore, for example, large-scale polynomial components and analyze singularities of higher order or small-scale variations of function f we may perform differentiation of either the analyzing wavelet or the function itself. This is a highly useful property, especially if we remember that the function f is often defined through a sequence of values while the analyzing wavelet is given by a formula.

For the wavelet transform there exists *an analogue of the Parseval theorem* and the identity

$$\int f_1(t) f_2^*(t)\, dt = C_\psi^{-1} \iint W_1(a,b)\, W_2^*(a,b)\, \frac{da\, db}{a^2} \tag{2.36}$$

Hence, it follows that the energy of a signal can be calculated in terms of amplitudes (coefficients) of the wavelet transform, in the same way it is computed through the components of the Fourier transform:

$$E_f = \int f^2(t)\, dt = \int |A(\omega) - iB(\omega)|^2\, d\omega \tag{2.37}$$

The definitions and properties of the one-dimensional continuous wavelet transform can be generalized to multidimensional and discrete cases.

Since the wavelet transform is a scalar product of the analyzing wavelet, of the given scale and signal being explored, coefficients $W(a,b)$ contain combined information on both the analyzing wavelet and the signal (similar to coefficients of the Fourier transform which bear imprints of both the signal and the sinusoidal wave).

The choice of analyzing wavelet is, as a rule, dictated by the type of information to be derived from the signal. Every wavelet has specific features in time and frequency domains and sometimes—by applying different wavelets—we may reveal more fully or emphasize one or other of the signal's characteristics.

Real-valued bases are frequently constructed from the derivatives of Gaussian functions:

$$\psi_m(t) = (-1)^m \, \partial_t^m \, [\exp(-t/2)] \tag{2.38}$$

$$\psi_m(k) = m(ik)^m \exp(-k^2/2) \tag{2.39}$$

where $\partial_t^m = \partial^m[\ldots]/\partial t^m$, $m \geq 1$. Higher derivatives have more zero moments and allow us to retrieve information about the features of higher orders contained in the signals.

Because of their shape, these functions came to be known, respectively, as the WAVE wavelet and the MHAT wavelet or the "Mexican hat" (looks like a sombrero).

The MHAT wavelet, with its narrow energy spectrum and two zero moments (zeroth and first), is well suited to analyze complex signals. Generalized to the two-dimensional case, the MHAT wavelet is frequently used to analyze isotropic fields. If the derivative is taken in one direction, an anisotropy basis can be obtained, with good angular resolution.

The wavelet transform, owing to its hierarchical basis, is well suited to analyze those cascade processes, fractal sets, and multifractal sets that have a hierarchical nature (Mandelbrot, 1977, 1982, 1989, 1998; Coniglio, 1987; Voss, 1989; Lovejoy and Schertzer, 1985; Lovejoy *et al.*, 1986).

We present an example concerning the analysis of a fractal set on the basis of a homogeneous triadic Cantor set. As is known, when constructing the first generation of this set, an interval is divided into three parts and the middle part is excluded; for the second generation the same procedure is excluded; for the third generation the same procedure is applied to the two remaining intervals, and so on at each subsequent stage up to infinity. Figure 2.27(a) displays the first stages of this construction.

Based on the set constructed, the Cantor dust (a numeric set) is created from zeros and ones (zeros correspond to excluded parts of the interval).

Figure 2.27(b) presents the patterns of coefficients and local maximum lines. They are reasonably detailed; however, linear scale sweep does not allow a broad-scale range to be presented. To demonstrate the general character of the process, Figure 2.27(c) shows the skeleton in logarithmic axes.

The pattern of coefficients displays the hierarchical structure of the set presented. It is seen even more clearly in the patterns of local maximum lines. The skeleton not only reveals the hierarchical structure, but also shows how the fractal measure—on which the set is formed—was constructed.

At every stage of the cascade process, every scale subdivision is marked off on the local maximum pattern by branching giving the appearance of a peculiar "fork": the line marking the local maximum position bifurcates into two independent local

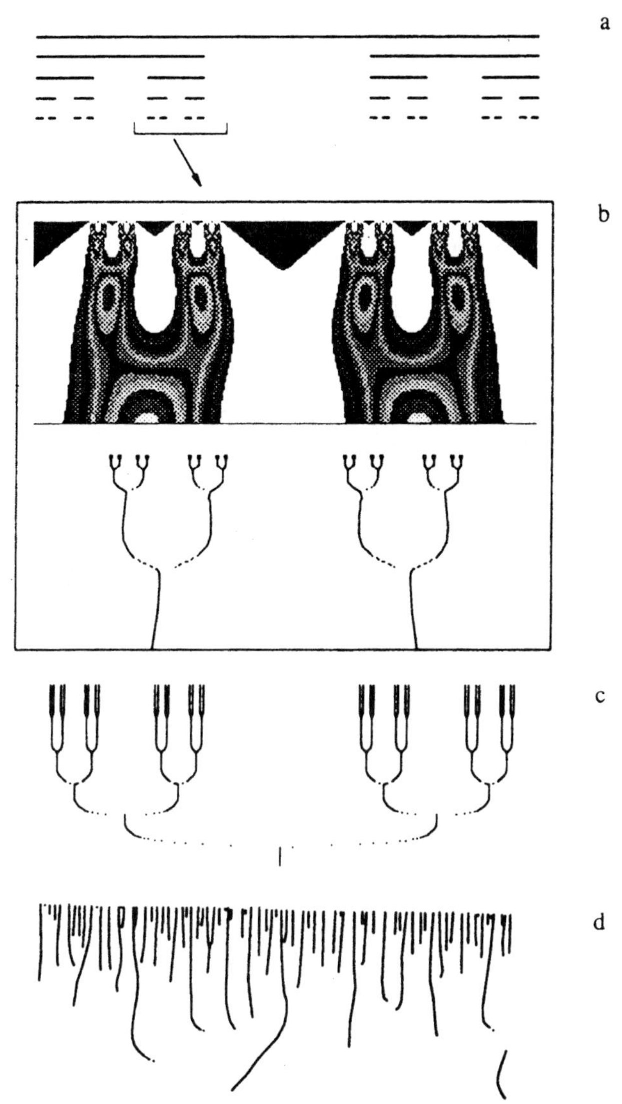

Figure 2.27. First generations of a homogeneous triadic Cantor series (a), fragments of the patterns of coefficients and local maximum lines (b), pattern of local maximum lines for a triadic homogeneous Cantor series (c), and a random process (d) on the logarithmic scale (from Astafyeva, 1996).

maximum lines. This is the invariably recurrent feature since the measure is self-similar and monofractal.

It is known that the fractal dimension or self-similarity dimension of the homogeneous Cantor set $D_f = \ln m / \ln s$, where m is the branching rate and s is the scale factor. In the case of the triadic set $D_f = \ln 2 / \ln 3$. Using wavelet transform coefficients as the limit (with the scale going to zero) we can also assess the dimension of the ratio $\ln N(a)/\ln a$, where $N(a)$ is the number of local maxima. The higher the order of generation used for the Cantor set, the more accurately can its dimension be determined; for the 10th–11th generations the value computed by wavelet transform coefficients practically coincides with that found analytically.

For comparison, Figure 2.27(d) pictures local maximum lines of a random process. We can see how different, even qualitatively, the tree-like structure of the skeleton of the cascade process and the grass-like skeleton of the random process are (they could be likened to periodic skeletons of harmonic functions and bushes of lines marking off the singularities of signals).

Numerous examples of using the wavelet transform are described by Astafyeva (1996). Let us look at one of them that relates to the transformation of the "devil's staircase"—a well-known and nearly differentiable function, built on the basis of the uniform Cantor set (Astafyeva *et al.*, 1994b). Figure 2.28 displays the typical "devil's staircase" (Figure 2.28a), the pattern of wavelet transforms $W(a, b)$ (Figure 2.28b), and a corresponding picture of local maximums of factors $W(a, b)$ (Figure 2.28c).

In Figure 2.28(b) several ranges of the values of factors $W(a, b)$ of wavelet transforms are marked by different tones. On the axis of abscissa a change to coordinate b is plotted and coordinate a is laid off on the axis of ordinates. So that details can be seen more clearly, only part of the picture showing the factors of wavelet transforms being changed is presented. The picture of the wavelet transform shows a hierarchical structure of the analyzing set (i.e., the "devil's staircase").

Nowhere else is it seen more clearly than in Figure 2.28(b) where the local maximums of wavelet transform coefficients are marked. The picture of local maximums reveals not only the hierarchical structure of the analyzing set, but also the way of building the fractal set to which it is connected. As is well known, on building the first generation of a uniform Cantor set, the length is divided into three parts, the average of which is rejected. On building the second generation such a procedure is performed using the two remaining lengths at each following stage, and so on.

Each such "crushing" scale is marked on the picture of local maximums by branching giving the appearance of a "fork" that notes the position of the local maximum, where the line bifurcates and separates into two independent local maximums. This can be repeated on all scales, since the process is fractal and possesses self-similar characteristics.

It should be noted that the "devil's staircase" has been successfully employed in describing the hierarchical oscillator system, showing evidence of the phase transition from aperiodicity to a chaotic regime (e.g., like ENSO; Jin *et al.*, 1994, 1996).

To summarize, the examination of wavelet patterns has much potential for yielding information about the possible type and quantitative characteristics of dynamical systems. By this expedient we can clearly identify the different types of dynamical systems like the "pure" random oscillator (e.g., Poisson type or Brownian motion model), the harmonic oscillator (e.g., van der Pol model), aperiodic and pulsed oscillators (e.g., multivibrators), the intermittent oscillator (e.g., trigger type and "irregular burst" type), the hierarchical system (e.g., Cantor set type and

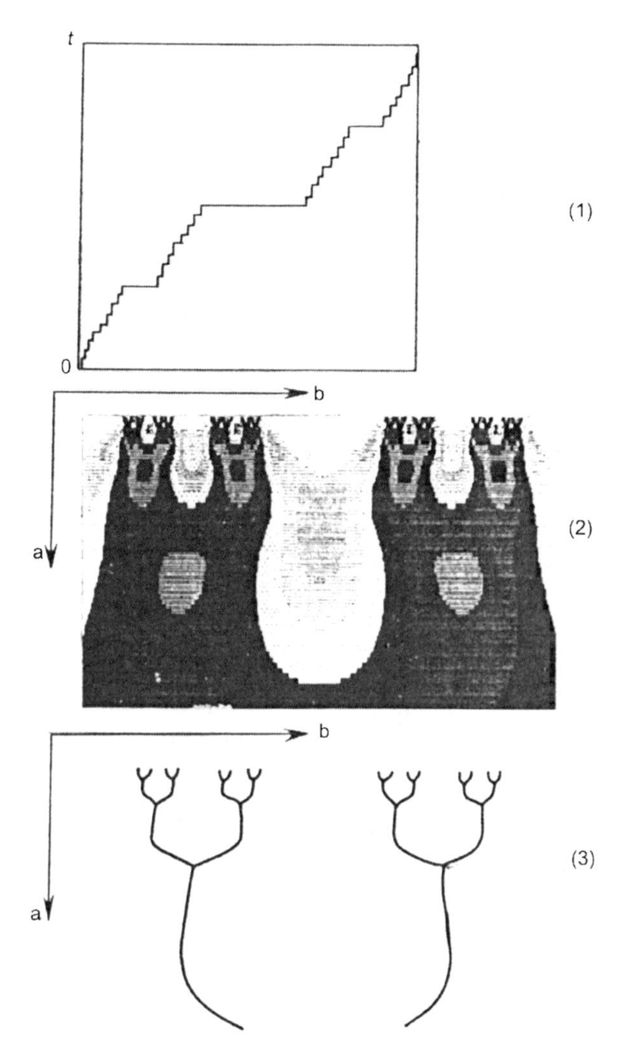

Figure 2.28. The "devil's staircase" function (1), the pattern of the wavelet transform (2), and corresponding local maximum lines (3) (from Astafyeva *et al.*, 1994b).

Sierpinski gasket type), and the dynamical system near the transition point from order to chaos (e.g., Lorenz's system and Duffing's system).

The useful information on the structure and the temporal behavior of the simulation of (theoretical) dynamical systems can be obtained from many publications (e.g., Schuster, 1984; Landa, 1996; Kadanov, 1993; Chen and Dong, 1996; Gilmore, 1998; Dubois, 1998).

Of course, the identification of contrasting types of dynamical system, like the harmonic oscillator and the multivibrator, can be performed more simply by examining the time series of these processes without resorting to the wavelet procedure.

However, naturally complex systems (e.g., physical, geophysical, biological, and economic) usually consist of many subsystems. While subsystems can interact with each other, they also interact with an external environment. These combined effects give rise to both the non-linear and chaotic behavior of the system as a whole (e.g., Jin *et al.*, 1996; Jensen, 1998; Schweitzer, 1997; Gohara and Okuyama, 1999; Boccaletti *et al.*, 2000). All of this means that the successful analysis of such dynamical systems can be performed only by a processing procedure that considers wide ranges of temporal scales and has good resolution at different scales simultaneously. One such method is the wavelet transform procedure. In Section 2.7.2 we will use this approach to study the hierachical structure of global tropical cyclogenesis as a complex natural system.

2.7.2 Wavelet patterns of global tropical cyclogenesis

The methodical basis of forming a time series of observational data of global tropical cyclogenesis offers up the idea of shaping and studying probabilistic features of the intensity of tropical cyclogenesis as signals haaving a complicated structure (type of telegraphic process) with the help of the mathematical device of the theory of random flows (see Section 2.1).

In accordance with this approach, the process being analyzed develops as follows. Leaving to one side the details and track record of individual tropical structures, we shall place each TD on the time axis as a pulse of single amplitude with a duration equal to the TC lifetime. The number of received pulses (indistinguishable events) in a single time interval (for us it is a day) is a natural physical parameter: the intensity of global tropical cyclogenesis.

The sequence of pulses formed by this method, with corresponding duration and amplitude, is simply the integer random time flow of indistinguishable events.

Figure 2.29a shows the time series formed by the method described from 5-year observational data, which reveals the changing intensity of global tropical cyclogenesis in the period 1988–1992.

Figure 2.29b, c presents the results of using the wavelet transform to obtain the time series (its length is 1,827 days).

As already noted, the wavelet transform of a one-dimensional signal shows us its two-dimensional expansion in temporal and spectral space $W(a,b)$. On the axis of abscissas changing time in days b is plotted and on the axis of ordinates the scale parameter a is presented on the linear scale. In order not to complicate the picture, obtained as a result of the wavelet transform, no different ranges of values of coefficient $W(a,b)$ are shown, just its positive values in black.

The pattern of the wavelet transform coefficient reveals the periodic behavior of analyzing the process on average scales: annual and semi-annual periods clearly stand out. Figure 2.29c presents the corresponding picture for focal maximums. It is proposed that semi-annual and annual features of the intensity of global tropical cyclogenesis are independent.

The part of the surface (three-dimensional image) described by coefficients $W(a,b)$ from 300 up to 1,500 days is shown in Figure 2.30. Six large-scale

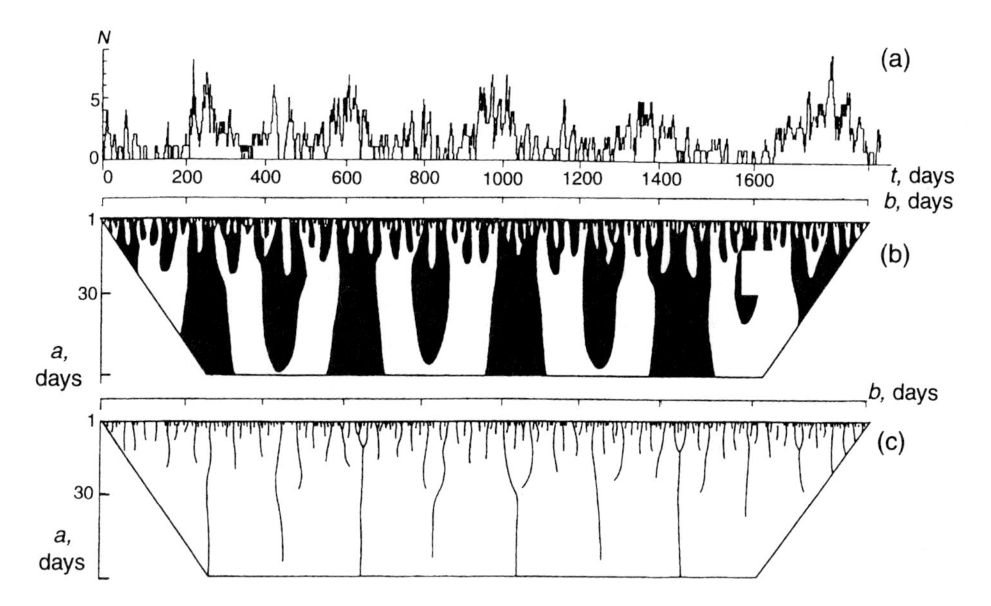

Figure 2.29. Time series of the intensity of global cyclogenesis (N) (a), the pattern of the wavelet transform (two-level) (b), and corresponding local maximum lines (c) for 1988–1992 (from Astafyeva *et al.*, 1994b).

"hunches" with approximately semi-annual period stand out sharply against the random noise: the three smaller ones are connected with increasing global tropical cyclogenesis in February and March, the three larger ones are connected with activity in August and September.

Figure 2.31 presents the changing intensity of global cyclogenesis (Figure 2.31a), the pattern of the wavelet transform (Figure 2.31b), and the pattern of local maximums (Figure 2.31c) for that part of the signal noted in Figure 2.29. These detailed results allow us to analyze the finer structure of the signal on small scales. Note the "forks"—which are an integration (bifurcation) of scales and the truth—are unlike the ones for the fractal set described above.

The pattern of the wavelet transform and the picture of local maximums, obtained with the fast-rising "hat" that reveals the properties of the process on greater timescales, do not allow us to reach an unambiguous conclusion about the appearance of "forks" resulting from the merging of annual local maximum lines. From the 5-year data we have been unable to ascertain whether a hierarchy of scales on the time order of several years exists.

We believe that the main result of scientific studies should be aimed at finding the hierarchical time structure of global tropical cyclogenesis. Such a result would allow a more goal-directed approach to building stochastic evolution models of tropical cyclogenesis (e.g., separate thermodynamic and kinetic physical factors that stipulate the developmental particularities of cyclogenesis).

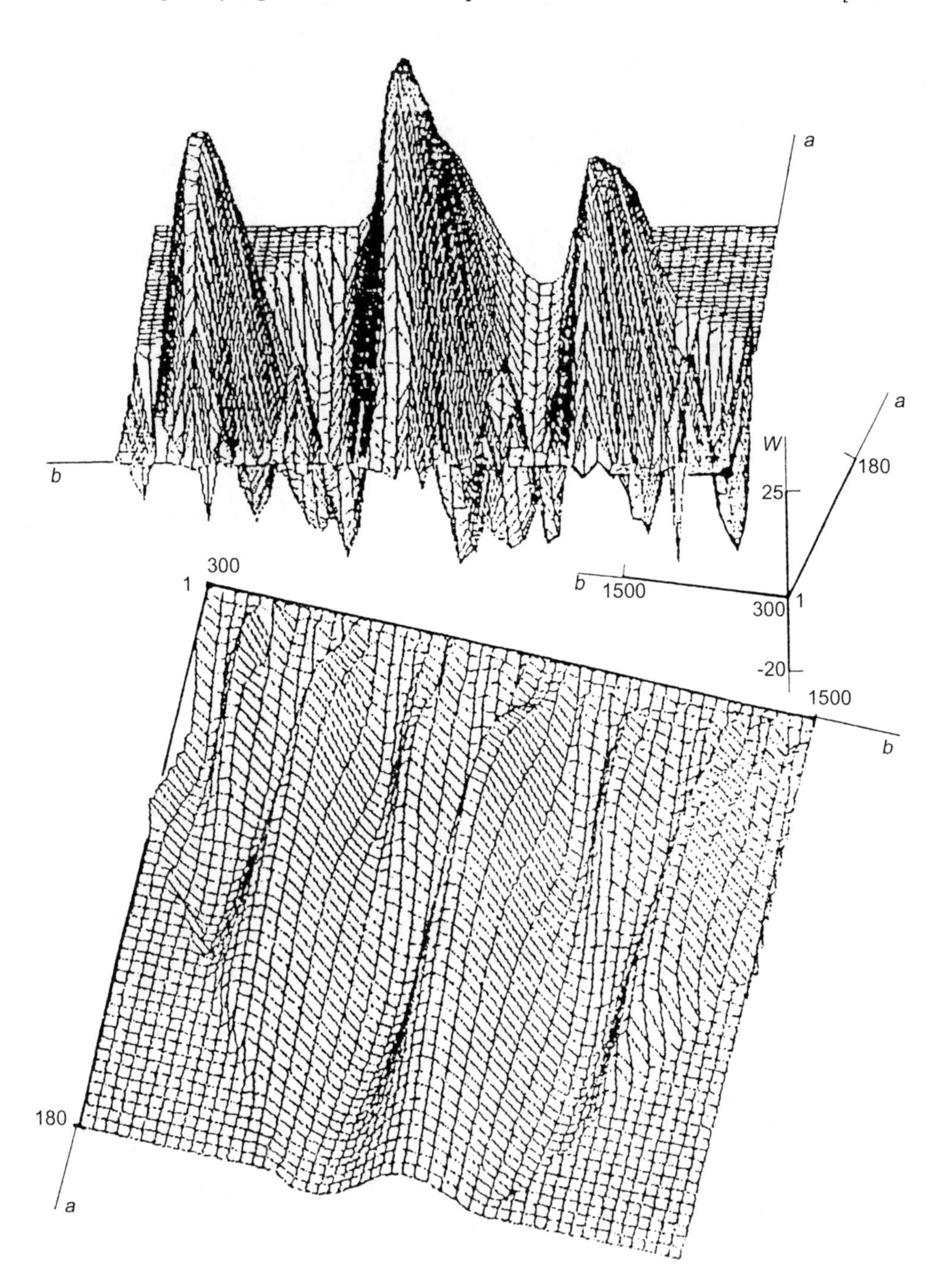

Figure 2.30. Three-dimensional pattern of the wavelet transform for time intervals from 300 to 1,500 days. The time series is given in Figure 2.29 (from Astafyeva *et al.*, 1994a).

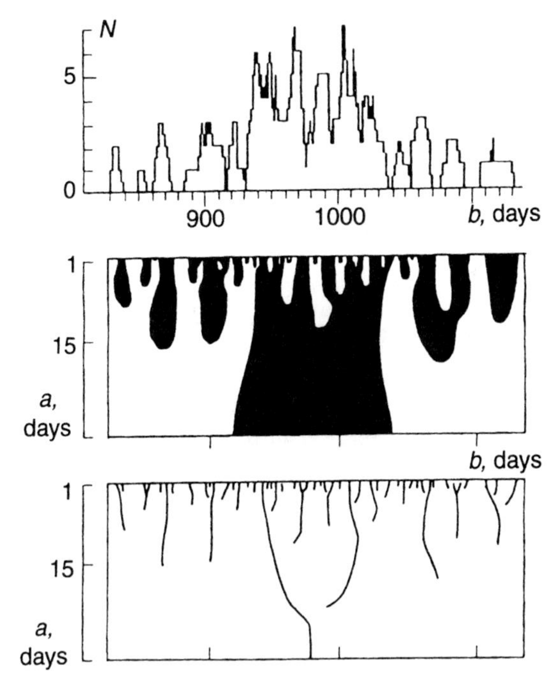

Figure 2.31. Time series of the intensity of global cyclogenesis (*N*) (a), the pattern of the wavelet transform (b), and corresponding local maximum lines (c) for the time interval marked in Figure 2.29 (from Astafyeva *et al.*, 1994b).

As is well known, natural systems form as a result of the interactions of a number of processes that have their own energy, timescales, and spatial sizes, which can noticeably differ. Accordingly, the hierarchical structure of a natural system is connected with the presence of several distinctive temporal and spatial scales.

In this case, tropical cyclogenesis can involve, at least once, three timescales, which are characterized by their own particularities. So, on timescales of the order of 1–6 days a system of global cyclogenesis presents itself (on first approach) as an object of Poisson type (e.g., volume of ideal gas by fluctuations of density). The main role here is played by the kinetics of the processes of birth and death of the elements of a structured system. In other words, there are processes leading to the generation, evolution, and dissipation of individual TCs that behave like non-interacting Brownian particles under a restoring interaction (see Section 2.2.4).

When turning to timescales of the order of 1 month, non-linear interactions are included in the mechanism of global cyclogenesis operation. These interactions form sorts of bifurcation "forks" with the integration of scales. Non-linear interactions are most likely caused by internal feedbacks in the global ocean–atmosphere system. In turn, internal feedbacks can be arranged by circulating the particularities of atmospheric macroflows.

On half-yearly timescales the system of global tropical cyclogenesis is obviously linearized—we nearly have a harmonic signal. The maximums of global cyclogenesis activity are connected with significant activity in August and September in the northern part of the Pacific Ocean and in the Atlantic, as well as in February and March in southern parts of the Indian and Pacific Oceans. Analysis of the pattern of the wavelet transform shows that cyclogenesis in different hemispheres (north and south) has an independent individual nature.

For greater timescales (of the order of several years) we can expect a new generation of bifurcation "forks" as a result of interannual interactions, which can turn out to be "north" maximums of activity of global cyclogenesis in August and September (the influence of "south" maximums on such greater timescales is not known). The physical reasons for such correlations will become clearer when the external influences are known. It may be possible to explain this by cosmic factors: solar–terrestrial relationships and magnetosphere–troposphere interactions.

Analysis leads us to expect global tropical cyclogenesis to behave, strictly speaking, like a non-equilibrium system with elements of kinetics, diffusions, and non-linear interactions.

2.8 HIERARCHICAL STRUCTURE OF POPULAR SERVICE SYSTEMS

Considerable recent attention has been focused on using the principles and methods of non-linear dynamics to resolve geophysical, physicochemical, and socioeconomic problems (Polak and Michailov, 1983; Coniglio, 1987; Kadanov, 1993; Jin *et al.*, 1994, 1996; Diaz and Margraf, 1993; Holdom, 1998; Dubois, 1998; Jensen, 1998; Roberts and Turcotte, 1998; Chen and Dong, 1996; Chang *et al.*, 1996; Ivanitskii *et al.*, 1998; Branover *et al.*, 1999). By applying the approaches of dynamic chaos, some interesting and important aspects of such physical systems have come to light. First, it should be pointed out that new auto-model features and hierarchical structures in the functioning of such physical systems have been detected. In other words, the heart of the problem is the internal multiscale interaction by virtue of both its physical reasons and structural topological features.

Recently, Astafyeva and Sharkov (1998) have demonstrated the striking fact that the hierarchical structures of two distinctly different processes—global tropical cyclogenesis and railway traffic—are closely analogous. It is normal to expect many natural objects, consisting of independent individual elements with weak "distant" correlations, to possess a hierarchical structure similar to the one considered above and to comprise the "Poisson" area, non-linear and linear areas of scales of interactions. In addition to natural objects, transport service systems (e.g., land, air, and sea), multicomponent socioeconomic processes (e.g., the correlation between demand and marketing, or market-related economic models at different stages of capital accumulation) (Samuelson, 1961; Travis, 1964; Fischer *et al.*, 1988; Holdom, 1998; Schweitzer, 1997; Grassia, 2000; Ponzi and Aizawa, 2000) fit into this group.

As the analogy between natural and social phenomena is not only of scientific but also of public interest, it would be well to consider the work by Astafyeva and Sharkov (1998) in more detail.

2.8.1 Critical parameter of traffic services

One of the important aspects of the study of a transport system (in particular, of railway traffic) is the consequences of natural or artificial malfunctions (interruptions, delays, and cancellations) of the traffic. The time it takes to get traffic back to normality (i.e., the reconstruction or correction period) and restore the timetables of the traffic will, to a considerable extent, be defined by how the traffic service is managed, and the exact details of the reconstruction period will be a factor (though possibly not directly) of such management. Knowledge of the objectives of the reconstruction period (in particular, its inerrancy, degree of non-linearity, hierarchy, etc.) will adequately influence the transport system and, in particular, give more reliable forecasts about volumes of transportation which is one of the most important factors in the efficient functioning of the whole transport system.

Despite significant efforts by train services to keep records of malfunctions (i.e., disruptions of services) of the traffic, the reconstruction period, particularly its temporal features, remains little explored. This is bound, on the one hand, by the absence of corresponding mathematical facilities for studying complex hierarchical systems and, on the other hand, by the absence of a united approach to the mathematical description of the process of malfunctions of the traffic and of the reconstruction period.

The mathematical description proposed in Astafyeva and Sharkov (1998) allows us to use a very effective goal-directed approach to the study of the transport system of rail traffic on different timescales.

We use wavelet analysis to study the timescale characteristics of the process of malfunctions.

2.8.2 Forming a time series for the traffic process

Studying probabilistic features of a signal of complex structure is conducted with the help of the mathematical technique of the theory of random flows.

In accordance with this approach, data about the analyzed process can be accumulated as follows. Notwithstanding the detailed structure and track record of individual malfunctions, we shall present each traffic malfunction on the time axis as a pulse of single amplitude with duration equal to the lifetime of the event (in other words, the whole reconstruction period before restoration of the service). It is possible to show that this simplification—at first glance at least—will not introduce principal changes to the general picture of the phenomena. The number of pulses (i.e., indistinguishable events) received in a single time interval (for our experimental data this value was formed after half an hour) is a natural physical parameter, a sort of instant intensity of traffic malfunctions (in other words, a break in traffic intensity).

In the language of statistical procedures, this approach is identified as a method of calculating the number of events with provision for their lifetime (Cox and Lewis, 1966; Apanasovich *et al.*, 1988).

A mathematically proposed procedure for accumulation of a signal can be described as:

$$N(t) = \sum \Theta(t - t_i; \tau_i) \tag{2.40}$$

where $N(t)$ is the "instant" intensity of traffic malfunctions (number of delays lasting half an hour); and $\Theta(t)$ is the Heaviside limited function:

$$\Theta(t - t_i; \tau_i) = \begin{cases} 1, & t_i \leq t \leq \tau_i \\ 0, & t_i + \tau_i < t < \tau_i \end{cases} \tag{2.41}$$

where τ_i is the time "existence" of the traffic malfunction (i.e., the length of disruption before restoration of the service; in other words, the rebuilding period); t_i is the time of origin of the traffic malfunction; $t_i + \tau_i$ is the time of liquidation of the traffic malfunction. The sequence of pulses formed by a similar image with corresponding duration and amplitude is the integer casual time flow of indistinguishable events. In this way, we have presented the time sequence of malfunctions as a statistical signal of complex structure.

Of course, this approach is greatly simplified (e.g., the danger of the event, material losses, the degree of financing involved in the reconstruction period, and other particularities are not taken into account). However, we can show that such an approach (and we will demonstrate this below) allows us to ascertain the important statistical regularities of traffic malfunctions to reveal its temporal variability, hierarchical, round-robin, or other structured particularities of the process under study. Moreover, particularities of the signal structure can turn out to be highly different on different timescales. Raw data on disruption to the timetables were obtained from the systematized archive of the Federation Traffic Service (Russia).

2.8.3 Wavelet patterns of disruption to traffic intensity

Figure 2.32 presents a time series of the intensity of traffic malfunctions (i.e., disruptions to services) that occurred on the North Railway Service (Russia) from 1992 till the middle of 1995. The time series of analyzing data involves 61,135 samples of the intensity of traffic malfunctions obtained by sampling every 30 minutes (Figure 2.32 presents 7.5 hours of data from 15 locations). In the lower part of the drawing the beginning and duration of each season (winter, spring, summer, fall) are defined. As expected, the intensity of traffic malfunctions is described by the signal peculiar to the telegraphic process.

The time series of the intensity of traffic malfunctions demonstrate the following distinctive particularities. The most powerful maximums of the intensity of traffic malfunctions are observed at the end of winter/beginning of spring (March) and in the summer. This phenomenon is associated with complex weather conditions at the beginning of spring and intensive transportation during summer. The maximum intensity of traffic malfunctions occurred in January 1995 and deserves further

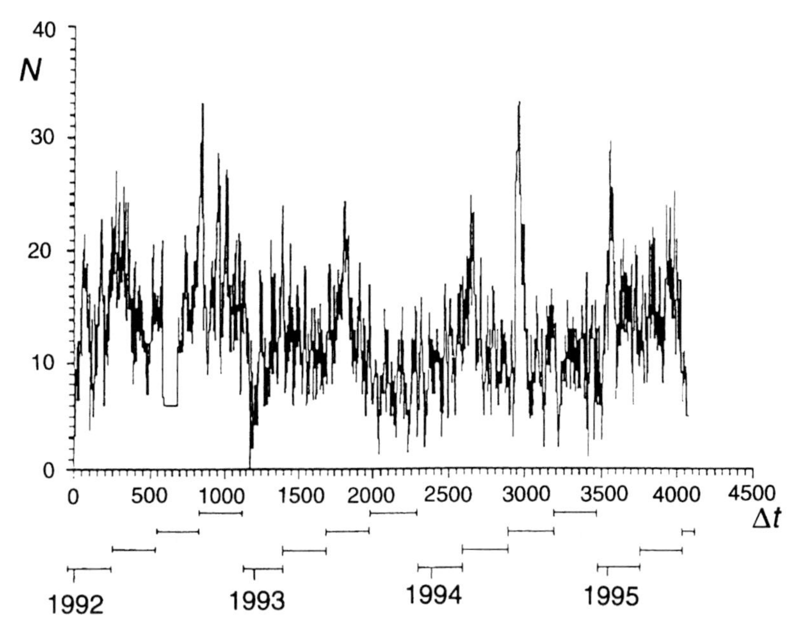

Figure 2.32. Time series of interruptions to railway traffic from January 1, 1992 to June 20, 1995 on a timescale (Δt) equal to 7.5 hours. Straight line segments represent quarters of the given year (from Astafyeva and Sharkov, 1998).

attention. Looking at the process for 1992–1994 this extremum is shown to be typical. We should also point out that the reduction in the intensity of malfunctions in the autumn–winter period is general for the whole analyzed set.

More fine particularities and regularities of the temporal flows obtained are shown by means of wavelet analysis. Figure 2.33 presents the results obtained by using the wavelet transform to describe the temporal set.

As already noted, the wavelet transform of a univariate signal provides a two-dimensional pattern in temporal and spectral spaces $W(a, b)$. On the axis of abscissas a change in time b is plotted and on the axis of ordinates the scale parameter a is presented (the scale decreases). To facilitate the collation of results, the time axes on Figure 2.32 (for the raw dataset) and on Figure 2.33 (for pictures of the factors of wavelet transforms) are alike. Dark areas on the pictures correspond to positive values of the coefficients of the wavelet transform and light areas negative values. The areas with different tones of gray reveal several ranges of the values of coefficients $W(a, b)$.

In Figure 2.33a the picture of factors of the wavelet transform is shown in such a range of scales as to compensate for the large-scale temporal dependence of the analyzing process. Here the scale changes every year. It is not difficult to see that the process demonstrates regular behavior on timescales from approximately 3 months up to 1 year (we call this range "large scale"). The round-robin annual cycle clearly stands out (particularly in the middle of the time range) with

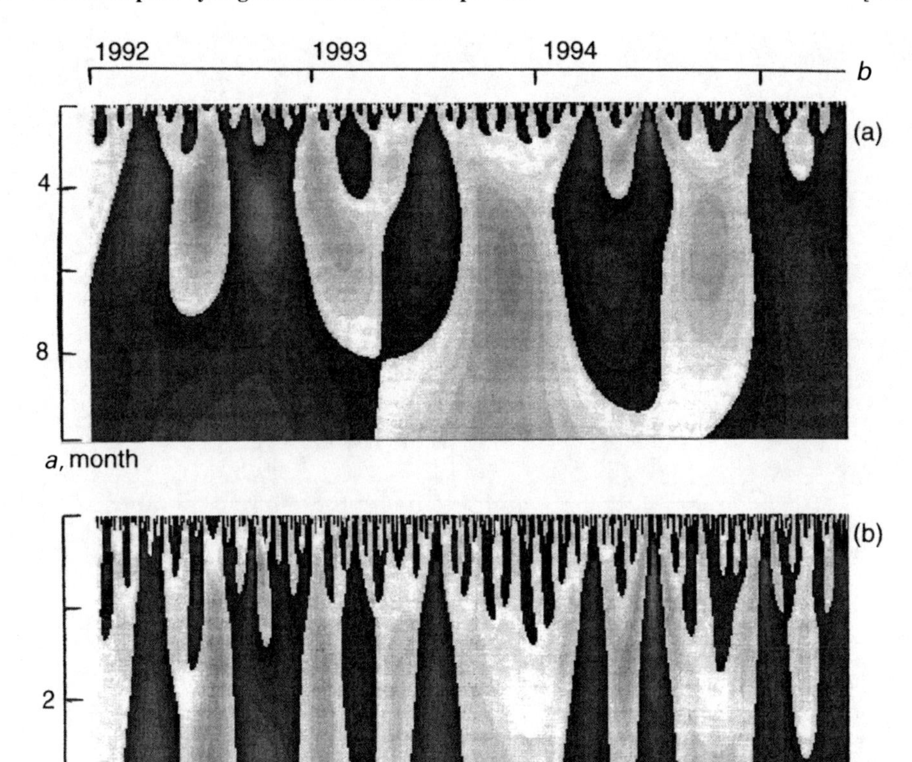

Figure 2.33. Wavelet transform two-dimensional patterns of a time series of railway traffic disruption (Figure 2.32) on timescales of up to 12 months (a) and up to 3.5 months (from Astafyeva and Sharkov, 1998).

intensity maximums in spring and summer and minimums in autumn and winter clearly defined.

In Figure 2.33b the small-scale part of the process (increased higher part of pictures shown in Figure 2.33b) is displayed using the most sophisticated treatment. Here the scale changes every 3 months for half of the months. We can see that the behavior of the process is sharply different from the large-scale one described above. A greater number of bifurcation "forks" are apparent, caused by non-linear internal mechanisms in the system on timescales of approximately 2 weeks up to 1 month (we call this range "average scale").

We consider more detailed analysis of the data for the first half of 1995 may be particularly revealing. Figure 2.34 presents a time series of the intensity of traffic malfunctions for this period. The lower part of the drawing shows (as steps) the duration of months (in the analyzed range of time) from January to May and 20 days in June.

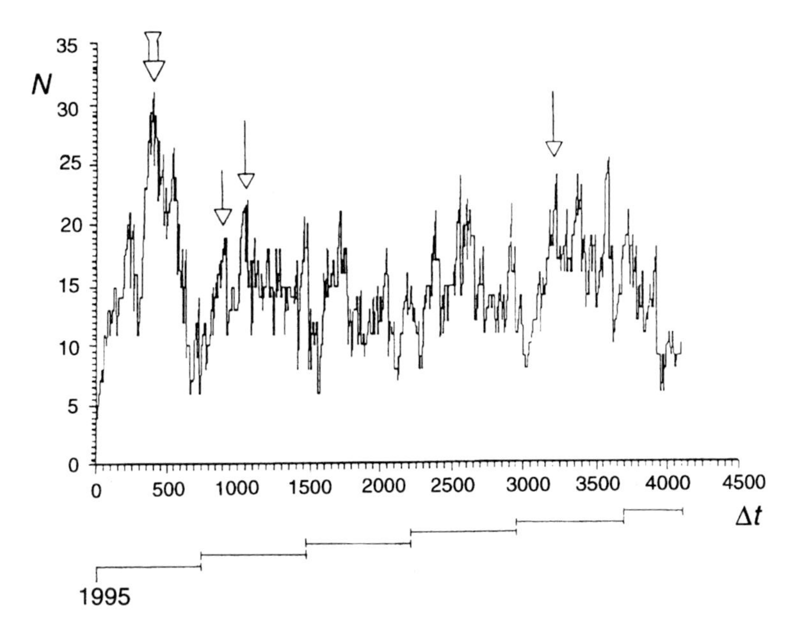

Figure 2.34. Time series of railway traffic disruption from January 1, 1995 to June 20, 1995 on a timescale (Δt) equal to 1 hour. Straight line segments represent months of the given year (1995) (from Astafyeva and Sharkov, 1998).

Note that the data demonstrate the presence of intensity maximums of the process of traffic malfunctions of at least one of the two types. Maximums having an accumulative period (i.e., indicated by arrows in Figure 2.34) can be seen, as can the maximum time of relaxation (indicated by the double arrow).

Figure 2.35 shows the pattern of the wavelet transform but this time with different scale values: here the scale changes every 1.5 months, whereas in Figure 2.34b it was every 11 days.

An important particularity of the process of traffic malfunctions follows from analysis of Figure 2.35. There is a whole family of small-scale events with very short times of relaxation (short "memory"). On the picture of coefficients they are presented as separate black lines. Relaxation processes can take (on evidence from wavelet analysis) from a fraction of an hour up to several (2–3) hours (we call this range "small scale"). On the other hand, events with distant time correlations exist (conditionally called "long memory" they occur every 1.5 months or so, as shown in Figure 2.33).

2.8.4 Possible physical models

The main result of using wavelet analysis to study the observational data of the process of railway traffic disruption is discovering the complex hierarchical time structure of the initial process. This allows a more goal-directed approach to

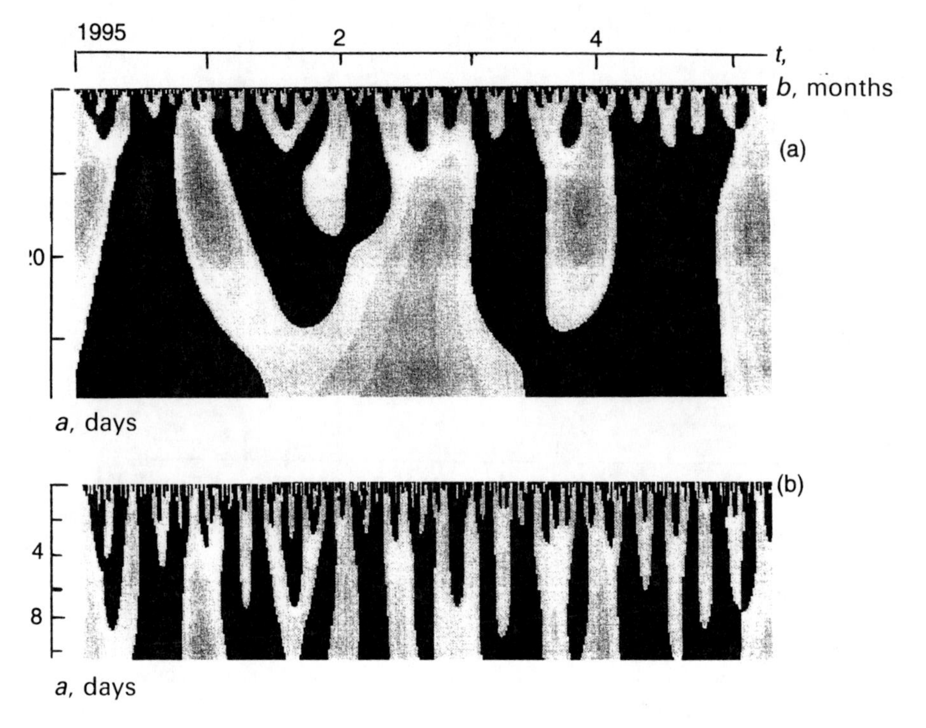

Figure 2.35. Wavelet transform two-dimensional patterns of a time series of railway traffic disruption (Figure 2.34) on timescales of up to 1.5 month (a) and up to 11 days (b) (from Astafyeva and Sharkov, 1998).

forming stochastic models of the process of traffic malfunctions (e.g., we can separate out the influence of technical, economic, and management factors that affect one or more particularities of the reconstruction period).

As is well known, the hierarchical structure of complicated systems is connected with the system having several different timescales, on the one hand, and having non-linear interactions affecting the time correlations of different scales, on the other hand.

As for the intensity of traffic malfunctions, we can choose at least one of three ranges of timescales, which are characterized by their own physical particularities. So, on small timescales of the order of several (2–3) hours the system of traffic malfunctions presents itself as a Poisson-type physical object (with independent events). Broadly speaking, this would be expected, because the reasons causing individual traffic malfunctions on these timescales clearly have no correlation with the traffic malfunction itself. In other words, the processes leading up to the events occurring (i.e., traffic disruption), their changes, and liquidations (i.e., the reconstruction period) in each concrete event behave like non-interacting Brownian particles in spring interaction models. For these timescales we can use the model of Brownian motion to move the particles from the viscous ambience into the system of finite sizes. The process is described by the well-known one-dimensional Langevin equation (Nicolis and Prigogine, 1977; Klimantovich, 1982; Polak and Michailov, 1983).

In the considered event it is necessary to note that a very quick reconstruction period for suppressing the majority of events exists. This can be a quantitative criterion of working a managerial system on primary levels.

When we consider average timescales (from 5–10 days up to a month) in the physical mechanism of how traffic malfunctions operate, we include the non-linear modes of operation of the managerial system, which refines the reconstruction period of traffic malfunctions. The analysis of wavelet diagrams shows non-linear modes forming bifurcation "forks" as timescales integrate. This can be interpreted as timescale integration of interactions in the managerial system. Most likely this occurs because of the presence in the managerial system of non-linear feedbacks which "work" at highly decelerated rates in both space and time.

On greater timescales (from a month up to a year) the physical system of traffic malfunctions is clearly linearized and presents itself almost as a harmonic signal. The activity maximums of this process prevail at the beginning of spring and summer every year.

The limited length of the time series does not allow us to pronounce on larger scales (e.g., interannual interactions) with any degree of confidence. However, it is highly probable that the interannual interaction (and connected bifurcation "forks") can be allied to standing management procedures (orders, instructions, etc.).

It follows from this that the process of traffic malfunction is a non-equilibrium system with elements of kinetics, diffusions, and non-linear interactions. On different timescales the sources of traffic malfunctions (disruption) and their liquidations (service restoration) are naturally determined by absolutely different physical processes: external influences (basically artificial) on timescales of sampling raw data (0.5 hour); inadequate management procedures on timescales of several hours up to 10 days; natural factors (floods, precipitation) and permanent management procedures and seasonal public–economic particularities (e.g., increasing intensity of transportation in summer) on timescales of several months up to a year.

On a broad range of timescales such factors as technical equipment of the railway network and non-linear feedback in the managerial system should be considered. We believe that the influence of these factors can be studied using wavelet analysis of the data for different railroads on different ranges of time and sampling.

2.9 MAGNETOSPHERE PROCESSES AND GLOBAL TROPICAL CYCLOGENESIS

The tropical zones of the global atmosphere–ocean system play a crucial role in the dynamics and evolution of synoptic and climatic meteorological processes on Earth. In this connection, any outer (Earth troposphere) interaction bringing additional influence on turbulent exchange in the tropical ocean–atmosphere system invites close investigation. Over the past 20 years, the findings of investigation into solar–terrestrial linkages demonstrate that statistical plausible components of tropospheric processes considered on the synoptic scale result from solar activity (Loginov, 1973; Danilov *et al.*, 1987; Raschke and Jacob, 1993; JSTC, 1995; King, 1999; WMO, 1999; Svensmark and Friis-Christensen, 1997; Bochnicek *et al.*, 1999; Gabis and Troshichev, 2000).

On the other hand, it is interesting to note that it was long proposed (at the end of the 19th century) that cyclic recurrences in solar activity can affect the frequency of individual TC generation in various ocean basins (see references in Tchijevsky, 1976 and the historical review in Elsner and Kara, 1999). Similar work has been carried out recently. In substance, the main experimental procedure in these investigations is called "the method of the superposition of epochs" (Tchijevsky, 1976). Despite this approach being used over a protracted period of the last 100 years, it is not mathematically valid. Alternatively, the results of an investigation into the cross-correlation between sunspots and North Atlantic hurricane activity (e.g., see fig. 10.15 in the book by Elsner and Kara, 1999) do not contain estimates of confidence levels of the correlation values observed. Thus, rigorous treatment for statistical procedures is lacking. Because of this, final conclusions cannot be reached re-garding the relationship between longer term solar activity and tropical cyclogenesis.

As for global tropical cyclogenesis, such observational attempts have long not met with success. This is primarily due to the fact that comprehensive data on global tropical cyclogenesis are lacking (see Chapter 5). It is only over the past 10 years that progress has been made in developing comprehensive datasets on global tropical cyclogenesis. This makes possible more solid investigations of solar–terrestrial links with "fast" (on the timescale of a day) interactions.

On the basis of what is currently known about climatic interactions, Bochnicek *et al.* (1999) point to the necessity of studying extraterrestrial–weather linkages on substantially shorter time intervals (i.e., a few days).

Gdalevich *et al.* (1994) present experimental evidence for the possible existence of new physical mechanisms providing "fast" interconnections between magnetosphere variability and the time evolution of global tropical cyclogenesis.

In this section, global tropical cyclogenesis is represented in simple form as a discrete Markov ("telegraph") process (see Section 2.2). By the "quantitative parameter of the activity of global tropical cyclogenesis" we mean the number of operating TCs in the World Ocean per day taking due account of the lifetime of individual TCs. So, the signal structure is the integer-numbered random time series of non-distinctive events (random impulse regime) (Figure 2.36). The characteristic, thus defined, determines the energy interaction in the ocean–atmosphere system and may be considered as the geophysical parameter needed to reveal the stability of the global ocean–atmosphere system (for the tropics).

Geomagnetic D_{st} variation is the global reduction in intensity of H-components in the terrestrial magnetic field, registered simultaneously by equatorial stations around the Earth. In accordance with the developing physical concept (see below), the quantitative features of geomagnetic variation reviewed for this study will be an integral of variation intensity within the current day and so $W = \int_{\Delta t} \Delta D_{st}\, dt$ where

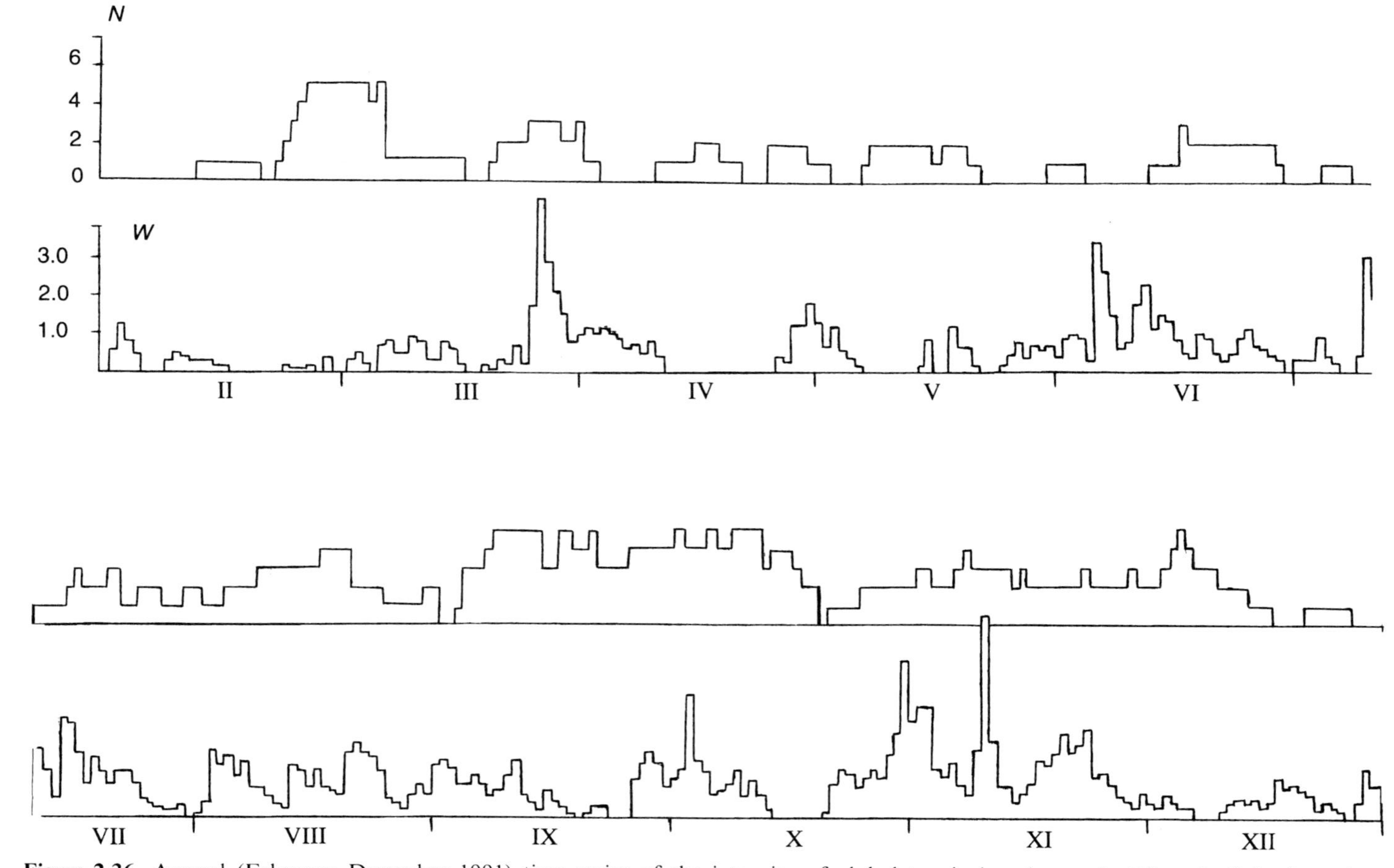

Figure 2.36. Annual (February–December 1991) time series of the intensity of global tropical cyclogenesis (N) and of the intensity of magnetospheric variability (W) (from Gdalevich *et al.*, 1994).

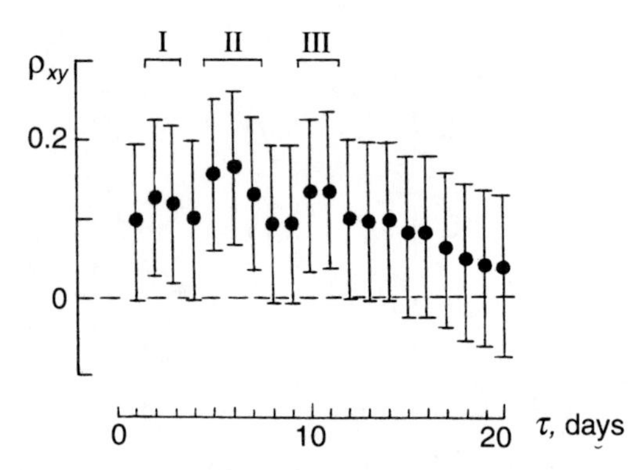

Figure 2.37. Cross-correlation coefficient ρ_{xy} of the intensity of magnetospheric variability and of the intensity of global tropical cyclogenesis. The $\log(\tau)$ is in days. Vertical line segments are 95% confidence levels. Black dots are sampling values of the normalized cross-correlation coefficient. The sample run is 365 days (Gdalevich *et al.*, 1994).

ΔD_{st} is the negative contrast between the non-perturbed and terrestrial magnetic field outraged by the condition of H-components of the terrestrial magnetic field; and $\Delta t = 24$ hours. In the physical sense W can be interpreted as a value proportional to the energy in the unit of time (power) of a geomagnetic storm.

The respective time series of parameter W are shown in Figure 2.36. Analysis of Figure 2.36b reveals the presence of all stages of a geomagnetic outbreak from the moment the main phase storm develops ("splash" of W) to phases of reconstruction (extended "tail" of the disturbance).

Experimental material was subjected to cross-correlation processing with a timestep (timelag) of 1 day, and, on the strength of supposed physically stipulated relationships between processes, temporal writing of the integral W "was shifted" onwards. Sampling values of the normalizing factor of cross-correlations with timelags from 1 up to 20 days are shown in Figure 2.37.

Statistical analysis of the validity of this probabilistic procedure, connected with the calculation of cross-correlations, was considered using two (generally speaking, independent) approaches. One considered the building of confidence-level intervals using a normalizing Fisher transformation with confidence-level probability of 0.95% (level of value 0.05). The other checked the truth of these statistical ("zero") hypotheses: $H_0[\rho_{xy} = 0]$, where ρ_{xy} is the truth factor of cross-correlations under different time shifts (from 1 up to 20 days).

The results of the first approach are shown in Figure 2.37 in the construction of confidence-level intervals of 0.95% for the truth factor value ρ_{xy}.

Figure 2.37 shows there are three time shifts (2–3, 5–7, and 10–11 days) when confidence-level intervals do not include $\rho_{xy} = 0$ and, hence, are responsible for the difference of $\rho_{xy} = 0$ in this temporal range from zero.

The second approach (i.e., checking the truth of hypothesis H_0) ends by building a selective function $t = r_{xy}(n - 2/1 - r_{xy}^2)^{1/2}$, where r_{xy} are the sampling values of cross-correlations, and n is the full volume of samples. Moreover, sampling statistics

t comply with the Student t-distribution with $f = n - 2$ degrees of freedom. Evaluation of selective functions t for shifts from the areas specified above gives a value of t equal to 2.7 here with the critical value of statistics of $t_{0.05;364} = 196$ (in the $\alpha = 0.05$ and $f = 364$ forms). We can see that $t > t_{0.05;364}$ and the H_0 hypothesis for specified temporal shifts must be rejected and, hence, the empirical value of factor cross-correlations differs substantially from zero. On the other hand, evaluation of functions t for other time shifts satisfies the inequality $t < t_{0.05;364}$ and, hence, hypothesis $H_0[\rho_{xy} = 0]$ can be accepted. In other words, deflections of the empirical factor r_{xy} from zero have a purely random nature.

This result can be interpreted as follows. During geomagnetic storms in the magnetosphere, energy (10^{23} erg) is released. This energy disperses on creation or reinforcement of the recirculating current—heating Arctic regions and exciting different waves. Variation of the magnetosphere current (World geomagnetic storms) disturbs ionospheric layers, which, in turn, generate infrasonic waves reaching the troposphere and disrupting the geostrophic balance. This can greatly transform spatial features of mesoscale turbulence (intensity and helicity) and, in accordance with the concept of the vortical dynamo (Chapter 4), will change the rate of generation and time of operation of tropical structures. Note that there may well be more than one interaction chain. Thus, it causes the presence of several timescales of efficient interactions. These considerations allow us to reach the non-trivial conclusion of the likely existence of a stochastic physical mechanism that is responsible for the generation of TDs by these teleconnections.

3

Regional tropical cyclogenesis

3.1 PACIFIC TROPICAL CYCLOGENESIS AS A STOCHASTIC PROCESS

Based on statistical processing of the time sequence of the number of active tropical cyclones (TCs) occurring in the form of a random stream of homogeneous events, we were able to show (see Sections 2.1 and 2.3) that a stochastic model of the intensity (amplitude characteristics) of global cyclogenesis may be represented in the form of two Poisson processes (in a yearly cycle) with stable characteristics. The characteristics of the geographical distribution of the centers of TC generation on global and regional scales over the last 10 years were studied in Section 2.6.

There is reason to suppose that the statistical models of regional tropical cyclogenesis are compatible with Poisson distributions, but have their own specific characteristics, including those of a seasonal nature. It is known (Neumann, 1993) that the Pacific Ocean Basin, which makes the decisive contribution to global cyclogenesis, may be divided into three active regions, the northwest Pacific (NWP), the northeast Pacific (NEP), and the southwest Pacific (SWP).

The purpose of this section is to give a representation of experimental results, based on the processing of time series of the intensity of Pacific Ocean cyclogenesis over 10 years (1983–1992), pointing to the possibility of using a Poisson model of intermittent nature (in time).

The methodological approach described in this chapter is based on representing the time series of the intensity of tropical cyclogenesis as a random flow of homogeneous events.

Since we are not interested in the detailed structure and dynamics of each individual tropical formation, on the time axis we shall represent each tropical disturbance as a pulse with a single amplitude and a random length (corresponding to the time over which the TC is active) with a random time of occurrence (generation of an individual TC). The number of pulses occurring (events) in a unit time interval

E. A. Sharkov, *Global Tropical Cyclogenesis* (Second Edition).

(24 hours in our case) is therefore a natural physical parameter, the "instantaneous" intensity of cyclogenesis, which determines the energetics of the ocean–atmosphere interrelationship. In the language of statistical procedures, this approach is a way of counting the number of events that takes into account event lifetimes (Cox and Lewis, 1966).

The raw data for 1983–1992 for regional cyclogenesis in the Pacific Ocean were taken from the systematized database Global-TC of remote observations of global tropical cyclogenesis (Pokrovskaya and Sharkov, 1994c, 1997a, 1999a, d).

3.1.1 Probability models of Pacific cyclogenesis intensity

Based on known rules, histograms of the integer-valued parameter $N(t)$ were constructed (these are the statistical analogues of sample probability densities) for each dataset for the three regions of the Pacific Ocean over a 3-month interval (for NWP and NEP June to August; for SWP January to March) and for each year from 1983 to 1992. In addition, sample means and variances were calculated for each dataset. The results of the processing arc are shown in Tables 3.1–3.3 and in Figure 3.1.

Table 3.1. Parameters of the distributions of the intensity of cyclogenesis in the NWP for 1983–1992. The number of days over which sampling was performed covered June–August 1992. The figures in parens give the tabulated values of the function $\chi^2(\alpha,f)$ for the Pearson criterion at the 95% confidence level for a corresponding number of degrees of freedom.

Year	*Total No. of occurrences* (n)	*Main numerical characteristics of histograms*		*Parameter of approximating Poisson law* (λ, day^{-1})	*Measure of deviation of Pearson criterion*
		Mean	*Variance*		
1983	58	0.62	0.61	0.6	0.49 (4.30)
1984	88	0.95	0.66	1.0	6.26 (12.7)
1985	94	1.02	1.04	1.0	8.58 (2.78)
1986	90	0.98	0.84	1.0	1.68 (4.30)
1987	86	0.93	0.58	1.0	6.45 (4.30)
1988	84	0.91	0.90	1.0	8.23 (3.18)
1989	115	1.24	0.82	1.2	2.81 (4.30)
1990	102	1.10	0.96	1.1	9.55 (4.30)
1991	71	0.75	0.91	0.8	0.08 (4.30)
1992	93	0.98	1.01	1.0	1.47 (3.18)

Table 3.2. Parameters of the distributions of the intensity of cyclogenesis in the NEP for 1983–1992. The number of days over which sampling was performed covered June–August 1992 (notation as in Table 3.1).

Year	*Total No. of occurrences* (n)	*Main numerical characteristics of histograms*		*Parameter of approximating Poisson law* (λ, day^{-1})	*Measure of deviation of Pearson criterion*
		Mean	*Variance*		
1983	70	0.76	0.42	0.8	8.00 (12.7)
1984	79	0.85	1.69	0.9	8.19 (4.30)
1985	110	1.20	0.86	1.2	17.36 (4.30)
1986	71	0.77	0.42	0.8	9.64 (12.7)
1987	74	0.79	0.63	0.8	5.33 (4.30)
1988	47	0.38	0.27	0.4	6.80 (12.7)
1989	76	0.82	0.77	0.8	4.78 (4.30)
1990	98	1.07	0.45	1.1	23.70 (4.30)
1991	56	0.60	0.68	0.6	14.00 (4.30)
1992	99	1.18	0.77	1.2	1.22 (4.30)

Comparison of sample histograms with possible approximating Poisson laws on linear and semilogarithmic scales (Figure 3.1) indicates the potential of studying hypotheses about the Poisson nature of fluctuations in the amplitude of the intensity $N(t)$ of tropical cyclogenesis in the Pacific Ocean Basin. Thus, according to Pearson's criterion, analysis of the measure of deviation of theoretical distributions and experimental histograms for data for the NWP for the period 1983–1992 (see Table 3.1) shows that the experimental samples are by and large compatible with the parent population having the Poisson distribution for the values of λ (see Section 2.2.4) indicated in the table. Data for 1985, 1988, and 1999 are exceptions, for which unsatisfactory agreement with theoretical approximations is observed.

The reasons for deviations are easily determined from analysis of the experimental histograms themselves. For the 1985 data the tail of the experimental histogram deviates from the Poisson branch (Figure 3.1c) because of an excessive grouping of four to five TCs. For the 1990 data, there are too many pairs of elements ($N = 2$) and too few single elements ($N = 1$) (Figure 3.1a).

As is the case for the NWP, there is satisfactory agreement by and large between the experimental histograms and the theoretical Poisson distribution observed for NEP data (Table 3.2), except for the data of 1985, 1990, and 1991. For 1985 and 1990 data, too many pairs ($N = 2$) and single ($N = 1$) elements are observed (Figure

Table 3.3. Parameters of the distributions of the intensity of cyclogenesis in the SWP for 1983–1992. The number of days over which sampling was performed covered January–March 1991 (notation as in Table 3.1).

Year	*Total No. of occurrences* (n)	*Main numerical characteristics of histograms*		*Parameter of approximating Poisson law* (λ, day^{-1})	*Measure of deviation of Pearson criterion*
		Mean	*Variance*		
1983	67	0.52	0.92	0.5	54.040 (3.18)
1984	51	0.56	0.41	0.6	0.400 (12.7)
1985	38	0.41	0.40	0.4	4.530 (4.30)
1986	13	0.14	0.20	0.1	31.400 (12.7)
1987	29	0.31	0.22	0.3	1.240 (12.7)
1988	37	0.40	0.61	0.4	27.650 (4.30)
1989	48	0.53	0.67	0.5	8.880 (4.30)
1990	41	0.45	0.40	0.5	0.030 (12.7)
1991	15	0.16	0.17	0.15	0.010 (12.7)
1992	54	0.56	0.39	0.6	0.048 (12.7)

3.1d) and the number of "zero" values of N is clearly low. On the other hand, the situation is reversed for the 1991 data, for which there are too many zero values of N and too few entries with ($N = 1$) (Figure 3.1e).

As far as the data for the SWP are concerned, there is considerable disagreement between the experimental histograms and theoretical distributions for 1983, 1986, and 1988 (Table 3.3). In this case, analysis of Figure 3.1f–h shows that the causes of these deviations are associated with the marked grouping of structural elements for $N = 3$, 4 in the tail area of the distribution.

By virtue of the large number of samples of events in the experimental data ($n > 50$, see Tables 3.1–3.3) an accurate estimate of the intensity of the stream from the Poisson distribution λ together with corresponding confidence levels may be obtained using a method based on the normal approximation of the Poisson distribution (Cox and Lewis, 1966) by means of the equation:

$$\lambda t_0 = n + \tfrac{1}{2} C^2_{\alpha/2} \pm C_{\alpha/2} n^{1/2} \tag{3.1}$$

where t_0 is the number of days over which sampling was performed (see Table 3.3); and C_α is the upper α-quantile of a unit normal distribution. Using this equation, for a probability of 95% we obtain confidence levels for λ. Accurate estimates and confidence-level intervals for the average value of λ for tropical cyclogenesis in the

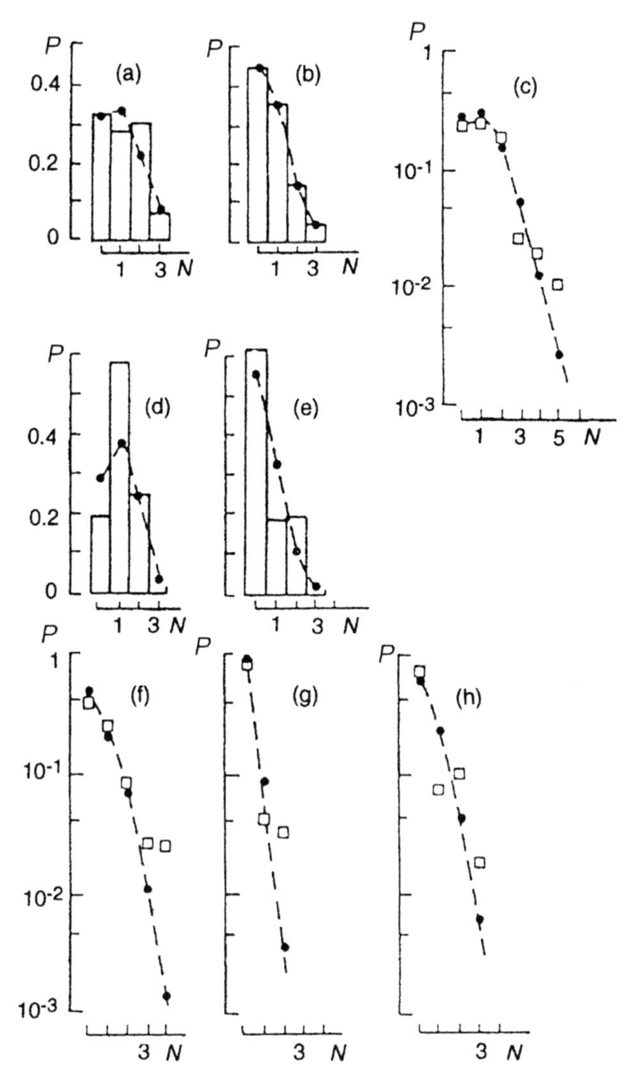

Figure 3.1. Sample probability densities for the intensity of cyclogenesis and theoretical Poisson distributions on linear (a)–(d) and semilogarithmic (e)–(h) scales. Stepped curves and open squares represent experimental histograms; full circles and dotted lines represent the theoretical Poisson distributions with the values given in Tables 3.1–3.3: (a) data for the NWP for 1990; (b) for 1991; (c) for 1985; (d) NEP for 1990; (e) for 1991; (f) SWP for 1983; (g) for 1986; (h) for 1988 (from Pokrovskaya and Sharkov, 1996a).

three regions of the Pacific Ocean studied over the period 1983–1992 are shown in Figure 3.2. Analysis of Figure 3.2 shows that over the 10 years studied the intensity of TCs exhibited significant variation; this applies to the NEP and the SWP in particular, although other trends may be detected (e.g., cyclogenesis in the NWP is most intense). The average intensity of TCs over the 10-year period is approxi-

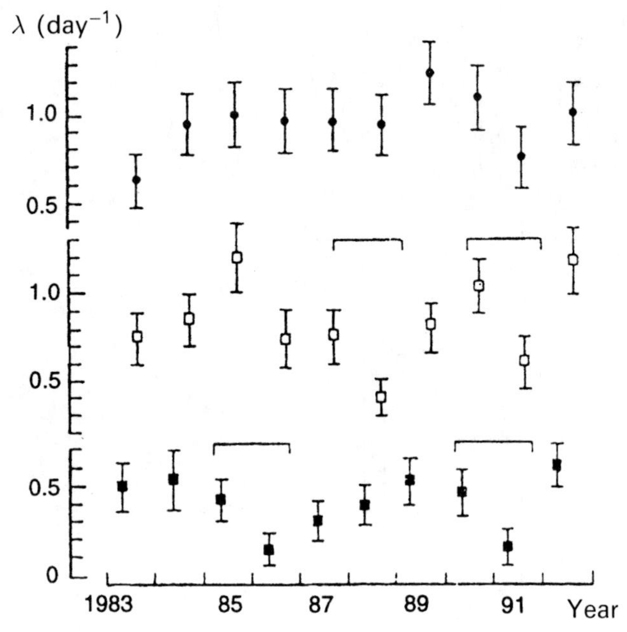

Figure 3.2. Experimental values of the intensity of cyclogenesis over periods of the year by region for 1983–1992. Vertical bars denote the 95% confidence limits for the average intensity value. Full circles denote data for the NWP, open squares denote data for the NEP, and full squares denote data for the SWP (from Pokrovskaya and Sharkov, 1996a).

mately $\bar{\lambda} = 1.0$; in other words, this implies the presence of one active TC per day in the region. The NEP has TCs with a somewhat lower intensity, for which $\bar{\lambda} = 0.9$ (over the 10-year period). The SWP has a low intensity $\bar{\lambda} = 0.45$. It is worth noting the marked decrease in the intensity of cyclogenesis in this part of the Pacific in 1986 and 1991 (by a factor of almost 4.5–5).

An analogous effect occurred in the NEP in 1988, when the intensity of cyclogenesis fell by a factor greater than 2. In this connection, cyclogenesis in the NWP was more stable and the differences remained less than 30%.

3.1.2 Interannual variability of Pacific cyclogenesis

We now consider the question of interannual variation in the intensity of the Poisson set simulating tropical cyclogenesis, which is important as far as the creation of prediction models is concerned. Essentially, we have to accept or reject the hypothesis that all the Poisson models identified (over 10 years) belong to the same parent population. For this, we use the method of the dual *t*-criterion (Feller, 1968), which is also applicable for asymmetric distribution.

The physical meaning of this statistical procedure is the following: there is a need to examine whether the physical reasons that form random year sample sets are common or not.

To test the hypothesis $H_0 : \bar{\lambda}_1 = \bar{\lambda}_2$, we consider the sampling function:

$$t = \frac{\bar{\lambda}_1 - \bar{\lambda}_2}{\sqrt{s_1^2 + s_2^2}} \sqrt{t_0} \tag{3.2}$$

where t_0 is the observed sample size $t_0 = 92$; and s_1^2 and s_2^2 are experimental values of the variance (see Tables 3.1–3.3). We compile the sampling function for values of parameters for 1985–1986 for the SWP (see Table 3.3) and obtain $t = 3.36$. For a given value of $\alpha = 0.01$ and $f = 182$ degrees of freedom, we obtain the value for the given quantile of the Student t-distribution, $t_{\alpha,f} = 2.33$. Comparison of t with $t_{\alpha,f}$ indicates that the hypothesis is false with a probability of 99% and samples for TC statistical processes for 1985 and 1986 should be assigned to different parent populations. An analogous procedure, carried out for the SWP for 1991 and 1990 and for the NEP for 1987–1988 and 1990–1991, showed that there is a significant difference in average intensity values. All four combinations are shown in Figure 3.2.

In all other combinations, no significant difference is observed; in particular, there is no significant difference over the whole 10-year period of tropical cyclogenesis in the NWP.

As previously mentioned, to identify the physical reasons for the significant differences in the average intensity values observed above, we need to carry out more detailed analysis of time variation in the set of events and construct the function $F(t)$ of the cumulative number of events (see Eq. 2.3).

Figure 3.3 shows the function of the cumulative number of tropical cyclogenesis events for the three studied regions of the Pacific for 1990 and 1991. Analysis of the data shown in Figure 3.3 and of the function $F(t)$ for other years (not shown in

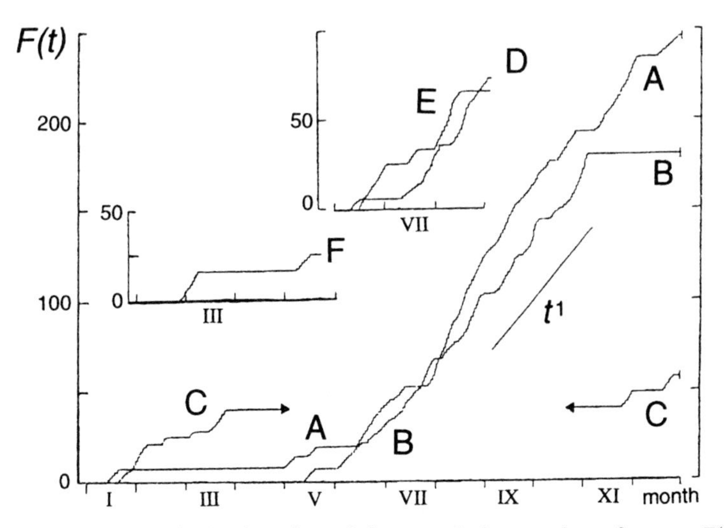

Figure 3.3. Time variation in the function of the cumulative number of events $F(t)$ for 1990 and 1991 in the three active zones of the Pacific Ocean: A, NWP, 1990; B, NEP, 1990; C, SWP, 1990; D, NWP, 1991; E, NEP, 1991; F, SWP, 1991 (from Pokrovskaya and Sharkov, 1996a).

Figure 3.3 for lack of space) shows that the behavior of $F(t)$ for 1990 and 1991 is fairly characteristic of the entire 10-year period studied. The main characteristic of time variation of the cumulative function is its discontinuous nature (or intermittency). In other words, periods in which TCs are generated and active—where the derivative of $F(t)$ with respect to time is non-zero—alternate chaotically with periods of "silence" —in which $dF(t)/dt = 0$—where these periods may be quite significant (up to a month).

It is known (Cox and Lewis, 1966) that the differential quotient of the cumulative function F for a Poisson process and the observation time t, $\Delta F/\Delta t$, is simply the average intensity value of the Poisson set of events $\bar{\lambda}$. If Δt is chosen sufficiently small, then we may talk about a kind of differential value of the stream intensity (λ_d) and its variations with time, $\lambda_d(t)$, and about its integral value $\bar{\lambda}$ if Δt is taken to be the whole period of observation. Detailed analysis of the time variation of $F(t)$, given in Figure 3.3 for 1990 and 1991, reveals an important characteristic, namely that in periods of generation of $F(t)$, the differential value $\lambda_d(t)$ is approximately constant ($\lambda_d \approx 1$ per day) for different areas (NEP, NWP, and SWP) and for different seasons (1990 and 1991). Differences only occur in the total generation time over the period of observation, and this is naturally apparent in the integral value of the stream intensity. Thus, the "sluggish" generation of cyclones in 1991 (and the correspondingly low value of $\bar{\lambda}$ for all three areas) was the result of a low total time for the generation of cyclones, as Figure 3.3 shows (inset for 1991 data). Here, the differential value λ_d remained approximately constant at $\lambda_d \approx 1$ per day, as for the 1990 data.

3.1.3 Intermittency coefficient and the true intensity of Pacific cyclogenesis

Two parameters are used to describe the intermittent process of generation and activity numerically. The first parameter is given by the ratio of the number of events (N_g) over the pure generation time (T_g) of the process (i.e., when $\Delta F/\Delta t$ is non-zero). Thus, $\lambda_0 = N_g/T_g$ per day is the true intensity of the Poisson set. The second parameter is the ratio of the pure generation time (T_g) to the time of a complete period of observations (t_0) or the number of days over which sampling took place. We define $\gamma = T_g/t_0$ to be the intermittency coefficient.

The values of the intermittency coefficient and the true intensity of the flows computed from experimental data for 1983–1992 for the three regions of the Pacific are shown in Table 3.4 and Figure 3.4. Special computations show that the intensity values of the Poisson flows $\bar{\lambda}$ found earlier (see Tables 3.1–3.3) are related to the true intensity λ_0 within a good degree of accuracy (better than 5%) by the equation $\lambda_0 = \bar{\lambda}/\gamma$. In other words, the physical sense of the true intensity is that it recovers the average value $\bar{\lambda}$ over a period of observation t_0 when the generation process is quasi-homogeneous during the period of observation. This is precisely what we expect for Poisson sets of quasi-homogeneous type (see Section 2.2.4).

Analysis of the data for λ_0 and γ in Table 3.4 and Figure 3.4 allows us to draw a number of important and fairly unexpected conclusions.

Table 3.4. Intermittency coefficient γ and the true intensity of cyclogenesis random flows (λ_0) for the three basins of the Pacific for 1983–1992.

	NWP		*NEP*		*SWP*	
Year	λ_0	γ	λ_0	γ	λ_0	γ
1983	1.38	0.46	1.18	0.64	1.45	0.50
1984	1.44	0.66	1.41	0.60	1.27	0.44
1985	1.59	0.64	1.59	0.75	1.15	0.36
1986	1.52	0.64	1.18	0.65	1.30	0.10
1987	1.38	0.67	1.48	0.54	1.16	0.27
1988	1.33	0.68	1.56	0.3	1.76	0.23
1989	1.62	0.77	1.41	0.58	1.50	0.35
1990	1.67	0.66	1.32	0.80	1.17	0.39
1991	1.39	0.55	1.55	0.39	1.66	0.10
1992	1.60	0.63	1.48	0.72	1.20	0.05

First, the average true intensity value $\bar{\lambda}_0$ over 10 years for all three areas of the Pacific Ocean is strikingly stable. For the NWP $\bar{\lambda}_0$ is 1.48, for the NEP it is 1.42, and for the SWP it is 1.37 per day. It is not hard to see that differences in the absolute value amount are smaller than 7%.

Second, interannual variability in the true intensity is also insignificant; the maximum differences for all three areas are smaller than 25%. The situation is rather different with the intermittency coefficient. For the NWP we observe (Table 3.4 and Figure 3.4) a clearly expressed homogeneity: the mean value $\bar{\gamma}$ over 10 years is 0.64 and maximum deviations range between 15% and 20%. For the NEP the situation is quite stable and similar to that for the NWP ($\bar{\gamma} = 0.64$) except for 1988 and 1991 where the decrease in γ was 100% (in both cases). Generation characteristics in the SWP are very different from those of the NEP and the NWP: the average value (over 10 years) of the intermittency coefficient is 0.32 (i.e., a factor of 2 smaller than the NWP and NEP). In addition, there is a clearly expressed sharp decrease in γ for 1986 and 1991. Recalling the previous relationship $\bar{\lambda} = \lambda_0\gamma$ it is not difficult to see that the marked variability in the values of $\bar{\lambda}$ observed by us (Figure 3.2 and Tables 3.1–3.3) is not the result of the internal generation properties of the ocean–atmosphere system (γ_0 remains constant for all three areas and over the 10-year period) but of particular external conditions which retard (or eliminate) the conditions for the generation of TCs. From the point of view of the generation properties of the system in this case, it is a matter of entering a "severe" cyclone generation regime.

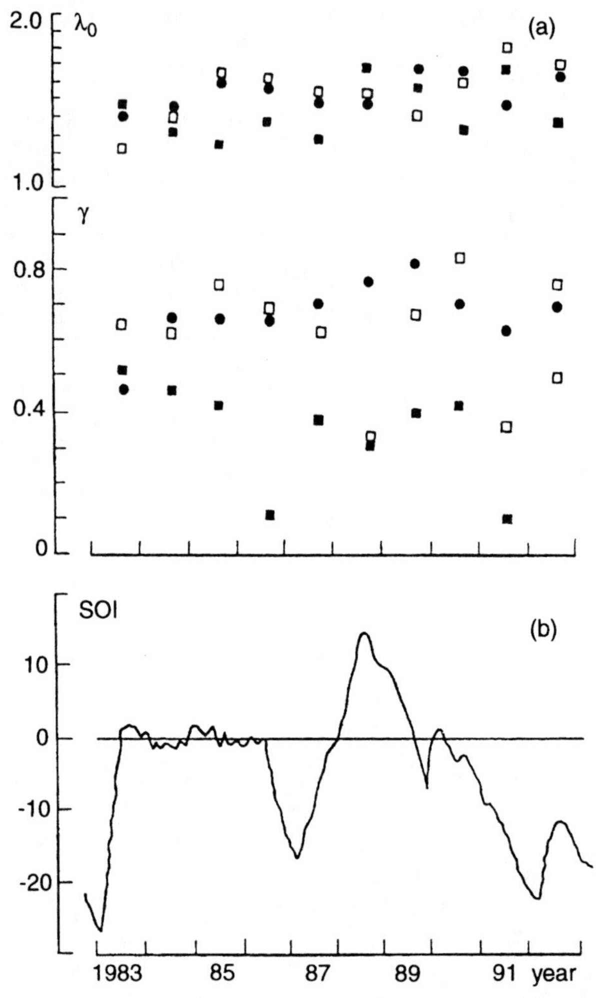

Figure 3.4. Experimental values of the true intensity λ_0 and the intermittency coefficient γ over the 10-year period: (a) see notation for Figure 3.2; (b) values of the SOI index for ENSO phenomena averaged over 5-month periods between 1983 and 1992 (from Pokrovskaya and Sharkov, 1996a).

This experimental fact constitutes positive evidence in support of the recently described (Chapter 4) model of global cyclogenesis showing gradual generation in an excited autowave medium of the ocean–atmosphere system (see Chapter 4).

The specific physical causes of the apparent sluggishness of cyclogenesis likely point to the possible effect of El Niño–Southern Oscillation (ENSO) phenomena. Attempts to link variation in the intensity of cyclogenesis in the Pacific Ocean Basin with characteristics of ENSO (Figure 3.4b) provide sound evidence in support of the hypothesis that ENSO phenomena affect cyclogenesis. For example, it follows from analysis of Figure 3.4b that during the intense ENSO episodes in 1983, 1986–1987, and 1991–1992 tropical cyclogenesis in the Pacific Ocean Basin experienced a sharp decrease in total generation time (intermittency coefficient) while, at the same time, the true intensity was practically unchanged. In this sense, standard (and repeated) assertions about the sluggishness of SWP cyclogenesis should be interpreted as referring to the severity of external conditions at the time of pure generation rather than to the weakness of the internal generation properties of the ocean–atmosphere system (these are the same for all areas of the ocean). We must bear in mind here that our research concerns the average daily intensity of cyclogenesis and consideration of other scales (monthly and annual) may in principle change the situation. For example, Revell and Goulter (1986) show that weak differentiation in the spatial zoning of centers of TC generation was identified (yearly averaged) in the SWP in both the presence and absence of ENSO phenomena. In such cases, the amplitude characteristics of cyclogenesis were not taken into account.

Thus, based on the statistical analysis of observational data, it has been shown that an intermittent-type Poisson process may satisfactorily describe the amplitude characteristics of all three active zones of TC cyclogenesis. It was found that the true intensity value of the stream is very stable (seasonally and across regions of the Pacific Ocean) while variation (decrease) in the integral values of intensity is the result of the short time of total cyclone generation.

The models described in this section and their numerical characteristics may be used in ballistic design procedures for potential satellite systems and in the operational design of space experiments.

3.2 THERMAL STRATIFICATION OF TROPICAL ATMOSPHERE AND PACIFIC CYCLOGENESIS

An important aspect in the study of the ocean–atmosphere interaction is the effect of intense TC-type eddy perturbations on the thermodynamics and kinematics of a tropical atmosphere (Riehl, 1979; Gray, 1979; Anthes, 1982; Merrill, 1988a, b). Pokrovskaya and Sharkov (1988, 1990, 1991, 1993b) carry out detailed analysis of the results of studies (using space and radiosonde data) of spatiotemporal variation in the thermal stratification of a tropical atmosphere over the Pacific Ocean for three very different synoptic situations: when conditions are calm (no TC effects); when there is continuous passage of a TC over the given geographical region (thermal "tracks" in the atmosphere); and during the thermal state of the tropical atmosphere over the Pacific Ocean at the location of future genesis of a TC at the time of its birth, during its intensification and transformation to the stage of a tropical storm (TS), and then after the TS leaves its birthplace.

3.2.1 Initial observational data and processing methodology

As initial data for the study we used the thermal profiles of a free (cloudless) tropical atmosphere over the Pacific Ocean in a belt with coordinates 10–25°N, 115–160°E

Table 3.5. Synoptic situations in the observation test basins of the Pacific.

Observation dates	*Test area coordinates*	*Synoptic situation*	*Data sources*
1–5.6.85	10–25°N, 115–160°E	Undisturbed atmosphere	Satellite data
18–24.6.85			
18–25.7.85			
10–14.9.85			
18–30.7.85	20–25°N, 130–135°E	TC "Jeff", square 4a	Satellite data
10–18.9.85	20–25°N, 130–135°E	TC "Val", square 4a	Satellite data
19–24.10.86	14°N, 113°E	TC "Georgia", square 1c	Radiosondes
12–16.11.86	13–25°N, 110–120°E	STS "Ida", squares 1a, 1b	Radiosondes

(known as test area A) over the periods June–September 1985 and October–November 1986, together with aerological soundings at observation stations in test area A taken by the Typhoon '86 Expedition on the 43rd voyage of the RV *Akademik Korolev* from September to December 1986 (Table 3.5). Detailed description of the synoptical situations are given by Pokrovskaya and Sharkov (1988).

Thermal profiles were recovered from a set of TOVS (TIROS[1] operational vertical sounder) vertical soundings by the TIROS satellite of the NOAA (National Oceanic and Atmospheric Administration) series at the following isobaric levels: 1,000, 850, 700, 500, 400, 300, 250, 200, 150, 100, 70, 50, 30, 20 mb with mean square error of recovery ±1–1.5 K. The spatial element of the resolution zone measured 110 × 100 km (Kidwell, 1988). The choice of geographical region for study was determined by a number of interconnected details. First, test area A is a zone of intense cyclogenesis. Second, this region provides suitable space thermal-sensing data. In addition, there was detailed information about the synoptic conditions in the studied region, radiosonde data, and satellite IR and TV images of cloud systems acquired by GMS (geostationary meteorological satellite) and NOAA satellites (MSC, 1986).

Detailed analysis of all the full-scale material acquired, including data recovered from space sensing and from radiosondes, showed that throughout the area of study there is a very stable (in both space and time) thermal height stratification of the atmosphere of the following form:

$$T(z) = T_0 - \gamma_k(z)z, \qquad (k = 0, 1, 2) \tag{3.3}$$

[1] TIROS is an acronym for "television and infrared observation satellite".

consisting of three height levels:

- between the isobaric surfaces of 1,000 mb and 850 mb with gradient $\gamma = \gamma_0$ (0–1,460 m);
- between 850 mb and 400 mb with $\gamma = \gamma_1$ (1,460–7,140 m); and
- between 400 mb and 150 mb with $\gamma = \gamma_1$ (7,140–19,580 m);

where γ is the temperature at height 100 m in degrees Celsius. A special graphical construction showed (Pokrovskaya and Sharkov, 1988) that mean square deviation in the values of the gradients found, for a single measurement, amounted to 0.002–0.003°C/100 m (Figure 3.5). The size of the zone for temperature profile

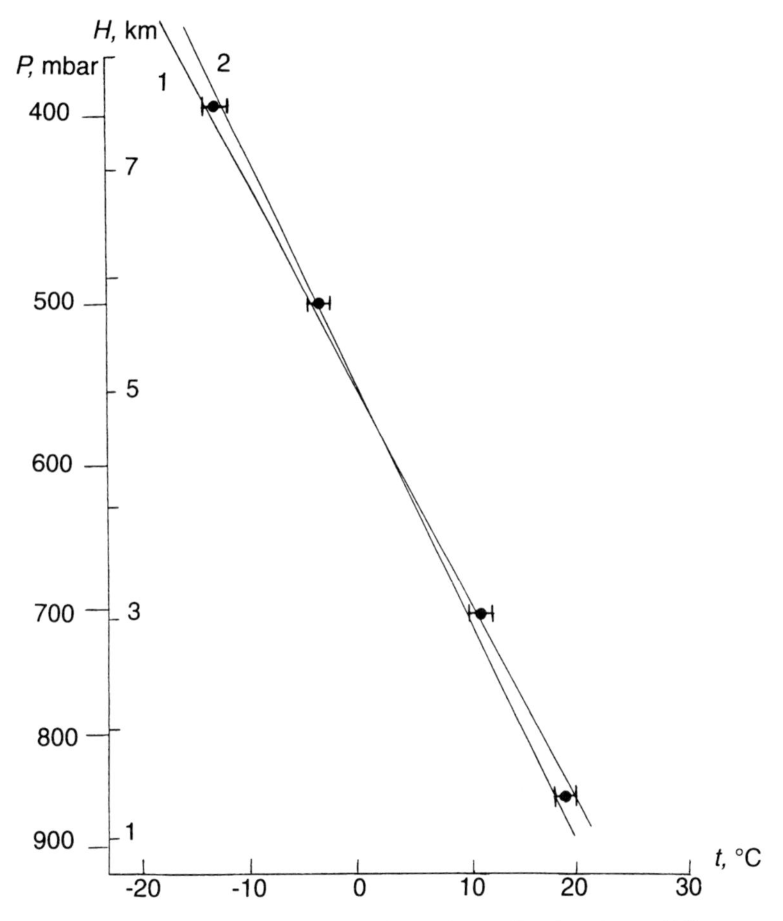

Figure 3.5. Linear approximation of the thermal profile in the middle troposphere (850–400 mb) at a point with co-ordinates 16°N, 140°E for June 1, 1985. Full circles represent temperature values recovered from NOAA data with restored errors. "1" and "2" represent the maximum and minimum temperature gradients, respectively (from Pokrovskaya and Sharkov, 1988).

Table 3.6. Parameters of the temperature profile of the atmosphere in squares of the main test area A for June 1–5, 1985. The upper number in each block denotes the number of sensing points (the sample size for the square). The lower two lines denote the ratios $\bar{\gamma}_i/\sigma(\gamma_i)$ $(i = 1, 2)$.

	1	*2*	*3*	*4*	*5*	*6*	*7*	*8*	*9*
a	4	6	5	12	12	22	15	5	3
	$\frac{0.57}{0.028}$	$\frac{0.55}{0.041}$	$\frac{0.56}{0.012}$	$\frac{0.55}{0.036}$	$\frac{0.58}{0.024}$	$\frac{0.57}{0.015}$	$\frac{0.58}{0.017}$	$\frac{0.57}{0.024}$	$\frac{0.60}{0.026}$
	$\frac{0.80}{0.044}$	$\frac{0.82}{0.024}$	$\frac{0.82}{0.020}$	$\frac{0.80}{0.048}$	$\frac{0.77}{0.016}$	$\frac{0.77}{0.018}$	$\frac{0.76}{0.017}$	$\frac{0.76}{0.017}$	$\frac{0.73}{0.007}$
b	3	5	2	5	18	19	13	8	2
	$\frac{0.56}{0.020}$	$\frac{0.60}{0.012}$	$\frac{0.50}{}$	$\frac{0.56}{0.036}$	$\frac{0.57}{0.018}$	$\frac{0.58}{0.018}$	$\frac{0.57}{0.019}$	$\frac{0.55}{0.029}$	$\frac{0.58}{}$
	$\frac{0.80}{0.010}$	$\frac{0.80}{0.022}$	$\frac{0.82}{}$	$\frac{0.82}{0.012}$	$\frac{0.79}{0.019}$	$\frac{0.78}{0.019}$	$\frac{0.76}{0.027}$	$\frac{0.77}{0.017}$	$\frac{0.76}{}$
c			1	4	6	12	8	6	
			$\frac{0.56}{0.015}$	$\frac{0.58}{0.013}$	$\frac{0.59}{0.025}$	$\frac{0.58}{0.022}$	$\frac{0.59}{0.026}$	$\frac{0.58}{}$	
			$\frac{0.88}{}$	$\frac{0.83}{0.017}$	$\frac{0.77}{0.023}$	$\frac{0.80}{0.021}$	$\frac{0.79}{0.026}$	$\frac{0.78}{0.027}$	

recovery (in our studies, a grid scale) was 100 × 100 km and was determined by angular characteristics of the field of view of the NOAA TOVS IR and microwave radiometer system (real time vertical sensing). Further studies have shown that it is possible to establish another approximation: the linear profile with a quadratic component.

The spatial and statistical characteristics of thermal features—gradients $\gamma_k(z)$—were studied for two spatial scales (mesoscale and synoptic scale): first, for test area A of dimension 5,000 × 2,000 km and, second, for the 27 squares (of size 500 × 500 km) into which the main test area A was fragmented (see Table 3.6).

The spatial scales were chosen based on the following considerations. While test area A covers practically the whole zone of tropical cyclogenesis in the NWP, the sizes of the squares (zones) are almost typical of the size of tropical cloud clusters in a tropical disturbance (TD).

The total number of recovery points in test area A (full sample set) was 188. Distribution of the recovery points across the squares (zones) is shown in Table 3.6. Each point of the temperature profile was recovered and the gradients were calculated at the previous height levels γ_k^n ($n = 1, \ldots, 188$, $k = 0, 1, 2$). The mean value $\bar{\gamma}_k^{ij}$ and root mean square (r.m.s) variation $\sigma(\gamma)$ were calculated in turn for each zone (for lines i, $i = 1, \ldots, 9$ and columns $j = a, b, c$) over all the recovery points in that zone (Table 3.6). The chart of large-scale spatial distribution of sea surface temperature (SST) showed weak gradients of 1°C over 1,000 km.

The main procedural problem of processing the TC onset situation is finding an appropriate synoptic situation and its geographical location in the zone of intense cyclogenesis in the Pacific Ocean; this could be carried out by selecting reliable remote-sensing data in the immediate area of future onsets of TSs.

3.2.2 Spatiotemporal statistics of gradient fields

In this section we give the detailed statistical characteristics of gradient fields under two different scale situations: (a) full test area A (sample size $n = 188$) and (b) 24 small squares.

Pokrovskaya and Sharkov (1988, 1991) showed—by forming the experimental differential distribution of the values γ_k^i ($i = 1, \ldots, 188$) for each group ($k = 0, 1, 2$) for the whole of test area A on a timescale of 5 days—that the hypothesis of experimental histograms approximating to the normal distribution can be accepted (Figure 3.6), where the means $\bar{\gamma}_k$ and the r.m.s.'s for this statistical object are given by:

$$\left.\begin{array}{lll} \bar{\gamma}_0 = 0.77 & \bar{\gamma}_1 = 0.57 & \bar{\gamma}_2 = 0.78 \\ \sigma(\gamma_0) = 0.094 & \sigma(\gamma_1) = 0.025 & \sigma(\gamma_2) = 0.038 \end{array}\right\} \quad (3.4)$$

Note that the temperature regime in the height regions between 850 mb and 150 mb is considerably more sensitive (see the r.m.s. value) than that of the lower driving layer of the atmosphere, which is very sensitive to different types of advective perturbing fluxes and motions.

Spatial fragmentation of the main test area A (5,000 × 2,000 km) into zones (500 × 500 km) leads to additional variation of γ_k^{ij}. It is important to clarify whether the sample means $\bar{\gamma}_k^{ij}$ and r.m.s.'s $\sigma(\gamma_k^{ij})$ belong to the general set of statistical parameters characterizing the main test area.

Special statistical analysis—using Pearson's criterion and the difference criterion (Cressie, 1993)—of all zones (Pokrovskaya and Sharkov, 1991) showed that the hypothesis H_0 (about the sample means $\bar{\gamma}_k^{ij}$ belonging to the general set of the full area) can be accepted because it is not contradicted by experimental data.

Daily variations were analyzed over 5 days (Table 3.7) in the most informative square (i.e., square 6a, Table 3.6).

Using sensing data, a time series was formed from the sample values γ_k^{6a}, where each term was averaged over a 1-day sample in a square of size 500 × 500 km. Table 3.7 shows these samples together with their calculated variances.

It is easy to see that, for the time series for γ_1 and γ_2, the difference between the means is insignificant; the values of the variances are uniform. Thus, the samples belong to the same general set (Eq. 3.4). However, the variation of γ_0 with time is

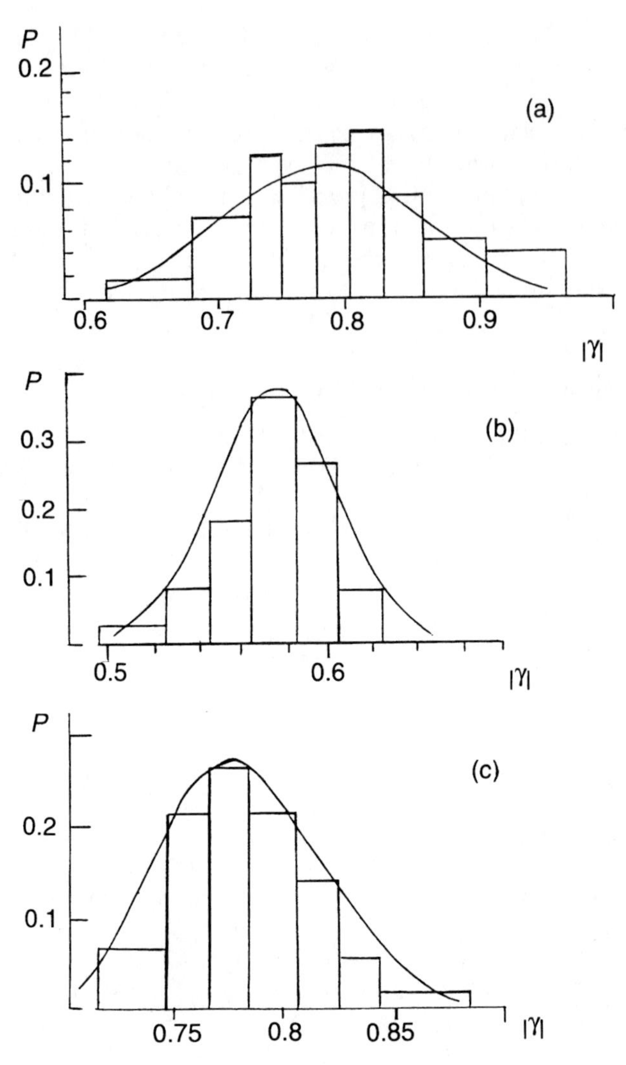

Figure 3.6. Experimental histograms of the thermal gradients and theoretical normal distributions for the whole of test area A on a timescale of 5 days. The run is 189. (a) Data of γ_0; (b) data of γ_1; (c) data of γ_2. Stepped curves represent the experimental histograms. The solid lines represent the theoretical normal distributions of the means and r.m.s.'s given by Eq. (3.4) (from Pokrovskaya and Sharkov, 1988).

characterized by large differences in mean values and a large r.m.s. around the daily average. In other words, the value of γ_0 varies over several days and also within a single day by up to 0.15°C/100 m. However, we already considered this fact when analyzing the distribution of γ_0 over the whole test area A. It is interesting to note that SST does not always determine this variation, since variation in SST within square 6a over this time period was less than 0.1°C.

Table 3.7. Parameters of the temperature profile of the atmosphere (square 6a) over 5 days (daily averaged).

Parameter	*Day 1*	*Day 2*	*Day 3*	*Day 4*	*Day 5*	*5-day average*
$\bar{\gamma}_0$	0.695	0.793	0.792	0.818	0.755	0.78
$\sigma(\gamma_0)$	0.03	0.063	0.068	0.056	0.055	0.067
$\bar{\gamma}_1$	0.572	0.574	0.572	0.564	0.54	0.57
$\sigma(\gamma_1)$	0.01	0.012	0.01	0.03	0.01	0.015
$\bar{\gamma}_2$	0.787	0.0774	0.77	0.774	0.775	0.77
$\sigma(\gamma_2)$	0.057	0.018	0.006	0.018	0.035	0.018
Sample size	4	5	6	5	2	22

Let us now turn to monthly variation. This characteristic can be determined from analysis of the mean values γ_k and their variances, obtained for 5-day samples of sensing data in June (two dates), July, and September in the tropical atmosphere over the same square 4a under calm conditions (Table 3.8). According to the Student t-test (for unknown variance in the general set), at a level of confidence of 0.05, the difference between γ_1 and γ_2 is insignificant and the samples may be assigned to the same general set. In other words, the thermal structure of the middle and upper troposphere is not subject to seasonal variation over the course of the 4 months. As

Table 3.8. Parameters of the temperature profile of the atmosphere (square 4a) for June, July, and September 1986.

Parameter	*Observation dates*				*4-month average*
	June 1–5	*July 18–24*	*July 18–25*	*September 10–14*	
$\bar{\gamma}_0$	0.87	0.63	0.58	0.63	0.64
$\sigma(\gamma_0)$	0.120	0.084	0.064	0.69	0.115
$\bar{\gamma}_1$	0.55	0.57	0.61	0.61	0.59
$\sigma(\gamma_1)$	0.036	0.029	0.010	0.015	0.033
$\bar{\gamma}_2$	0.80	0.80	0.77	0.74	0.79
$\sigma(\gamma_2)$	0.048	0.032	0.016	0.026	0.041
Sample size	5	15	15	13	48
SST	25.5	28.5	29.4	30.0	

for the lower troposphere, as we might expect, the differences in values of γ_0 are significant, although they are not determined by SST.

Thus, the thermal structure of the tropical atmosphere, under undisturbed conditions for the height band between 850 mb and 150 mb and spatial scales of 500 km to 5,000 km for periods from 1 day to 4 months, is a statistical object with systematic constraints, which forms a single general object with the parameter values given in Tables 3.7 and 3.8. The thermal structure of the lower layer of the troposphere (1,000–850 mb) under undisturbed conditions was subject to considerable variation both over several days and within a day (also within a month) and cannot be viewed as a single general set in the statistical sense.

3.2.3 Thermal stratification of the atmosphere by the action of tropical cyclones

In what follows (Pokrovskaya and Sharkov, 1991) we consider the properties of thermal stratification of the atmosphere in square 4a of test area A for continuous perturbation by a TD in this part of the atmosphere in the form of two tropical storms—"Jeff" (8507) and "Val" (8517)—using remote-sensing results. A special set of experimental data enabled us to track the evolution of the state of the atmosphere over square 4a, beginning in a calm state (the TS was more than 1,500–2,000 km away), including the approach of the TS and the effect of its edges, the continuous drawing in of the atmosphere into the tropical storm eddy, and the departure of the TS from the given geographical zone (1,000 km away).

This actively affected square 4a from July 25 to 27, 1985. Figure 3.7a shows the space-sensing data and the synoptic conditions (schematically) at the time the TS affected the region. At 21°N, 148°E, on July 21 the initial TD formed in a mass of convection clouds. On July 22 it became a depression and on July 23 it became TS "Jeff" (8507). The storm moved northwards, on July 25 its path turned to the north-

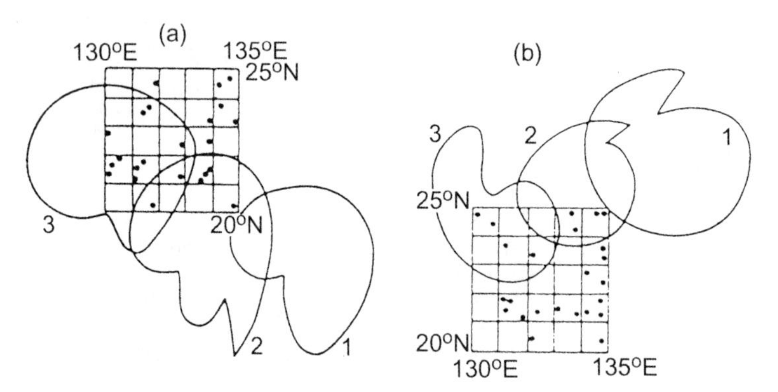

Figure 3.7. Characteristics of square 4a for spatial data (dots are the points of recovery of thermal profiles) and the position of the edges of the cloud mass: (a) TS "Jeff" for 1 on July 26, 1985 at 12:00 GMT, 2 on July 27, 1985 at 00:00 GMT, 3 on July 27, 1985 at 12:00 GMT; (b) TS "Val" for 1 on September 14, 1985 at 12:00 GMT, 2 on September 15, 1985 at 00:00 GMT, 3 on September 15, 1985 at 12:00 GMT (from Pokrovskaya and Sharkov, 1991).

east and on July 26 it began to affect the given region. On July 27 the storm passed over the region and between July 28 and 29 the effects of its eastern region continued to be felt.

The results of processing space-sensing data are shown in Figure 3.8a. From these results it is easy to see that samples γ_1 and γ_2 from July 18 to 26, 1985 (outside the TS), those from July 26 to 29, 1985 (within the TS and on its periphery, individual measurements obtained in broken cloud cover), and those of July 30, 1985 belong both to the general set for the whole test area A (5-day model) and to the general set for square 4a (4-month model). While no significant differences were observed between values of γ_1 over all the given dates, the differences between the values of γ_2 for July 25 to 27 and July 30 could possibly be significant.

The physical meaning of the statistical procedure (Student t-test) used below is as follows: there is a need to ascertain whether the difference between the two values of the parameter being studied is significant (i.e., governed by fundamental physical reasons) or random (i.e., associated with a limitation in sample size).

To determine whether there is significance, we form the dimensionless coefficient of the Student t-test (Cressie, 1993):

$$t = \frac{\bar{\gamma}_1' - \bar{\gamma}_2''}{\sqrt{\dfrac{(n_1 - 1)\sigma_1^2 + (n_2 - 1)\sigma_2^2}{n_1 + n_2 - 2}}} \sqrt{\frac{n_1 n_2}{n_1 + n_2}} \tag{3.5}$$

where the overbars and indices 1 and 2 relate, respectively, to July 25 to 27 and July 30; and σ and n are the r.m.s. and sample size, respectively. In our case, $t = 2.69 < t_{\alpha;f} = 3.25$ for $\alpha = 0.01$ and $f = 9$. Thus, the differences in the mean values of γ_k $(k = 1, 2)$ before and after the passage of the TS are not significant. The same holds for the period when the TS directly affected the region.

In the lower troposphere, considerable variation is observed both inside the TS and outside the area of its effect.

Starting on September 10 the given region of the Pacific Ocean was affected by the crest of a Pacific Ocean subtropical anticyclone, on the southern edge of which, as a result of of clearly visible near-equatorial troughs sharpening, large masses of compact cumulus cloud formed in which the initial TD was born at 12°N, 145°E on September 12. On September 13 the disturbance became a tropical depression, which moved slowly in a northwesterly direction. On September 14 the cloud mass associated with the tropical eddy strengthening began to affect the region (square 4a) and on September 15 TS "Val" (8517) passed through the region with windspeeds of 20 m/s to 22 m/s. Moving in a westerly direction, by September 16 the TS cloud system no longer affected square 4a.

Figure 3.7b shows the thermal sensing data for square 4a and a schematic view of the synoptic conditions under which TS "Val" affected the region of square 4a.

The dense cloud mass between September 14 and 15, 1985 over square 4a prevented us from acquiring sufficient samples to recover temperature profiles, except for a single recovery in this region on September 14. As in the case of TS

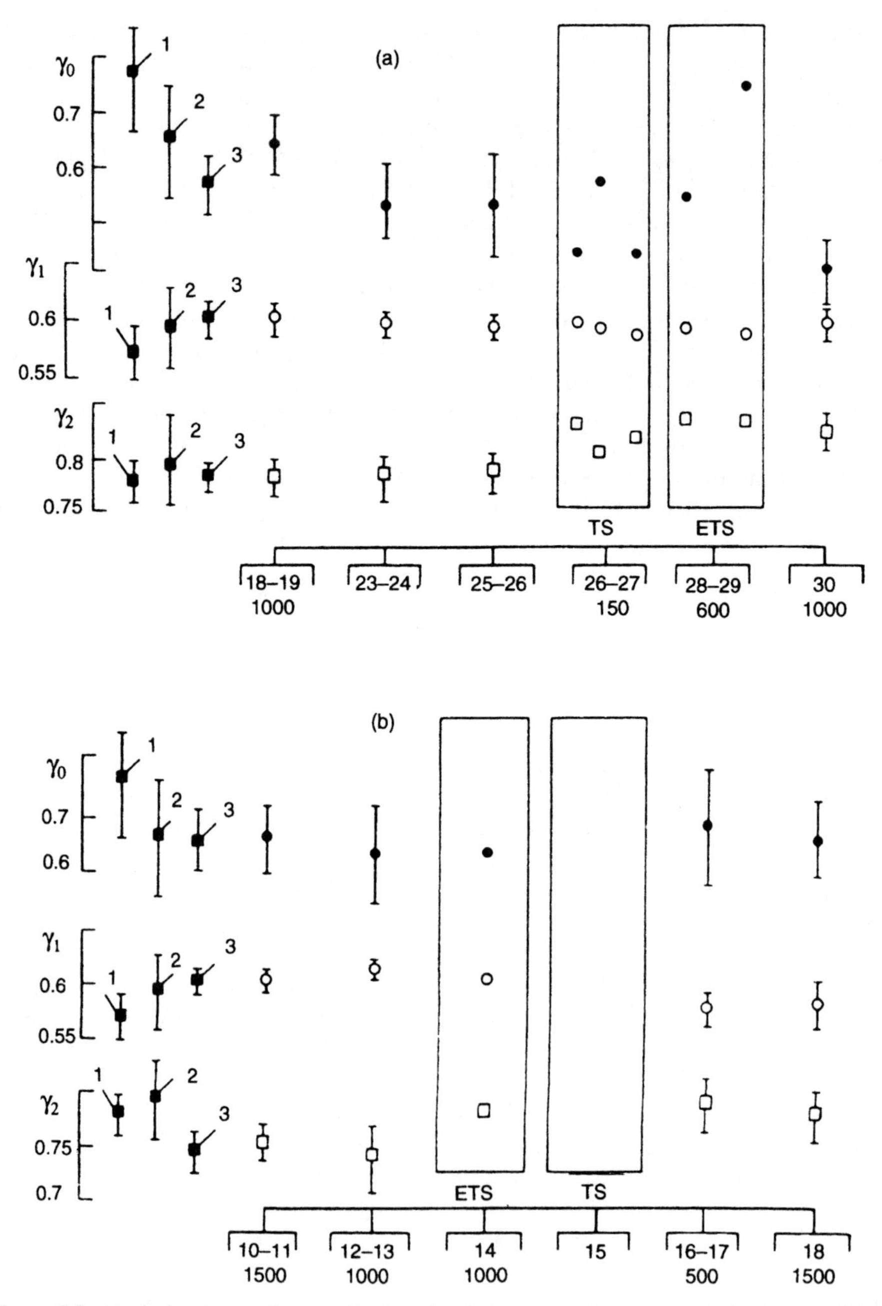

Figure 3.8. Variation in gradients with time for 2-day averaging under the influence of (a) TS "Jeff" (TS) and its edges (ETS) between June 18 and 30, 1985 in square 4a; (b) TS "Val" 1 to 5-day model for test area A; 2 to 4-month model for square 4a; 3 to 5-day model of γ between July 18 and 25, 1985. The lower numbers indicate the distance from the center of the TS (from Pokrovskaya and Sharkov, 1991).

"Jeff", the differences in the values of γ_1, γ_2, and γ_0 before and after the passage of the TS over the region was insignificant (Figure 3.7b).

Thus, the passage of two TSs (in the course of 2 months) through the same geographical square was not apparent in the state of thermal stratification of the middle and upper troposphere from either mean gradients or their variances.

3.2.4 Thermal stratification of a disturbed atmosphere (resulting from radiosonde data)

In this section we analyze the state of stratification of the atmosphere using the previous method for radiosonde data from the RV *Akademik Korolev* (at drift), acquired sequentially in different zones affected by the TS (details of the trajectories are given by Pokrovskaya and Sharkov, 1988). In view of the much reduced (in comparison with the space variant) spatial scales, when averaging individual measurements for radiosonde data we would expect more significant variation in estimates of the characteristics of thermal stratification under the influence of TDs. However, we shall show below that this was not the case.

TS "Georgia" (8622) was born in the form of a TD on October 14, 1986 at 8.5°N, 132°E in the zone of a sharpening monsoon trough, oriented from the South China Sea to the northeast. Spreading westwards, the disturbance developed slowly and on October 17 reached the stage of a TD. The formation of a high anticyclone over South East Asia on October 19 led to an increase in windspeed in an easterly direction of up to 20 m/s and to an intensification in the processes of cyclogenesis. This in turn led to the TD transforming into a TS on October 18. Spreading westwards on its southern periphery on October 19, the storm passed through the Philippine Islands, which led to dissipation of the general energy of the system.

As the eddy moved farther westwards over the warm water of the South China Sea on October 20 and 21 cyclogenesis occurred once more and the cloud mass increased in diameter to 750 km to 850 km. On October 22, the TS reached the continent where it stopped rapidly.

Figure 3.9 shows the geographical position of the RV (point of radiosonde observation) according to its own navigational equipment together with the temporal evolution of the edge of the main cloud mass of the TS according to IR soundings from the GMS geostationary satellite. Note the abrupt displacement at the center of the barometric formation (the eye of the TS) towards the edge of the cloud mass, which usually occurs during activation of a TS. Comparison of the RV's navigational data with space-sensing data showed that the center of the barometric system moved to within 110 km to 120 km of the RV. Figure 3.10 shows detailed data about stratification computed from radiosonde data. Unlike the previous figures which show spatial r.m.s.'s (on a 500 km scale), Figure 3.10 shows individual errors in the gradients of γ_0, γ_1, and γ_2.

Analysis of the temporal evolution of the parameters for thermal stratification of the atmosphere under the influence of the TS, as determined from space (Figures 3.7 and 3.8) and radiosonde data (Figures 3.9 and 3.10), show that no qualitative differences were observed in these situations, despite large numerical differences in

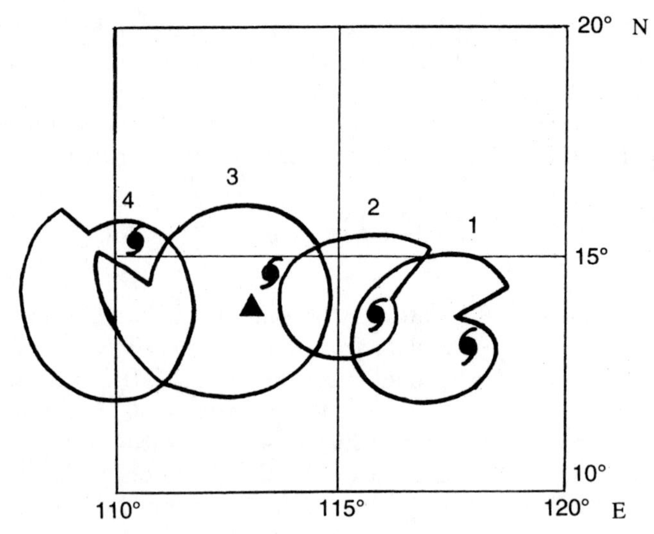

Figure 3.9. Position of the RV (denoted by triangle) and the edges of the continuous cloud mass of TS "Georgia" (position of the eye marked): 1 on October 20, 1986 at 00:00 GMT; 2 on October 20, 1986 at 12:00 GMT; 3 on October 21, 1986 at 00:00 GMT; 4 on October 21, 1986 at 12:00 GMT (from Pokrovskaya and Sharkov, 1991).

the spatial scales for these measurements. In both observational situations, the stratification of the lower layer of the troposphere (1,000–850 mb) appeared very sensitive to the dynamic state. One characteristic here is the sharp decrease in gradient γ_0 at 00:00 hours on October 21, 1986 from $\gamma_0 = 0.65$ to $\gamma_0 = 0.41$, which is associated with the sharp decrease in agitation in the zone near the eye of the TS and the significant decrease in temperature of the driving layer (Figure 3.10). It is well known that the temperature of the driving layer decreases from the edge to the center of TSs.

It is interesting to note that the values of gradients computed from radiosonde data do not show traces of the warm core, which is usually found at heights of 500 mb to 200 mb. This fact is observed in the case of space data where they are physically more intelligible and explained, since the spatial resolution exceeds 100 km and, moreover, the temperature profile is recovered outside the cloud cover zone.

In the following, we consider the time series of gradient values for October 19 to 24, 1985 as a statistical object and determine the degree of independence of the sample elements together with the relationship of the statistical characteristics of this local (in the geographical sense) system to the characteristics of the general set γ_k, the statistical properties of which reflect the state of an undisturbed tropical atmosphere (from space data).

Thus, we form three samples of 15 elements γ_k^i ($i = 1, \ldots, 15$; $k = 0, 1, 2$). The mean values $\bar{\gamma}_k$ and their r.m.s.'s are given by:

$$\left.\begin{array}{lll} \bar{\gamma}_0 = 0.60 & \bar{\gamma}_1 = 0.54 & \bar{\gamma}_2 = 0.76 \\ \sigma(\gamma_0) = 0.049 & \sigma(\gamma_1) = 0.021 & \sigma(\gamma_2) = 0.038 \end{array}\right\} \quad (3.6)$$

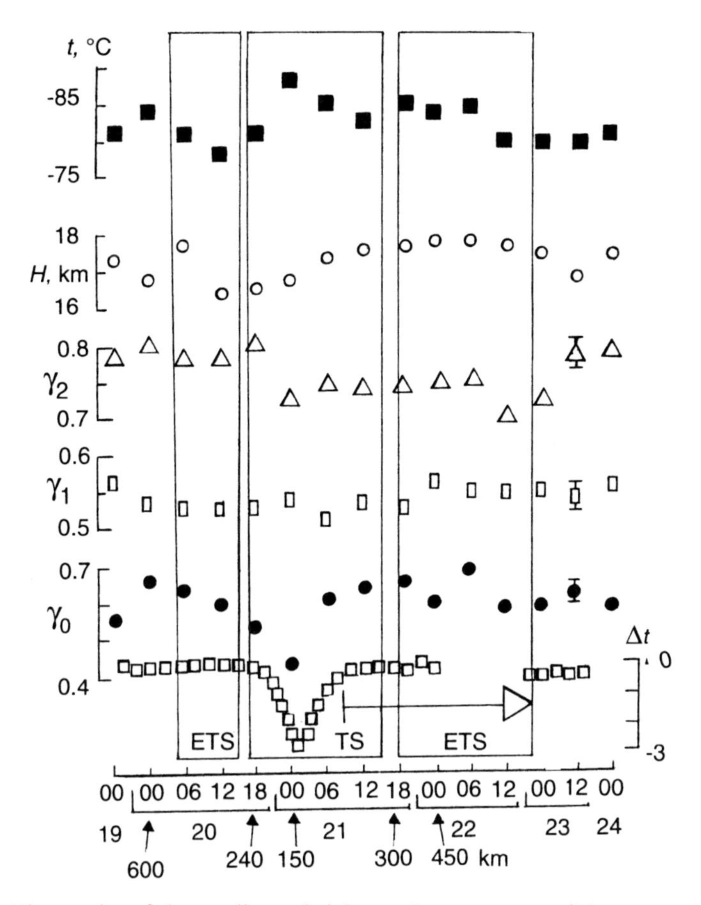

Figure 3.10. Time series of the gradients, height, and temperature of the tropopause and the difference between air and water temperatures (Δt) over time, during the passage of TS "Georgia" over the position of the RV. ETS denotes the edge of the TS and TS refers to the center of the storm (from Pokrovskaya and Sharkov, 1991).

Moreover, using the criterion of the ratio of the squares of sequential differences, we compute the dimensionless parameters $\beta(\gamma_k)$ for samples γ_k: $\beta(\gamma_0) = 1.15$, $\beta(\gamma_1) = 0.57$, $\beta(\gamma_2) = 0.73$.

At the confidence level of $\alpha = 0.05$ we have $\beta^{\min}_{0.05} = 1.4$. Comparing $\beta\gamma_k$ and $\beta^{\min}_{0.05}$ we deduce that, for all three samples, hypothesis H_0 is untenable (the sample terms are internally related and are not independent).

We consider the relationship of the current samples with the general set γ_k for May 4, 1985 for space data. We define the dimensionless parameter as $Z_k = (\bar{\gamma}_{ok} - \bar{\gamma}_k)\sqrt{n}/\sigma(\gamma_{ok})$ where $\bar{\gamma}_{ok}$ is the mean value of the general set for test area A: $Z_1 = 1.38$, $Z_2 = 5.77$, $Z_3 = 2.7$.

Comparison with the critical value $Z_{0.01} = 2.32$ shows that hypothesis H_0 for γ_0 at this confidence level is confirmed but that it is unacceptable for γ_1 and γ_2. This last fact is associated with the systematic bias of the mean values of γ_1 and γ_2

($\Delta\gamma = \bar{\gamma}_{ok} - \bar{\gamma}_k = 0.02$–$0.03$) for radiosonde measurements (October 1986) (easily seen from the specific form of the sample itself). These values and the signed systematic biases are also observed when we compare $\bar{\gamma}_k$ with models having 5-day (Table 3.7) and 4-month (Table 3.8) averaging (spatial scale of 500 km).

It is interesting to note that 1 month later (November 12 to 16, 1986) the synoptic situation was almost exactly repeated in the given geographical region (the region of drift of the RV *Akademik Korolev*) when a stronger TS than TS "Georgia"—super-tropical storm (STS) "Ida" (8624)—passed over (for more details see Pokrovskaya and Sharkov, 1988).

Analysis of time samples of the stratification parameters (14 sample elements for November 12 to 16) showed that qualitatively there was no difference in patterns. The mean values $\bar{\gamma}_k$ and their r.m.s.'s for the given samples were:

$$\left.\begin{array}{lll} \bar{\gamma} = 0.58 & \bar{\gamma} = 0.53 & \bar{\gamma}_2 = 0.78 \\ \sigma(\gamma_0) = 0.075 & \sigma(\gamma_1) = 0.033 & \sigma(\gamma_2) = 0.028 \end{array}\right\} \tag{3.7}$$

Comparing these parameters with the data (Eq. 3.6), using the Student *t*-test (see its physical meaning in Section 3.2.3), showed that the difference in statistical characteristics of samples acquired in the same geographical region at a monthly interval under the influence of the two very different TSs (two combinations of factors) was insignificant.

3.2.5 Thermal stratification of the tropical atmosphere by the action of tropical cyclone formation

Experimental study of the condition of thermal stratification in the spatial basin where TCs will develop in the future is important for derivation of the remote precursors that predict the onset of TCs. On the basis on satellite data, Pokrovskaya and Sharkov (1990, 1993b) study the thermal state of the tropical atmosphere over the Pacific Ocean at a point of future genesis, intensification, and death of TCs.

The test area (15–25°N, 145–155°E) of scale 1,000 × 1,000 km (squares 7a, 6, 8a, and 8b of test area A) was studied in Section 3.2.2.

We now give a brief description of the large-scale atmospheric processes over this area between July 15 and 30, 1985. Figure 3.11 shows simplified maps as a result of cloud analysis of the state of the cloud cover over this area between July 15 and 17 during which the region was affected by the ridge of a Pacific Ocean anticyclone on the southern edge of which masses of thick cumulus cloud 15 km to 17 km in height formed periodically. On July 18 (Figure 3.11a) an initial tropical depression (TDE) formed in one of the cloud masses at 20°N, 150°E; during the next 3 days this developed gradually, so that by July 22 (Figure 3.11b) it became a TDE at the same position. However, by July 23 (Figure 3.11c, d) it reached the stage of a TS. TS "Jeff" continued to strengthen and moved in a northerly direction. By July 28 it no longer affected the studied region. A more detailed cloud analysis of this synoptic situation is performed by Pokrovskaya and Sharkov (1990).

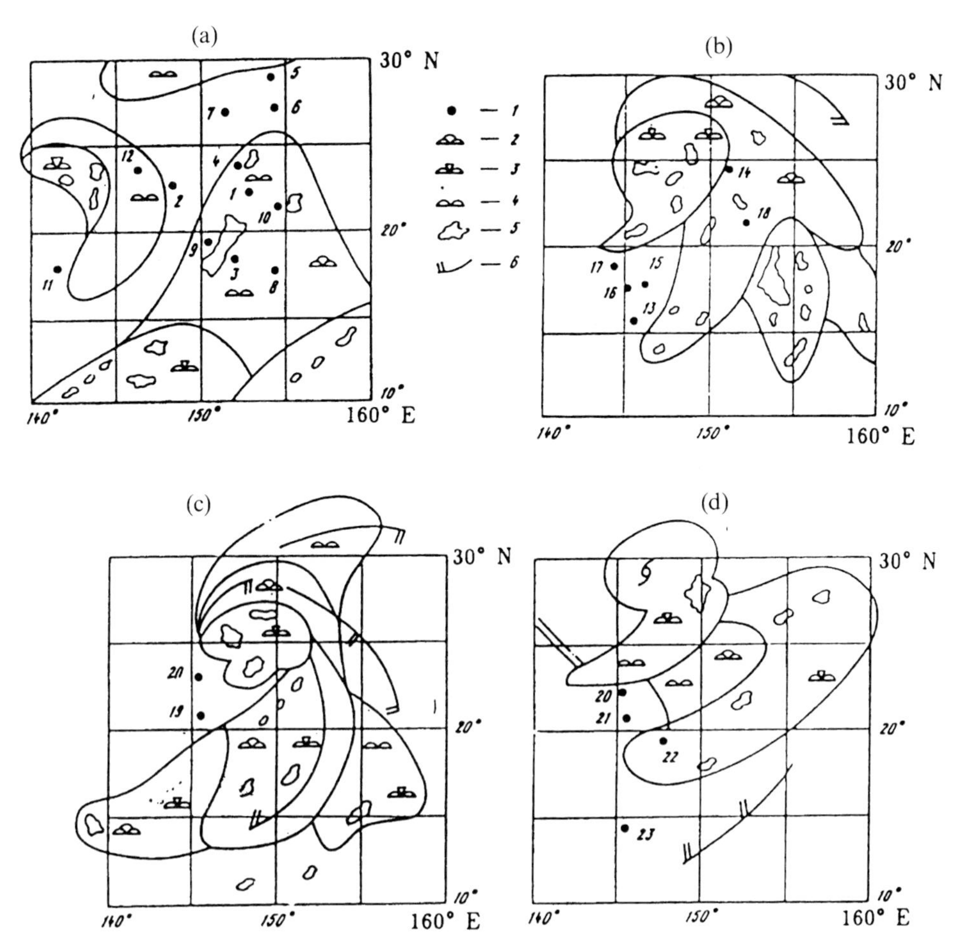

Figure 3.11. Maps of the state of cloud cover for the tropical atmosphere over the NWP: (a) on July 18, 1985; (b) on July 22, 1985; (c) on July 23, 1985; (d) on July 28, 1985; 1, points of recovery of thermal stratification from NOAA data (these points are numbered as in Table 3.1); 2, thick cumulus cloud; 3, cumulonimbus cloud; 4, cumulus cloud; 5, position of the accumulation of thick and active cumulus clouds; 6, bands of strata of cirrocumulus clouds (from Pokrovskaya and Sharkov, 1993b).

Figure 3.12 shows average values of the field γ_k and their mean square deviations throughout the evolution of cloud systems, from an undisturbed atmospheric state (UA 1), through the birth of the initial tropical disturbance (ITD), the stages between a TD and TDE, and finally to the development stage of a TS. After its gradual departure between July 27 and 30 there was again a return to UA 2.

The vast dense cloud mass between July 24 and 27 made it impossible to acquire reliable space data and to recover values of the profile of thermal stratification. Pokrovskaya and Sharkov (1990) construct the following statistical models:

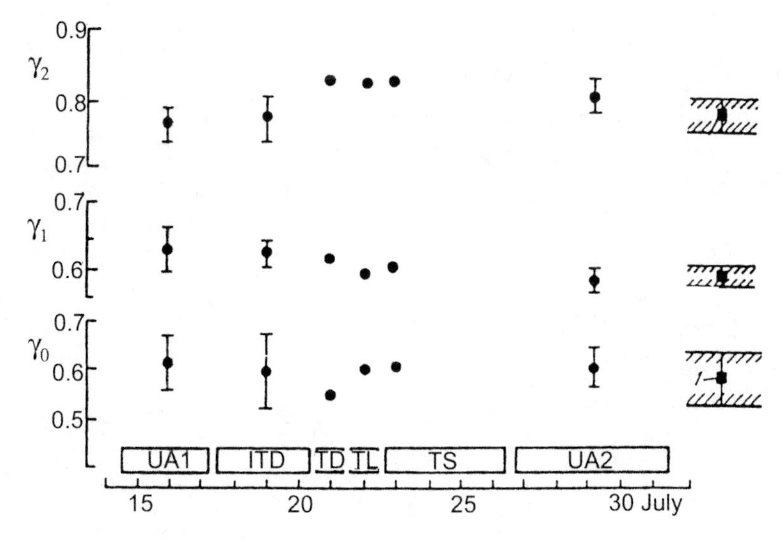

Figure 3.12. Time dependence of the mean values of the gradients and their r.m.s.'s at three height levels in the atmosphere in the studied test area during the recovery of TS "Jeff" data between July 15 and 31, 1985 (see the text for an explanation of the abbreviations): 1, statistical model of an undisturbed atmosphere with spatiotemporal scales of averaging of 500 km and 7 days (Pokrovskaya and Sharkov, 1993b).

- for stage UA 1 (July 15 to 17, 1985)

$$\left.\begin{aligned} \bar{\gamma}_0 &= 0.61 \qquad \sigma(\bar{\gamma}_0) = 0.045 \\ \bar{\gamma}_1 &= 0.63 \qquad \sigma(\bar{\gamma}_1) = 0.023 \\ \bar{\gamma}_2 &= 0.76 \qquad \sigma(\bar{\gamma}_2) = 0.028 \end{aligned}\right\} \tag{3.8}$$

- for stage ITD (July 18 to 20, 1985)

$$\left.\begin{aligned} \bar{\gamma}_0 &= 0.60 \qquad \sigma(\bar{\gamma}_0) = 0.075 \\ \bar{\gamma}_1 &= 0.62 \qquad \sigma(\bar{\gamma}_1) = 0.013 \\ \bar{\gamma}_2 &= 0.78 \qquad \sigma(\bar{\gamma}_2) = 0.049 \end{aligned}\right\} \tag{3.9}$$

- for stage UA 2 (July 15 to 17, 1985)

$$\left.\begin{aligned} \bar{\gamma}_0 &= 0.61 \qquad \sigma(\bar{\gamma}_0) = 0.063 \\ \bar{\gamma}_1 &= 0.58 \qquad \sigma(\bar{\gamma}_1) = 0.015 \\ \bar{\gamma}_2 &= 0.82 \qquad \sigma(\bar{\gamma}_2) = 0.025 \end{aligned}\right\} \tag{3.10}$$

Note that the thermal regime between 850 mb and 150 mb is much more stable (in the sense of variance values) than in the lower troposphere, which is sensitive to a different type of convective fluxes and agitation.

To test the uniformity of the fields γ_k at various stages of the evolution of a TS, let us carry out a number of statistical procedures (Feller, 1968; Cressie, 1993).

First, let us consider the question of statistical distinguishability of samples of fields γ_k acquired at the same geographical location (birthplace of future TDs) at different dates:

(a) when there is a free undisturbed atmosphere over the 5 days before the onset of the TD (UA 1);
(b) at the stage of the ITD;
(c) after departure of the TD from its birthplace (UA 2).

To do this we determine the dimensionless coefficient of the Student t-test (see physical meaning in Section 3.2.3):

$$t_k = \frac{\bar{\gamma}'_k - \bar{\gamma}''_k}{\sqrt{\dfrac{(n'-1)\sigma_k'^2 + (n''-1)\sigma_k''^2}{n'+n''-2}}} \sqrt{\frac{n'n''}{n'+n''}} \tag{3.11}$$

where $\bar{\gamma}_k$ and σ_k are the values of the means and r.m.s.'s of the gradients; overbars denote the different times of observation of the phenomena; and n is the size of the corresponding sample.

Thus, for three combinations of this stage, we can construct values of coefficient t as shown in Table 3.9. The values of the critical coefficient $t_{\alpha,f}$ corresponding to a significance level of 0.05 (probability 0.95) are given in Table 3.9. Comparison of these values shows that for this significance level the differences between the mean values of γ_k fields in situations UA 1 and ITD are insignificant and the samples belong to the same general set. For situations ITD and UA 2 the difference in the lower and upper layers of the troposphere is statistically insignificant (significance level 0.05). At the same time, a significant displacement (−0.04) is observed in the field γ_2 of the middle troposphere. This also occurs for stages UA 1 and UA 2, although the difference between the means has the opposite sign (+0.06). However, it would be inaccurate to treat these characteristics as direct effects of the departing TS.

We now consider variance uniformity in the combinations of situations studied using the Fisher criterion. For this, we form the sampling function $F_k = \sigma_k'^2/\sigma_k''^2$ from the variance in γ_k fields in different situations (denoted by overbars). The hypothesis that the variance is equal $H_0(\sigma'^2 = \sigma''^2)$ is confirmed provided the condition

Table 3.9. Values of the coefficients t and F of the Student t-distribution and the Fisher criteria for three situations.

Situation	t_0	t_1	t_2	t_{cr}	F_0	F_1	F_2	F_{cr}
UA 1–ITD	0.32	0.95	1.02	2.14	1.65	1.77	1.74	4.77
ITD–UA 2	0.31	3.45	2.10	2.14	1.20	1.18	1.94	4.77
UA 1–UA 2	0.10	5.50	5.02	2.10	1.38	1.49	1.09	3.30

$F_k < F_{\alpha;f_1;f_2}$ is a quantile of the Fisher distribution at the significance level α and the two degrees of freedom are $f_1 = n' - 1$ and $f_2 = n'' - 1$. It can be seen from Table 3.9, which shows the values of the coefficient F_k for the three situations and values of $F_{\alpha;f_1;f_2}$ for $\alpha = 0.05$, that at this level of significance the variance values acquired by sampling the fields γ_k does not contradict the assertion that the variance in γ_k fields at different stages of the evolution of the TS is equal.

In other words, the thermal structure of the troposphere (field of mean values $\bar{\gamma}_k$ and variance) does not exhibit any statistically significant changes (other than a shift of means in the middle troposphere) at the stage of formation of the TD, at the stage of the ITD, or after the departure of the TS from the region of its birth.

Second, let us now consider the statistical hypothesis H_0 that the values of γ_k at the individual points of recovery of the field γ_k, acquired at the TD, TDE, and TS stages (Figure 3.12), belong to the statistical model $\gamma_k^{\rm ITD}$ at the ITD stage, which we shall apply beyond the general set whose variance is known (see Eq. 3.9). According to statistical rules, the hypothesis is confirmed at significance level α if the sampling function $Z_k = (\gamma - \bar{\gamma}^{\rm ITD}/\sigma_k)\sqrt{n}$ (where $\bar{\gamma}$ is the mean and n is the size of the studied set) satisfies the inequality $|Z_k| < u_\alpha$, where u_α is a quantile of the normal distribution. The computed values of Z_k are $|Z_0| = 0.38$, $|Z_1| = 2.58$, and $|Z_2| = 1.74$. Analysis of these values at the significance level 0.01 ($u_{0.01} = 2.32$) shows that the hypothesis that γ_k at stages TD–TDE–TS belongs to the statistical model for the ITD stage is not contradicted by the experimental data; thus, in what follows we shall operate with this model for stages ITD, TD, TDE, and TS.

Third, let us now consider the statistical hypothesis H_0 that the values of γ_k at the various stages of formation of the TD (UA 1, ITD, TD–TDE–TS, UA 2) belong to the statistical model γ_k^0 acquired between July 18 to 25, 1985 in a sea area 1,000 km
away in which no disturbances were recorded over a month (known as the reference region or RR). This model was constructed in Section 3.2.2 and is given by:

$$\left.\begin{array}{ll} \bar{\gamma}_0 = 0.58 & \sigma(\bar{\gamma}_0) = 0.064 \\ \bar{\gamma}_1 = 0.61 & \sigma(\bar{\gamma}_1) = 0.010 \\ \bar{\gamma}_2 = 0.77 & \sigma(\bar{\gamma}_2) = 0.016 \end{array}\right\} \tag{3.12}$$

According to the rules, hypothesis H_0 is confirmed at the given significance level if the sampling function $Z = (\bar{\gamma}_k - \bar{\gamma}_k^0/\sigma_k^0)\sqrt{n}$ (where n is the sample size) satisfies the condition $|Z| < u_\alpha$ where u_α is a quantile of the normal distribution. Comparing the computed values with the critical value $u_\alpha^{\rm cr} = 2.32$ for $\alpha = 0.01$ (Table 3.10), we deduce that hypothesis H_0 is not contradicted by the experimental data at the given significance level. Thus, differences in the field of temperature gradients at all stages of development of the TD are insignificant (except in four cases) and the samples may be assumed to belong to the same general set that characterizes an undisturbed zone of the atmosphere. In other words, the thermal structure of the atmosphere in the zone of TD generation has the same statistical properties as undisturbed zones. The four questionable cases in which $|Z| > u_{0.01}$ are associated

Table 3.10. Values of the sampling function Z for three situations, relative to the reference area.

Situation	Z_0	Z_1	Z_2
UA 1–RR	1.47	6.0	1.90
ITD–RR	0.76	2.4	1.52
UA 2–RR	1.47	9.0	9.60
TD–TL–TS–RR	0.14	1.7	11.60

with the fact that the RR had very uniform spatial statistics, which is reflected in the low r.m.s. values (see Eq. 3.12).

Thus, it has been shown experimentally that the temperature gradients of the middle and upper troposphere (in the zones of spatial variation) are of a stable nature and are not subject to the effect of nascent TDs of various intensity, up to the strongest TSs. The radiation balance (its longwave component) of the planet–space system as a physical factor underlying this stability has also been shown to exist (Kondratyev, 1980).

As far as situations directly associated with the origin of TDs are concerned it is commonly believed that serious (and sharp) variation in thermal stratification (and also in humidity) of the atmosphere occurs in zones of cyclogenesis on the birth of TDs. However, such discussions are moot.

Our results show that in zones of intense cyclogenesis the field of temperature gradients has no significant anomalies regarding natural spatial variation in time intervals preceding the generation of TDs (up to 5 days), directly after their generation, and after their departure from their area of birth.

Specific experimental confirmation of the stability of thermal stratification in zones of intense cyclogenesis has been obtained from radiothermal measurements taken on board research vessels (Rassadovsky and Troitsky, 1981). From the point of view of the methodology of our experiments these are localized spatial measurements within the mass of TD cloud. Nevertheless, the conclusions reached by Rassadovsky and Troitsky (1981) are analogous with our own.

In view of the high stability of thermal stratification in zones of cyclogenesis, it is clear that little information is to be obtained by using the values of temperature profile gradients as standalone features for prediction purposes. It is interesting that an analogous conclusion was obtained for large-scale spatial fields of humidity (bearing in mind that the content of water vapor was averaged over a complete column of the atmosphere and not over its profile) (Petrova, 1987; Pokrovskaya and Sharkov, 1997c).

However, we should bear in mind that the spatial indistinguishability of situations preceding the creation of large-scale eddy structures does not mean that variation in temperature gradients and humidity does not contribute critically to metastable conditions leading to the creation of these structures. Clearly, this is exactly the reason remote observation and identification of conditions for generation of large-scale structures is complicated since critical conditions are, as it were, "buried" in the spatial variation of hydrometeorological parameters.

To avoid misunderstanding, it will be noted that these conclusions are based on the results of data recovered using existing satellite IR and microwave instrumentation—namely, the HIRS (high-resolution infrared radiation sounder) and the MSU (microwave sounding unit) of the TOVS system—which have an accuracy of recovery of $\pm 1°C$ to $1°C$ at each height level. The application of highly sensitive radiothermal and IR systems to this problem and the use of more accurate statistical and fractal (or wavelet) analysis (e.g., Astafyeva, 1996) may change the strategy for remote experiments on this topic in a fundamental manner.

3.3 SPATIOTEMPORAL FEATURES OF ATMOSPHERIC MOISTURE AND PACIFIC CYCLOGENESIS

In the last 30 years there has been an intense search for the key geophysical parameters, which would help resolution of the problem of the occurrence and dynamics of crisis situations in the Earth's atmosphere. This direction of research has stimulated the development of theoretical concepts regarding the "self-organization" of mesoscale spiral turbulence in large-scale eddy structures (e.g., see Moiseev *et al.*, 1988; Zimin *et al.*, 1992 and the references cited in both). According to these theoretical concepts, the conditions for thermal/moisture content stratification—nonlinearity of the temperature profile (Rutkevich, 1994) and precipitable water content (Moiseev *et al.*, 1990)—play crucial roles in the advent of critical conditions under which organized eddy structures occur (together with the turbulent energy reserves and the spiral nature of the velocity field). But, first, we must mention total water vapor content as one of the most important thermodynamic parameters of the tropical atmosphere.

From the point of view of these thermodynamic concepts and how they relate to the occurrence of a TC—for example, Kuo's model (Kuo, 1965), the convection hypothesis (Palmen and Newton, 1969), the Karno cycle model (Emanuel, 1987)—the manifestation of excessive water vapor concentration in comparison with the natural background plays a decisive (essential) role in formation of the warm core as a result of the condensation of vapor into droplets and the generation of internal baroclinicity, as a result of which a TC forms and its existence is supported.

Thus, from the point of view of both concepts (i.e., the self-organization and thermodynamic approaches), we require detailed knowledge of the spatiotemporal statistical characteristics of the moisture content field of an unperturbed (normal background) and a perturbed (where an eddy structure has formed) atmosphere on both the mesoscale and synoptic scale.

Various aspects of the effect of this parameter on tropical cyclogenesis have been considered in a number of experimental works (see Anthes, 1982; Minina, 1987; Petrova, 1987; Merrill, 1988a, b; Liu *et al.*, 1994, 1995 and the references therein). The conclusions of these works about the effect water vapor content has on cyclogenesis vary, although the notion proposed by Petrova (1987) that the values of water vapor content for mature forms of TSs and for initial forms, which have not developed into well-formed structures (TC), are by and large statistically indistinguishable is predominant. This has led to the conjecture that the natural background for the moisture content field is stable (in the statistical sense) during the immediate creation (formation) of a TD and during the transition from this form to mature structures (TS and beyond). The importance of experimental proof of this hypothesis for the development of systems to predict remote geophysical features of the occurrence of a TC is clear.

The purpose of this section is to analyze the results of studying (based on satellite data) the spatiotemporal variability in water vapor content (precipitable water) over the Pacific Ocean at a point of future genesis of TCs on different dates over the months leading up to formation of the TC, directly before the start of active cyclogenesis, during the creation of the initial form (TDE), during transition from the TDE to the mature form of a TS, and after the latter leaves the region in question (TC origination).

3.3.1 Observational data and processing procedure

The main technical difficulty in carrying out the research lies in finding a suitable synoptic situation at a location in a zone of intense cyclogenesis within the Pacific Ocean that would provide a set of reliable space-sensing data in the immediate zone of future TC formation and throughout the whole period of observation (before birth, at the time the structure is born, and after the TS leaves the birthplace).

The raw data used to form the spatial mesoscale fields of moisture content were based on results from the recovery of total water vapor in a column of the tropical atmosphere over the Pacific Ocean for a latitudinal belt ($1,000 \times 1,000$ km) at 15–25°N, 145–155°E (test area A) from June 1 to August 10, 1985 (see Section 3.2.1). Total water vapor content in the column of atmosphere from the sea surface (1,000 mb level) to the 400 mb level were recovered under cloud-free conditions from systematic vertical sensing by TOVS of the NOAA series (Kidwell, 1988). The spatial element of the resolution area measured 110×110 km. Taking into account the gaps in space and in time, the authors of the *NOAA Polar Orbiter User's Guide* (Kidwell, 1988) recommend maximum errors in the absolute recovery of the moisture content should be estimated as ±30%. However, for relative recovery the error is considerably less and may amount to 0.5% (MSC, 1986). We shall orient ourselves towards the latter figures, since our special analysis of a large volume of recovered data published in the dataset of MSC (1986) showed that the results of relative recovery were very stable, while absolute comparison of recovered data with data from subsatellite measurements revealed a small but constant bias. The spatial fields of total water vapor were formed using the Global-TC systematized database (see Chapter 5 and Pokrovskaya *et al.*, 1994).

The choice of region to be studied and the dates of observation were determined, on the one hand, by the fact that the given region is a zone of intense cyclogenesis and, on the other hand, by the fact that the period of observation was characterized by highly contrasting meteorological conditions, which we consider in detail on

p. 134. In addition, we had previously carried out a detailed spatial and statistical analysis of the gradient fields of the atmospheric temperature stratification for this area (see Section 3.2.1), in which it was shown that the temperature gradient field did not have significant features within the range of natural spatial variation during the cyclogenesis of TC "Jeff" (8507) from July 20 to 28, 1985.

We now give a brief description of the nature of large-scale atmospheric processes over the given waters from June 1 to August 10, 1985.

Period I (June 1–5, 1985)

In episode 1 (Table 3.11) the region studied was affected for a protracted period by the ridge of a Pacific Ocean subtropical anticyclone. Clear weather with few clouds predominated.

Table 3.11. Meteorological conditions and the statistical characteristics of total moisture content q in the test area.

Episode No.	*Period of observation*	*Meteorological conditions*	*Sample size*	*Average value of q* ($g\,cm^{-2}$)	*r.m.s., $\sigma(q)$* ($g\,cm^{-2}$)
1	June 1–5	Anticyclone	39	3.78	0.34
2	June 6–8	Frontal division	22	4.39	0.44
3	June 11–17	ITCZ active	51	4.24	0.54
4	June 22–24	Effect of periphery of TS (8505)	38	4.08	0.41
5	June 25–30	Anticyclone	42	3.81	0.55
6	July 1–3	Effect of marginal frontal divisions	26	4.12	0.56
7	July 4–10	Anticyclone	44	3.81	0.60
8	July 11–13	TD generation	25	4.03	0.42
9	July 14–18	ITCZ active	23	4.35	0.45
10	July 19–23	TD generation, TL → TS	7	4.71	0.35
11	July 28–31	New TD generation	11	4.86	0.25
12	August 1–5	Effect of periphery of TS (8508)	23	4.43	0.43
13	August 6–10	Effect of distant periphery of TS	29	4.48	0.44

Period II (June 6–24, 1985)

In episodes 2 and 3 we observed the effect of northeasterly frontal divisions, activity of the intertropical convergence zone (ITCZ), and the formation of thick cloud clusters from June 22 to 24, 1985. In episode 4, the test area was affected by the marginal region of TC "Hal" (8505) (distance from the center of the TC was 2,000 km) and the birth of powerful TC "Irma" (8506) (distance from the center of the TC was 1,000 km).

Period III (June 25 to July 10, 1985)

In episodes 5–7 the ridge of a Pacific Ocean subtropical anticyclone moved across the test area and frontal divisions had a weak effect. Figure 3.13a shows a simplified cloud analysis map (episode 7).

Period IV (July 11–18, 1985)

In episodes 8 and 9 the test area was affected by the marginal regions of northeasterly frontal divisions, a TDE formed (Figure 3.13b) and moved northeast to merge with the system of a frontal division (Figure 3.13c). At the same time, the ITCZ was active to the south of the test area.

Period V (July 19–23, 1985)

In episode 10, an initial TD formed (July 19 at 12:00 GMT) in one of the cloud clusters at 20°N, 150°E; during the next 3 days this developed slowly and on July 22 it reached the TL stage before becoming TS "Jeff" (8507) on July 23. The TS moved in a northerly direction and affected the test area for 4 days.

Period VI (July 28 to August 10, 1985)

In episodes 11 and 12, under the effect of frontal divisions within the test area, a new TDE formed and moved rapidly in a westerly direction crossing the boundary of the test area on July 29 (Figure 3.13e). On August 3 the TDE became the strong STS "Kit" (8508), which had a weak effect on the test area (Figure 3.13f) and lasted until August 10 (episode 13).

3.3.2 Precipitable water fields in an undisturbed atmosphere

At the beginning of June and July 1985 (observation periods I and II) the basin of the northwestern area of the Pacific Ocean was affected for a long time by the ridge of a Pacific Ocean subtropical anticyclone. Under conditions of little cloud cover and a long period during which the tropical atmosphere was exposed to solar radiation, the region in question acquired the properties of a dry air mass, resulting from descending currents that prevail under anticyclonic conditions and the presence of retentive temperature inversion layers. The adverse (5-day) total content of moisture precipitated in the form of vapor, $q\,(\mathrm{g\,cm^{-2}})$ in the atmospheric column throughout

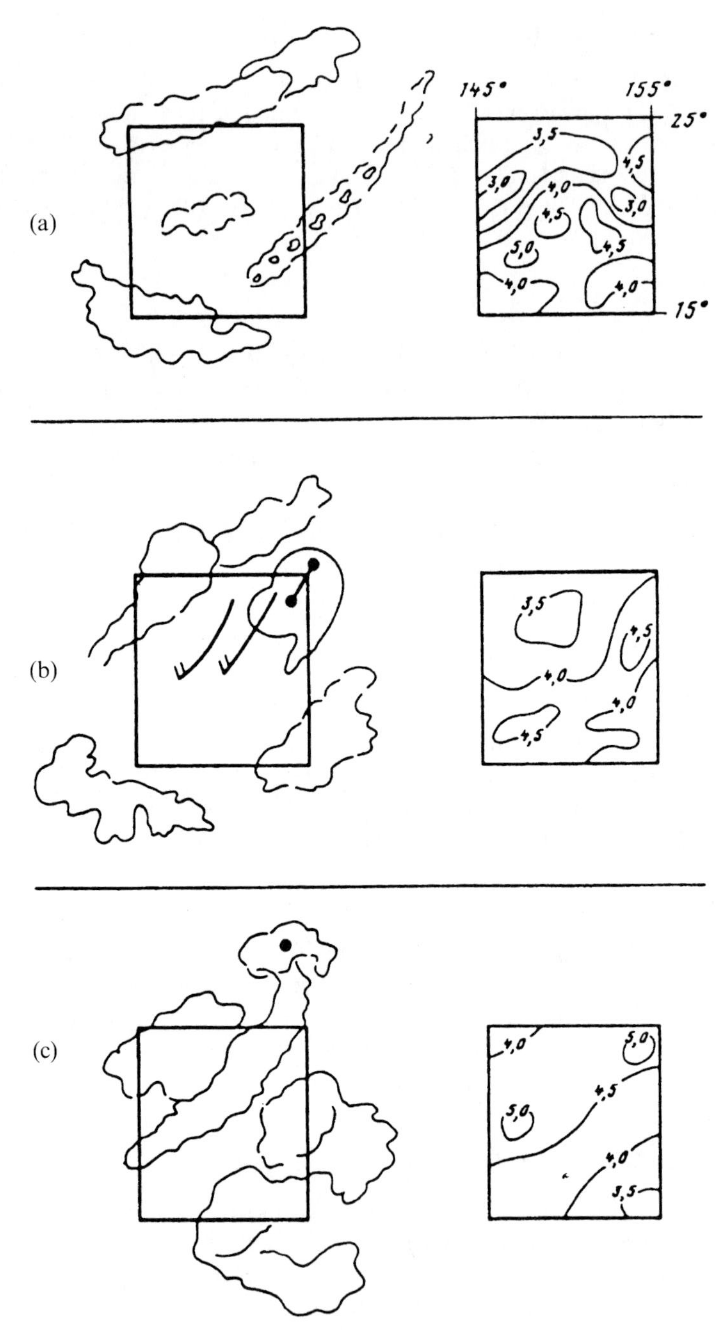

Figure 3.13. Cloud analysis map of the state of cloud cover and the moisture content map of the tropical atmosphere over test area A. Left-hand column—cloud analysis: (a) on July 8 at 00:00 GMT; (b) on July 13 at 12:00 GMT; (c) on July 14 at 12:00 GMT; (d) on July 22 at 12:00 GMT; (e) on July 31 at 12:00 GMT; (f) on August 2 at 12:00 GMT. Right-hand column—

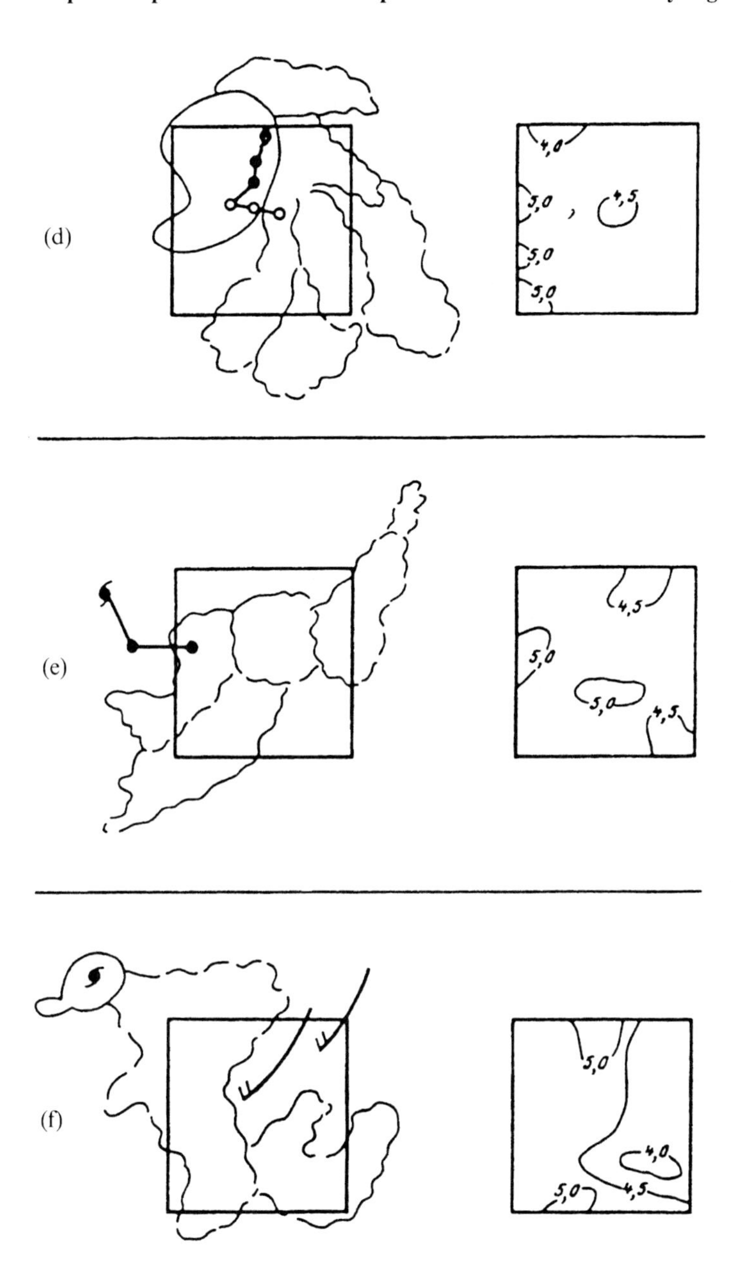

moisture content: (a) July 4–10; (b) July 11–13; (c) July 13–18; (d) July 19–23; (e) July 28–30; (f) August 1–5. Black dots and circles denote the geographical centers of TDs (see text for details) (from Pokrovskaya and Sharkov, 1997c).

the region was (period I) $q_{\text{ave}} = 3.78 \pm 0.347$ (here and in what follows we give the r.m.s. deviation). For period III, the average r.m.s. was practically unchanged (see Table 3.11). To the south of the test area, under the effect of the near equatorial trough of the ITCZ more moist points are recorded in the moisture content field (Figure 3.13a). We should also mention the important structural property of the moisture content field. The structure is by and large very "variegated" and spatial features of 100 km to 300 km in size are observed with marked spatial gradients $(\Delta q/\Delta x) \cong 0.03$–$0.02\,\text{g cm}^{-2}\,\text{km}$.

3.3.3 Precipitable water fields in a disturbed atmosphere

In the second 10-day period in July 1985 the test area was actively affected by the near equatonal trough of the ITCZ (episodes 8 and 9), which led to considerable transformation of the structure of the moisture content field: it became smoother, the spatial features expanded to 1,000 km, and the spatial gradients were significantly decreased to values between $0.001\,\text{g cm}^{-2}\,\text{km}$ and $0.0055\,\text{g cm}^{-2}\,\text{km}$. A TDE came into being on July 19 in a very dry region of the test area with $q \cong 3.5\,\text{g cm}^{-2}$ (Figure 3.13b, c). The next 10 days of the month (episode 10) were of particular interest to us, since an initial form of tropical disturbance was born immediately in the test area, reached mature form, and intensified. We shall analyze this episode more thoroughly in Section 3.3.4. Here, we only mention the weak growth of q to $q = 4.86$ and further smoothing of the gradients of the field, which decreased to 0.0005 and less (Figure 3.13d). On the whole, the q field still had low gradients (Figure 3.13e). After the storm left the region in question and right up to August 10 the atmosphere remained moist ($q = 4.43$) with low gradients.

3.3.4 Model of precipitable water fields and their properties under various synoptic conditions

Experimental datasets for average values of the q field and their mean square deviations obtained over 71 days with spatial averaging over $1{,}000 \times 1{,}000$ km and temporal averaging over 36 days are presented in Table 3.11 and Figure 3.14. During the interval in question, various synoptic conditions were recorded from anticyclonic to pronounced activity of the ITCZ. Analysis of the data of Table 3.11 and Figure 3.14 shows that conditions at the time of ITCZ activity led to a comparatively small shift (increase) in the average value of g for a practically constant value of the r.m.s.

Let us consider the differences between models of q under different synoptic conditions using a number of well-known statistical procedures (the Student t-test, the Fisher criterion). Let us choose the models of q obtained at the same geographical point but under very different synoptic conditions—episodes 3, 6, 10, and 13. In the procedure we define the dimensionless Student coefficient:

$$t_{ij} = \frac{\bar{q}_i - \bar{q}_j}{\sqrt{\dfrac{(n_i - 1)\sigma_i^2 + (n_j - 1)\sigma_j^2}{n_i + n_j - 2}}} \sqrt{\frac{n_i n_j}{n_i + n_j}} \tag{3.13}$$

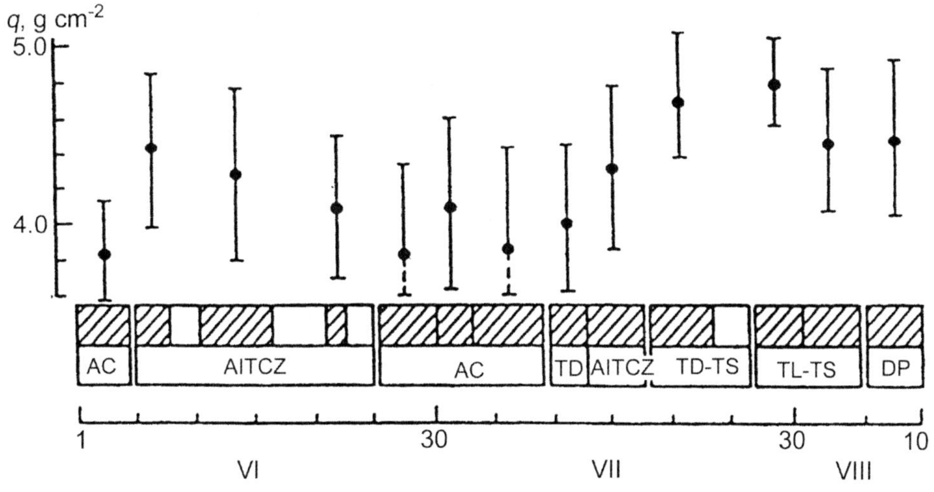

Figure 3.14. Time dependence of the average values of total water vapor in test area A and their r.m.s.'s from June 1 to August 10, 1995. The variability in the general synoptic situation is reflected by the division of the time axis into episodes (rectangles). The shaded area of a rectangle corresponds to the time over which the information was accumulated and averaged for the given value of the parameter q (for episode numbers, see Table 3.1): AC, anticyclone; AITCZ, active ICTZ; TD–TS–STS, generation of TD followed by formation of TS and then of STS; TL–TS, formation of TS from TL; DP, effect of the distant periphery of TS (from Pokrovskaya and Sharkov, 1997c).

Table 3.12. Values of the coefficients t and F of the Student and Fisher criteria for four combinations of episodes.

Situation (episode number)	t	t_{cr}	F	F_{cr}
AITCZ–AC (3–6)	0.92	2.00 / 2.65	1.07	1.78 / 2.26
AC–TS (6–10)	2.52	2.04 / 2.75	2.58	2.51 / 3.70
TS–DP (10–13)	1.31	2.04 / 2.75	1.58	2.50 / 3.70
AC–DP (6–13)	2.05	2.00 / 2.66	1.63	2.10 / 2.90

Note: The numbers in the numerators correspond to critical values of the parameters at the 0.05 significance level, those in the denominators correspond to the 0.01 level. AC stands for anticyclone and DP for distant periphery.

where $\bar{q}$ and σ are values of the average moisture content and its r.m.s.; and n is the size of the corresponding sample $i, j = 3, 6, 10, 13$ (Table 3.11). For the four situations, we construct the values of coefficient t in Table 3.12. For significance levels of 0.05 and 0.01 (probabilities of 0.95 and 0.99) the values of the critical coefficient $t_{\text{cr}} = t_{\alpha,f}$ are also given in Table 3.12. Comparison shows that at the 0.01 level of significance the differences between the average values of the q field in all situations are insignificant and the samples may all belong to the same general population. Let us consider the uniformity of variance in q fields in the given observations using the Fisher criterion. For this we define the function $F_{i,j} = \sigma_i^2/\sigma_j^2$ from the variance in q fields under different synoptic situations (i, j). The hypothesis that the variance is equal $H_0(\sigma_i^2 = \sigma_j^2)$ is accepted if the condition $F_{i,j} < F_{\alpha,f_i,f_j}$ is satisfied, where F_{α,f_i,f_j} is a quantile of the Fisher distribution at the significance level for the two degrees of freedom $f_i = n_i - 1$ and $f_j = n_j - 1$. From analysis of data in Table 3.12, which gives the experimental values of the coefficient $F_{i,j}$ for four combinations and values of F_{α,f_i,f_j} for $\alpha = 0.05$ and 0.01, it is not difficult to see that under the given conditions the significance of the variance values obtained for the samples of q fields does not contradict the assertion that variance is the same for different synoptic situations.

In other words, the statistical spatial structure of the moisture content field is invariant (either in terms of the average or in terms of variance) in very different synoptic situations (at least under the conditions in which this experiment was carried out).

3.3.5 Mesoscale variability in precipitable water fields in the process of cyclogenesis

Let us consider episode 10 in more detail, dividing it into four stages: the unperturbed situation (July 17–18), the creation of a TD (July 19–20), the formation of a TDE (July 21–22), and transition to a mature TC structure (July 23–24). The spatial q fields for these four stages are shown in Figure 3.15 where not only the area but also the surrounding waters are shown. In the two days before creation of the TD, the q field in the test area had weak gradients ($\Delta q/\Delta x < 0.003\,\text{g}\,\text{cm}^{-2}\,\text{km}$). The maximum value of q was found outside the test area at a distance of 500 km (Figure 3.15a). At the time the TD was created no serious transformations in the q field affected the local maximum ($q = 5.0$) found outside the test area. Formation of the TDE (July 21) took place on the boundary of the 4.5 $\text{g}\,\text{cm}^{-2}$ isoline, and the TDE as a formation moved in a northerly direction from a moist ($q = 5.0\,\text{g}\,\text{cm}^{-2}$) to a very dry region ($q = 3.5\,\text{g}\,\text{cm}^{-2}$). Transition of the TDE to mature form occurred on the northern boundary of the test area with an instantaneous value of $q \cong 4.3\,\text{g}\,\text{cm}^{-2}$. A vast region with high values $q = 5.0\,\text{g}\,\text{cm}^{-2}$ remained outside the area toward the west.

During the four stages the average value of q in the test area was around 4.5 $\text{g}\,\text{cm}^{-2}$ to 4.6 $\text{g}\,\text{cm}^{-2}$, local maxima were not recorded, and the field continued

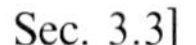

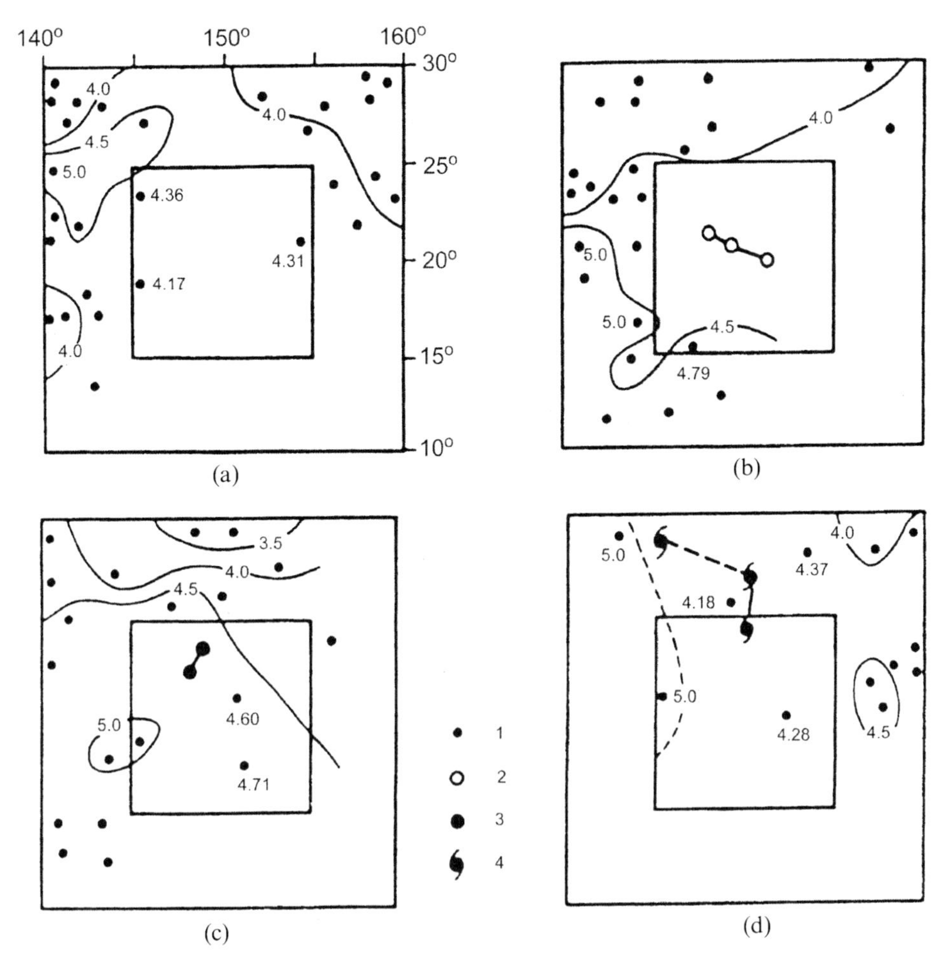

Figure 3.15. Maps of the moisture content of the tropical atmosphere over test area A and the surrounding atmosphere for episode 10 (internal square represents test area A): (a) July 17–18; (b) July 19–20; (c) July 21–22; (d) July 23–24; 1, points of recovery of moisture content stratification from NOAA data for the corresponding dates; 2, initial TD; 3, TL stage; 4, TS "Jeff" (8507). Numbers next to the recovery points denote the total water content of the atmospheric column (from NOAA data) (Pokrovskaya and Sharkov, 1997c).

to have weak gradients although the specific arrangement of the isolines in the q field varied noticeably. In this sense, the q field was highly variable.

3.3.6 Correlation with "one-point" measurements of precipitable water content

Earlier, we showed experimentally (Section 3.2) that the stable nature (within the range of natural spatial variation) of thermal stratification of the middle and upper troposphere was not affected in any considerable way by TCs of different intensities. On the other hand, with the direct formation of a TC, the field of temperature gradients does not contain significant variation in zones of intense cyclogenesis. As far as the moisture content field is concerned, according to a number of thermodynamic concepts (and, in particular, according to the convection hypothesis; (Palmen and Newton, 1969) the creation of a TC is based on a view that is widely held (Minina, 1987; Dobryshman, 1995) that the concentration of water vapor in the places where TCs form is very significant immediately prior to formation of their initial forms (TDs) in the field of an unperturbed atmosphere.

Our results do not contradict this view (under the given conditions) and indicate that in zones of intense cyclogenesis the spatial field of moisture content is very homogeneous and has no significant characteristics within the range of natural variation (spatial scale $1{,}000 \times 1{,}000$ km, timescale of 2 days). As we have shown, TDs form under very dry unperturbed atmospheric conditions.

It is of interest to correlate our spatiotemporal measurements with the results of the one-point study performed using a multiwavelengh radiometer (1.35 cm, 8 mm, 6 mm, 5.6 mm) on board an RV in the Pacific during September–October 1978

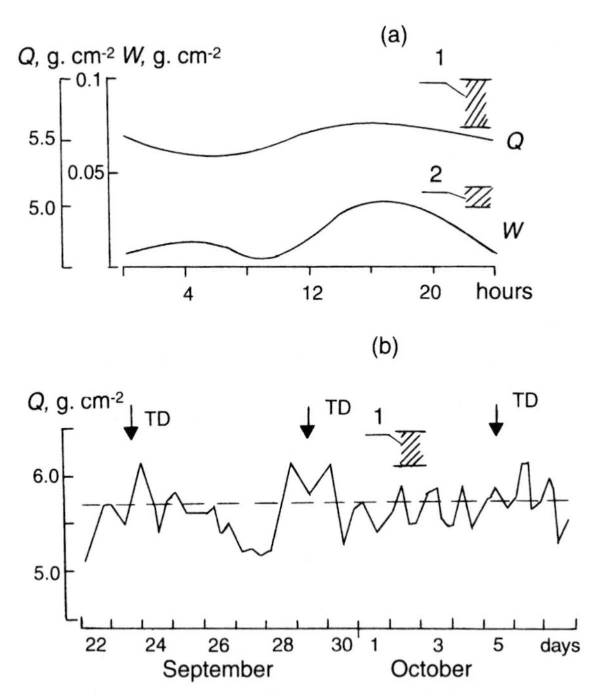

Figure 3.16. Daily (a) and monthly (b) time series of recovered values of precipitable water content (Q) and cloud liquid water content (W) at about 11°N, 145°E during between September 23 and October 7 1978. 1 and 2 are recovery errors of Q and of W. Full arrows represent the points in time at which tropical disturbances (TD) occur. The plot was reproduced by the author from data reported by Rassadovsky and Troitsky (1981).

(Rassadovsky and Troitsky, 1981). The observation point was located at about 11°N, 145°E. Figure 3.16 shows a time series—inferring precipitable water content and liquid water content—obtained under distinct meteorological conditions: without non-convective cloud (a) and with convective activity (b).

Under non-convective conditions (Figure 3.16a) daily cyclonicity was obvionsly found in the time series of Q that had daily-averaged variance of $\Delta Q = 0.35\,\mathrm{g\,cm^{-2}}$. In this case, the relative liquid water content in the tropical atmosphere accounts for 0.8% of total precipitable water content. Under convective conditions (Figure 3.16b), experimental results have shown that the maximum difference of daily precipitable water content may amount to $1\,\mathrm{g\,cm^{-2}}$ (about 20%) and relative liquid water content accounts for 3.5% of total precipitable water content.

It is interesting to note that no significant variability in total precipitable water content was found when TDs occur in this location (full arrows represent the points in time).

The absence of significant differences in the characteristics of the moisture content field in different synoptic situations may be a consequence of the statistical criteria used being weak and only relating to general "coarse" differences and trends (of which there were none in this case).

It is clear from study results that the use of characteristics of the moisture content field as features predicting the birth of TDs or trends in their evolution is not very informative when they are the only criteria used or when weak statistical criteria are applied.

The use of highly sensitive radiothermal and IR systems for recovery purposes and the application of a fine fractal and wavelet analysis may alter a remote experimental strategy greatly in this respect.

Thus, it has been shown that the moisture content field is statistically homogeneous and we have constructed a statistical model on spatiotemporal scales of 1,000 km and 3–6 days; it is not subject to significant changes in different synoptic situations. In anticyclonic weather the field has a very uniform structure with a predominance of sharp gradients; when the ITCZ is active the field has low gradients. We have not identified any significant features (within the range of natural variation) in the moisture content field at any stage of cyclogenesis.

3.3.7 Atmospheric water balance under pre-typhoon and typhoon conditions

In this section we present results on the atmospheric water balance of typhoon "Nina", which formed near 5°N, 160°E on November 18, 1987 and moved northwestward during its development (Liu *et al.*, 1994, 1995). The water vapor path, liquid water path, ice index, and precipitation amount are determined in the vicinity of the typhoon using data from the SSM/I (special sensor microwave/imager). The water balance in the typhoon cloud is then examined during its different development stages. An ice index is derived from SSM/I data that are used to investigate the overall ratio of ice/liquid water change in the typhoon during its development. By comparing the ice/water ratio of different mesoscale convective cells in the typhoon, attempts were made to interpret the different cloud structures and development

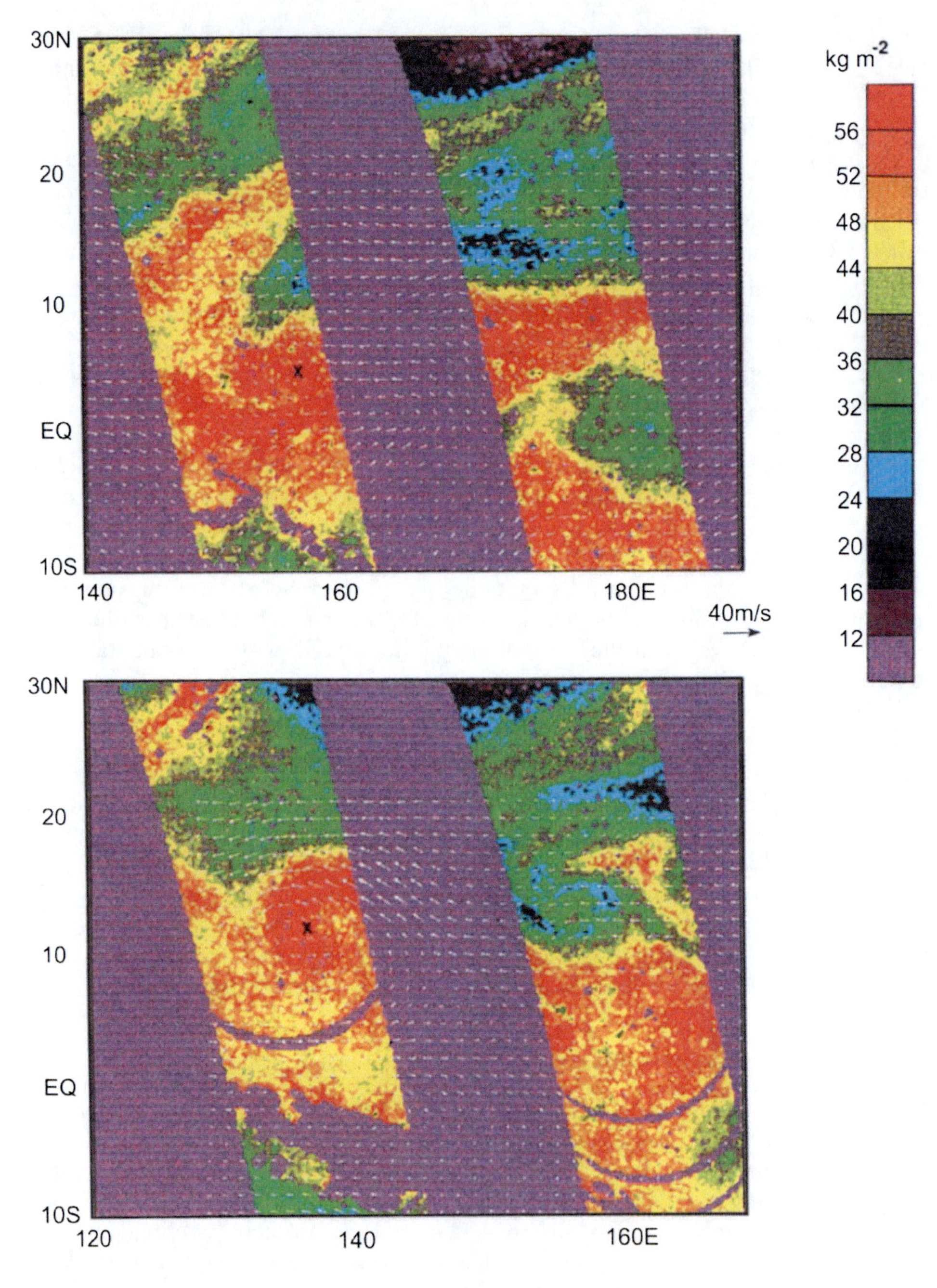

Figure 3.17. Horizontal distribution of vertically integrated water vapor retrieved from SSM/I data with superimposed ECMWF 850 mb wind field for: (a) November 18, 1987 at 19:00 PM (formation stage of typhoon "Nina") and (b) November 18, 1987 at 22:00 PM (developing stage) (Liu *et al.*, 1994).

stages of individual mesoscale cloud cells relative to their position from the typhoon center.

The horizontal distribution of the vertically integrated water vapor amount (WV) (or precipitable water content) in the vicinity of the storm is shown in Figure 3.17 for 19:00 UTC (formation stage). Superimposed on this figure is the 850 mb wind field at 18:00 UTC from the ECMWF (European Center for Medium-Range Weather Forecasting) analysis at a resolution of 1.25°. Only large-scale features can be seen from the wind field, which shows smaller windspeeds than the actual maximum wind near the storm. Nevertheless, it is clear there is strong convergence near the storm center (5°N, 157°E). Two major bands of water vapor convergence in the storm are seen: one has a very long tail originating from the northeast and wrapping around from the northwest; and the other from the east. Low amounts of WV are seen on both sides of the high WV bands. This indicates that water vapor is not being supplied to the storm by means of uniform convergence, but in bands of water vapor convergence. Figure 3.17b shows the WV and the 850 mb wind fields at 22:00 UTC when the storm was developing. The storm center at this time was located at about 11°N, 137°E. Low-level cyclonic circulation was stronger than in the formation stage (Figure 3.17b). To the north and northeast of the storm, there is a widespread low WV zone. Major water vapor supply to the storm is delivered by the southeasterly flow to the east of the storm while the northwesterly flow to the west of the storm may also have made some contribution.

It was found that the atmospheric water budget in the typhoon was mainly balanced by the horizontal transport of water vapor into the region, evaporation from the ocean, and precipitation. Of the two source terms, horizontal transport played the major role with a contribution of more than 65% in all storm stages for a 1° radial area or larger. In addition, the horizontal transport of water vapor seemed to converge through several "bands" instead of uniformly. Mesoscale convective cells, which may consist of several cumulonimbus clouds in each, developed in the bands, with convectively more active cells occurring upwind and dissipating ones downwind. It was also found that maximum latent heat release preceded maximum storm intensity.

During the week prior to the genesis of "Nina", several cloud clusters were observed in the region of subsequent genesis (near 5°N, 170°E).

Figure 3.18 shows the GMS IR sequence from 00:00 UTC on November 12 to 18:00 UTC on November 19 in the domain 0–11°N and 113–200°E every 6 hours (Liu *et al.*, 1995). Although it is hard to determine when the genesis process started, the cloud cluster (F) that appears to be most directly related to the eventual typhoon first appeared near 5°N, 170°E early on November 16. Before this date several cloud clusters were also observed in this region (A, B, C, D, and E as shown in Figure 3.18), of which clusters A and D were in almost the same location as where the cyclogenesis occurred. Clusters B and E initiated to the west side of A and D, respectively, and moved to the west while cluster C initiated to the east of A and moved to the east. None of these cloud clusters directly developed to the TS stage, although they all had a large area where the cloud-top temperature was cold. During November 16 to 19 the pre-tropical storm cloud cluster (F) underwent a major

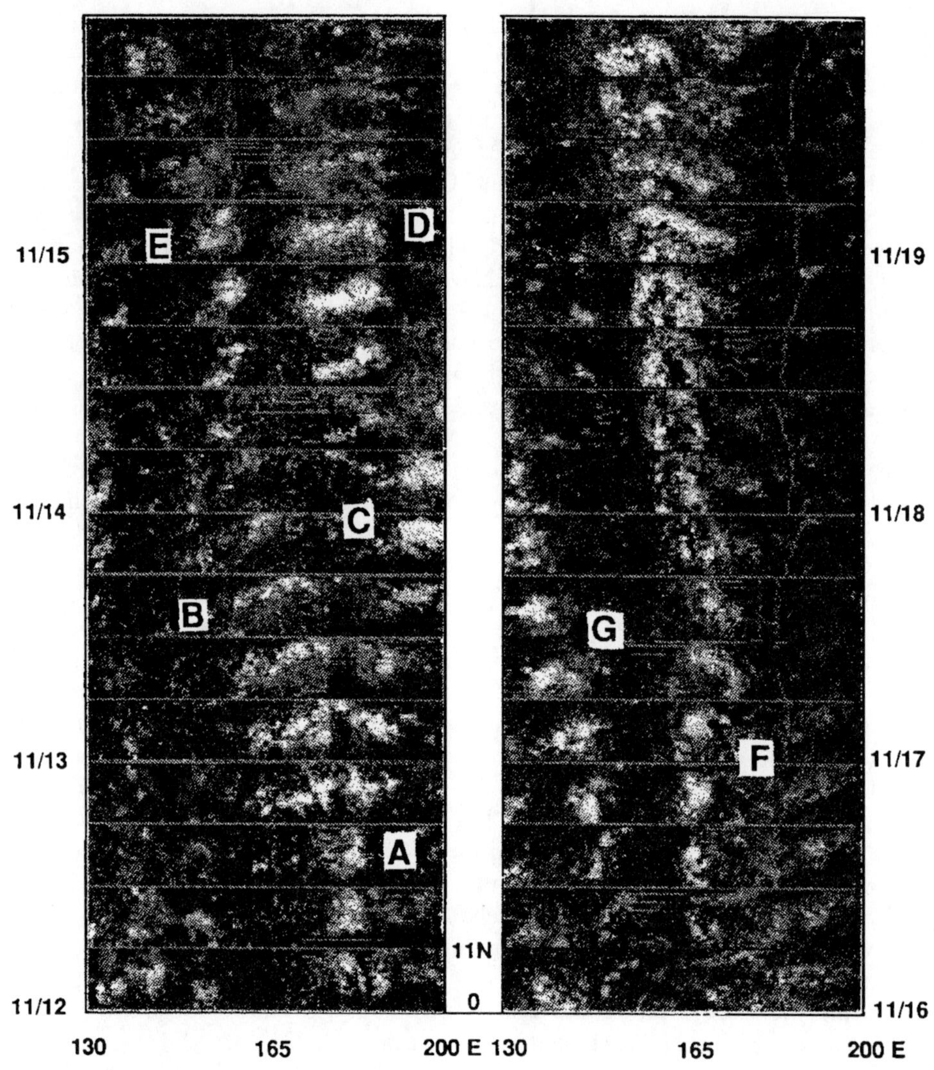

Figure 3.18. GMS infrared image taken every six hours from November 12 00:00 UTC to November 15 00:00 UTC (left) and November 16 00:00 UTC to November 19 18:00 UTC (right) in the region 0–11°N, 130–180°E. Letters A, B, C, D, E, F, and G indicate the cloud clusters mentioned in the text (Liu *et al.*, 1995).

active–inactive–active variation as shown by the horizontal extent of the cluster. Figure 3.19 shows the percentage of the area with cloud-top temperature CTT $<-70°$C (in a 0–2° radial area) that varied from the disturbance center. Similar to Zehr (1992) there was clearly an "early convective maximum" late on November 16. Convective activity temporarily weakened following the early convective maximum before it regained strength late on November 18. The area of cold

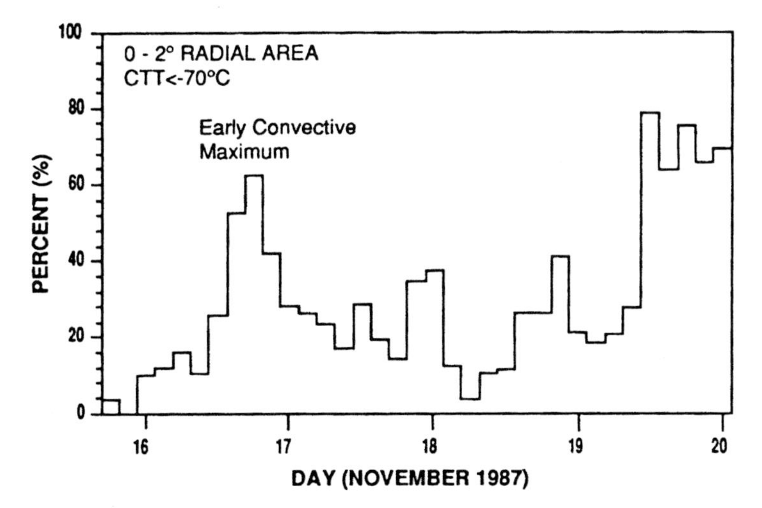

Figure 3.19. Variation of the percentage of cold cloud-top temperature (<−70°C) pixels from total pixels in the 0–2° radial area around the storm center (from Liu *et al.*, 1995).

cloud-top temperature from late on November 18 to early on November 19 does not show a large increase in Figure 3.19 although the apparent increase of deep clouds can be seen in Figure 3.18. This is because deep clouds did not gather at the center, but rather spread in the outer region.

Cloud clusters in the region where "Nina" developed into a TS were studied by comparing the structure of the cloud clusters with the surface wind field. The cloud clusters had a lifecycle of less than 2 days except for one which finally became the TS. Although there were similarities in cloud structure between cloud clusters, the wind fields associated with them showed a variety of patterns. One cloud cluster (B) occurred in a strong wind stream, but dissipated as the wind weakened. A cluster (E) developed in a weak vortex, and also dissipated about 1 day later. Other cloud clusters (D and F) occurred in the stronger wind region of a strengthening cyclonic circulation. Cluster D lasted only about 1 day; but cluster F, immediately west of D, maintained itself for several days as the cyclonic wind strengthened and finally developed into a TS.

Comparison of evaporation from the ocean surface with precipitation showed a large water vapor deficit in the cloud cluster area when there was no water vapor supply by horizontal transfer. Evaporation generally did not exceed 1 mm/h while precipitation could be larger than 5 mm/h for a $1 \times 1°$ area average. Additionally, the apparent decrease in total water vapor around the cloud clusters (0–2° radial area) occurred after their mature stage, except for cluster F. By considering the water vapor balance in cloud clusters, cyclogenesis could be interpreted as the transition from an unbalanced water budget (short-lived cloud cluster) to a balanced water budget (typhoon). The horizontal transfer of water vapor in a water vapor–unbalanced cloud cluster is not large enough to overcome the deficit caused by precipitation over evaporation. This cluster might have accumulated a large amount of water (water vapor, liquid water, and ice) around its center during its premature stage. Once a cluster reaches maturity, the shortage of water vapor will result in dissipation of the cluster. When the horizontal advection of water vapor is large enough, a cloud cluster may turn into a typhoon. In such a case, the combined contribution of evaporation and horizontal transfer can provide enough water vapor to balance the removal by precipitation so that precipitation does not "dry up" the atmosphere as in the unbalanced case. This appears to be the necessary condition for cyclogenesis. In the case studied here, the increase in horizontal transfer of water vapor is associated with the increase in surface cyclonic wind. It is not clear what the cause of the increase in surface wind is. It appears that latent heat release alone is not likely; the presence of an equatorial westerly wind burst at this time contributed to the production of low-level cyclonic vorticity. It should also be mentioned that the role played by low-level convergence emphasized in this study is different from what was discussed by McBride and Zehr (1981), Merrill (1988a, b) and Zehr (1992). While they focus on its dynamical importance, Liu *et al.* (1995) emphasize the role played by low-level convergence as a source of water vapor, although it may have played both roles.

3.4 INITIAL STAGE OF TROPICAL CYCLOGENESIS IN THE PACIFIC: DYNAMICS, INTERACTIONS, HIERARCHY

In the last 40 years remote-sensing investigation has been used. Its purpose is to study all temporal phases in the evolution of natural atmospheric catastrophes (McBride, 1981a, b; Zehr, 1992; Liu *et al.*, 1995; Ritchie and Holland, 1997; Zimin *et al.*, 1992). Remote sensing is especially useful in the study of initial forms of TDs. Remote monitoring of the ocean–atmosphere system that transfers from turbulent chaos to an organized (coherent) structure is of great interest both from the physical point of view and that of the solution of prediction problems.

The new aspect here, in the remote study of the genesis of eddy systems in the terrestrial atmosphere, is the development of a methodology for remote monitoring of initial forms of TDs, including those that transfer into developed forms and those that do not. Remote monitoring of mature stages is now well established.

When carrying out detailed remote aerospace observations of the initial forms of TCs, it is important to find systematic data about the sites at which disturbances occur and evolve in basins where intense cyclogenesis occurs. However, when dealling with this problem, we are faced with certain difficulties associated, first, with the systematic accumulation of remote data about initial forms of TDs in all basins where cyclogenesis occurs and, second, with methodological problems relating to the analysis of remote aerospace observations arising from the difficulties of identifying and differentiating different forms of the initial stages of TCs (see Zehr, 1992; Liu *et al.*, 1995).

Careful analysis (Pokrovskaya and Sharkov, 1994c, 1997a, 1999a, d) of geographical information systems (GISs) and the datasets relating to large-scale TDs now in use around the world (and modifications of them) shows that the development of databases on tropical cyclogenesis takes place at the regional level and is largely based on data for mature forms of TDs. The initial stages are deliberately not recorded and are not gathered.

Work on observations of the first (non-developed) forms of TDs (McBride, 1981a, b; McBride and Zehr, 1981) essentially compares those that become TCs with those that do not, based on the microthermal characteristics of individual cloud systems, and quite deliberately does not consider the spatial characteristics (on macroscales) of tropical formations at the pre-crisis stage or the numerical characteristics of these as structural elements of a complex evolving barometric field.

It is necessary once more to stress that the problems resulting from investigating transition stages are very complicated. Up to now there has been little detailed work done (apart from the authors cross-referenced here) to identify such transitional forms as Mediterranean Sea hurricanes (Blier and Ma, 1997), and cyclogenesis mesostructures over the eastern Mediterranean (Irisova, 1974), or polar mesolow systems (Lilly and Gal-Chen, 1983).

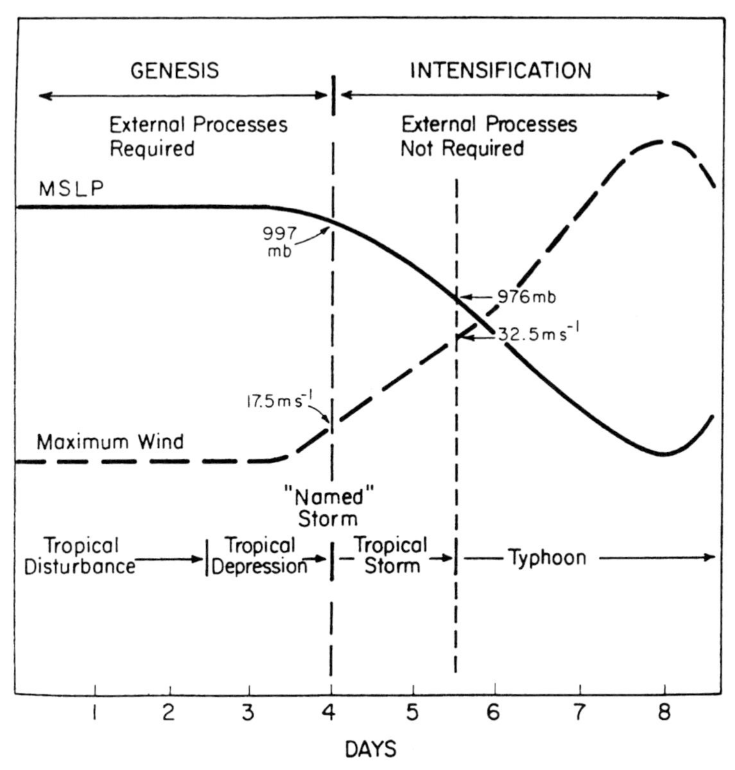

Figure 3.20. Depiction of typical genesis and intensification periods and their associated minimum sea level pressure (MSLP) and maximum surface windspeed (from Zehr, 1992).

3.4.1 Two-stage conceptual model of tropical cyclogenesis

The commonly accepted model of the genesis and intensification periods (stages) of tropical systems is presented in Figure 3.20. In the study of these systems, by TD we mean all forms of tropical formation, characterizing both non-stationary wave regimes and weakly organized structures. The main purpose of this work is to identify the several types of original TDs that arise.

The genesis stage is the original form of tropical system development including the tropical disturbance (TDI), areas where there are lows (Neumann, 1993), and the TDE. As a rule, at this stage, the TDI and TDE are zones of relatively low pressure (2–4 mb lower than the background). These disturbances may either become mature TCs or remain in their original non-developed form. The detailed spatiotemporal numerical characteristics of these disturbances, the regions in which they occur, and the trajectories along which they move are inadequately known today, although from preliminary (and very incomplete) data it is known that in active seasons TDs and TDEs are born and exist practically on a day-to-day basis.

Satellite infrared and visible data have been used in TC studies in the 1980s and 1990s (e.g., Dvorak, 1984: Zehr, 1992). Dvorak has developed a technique to estimate TC intensity based solely on satellite imagery that is widely used operationally. Zehr has investigated the variation in CTT using digitized infrared satellite data. He found a strong diurnal cycle of cyclone cloud: cold cloud tends to have maximum cloud amount in early morning while the maximum warm-cloud amount occurs in late afternoon. By analyzing GMS IR data in the area within 2° latitude radius from the cloud system center, Zehr conceptually divided tropical cyclogenesis into the pre-genesis stage and two active stages (called Stage 1 and Stage 2), each corresponding to a maximum of deep convection (CTT $< -70°C$). The two active phases are considered to be induced by environmental forcing (wind surges, low-level convergence, etc.). Zehr also emphasized two factors necessary for tropical cyclogenesis: (1) strong mesoscale convection in Stage 1, which allows initiation of the mesoscale vortex; and (2) coincidence of the location of deep convection in Stage 2 with the location of the mesoscale vortex initiated in Stage 1 in order to decrease the surface pressure sufficiently.

The characteristics of the two-stage conceptual model of tropical cyclogenesis are illustrated in Figure 3.21. As discussed previously, the time periods associated with the genesis period are highly variable. The timescale in Figure 3.21 represents average time rates of change. The plot of deep convective clouds (Cb) is an idealized depletion of cold IR areas with diurnal variation removed. Similarly, the associated trends of minimum sea level pressure (MSLP) and maximum surface wind speeds (V_{max}) do not include diurnal and short-term small variation, which may take place. Stage 1 is characterized by V_{max} and MSLP that are not much different from those associated with a pre–Stage 1 disturbance or with a non-developing TD. During Stage 2, the MSLP begins to decrease faster, lowering to an average value of 997 mb at the end of this stage. This coincides with V_{max} increasing to 17.5 m s^{-1} (35 kt), according to the average pressure–wind relationship for the western North Pacific. This represents a minimum sea level pressure about 10 mb lower than the

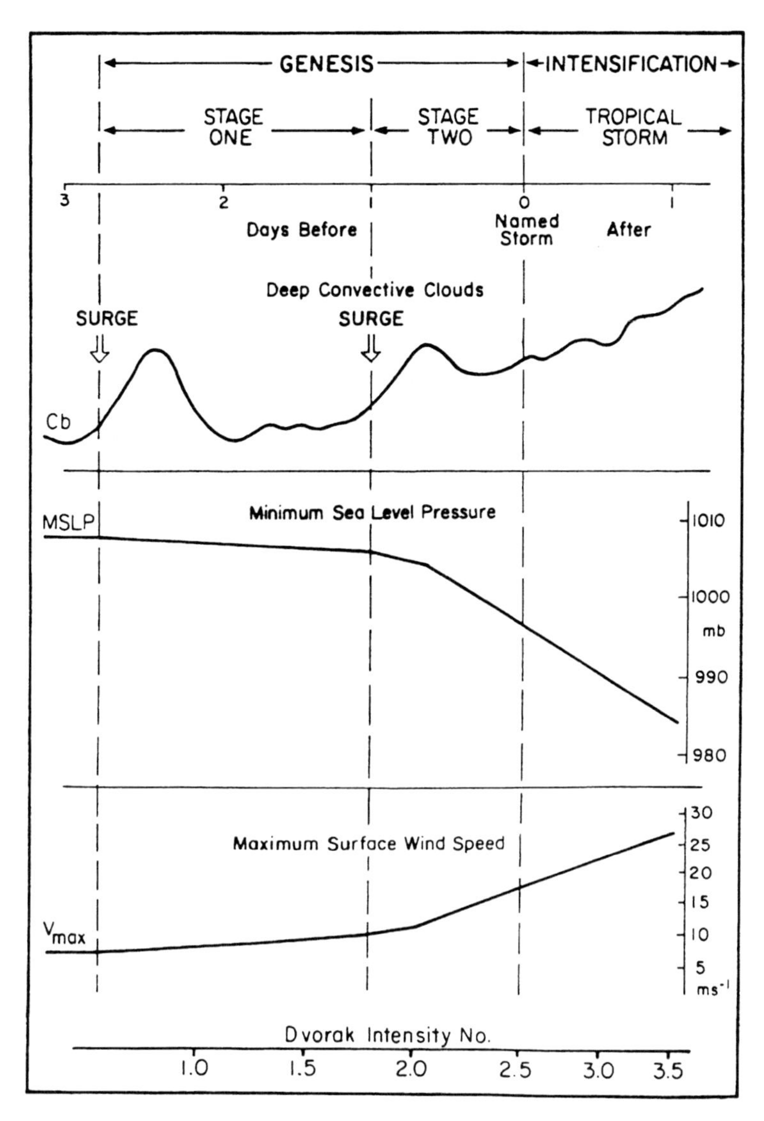

Figure 3.21. Conceptual model of important changes in deep convective clouds (Cb), minimum sea level pressure (MSLP), and maximum surface windspeeds (V_{max}) during tropical cyclogenesis and estimates of the associated numerical values. The associated Dvorak intensity T-numbers are also shown (Dvorak, 1984). The plot of Cb is a depiction of IR satellite data illustrated by areas colder than −70°C (from Zehr, 1992).

average 1,007 mb associated with non-developing TDs. Almost all of this drop in pressure occurs during Stage 2. These idealized changes in MSLP and V_{max} during tropical cyclogenesis are also supported by previous studies.

The time evolution of deep convective clouds (Cb) in the conceptual model includes two convective maxima that mark the beginning of the genesis process.

Stage 1 also includes an inactive period in which Cb activity is considerably less than that associated with either the Stage 1 convective maximum or with Stage 2. Stage 2 is characterized by increasing Cb clouds, which may occur as a secondary convective maximum. Stage 2 Cb activity may also occur as part of a general increasing trend, which continues into the intensification period.

The main differentiating feature of Stage 2 convection is that it is organized in a pattern rotated about a cyclonic circulation center. It also becomes more concentrated around that center with time. These observations are closely related to the widely used current intensity numbers (I-numbers) of the Dvorak technique (Dvorak, 1984) for the analysis of TC intensity (Figure 3.21). Specific cloud pattern characteristics must be observed before an initial Dvorak classification of T1.0 is assigned to a particular disturbance. This is typically followed 1–2 days later by the first designation as a TS (T2.5). Stage 1 in Figure 3.21 corresponds to Dvorak's T1.0–1.5 and Stage 2 with T2.0–2.5. However, as noted by Dvorak (1984), cloud patterns during the early stages are highly variable and objective assignments of the extent to which cyclogenesis has occurred, are very difficult to interpret, and sometimes unreliable.

Figure 3.21 also denotes the occurrence of a surge associated with the onset of increasing deep convection. A surge is defined here as a low-level windspeed maximum typically appearing in the 850 mb analysis. An area of low-level mass convergence is associated with a surge. When the surge interacts with a TD or a TDE, it will enhance low-level convergence and, therefore, result in deeper convection; the conceptual model in Figure 3.21 includes a surge interaction immediately preceding enhanced deep convection.

Figure 3.22 is an expanded illustration of the tropical cyclogenesis conceptual model introduced by Zehr (1992). In addition to deep convective clouds (Cb), MSLP, and $V_{\max}$, several other quantities have been added. They are maximum low-level relative vorticity ($VOR_{\max}$), mean low-level relative vorticity (VOR_{mean}), and mean low-level convergence ($-DIV_{\text{mean}}$). VOR_{mean} and $-DIV_{\text{mean}}$ represent average quantities over a 2° latitude radius area.

While $VOR_{\max}$ shows a substantial increase during Stage 1, the increase in VOR_{mean} is much smaller. This is because the increase in $VOR_{\max}$ is primarily due to a concentration of cyclonic circulation rather than an increase in windspeed. The difference in VOR_{mean} between the pre-genesis and Stage 1 periods is due primarily to synoptic-scale characteristic differences between non-developing and pre–tropical disturbances. The large increase in $VOR_{\max}$ during Stage 1 is due to the formation of the initial mesoscale vortex which has only a small effect on VOR_{mean}.

VOR_{mean} shows a larger increase during Stage 2 (from $1\text{–}2 \times 10^{-5}\,\text{s}^{-1}$ to $9 \times 10^{-5}\,\text{s}^{-1}$). This is due primarily to an increase in outer cyclonic circulation just prior to becoming a named storm. The inner mesoscale circulation also winds itself up during this period contributing to the increase in VOR_{mean}.

Mean low-level convergence ($-DIV_{\text{mean}}$) shows two sharp increases that coincide with the surges indicated in the plot of deep convective clouds. This occurs both during Stage 1 and again during Stage 2, typically about 2 days later. The enhanced low-level convergence is forced by environmental influences external

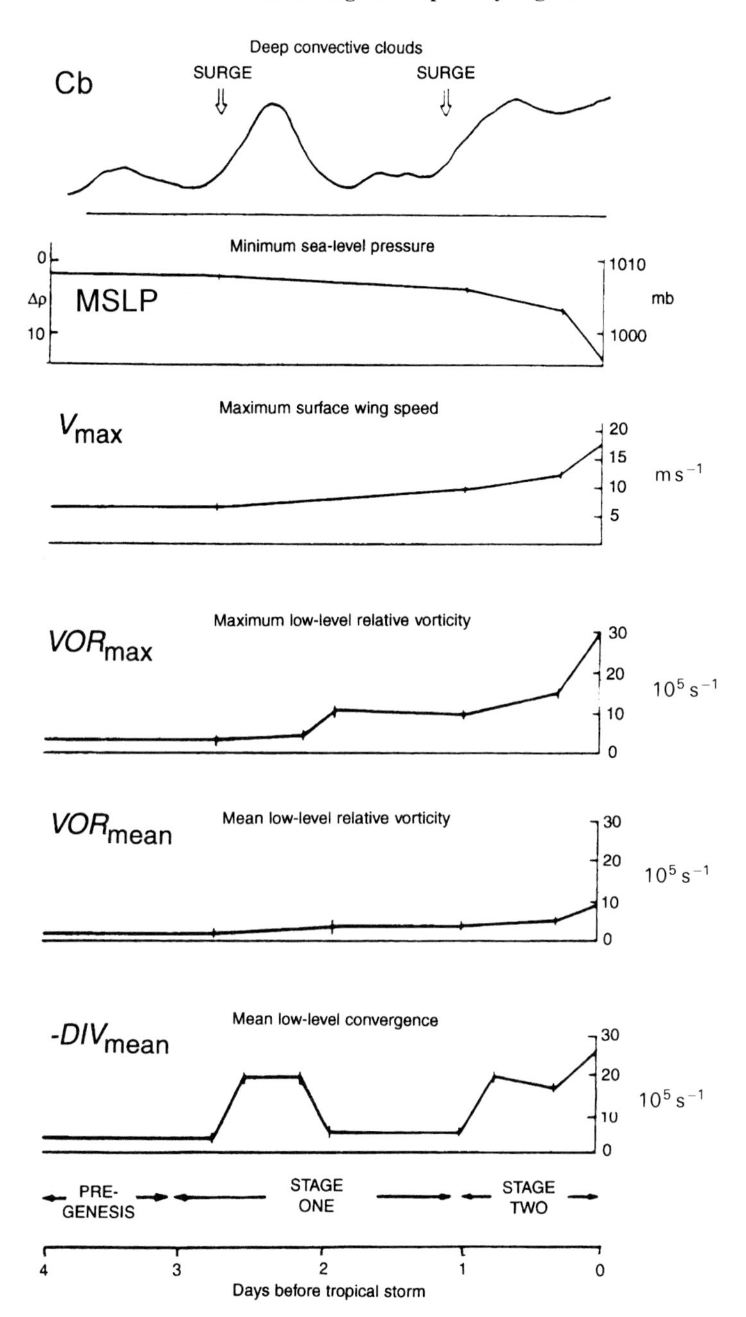

Figure 3.22. Changes in the important quantities of a pre–tropical storm disturbance according to the detailed conceptual model of tropical cyclogenesis. Cb represents a quantitative time-averaged satellite measurement of cold IR cloud area. The mean low-level relative vorticity (VOR_{mean}) and the mean low-level convergence ($-DIV_{\text{mean}}$) are averages over an $R = 0$–$2°$ latitude area (from Zehr, 1992).

to the disturbance itself. Such influences appear as maxima of inward radial wind, analyzed relative to the center of the disturbance and its motion. They are referred to as surges and act to enhance $-DIV_{\text{mean}}$. The temporarily enhanced low-level convergence is the forcing responsible for the convective maxima. Deep convection is essential for cyclogenesis to occur. However, it may only be required at key periods, on timescales less than 24 h, to allow important structural changes to take place. Substantial portions of the genesis period may be inactive with regard to deep convection located near the circulation center. Mean low-level convergence averaged over a 3 to 4-day period may be sufficient to maintain a persistent TD, but it is likely inadequate at forcing deep convection of the magnitude needed for tropical cyclogenesis. Therefore, external forcing in the form of low-level wind surges is required both at Stage 1 for initial vortex formation and at Stage 2 for initial vortex deepening.

Conceptual models of the observable quantitative structural changes that take place during tropical cyclogenesis are presented in Figure 3.22. This view of cyclogenesis has several unique features that distinguish it from previous explanations of genesis:

1. Rather than a gradual transition from a pre-existing disturbance to a TC, cyclogenesis is described by two distinct phases or events (Stage 1 and Stage 2). Both are observed in association with the enhanced intensity and area coverage of deep convective clouds. They occur on a timescale of 6–24 h and may be separated by variable periods of up to several days, often characterized by inactive convective periods. An important structural change is noted with each of the two stages. During Stage 1 a mesoscale vortex is initiated, which is embedded within the pre-existing disturbance circulation. During Stage 2 the central pressure of that vortex decreases, and the tangential wind increases in response, resulting in a minimal TS. This process is illustrated qualitatively in Figure 3.23.
2. The enhanced convection at both stages is forced by external low-level wind influences, referred to as surges.
3. In addition to having deep convection, three necessary conditions must be present in order for cyclogenesis to occur: (a) vertical wind shear near the circulation center must be small; (b) low-level vorticity associated with the pre–tropical storm disturbance must be sufficiently high; (c) low-level convergence must be sufficient not only to maintain the disturbance but also to provide the additional forcing required for enhanced convection (No. 2 above). If any one of the three conditions is unfavorable, cyclogenesis will not proceed. The resultant weather system may persist as a non-developing TD. If Stage 1 of tropical cyclogenesis has occurred and deep convective clouds are located near the circulation center, it may be designated a TDE. However, it does not become a named storm unless Stage 2 of tropical cyclogenesis occurs.

The physical mechanisms that can theoretically induce tropical cyclogenesis, as described by this conceptual model, can be discussed within the framework of the

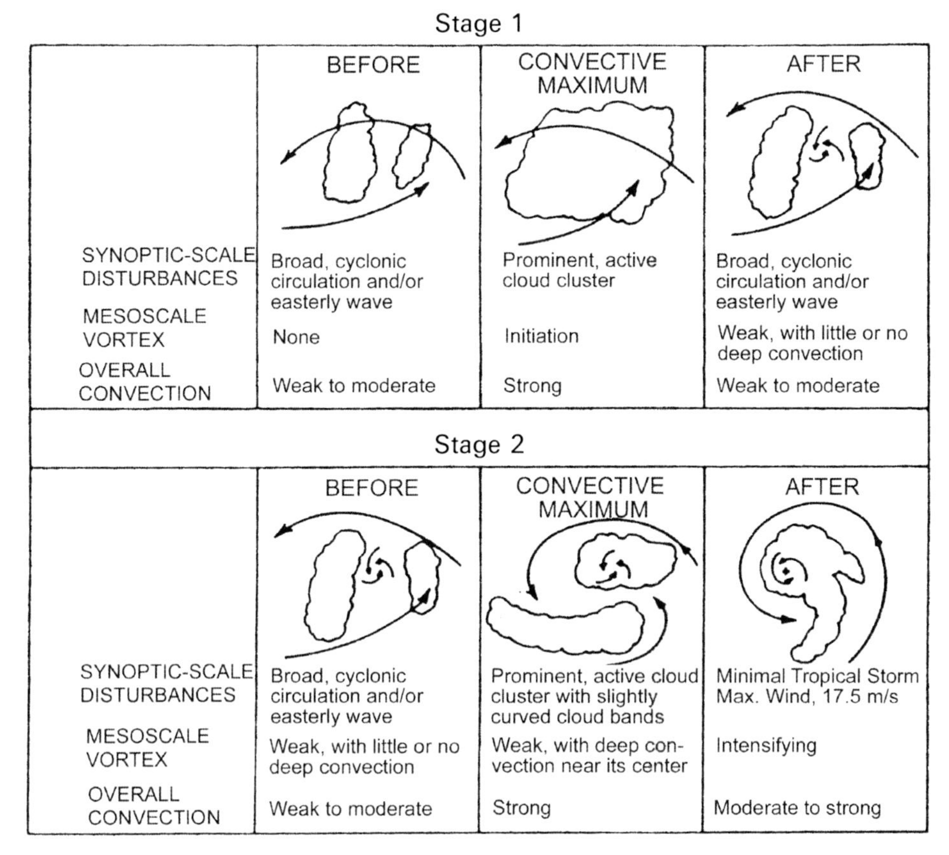

Figure 3.23. Conceptual model of tropical cyclogenesis with illustrations and descriptions of characteristics that are observable using satellite images (from Zehr, 1992).

self-organization concept (Moiseev *et al.*, 1983a, 2000; Moiseev, 1990; Zimin *et al.*, 1992). Two features of this concept require physical explanation: initial mesoscale vortex formation and the decrease in initial significant central pressure (MSLP).

The physical processes in a fully developed typhoon are considered to be different from the genesis stage of a TC. It is believed that a mature TC can be thought to be in steady state with a primary (azimuthal) circulation and an in–up–out secondary circulation, while cyclogenesis, in contrast, is the transition of the disturbance from a non-steady state to a steady state.

3.4.2 Dynamics and interactions of original forms of TCs

The purpose of this section is to present the results of analysis of remote satellite observations of cloud systems (from GMS and NOAA satellites) and charts of the barometric situation in the northwestern area of the Pacific Ocean in a period of active tropical cyclogenesis, with a view to identifying the original forms of tropical cyclogenesis, observing the process of their development and the special dynamics in this process, and obtaining some numerical characteristics. In this section we use experimental operational data acquired directly from the Japanese meteorological center (Tokyo) on board the RV *Akademik Shirshov* during maritime experiments which took place in the NWP and the South China Sea from July 2 to October 10, 1988 and from April 28 to August 11, 1989 (Pokrovskaya and Sharkov, 1997b).

Our purpose here is to identify the several types of original forms of TCs that arise on the basis of the dynamic concept described in Section 3.4.1.

One of these (type A) is the original form of tropical system that leads up to TC genesis (Stage 1 and Stage 2).

Type B is a tropical system at the stage of its transition to a TS—essentially, the moment at which a TC is generated (formed). The spatial characteristics of the fields in which TCs are generated have been determined. They have a pronounced non-uniform structure that persists right up to the formation of active centers in the regions of their generation (northern hemisphere) (Section 2.6.3). The physical conditions under which TDEs evolve into an organized TC structure are currently based on the concept of self-organization.

These are active TCs (type C) with a marked inward-spiraling tropospheric circulation and outward-spiraling stratospheric circulation. According to theoretical ideas (linear stage) the TC is fundamentally a three-dimensional structure in the form of coupled toroidal and poloidal windspeed fields and is a typical example of the formation of a large-scale coherent structure from the spiral of turbulent chaos. The remote features for recognition and identification of the stage of TC evolution, the seasonal characteristics of cyclogenetic activity, the geographical regions in which TCs are active, and the trajectories of TCs in the northern hemisphere are well known (Neumann, 1993).

In July 1988 in the troposphere over the NWP a latitudinal movement of air masses was observed that was not conducive to the exchange of these air masses between low and high latitudes. The circulation in the upper troposphere had anomalous properties which gave rise to unfavorable conditions for the development of vertical-penetrating convection. Because of this, tropical cyclogenesis in July was weak, as a result of which a total of 12 TDs were born in the NWP, only three of which developed to become mature TCs.

In August the atmospheric circulation in the region did not change significantly, cyclogenesis continued at low intensity, and during that month 18 disturbances formed but only 7 developed to become mature TCs.

In September 1988 the Asian monsoon trough became more intense. The paucity of clouds facilitated the heating of oceanic water, which ultimately led to an intensification in cyclogenesis. As a result of this, 18 TDs occurred in these basins, only half of which reached the stage of mature TCs.

Thus, as a consequence of the regional nature of high and surface barometric fields, tropical cyclogenesis from July to September 1988 was of low intensity. Of the 53 TDs that occurred during these 3 months, 19 (or 35%) became mature TCs, while more than half (34 or 65%) broke up (dispersed) without reaching the stage of TS.

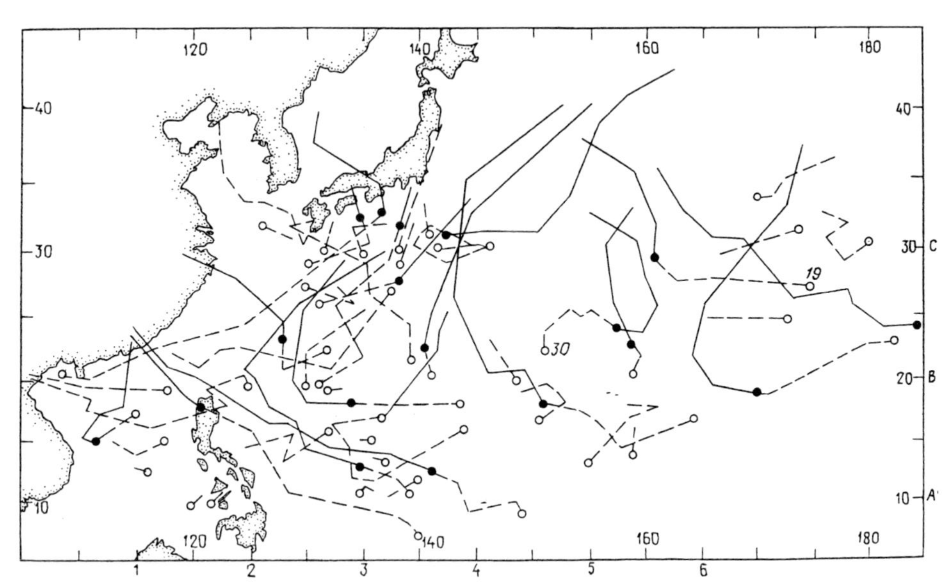

Figure 3.24. Spatial and geographical distribution of the original forms of TDs and their subsequent evolution (data for July–September 1988). Open circles, the birthplace of disturbances; full circles, the point at which the disturbance became a mature TC; dashed and continuous lines, trajectories of the movement of original forms and the TC, respectively. The numbers 19 and 30 denote TDs that developed into TC "Clara" and TC "Elsa", respectively (from Pokrovskaya and Sharkov, 1997b).

The spatial and geographical distribution of points at which the original TDs (both those that matured and those that did not) were born, together with the paths on which they moved, are shown in Figure 3.24.

In order to obtain a sufficiently clear representation of the main spatiotemporal characteristics of the evolution of TDs in the region in question, we shall divide the spatial field of the region of cyclogenesis into squares of size $1{,}000 \times 1{,}000\,\text{km}^2$ (Figure 3.24) and generate the following numerical parameters which will characterize different stages in the spatiotemporal evolution of TDs:

- the birth rate (rate of occurrence) of TDs of type A in one month in the active season (parameter G);
- the rate of transition (formation) of TDs of type A to an organized TC structure (type B) (parameter T);
- the number of original forms (type A) and TCs occurring in (or with trajectories crossing) the square in question (parameter S).

Depending on the purpose of specific space experiments, it will be necessary to use spatial charts of different parameters: charts of G when studying the occurrence of initial forms; charts of T when studying the physical mechanism governing the formation of TCs; charts of S when studying the general properties of TDs.

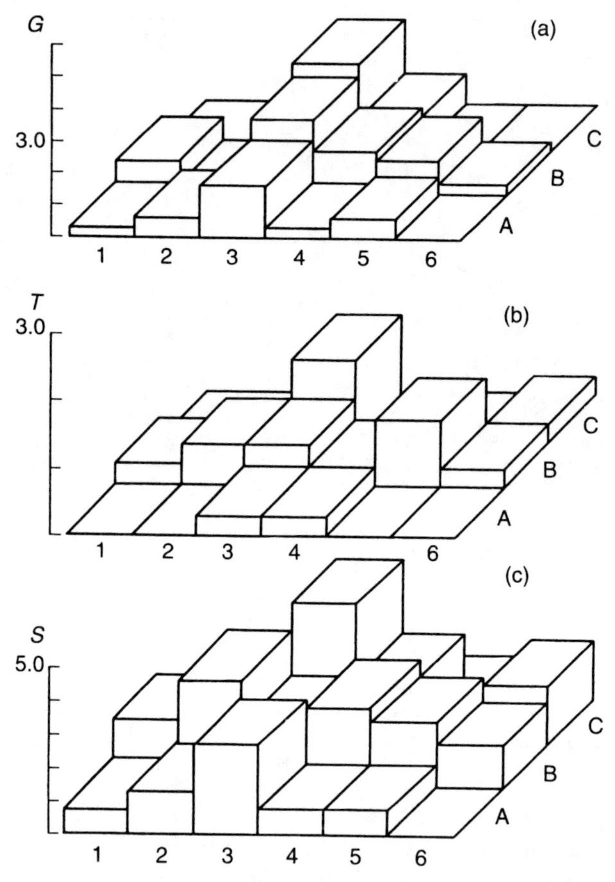

Figure 3.25. Spatial relief maps of the values of evolution-related parameters for cyclogenesis in 1988. (a) Birth rate of original forms (G); (b) rate of transition from original forms to mature TC (T); (c) number of original forms and TC in a given square (S). The geographical attachment of the coordinates is shown in Figure 3.24 (from Pokrovskaya and Sharkov, 1997b).

Spatial charts of these evolutionary parameters (for 1988) are given in Figure 3.25a–c in the form of relief maps of G, T, and S. Analysis of Figure 3.25a reveals that the spatial distribution of G (TD birth rate) has the form of a hill with a single peak at maximum value ($G_M \cong 3$) in square 3C, decreasing toward the south and with a marked drop to $G \cong 0.5$–0.7 in the east and west. In the northern squares (above 35°N) no original stages formed ($G = 0$). Note once again that, when we choose to analyze a limited basin (e.g., as in our case, a basins $1{,}000 \times 1{,}000\,\text{km}^2$ in size), the birth rate of TDs is very low (no more than three per month).

Analysis of Figure 3.25—which shows the spatial field of the rate of transition from a TD, in its original stage, to the form of a mature TC—it follows that the distribution has a structure with two peaks with maxima in squares 3C and 5B, falling off gradually to the south, west, and east. Above 35°N, as expected, $T = 0$.

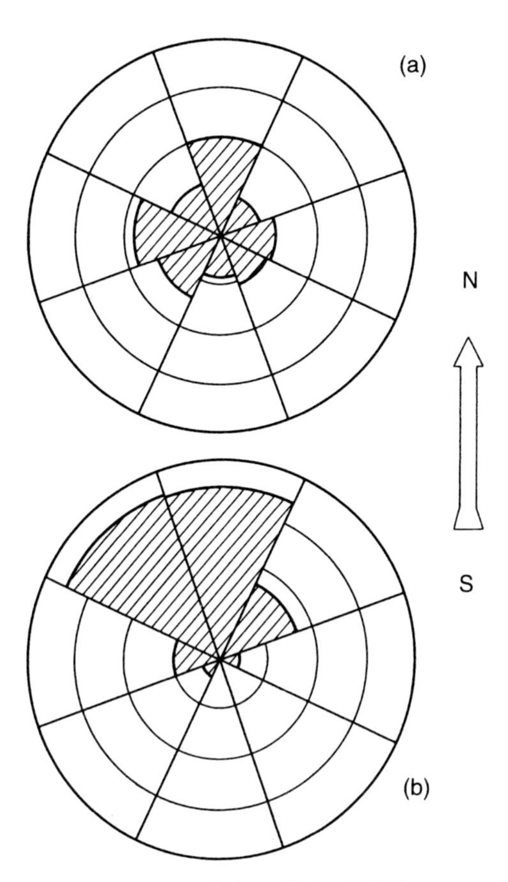

Figure 3.26. Azimuthal starting characteristics of the initial stages of TD trajectories (for 1988): (a) original forms, (b) TCs. Each radial division corresponds to 45° (from Pokrovskaya and Sharkov, 1997b).

Differences from the G-field are essentially associated with the very complicated trajectories of original forms and with the finite and (generally speaking) random lifetime in this state for an individual TD of type A. The maximum value is $T_M \cong 1$, giving the ratio $T_M/G_M = \frac{1}{3}$. In other words one third of all TDs become mature TCs.

The spatial field of S (Figure 3.25c) is characterized by a complex multipeaked relief with local maxima in squares 3C, 2B, 3A, and 3B and deep local minima in square 2B. Maximum values are $S_M \cong 3$–4 and minimum values are in the range 0.5–0.6. The complex relief of the field of S is largely determined by known characteristics of the trajectories of mature TCs (movement in a northwesterly direction, with a shift in movement to a northeasterly direction).

To numerically estimate the trajectories taken by TDs of type A, we consider the azimuthal characteristics of the movement of TDs in the initial stages of evolution (the relative number of "starting" azimuths in a given azimuthal sector) compared with the azimuthal characteristics of the trajectories of mature TCs formed by a fraction of these original forms. Figure 3.26 shows that, in the case of original forms, we can detect the more or less stable isotropy of the "starting" azimuths of trajectories. Thus, approximately half the total number of disturbances that arose corresponded to a traditional (northerly, northwesterly, or westerly directions) path of movement for this region for mature TCs. The remaining half clearly moved in a non-traditional manner in northerly, easterly, or even southerly directions, on so-called "unconventional" trajectories (Figure 3.26a).

The "starting" azimuths of trajectories of a developing TC (transition to mature form) are characterized by a clear anisotropy (Figure 3.26b)—more than 80% of them lie in the north or northwest sectors.

Another revelation from remote sensing shows an original form moving along an unconventional trajectory, but after transition to a mature form (TC) its trajectory changed abruptly, bringing it on to a traditional path for TCs. When a TD moved in a traditional manner, there was no such change in trajectory. Typical examples of this include original forms 19 and 30 in Figure 3.24 and the corresponding TC "Clara" (8810) and TC "Elsa" (8814).

Since the trajectory characteristics of the objects in question are closely related to their kinematic structure, the characteristics we have identified evidently reflect the large-scale reconstruction of the kinematic structure of an original form on transition to a mature form: namely, the formation of a spiral structure for the object with coupled inward-spiraling (tropospheric) and outward-spiraling (stratospheric) circulations.

In the region studied, there was a predominant regional transport of air masses typical of spring in the first half of May (1989). Thus, the weak TD that was created evolved no further.

In the second half of the month there was a change in meridianal circulation, which led to intensification of the ITCZ and the development of TDs. During the month seven original TDs were born in this area, but only one developed further to become a TC.

In mid-June the monsoon trough became active. This led to the occurrence of nine original TDs, two of which became mature TCs. In July cyclonic activity over the Asian continent, associated with the monsoon trough, reached its maximum. As a consequence of these processes, cyclonic activity in the tropical region grew, as a result of which 20 original TDs occurred during the month, 7 becoming mature TCs.

Thus, from May to July 1989 36 original TDs formed in the NWP, 10 of which (or 28%) became mature TCs and 26 (or 72%) evolved beyond the original stage before breaking up.

The spatial and geographical distribution of points at which original TDs (both developing and non-developing) formed and the paths of their movement are shown in Figure 3.27. The spatiotemporal properties of 1988 are also characteristic of the process of cyclogenesis in 1989 (see Figure 3.27). In view of the low intensity of cyclogenesis in general, it is only possible to identify one active region, in which the rate of occurrence (G) is comparable with the value for the 1988 data at 2; in the remaining regions the rate was considerably lower at 0.3–1.

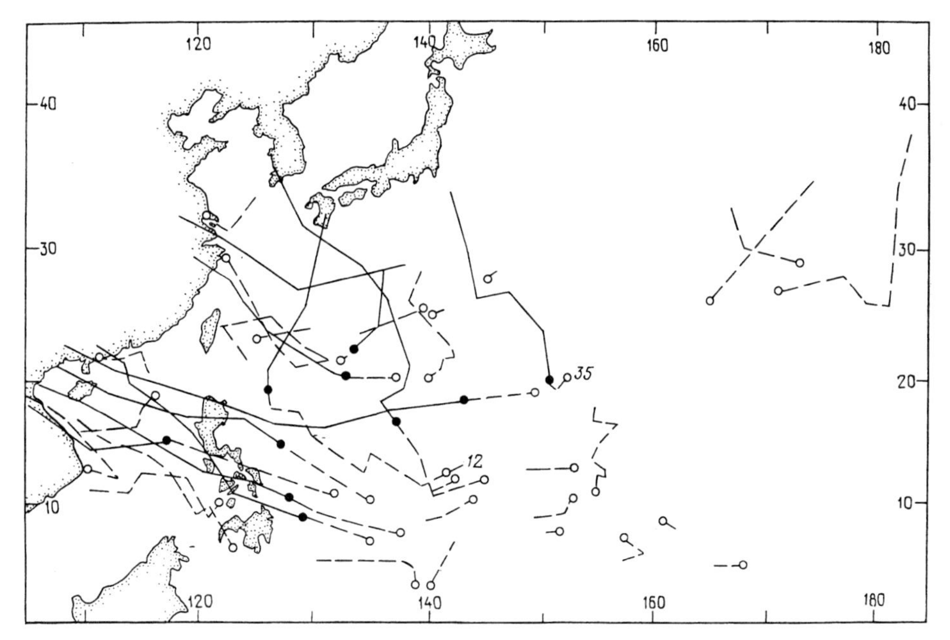

Figure 3.27. Spatial and geographical distribution of original forms of TDs and their evolution (data for May–July 1989). Notation as in Figure 3.25. The numbers 12 and 35 denote TDs that developed into TC "Elsa" and TC "Mac" (from Pokrovskaya and Sharkov, 1997b).

As was the case in 1988, around half the original forms moved along "unconventional" trajectories and later, if and when they subsequently formed TCs, an abrupt change (so-called "breaking") of the trajectories occurred. Typical examples of this include original forms 12 and 35 in Figure 3.27 and the corresponding TC "Elsa" (8906) and TC "Mac" (8913).

Analysis of the surface barometric fields in which original forms occur shows that a preliminary classification can be made by dividing the barometric conditions for occurrence into three types:

- type I is a gradientless almost uniform barometric field (without frontal effects) with a spatial gradient of $10^{-3}\,\mathrm{mb\,km^{-1}}$ or less;
- type II is a barometric field with a marked spatial gradient of the order of $10^{-2}\,\mathrm{mb\,km^{-1}}$ (low-gradient field);
- type III is a barometric field with a gradient of the order of $\geq 10^{-2}\,\mathrm{mb\,km^{-1}}$ or a barometric field in the vicinity of active TCs and "on the tail" of cold fronts and occlusions.

Typical examples of specific barometric situations are shown in Figure 3.28. For example, on September 5, 1988 (Figure 3.28a) in a low-gradient barometric field (spatial gradient less than $1 \times 10^{-3}\,\mathrm{mb\,km^{-1}}$) three original TDs were detected.

One, a TDE, moved eastward and became TC "Hal" (8818); the other two, a TD and a TDE, dispersed. Figure 3.28b shows a barometric situation of type II when a disturbance in the form of a TD (No. 34) was born in the gradient field (gradient $\cong 9 \times 10^{-3}$ mb km^{-1}); this later became TC "Hal" (8818). Figure 3.28c shows a barometric situation of type III where a single barometric system (gradient $\cong$ 6–4 $\times 10^{-3}$ mb km^{-1}) contained a TC—"Agnes" (8807)—at the super-tropical storm (STS) stage and an original disturbance in the form of a TD, occurring in a cloud cluster in the tail of "Agnes" (No. 11), as a component of it (satellite), which subsequently developed into a mature TC (8808).

The distinction between the barometric conditions of types II and III is provisional since, we mainly encounter situations of a hybrid type, when the large-scale barometric field is perturbed by the influence of frontal divisions and active TCs. Often, a peculiar effect of "self-reproduction" of an original form of a disturbance is observed in a single barometric system, when the original disturbance divides into two or three components (satellites) and the whole system continues to operate as one unit. Subsequently, either the components merge into a single component or they disperse. A typical example of the "fusion" effect is shown in Figure 3.29a–c which shows the evolution of three original TDs. Two disturbances in the form of a TD and a TDE—observed on August 8, 1988 at 12:00 GMT (Figure 3.29a) under

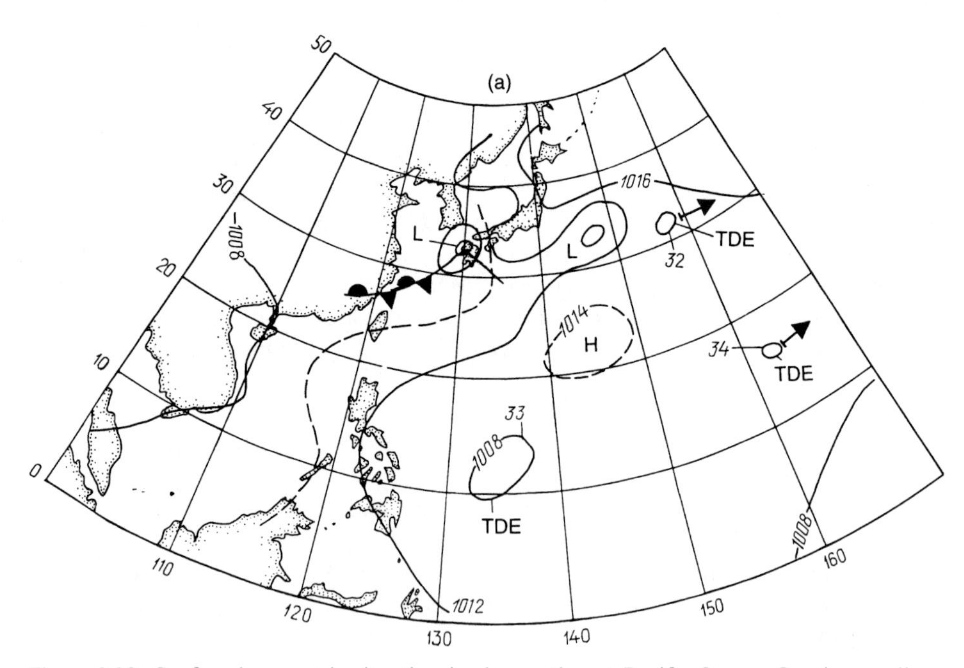

Figure 3.28. Surface barometric situation in the northwest Pacific Ocean. Continuous lines, isobars; numbers, values of the pressure in millibars; L and H, zones of low and high pressure, respectively; (a) barometric situation of type I, data of September 5, 1988 at 12:00 GMT. Arrows, "starting" azimuths of the trajectories of disturbances.

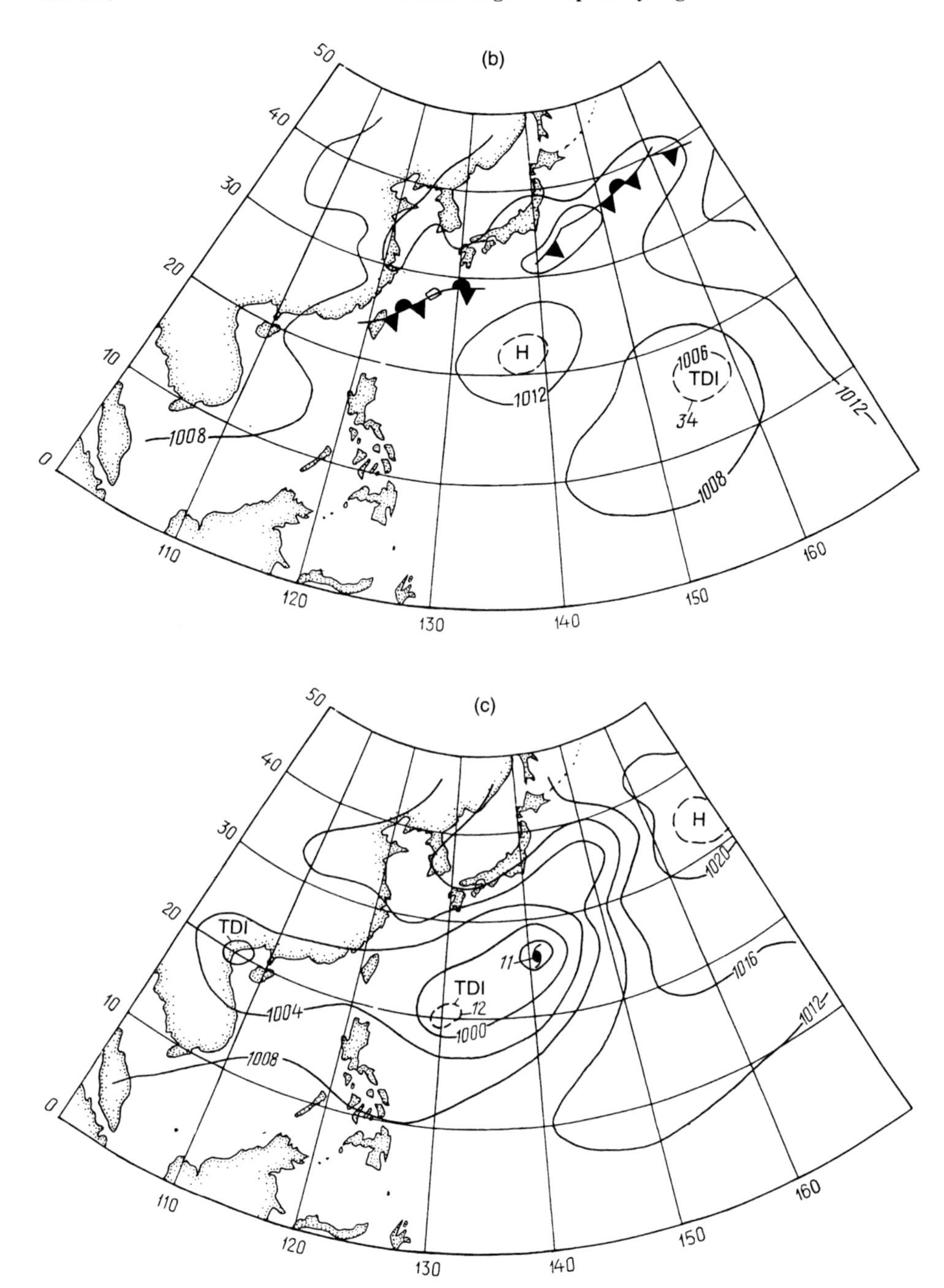

Figure 3.28. (*continued*) (b) Barometric situation of type II, data of September 7, 1988 at 12:00 GMT; (c) barometric situation of type III, data of July 30, 1988 at 00:00 GMT. TDEs (see facing page) and corresponding figures are reported tropical depressions. TDIs and corresponding figures are reported tropical disturbances (Pokrovskaya and Sharkov, 1997b).

different barometric conditions—merged in the course of a day (August 9, 1988 at 12:00 GMT) into a single barometric system which "reconnected" at the barometric level of 1,004 mb and deepened to the level of 1,000 mb. By 12:00 GMT on August 10, 1988 (Figure 3.29c) the two objects had merged and filled up to the 1,004 mb level, with subsequent dissipation of the single object. At the same time, the other tropical depression continued to deepen rapidly and subsequently became TC "Clara" (8810).

It follows from analysis of this example that it is impossible to track the full evolution of this process from surface barometric charts generated every 6–12 hours because of the highly dynamic nature of the objects.

The percentage ratio of developing and non-developing forms of original TDs under given barometric conditions is in our opinion an important parameter, which can characterize the activity of the processes of cyclogenesis in different barometric systems. Table 3.13 shows the values of this parameter for the barometric situations studied and the different ratios. Despite the somewhat limited nature of the statistics, we can still identify an important trend: the ratio of developing to non-developing forms of original TDs remains practically constant regardless of the barometric situation; of the total number of original forms around one third (30%) develop.

Analysis of the results obtained raises a number of questions for discussion. The main question is whether a TD that does not develop to become a mature TC may be

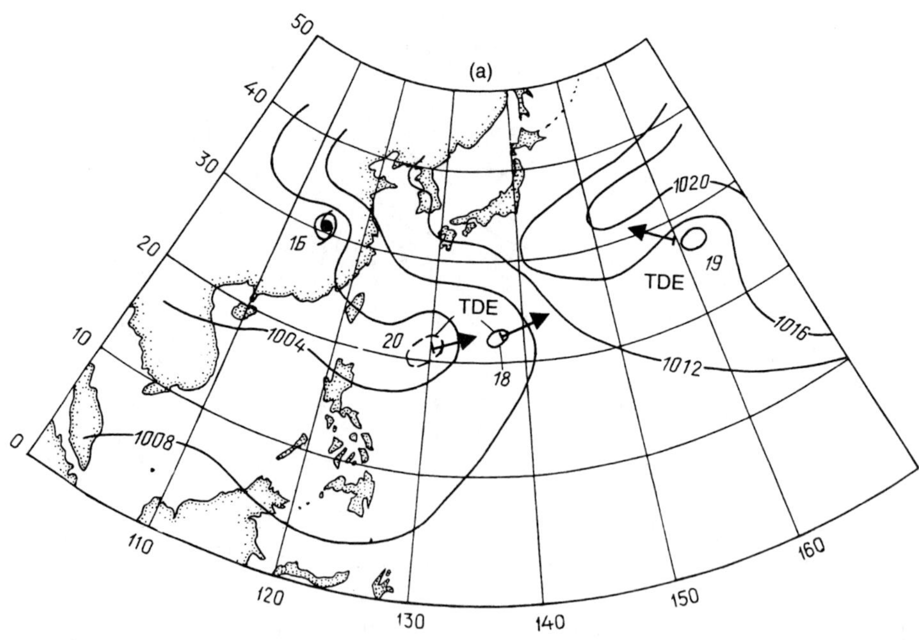

Figure 3.29. Surface barometric situation in the northwest Pacific: (a) on August 8, 1988 at 12:00 GMT; (b) on August 9, 1988 at 12:00 GMT; (c) on August 10, 1988 at 00:00 GMT. Notation as in Figure 3.28 (from Pokrovskaya and Sharkov, 1997b).

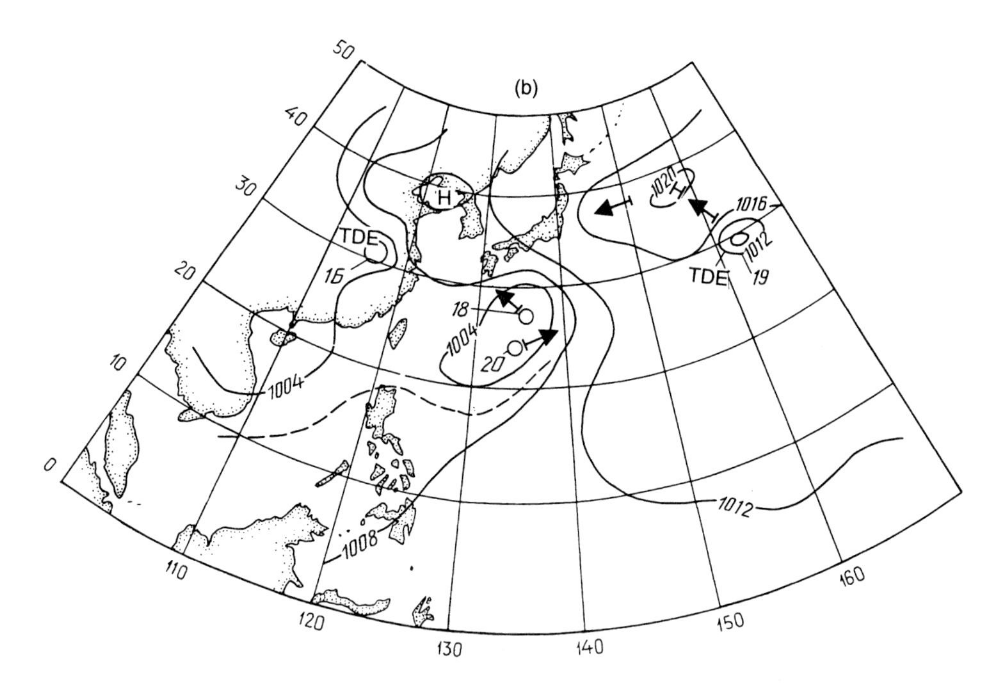
(b)
50
40
30
20
10
0
H
TDE
16
18
20
1004
1004
1008
1012
1020
1016
1012
19
TDE
110
120
130
140
150
160

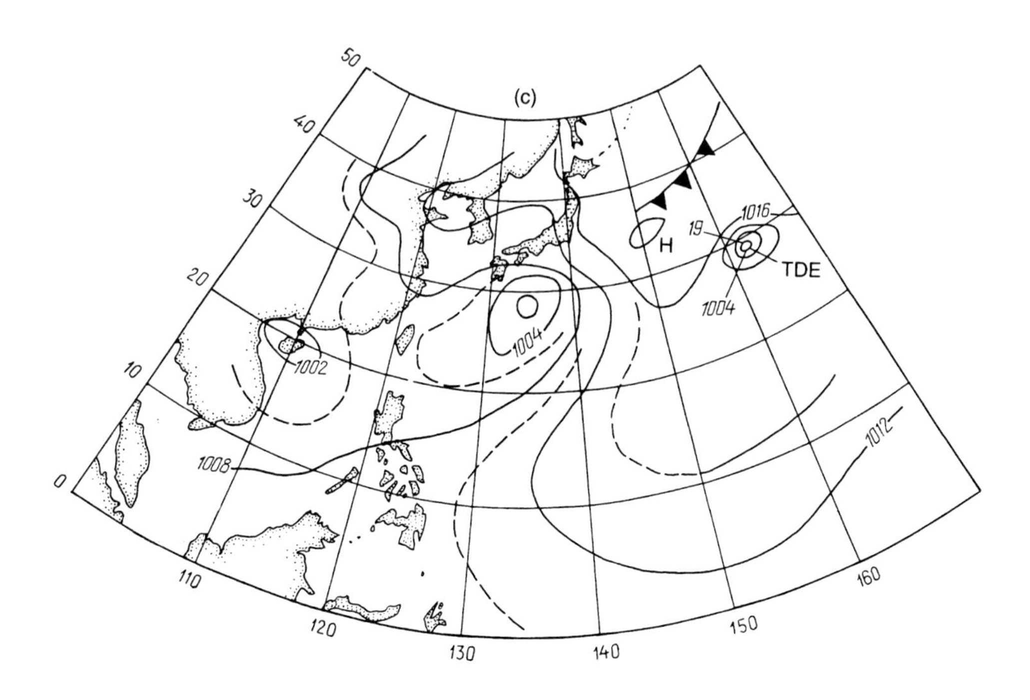
(c)
50
40
30
20
10
0
H
1016
19
TDE
1004
1004
1002
1008
1012
110
120
130
140
150
160

Table 3.13. Comparison of developing and non-developing TD forms under different barometric conditions for the active seasons of 1988 and 1989.

Type of barometric situation	*Number of initial forms*	*Initial forms developing into TCs* (%)	*Initial forms not developing into TCs* (%)
I	31	27	73
II + III	61	31	69
II	22	30	70
III	39	27	73
I + II + III	92	30	70
I + II + III (1988)	53	35	65
I + II + III (1989)	36	28	72

assumed to be an individual structural element (coherent structure in the sense of chaotic dynamics) with ascribable structural and kinematic characteristics (boundaries, velocity, trajectory, etc.), as in the case of mature TCs. Or should their existence be viewed as an element of the wave process in the barometric field of a tropical atmosphere? This is closely related to the question about individual identification of an original disturbance. Because of the highly dynamic nature of the evolution of these objects and the "coarse" representation of barometric fields (isobars with a discreteness of 4 mb) on surface charts giving a barometric analysis at synoptic dates, it is often difficult to identify disturbances of type A individually, even using satellite information about tropical zone cloud systems. In a number of cases, it is basically impossible to identify disturbances of type A individually; to do this properly requires more detailed (with a discreteness of 1 mb) and more frequent 1 to 2-hour representations of the barometric fields on surface charts.

The independence of the ratio between developed and non-developed forms under differing barometric conditions could be explained in the context of the notion of "self-organization" (Moiseev *et al.*, 1983a, 2000; Moiseev, 1990; Zimin *et al.*, 1992). According to this notion the physical nature of the formation of structures from chaos involves particular geometrical properties of mesoscale turbulence whose intensity does not depend directly on the type of barometric situation. However, the latter can serve as a trigger mechanism for formation of a structure. Thus, our results can be interpreted as indicating the existence of a mechanism for the formation of a structure that is independent of the "trigger" mechanism.

For these reasons the material in our work is, in a certain sense, of a preliminary nature. Nevertheless, even our first results indicate the need to define the experimental work that needs to be done regarding the remote study of initial forms of atmospheric catastrophes. In fact, the problem of detecting and identifying original forms is central to experimental work within any program monitoring natural atmospheric catastrophes.

Analysis of hydrometeorological and satellite data on original tropical eddy disturbances in the NWP (June–September 1988 and May–July 1989) has enabled us to draw the following conclusions:

- we have identified three types of barometric conditions for the occurrence of original forms of TD; we have noted the effects that "reproduction" and "fusion" of the original forms have among themselves and in interaction with developed forms;
- the spatial and geographical regions in which original TD forms occur by and large correspond to general areas of tropical cyclogenesis; we have recorded the centers in which original forms are concentrated;
- the percentage ratio of developed and non-developed TD forms remains practically constant and is independent of the processes of cyclogenesis and of the barometric situation; of the total number of original TD forms occurring 30% develop;
- we have identified the isotropy of the "starting" azimuths for the movement of original TD forms.

The spatial and geographical characteristics of original TD forms presented in this book can be used both for the execution and planning of ballistic procedures for potential satellite systems and for the operational planning of space experiments.

3.4.3 Hierarchy and clusterization of tropical convective systems

As noted above (Section 3.4.1), mesoscale convective systems are major dynamical components when tropical cyclogenesis occurs. It is for this reason that considerable attention has been given to the study of various evolutionary forms and the turbulent structure of convective cloud systems in the context of global climate change and atmospheric catastrophes (Marchuk *et al.*, 1986; Matveev *et al.*, 1986; Houze, 1993; Emanuel, 1996; Hansen *et al.*, 1997).

Satisfactory analysis of multiscale convective systems can be performed by thermal IR and radiophysical Doppler instruments. The first type of remote-sensing measurements provides the pure thermal signatures of the natural objects under study, whereas the second method yields pure dynamic patterns. It is self-evident that the two types of remote-sensing instruments complement each other.

A number of investigators (Webster and Lukas, 1992; Mapes and Houze, 1993; Chen *et al.*, 1996; Chen and Houze, 1997a, b) have obtained dramatic results from the remote-sensing study of the hierarchy, time series, and clusterization effects of mesoscale convective systems. A brief review of the main results now follows.

A cloud cluster is defined as a closed contour of a threshold IR temperature in a GMS image. An objective technique to identify and track connected cloud clusters of a specified IR temperature threshold was developed by Williams and Houze (1987) and further applied and extended by Mapes and Houze (1993). To identify a cluster at a particular time, connected areas of cold cloudiness (called "line clusters") are found within each line (row) of the data array containing one satellite image. The line clusters on two successive lines must share a column (not merely touch diagonally) to be considered connected. Each cloud cluster is summarized by its area and centroid position. An IR temperature threshold of 208 K defines the cloud clusters identified in this study. This method was used to identify all such cloud clusters in the field of view of the GMS and the COARE (Coupled Ocean–Atmosphere Response Experiment) domains (Pacific basin).

In what follows we use several terms to describe the satellite-observed cloud patterns associated with deep convection (Chen *et al.*, 1996):

- *Cloud cluster*. A cloud cluster is a region of cold cloud tops surrounded by a single closed contour of a threshold IR temperature in a GMS image. For a low-threshold IR temperature, which roughly outlines a precipitation area, a cloud cluster corresponds to what is commonly called a "mesoscale convective system" (MCS) (Houze, 1993).
- *Time cluster*. A time cluster is a set of cloud clusters that exhibits temporal and spatial continuity across at least two frames of satellite imagery. Because cloud clusters can split and merge, a single time cluster may consist of several cloud clusters in any given frame. If the time cluster consists of a group of cloud clusters, the overall group must exhibit a specified amount of close proximity in space and continuity in time. The members of a time cluster may form, dissipate, merge, or split during the lifetime of the time cluster.
- *Supercluster*. The term "supercluster" is used to describe eastward-propagating cloud ensembles with a horizontal scale of several thousand kilometers, within which are embedded westward-moving individual cloud clusters. Mapes and Houze (1993) suggest a quantitative definition of a supercluster as a time cluster exceeding 2 days in duration.
- *Superconvective system*. A superconvective system (SCS) is defined as a time cluster whose maximum instantaneous size (possibly comprising several cloud clusters, as noted above) exceeds a total area of $9 \times 10^4\,km^2$. No minimum lifetime criterion is imposed. The lifetimes of SCSs range from several hours to a few days. The supercluster, defined above, is a special long-lived case of the more general SCS.
- *Tropical intraseasonal oscillation (ISO) cloud ensemble (ICE)*. An ISO has a convectively active phase and a convectively suppressed phase. We call the cloud population constituting the convectively active phase of the ISO the ISO cloud ensemble (ICE).

Representing high cloudiness as the percent high cloudiness (PHC) is a good way to indicate the overall *amount* of cloudiness in a given area and/or time period (e.g., see Marchuk *et al.*, 1986; Matveev *et al.*, 1986; Houze, 1993). However, it does not indicate how the cloudiness arose. For this purpose, we must delineate and characterize the phenomena producing the cloudiness. Most high cloudiness in the

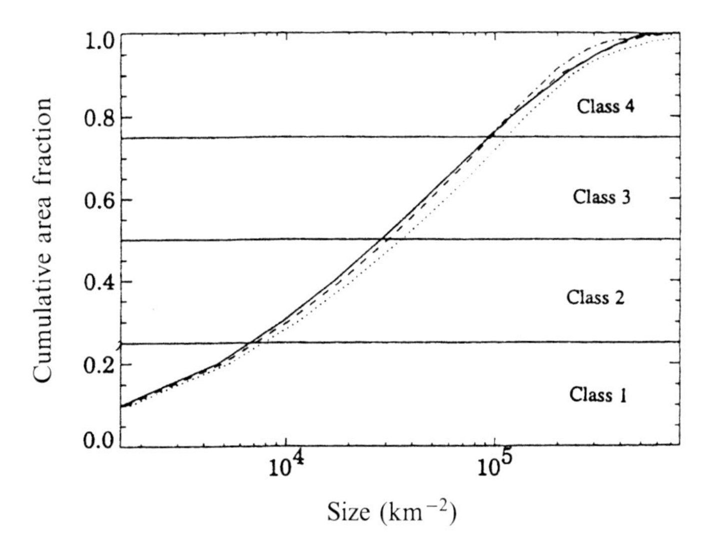

Figure 3.30. Cumulative size distributions of 208 K cloud clusters. The fraction of the total sampled cloud area that is accounted for by 208 K cloud clusters up to the indicated size for 1986–1987 (dotted line), 1987–1988 (dashed line), 1988–1989 (dot-dashed line), and 1992–1993 (solid line). Thin horizontal lines divide the size distribution into size quartiles, each quartile contributing an equal amount to the total area of cloud top colder than 208 K (Chen *et al.*, 1996).

COARE region is produced by discrete mesoscale convective systems, which we represent as cloud clusters. The ensemble properties of cloud clusters are thus an important element of a complete climatology of tropical convection.

Figure 3.30 shows the cumulative fraction of total sampled cloud area in the GMS domain accounted for by the 208 K cloud clusters up to the indicated size for 1986–1987 (dotted line), 1987–1988 (dashed line), 1988–1989 (dot-dashed line), and 1992–1993 (solid line). Thin horizontal lines divide the size distribution into size quartiles, each quartile contributing an equal amount (25%) to the total area of cloud top that is colder than 208 K. The overall size distribution of cloud clusters for the intensive observing period (IOP) (November 1992 through February 1993) is similar to the previous 3-year climatology of Mapes and Houze (1993). The boundaries for the quartiles are class 1 ($<6{,}800\,km^2$ in area or $<\sim 80\,km$ in dimension), class 2 ($6{,}800$–$28{,}300\,km^2$ or ~ 80–$170\,km$), class 3 ($28{,}300$–$92{,}800\,km^2$ or $>\sim 170$–$300\,km$), and class 4 ($>92{,}800\,km^2$ or $>\sim 300\,km$). Previous studies of tropical cloud clusters (Williams and Houze, 1987; Mapes and Houze, 1993) show that cluster sizes are nearly log-normally distributed but deviate from log-normality at both the smallest and largest ends of the distribution.

Figure 3.31a shows the frequency of occurrence of tracked time clusters (or convective systems) as a function of their lifetime and maximum size. The distribution includes all the time clusters occurring in the GMS domain during TOGA (Tropical Oceans and Global Atmosphere) COARE. The lifetime is the

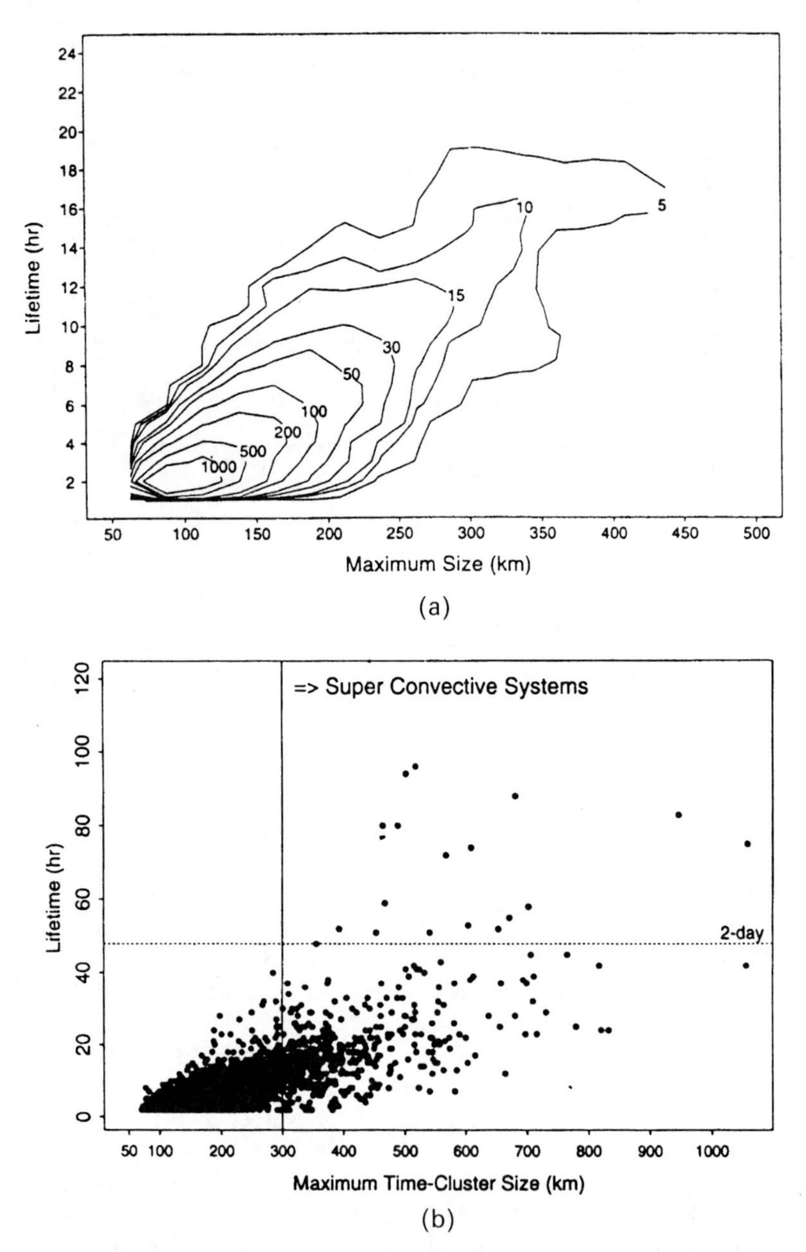

Figure 3.31. (a) Frequency distribution of time cluster (convective system) occurrences (per 25 km size interval per hour) as a function of the maximum size (abscissa) reached by a convective system during its lifetime (i.e., from the start to the end of the lifecycle) for all time clusters that occurred over the Warm Pool during TOGA COARE. (b) Scatterplot of time cluster sizes versus lifetime over the Warm Pool during the COARE IOP. Each dot represents the maximum size reached by a particular time cluster during its lifetime (from Chen *et al.*, 1996; Chen and Houze, 1997a, b).

time period from the beginning to the end of a trackable convective system. The maximum horizontal size is the square root of the net area of IR temperature <208 K, which is the sum of the area of all the individual clusters making up the time cluster, at the time of the maximum areal extent of the time clusters. The tendency for convective systems with a greater horizontal dimension to have a longer lifetime is clearly evident in Figure 3.31a. The peak frequency of time clusters exceeding a particular lifetime increases with increasing maximum size more or less linearly; however, the distribution has considerable variability (breadth in the envelope of contours in Figure 3.31a). The overwhelming majority of time clusters whose maximum size was small lasted only 1–3 h. Most of the largest group of time clusters (>300 km in horizontal dimension, the SCSs) have a lifetime between 8 h and 20 h. There were some very long-lasting SCSs (less frequent than the lowest plotted frequency contour 5 in Figure 3.31a which lasted more than a few days).

In the case of time clusters whose horizontal dimensions are very large, the times obtained from Figure 3.31a are a considerable underestimate of the true convective system lifetime since the area of cloud tops <208 K defined time clusters by the area of cloud tops <208 K.

Figure 3.31b shows the lifetime versus maximum size achieved by each time cluster during the COARE IOP over the entire GMS domain. The maximum size is expressed as a characteristic linear dimension (the square root of the area of IR temperature <208 K of the cloud extent at the time of maximum extent). Although maximum size is not very well correlated with lifetime, the likelihood that a time cluster will exceed a particular lifetime increases with increasing maximum time cluster size (the upper limit of the spread of dots in Figure 3.31b increases more or less linearly with increasing maximum time cluster size). Small short-lived time clusters are the most numerous, but the few time clusters with very long duration and very large maximum size contain a sizable fraction of the total 208 K and colder cloud top areas. In particular, time clusters lasting over 48 hours (including five TCs) constituted some 15% of total time cluster cloud coverage. Mapes and Houze (1993) term these long-lived (>2 days) trackable entities *superclusters* (see above). Time clusters lasting over 24 hours constituted ~30% of total time cluster cloud coverage. Note that small cloud clusters with 208 K area less than 5,000 km^2, which constituted about 20% of total 208 K cloud coverage, are not included in the cloud coverage of time clusters in Figure 3.31b.

Figure 3.31 includes many time clusters with a maximum size exceeding 300 km, which is the threshold that distinguishes the upper quartile (class 4) of cloud clusters (Figure 3.30). However, many of these large time clusters were of relatively short duration (12 h or less). To describe these large time clusters we define, without regard to duration, an SCS as any time cluster that achieves a maximum size >300 km (right-hand portion of Figure 3.31b). This size criterion is the same as that used to define a class 4 cloud cluster (exceeding 9×10^4 km^2 in area, see Figure 3.30). Note that every class 4 cluster (largest size quartile) is part of an SCS. The superclusters that had lifetimes exceeding 2 days were a subgroup of SCSs (upper right portion of Figure 3.31b).

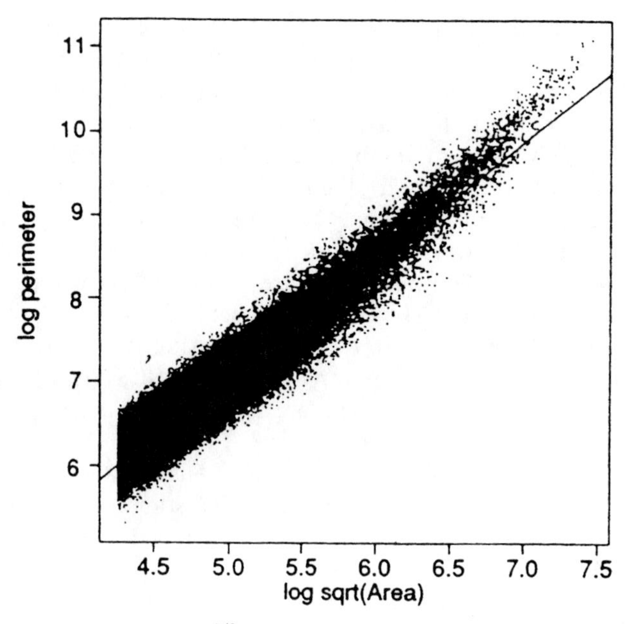

Figure 3.32. Perimeter versus $(\text{area})^{1/2}$ for 235 K cloud clusters. The ordinate is the log of the square root of cloud cluster size (km), and the abscissa is the log of perimeter length (km). The fitted line has a slope ("fractal dimension") of 1.4 (from Mapes and Houze, 1993).

Williams and Houze (1987) and Mapes and Houze (1993) find that giant cloud clusters start to deviate from a log-normal distribution when they reach 300 km in dimension. The deviation from log-normality suggests that 300 km is the typical limit of the size of upper-level cloud tops of an individual cloud system. A scale of 300 km is approximately the maximum horizontal distance that ice particles can travel from their convective cell sources before falling out. SCSs whose maximum size exceeds 300 km were invariably composed of multiple MCSs.

Another structural parameter of interest is the perimeter of each cloud cluster. Figure 3.32 shows a log-log plot of perimeter versus $(\text{area})^{1/2}$ for $-38°\text{C}$ cloud clusters. The slope of the fitted line, which corresponds to the fractal dimension of the perimeters, is 1.4, in agreement with the 1.4–1.5 ranges quoted by Lovejoy (1982) and Cahalan and Joseph (1989) for their ITCZ deep-convective clouds. The linear fit is not especially good: the larger clusters are increasingly ragged, with very long perimeters for their area.

This result is noteworthy. The point is that the manifestation of fractal features in the physical system arises from multiscale interactions and possible self-organization effects in the dynamic evolution of the system under study (see Section 2.7.1). The fractal characteristic used by Mapes and Houze (1993) is no more than a rough approximation for describing self-similar properties of the self-organized system. In this case more sophisticated analyses, based on complicated processing procedures, are required (see Section 3.4.4 for details).

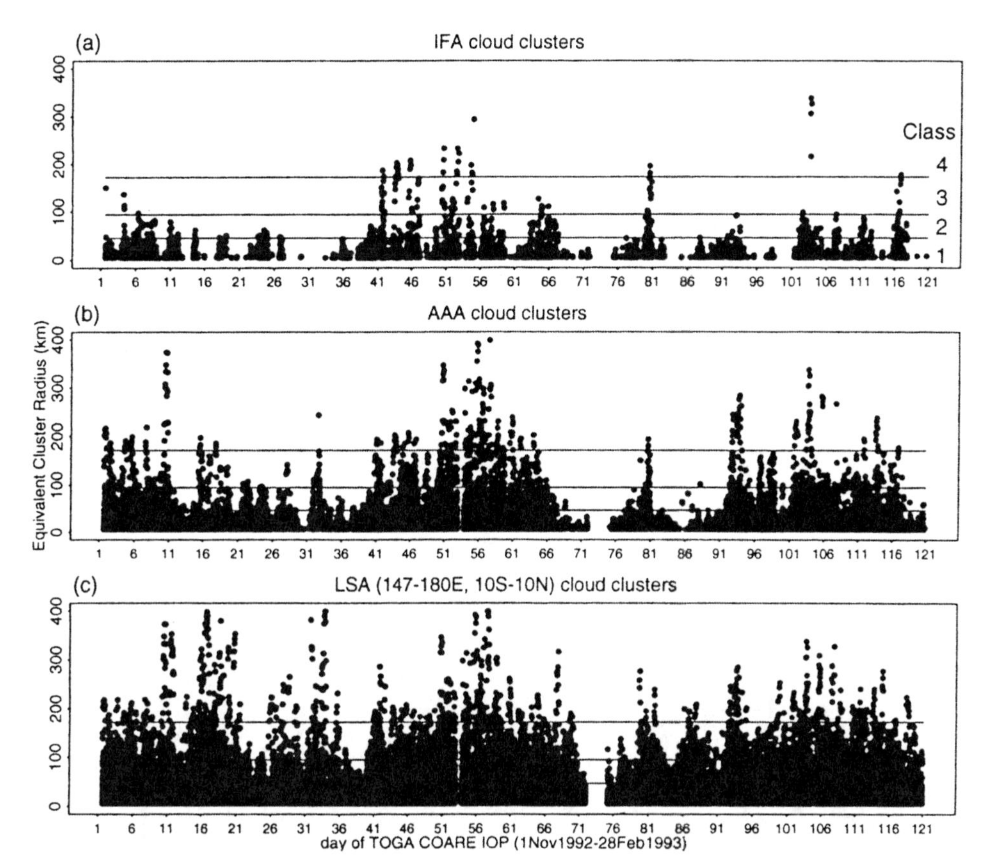

Figure 3.33. Time series of occurrence of cloud clusters (closed contours at the infrared temperature of 208 K) as a function of their size (January 11–14 missing data). Horizontal lines are boundaries of size classes 1–4 indicated in Figure 3.31. (a), (b) and (c) Clusters in the three COARE domains (Chen *et al.*, 1996).

The population of cloud clusters in the three COARE domains—intensive flux array (IFA), aircraft-accessible array (AAA), and large-scale array (LSA) in fig. 1 of the paper by Chen *et al.* (1996)—is shown in the time series in Figure 3.33. Each dot represents the occurrence of a cloud cluster of specific size, expressed as equivalent radius $R_e(A/\pi)^{1/2}$, where A is the cluster area on an hourly GMS IR image. Horizontal lines mark the boundaries of the quartiles for each class defined by Figure 3.30, which are slightly different from the classification used for the COARE IOP aircraft missions. Small class I cloud systems were present on more than 80% of the IOP days in the IFA (Figure 3.32a), whereas class 4 systems occurred within the IFA on only ~10% of the IOP days.

The different large-scale regimes—including two suppressed phases (from mid-November to early December and mid-January) and two active phases (mid-December to early January and February)—of deep convection associated with the ISO are apparent in the total number of cloud clusters shown in Figure 3.30.

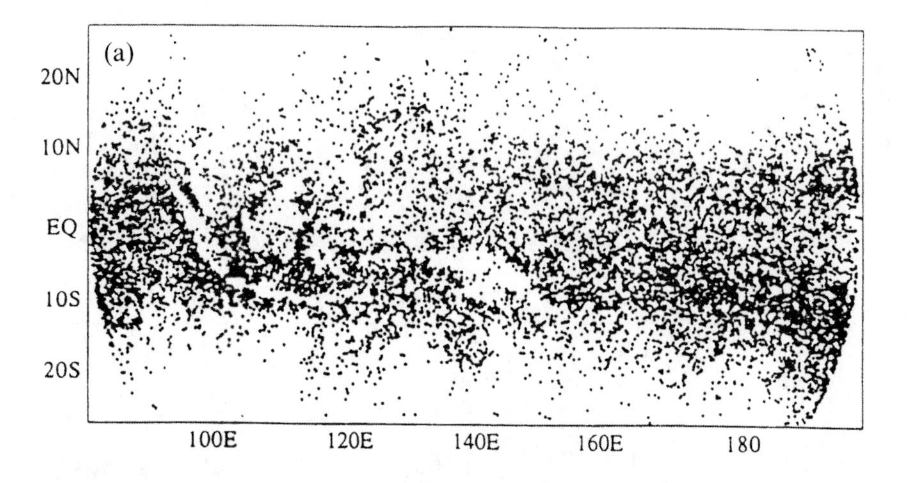

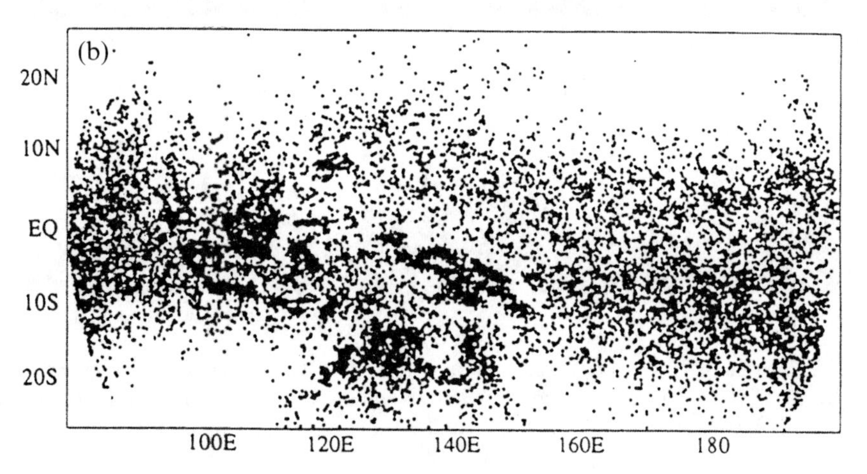

Figure 3.34. Maps of all very cold (<208 K) cloud clusters 1,000–3,000 km^2 in area observed in images at (a) 23:00 UTC (09:00 LST at subsatellite point) and (b) 08:00 UTC (18:00 LST at subsatellite point) (from Mapes and Houze, 1993).

This variation is much more pronounced in larger clusters than smaller clusters. Most of the very large clusters were found only during active phases of the ISO. The cloud population in other years has a similar behavior (Mapes and Houze, 1993).

The diurnal cycle of deep convection over tropical oceans is a surprisingly rich and subtle question. There are a few situations in which the diurnal cycle is strong and easy to understand. For example, maps of the incidence of small (1,000–3,000 km^2) very cold cloud clusters in the morning and afternoon need no geographical overlay (Figure 3.34). The small short-lived convective clouds shown in Figure 3.34 occur over large islands of the maritime continent during the afternoon, and over the surrounding seas during the night and morning.

Similar maps of incidence of large (>30,000 km^2) very cold (208 K) cloud clusters, at the minimum and maximum of their diurnal cycle, do not show the geography quite so clearly. Instead, a pronounced diurnal variation in the *total number* of large clusters is apparent, even over the open ocean, with numerous clusters observed during early morning and fewer in the evening hours.

Figure 3.35 shows time–longitude plots of all 208 K cloud clusters observed during the 1986–1987 season within the entire tropical latitude belt, in three size ranges. The three size ranges represent comparable fractions of total 208 K cloud cover, and the plotting characters have been scaled so that a similar amount of ink appears in each panel. Westward-moving TCs and other individual cloud clusters are clearly evident. An eastward-moving intraseasonal envelope modulating cloudiness can also be detected.

Figure 3.36 is a time–longitude plot of all cloud clusters between the equator and 10°S (except in Figure 3.36b and 3.36d which include clusters between 1°N and 10°S) during December 1992. Boxes in Figure 3.36a indicate regions that are examined in greater detail in Figure 3.36b–e. Each cloud cluster is indicated at the position of its centroid by an oval. The size of each oval is proportional to the actual size of the cloud cluster (cloud clusters smaller than 2,000 km^2 are not plotted). Chains of cloud clusters show the lifecycles of particular convective systems that consist of these cloud clusters. The east–west progression of the chains indicates the zonal component of their motions. Note that the zonal motion of these satellite-observed cloud clusters may differ from the motion of intense convective components within each cloud cluster.

The general band of enhanced cloudiness from the upper left to lower right indicates the progression of ICE2, which became particularly intense after December 20 near 160–180°E. As ICE2 migrated discontinuously eastward from the eastern Indian Ocean to the central Pacific, the patterns formed by the cloud clusters within ICE2 were highly variable. However, Figure 3.36 reveals certain systematic features within this large-scale ensemble. The myriad of individual clusters in Figure 3.36a is not a random distribution. Within the envelope of ICE2, the westward-propagating streaks of cloud clusters, with a periodicity of ~2 days, are apparent, especially during middle to late December. These wavelike disturbances consisted of *groups* of westward-propagating (and some eastward-propagating) cloud clusters.

Figure 3.36b, c shows that these disturbances propagated westward at 10 m s^{-1} to 15 m s^{-1} over a zonal distance of 2,000 km to 4,000 km. At a fixed longitude, the passage of these disturbances accounts for the 2-day periodicity of cloudiness. Figure 3.36b, c shows that a westward-propagating 2-day disturbance was typically made up of a fluctuating group of separate convective systems whose lifetimes were individually much less than the 2-day disturbance in which they occur. Cloud clusters within the 2-day disturbances underwent sequential growth, splits, mergers, dissipation, and redevelopment within the same zonal disturbance with a horizontal scale of thousands of kilometers.

Figure 3.36b and its further enlargement Figure 3.36d show that a series of westward-propagating cloud clusters, or groups of cloud clusters, affected the

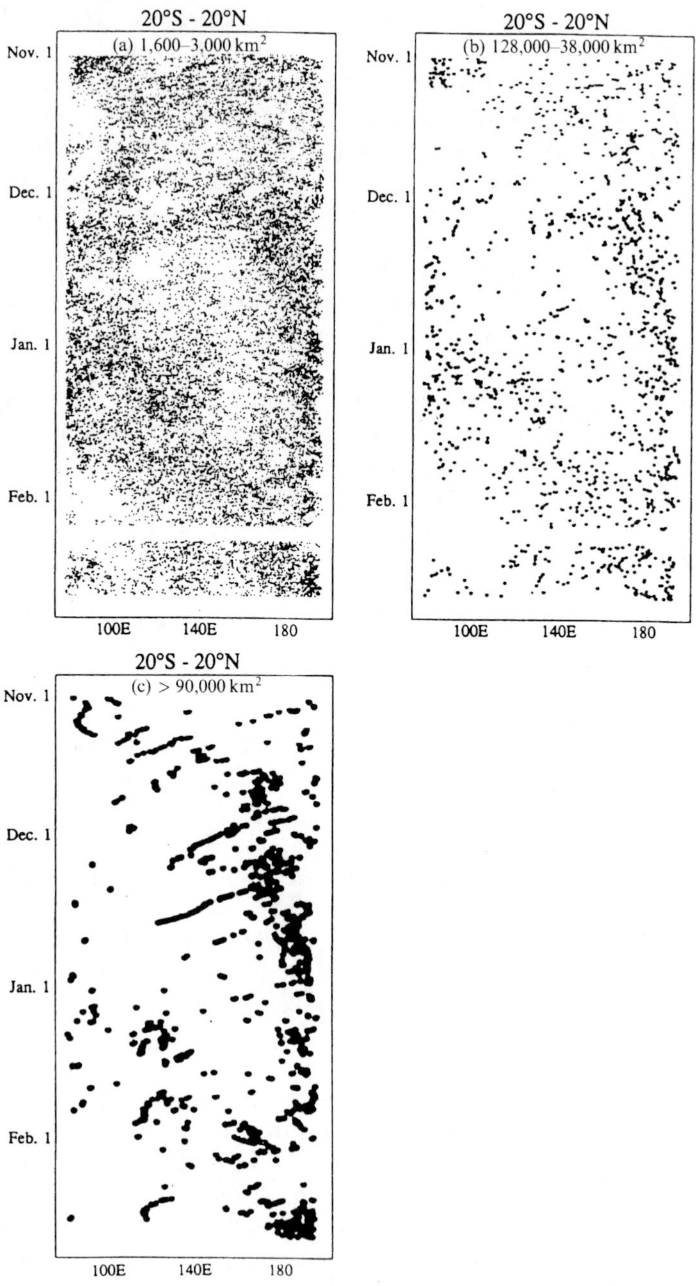

Figure 3.35. Time–longitude sections of very cold (<208 K) cloud clusters within the indicated size categories between 20°S and 20°N during the 1986–1987 season. The three size categories account for similar fractions of total cloud, and plotting characters are scaled to put a similar amount of ink in each panel. (a) 1,600–3,000 km^2; (b) 28,000–38,000 km^2; (c) greater than 90,000 km^2 (from Mapes and Houze, 1993).

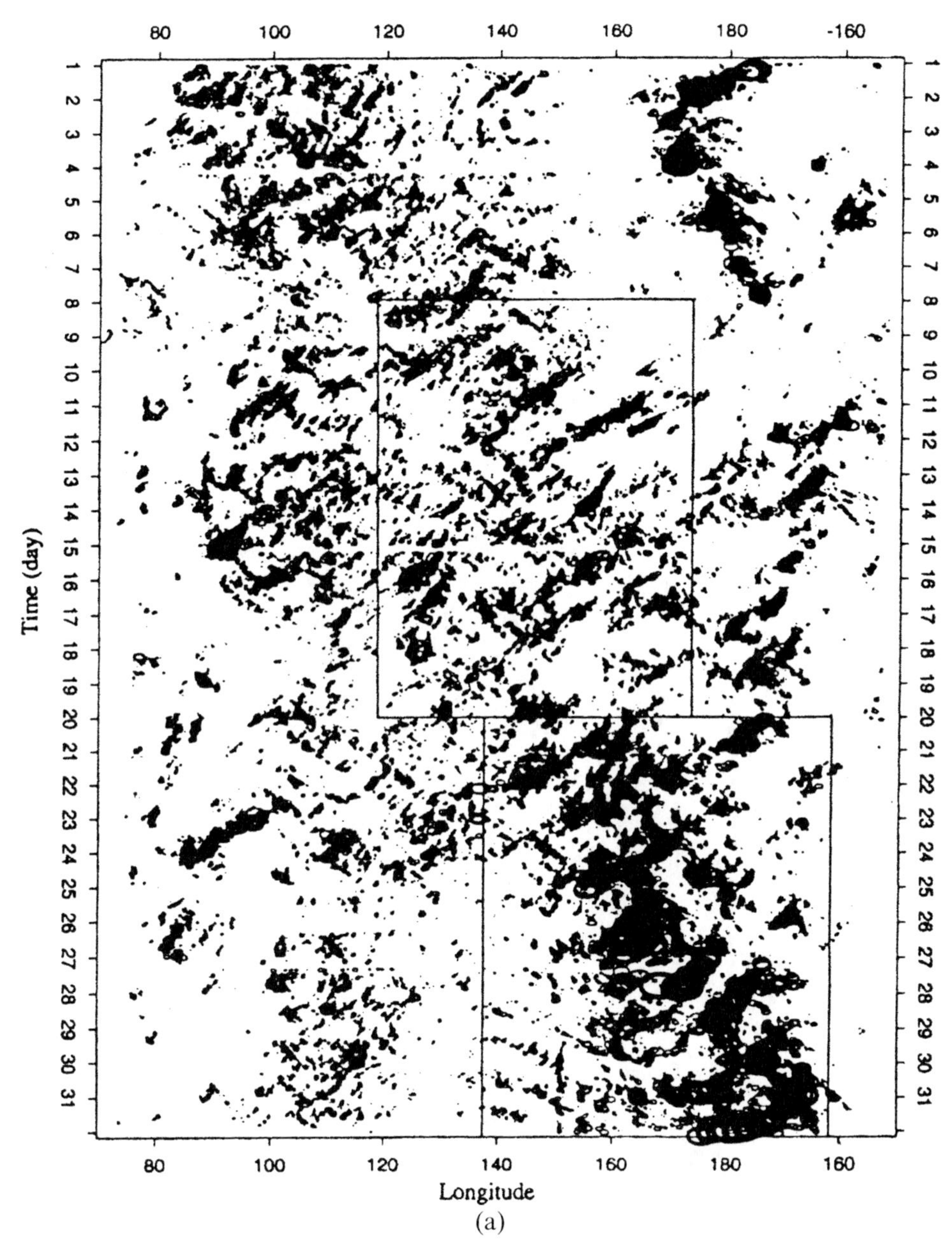

(a)

Figure 3.36. Time–longitude diagrams of cloud clusters (IR temperature <208 K) centered between 0° and 10°S and between 1°N and 10°S for (b) and (d). (a) December 1992; (b) and (c) enhanced windows for December 8–19 and December 20–31; and (d)–(e) enhanced windows for December 11–17 and December 22–28. The date marker is at 00:00 UTC (or 11:00 LST at 156°E). Sizes of each oval are proportional to the sizes of actual cloud clusters. Arrows and numbers in (d) indicate the location and date of each aircraft mission during December 12–15, 1992. A, B and C in (e) indicate three convective systems (Chen *et al.*, 1996).

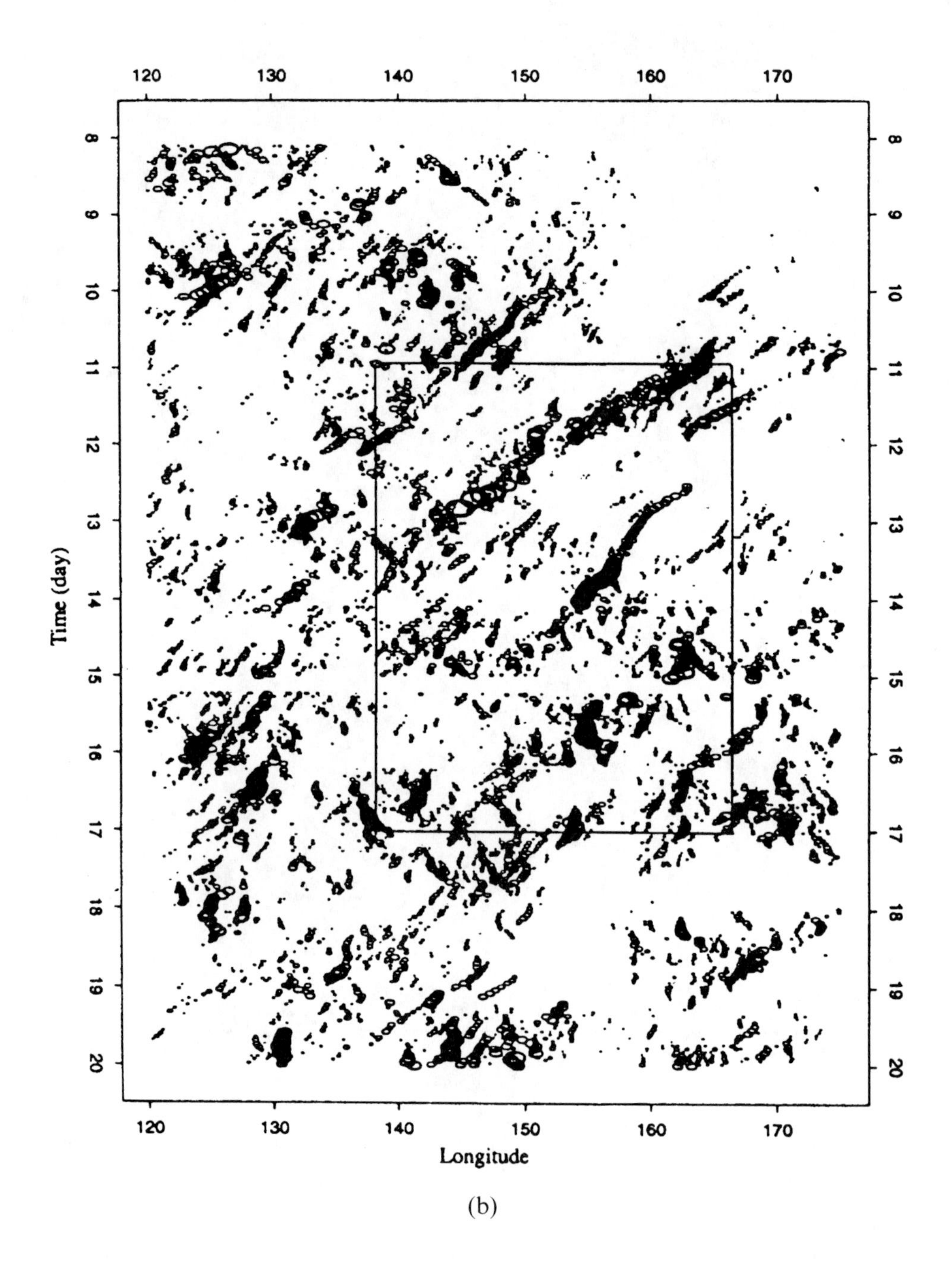

(b)

Figure 3.36 (*continued*)

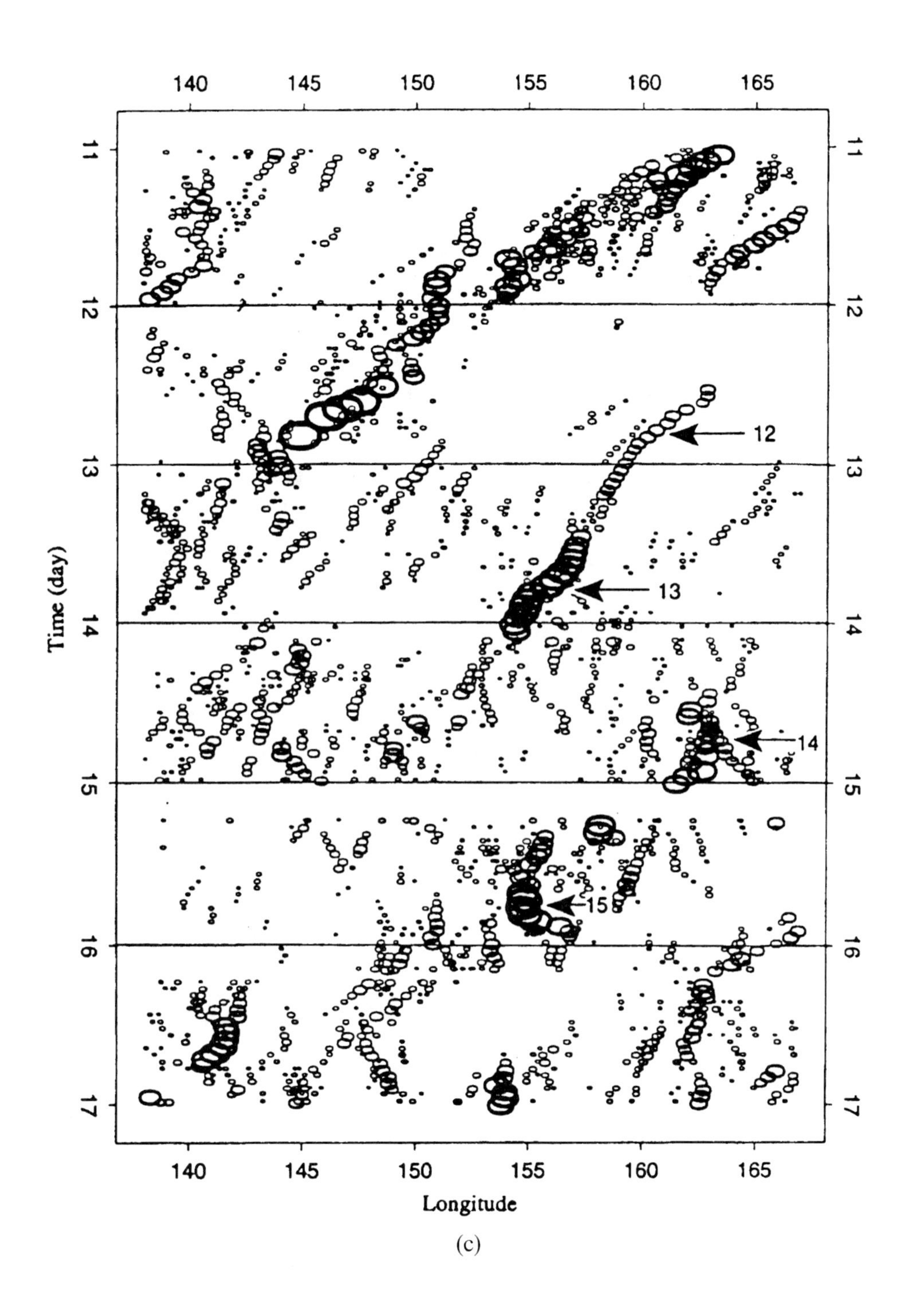
140
145
150
155
160
165
11
12
13
14
15
16
17
Time (day)
Longitude
12
13
14
15

(c)

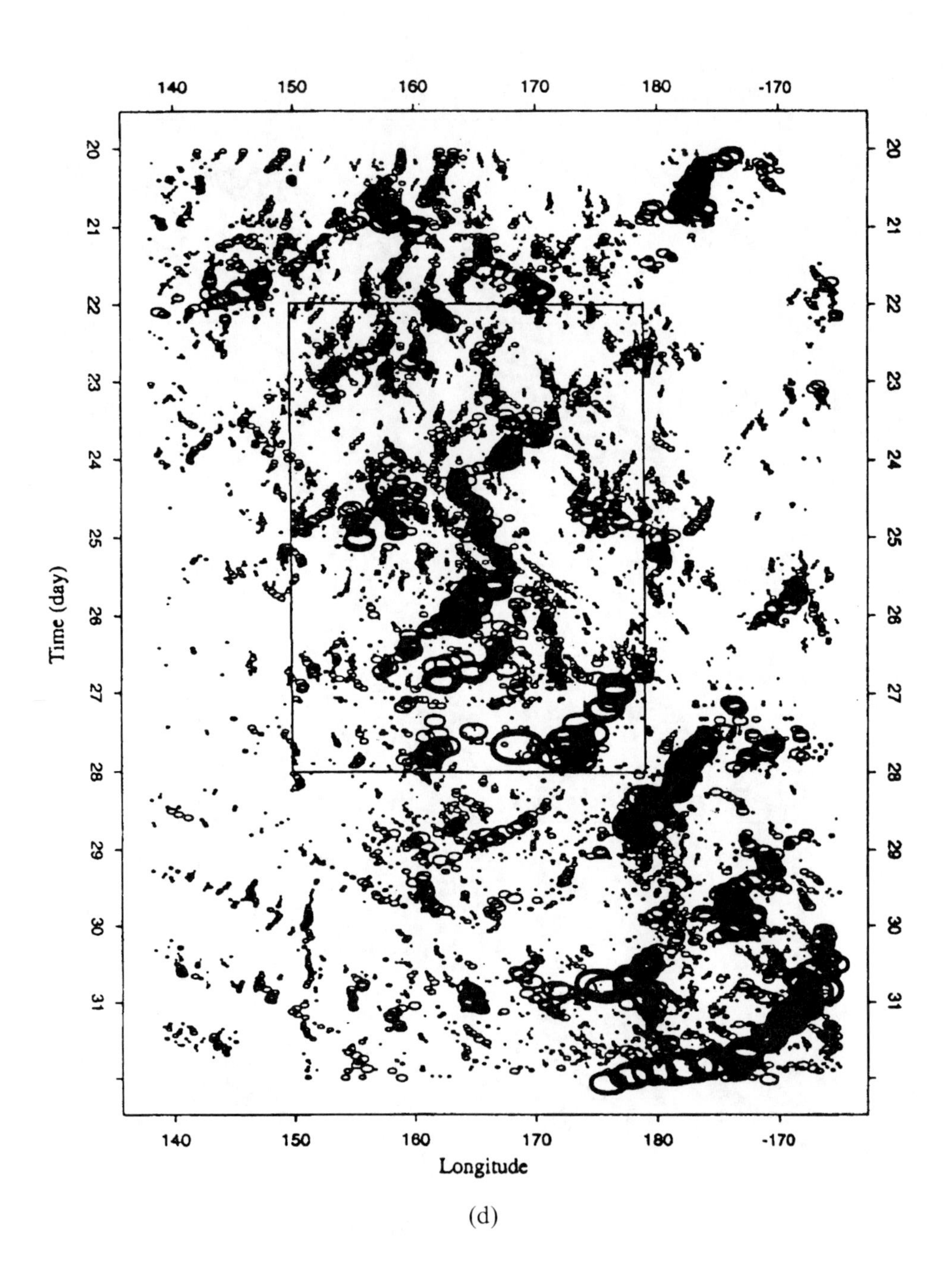

(d)

Figure 3.36 (*continued*)

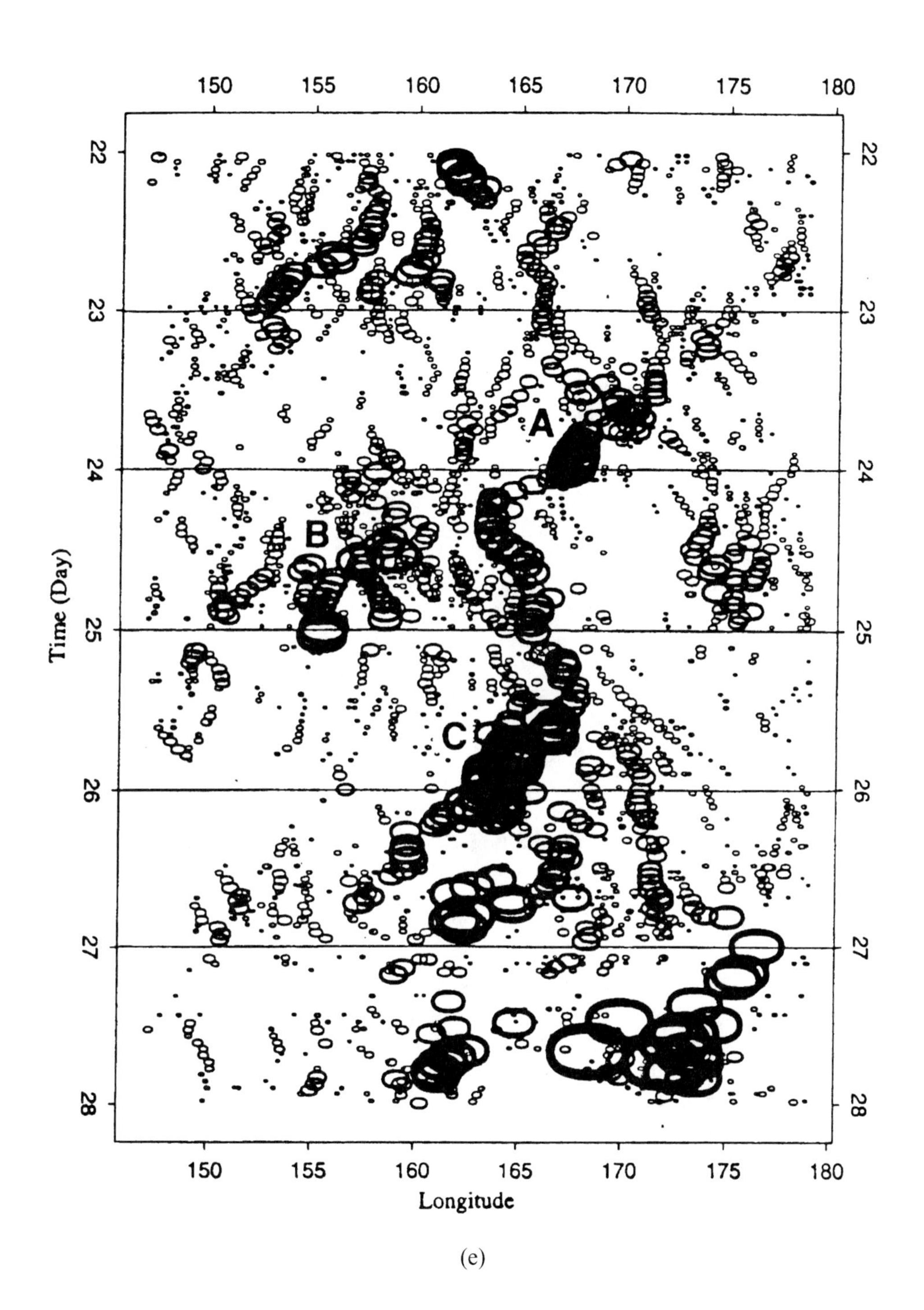

150
155
160
165
170
175
180
22
23
24
25
26
27
28
A
B
C
Time (Day)
Longitude

(e)

COARE longitudes during mid-December in the early stage of development of intense convection and strong low-level westerlies of the December wind burst over the COARE domain. Figure 3.36d indicates that, in some cases, cloud clusters comprising the 2-day westward-propagating disturbances moved at the same speed as the overall disturbance, while at other times the centroids of the cloud clusters were zonally stationary or even moved eastward (e.g., on December 14 and 15). These observations suggest that the 2-day wave has an existence and intrinsic scale beyond its embedded deep convection structures or, in other words, that it contains a clear air signal in addition to its convective cloudiness signal with comparable time and space scales for cloudy and clear regions. Perhaps a very careful examination of thermodynamic data from this period could illuminate the mechanism by which the 2-day wave organizes convection.

Three well-defined examples of westward-propagating 2-day disturbances occurred between 140°E and 165°E (in the vicinity of the IFA) during December 11–15. Four COARE aircraft missions sampled two of these disturbances. Arrows in Figure 3.36d show the location of the aircraft at the approximate mid-times of the four flights. The flights were in four separate convective systems.

Figure 3.37a, b shows both satellite and radar views of two of the cloud clusters sampled by the COARE aircraft. Airborne radar data obtained using the NOAA WP3D aircraft lower-fuselage radar on December 12 (Figure 3.37a) showed that the cloud cluster in Figure 3.36d represented a typical large tropical oceanic cloud cluster. This cloud cluster had a leading region of convective precipitation (more intense echo spots on its southwest side in Figure 3.37a) and a trailing region of stratiform precipitation (relatively uniform weak–moderate echoes on the northeast side in Figure 3.37a). The leading line–trailing stratiform echo structure was maintained by discrete redevelopment—not continuous propagation. The December 12 system died during the time of diurnal minimum in the early afternoon of December 13. New convection broke out along a convective cloud line in the vicinity of the dissipated December 12 convective system and tracked slowly northwestward as it grew and intensified (Figure 3.37a–c). The precipitation patterns of the December 13 system were more complex with multiple convective and stratiform regions (Figure 3.37b). In three panels the radar indicates large contiguous rain areas comparable with $T < 208$ K areas, which are enhanced in red.

The cloud clusters in the next 2-day disturbance in the vicinity of the IFA (150–165°E) during December 14–15 formed a very complex pattern in the time–longitude plot shown in Figure 3.36d. Aircraft sampled two separate connective systems during December 14–15 (arrows in Figure 3.36d), respectively. The cloud clusters on these two days did not form as a continuous pattern in time–longitude space as did the cloud clusters during December 12–13.

The airborne radar data obtained on all four flights (December 12–15) repeatedly showed precipitation patterns typical of mesoscale precipitation features. The mesoscale precipitation features were individually between about 100 km and 200 km in dimension. Often these mesoscale precipitation features were juxtaposed so closely that they combined to form contiguous or nearly contiguous precipitation over a

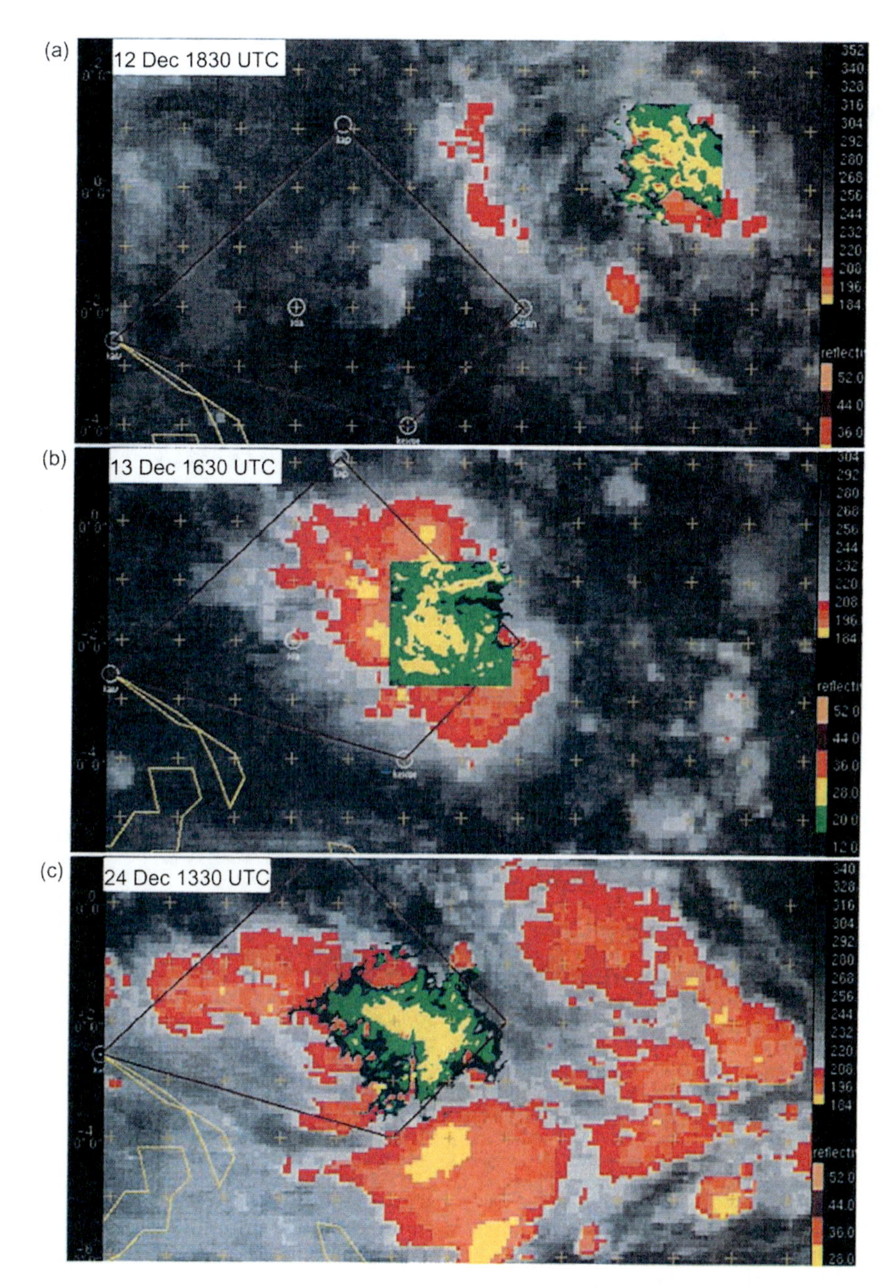

Figure 3.37. NOAA WP-3D N43RF aircraft lower-fuselage radar image overlaid on GMS IR images at: (a) 18:30 UTC on December 12, 1992; (b) 16:30 UTC on December 13, 1992 and (c) MIT radar image from RV *Vickers* overlaid with GMS IR image at 13:30 UTC on December 24, 1992. Both radars are C-band (5 cm wavelength). These images illustrate part of the precipitation region of cloud clusters shown in Figure 3.36d, ee. Color scales indicate radar reflectivity (dBZ) and IR temperatures (K). The COARE IFA is outlined (Chen *et al.*, 1996).

region several hundred kilometers in dimension (Figure 3.37b). As in the cloud clusters observed by aircraft, the satellite-observed cloud shields with $T < 208$ K reached class 4 in size (Figure 3.30) and corresponded closely to the large precipitation area formed by the juxtaposed mesoscale precipitation features. Two of the four convective systems (December 13 and 15) investigated by aircraft were class 4 in size. Note that, even at this extreme size, these convective systems all had lifetimes of less than 1 day.

The effect of the diurnal cycle on cloud clusters is clearly evident. Each of the four aircraft-sampled convective systems died out during the time of diurnal minimum. Most were enhanced with a large areal extent at ~18:00 UTC (04:00 LST). Such diurnal behavior is found in almost all large convective systems (Chen and Houze, 1997a).

Figure 3.36c, e focus on the intense convection of late December. During this period ICE2 became very intense and lower-tropospheric westerlies spanned over 40° of longitude. The ovals making up ICE2 in Figure 3.36c, e are larger and much more numerous than those in Figure 3.36d. Despite the strong westerly flow, the chains of ovals in Figure 3.36c, e indicate that a great many cloud clusters (more than 60%) tended to track westward, presumably by discrete propagation or successive redevelopment. Again, the diurnal effect is apparent and the clusters are more pronounced around 18:00 UTC (early morning) than at 06:00 UTC (afternoon) on each day. Four 2-day westward-propagating disturbances (i.e., December 21–22, 23–24, 25–26, and 27–28) were still evident, though less clear than those in Figure 3.36d, during this period.

During the most convectively active period, shown in Figure 3.36e, the ovals representing cloud cluster centroids tend to jump around rather erratically (especially system B on December 24). The reason is that—as cloud clusters merge and split from one satellite image to the next—centroids can move discontinuously. This frequent merging and splitting is an early indication that the convection of late December was so active that the scale separation between MCSs, with typical scales of between about 100 km and 200 km, and the larger scales (thousands of kilometers) began to break down.

The horizontal structures of three convective systems A, B, and C (as indicated in Figure 3.36e) are illustrated in Figure 3.38, which shows satellite IR images every 6 hours (except panels a, b) for 2 days. The three convective systems were extremely large, tangled, and complex, and there is ambiguity about where one ended and the next began. The colder cloud tops suggest a considerable substructure on the 100 km to 200 km scale, the size of mesoscale precipitation features. Figure 3.37c shows a radar echo in a portion of convective system B two hours after the time of Figure 3.37c. This shows that a large contiguous region of precipitation underlay the region of CTT < 208 K, extending to and presumably beyond the range of the MIT radar on RV *Vickers*. Viewing the ship radar data in time lapse reveals that convective system B contained several mesoscale precipitation features between about 100 km and 200 km in dimension. At some times these precipitation features were separated. At others they merged and interacted, like that shown in Figure 3.37c. The rain area covered a region >350 km in dimension, corresponding roughly to the region of IR temperature <208 K. However, close inspection of sequences of radar images reveals that the approximately 100 km to 200 km features were present as regions of higher reflectivity within the general large rain area.

The diurnal cycle of various CTTs is well demonstrated in Figure 3.38. There were many more large blue areas (220–235 K) than red (<208 K) during the local afternoon (Figure 3.38b, f). The 208 K area was dominant during pre-dawn hours (Figure 3.38d, h). Convective system B developed 10° to the west of A (Figure 3.38b) and intensified to maximum areal extent at the time of diurnal maximum in the early morning (Figure 3.38d). Convective system C developed in a relatively clear region where convective system A was active two days earlier and dissipated near 168°E (Figure 3.38b). Such behavior of convective systems was quite common within ICE2.

Satellite data (Chen and Houze, 1997b) show that the total amount of deep convection was nearly constant from year to year over the Warm Pool region of the tropical Indian and Pacific Oceans, but the spatial distribution varied. In the 1986–1987 warm ENSO event, the maximum of deep convective activity was over the high SST of the west central Pacific, with a local minimum over the eastern Indian Ocean. In the 1988–1989 cold ENSO event the maximum convective activity switched to the eastern Indian Ocean, with a minimum over the west central Pacific. SST changed very little over the eastern Indian Ocean from year to year and the centers of convective activity were always co-located with the 4-month mean low-level westerlies in each year. The latitude of the cloudy region of the ISO varied interannually. The interannual differences in deep convective activity over the Warm Pool were accounted for almost exclusively by the occurrence or non-occurrence of extremely large long-lasting deep mesoscale cloud systems (cloud tops <208 K with horizontal dimensions ~300–700 km). Such cloud systems occurred more frequently than normal over the western Pacific during the warm ENSO event and over the eastern Indian Ocean during the cold event. They occurred more frequently over the tropical eastern Indian and western Pacific Oceans than over the maritime continent, thus accounting for the see-saw pattern of observed cold cloudiness between the two oceanic regions on both interannual and intraseasonal timescales.

The total domain-averaged PHC and MSU-estimated precipitation as well as the fractional coverage of the PHC and precipitation, however, remained nearly constant for all three years over the Indo-Pacific Warm Pool region. It suggests a near conservation of total amount of cold cloudiness and precipitation over the Warm Pool region under the extreme conditions of the ENSO cycle of 1986–1989. This result indicates that the mechanism(s) controlling the total amount of deep convection over the Warm Pool may be independent of the ENSO cycle. However, spatial shifting of the location of the convective activity center within the region, between the western central Pacific and the eastern Indian Ocean, and between the equatorial region and the SPCZ, is still the main reason for the global response to the convective variability related to the ENSO cycle.

In closing it should be noted that many of the physical systems show a capacity for clustering its elements or, in other words, for self-organizing effects (Pietronero and Tosatti, 1986).

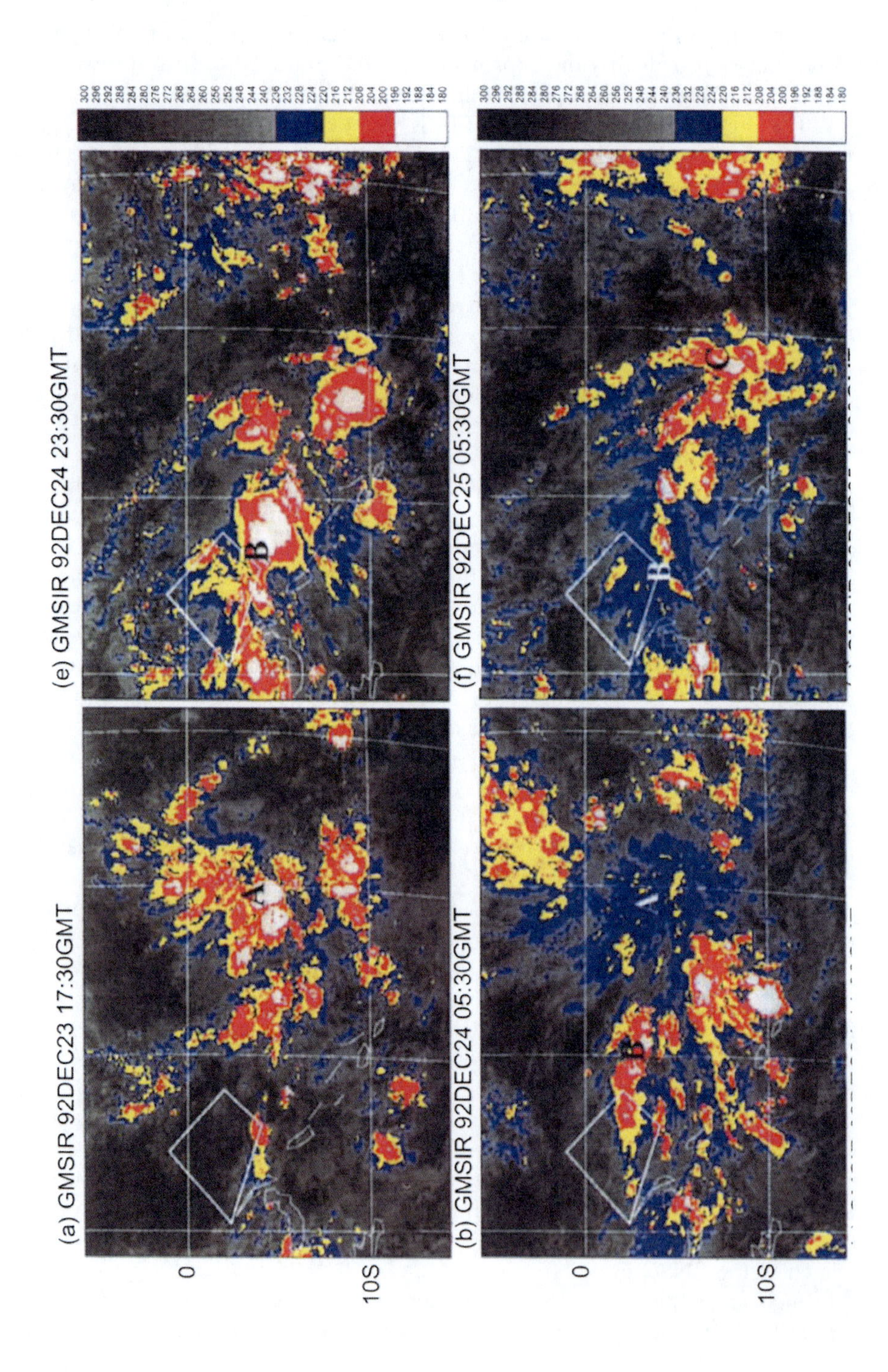

(a) GMSIR 92DEC23 17:30GMT
(b) GMSIR 92DEC24 05:30GMT
(e) GMSIR 92DEC24 23:30GMT
(f) GMSIR 92DEC25 05:30GMT
0
10S
A
B
C

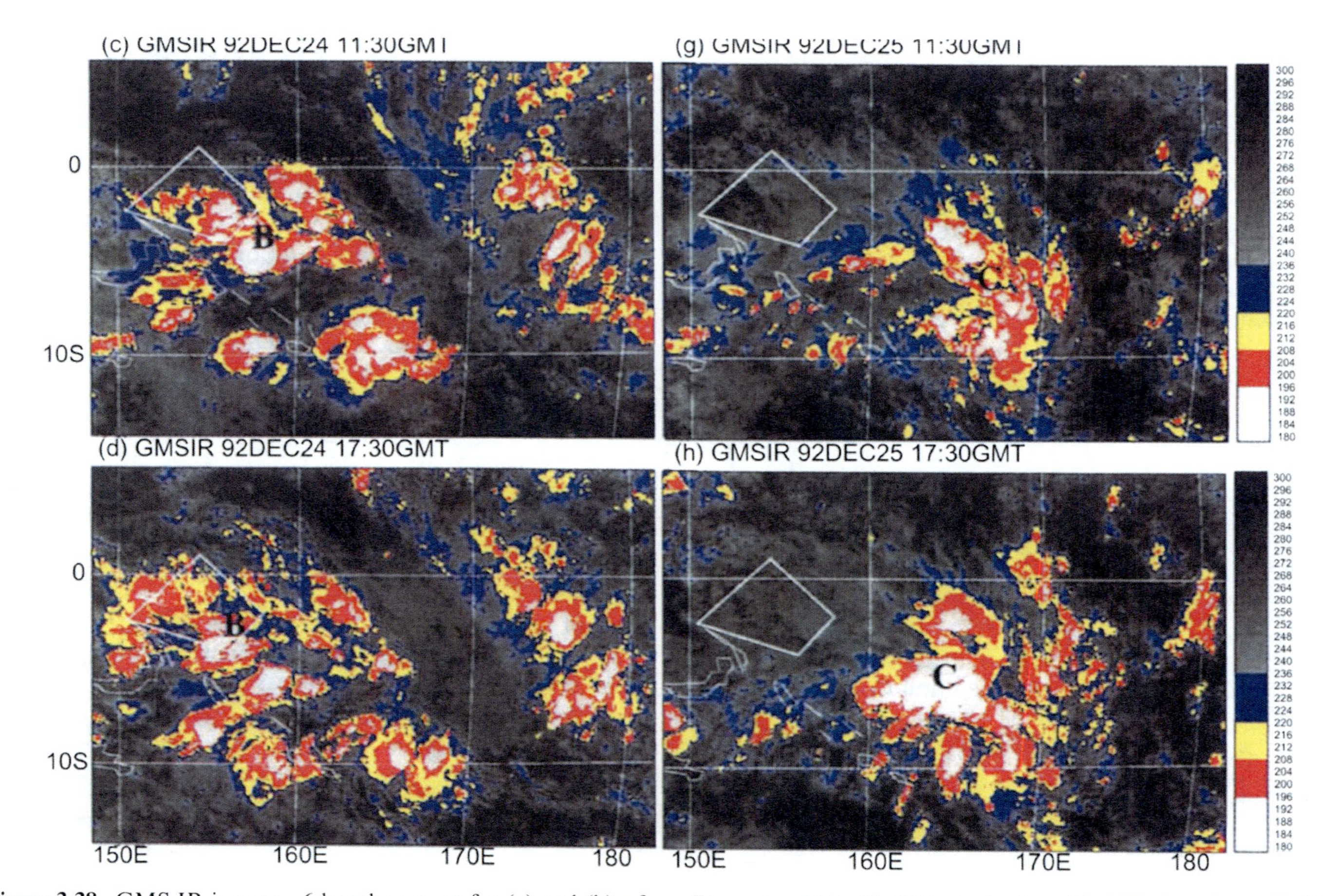

Figure 3.38. GMS IR images—6 hourly, except for (a) and (b)—from December 23 to 25, 1992. A few key IR temperature ranges (K) are highlighted. A, B, and C indicate the locations of each convective system. The COARE IFA is outlined (Chen *et al.* 1996).

3.4.4 Fractal features of superconvective clusters

As mentioned in Section 3.4.3, in view of the presence of multiscale processes in the clusterization of tropical convective systems the fractal features of such systems must clearly be of great variety. The early sophisticated works of Baryshnikova *et al.* (1989a, b) have shown that this is the case.

Baryshnikova *et al.* (1989a, b) undertake the task of describing and measuring convective turbulent conditions in complicated pre-hurricane situations using the concept of fractal dimension (Mandelbrot, 1982; Lovejoy and Schertzer, 1985; Voss, 1989). They study the fractal dimensions of constant temperature lines on satellite IR (NOAA) images of convective cloud fields and the link that this characteristic has with the turbulent region in the cloud system.

It is known that air temperature at the cloud boundary depends on the humidity and temperature inside a cloud. Moreover, Baryshnikova *et al.* (1988) show that temperature T measured at the cloud boundary at some initial moment $t = t_0$ is given by:

$$T(x, y) = \alpha\langle\kappa(x, y, z_b)\rangle + \beta\langle\eta(x, y, z_b)\rangle + \sigma \tag{3.14}$$

where κ is the deviation of humidity from the equilibrium value; η is the deviation of temperature from the linear vertical distribution taken on the cloud boundary z_b; α, β, and σ are certain constants; and $\langle\cdots\rangle$ denotes averaging over the resolution element. Note that temperature is determined by the above-mentioned deviations themselves rather than by functions of them.

In a turbulent atmosphere κ and η are random and behave like a scalar admixture (i.e., their values are governed by the transfer equation). In the case of developed Kolmogorov turbulence we have κ, $\eta \cong \lambda^{1/3}$ where λ is the characteristic scale of pulsations, while the surface $T(x, y)$ has fractal properties with dimension $D_2 = 2.34$. When the properties of turbulence change, the fractal dimension of this surface also changes (e.g., development of atmospheric large-scale instabilities can be expected to result in a decrease of D_2 towards 2).

In the case of homogeneous isotropic turbulence, which is typical of atmospheric conditions in a wide range of scales, the fractal dimensions of the surface $T(x, y)$, D, and the isothermal line D_1 are related: $D_2 = 1 + D_1$ (Mandelbrot, 1982).

For inhomogeneous turbulence the relation between D_2 and D_1 is more complicated. However, changes in the nature of atmospheric motions lead to changes in D_1 as well. Therefore, estimation of D_1, which is more easily achieved, can be made instead of D_2 when addressing problematic atmospheric issues.

Thus, the fractal dimension of isothermal lines in cloud images should depend on the state of atmospheric turbulence (e.g., on the development of instabilities). Such dependence may take place when large-scale structures develop in the atmosphere (e.g., hurricanes). In this case the fractal and spectral properties of wind flows should change (Levich and Tzvetkov, 1985). Thus, the fractal dimension of isothermal lines on IR images of clouds can serve as an indicator of large-scale structure development in the atmosphere.

In order to verify this presumption Baryshnikova *et al.* (1989a, b) consider a series of IR images of different stages of typhoon "Nelson" (August 1985) (as well as initial superclusters) obtained by the Japanese geostationary satellite GMS. For data processing Baryshnikova *et al.* (1989a, b) choose regions from the central part of each image, restricted by tropical zone (from 15–25°N and from 130–150°E).

Definition of the fractal dimension and its possible applications in physical problems is considered in many papers (see Mandelbrot, 1982; Lovejoy and Mandelbrot, 1985 and references therein). The investigation of atmospheric turbulence and, particularly, analysis of the atmospheric state are possible applications of this approach. The fractal dimension d of an object embedded in n-dimensional space is usually defined as:

$$d = \lim_{r \to 0} \frac{\ln N(r)}{\ln(1/r)} \tag{3.15}$$

where $N(r)$ is the minimal number of n-dimensional spheres of radius r that cover the object. Note that for real objects the limit $r \to 0$ is often not reached. However, for many physical processes that are self-semilar over a wide scale range ($r_{\min} < r < r_{\max}$), the number of covering surfaces $N(r)$ scales as r^{α}, where α is a certain constant. Therefore, the value $\ln N(r)/\ln(1/r)$ characterizes this process in the given range of r and the process itself can be considered fractal. An important example of such a process is Kolmogorov turbulence.

Fractal dimension has proved to be a convenient characteristic of turbulent processes (Frisch *et al.*, 1978; Zeldovich and Sokolov, 1985; Levich and Tzvetkov, 1985). When turbulence is homogeneous, the fractal dimension is uniquely related to the spectrum slope in those frequency intervals where the spectrum is a power law. When turbulence intermittency appears, the fractal dimension of the velocity field of such a system changes. This is also true for regular motion emergence on a turbulent background (Levich and Tzvetkov, 1985).

The results of data processing are as follows. The dependence of the logarithm of the covering area on the covering radius is sufficiently close to the linear dependence (Figure 3.39). Thus, a certain fractal dimension can be attributed to isothermal lines in the investigated range of scales (the covering radii varied from 5–100 km). This fractal dimension is practically independent of the temperature (see Figure 3.39) of the line considered, except in cases of high temperature (when the data are constrained by the underlying surface and transition layer) and low temperature (when the signal-to-noise ratio is too low). Therefore, the properties of turbulence do not vary much, with height ranging from 2 km to 15 km, from where measurements were made. When considering the cloud images that correspond to different stages of typhoon development, from TDE to TS, we derived an unexpected result: the fractal dimension is practically the same for all images and equals 1.35 ± 0.05. We also analyzed data for a midlatitude cyclone in the Okhotsk Sea, and the derived fractal dimension does not differ from that of tropical cloud systems. The fractal dimension is a stable characteristic of various cloud structures, independent of the presence of developed convection, large-scale vortices, and wind flows. Similar results were obtained in other studies; for example, Lovejoy (1982) shows that $D_{\mathrm{l}} \approx 1.35$ irrespective of cloud type. This value of fractal dimension is close to that typical of homogeneous Kolmogorov turbulence, $D_{\mathrm{l}} \approx 1.34$. We are therefore

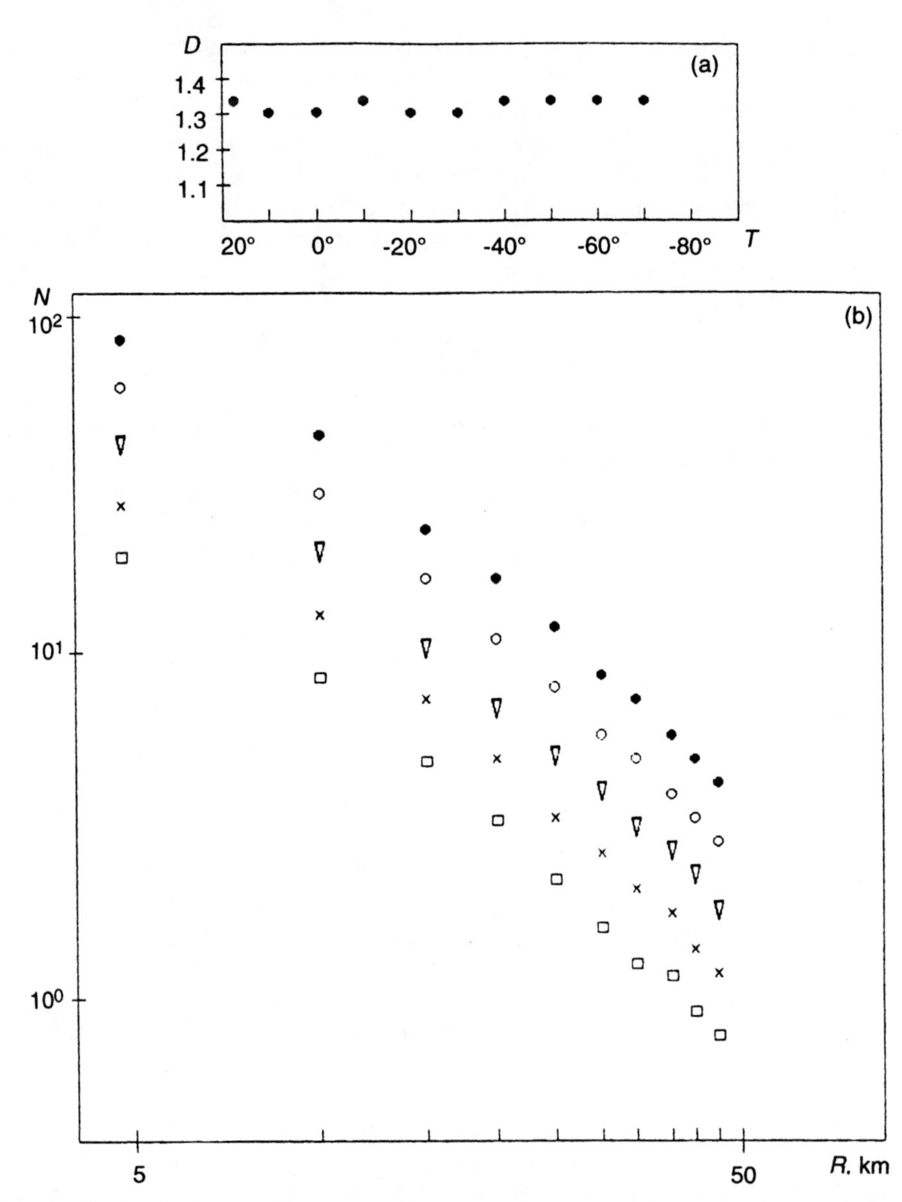

Figure 3.39. Typical plots: (a) fractal dimension D versus temperature T of the chosen isotherm; (b) number of circles in N coverage versus radius r (plots for different temperatures are shown and each curve's own scale was chosen for N) (from Baryshnikova *et al.*, 1989b).

tempted to say that atmospheric turbulence over the investigated scales (from 5 to 10 km) is homogeneous and of Kolmogorov type. Such a conclusion seems, however, to be premature since in the analyzed regions both the zones of intense convection (small-scale instabilities) and cyclones (large-scale regular structure) are present. This

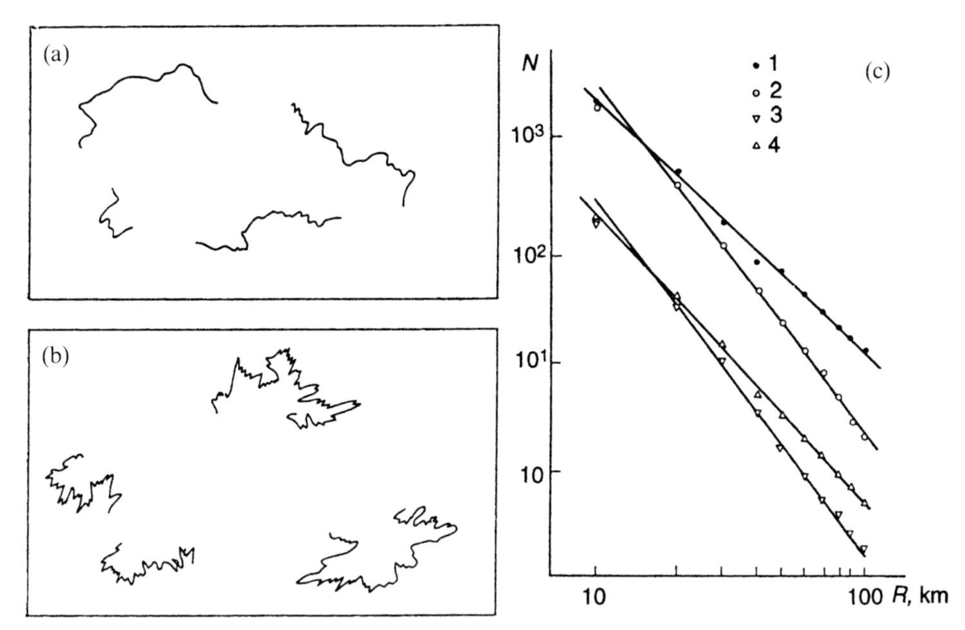

Figure 3.40. Sample images (a) and (b) of those portions of constant temperature curves (IR images) with distinct temperature gradients: (a) with large values of gradients; (b) with small values of gradients. (c) The number of circles in N coverage versus radius R for curves with a distinct gradient. 1, $\partial T/\partial z > 6$; 2, <6; 3, $>4, 5$; 4, $<4, 5$ gradients/km (from Baryshnikova *et al.*, 1989a).

should inevitably result in different turbulence regimes (Levich and Tzvetkov, 1985) and deviations from homogeneity and Kolmogorov's law.

Another possible explanation is the following. Suppose that the isothermal lines investigated pass through two different regions, one with Kolmogorov turbulence and another with regular motions. In this case the fractal dimension of the first segment of the line is $D_l = 1.34$ while that of the second segment is of the order of unity. Thus, different segments can have different dimensions. When estimating the dimension of such a line we always obtain the largest value present (i.e., 1.34).

Therefore, it seems reasonable to divide the lines into segments and calculate the fractal dimension individually (Figure 3.40a, b). For division into segments we chose the transverse temperature gradient with its suitably chosen threshold value. This value varied between 10°C/km and 30°C/km in different runs. The calculated fractal dimension of large-gradient segments turns out to be of the order of unity while in the case of small-gradient segments $D_l \approx 1.3$. Note that regions with large temperature gradients are usually associated with developing clouds forming in intense upstream flows (e.g., in convection). Regions with small temperature gradients, in turn, correspond to decaying clouds and usually the absence of regular motions. Thus, we see that regions with different types of air motion are indeed characterized by different values of the fractal dimension of the upper cloud boundary. Therefore,

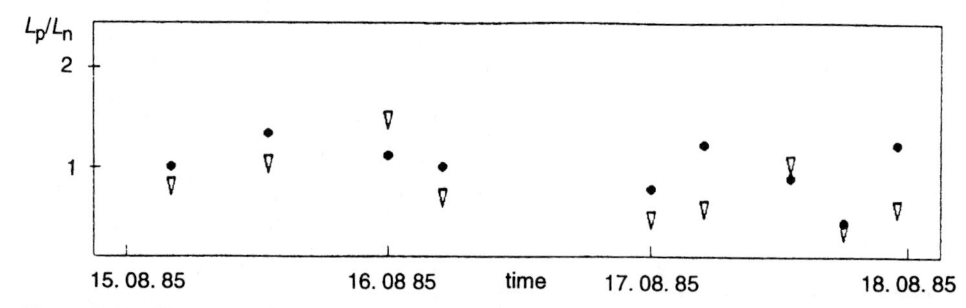

Figure 3.41. The ratio of regular interval L_p total length to irregular interval L_n total length on a temperature isoline versus the observation time for specific atmospheric regions: •, depression region where cloudiness was studied which eventually developed into a typhoon. ∇, cloudiness was studied but there were no regular motions (Baryshnikova *et al.*, 1989b).

estimation of the fractal dimension of different segments of isothermal lines allows determining whether or not regular motions (e.g., convection) are present there. In large areas (500×500 km) lines with both small ≈ 1 and large ≈ 1.3 fractal dimensions can always be found simultaneously, which seems to indicate the intermittent nature of turbulence with alternating regions of either enhanced turbulence or regular flows. We note that the ratio of total areas of such regions can serve as an approximate measure of the energy stored in wind flows. It is expected that the area fraction of regular motions is larger when such structures are present than with cases of purely chaotic motions. This is substantiated by the data given in Figure 3.41.

The main result obtained is that tropical turbulent cloud fields represent a complicated "mixture" of zones with intensive convection (mesoscale instabilities) and more "quiet" zones with a large-scale regular structure. Using the fractal method the authors separated the region with regular motions (where the fractal dimension is about unity) from the region with completely turbulent flows (where the fractal dimension is about 1.34 indicating a Kolmogorov type of turbulence). They found that the number of zones with regular motions in structures developing into TCs is greater than in structures from which no TC develops.

Thus, the fractal dimension of cloud structures reflects the state of atmospheric turbulence and can be used for diagnostic purposes. One should be careful when considering small regions where turbulence can be considered homogeneous. This requires sufficiently good resolution both in space and in temperature. We hope that more detailed application of this approach to regions with different turbulence regimes will yield additional information about atmospheric processes.

3.4.5 Spatiotemporal evolution of convective cluster mesoturbulence with high-precision Doppler radar

As noted above (Section 3.4.3), the achievements of the remote-sensing study of mesoscale and large-scale convective tropical systems using IR satellite instruments are impressive. However, it should be noted that this type of remote-sensing measurement provides the pure thermal "frozen" signature of the cloud system of interest. Using IR signatures it is difficult to obtain observational data about the detailed "inner" structure of the turbulent fields of wind in cloud volumes: namely, the turbulent field kinematic characteristics of convective scales and mesoscales are primarily under consideration because it is the processes of such scales that determine (within the framework of the models under discussion, see Section 3.4.1) the energy potential of an incipient catastrophic vortex.

According to the self-organized concept (Chapter 4) the dynamic parameters of a turbulent air mass (i.e., its turbulent store of energy and spiral and helical properties) play the main part in forming large-scale critical states (like TCs at birth).

In addition, the entropy and helical properties of turbulence are responsible for such important characteristics of turbulent flows as the variability of intermittency, the onset of unstable flow (transition to chaos), and anomalous diffusion (Levich and Tzvetkov, 1985; Hussain, 1986; Sanada, 1993; Tsinober, 1994; Lesieur, 1997).

Klepikov *et al.* (1990, 1995) and Sharkov (1996b, c; 1998) present the results of satellite observations and full-scale experimental explorations of the spatial characteristics of convective turbulence, which were performed aboard a research ship (RS) in the NWP by means of high-precision Doppler radar providing a high-accuracy estimation of wind flow velocity (the r.m.s. instrumental error for determining wind flows is a resolution cell of 15 cm/s).

During October 1–2, 1988 the weather where the RS was sailing was dominated by multilayered cumuli and short-term precipitation, with the velocity of the wind (from the north and northeast) of 8 m/s to 9 m/s, and SST varying between 28°C and 29°C. A TD, with a pressure of 1,002 mb at its center accompanied by a weak southeast wind, existed to the west of the sailing region near Hainan Island. It moved slowly toward the Chinese coast. In the waters of the Philippine Sea the TD became STS "Nelson" (8824) with a pressure of 985 mb at its center and with a wind velocity of 26 m/s. The velocity of the northwest-propagating storm was equal to 10 knots. Doppler sounding from the RS was carried out in an evolutionary cloud cluster in that part of the atmosphere that was absorbed by the cloud system that would become a TC within 24 hours. Within the next 24 hours the central part of the typhoon traveled straight through the geographical location where the Doppler measurements were conducted.

The initial data of the measuring system (for this experiment) consisted of a set of Doppler velocities (in a polar coordinate system) of fine water drop clouds entrained in the kinematic dynamics of turbulent wind flows. The one-dimensional longitudinal structure function (SF) was measured, using a Doppler velocity vector (velocity vector projection on the sounding direction), for a two-dimensional frame of cloud formation obtained as a result of azimuthal scanning with a step of 3° at an elevation angle of 2° from the horizon. The "instantaneous" pulse volumes were equal to $3{,}000\,\mathrm{m}^2 \times 600\,\mathrm{m}$ for small ranges and $0.75\,\mathrm{m}^2 \times 0.6\,\mathrm{km}$ for large ranges ($\approx 30\,\mathrm{km}$).

In the course of investigations the spatiotemporal behavior of the one-dimensional SF (up to sixth order) was analyzed during the entire lifetime of cloud meteoformation from its generation to its complete decay (1.5–2 h). The corresponding statistical average, which is necessary for finding the longitudinal SF, was constructed as follows:

$$D_n = \frac{\sum_{k=1}^{M}\sum_{i=1}^{N}\sum_{j>1}^{N} B_{ij}|V_{ik} - V_{jk}|^n}{\sum_{k=1}^{M}\sum_{i=1}^{N}\sum_{j>1}^{N} B_{ij}} \tag{3.16}$$

where $R = r_i - r_j$ is the turbulent scale; i and j are the indices over the range of resolution cells; N is the maximum number of resolution cells; M is the number of independent azimuthal directions; and $B_{ij} = 1$ (when velocity data obtained at both points i and j are available) and $B_{ij} = 0$ (otherwise).

The technique used consists of summing and averaging the differences of the Doppler velocities of pulse volumes for different separations (turbulent scale) between the volumes. The calculation is carried out both along a unit ray and in the azimuthal direction. The calculation of the SF for a two-dimensional velocity field by the technique suggested is complicated because of the stochastic two-dimensional structure of the field of wave scattering on whose background the kinematic field itself is revealed. The efficiency of this technique at calculating the structure function was demonstrated by calculating the SF of two-dimensional "white" noise as an example.

As is known, the first-order SF corresponds to the correlation function of a process (with a minus sign), the second-order SF determines the behavior of the energy spectrum of turbulent flow velocity fluctuations, and the SFs of the next orders are associated with higher moments of velocity fluctuations. Analysis of the functions $D_n(R)$ shows the behavior of the SFs of different orders to be similar on the whole. For example, all the functions $D_n(R)$ have pronounced power law branches, although they differ essentially in details—the boundary between two power law branches is noticeably shifted in the direction of larger scales with SF order. The behavior of higher order SFs for convective turbulence is the subject of a separate investigation.

Figure 3.42 shows the spatiotemporal evolution of second-order SFs (spectral energy) during radar observation from 13:22 LT to 15:24 LT. First, we note that the SFs had certain characteristic properties during the observation time: namely, the presence of two pronounced power law branches. One of them (with lesser gradient) was observed on scales of the order of several kilometers, and another (with greater gradient) was observed on scales ranging from 5 km to 20 km. It should be noted, however, that these branches were dynamic. For example, in four minutes (from 13:22 LT to 13:26 LT) the boundary between the two ranges shifted from 4 km to 8 km, although the character of the turbulent regime itself (the presence of two power branches) held on the whole. At 13:26 LT a new feature of the turbulent regime was observed: a descending branch of the SF on scales exceeding 21 km. At 13:33 LT an essential transformation in the turbulent regime was revealed: there

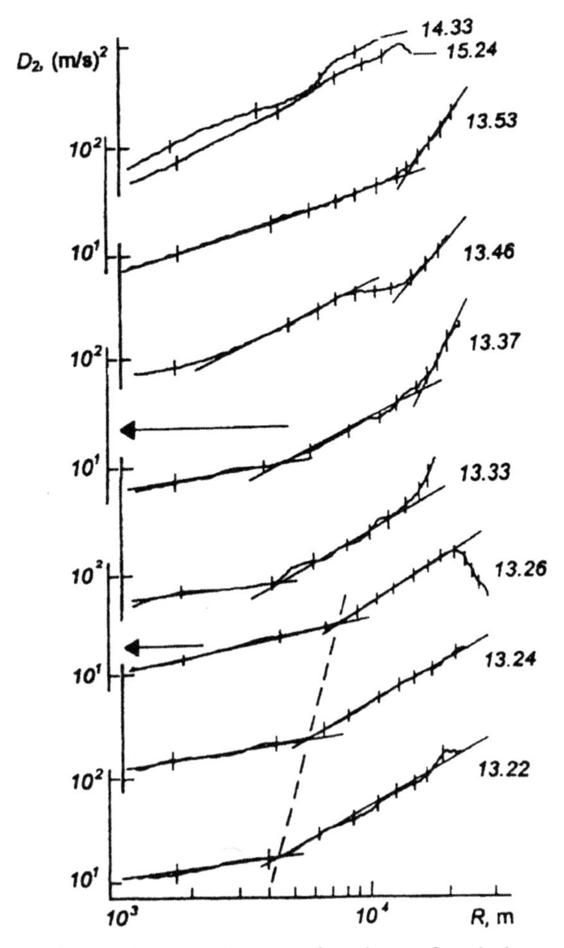

Figure 3.42. Experimental second-order structure functions of turbulent scale for different cloud cluster lifetimes. Numbers near the curves indicate the local time of cloud system sounding (from Klepikov *et al.*, 1995).

was an obvious growth in the SF on scales of 5 km to 12 km, forming a peculiar kind of "plateau" on scales of 10 km to 15 km. These features can be interpreted as turbulent fluctuation energy transfer, accumulation, and downward release on scales that by 13:53 LT had formed a pure Kolmogorov regime (the gradient of the power branch was about 0.7) on scales of 1 km to 17 km and a power branch with gradient magnitude ≈ 2.78 on scales of 17 km to 21 km. By 14:33 LT the spatial form of the SF had essentially changed such that irregular behavior (on scales of 5–6 km) appeared in the background of a power branch (with a gradient of 1.17).

The next survey frame was taken at 15:24 LT but took an hour to complete. The frame demonstrates the absence of this feature. In this case, a gradient estimate over the entire range of scales (1–17 km) gives a value of 1.3 which is substantially greater

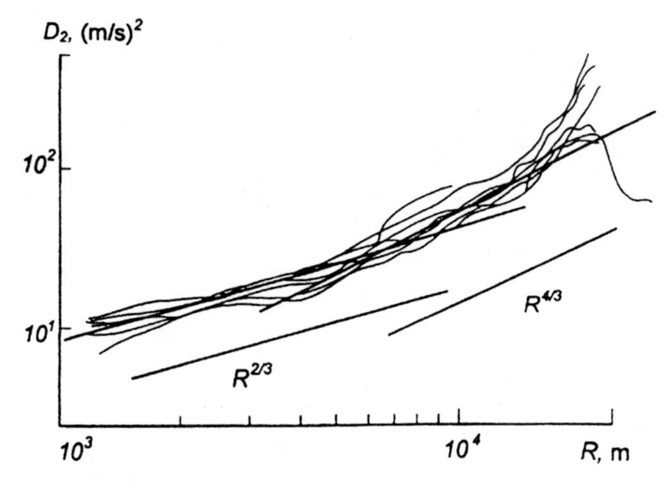

Figure 3.43. Experimental second-order structure functions of turbulent scale averaged over a time equal to 1 h 10 min (from Klepikov *et al.*, 1995).

than the value corresponding to Kolmogorov branches and close to the known power spectrum law of −7/3.

The following survey frame was taken at 16:16 LT but once again took an hour to complete. Analysis of SFs points to the presence of the Kolmogorov branch within the range of 1 km to 5 km. About 10 km away, the SF resembled a "noise-like" signal. Visual inspection showed dissipation of the cloud cluster.

Figure 3.43 shows experimental SFs averaged over 1 h 10 min (with a preserved spatial grid scale of ~1 km). Here, Kolmogorov power branches can be clearly recognized on scales of 1 km to 7 km, and branches of −7/3 type can be seen on scales of 8 km to 15 km.

It is known (Lesieur, 1997) that—to reveal the type of turbulence model—both the exponent of a power law approximation for a structure function:

$$D_n(R) = \langle |\Delta U|^n \rangle \cong AR^{\zeta_n} \tag{3.17}$$

and its dependence on the order of the SF (i.e., $\zeta_n = f(n)$) are of basic importance.

To specify the performance data of potential space Doppler systems, it is first necessary to form a spatiotemporal model of the second-order SF responsible for the spectral characteristics of the store of turbulent energy. On the basis of the experimental data presented it is possible to form a spatiotemporal model of the "second-order SF of type (3.17)" (Sharkov, 1996c). However, we should remember that model parameters are highly dependent on the spatiotemporal grid scale chosen. For example, for a spatiotemporal scale of order 1 km and 1 min, ζ_2 may be represented by three power branches—with a small exponent $\zeta_2 \cong 0.3$–0.6, an intermediate exponent $\zeta_2 \cong 1.2$–1.7, and a large exponent $\zeta_2 \cong 2$–3.5—for $R > 15$ km. The boundary between the first two branches is highly variable and, hence, final time dependence cannot be determined. Estimates show the time "pixel" of the model to be about 1 min.

These features can be interpreted (Sharkov, 1999b) as turbulent fluctuation energy transfer, accumulation, and downward release on scales of 1 km to 15 km of a "pure" Kolmogorov regime (the power branch gradient was about 2/3), which "existed" for 20 min, then transformed into the branch with a gradient of 4/3, and then into the known power spectrum law of −7/3 (Tsinober, 1994). It is interesting to note that a branch with zero gradient also existed, corresponding to the famous $1/f$ power spectrum. An important connection between $1/f$ noise, self-similarity, and self-organized criticality (SOC) has been emphasized by Mandelbrot (1977) and Bak *et al.* (1987). $1/f$ noise is as basic to off-equilibrium response as equipartition is to equilibrium. Note should be taken that the energy spectrum ω^{-1} corresponds to the isotropic inverse cascade because of the helicity invariant (Levich and Tzvetkov, 1985). We managed to find branches with gradients that were equal to about 1. It is known that this is the situation of the interesting power spectrum (ω^{-2}) of the random walk model. Note that SF branches with negative gradients (and hence with equilibrium power spectrum) were extremely seldom found and then only within large scales (15–30 km). It does seem likely that, on the one hand, all the above-mentioned regimes peacefully co-exist and, on the other hand, are antagonistic to each other.

When a spatiotemporal grid scale of order 1 km and 1 hour is chosen, the situation of the model becomes somewhat more stable. The Kolmogorov power branch, recognized on scales of 1 km to 7 km and a branch of −7/3 type, can be seen on scales of 8 km to 15 km. This is the means by which fast-acting interactions cause, to some degree, the chaotic structural background, from which Kolmogorov and non-Kolmogorov stable branches are formed (Figure 3.43). The following approximation is possible here:

$$D_2 = \begin{cases} A_1 R^{2/3} & R \in (1\text{–}7\,\text{km}) \\ A_2 R^{3/4} & R \in (7\text{–}15\,\text{km}) \end{cases} \tag{3.18}$$

where $A_1 \cong 0.1\,\text{m}^{3/4}\,\text{s}^{-2}$; and $A_2 \cong 3 \times 10^{-4}\,\text{m}^{2/3}\,\text{s}^{-2}$. Further experimental studies are necessary to elucidate the spatiotemporal "stability" of the model suggested.

Using experimental results, Sharkov (1999b) performs an assessment of the main parameters involved in convective tropical turbulence in a pre-hurricane situation (supercluster). They are the following:

Reynolds number	3×10^9
Mesoscale Richardson number	$-5 \div -20$
Kolmogorov scale (η)	0.5 mm
Energy dissipation rate (ε)	$1.2 \times 10^{-2}\,\text{m}^2\,\text{s}^{-3}$
Taylor microscale	14 cm
Taylor microscale Reynolds number (R_λ)	10^5

The other important problem of turbulent field observations is to find the structure elements (if they exist) and their interactions in chaotic natural velocity patterns. The one-dimensional chaotic time series can be studied by wavelet analysis (Section 2.7). Voiskovskii *et al.* (1987, 1989) and Sharkov (1996c) demonstrate other approaches for detecting morphological elements, on the basis of two-dimensional Doppler radar velocity field data. These papers embrace the results of the special processing of two-dimensional Doppler radar velocity fields (cloud areas observed were about 5,000–7,000 km^2) that were used in the above-mentioned process. Digital processing was designed to detect the discrete eddying complexes embedded in random fields with only radial components. The morphological structure of the eddying complexes to be detected includes both rotatable and divergent (or convergent) components. The approach was proposed and developed by Klepikov and Sharkov (1987). A series of discrete complexes was resolved from the chaotic turbulent fields. Typical radii of the detected eddies were observed within the limits of 2 km to 20 km. The discrete divergent complexes had the property of relatively temporal stability. The reciprocal relationship between vorticity and the typical radius of the detected eddy was found, at the same time, for the Kolmogorov cascade to be equal to (−2/3). Special techniques for estimation of vertical velocity components and, hence, the helicity of complexes were proposed and developed by Sharkov (1996c). However, a statistically determinate relationship between the helicities and typical radii of the detected complexes was not found.

Note that detection of the fine spatiotemporal structure of natural turbulence can only be obtained by means of high-precision Doppler radar sets and high-duty data-handling systems.

This is the reason airborne Doppler radar observations were made (Chong and Bousquet, 1999) to investigate the internal structure of a mid-level mesovortex that developed within the rear part of the stratiform precipitation region of a mature-to-decaying mesoscale convective system. This system, composed of several convective elements, occurred on December 13, 1992 on the eastern side of the intensive flux array of TOGA COARE near the equator (2°S).

Doppler radar observations showed that the closed wind circulation at mid-levels had a horizontal dimension of 150 km and was coincident with a marked inflow at the central rear of the stratiform region where a notch pattern clearly identified the intrusion of dry air. It was associated with positive vertical vorticity (anticyclonic for the southern hemisphere) and could be classified as a cooling-induced vortex. Maximum positive vorticity was concentrated along the converging interface between this flow and in-cloud rearward flow. Mesoscale downdraft air was the primary source of the rearward outflow observed at low levels. The vorticity budget performed within the mesovortex reveals tilting of the horizontal component of vorticity to be the prominent dynamical process influencing vortex development, although a part of the vortex amplification at mid to high levels was due to stretching. At lower levels, tilting tended to inhibit the mesovortex, by converting horizontal vorticity to negative vertical vorticity. Close examination of system-induced vertical wind shear as a result of mesoscale momentum transport reveals an evident correlation with the tilting mechanism. Overall, vorticity changes as a result of tilting and stretching were negatively correlated with vertical and horizontal advection, respectively. Vertical advection redistribution was accomplished by means of vertical motion, which is found to transport positive vorticity from the higher part of the mesovortex down to low levels.

It is important to note that the Doppler radar–derived fields of wind and reflectivity have been deeply involved in retrieving temperature perturbations, water vapor levels, and cloud water content (Bielli and Roux, 1999).

3.5 VARIATION IN TROPICAL CYCLONE ACTIVITY AND EL NIÑO–SOUTHERN OSCILLATION

As indicated in Section 2.6, and in works by Neumann (1993) and McBride (1995), most cyclone-active basins have relatively large interannual variabilities of cyclone activity. The ratio between the standard deviation in cyclone numbers and the mean ranges from 25% to 40%, while the corresponding ratio for severe TCs ranges from 34% to 69%. The exception is the western North Pacific basin, which has (relatively) small variability with a ratio of only 16% for cyclone numbers and 23% for severe TC numbers (McBride, 1995).

A number of authors have demonstrated relationships between the interannual fluctuations of cyclone numbers and the slowly varying aspects of large-scale tropical circulation and environment. The significance of the term "slowly varying" is that such relationships can then be used to forecast TC activity in advance by monitoring the large-scale atmospheric (and oceanic) structure at the beginning of the cyclone season.

It is important to reiterate (see Section 2.6) the fact that initial data, obtained under various averaged conditions, can lead to dissimiliar results (see below). The majority of authors lean toward year-averaged cyclone activity (number of TCs throughout the year), whereas the changeover to month-averaged data can radically alter the results (see Nicholls, 1985, 1992; McBride, 1995).

The pioneering work in this field was by Nicholls (1979, 1984, 1985) for cyclones in the two Australian basins. He demonstrated an association between the Southern Oscillation Index (SOI) during the southern hemisphere winter and the number of TCs close to Australia (from 105–165°E) during the subsequent cyclone season (i.e., from October to April). The number of cyclone days over a season is correlated with mean sea level pressure for the preceding July–September. The linear correlation coefficient for the two series over this 25-year sample is −0.68. Nicholls (1985) demonstrates the robustness of the relationship by calculating separate lag correlations for each 10-year dataset from 1909 to 1982. For all seven 10-year subsets the correlation coefficient between July–September pressure and subsequent October–April cyclone numbers ranged from −0.41 to −0.72 (provided he removed one "bad" data point for the 1943–1944 cyclone season).

However, Nicholls (1992) detects a sudden decrease in cyclone numbers within the region after the end of the 1985–1986 season that was not accompanied by a corresponding decrease in the SOI. Thus, his method would have consistently over-predicted cyclone activity during the 1986–1987 through to the 1990–1991 seasons. This sudden change in the relationship between the SOI and cyclone numbers may have been a real physical change, perhaps a result of changes in satellite imagery interpretation, or maybe inadvertent changes in the SOI. Nicholls (1992) suggests that a possible remedy may be to correlate the trend in the SOI versus the change in cyclone numbers from one season to the next.

It is believed (McBride, 1995) that the physical reason for the association is that the number of TCs during the season has a simultaneous high negative correlation with the large-scale surface pressure in the region (i.e., low surface pressure is consistent with a large number of TCs). Since northern Australia is close to one of the activity centers of the Southern Oscillation, variation in this large-scale pressure is effectively equivalent to variation in the SOI. Predictability (or the lag relationship) comes about through slow variation (or large serial correlation) in the SOI at the turn of the year preceding the southern hemisphere cyclone season.

During an ENSO warm event in the eastern South Pacific the pressure over Australia is high, which leads to a reduced number of cyclones in that region. Revell and Goulter (1986) point out that the frequency of cyclone formation at the eastern end of the Australian/South Pacific basin (i.e., east of 170°E) actually increases during an ENSO warm event. Although the relationship between the formation longitudes in the region and the SOI is weak, it is statistically significant. However, the relationship appears to be dominated by extreme events (i.e., warm events). If the relationship is real, the eastward movement of formation locations can be explained in terms of favorable factors for cyclone formation. During an ENSO warm event the region with SST exceeding 26°C extends much farther eastward across the South Pacific. Of probably more importance is the fact that southern hemisphere monsoon westerlies extend to the dateline, which means that the monsoon shearline extends much farther eastward.

The work by Hastings (1990) and others has clearly demonstrated an eastward displacement of the primary centers of TC activity during El Niño years. TCs extend a considerable distance eastward during strong El Niño episodes and there is a reduced frequency in the Coral Sea and the eastern Australian region. These results agree with the earlier findings of Revell and Goulter (1986).

TC activity during the intense 1982–1983 El Niño provided an extreme example of this eastward shift in TC occurrence. Many more TCs (eight systems) formed east of 180° in the South Pacific in 1982–1983 than in any previous year on record, and activity in the Australia region was much reduced. Hastings (1990), Hanstrum *et al.* (1999), and Shaik and Bate (1999) show that a later than normal start of the South Pacific TC season typically occurs during the year following an El Niño.

A similar but less dramatic change occurs in the western North Pacific basin. A number of authors have studied the association between the SOI and cyclone activity in the western North Pacific basin (Chan, 1985; Zhang *et al.*, 1990; Lander, 1994). In each study, the simultaneous SOI relationship with the total number of cyclones over the basin has been quite weak. However, Chan (1985) reports that the number of cyclones east of 150°E increases when large-scale pressure is high (i.e., during an ENSO warm event). The reason (given by Chan) is the same as given above for the

South Pacific. During an ENSO warm event, the monsoon shearline extends farther eastward than normal, which is a condition conducive for cyclone formation. This was extended by Lander (1994) who show that a large number of monsoon shearline-type cyclone formations occur late in the season of a warm event in a region east of 160°E and south of 20°N. Conversely, during a cold event no formations occur in that southern and eastward part of the basin.

Some limited potential for seasonal prediction has been found in the western North Pacific. Zhang *et al.* (1990) study interannual variation in cyclones in this basin during the peak months of July–October for 1959–1979. They find that 40% of the variance in this series could be explained through regression with the SOI and the trend in the SOI series. However, no significant trend for this basin can be detected in the numbers (McBride, 1995). Inspection of Zhang *et al.*'s data reveals their observed trend is heavily influenced by large values in the first 9 years of their series. Thus, any firm conclusions on the predictability of annual TC occurrence for this basin must await analysis with a longer time series.

As for the Central Pacific, Chu and Wang (1998) examine historical records (1949–1995) of cyclones and classified them into El Niño and non–El Niño batches. A bootstrap resampling method is used to simulate sampling distributions of the annual mean number of TCs for the above two batches individually. The statistical characteristics for the non–El Niño batch are very different from the El Niño batch.

A two-sample permutation procedure is then applied to conduct statistical tests. The results of a testing hypothesis indicate that the difference in the annual mean number of cyclones between El Niño and non–El Niño batches is statistically significant at the 5% level. Therefore, we can say with statistical confidence that the mean number of cyclones in the vicinity of Hawaii during an El Niño year is higher than that during a non–El Niño year. Likewise, the difference in variances between El Niño and non–El Niño batches is also significant. Cyclone tracks passing Hawaii during the El Niño batch appear to be different from those of the non–El Niño composite. A change in the large-scale dynamic environment or the thermodynamic environment is believed to be conducive to the increased cyclone incidence in the vicinity of Hawaii during an El Niño year.

Irwin and Davis (1999) study the relationship between the SOI and TC tracks in the eastern North Pacific (ENP). It is known that most ENP TCs are generated between the Mexican western coast and Clipperton Island (at 110°W), but storms have developed west of 140°W in the ENP.

Previous work on western Pacific storms has noted a shift in the genesis region toward the central Pacific during El Niño events. If the response of ENP TCs to the ENSO cycle is similar to those in the western Pacific basin, Irwin and Davis hypothesize that they would originate and track significantly westward of the mean point of origin.

Irwin and Davis investigate points of origin and downgradation of TCs in the ENP east of 160°W related to the SOI during the hurricane season for 1966–1997. All ENP TSs and hurricanes from 1966 through 1997 were grouped into SOI-based categories: those occurring during strong El Niño events (mean SOI < -0.6), strong

La Niña events (mean SOI >0.6), or near-zero periods ($-0.6 <$ mean SOI < 0.6). During El Niño storm season, ENP TCs originated approximately 5.7° (617 km) west and downgraded 7.5° (780 km) west of the long-term mean longitudes for the positive SOI group. Near-zero group storms also followed more northerly tracks than the negative SOI group storms. However, no significant differences in storm track are evident between the positive SOI and near-zero groups, and the track length is not significantly different for any storm group.

Statistical analysis of the records for the last 40 years of TCs in the Indian Ocean indicates no obvious systematic ENSO-related variation in seasonal TC frequency or location in the North and South Indian Oceans (Gray, 1993). However, more careful studies of Indian Ocean cyclones are needed.

It is a striking fact that TC activity in the North Atlantic is more sensitive to El Niño influences than in any other ocean basin. Gray (1993) indicates that, comparing a non–El Niño year to one with moderate to strong El Niño, the North Atlantic basin experiences:

(1) a substantial reduction in cyclone numbers, especially at low latitudes;
(2) a 60% reduction in the number of hurricane days (defined as the number of days that a hurricane was present in the basin);
(3) an overall reduction in system intensity.

Gray (1993) attributes this reduction in TC activity to the anomalously strong westerly winds that develop in the western North Atlantic and Caribbean region during El Niño years.

Major developments in documenting the predictability of this basin have been achieved (Gray, 1979, 1984a, b, 1993; Shapiro, 1982a, b). The number of TCs in a season has been related to various aspects of large-scale flow, including sea level pressure, the patterns of SST, upper-tropospheric zonal winds, and seasonal rainfall in the Sahel of West Africa. The strongest relationships have been with the Southern Oscillation and with the stratospheric Quasi-Biennial Oscillation (QBO), which is a quasi-periodic reversal of zonal winds over the equator. TC activity in seasons during the west phase of the QBO is a factor of 1.4 greater than during the east phase. Ascribing a simple index of $+1$ for a season in the west phase, 0 for transition seasons, and -1 for the east phase of the QBO, the correlation of this index with the cyclone numbers accounts for 33% of the variance. Similarly, large correlations are found between cyclone numbers and an index of ENSO activity based on SST anomalies over the equatorial eastern Pacific.

The relationship between ENSO and cyclone activity in the North Atlantic basin apparently is quite different. Shapiro (1987) reports a correlation of -0.34 (only 12% of the variance) between cyclone numbers and the warm-water anomaly in the equatorial eastern Pacific. Both Shapiro (1982a, b) and Gray (1984a) provide evidence of a physical link that higher equatorial SST values increase the activity of tropical convection, which increases upper-level westerly zonal winds and the vertical wind shear downstream over the primary formation region of Atlantic cyclones. As discussed above, large vertical shear represents an unfavorable condition for TC formation (on this seasonal timescale).

Important results on the statistical features of cyclone activity in the North Atlantic basin have been obtained in a number of works (Walsh and Kleeman, 1997; Elsner, 1997; Elsner *et al.*, 1999; Elsner and Kara, 1999; Landsea *et al.*, 1999).

Walsh and Kleeman (1997) demonstrate significant relationships on timescales longer than 5 years between northern Pacific SSTs, North Atlantic TC numbers and north-eastern Australian rainfall. In agreement with previous work, the global nature of these correlations suggests that they are indications of an intrinsic inter-decadal mode of oscillation in the global ocean–atmosphere system. The persistence in time of SST variation may enable forecasts to be made of North Atlantic TC numbers several years in advance. A similar capability for the prediction of north-eastern Australian rainfall is also shown. Thus these relationships may be useful as an aid in long-term planning decisions.

The temporal relationship between northern Pacific SSTs and Atlantic TCs is shown in Figure 3.44 where SST values are taken at the point in the North Pacific (42.5°N, 155°E) which exhibited the highest correlation with Atlantic TC numbers (0.76, about 58% of the variance) over the period 1948–1989, using available SST data. Analysis of observations suggests that a similar mode of oscillation of the ocean–atmosphere system has persisted for more than 100 years. In Figure 3.44 both time series show clear variation on interdecadal timescales. The amplitude of

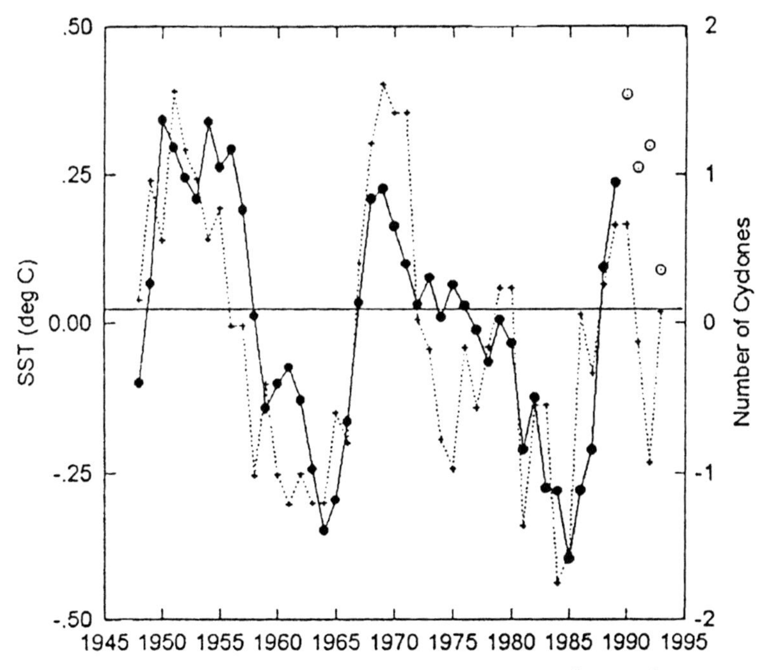

Figure 3.44. Five-year running means of SST anomalies at 42.5°N, 155°E (solid line) and anomalies in the number of Atlantic tropical storms and hurricanes relative to the long-term average (dashed line). Time series are shown over the period 1948–1993; the years 1990–1993 for the SST time series are designated by circles (from Walsh and Kleeman, 1997).

SST variation is not large, about 0.8°C from peak to trough, whereas TC numbers vary from peak to trough by about 30% of their long-term mean of about ten TCs/hurricanes per year.

The main mechanism previously proposed to relate Atlantic SST anomalies to TC numbers in that basin is through the influence of SST on tropospheric vertical wind shear (i.e., the difference between the wind strength and direction in the upper and lower troposphere), with higher shear known to inhibit the formation of TCs (McBride, 1995).

The high correlation between northern Pacific SSTs and total TC numbers in the North Atlantic on a 5-year timescale, combined with the apparent almost regular variability in SST in this region over the last 50 years or longer, suggests that a preliminary forecast of mean TC activity over the Atlantic during the next few years can be made based upon the observed SST anomalies over the last 5 years in the North Pacific region. Figure 3.44 shows an extension of both time series up until 1993, using SST data from 1990 to 1993, although these data are not yet included in the correlation analysis because of unresolved issues regarding homogeneity between the two SST datasets. The peak values of the 5-year running mean of SST in the chosen location were reached in 1990 and have begun to decline since. A return to generally more quiescent conditions (at least in terms of total numbers of TSs and hurricanes) is likely towards the end of this decade and into the next.

The annual record of hurricane activity in the North Atlantic basin for the period 1886–1996 is examined by Elsner *et al.* (1999) from the perspective of time series analysis. Single-spectrum analysis combined with the maximum entropy method is used on the time series of annual hurricane occurrences over the entire basin to extract the dominant modes of oscillation. The annual frequency of hurricanes is modulated on the biennial, semidecadal, and near decadal timescales. The biennial and semidecadal oscillations correspond to two well-known physical forcings in the local and global climate. These include a shift in tropical stratospheric winds between an east and west phase (QBO) and a shift in equatorial Pacific Ocean temperatures between a warm and cold phase (ENSO). These climate signals have previously been implicated in modulating interannual hurricane activity in the North Atlantic and elsewhere. The near decadal oscillation is a new finding. Separate analyses on tropical-only (TO) and baroclinically enhanced (BE) hurricane frequencies show that the two components are largely complementary with respect to their frequency spectra. The spectrum of TO hurricanes is dominated by timescales associated with the ENSO and the QBO, while the near decadal timescale dominates the spectrum of BE hurricanes. Speculation as to the cause of near decadal oscillation of hurricanes is centered on changes in Atlantic SST possibly through changes in evaporation rates. Specifically, cross-correlation analysis points to solar activity as a possible explanation.

We will now show that the method of cumulative events proposed and developed in Section 2.2 is acutely sensitive to ENSO variation. The results are presented in Figure 3.45. We performed the calculations on the basis of initial data for Atlantic tropical activity (Pokrovskaya and Sharkov, 1997a, 1999d). The temporal relationship between the North Atlantic cumulative function (CF), found

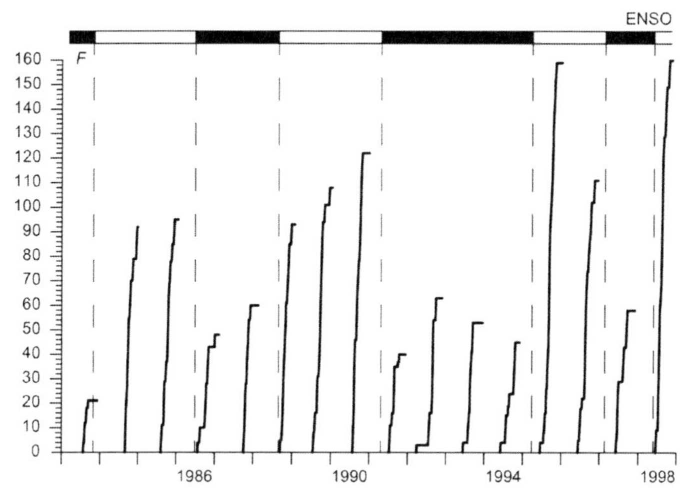

Figure 3.45. Time series of the cumulative functions of Atlantic tropical cyclone activity during 1983–1998. Solid and open bars indicate active and suppressed periods of ENSO, respectively.

from 1983 to 1998, and ENSO episodes is shown in Figure 3.45, where both time series show clear variation on inter-year timescales. The amplitude of the cumulative function (for the given years) for the active ENSO episode is small, whereas for the non-active (suppressed) ENSO episode the amplitude of the CF rises sharply (see non-ENSO episodes of 1995–1996 and of 1998).

Analyses were also performed to contrast the impact that various environmental factors have upon Atlantic TC variability (Landsea *et al.*, 1999). Figure 3.46 is a schematic summarizing the interannual and interdecadal forcing of Atlantic TCs. The largest interannual variation appears to be associated with Caribbean sea level pressures (SLPs) and 200 mb zonal winds: years of low pressures and easterly 200 mb wind anomalies corresponded with more frequent and more intense TCs. Of course, the existence of more intense TCs will directly contribute to lower surface pressures and easterly wind anomalies.

The other environmental factors considered here—ENSO, the stratospheric QBO, West Sahel rainfall, and Atlantic SST—also showed moderate to strong influences on Atlantic TC activity, confirming previous studies. A new finding is that some of the environmental factors including Caribbean SLP, Atlantic SST, and West Sahel rainfall did induce a weak opposite forcing of increased activity in the northerly portion of the basin during years of high pressure, cool SST, and dry West Sahel seasons. In contrast, two of the environmental factors appeared to cause consistent basinwide alterations—ENSO and the QBO—though their effects in southerly latitudes were strongest. Thus, the hypothesis that there is an "out-of-phase" relationship between southerly and northerly latitude storm formations because of vertical shear proves true for West Sahel rainfall, but not for ENSO.

The analysis is not meant to imply that these various environmental factors independently affect Atlantic TCs without interacting with one another. Of the six

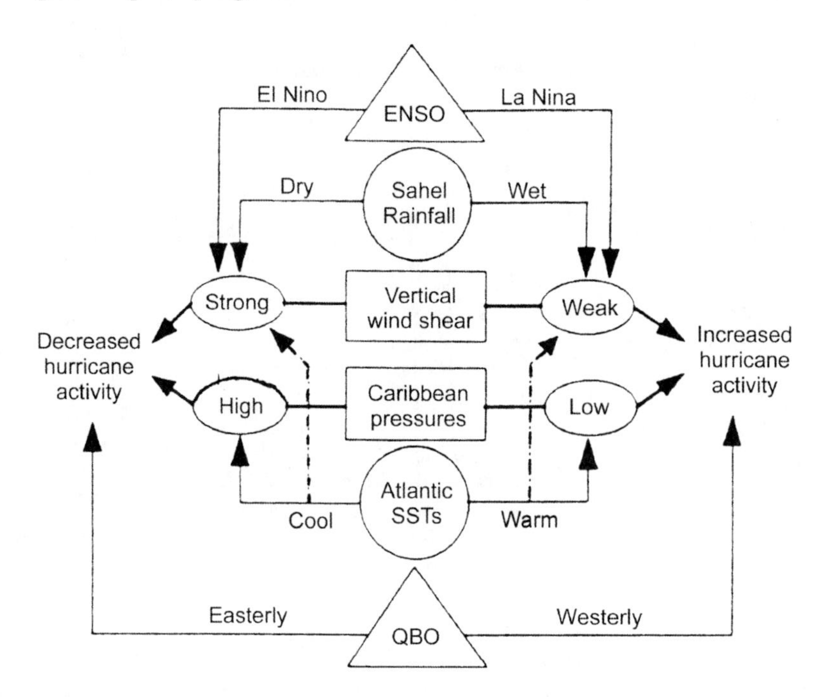

Figure 3.46. Schematic of environmental forcing of Atlantic TCs on primarily interannual (triangles) and interdecadal (circles) timescales. The rectangles indicate the physical factors directly responsible for TC variations (from Landsea *et al.*, 1999).

environmental factors considered here, only five pairs (out of fifteen possible) have a covariance exceeding 25% of the variability: West Sahel rainfall and Caribbean SLP (27%), Caribbean SLP and Caribbean 200 mb zonal wind anomalies (29%), West Sahel rainfall and Caribbean 200 mb zonal wind anomalies (32%), tropical North Atlantic SSTs and Caribbean 200 mb zonal wind anomalies (34%), and West Sahel rainfall and tropical North Atlantic SSTs (38%). The remaining ten pairs of combinations had relationships explaining less than one fifth of the variance between the two environmental factors. While most of these relatively small values of covariability suggest that environmental factors may be independently related to Atlantic TCs, further study is certainly warranted to investigate the interdependence of these conditions.

While Atlantic SST was a weak interannual environmental factor for intense hurricanes, the multidecadal mode of Atlantic SST in contrast corresponds strongly to observed decade-to-decade changes in intense Atlantic hurricanes. In particular, the quiet regimes of 1899–1925 and 1971–1994 are well related to an SST regime of cold North Atlantic conditions. The years 1926–1970, which were distinctly warm in the North Atlantic, correspond to active conditions for Atlantic hurricanes. There was an 80% increase in intense hurricanes during warm North Atlantic decades compared with cold North Atlantic years. These changes have direct impacts on the hurricanes that strike the Caribbean islands and the United States, especially the

Table 3.14. Linear correlation coefficients between the series of TC numbers and the Troup Index of the Southern Oscillation (anomalies of monthly Tahiti minus Darwin SLP differences normalized by the standard deviation of the Tahiti minus Darwin series). The upper row is for the entire 22-year series and the second row is for the series omitting the extreme values (upper and lower) of the SOI series. For the four northern hemisphere basins the SOI has been averaged from July to September, while it is averaged over January to March for the three southern hemisphere basins.

Correlation coefficient	*North Atlantic*	*Eastern North Pacific*	*Western North Pacific*	*North Indian*	*South-west Indian*	*Australia/ Southeast Indian*	*Australia/ Southwest Pacific*
22-year series	+0.31	+0.00	+0.13	−0.04	+0.19	+0.37	−0.01
20-year series	+0.29	+0.20	+0.35	−0.17	+0.07	+0.19	+0.01

intense hurricanes making landfall from the Florida peninsula to New England. U.S. normalized damage between the two SST regimes is striking, with a factor of 4 increase in median damage during warm North Atlantic decades.

There are noticeable differences in the physical causes of ENSO-induced variation in TC activity between the Australian and North Atlantic regions (Gray, 1993). Whereas the strength of upper-level westerly winds and the vertical shear mechanism appears to account for TC reduction in the Atlantic in El Niño years, differences in SST and surface pressure seem to provide the primary physical linkage in the Australian region. Here, cool SST anomalies and associated high barometric pressure accompany El Niño events and are associated with the diminished frequency of Australian Coral Sea cyclones. By comparison, El Niño events appear to cause no significant alteration in Atlantic SST or SLP.

For completeness, the time series of TCs in each active basin in Section 2.6 and in work by McBride (1995) have been correlated with the simultaneous SOI (Table 3.14). Positive correlations are calculated for four of the seven basins using the 22-year series. Although the largest values are for the Australian/southeast Indian basin and for the North Atlantic, the percentage of explained variance is small (14% and 10%, respectively). Inspection of the SOI time series for January–March showed that the 1982–1983 El Niño value was well outside the range of values for the remainder of the series, which may have an undue influence on linear correlations. To test the robustness of the correlations, the years with extremely positive and extremely negative SOI values were excluded from the time series and the correlation analysis was repeated. The large change in correlation coefficients between the 22-year series and the 20-year series indicates some caution is necessary when inferring physical relationships based on this sort of time series.

Using consistency between the two rows of Table 3.14 as a criterion for robustness, three basins appear to have consistent correlations between cyclone numbers and the SOI: the Australian/southeast Indian, the North Atlantic, and the western North Pacific. During an ENSO warm event, when the pressure is high at Darwin, the number of cyclones is suppressed near Australia and in the North Atlantic. For most climatic parameters, the relationship with the SOI has the opposite sign in the West Pacific from that in the West Atlantic. The reason for the consistent sign with TCs is because of the two very distinct physical mechanisms explaining the relationships (as described above).

Longitudinal displacements in formation areas are not shown in Table 3.14 because of conflicting basin boundary definitions. In particular, Nicholls (1984, 1985) indicates a correlation coefficient of approximately 0.7 between the SOI and the combined number of TCs in the Australian/southeast Indian basin and the western end of the Australian/southwest Pacific basin. Although a smaller positive correlation is calculated for the Australian/southeast Indian basin (Table 3.14), a zero correlation is found for the entire Australia/southwest Pacific basin, which is sufficiently large that a decrease in cyclone numbers at the eastern end of the basin (following Nicholls, 1992) offsets an increase in cyclone numbers at the western end (following Revell and Goulter, 1986) during an ENSO warm event.

The ENSO modulation of total TC frequency and intensity is strongest in the North Atlantic basin. Strong zonal displacements in TC location occur in the central and western Pacific and a later start of the TC season typically occurs in years following an El Niño. Table 3.15 summarizes ENSO influences by region and provides a basin–forecast relationship (Gray, 1993). If a significant El Niño (or

Table 3.15. Recommended seasonal TC activity forecasts by region during a moderate or strong El Niño event or a moderate or strong non–El Niño event.

	El Niño years		*Non–El Niño years*	
Cyclone basin	*Frequency*	*Intensity*	*Frequency*	*Intensity*
North Atlantic basin	Large decrease	Small decrease	Small increase	Small increase
Eastern North Pacific basin	Slight increase	Increase	Slight decrease	Decrease
Western North Pacific basin				
Eastern part	Increase	No change	Decrease	No change
Western part	Decrease	No change	Increase	No change
North Indian Ocean	No change	No change	Increase	No change
South Indian Ocean	No change	No change	No change	No change
Australian region				
Western	Slight decrease	No change	Slight increase	No change
Central and east	Decrease	Slight decrease	Increase	Slight increase
South and Central Pacific (>160°E)	Increase	Increase	Decrease	Slight decrease

non–El Niño) event is occurring or is expected to occur, then the anticipated seasonal TC activity and intensity can be qualitatively altered as specified in this table. Care needs to be taken, however, as it is to be expected that some El Niño and non–El Niño years will not fit these guidelines. The forecasts should be used for general guidance only.

4

Global tropical cyclogenesis and global change

The time rate of generation and evolution of atmospheric catastrophes—tropical cyclones (TCs)—on climatic scales represents a serious and still unsolved problem. The present chapter outlines a series of important questions concerning global tropical cyclogenesis and telecommunication links on the climatic scale in the ocean–atmosphere system. First, the stable integrated mode of multiple cyclogenesis generation for the 25-year period (1983–2007) is revealed both in the cyclone-generating water areas of the World Ocean and in the water areas of the northern and southern hemispheres where there are universal generation constants. These constants were found to be not dependent on the state of internal telecommunication links in the ocean–atmosphere system. The dependence on El Niño–Southern Oscillation (ENSO) episodes for regional cyclogenesis (the North Atlantic) in the annual accumulation mode is considered. Second, statistical features of the tropical cyclogenesis of primary and advanced (tropical cyclones) forms of tropical disturbances that have arisen in World Ocean water areas during their formation in the ocean's surface temperature field are analyzed—this temperature being considered as a long-term monthly average and three-monthly average in each particular analyzed year. Detailed analysis of the generally accepted hypothesis of "critical temperature", as the first necessary condition for TC onset, is carried out. The third set of questions considered in this chapter concerns studying the vertically integrated water vapor amount in the equatorial zone as an energy source for the genesis and intensification of individual tropical cyclones, and also as a spatiotemporal basis for the trajectory evolution of TCs.

4.1 UNIVERSAL GENERATION CONSTANT FOR A STOCHASTIC MODE OF GLOBAL TROPICAL CYCLOGENESIS

4.1.1 Statement of the problem

We have already addressed the fact that the tropical zone of the global ocean–atmosphere system plays a key part in the dynamics and evolution of synoptic

E. A. Sharkov, *Global Tropical Cyclogenesis* (Second Edition).

and climatic meteorological processes on the Earth. The ocean–atmosphere system of Earth's tropical zone possesses the unique property of being able to generate well-organized and stable mesoscale vortical structures—tropical cyclones (TCs)—from the atmospheric spatiotemporal chaos in the global circulation system. Purposeful remote sensing of such complicated structures requires, first of all, sufficiently clear understanding of the spatiotemporal picture of this phenomenon as a multiple process. However, when considering these structures as a temporal flux of events, they themselves represent the spatiotemporal chaotic signal with a rather complicated internal multi-correlation structure, that was revealed for the first time in the work of specialists at the Space Research Institute of the Russian Academy of Sciences (SRI RAS) (Sharkov, 1996a–d, 1997, 1998, 2000, 2006, 2010). Simultaneously, attempts were undertaken to describe the statistic regularities of tropical cyclogenesis by means of more general theoretical-probabilistic consideratons (Golitsyn, 1997, 2008). The reason such systems are studied—in particular, the attempts to reveal possible determinate components on both global and regional scales—is explained by many circumstances. First of all, these atmospheric processes represent a direct physical hazard to humankind and often cause considerable material damage, not to mention the administrative–social problems that become more complicated as humankind develops (Elsner and Kara, 1999) Humankind has long considered tropical cyclones the most destructive components of the ocean–atmosphere system causing significant material losses and human victims. Serious efforts have been made (mainly in the U.S.A.) to suppress this form of ocean–atmosphere system activity by any technical means available. All such efforts have completely failed (Willoughby *et al.*, 1985; Gray, 1997). As a result of this, the basic attention of the Western scientific community has switched to the study of the paleoactivity (paleoclimate) of tropical cyclogenesis (among the recent works in this direction are Mann *et al.*, 2009; Fedorov *et al.*, 2010). On this basis, active attempts have been made recently to forecast cyclogenesis from the viewpoint of both individual cyclogenesis and the regional components of global cyclogenesis (mainly in the North Atlantic water area for obvious administrative and geographical reasons). The flux of events here is represented in its simplest form (and, as we shall show later, least informative form)—namely, as a set of individual (delta-shaped over time) structural elements for the time interval under study: tropical cyclones (Elsner and Kara, 1999; Chylek and Lesins, 2008; Mock, 2008; Knutson *et al.*, 2008; Semmler *et al.*, 2008).

In contrast to the Western scientific community, however, the wider scientific community has come up with a different concept. This was initially based on the naive notion that nature has no wish to intentionally "harm" the human community, and that the stable and now proven to be historically long (at least since the time of the last glacial age) functioning of tropical cyclogenesis was caused by something else. It is only through recent investigations using space remote-sensing data and the latest achievements in the theory of complex systems which point to this principally different view of tropical cyclogenesis that we can state with a high probability that the determining role of tropical cyclones in carrying out global mass transfer and energy transfer in the global ocean–atmosphere system and in regulating the greenhouse effect on Earth is favorable for biological life (including humankind). Thus, global tropical cyclogenesis is most likely a necessary and, probably, determining factor in the ecological equilibrium (understood in a broad sense) both in the geophysical ocean–atmosphere system and in the ecological systems of the Earth. Catastrophic atmospheric vortices represent a peculiar mechanism to effectively "dump" excessive heat in the atmosphere under conditions in which the actions of the usual mechanisms—turbulent convection and global circulation—for some reason clearly become insufficient. Thus, catastrophic phenomena play an important (and, as paradoxically as it may seem, useful to humankind) role in establishing the climatic temperature regime of the Earth (the greenhouse effect), removing excessive heat and promoting the prevention of excessive overheating of the planet in the tropical zone.

The purpose of this section is to reveal—on the basis of using our approach to form a time flow of tropical events over 25 years (from 1983 to 2007)—the degree of stability of integrated and differential regimes of multiple cyclogenesis generation both in the cyclone-generating water areas of the World Ocean and in the water areas of the northern and southern hemispheres, as well as the dependence of the generation rate of cyclones on the features of global circulation—mainly, on El Niño–Southern Oscillation (ENSO) episodes.

4.1.2 Modern approaches and signal formation principles

At present, study of the genesis and time evolution of stable vortical systems as a flux of events within the framework of global circulation and the turbulent chaos of the tropical atmosphere is developing in two principal directions (Sharkov, 2000):

- *the "local" approach (individual cyclogenesis)*—used to study the formation of a single (individual) vortical structure from wave motions in the atmosphere and turbulent chaos under conditions when there is a local and strong imbalance in ocean–atmosphere systems; and
- *the "global" approach (multiple cyclogenesis)*—which considers the formation of vortical systems in World Ocean water areas as a set of centers of generation of vortical systems in the active medium of the natural ocean–atmosphere system (in this case the atmosphere is considered globally). This approach was proposed by co-workers at the Space Research Institute of the Russian Academy of Sciences (SRI RAS) in 1993 (Pokrovskaya and Sharkov, 1993a, b) and is currently being successfully developed (see Chapters 2 and 3 for more details).

Tropical cyclogenesis considered globally still remains a rather poorly studied physical process, nevertheless, some serious and non-trivial results have already been obtained on the basis of the multiple cyclogenesis concept offered by SRI RAS specialists (Sharkov, 2000). Obviously, the structural basis of multiple cyclogenesis studies should be the technique used to construct the time series of global tropical cyclogenesis—the physical process considered simultaneously over the whole water area of the World Ocean (or over the water areas of both hemispheres).

However, this seemingly simple question—how can the time sequence of a TC be simulated?—is not as trivial as it may seem; what is more, it is of principal importance, because its solution influences the physical significance of the final result.

Once again, we shall briefly repeat the methodology and principles of the signal formation, because these circumstances are very important in the given approach. Disregarding the detailed structure and dynamics of each individual tropical formation, we shall present on the time axis each tropical disturbance as a pulse of unit amplitude with random duration (corresponding to the TC functioning time) and with random instants of appearance (individual TC generation). The number of arrived pulses (events) per unit time interval (a day in our case) is, in such a case, the natural physical parameter—the intensity of global cyclogenesis that determines the energy of the ocean–atmosphere interaction.

Mathematically, the proposed signal formation procedure can be written as follows (Sharkov, 2000):

$$I(t) = \sum_{t} \theta(t - t_i; \tau_i), \tag{4.1}$$

where $I(t)$ is the instantaneous cyclogenesis intensity (the number of active TCs in a day); and θ is the restricted Heaviside function:

$$\theta(t - t_i, \tau_i) = \begin{cases} 1, & t_i < t < t_i + \tau_i, \\ 0, & t_i + \tau_i < t < t_i. \end{cases} \tag{4.2}$$

where τ_i is the lifetime (the time of existence) of an individual TC; t_i is the time of its formation (generation); and $t_i + \tau_i$ is the time of its dissipation.

A sequence of pulses generated in such a manner is little more than an integer random temporal flux of indiscernible events. Thus, our analysis is based on the notion of a time sequence of global tropical cyclogenesis intensity as a statistical signal of complicated structure (Sharkov, 2000).

In our study of possible systematic variations of the time course of the intensity of a flux of events (which can also be non-stationary in the majority of real geophysical natural objects) we shall use a graphical construction of the function of the accumulated number of events during the observed interval (Cox and Lewis, 1966):

$$F(t) = \sum_{k=1}^{M} I_k(t) \vartheta_0(t - t_k) \tag{4.3}$$

where $I_k(t)$ corresponds to Eq. (4.1); t_k is the time instant at which the accumulation function underwent positive variation; M is the number of days in the considered period; and $\vartheta_0(t - t_k)$ is the Heaviside function that is equal to

$$\theta_0(t - t_k) = \begin{cases} 1, & t \geq t_k, \\ 0, & t < t_k. \end{cases} \tag{4.4}$$

An important property of the accumulation function is the fact (Cox and Lewis, 1966) that the ratio between an incremental function for the Poisson process and the time of observation is none other than the average value of the intensity of the

Poisson flux of events (for the given observation interval). Thus, study of the accumulation function on various scales (observation times) allows us to obtain the quantitative characteristics of the intensity of the Poisson flux of events on various observational scales.

As the unit interval increases up to a week or month, the contribution of an individual tropical cyclone's lifetime to the formed signal sharply decreases (because of the cyclone's own limited lifetime). If the unit interval is chosen to be one year (or even a decade when studying the cyclogenetic paleoclimate), the contribution of the lifetime of an individual formation to signal formation virtually disappears. However, it is obvious from analysis of the signal structure (Eq. 4.1) that, depending on the scale of the phenomenon studied, the contribution of the lifetimes of individual cyclones may have an essential effect on the intra-correlation properties of the process (Eq. 4.1). Nevertheless, in the overwhelming number of studies—among recent works we recommend Elsner and Kara (1999); Chylek and Lesins (2008); Mock (2008); Knutson *et al.* (2008); Semmler *et al.* (2008)—devoted to studying the time structure of cyclogenesis, the lifetime of an individual cyclone is accepted as a delta function. Therefore, cyclogenesis intensity (Eq. 4.1) is essentially the quantity of cyclones in the considered interval (the re-calculating scheme). As is well known from the statistical theory of radiophysical signals (Rytov, 1966), in such a construction of a signal's model structure the internal correlation properties of a system will initially be lost, and the cross-correlation dependences between cyclogenesis and other physical parameters cannot be obtained. This mainly concerns the numerous (but, note, unsuccessful) attempts to find a stable cross-correlation link between cyclogenesis and solar activity, which in turn is also determined by means of a delta-shaped signal in the form of a Wolf numbers (Elsner and Kara, 1999). However, the use of cyclogenesis intensity in the form of a quasi-continuous function (Eq. 4.1) and the solar activity parameter in the form of a continuous function of radioemission intensity at a frequency of 10.7 GHz has led to principally different results from those of cross-correlation wavelet analysis (Afonin and Sharkov, 2003) (see Chapter 5).

For our purposes, initial data on the global cyclogenesis of mature TC forms for 1983–2005 were taken from the systematized database (DB) of remote observations of global tropical cyclogenesis: the Global-TC database. This database was generated and developed in the SRI RAS by applying the scenario principle of exploring both satellite and ground-based data (Pokrovskaya and Sharkov, 2006; Pokrovskaya *et al.*, 2004) (see Section 8.6). A database of mature forms of tropical disturbances over all World Ocean water areas for 2006–2008 was set up on a computer platform of the Global-TC DB using object analysis (i.e., pre-processing) (see Section 8.4 for more details) by systematizing and archiving primary information on the tropical zone state that has been regularly (daily) obtained throughout the observation period via the Internet from regional centers of typhoon hazard forecasting that collate satellite data and water surface measurements.

Note that Sharkov (2000) shows that a random process generated in such a manner represents a typical telegraph process. Moreover, the same work shows that the probabilistic structure of fluctuations in the amplitude of the studied flux is close to the structure of a Poisson-type flux in the presence of certain deviations from the Poisson distribution. The latter circumstance, as is well known (Sharkov, 1996a–d), plays a principal part in the analysis of components of a studied stochastic process, revealing the relationship (and competition) between the kinetic and diffusion components of a process.

As tropical cyclogenesis clearly represents a manifestation of the complicated, nonlinear behavior of a unified thermo-hydrodynamic system (i.e., the Earth surface–atmosphere system) that possesses its own intrinsic dynamic properties, characteristic timescales, and (possibly) resonant frequencies (Afonin and Sharkov, 2003), we shall try to compare those features of cyclogenesis time variation that have the brightest and strongest manifestations of telecommunication links in the Earth climate with the ENSO phenomenon (McPhaden *et al.*, 2006).

4.1.3 Universal constant for global and hemisphere cyclogeneses

To reveal the scaling features in the time variation of a flux of events, Figure 4.1 presents in graphical form the signal accumulation function for global cyclogenesis considered for all active water areas of the World Ocean (WO) and for cyclogenesis in the active water areas of the northern hemisphere (NH) and southern hemisphere (SH) for three timescales: 25-year period (1983–2008), 5-year period (1983-1987), and 1-year period (1983). Analysis of the data in Figure 4.1 shows that the behavior of $F(t)$ on various timescales changes radically. So, global cyclogenesis for the 25-year time cycle basically represents a strictly homogeneous process with the universal constant of process intensity both on the global scale ($dF/dt = 1.64$ L/day) and for cyclogeneses developing in the water areas of both hemispheres. So, the intensity of cyclogenesis in the northern hemisphere will be 1.14 L/day and 0.5 L/day for cyclogenesis in the southern hemisphere. Small wavy "ripples" in the basic course of the accumulation function are associated with semi-annual variations of cyclogenesis in both hemispheres. Note that telecommunication links in the Earth's climatic system have no effect on the running of global cyclogenesis, and the universal constant of cyclogenesis remains a constant quantity. A similar construction performed earlier by the author during the 10-year period (Sharkov, 2000) has shown that the integrated intensity of global cyclogenesis remained a strictly constant quantity, equal to 1.64 L/day on this time interval as well.

When considering the 5-year time cycle of cyclogenesis within the basic process framework, the alternating character of cyclogenesis generation becomes explicit (i.e., "silence" intervals alternating with generation intervals; Sharkov, 2000). When considering the 1-year time cycle (Figure 4.1) the well-known features of cyclogenesis in the both hemispheres are clearly revealed. To describe cyclogenesis rates quantitatively one should introduce both a differential description of generation intensity and average annual intensity (Sharkov, 2000).

To demonstrate the poor efficiency of the popular re-calculating scheme, Figure 4.2 presents diagrams constructed in this scheme's format of a quantity of TCs functioning in all the active water areas of the World Ocean (W) and in the active water areas of the NH and SH for the 26-year period (1983–2008). This cyclogenesis

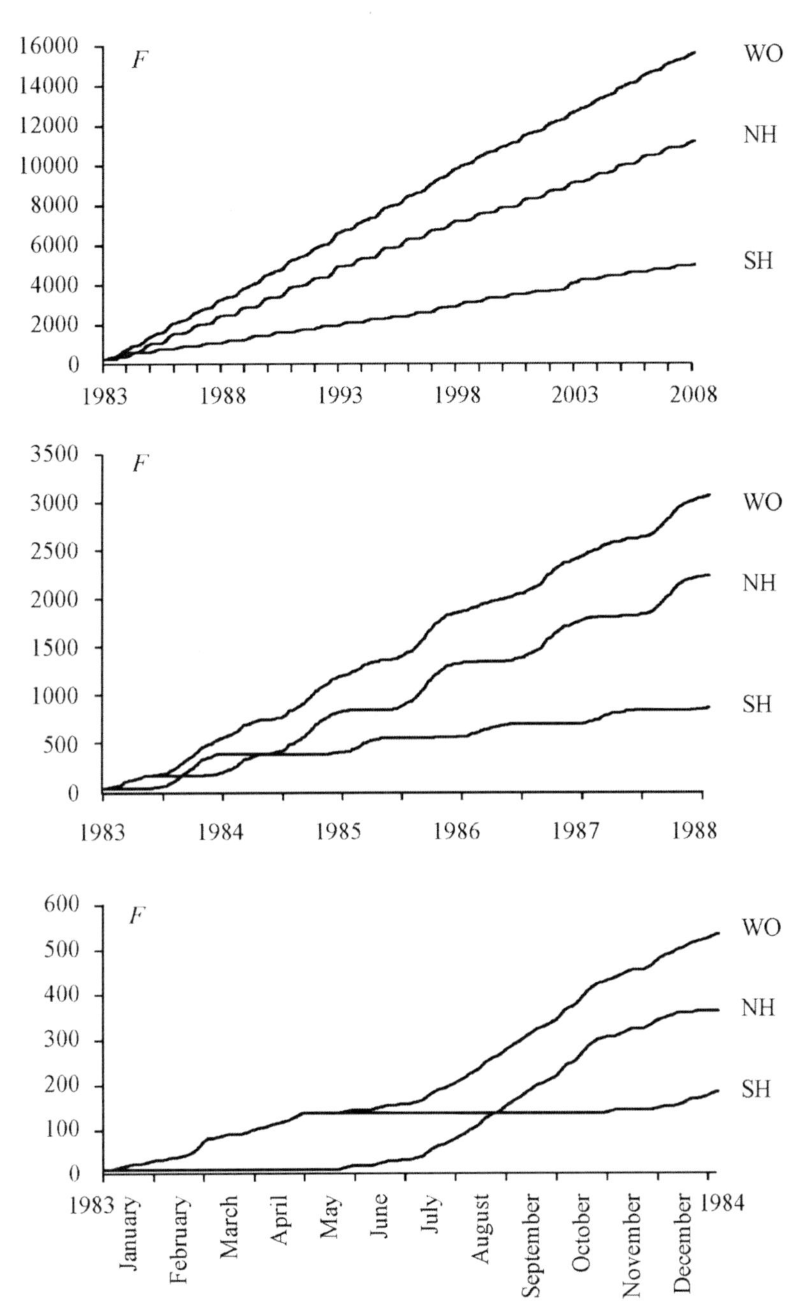

Figure 4.1. Signal accumulation function for global cyclogenesis, considered for all active basins of the World Ocean (WO), and for cyclogenesis in active basins of the northern and southern hemispheres (NH, SH) over three time periods. The upper diagram corresponds to the 25-year period (1983–2007), the middle diagram corresponds to the 5-year period (1983–1987), and the lower diagram corresponds to a single-year period (1983).

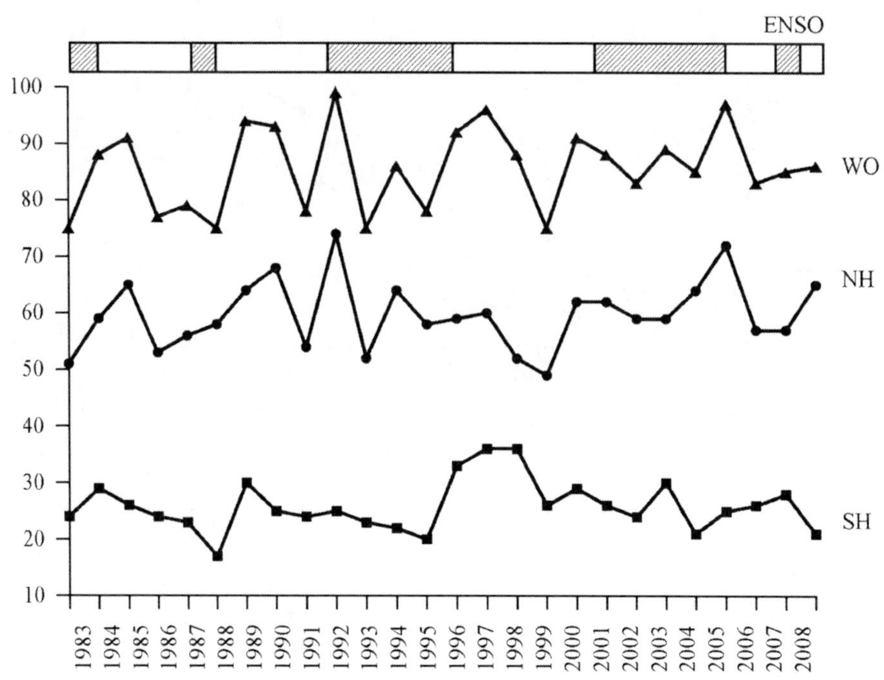

Figure 4.2. Global cyclogenesis intensity, considered for all active basins of the World Ocean (WO), and for cyclogenesis in active basins of the northern and southern hemispheres (NH, SH) during a 26-year period (1983–2008). Hatched and clear rectangles correspond to active and depressed time periods of the ENSO phenomenon (McPhaden *et al.*, 2006).

parameter possesses rather high variations both on the global scale and the scale of Earth's hemispheres. No clear evidence of the influence of telecommunication links in the Earth's climatic system on the course of global cyclogenesis (in the recalculating scheme format) were found. Active attempts to form by any means linear trends in these dependences have been undertaken in many works (among the most recent are Chylek and Lesins, 2008; Mock, 2008; Knutson *et al.*, 2008; Semmler *et al.*, 2008). However, even the diagrams presented for the 25-year term are evidence that no linear trends could in fact be constructed. The construction of linear trends will depend highly on the particular time interval to be analyzed and, hence, they cannot be considered a reliable forecasting element in the long run (i.e., 25–100 years).

Figure 4.3 presents the average annual rates of cyclogenesis generation for the 25-year term and the maximum values of differential values of cyclogenesis intensities for the active water areas of both hemispheres. Such a choice of parameters is brought about by the presence of high intra-year variability in the rates of generation of hemisphere cyclogeneses (Figures 4.1, 2.8, and 2.10). The average annual rates of TC generation in the World Ocean vary little around the long-term average value of the global cyclogenesis generation rate (the universal constant). The maximum

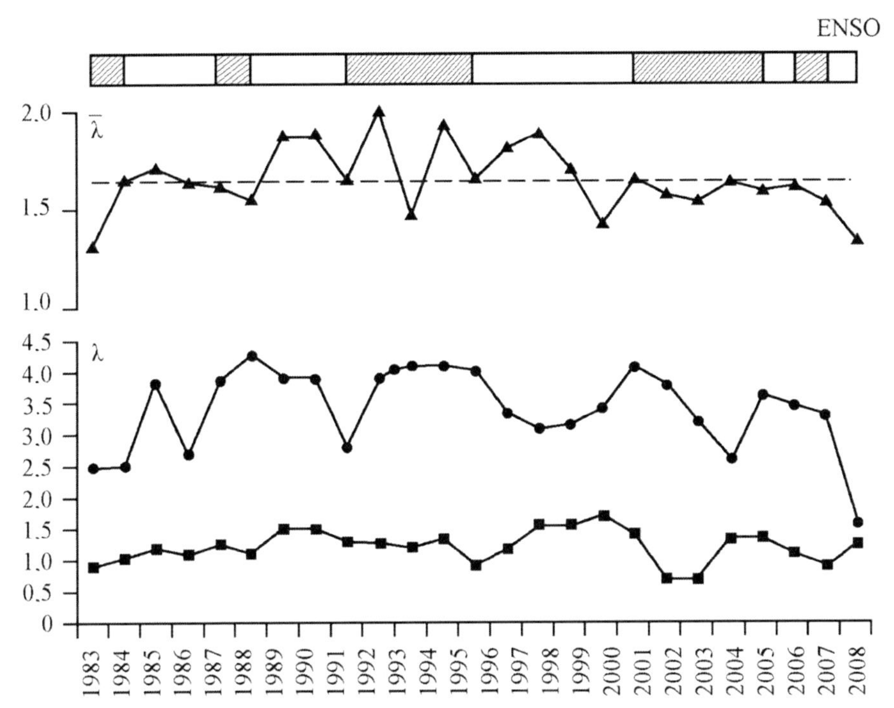

Figure 4.3. Time series of the annual mean of global cyclogenesis intensity and maximum values of differential intensities for cyclogenesis in active basins of the northern and southern hemispheres during a 25-year period (1983–2007). The dashed line is the multiyear mean of global cyclogenesis intensity (1983–2007). Hatched and clear rectangles correspond to active and depressed time periods of the ENSO phenomenon (McPhaden *et al.*, 2006).

values of differential cyclogenesis intensities for the active water areas of both hemispheres also vary little. No clear evidence of the effect of telecommunication links in the Earth's climatic system on the parameters involved in the course of global cyclogenesis were found. Moreover, there is a general feeling that the effect of global circulation on tropical cyclogenesis is virtually unnoticeable. However, we shall show in the following that this is not the case.

The effect of telecommunication links in the Earth's climatic system on tropical cyclogenesis is not only noticeable, it may also be a key factor for regional cyclogeneses. The problem consists in correct selection of parameters and in representing them adequately. Since the telecommunication links of the ENSO phenomenon and circulation features of the atmosphere over the North Atlantic are known (McPhaden *et al.*, 2006) to be prominent, one should expect a similar situation for regional North Atlantic cyclogenesis as well. Figure 4.4 presents time variation in the annual value of the function of tropical cyclogenesis accumulation in North Atlantic water areas for the 16-year term of observations (1983–1998). The same figure shows the active and passive phases of ENSO for the same term. Analysis of the figure's data indicates that active ENSO phases suppress very highly the values of the

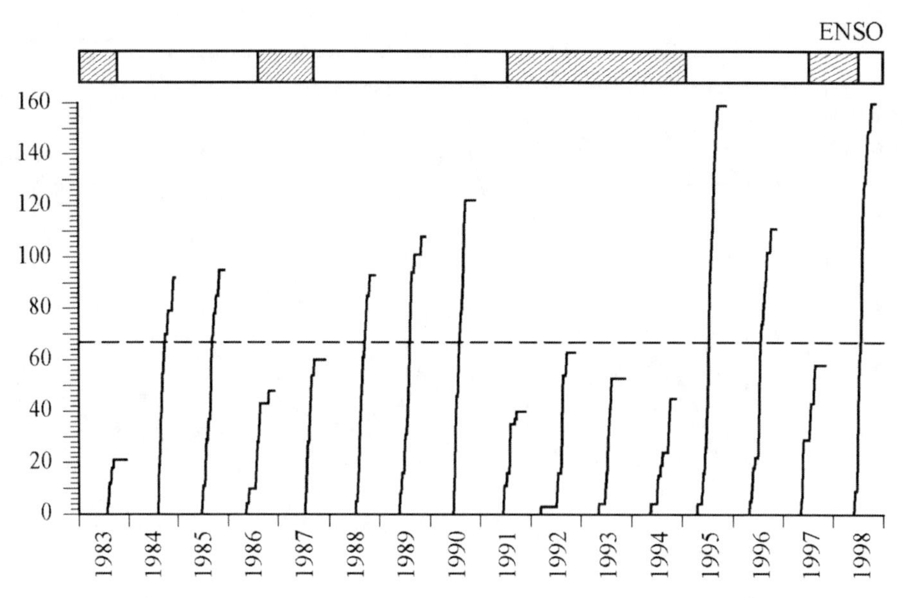

Figure 4.4. Time series of the annual mean of the accumulation function of North Atlantic cyclogenesis during a 16-year period (1983–1998). The dashed line is the critical value of the accumulation function ($F = 66$) during active ENSO periods. Hatched and clear rectangles correspond to active and depressed time periods of the ENSO phenomenon (McPhaden *et al.*, 2006).

accumulation function; moreover, one can even indicate the critical value of the accumulation function ($F = 66$) above which the accumulation function does not assume their values in the active phase of ENSO.

The results presented in this section make it possible to apply a principally new approach to solving the problem of ascertaining the time rate of generation and evolution of atmospheric catastrophes—tropical cyclones (TC)—on climatic scales.

Thus, on the basis of this approach—related to forming the time flux of tropical cyclones as a pulse of unit amplitude with random duration and random moments of appearance and considered for the 25-year period (1983–2007)—the stable integrated mode of multiple cyclogenesis generation is revealed both in the cyclone-generating water areas of the World Ocean and in the northern and southern hemisphere water areas. The intensities of the cyclogenesis process considered both on the global scale and on the hemisphere scale represent the universal generation constants, which do not depend on telecommunication links in the Earth's climatic system. Explicit dependence on ENSO episodes is only revealed for regional cyclogeneses (the North Atlantic) and in a certain mode of annual accumulation. But this telecommunication link is completely absent for global cyclogenesis, where the integrated generation mode for the 25-year period is considered.

4.2 TROPICAL CYCLOGENESIS AND SEA SURFACE TEMPERATURE FIELDS (GLOBAL AND REGIONAL SCALES)

Analysis of the results of remote and hydrological observations for 1983–2003 shows the basic characteristic structural features of the genesis of those tropical cyclones that arose in the cyclogenesis basins of the World Ocean (global and regional cyclogenesis). These are considered in this section. A search for the rigid boundary at which cyclones form in the field of sea surface temperature (SST) is carried out. This temperature is considered as both a monthly averaged long-term temperature (in situ measurements of temperature in the upper quasi-homogenous layer of the ocean) and as a 3-monthly averaged temperature (remote measurements of the ocean's surface skin layer temperature) for each analyzed year. Regional cyclogenesis is shown to possess some peculiar ranges of surface temperatures at which the processes involved in the generation of primary TC forms ("smeared" ranges, extrema with tails, delta-shaped forms of histograms) take place. The stable character of statistical histograms of the intensity of TC genesis distributions on the ocean's surface temperature at the instant of transition into mature forms of tropical disturbances (SST monthly averaged long-term values), both in a 21-year sample (1983–2003) and in a 5-year sample (1999–2003), in all active basins of the World Ocean is demonstrated. Thus, the use of a single criterion for the cutoff temperature, determined by remote IR measurements for cyclone-generating basins of the World Ocean, is non-productive.

4.2.1 Statement of the problem

In recent years, efforts to resolve the Earth remote-sensing problem have taken the direction of studying parameters of the geophysical environment that relate to the various time phases of evolution of natural accidents. This mainly concerns natural atmospheric catastrophes, such as TCs, whose effect involves considerable material damage.

The remote study of primary forms of TCs and their geophysical environment occupies a special place in programs designed to monitor tropical disturbances remotely. One should mention especially the problems involved in forecasting the appearance of primary forms of a disturbance and subsequent transition of an individual primary tropical disturbance into a well-developed form of TC, as well as detailed remote study of the structural, dynamic, and thermodynamic features of a tropical disturbance at the instant one of its primary forms suddenly—and unexpectedly for observers—intensifies. TC "Katrina" is an impressive example of such rapid intensification—one that entailed a loss of lives and great material damage.

However, attempts to study the primary forms of tropical disturbances remotely face a number of difficulties, the principal one of which is the absence of a generally accepted physical model of this complicated geophysical phenomenon and, accordingly, the necessary geophysical parameters subject to measurement. Despite researchers making considerable efforts to observe and record individual (and fragmentary) optical and IR images of tropical vortical disturbances in various phases (see, e.g., Sharkov, 1997, 1998, 2000, 2006), the generally accepted remote criteria of the geophysical environment's "proximity" to generating an individual tropical disturbance and to the critical moment of transition to the well-developed form are still missing. One should consider the results of complex multichannel optical microwave satellite investigations of the evolution of an optical TC image in the field of integral water vapor as a principally new step in studying remote criteria of TC genesis, whose analysis resulted in discovering the fundamental contribution of the horizontal transfer of water vapor by jet fluxes (Sharkov *et al.*, 2008a, b).

In contrast, Gray (1979) promulgates the idea of a set of so-called "necessary" (and, substantially, phenomenological) geophysical parameters that mesoscale vortical steady systems should generate in the tropical atmosphere (in the climatological aspect). This set is considered to be the classical set and represents an indispensable attribute of the majority of publications concerning the discussion of TC generation (see, e.g., Palmen and Newton, 1969; Tarakanov, 1980; Golitsyn, 2008). One of the major points of this set—which is often called "the first necessary condition for the appearance of typhoons" (see, e.g., Tarakanov, 1980; Golitsyn, 2008; Dobryshman and Makarova, 2004)—is the high value of surface temperature, which should exceed (necessarily) 26°C and fall between 26.3°C and 26.8°C (the so-called critical temperature, or the "cutoff" temperature) at the deep upper quasi-homogenous layer of the ocean (a deep thermocline). However, it needs to be said that authors do not present any serious experimentally substantiated proofs, referring to Palmen's theoretical instruction made in 1948 and reproduced in 1969 (Palmen and Newton, 1969). In addition, authors generally do not explain what temperature field they have in mind: that measured by standard oceanological techniques at a depth of 1 m to 2 m or that measured by remote (IR thermal and microwave) techniques—the so-called temperature of the surface skin layer. The quantitative value of a "sharp cutoff" temperature and the physical reasons for the appearance of a critical temperature have been discussed at different levels: from scientific literature to popular scientific publications and Internet sites intended for the general public.

But, there are other points of view. Some authors (Henderson-Sellers *et al.*, 1998) call the idea of a critical temperature a "fallacy", but fail to give any appropriate strict experimental proofs in support. Golitsyn (2008) supposes it expedient to consider "the energy flux from the ocean to the atmosphere of value 700 W/m^2 and higher" as "the first necessary condition for TC occurrence"—not the ocean surface temperature. However, the author obtains this value directly under equilibrium conditions (Clapeyron–Clausius equations) and for mean tropical climatological parameters.

The search for a critical temperature is topical in the field of remote sensing since, when proving the presence of a sharp cutoff in the World Ocean's surface temperature, it is possible to use automatic remote detectors that could essentially simplify the spatiotemporal predictability of crisis situations.

In previous publications (Sharkov and Pokrovskaya, 2006) the authors have shown by analyzing the results of remote and hydrological observations for 1983–2003 that the basic characteristic structural feature of the cyclogenesis of primary and well-developed (TC) forms of tropical disturbances, which arose in the water areas of the World Ocean (considered both globally and on the hemisphere scale) for this period, was the absence of a strict boundary (at TC formation) of the ocean's surface temperature when considered both as a monthly averaged long-term temperature (in situ measurements of temperature in the upper quasi-homogenous layer of the ocean) and as a 3-monthly averaged temperature (remote measurements of the ocean's surface skin layer temperature) for each particular analyzed year. Statistical histograms of the distributions of ocean surface temperatures at the instant of transition into mature (tropical storm or TS) forms of tropical disturbances (SST monthly averaged long-term values), both in a 21-year sample (1983–2003) and in a 5-year sample (1999–2003), in the basins of the World Ocean are demonstrated to be stable. In addition, when considering the situation in regional cyclone-generating water areas, the influence of a temperature field will be significant (depending on hydrological and geographical conditions) and, undoubtedly, reflected in the form and quantitative characteristics of histograms.

The purpose of this section is to present—by comparing the spatiotemporal fields of generation of TC primary forms in the surface temperature field determined by means of standard oceanologic in situ measurements with the fields of tropical disturbance transition from the initial stage to the mature stage determined using remote IR satellite measurements over six regional cyclone-generating water areas of the World Ocean for 1983–2006—experimental results (expressed in detailed histograms) indicating the presence of the wide (and unique to each region) range of surface temperatures at which the processes involved in mature form generation occur and the absence of a critical (threshold) temperature on the regional scale of tropical cyclogenesis.

4.2.2 Observational data and their processing

On close analysis, comparison of TC genesis and SST fields appears to be both complicated and ambiguous, because it is necessary to ascertain correctly the spatiotemporal variability of two stochastic processes: surface temperature and global tropical cyclogenesis, which possess essentially different time variability (Sharkov, 2000; Timmermann, 2003; Semmler *et al.*, 2008). This is the reason the approaches should be completely different and, accordingly, the results of investigations are scarcely comparable. There are approaches to group (multiple) cyclogenesis or to individual cyclogenesis that are dated to SST fields on various spatiotemporal scales (beginning with a week or a season and finishing with 1 or 5years) (Sharkov, 2000; Webster *et al.*, 2005; Semmler et al, 2008). Such results receive, as expected, a brisk (and, one should say, strict) polemic in the Western scientific literature (Webster *et al.*, 2005; Trenberth, 2005; Vecchi and Soden, 2007; Hassim and Walsh, 2008).

In addition, we note that some theorists (see, e.g., Palmen and Newton, 1969; Golitsyn, 2008) operate with abstract temperatures that could be attributed with difficulty to temperature fields observed remotely or obtained from in situ measurements. Oceanologists, in turn, mean by "ocean surface temperature" the average temperature over depth in the upper intermixed (quasi-homogenous) layer of water (up to the thermocline) and exclude from consideration the uppermost near-surface layer with a high temperature gradient because, in their opinion, the contribution of this layer to the heat stock of an active ocean's layer is negligible (Fedorov and Ginzburg, 1988).

For the present work, initial data for 1983–2006 on the global cyclogenesis of mature TC forms were taken from the systematized database (DB) of remote observations of global tropical cyclogenesis—Global-TC—that was formed and developed at the Space Research Institute of the Russian Academy of Sciences (SRI RAS) under the scenario principle of utilizing the satellite and ground-based data (Pokrovskaya and Sharkov, 2000, 2001; Pokrovskaya *et al.*, 2004). The database of mature forms of tropical disturbances throughout the water areas of the World Ocean for 1999–2003 was generated on the Global-TC computational DB platform by means of object analysis (pre-processing) (see Pokrovskaya and Sharkov, 2001, 2006 for more details), systematizing and archiving primary information on the tropical zone state, which has been regularly (daily) received throughout the observation period via the Internet from regional centers that forecast typhoon hazards using satellite data and water surface measurements.

Initial data on the spatiotemporal characteristics of the World Ocean's surface temperature for 1983–2006 were determined using data obtained by means of standard oceanologic measurements from research vessels (in situ at a depth of 1–2 m). In this case the spatial fields of ocean temperature that were obtained were formed as monthly averaged long-term fields. The data were taken from atlases of oceans (*Atlas* ..., 1977) issued in the U.S.S.R. in 1977. Because we are dealing here with monthly averaged long-term values of the SST field, the particular (absolute) time of their use does not matter. Comparison of the information on SST fields with the fields of mature form cyclogenesis was performed in conformity with the spatiotemporal features of temperature fields. So, time samples of the moments of genesis of mature forms corresponded to the monthly averaged long-term value of the temperature field with subsequent geographical attribution of genesis locations to a particular month of the year for two observation terms—from 1983 to 2003 (accumulated over 21 years) and from 2002 to 2006 (accumulated over 5 years). The second technique was based on using remote IR thermal data (the temperature field in the surface skin layer was about 10 μm). In this case the spatial fields of ocean temperature obtained were the result of 3-monthly averaging of each particular observed year for the term of 1999–2003. The data were taken from the website of the NOAA-CIRES Climate Diagnostics Center in the U.S.A., *http://www.cdc.noaa.gov/forecast1/histfrcstatlcl.html*

In both the first and the second technique the initial information on temperature fields was represented as a field of isotherms with a temperature step of 10°C. It is this value that can be accepted as the maximum error in OST fields when spatiotemporally attributing studied objects (TC) to SST fields.

Of course, further approaches with other spatiotemporal scales of the intensity parameters of cyclogenesis and SST fields are possible, as well as various processing techniques in the form of a combined time series or in the form of scattering diagrams (see, e.g., Webster *et al.*, 2005; Vecchi and Soden, 2007; Hassim and Walsh, 2008; Semmler *et al.*, 2008).

4.2.3 Statistical processing results for global cyclogenesis

The results of statistical processing are presented in the form of histograms of cyclogenesis intensity distributions at the moment of genesis of initial (TC) forms and their transition into mature (TS) forms of tropical cyclones in World Ocean water areas (global cyclogenesis), in the water areas of NH cyclogenesis and in the water areas of SH cyclogenesis (hemisphere cyclogenesis) for 1999–2003. The statistical sample (in other words, the number of tropical disturbances used to form the statistical histogram) of World Ocean tropical disturbances was 2,058 for primary forms and 425 for mature forms. Similarly, the statistical sample for NH water areas was 1,613 and 291 and for SH water areas it was 445 and 134, respectively. In Figure 4.5 SST means the long-term monthly averaged temperature, whereas in Figure 4.6 SST means the 3-monthly averaged temperature of a particular year.

Analysis of Figure 4.5a clearly shows that the histograms of primary and mature forms in World Ocean water areas represent typical unimodal distributions with a sharp break at high temperatures (higher than 30°C) and long-tailed distributions at average temperatures, up to 24°C, and with a maximum at 28°C to 29°C. At these temperatures a certain fraction (more than 40%) of both primary and mature TC forms are born. What is important about these results is the absence of a sharp break in the shape of histograms at temperatures of 26°C to 27°C, which would follow from theoretical constructions. Nevertheless, at a temperature lower than 26°C a significant quantity of tropical systems are born: about 5% to 6% of the total number of TCs and primary forms. An overwhelming fraction (more than 90%) of primary and mature TC forms are born when the surface temperature exceeds 27°C. No less interesting is the practical similarity of histograms for the genesis of both mature and primary forms. Similar conclusions can also be drawn for the water areas of both hemispheres, where tropical cyclogenesis takes place in the explicit form (Figure 4.5b, c).

Analysis of Figure 4.6, which is constructed according to a principally different technique, clearly shows that histograms of primary and mature forms in World Ocean water areas (3-monthly averaged over a particular year) represent typical unimodal distributions with a sharp break at high temperatures (higher than 30°C) and long-tailed distributions at average temperatures, up to 24 °C, and with a maximum at 28°C to 29°C. Comparison with Figure 4.6a shows virtually full conformity with respect to qualitative and quantitative characteristics. Similar conclusions can also be drawn for the water areas of both hemispheres, where tropical cyclogenesis takes place in the explicit form (Figure 4.6b, c).

It is important to note that cyclogenesis in both hemispheres with respect to the temperature field both qualitatively and quantitatively is very close to each other, though this in no way can be said about the intensity of the stochastic process as such (the intensity of cyclogenesis in the SH is three times weaker than that of the NH) (see Chapter 2).

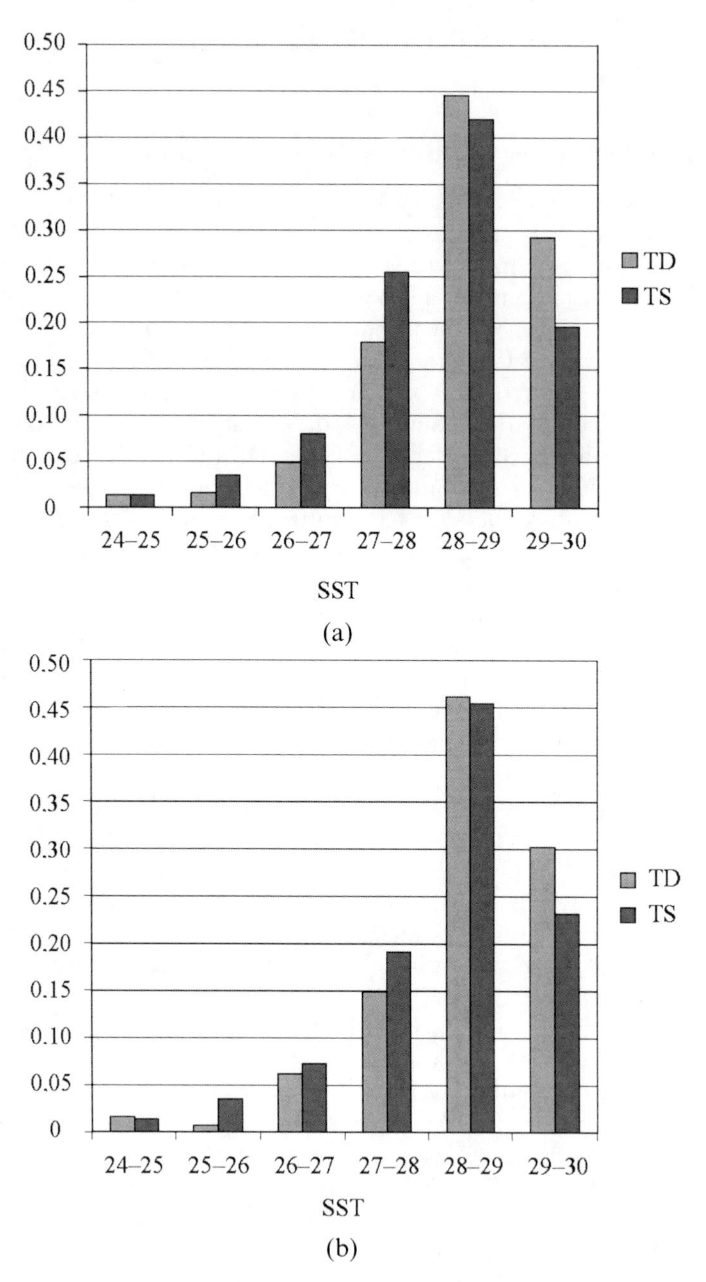

Figure 4.5. Density of probability of tropical disturbance transition to the tropical storm stage in cyclone-generating World Ocean basins for 1999–2003. SST is sea surface temperature as long-term monthly average values from in situ measurements. Rem is remote satellite IR data on measurement of SST (3-monthly averaged values in one year of observation of the given disturbance): (a) World Ocean basins; (b) northern hemisphere basins; (c) southern hemisphere basins.

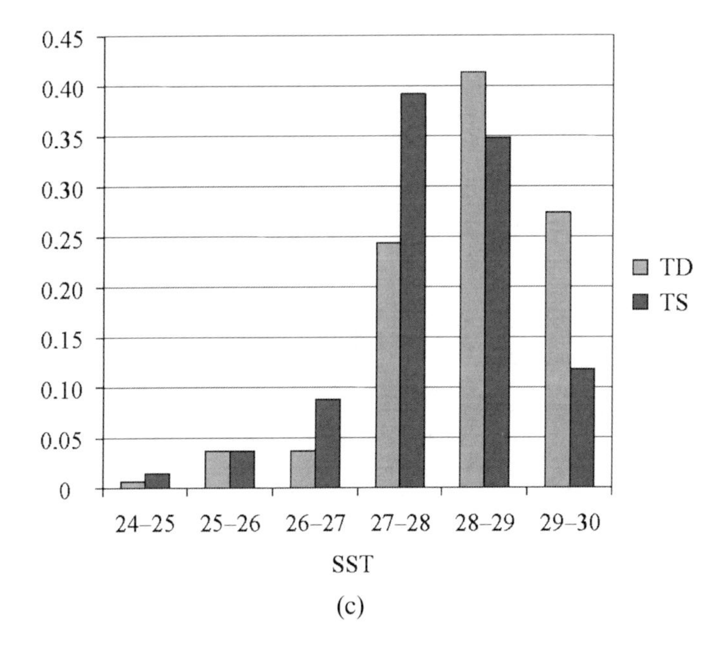

(c)

An important issue when forming statistical histograms is the minimum value of a sample at which the histogram is stable (i.e., unequivocally determining the process statistics). For this purpose we shall make use of the genesis data for mature forms of tropical disturbances during two principally different terms: 1983–2003 (21 years) and 1999–2003 (5 years). Note that accumulation and formation of these databases has proceeded according to a strictly invariable pre-processing technique (Pokrovskaya and Sharkov, 2006). The statistical sample (in other words, the number of tropical disturbances used to form the statistical histogram) for World Ocean tropical disturbances was 1,787 for primary forms during the 21-year term and 425 during the 5-year term. Similarly, the statistical sample for NH water areas was 1,238 and 291 and for SH water areas 549 and 134, respectively. The results of statistical processing are presented in the form of histograms of the distribution of SST (Figure 4.6) at the moment of genesis of mature (TS) forms of tropical cyclones in World Ocean water areas (global cyclogenesis), in the water areas of NH cyclogenesis and in the water areas of SH cyclogenesis (hemisphere cyclogenesis) for 1983–2003 and 1999–2003. In Figure 4.6, OST means the long-term monthly averaged temperature (the data were taken from *Atlas* ..., 1977).

Analysis of Figure 4.7 clearly shows that surface temperature histograms—in which TS genesis in basins of the World Ocean basins takes place for the two terms considered—represent typical unimodal distributions with a sharp break at high temperatures (higher than 30°C) and long-tailed distributions at average temperatures, up to 24°C, and with a maximum at 28°C to 29°C. Note that global cyclogenesis and cyclogenesis of basins in the NH and SH with respect to the

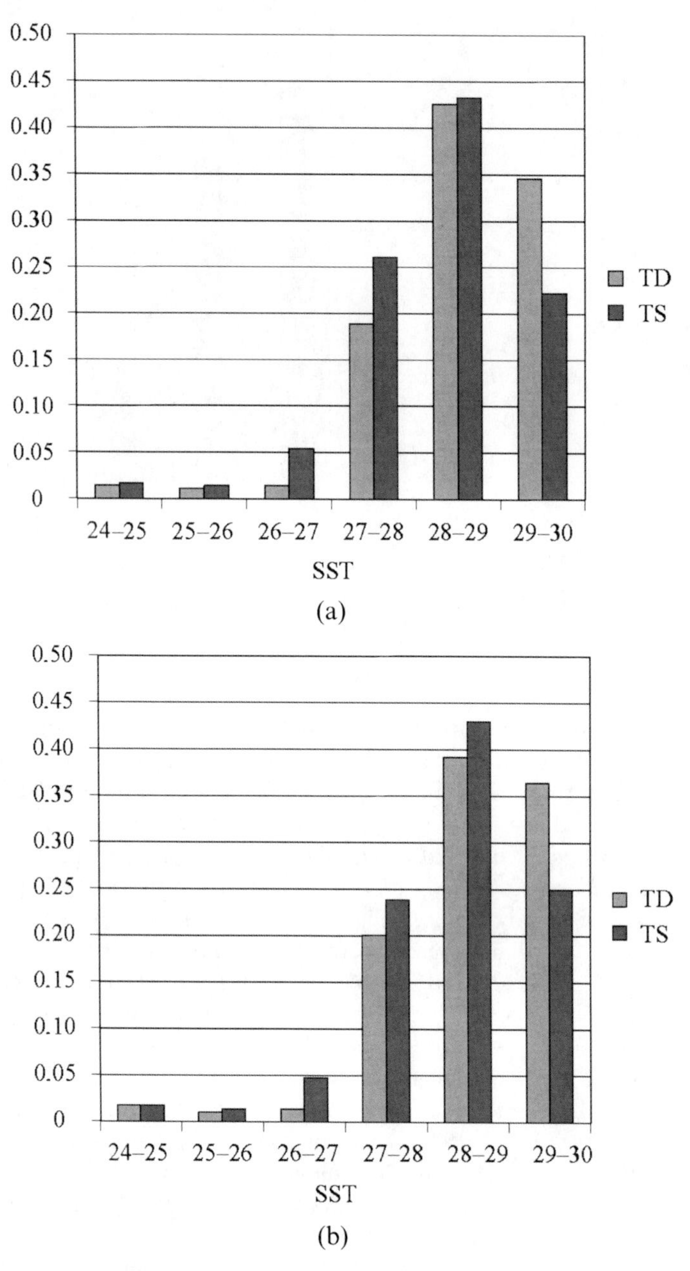

Figure 4.6. Density of probability of tropical disturbance transition to the tropical storm stage in cyclone-generating World Ocean basins for 1999–2003. SST is sea surface temperature as long-term monthly average values from in situ measurements. Rem is remote satellite IR data on measurement of SST (3-monthly averaged values in one year of observation of the given disturbance): (a) World Ocean basins; (b) northern hemisphere basins; (c) southern hemisphere basins.

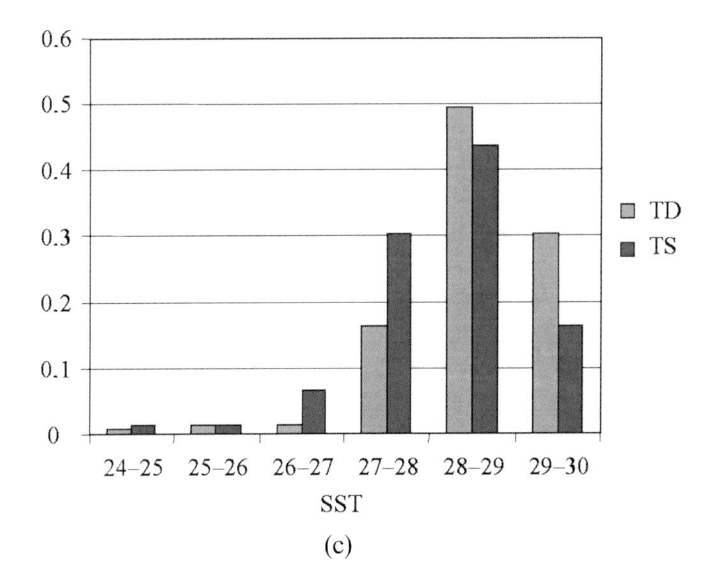

(c)

temperature field are, qualitatively and quantitatively, very close to each other for the two studied observation terms. Thus, it is reasonable to assert that the accumulation of statistical cyclogenesis data within the limits of 21 years to 5 years reveals a very stable general picture and representative results.

4.2.4 Statistical properties of global cyclogenesis

Despite the fact that the spatiotemporal characteristics of the World Ocean's surface temperature were determined by two principally different techniques and were taken principally from different literature and Internet sources, cyclogenesis in the NH and SH with respect to the temperature field nevertheless turned out to be very close to each other both qualitatively and quantitatively,. As mentioned earlier, this can in no way be said about the intensity of cyclogenesis as a stochastic process (the differential intensity of cyclogenesis un SH water areas is three times weaker than that in NH water areas). This circumstance is most likely due to the fact that the temperature field is much more inertial than the stochastic field of cyclogenesis.

Analyzing the appearance of 5% to 6% of TCs at temperatures lower than 26°C, it is reasonable to suppose that these disturbances are some kind of transition from disturbances at moderate latitudes to tropical disturbances, because they arise in the boundary zone between the tropical zone and the moderate latitude zone where there is a considerable spatial latitude gradient in surface temperature. This, in its turn, can result in considerable errors when determining the temperature at the location of tropical disturbance genesis. Here, depending on which processes prevail in the atmosphere in the given region, further development of the disturbance will proceed either according to the moderate latitude scenario or to the purely tropical scenario (the genesis of primary and mature forms of TCs).

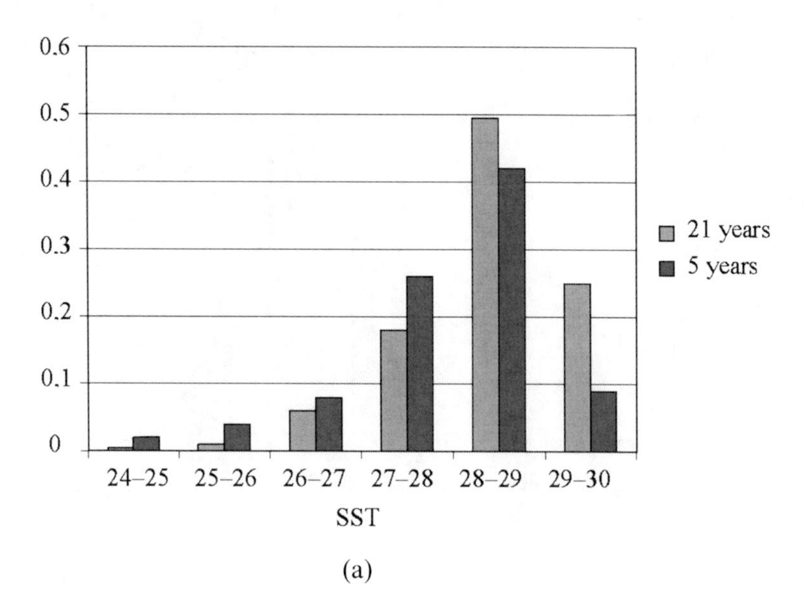

(a)

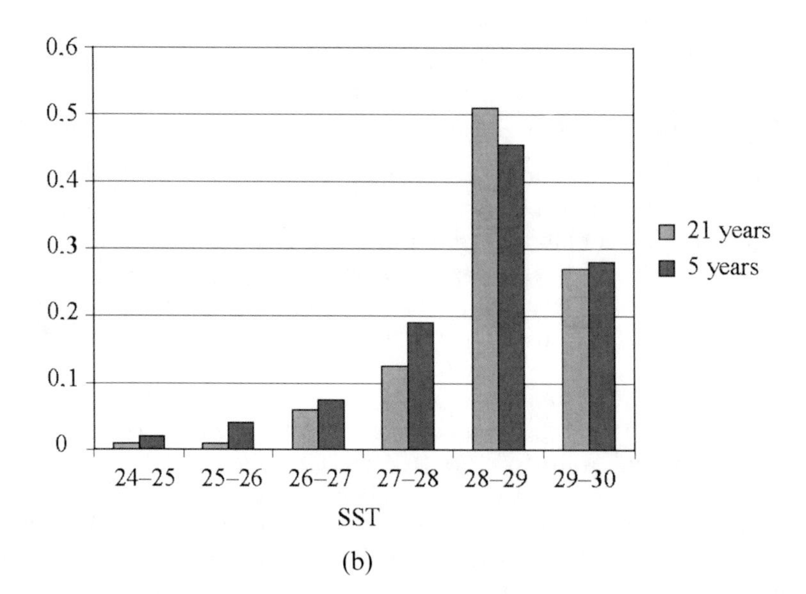

(b)

Figure 4.7. Density of probability of occurrence of tropical disturbances in cyclone-generating World Ocean water areas for 1983–2003 (accumulated over 21 years) and 2002–2006 (accumulated over 5 years). SST is sea surface temperature as long-term monthly averaged values from in situ measurements and archive data: (a) World Ocean basins; (b) northern hemisphere basins; (c) southern hemisphere basins.

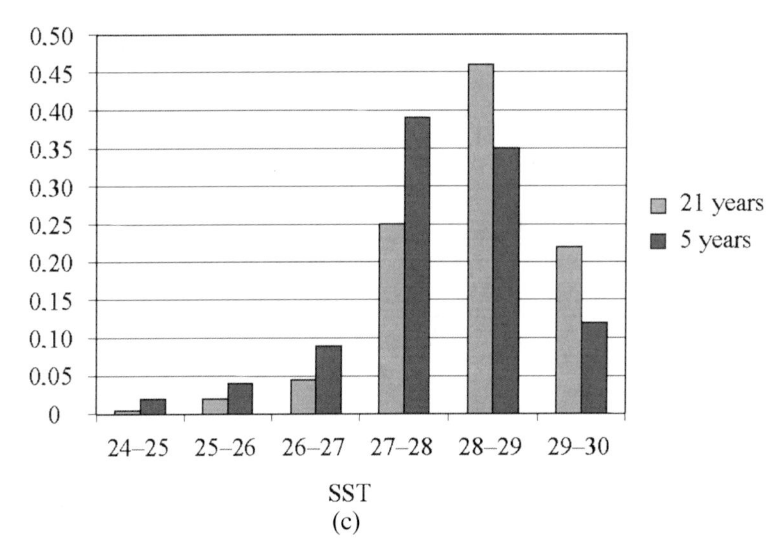

(c)

It is also important to note that the differential intensity of the stochastic process of cyclogenesis (of both hemispheres and globally) is apparently not associated directly with variations of the temperature field in cyclone-forming water areas, though seasonal variations take place both for cyclogenesis in NH water areas and in SH water areas. However, if we consider each cyclone-forming water area individually, the effect of the temperature field will be significant (depending on hydrological and geographical conditions), which will undoubtedly be reflected in the shape and quantitative characteristics of the histograms.

Thus, spatiotemporal comparison of the spatiotemporal fields of the generation of primary forms and the cyclogenesis of mature forms in the surface temperature field—determined (a) by means of standard oceanologic measurements (in situ at a depth of 1 m) and (b) from remote IR data (the temperature field in the surface skin layer) for the oceanic water areas of Earth's two hemispheres—presents experimental results that indicate the presence of a wide range of surface temperatures at which the processes of generation of primary forms and their transformation into mature forms occur and the absence of a critical (threshold) temperature and, accordingly, the absence of a strict boundary at the time of their generation in the ocean's surface temperature field, considered as a monthly averaged long-term value and as a 3-monthly averaged value of each year observed. The stable character of statistical histograms of the distributions of ocean surface temperatures at the moment of the transition to the mature (TS) forms of tropical disturbances (monthly averaged long-term OST values)—both in the 21-year sample (1983–2003) and in the 5-year sample (1999–2003) in World Ocean water areas—is demonstrated. It is important to note that cyclogenesis in the water areas of the NH and SH with respect to the temperature field are, both qualitatively an quantitatively, close to each other, though once again this can in no way be said about the intensity of the stochastic process (the differential intensity of cyclogenesis in SH water areas is three times weaker than that in NH water areas).

4.2.5 Statistical features of mature form genesis in the SST field based on in situ data

The results of statistical processing are presented in the form of histograms of the distributions of cyclogenesis intensity (the relative rate of occurrence) of the first mature TC form—the tropical storm (TS) (Pokrovskaya and Sharkov, 1999d, 2006)—from SSTs in six cyclone-forming water areas of the World Ocean: namely, the northwest part of the Pacific Ocean (NWPO), the northeast part of the Pacific Ocean (NEPO), the Atlantic Ocean (AO), the northern Indian Ocean (NIO), the southern Indian Ocean (SIO), the southwest Pacific Ocean (SWPO). These are the areas where so-called "regional cyclogenesis" took place between 1983 and 2006 (Figures 4.8a–f).

The statistical sample (in other words, the number of mature TC forms from which the statistical histogram is formed) for World Ocean tropical disturbances for mature forms was 570 in the NWPO water area, 213 in the NEPO water area, 213 in the SWPO water area, 225 in the AO water area, 93 in the NIO water area, and 338 in the SIO water area. Similarly, the statistical sample for NH water areas was equal to 1,613 primry forms and 291 mature forms and for SH water areas 445 and 134, respectively. In Figure 4.8, SST means the long-term monthly averaged values of temperature. Analysis of Figure 4.8a–f clearly shows that the histograms of primary forms in the active water areas of the World Ocean possess (as expected; Sharkov and Pokrovskaya, 2006) a high diversity of forms. So for cyclone-generating NWPO, NEPO, SIO, and SWPO water areas the histograms represent typical single-apex distributions with sharp breakage at high temperatures (higher than 30°C) and elongated tails at average temperatures, up to 24°C, and with a maximum at 28°C to 29°C. At this temperature a considerable fraction (about 50%) of mature TC forms are born. The only feature characteristic of all diagrams is the presence of a sharp drop in histograms with a surface temperature higher than 30°C. This feature is well known (Tarakanov, 1980). It is associated with the absence of considerable water areas with temperatures higher than 31°C. Note that—for the AO water area—the histogram has an essentially "smeared" character to the side of low temperatures, up to 24°C. Moreover, one should emphasize that the analysis of these results shows the absence of a sharp breakage in the form of histograms at temperatures of 26°C to 27°C, which would follow from various theoretical constructions. Nevertheless, at temperatures lower than 26°C a significant number of mature forms of tropical systems are born—about 10% of the total number of TCs. An overwhelming fraction (more than 90%) of primary and well-developed TC forms occur in the AO water area, even when the surface temperature exceeds 27°C. The NIO water areas differ from the other cyclone-generating water areas in that the shape of the histogram of mature form genesis is more or less delta-like—all mature forms are generated in the range of surface temperatures from 28°C to 30°C.

An important issue when forming statistical histograms is the minimum value of a sample at which the histogram is stable (i.e., unequivocally determining the process statistics). For this purpose we use the genesis data of mature forms of tropical

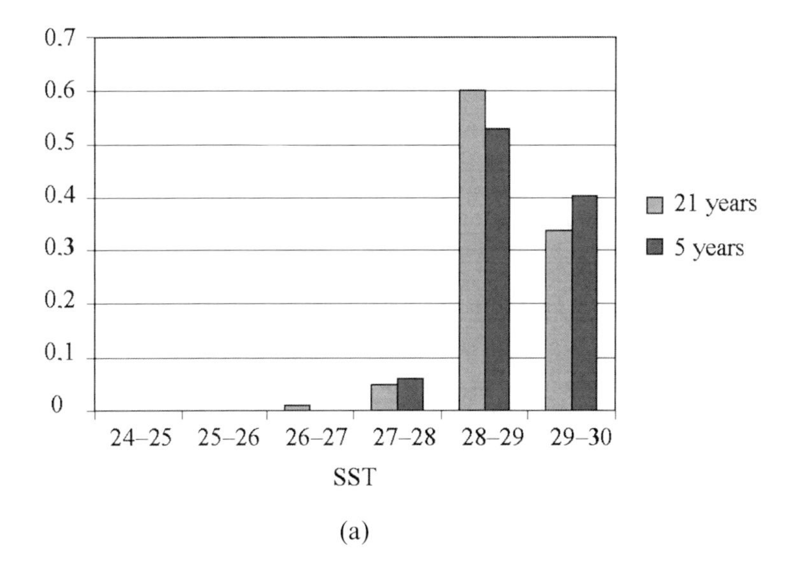

(a)

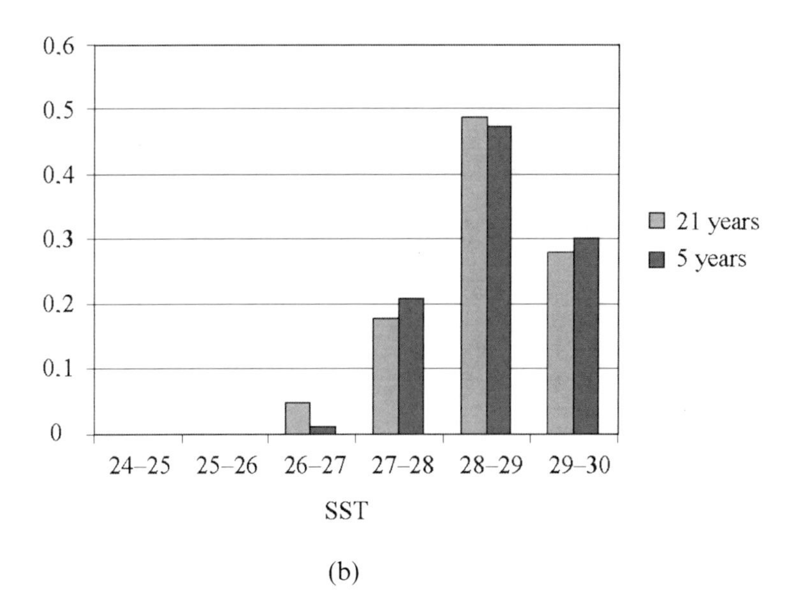

(b)

Figure 4.8. Density of probability of occurrence of tropical disturbances in cyclone-generating World Ocean water areas for 1983–2003 (accumulated over 21years) and 2002–2006 (accumulated over 5 years). SST is sea surface temperature as long-term monthly averaged values from in situ measurements and archive data: (a) water areas of the northwest Pacific Ocean; (b) water areas of the northeast Pacific Ocean; (c) water areas of the northern Atlantic Ocean; (d) water areas of the northern Indian Ocean; (e) water areas of the southern Indian Ocean; (f) water areas of the southwest Pacific Ocean.

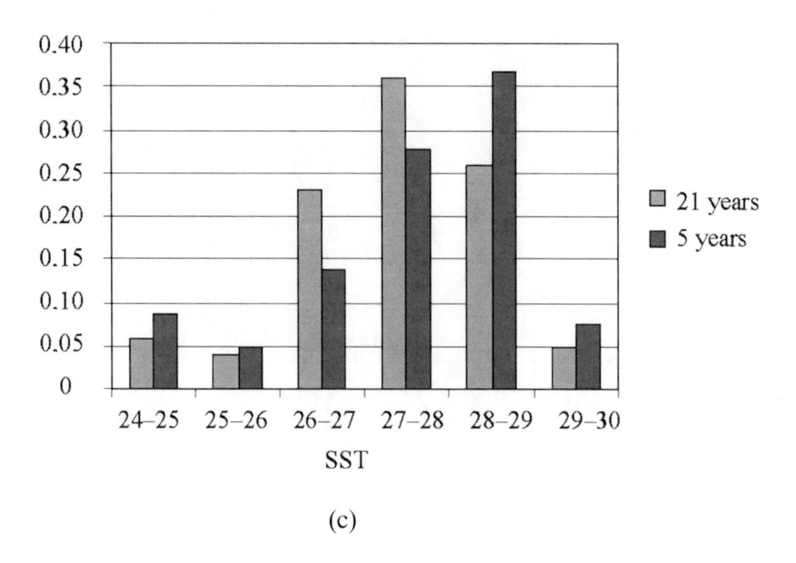

(c)

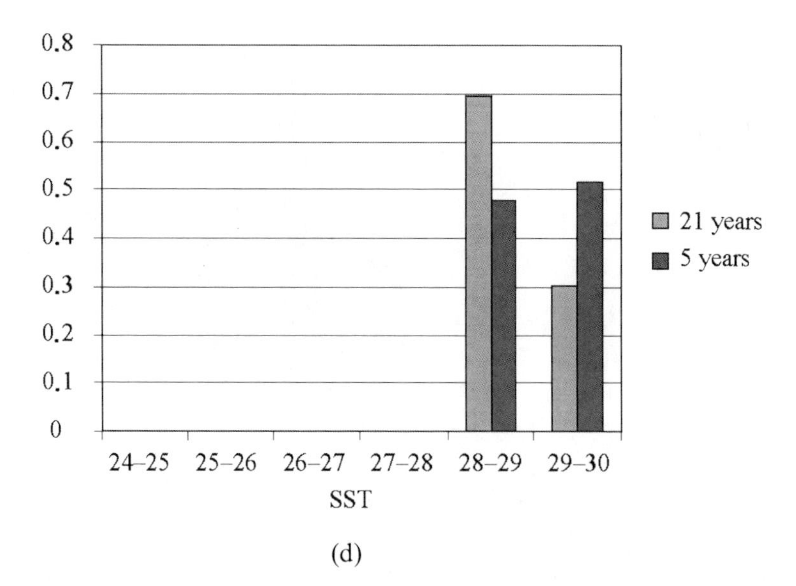

(d)

Figure 4.8 (*cont.*). Density of probability of occurrence of tropical disturbances in cyclone-generating World Ocean water areas for 1983–2003 (accumulated over 21years) and 2002–2006 (accumulated over 5 years). SST is sea surface temperature as long-term monthly averaged values from in situ measurements and archive data: (c) water areas of the northern Atlantic Ocean; (d) water areas of the northern Indian Ocean; (e) water areas of the southern Indian Ocean; (f) water areas of the southwest Pacific Ocean.

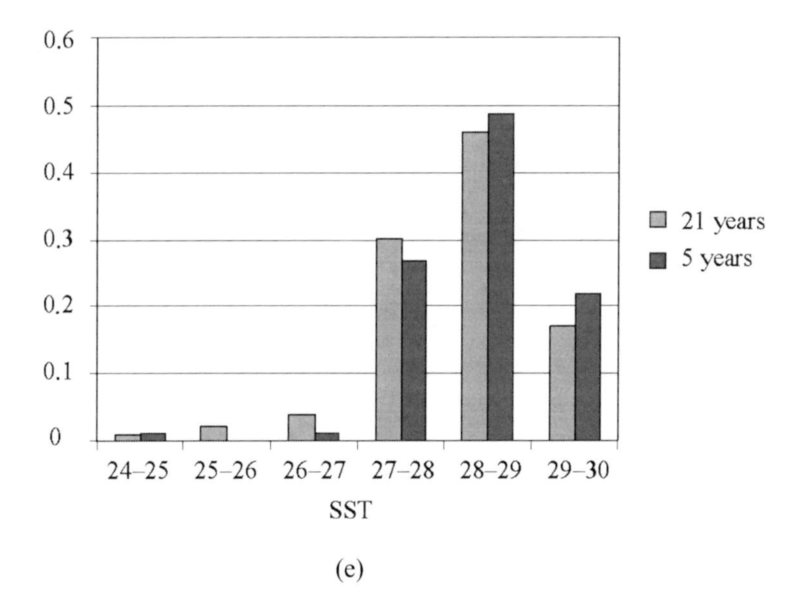

(e)

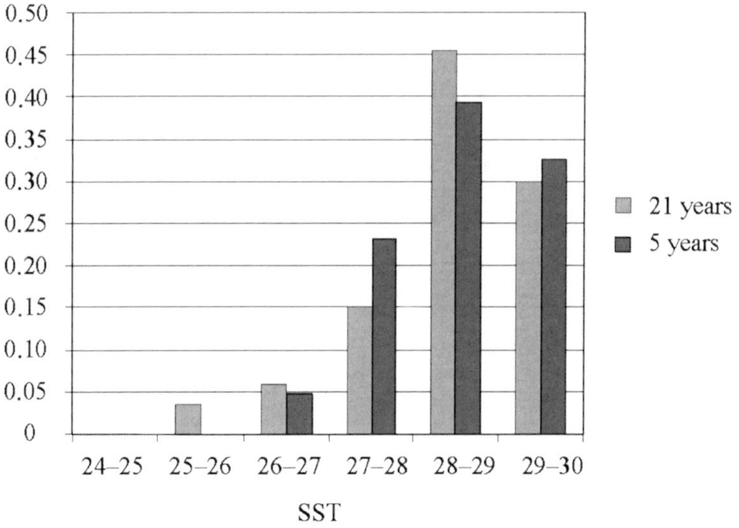

0.50
0.45
0.40
0.35
0.30
0.25
0.20
0.15
0.10
0.05
0
24–25 25–26 26–27 27–28 28–29 29–30
SST
21 years
5 years

(f)

disturbances for two principally different terms: 1983–2003 (21 years) and 1999–2003 (5 years). Note that the accumulation and formation of these databases has proceeded according to a strictly invariable pre-processing technique (Pokrovskaya and Sharkov, 2006). The results of statistical processing are presented in the form of histograms of the distributions of the genesis intensity of well-developed (TS) forms of tropical cyclones in World Ocean water areas for the period 2002–2006 in Figure 4.8a–f along with the period 1983–2003. Analysis of these combined histograms in Figure 4.5 clearly shows that histograms of the genesis intensity of mature forms in cyclone-generating World Ocean water areas for the two considered terms are close to each other; moreover, the features of each cyclone-generating water area are more or less similar qualitatively and quantitatively. Thus, it is reasonable to assert that—within the limits of 21 years to 5 years—the accumulation of statistical cyclogenesis data reveals a stable general picture and representative results.

4.2.6 Statistical features of mature form genesis in the SST field based on remote and in situ data

As already noted, an important aspect of investigating the SST field is consideration of the proximity of the results of studying TC genesis between in situ SST measurements and the results of remote satellite observations of SST.

Figures 4.9a–f present the probability densities of transition from the tropical disturbance stage to the tropical storm (TS) stage in six cyclone-generating World Ocean water areas for 1999–2003 (accumulated over 5 years) from two SST field presentations: long-term monthly averaged values (in situ measurements) and remote satellite IR data of SST measurements (3-monthly averaged values in the year of observation of the given tropical disturbance).

Analysis of Figure 4.9, in which the SST data are constructed according to a principally different technique, shows that histograms of the intensity of the transition of primary forms to mature forms in six cyclone-generating World Ocean water areas demonstrate more or less satisfactory qualitative conformity both in the form and character of histograms. However, from the viewpoint of the quantitative estimation of histograms there are serious distinctions. So, for the AO basin the histograms represent single-apex distributions with sharp breakage at high temperatures (higher than 30°C) and elongated tails at average temperatures, up to 24°C, and with a maximum at 28°C to 29°C. Quantitatively, at a temperature of 26°C a considerable drop is observed, which is compensated by a sharp increase at a temperature of 27°C. For the SIO water area the delta-shaped form is characteristic of the form of histograms (Figures 4.8d, 4.9d); however, the quantitative relation between temperatures of 27°C and 28°C is sufficiently noticeable (about 12%) (Figure 4.9d). A similar situation is observed for SIO, SWPO, NWPO, and NEPO water areas as well.

Note that the histograms of regional cyclogenesis in the World Ocean with respect to the temperature field do not qualitatively correspond to differences in the intensity of the stochastic process as such. So, the hottest cyclone-generating SIO water area (with an average temperature of 29°C; Figures 4.8d, 4.9d) has the

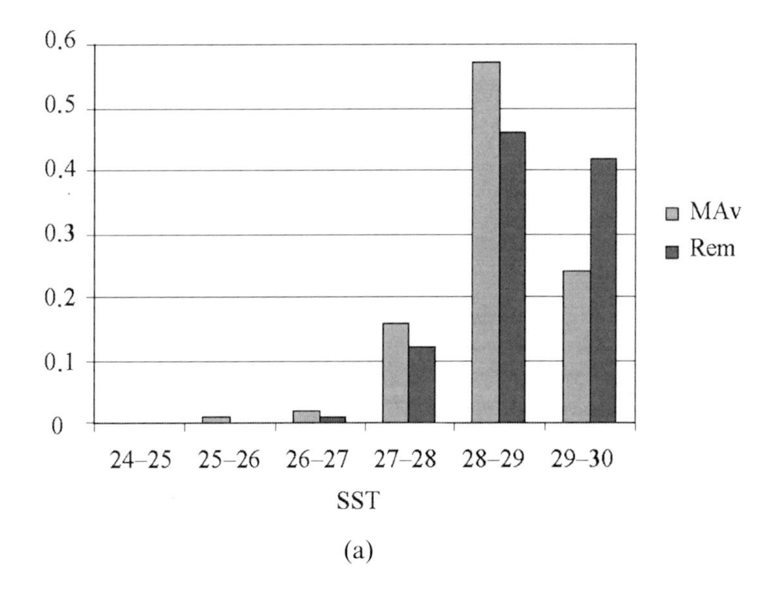

(a)

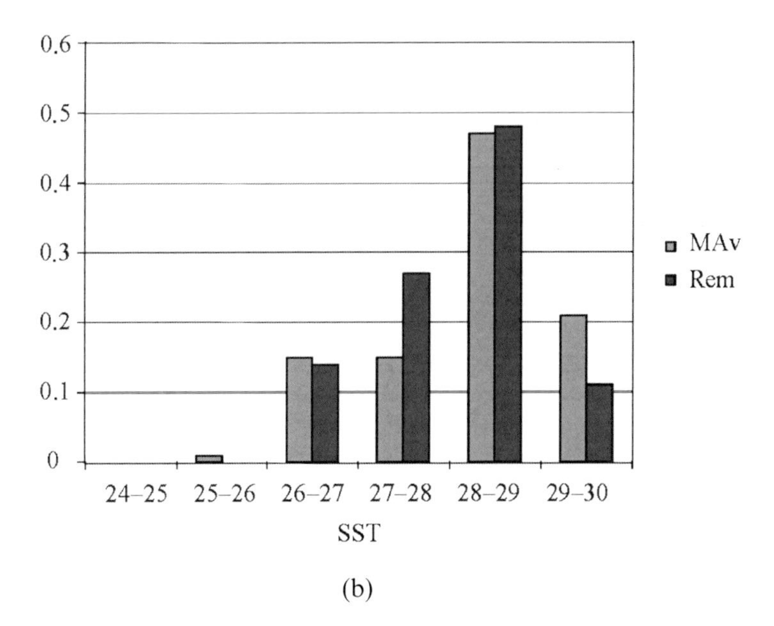

(b)

Figure 4.9. Density of probability of tropical disturbance transition to tropical storm stage in cyclone-generating World Ocean water areas for 1999–2003. MAv is sea surface temperature long-term monthly averaged values from *in situ* measurements. Rem is remote satellite IR data on sea surface temperature (3-monthly averaged values in a single year of observation of the given disturbance): (a) water areas of the northwest Pacific Ocean; (b) water areas of the northeast Pacific Ocean.

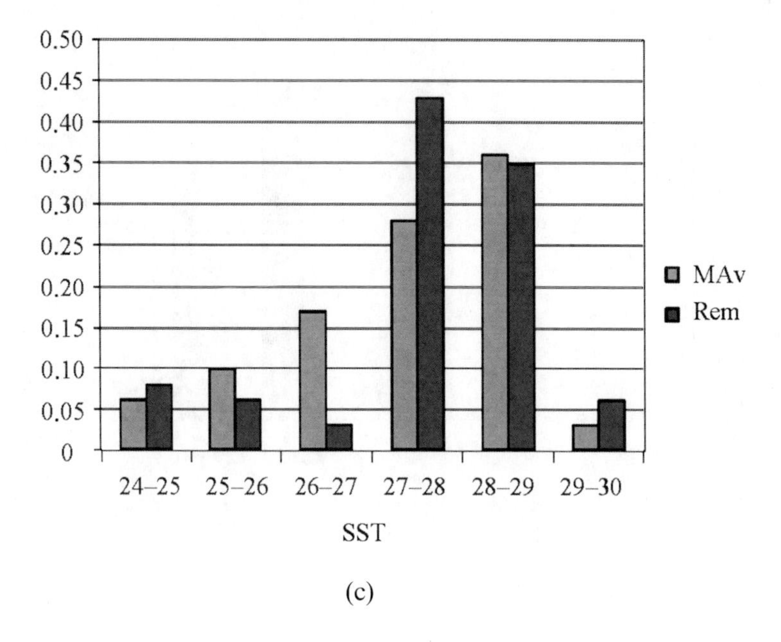

(c)

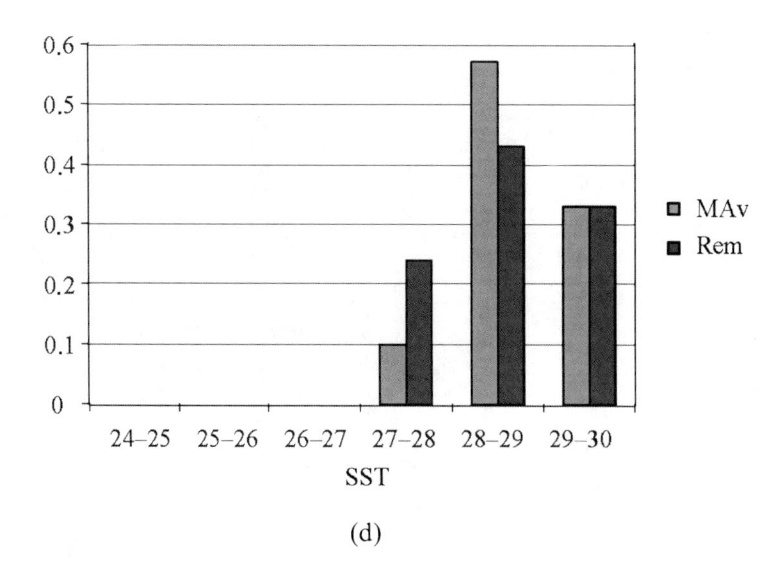

(d)

Figure 4.9 (*cont.*). Density of probability of tropical disturbance transition to tropical storm stage in cyclone-generating World Ocean water areas for 1999–2003. MAv is sea surface temperature long-term monthly averaged values from *in situ* measurements. Rem is remote satellite IR data on sea surface temperature (3-monthly averaged values in a single year of observation of the given disturbance): (c) water areas of the northern Atlantic Ocean; (d) water areas of the northern Indian Ocean; (e) water areas of the southern Indian Ocean; (f) water areas of the southwest Pacific Ocean.

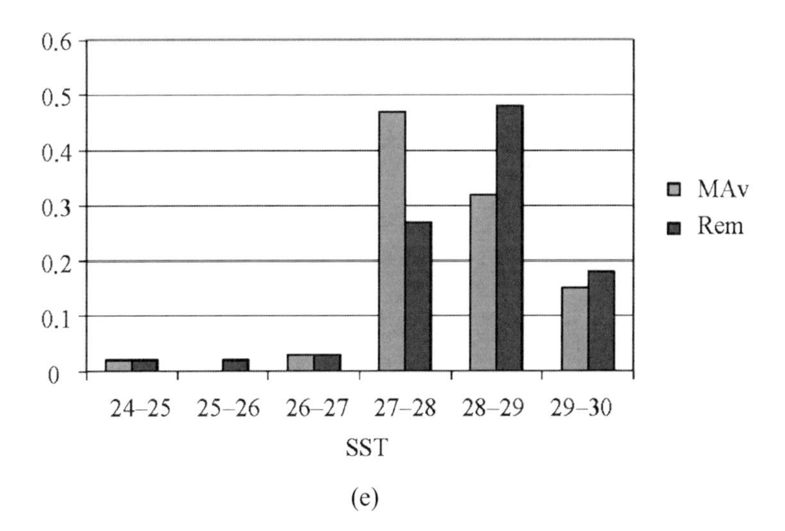
0.6
0.5
0.4
0.3
0.2
0.1
0
24–25
25–26
26–27
27–28
28–29
29–30
SST
MAv
Rem

(e)

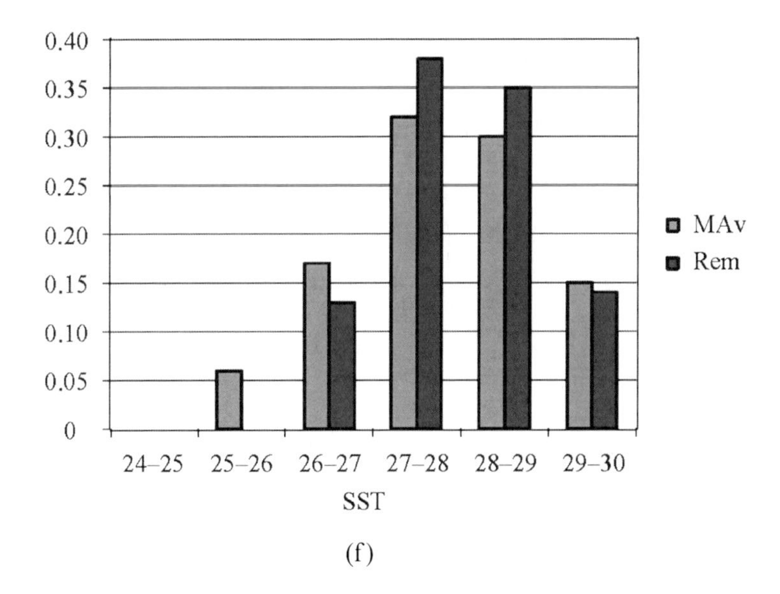
0.40
0.35
0.30
0.25
0.20
0.15
0.10
0.05
0
24–25
25–26
26–27
27–28
28–29
29–30
SST
MAv
Rem

(f)

weakest intensity among all World Ocean cyclogeneses: its intensity is seven times weaker than that of the NWPO water area (Sharkov, 2000).

4.2.7 Features of the statistical properties of regional cyclogenesis

Despite the fact that the spatiotemporal characteristics of the World Ocean's surface temperature were determined by two principally different techniques and were taken from principally different literature and Internet sources, regional cyclogenesis with respect to the temperature field nevertheless turned out to be very close to each other both qualitatively and quantitatively. Though this in no way can be said concerning the intensity of cyclogenesis as a stochastic process—the differential and integral intensity of cyclogenesis varies within very wide limits (Sharkov, 2000). This circumstance is most likely associated with the fact that, on one side, the temperature field of cyclogenesis is considerably more inertial than the stochastic field and, on the other side, the direct role of the ocean's surface temperature is not the main factor in TC genesis.

Analyzing the appearance of 5% to 6% of TCs in the AO water area at temperatures lower than 26°C (Figures 4.8, 4.9), it is reasonable to suppose that these disturbances are some kind of transitional form between disturbances at moderate latitudes and those at tropical disturbances, because they arise in the boundary zone between the tropical zone and the zone of moderate latitudes where there is a considerable spatial latitude gradient in surface temperature, which, in its turn, can result in considerable errors when determining the temperature at the location of tropical disturbance genesis. Depending on what processes prevail in the atmosphere of the given region, further development of the disturbance will proceed either according to the moderate latitude scenario or according to the purely tropical scenario (the genesis of primary and mature forms of TCs).

Note also that the differential intensity of a stochastic process of regional cyclogenesis is apparently not associated with variations in the temperature field of cyclone-forming water areas at all, though seasonal variations take place both for cyclogenesis in NH and SH water areas. However, if we consider each of the cyclone-forming water areas individually, the effect of the temperature field will be significant (depending on hydrological and geographical conditions) and reflected in the form and quantitative characteristics of the histograms.

Thus, the SST field measured by remote satellite instruments cannot serve directly as a strict remote criterion for forecasting TC generation on regional scales. Moreover, the results of statistical analysis indicate that even statement of the problem to search for a critical temperature over all cyclone-generating World Ocean water areas and regionally is unproductive physically. Designing and manufacturing automatic remote detectors in the IR thermal range, which could essentially simplify the predictability of crisis situations, are not topical because the absence of a sharp cutoff in the surface temperature field at TC generation can be proven experimentally.

As already noted, Golitsyn (2008) is at variance with the viewpoint that achievement of a critical temperature "should be considered as the first necessary condition for TC onset". This statement is essentially the result of the present work regionally and the result of earlier works of the authors (Sharkov and Pokrovskaya, 2006a, b) globally. However, instead of a critical temperature Golitsyn (2008) introduces the notion of a critical energy flux (basically, latent heat) from the ocean to the atmosphere via convection. However, the study of TC evolution by means of satellite microwave instruments on various scales and frequencies (Sharkov *et al.*, 2008a, b, 2009, 2010, 2011a, b; Kim *et al.*, 2009, 2010) has shown that power supply by latent heat should occur in an absolutely different manner—not by convective fluxes. This is due to horizontal transfer in jet-type fluxes into integral water vapor fields. In such a case, the direct role played by surface temperature is not the major factor in TC genesis—moreover, it seems to be indirect, which just goes to show the diversity of the experimental histograms revealed in our work (Figures 4.8, 4.9).

So, spatiotemporal comparison of the spatiotemporal fields of generation of mature TC forms in the SST field—determined both by means of standard oceanological measurements (in situ at a depth of 1 m), and 3-monthly averaged measurements (remote satellite IR measurements of the ocean's surface skin layer temperature) of each analyzed year over cyclone-generating water areas of the World Ocean—reveals experimental results that point to the fact that regional cyclogenesis possesses unique ranges of surface temperatures, at which the processes of generation of primary TC forms (smeared ranges, extrema with tails, delta-shaped forms of histograms) take place. The presence of a wide range of surface temperatures at which the processes of generation of mature TC forms take place, and the absence of a critical (threshold) temperature and, accordingly, of a rigid boundary at their generation in the SST field, considered on a monthly averaged long-term basis and on a 3-monthly averaged remote basis, are shown. The stable character of statistical histograms of SST distributions at the moment of transition into mature (TC) forms of tropical disturbances (monthly averaged OST long-term values), both in a sample for 21 years (1983–2003) and in a sample for 5 years (2002–2006), for regional cyclogenesis in World Ocean water areas is demonstrated.

4.3 GLOBAL RADIOTHERMAL FIELDS FOR STUDY OF THE ATMOSPHERE–OCEAN SYSTEM

Large-scale thermodynamic interactions in the ocean–atmosphere system are among the most important factors of climate variability. The processes affecting the transport and dissipative properties of the atmosphere and influencing climate formation have a great spatial extension, occur on long timescales, are characterized by a complicated spatiotemporal structure, and evolve within a very wide range of temporal and spatial scales. This section shows that existing space microwave complexes, by virtue of the orbital and technical characteristics of their carriers, do not provide sufficient regularity and density of Earth surface coverage to provide the initial data necessary to form the global field or the time sequence necessary to reveal the highly dynamic atmospheric processes that are at work. A specialized technique for constructing long-term global radiothermal fields, suitable for the scientific analysis of highly dynamic atmospheric processes such as tropical cyclones, is considered in this section.

4.3.1 The role of microwave sounding in space monitoring

The urgency of studying the large-scale thermodynamic processes in the ocean–atmosphere system of Earth is undoubtedly one of the most important tasks of the scientific community, because it not only has fundamental scientific but also important practical significance. As data accumulate about the components that make up this complicated system, scientific views on the global climatic system of the Earth as a whole and on possible climate variability are essentially undergoing change.

Large-scale thermodynamic interactions in the ocean–atmosphere system, taking place as a result of energy exchange and mass transfer processes evolving within a wide range of intensities and spatiotemporal scales, are among the most important factors of climate variability. The complexity of the structure of natural processes is also reflected in the structure of the geophysical parameters describing them. As a result, the necessity arises to measure these parameters at a given spatiotemporal resolution. This determines the specifications of the instruments performing measurements and the carriers on which they are mounted. There exists a wide scope of technical possibilities for observational data acquisition: the advanced network of meteorological fixed stations and balloon-borne sounders, weather research vessels and buoys, aircraft, satellites, and others. Observational data obtained by these means use essentially different spatiotemporal digitization techniques and can be used to solve global scientific problems by means of very complicated (but sometimes ambiguous) procedures for the adoption of data—satellite monitoring data in particular.

The processes affecting the transport and dissipative properties of the atmosphere and influencing climate formation cover very extensive areas, have long timescales, are characterized by a complicated spatiotemporal structure, and evolve on a very wide range of temporal and spatial scales. To ascertain the general regularities of such processes it is necessary to analyze the observational data characterizing their energy and dynamics: for example, the size of their spatiotemporal extent, their frequency, and the density of global surface coverage observations required.

Ground-based observational data are usually acquired as local (point) measurements and (much less often) as fields of spatiotemporal measurements since the latter face major technological difficulties and need to use model views of the object of interest. Only the instruments installed on satellites can now provide global observations of geophysical parameters in the form of fields at the spatial frequency and temporal regularity required for further analysis.

The inclusion of microwave diagnostics techniques in air and space observations in the 1960s and 1970s represented a major landmark in the development of Earth remote sensing (Sharkov, 2003, 2005). Studying and understanding microwave images of the Earth surface–atmosphere system have provided the essentially new physical information capacity of microwave sounding to study terrestrial objects (compared with the case of using optical and infrared ranges only). This cardinally changed both the configuration of potential satellite systems designed for Earth sensing, and the character and information filling of Earth remote sensing as a whole. The obvious advantages of microwave diagnostics—obtaining information at any time of a day, the wide weather range, the lack of dependence on solar illumination—have drawn the attention of a great number of researchers. These were decisive factors in the early stages of introducing radiophysical methods into remote-sensing tasks. However, further development of these methods has shown that the principal significance of introducing radiophysical methods into remote sensing lies on an absolutely different plane: namely, in the diffractive nature of the interaction of electromagnetic waves in the microwave range with rough elements of the Earth surface, with meteostructures in the Earth atmosphere, with features of quantum radiation and absorption of a gas phase of the Earth atmosphere in the microwave range, and with using the Rayleigh–Jeans approximation in the microwave range, which essentially simplifies the formulation of inversion procedures (Sharkov, 2003, 2005).

Modern microwave radiothermal instruments installed on satellites allow global data of radiobrightness observations to be gathered at the desired spatial resolution, extension, and temporal regularity. In recent years, a great deal of experience has been built up in using satellite microwave and IR radiometry (Sharkov, 2003) to analyze the thermal and dynamic interaction of the ocean and atmosphere on the basis of data provided by multichannel radiometric instruments installed on satellites: the radiothermal instruments SSMR (the NIMBUS 7 mission), SSM/I (the DMSP mission), AMSR (the ADEOS-II mission), AMSR-E (the Aqua mission) in the microwave range and AVHRR (the NOAA mission) in the IR range (see Chapter 9).

Microwave data are widely used to construct surface wind velocity, to determine the (integral) moisture content in the atmosphere over the World Ocean and rainwater in clouds, to reveal the zones of maximum rainfall, and to estimate heat and momentum fluxes. Most of the parameters needed to construct these factors cannot be obtained, in principle, from monitoring data in the optical and IR ranges.

Microwave satellite data are now used both indirectly (i.e., for constructing the values of standard meteorological characteristics on the basis of semi-empirical formulas) and directly (as the direct characteristics of thermal and dynamic interaction in the ocean–atmosphere system). Estimates have shown the legitimacy of using satellite microwave radiothermal fields to study thermodynamic processes in the ocean–atmosphere system on short-period (synoptic and meso-meteorological), intra-annual (monthly averaged, seasonal) and larger (inter-annual) timescales.

However, direct formation of the global radiothermal fields of the ocean–atmosphere system on the basis of satellite data is a complicated matter. This is due to the fact that existing space microwave complexes, by virtue of the orbital properties of carriers and the technical characteristics of complexes (mainly, swath widths) do not provide sufficient regularity and density of Earth surface coverage by the initial data for global field formation on the timescale necessary to reveal highly dynamic atmospheric processes.

In the present section, data from the multichannel radiometric instruments SSM/I (special sensor microwave/imager) installed on the F10/F15 spacecraft of the DMSP mission are used to construct the global radiothermal field of the Earth. The specialized supplementation techniques developed by Ermakov *et al.* (2007) make it possible to construct the global radiothermal fields (two fields per day) using data from all spacecraft in the DMSP mission. The first versions of the global radiothermal fields constructed according to the offered technique were demonstrated in Astafyeva and Sharkov (2008) and Astafyeva *et al.* (2006a, b, 2008). This section also indicates some possibilities of using the global fields of radiobrightness temperature for detailed study of the heat–mass exchange processes in the ocean–atmosphere system, which influence the general circulation of the atmosphere and participate in forming the planetary climate.

4.3.2 Representation of the global radiothermal field as a computer animation

As already noted, observational data about changing geophysical parameters are usually adapted in the form of local (point) measurements and (much less often) in the form of fields of spatiotemporal variations of geophysical parameters. At the same time, natural processes, which can essentially influence planetary climate formation, have a rather large spatial extension, take a long time, and are characterized by a complicated spatiootemporal structure. To reveal the general regularities of the most influential thermodynamic processes in the ocean–atmosphere system and to understand the physical mechanisms controlling them, the analysis of parameters (observational data) that represent and describe their power capacity and dynamics is necessary. To adequately study the spatial structure and time variability of the processes we are interested in, observational data should be presented in the form of fields with sufficient spatial resolution, extension, and temporal regularity. Moreover, as studies of such fields have shown, animating the spatiotemporal evolution of global fields is of principal significance in studying the dynamic processes on various scales.

In the present section, data from the SSM/I multichannel radiometric instruments installed on the F10/F15 spacecraft of the DMSP series of meteorological satellites are used to construct the global radiothermal field of the ocean–atmosphere system. We shall briefly describe the structure of the satellite data used and the principles involved in constructing the global radiothermal field of the Earth (see Sections 8.7 and 8.8 for more details).

The DMSP series spacecraft with SSM/I radiometric instruments onboard were launched within the framework of the Defense Meteorological Satellite Program of the U.S. Department of Defense. Long-term Earth monitoring within the DMSP program framework is directed at providing global meteorological, oceanographic, and solar-geophysical operational information. The microwave data supplied by the DMSP series meteorological satellites have been generally accessible to the scientific community since 1992. In the present section, data are used from the continuously updated electronic SRI RAS database (Ermakov *et al.*, 2007), which currently contains satellite information from 1995 to 2011.

The seven-channel four-frequency VHF radiometric SSM/I complex, installed on the satellites of this series, receives linearly polarized radiation at frequencies of 19.35 GHz, 22.24 GHz, 37.0 GHz, and 85.5 GHz. At the 22.24 GHz frequency only vertically polarized radiation is measured, whereas at other frequencies both horizontally and vertically polarized radiation is measured.

The DMSP series satellites allow data from radiobrightness observations in the form of fields (the scanning pattern of subsatellite swath widths) to be obtained with excellent temporal regularity and spatial resolution. The DMSP series satellites have a solar-synchronous nearly polar low-altitude circular orbit. The orbital parameters are: inclination is 98.8°, altitude is about 850 km, and the period is 102 minutes. Each satellite of this series makes 14.2 revolutions per day. Global coverage of the Earth is carried out for approximately three days and incomplete coverage takes place on another day. The subsatellite swath widths are strictly repeated twice per month, approximately every 16 days. The instrumental swath width equals 1,400 km, the size of the resolution element on the Earth surface depends on the instrument's range and latitude and varies from 12.5 km to 25 km. The total error in determining the resolution element's coordinates is between 20 km and 30 km, but reduces to 5 km after using a special correction.

To construct global radiothermal fields we have used data representing the radiobrightness temperature recalculated for the Earth surface with steps of 0.5×0.5 degrees. The trajectories chosen for the vehicles of this series and the instruments installed on them is such that the subsatellite swath widths do not completely cover the planet's surface in a single day. As a result, coverage is excessively thick at high latitudes, but leaves large lacunae at midlatitudes and, especially, in the near equatorial region. What is more, the information can be lost because of failures in the operation of instruments and information-downloading systems.

Radiothermal fields constructed from data from one of instruments for approximately a day, during the ascending and descending orbits of a satellite, have the appearance shown in Figure 4.10 (here and in the following figures the global radiothermal fields are presented as Mercator projections). Apart from the standard lacunae, caused by satellite trajectory features (the blackened areas on the left side of the figure), loss of information also occurs because of failures in the operation of transmitting or receiving instruments, which can be seen on the right side of Figure 4.10, obtained in a less successful day. The lacunae that form because of failures in the instrument's operation, can be much more extensive than those shown here. Note that the presence of lacunae in the initial information seriously complicates the formation of dynamic pictures at the mesoscale and microscale and, in some cases, simply spoil the dynamic picture.

The lacunae should be filled with data, but it is difficult to do this directly (Ermakov *et al.*, 2007) in light of the fact that the satellites pass over the planet areas of interest at different times and look at them from different angles and the fact that processes in the atmosphere are dynamical. Specialized techniques developed in the Space Research Institute (IKI) of the Russian Academy of Sciences allows

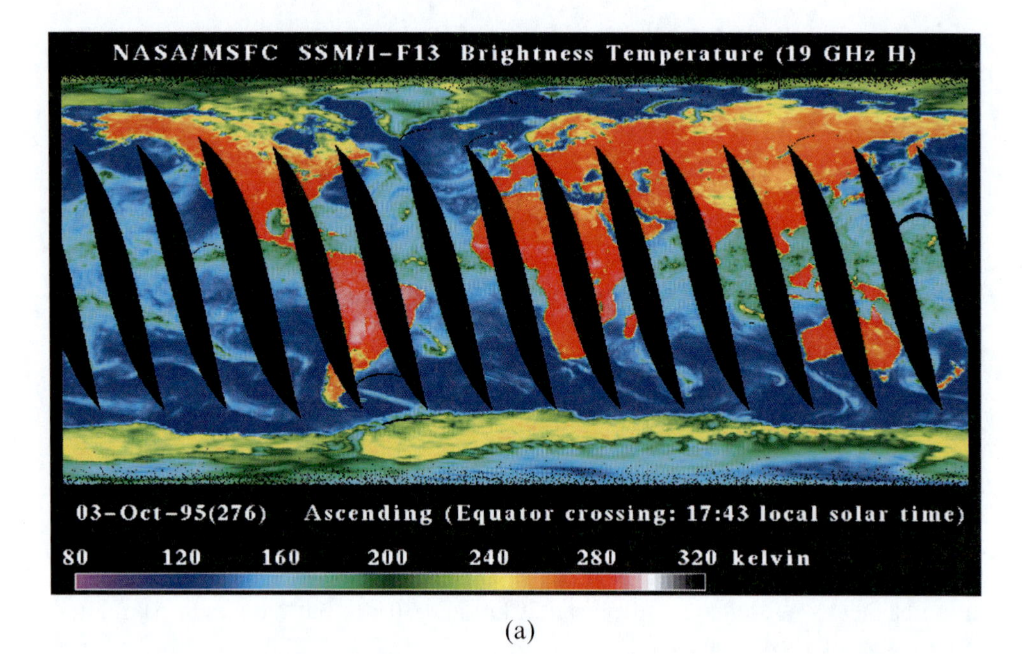

(a)

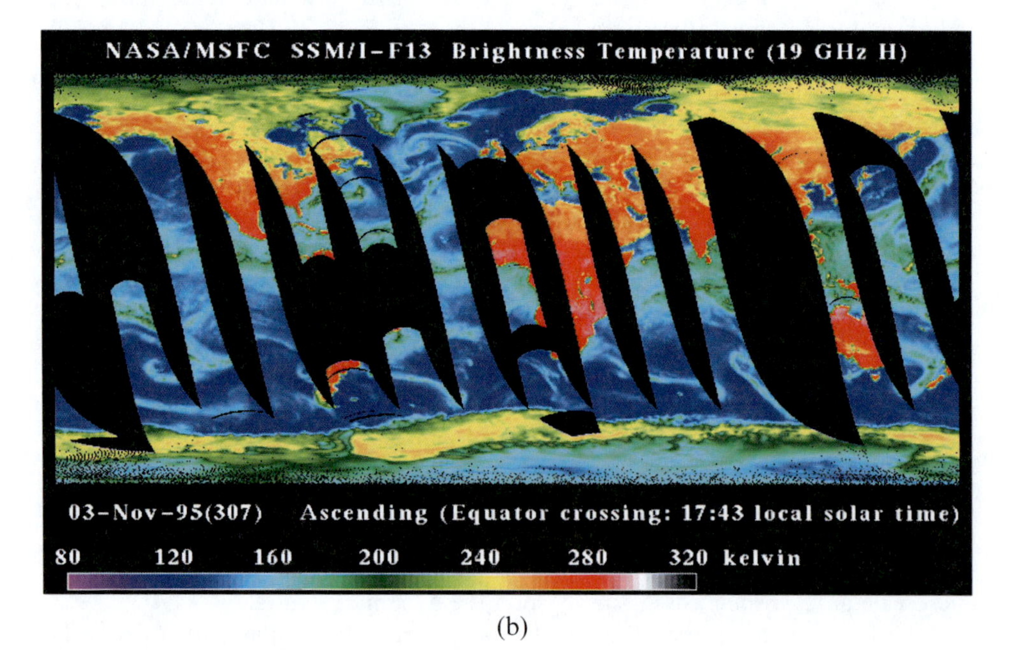

(b)

Figure 4.10. Typical patterns of the global radiothermal field constructed from satellite data from a DMSR mission satellite during a single day at 19.35 GHz. Black fields on patterns are the standard and additional lacunas associated with (a) satellite track features and (b) instrument malfunctions.

missing information to be filled in completely or with a very small percentage of blank areas and allows the global radiothermal fields to be constructed. Data from the F10/F15 spacecraft of the DMSP series were used to fill in the lacunae. The passage time of each satellite is contained in the initial information, and this allows the data to be constructed (obtained during ascending and descending orbits) according to the satellite passage time. After calibrating the initial data, which is required in some cases, the long-term (1995–2005) global radiothermal fields are generated, two fields per day, which are suitable for further scientific analysis (see Section 8.7 for more details).

On the basis of the global radiothermal fields obtained, a computer animation was produced—two fragments of which can be seen on the SRI RAS website (*http://www.iki.rssi.ru/asp/lab_555.htm*)—by the Climatic Research Laboratory, which is part of the Earth Research from Space Department at SRI RAS. Note that the computer animation's representation of dynamic processes possesses many advantages when analyzing dynamic processes: in particular, the ease with which analysis of the phenomenon under study can be controlled. The computer animation allows processes occurring in the ocean–atmosphere system to be observed as well as its dynamics and during long intervals of time. In such an approach the detailed structure of atmospheric processes can readily be seen simultaneously on a wide range of scales: from hundreds of kilometers (the tropical cyclone structure) to the planetary scales of quasi-stationary atmospheric structures (the basic zones of depressions, the northern and southern tropics, the basic energy active zones). Computer animation also allows analysis of the processes at the timescale of interest and enables repetition of the separate spatial fragments as often as required.

4.3.3 Possibilities of using the radiothermal data of satellite monitoring

There are many works devoted to analyzing the ways in which satellite microwave data can be used to estimate characteristics of the thermodynamic processes involved in the ocean–atmosphere system (see the references in Sharkov, 1998, 2003, 2005). We will now present some general ideas concerning possible uses for the radiothermal data of satellite monitoring.

The SSM/I radiothermal complex records the intrinsic radiation of the ocean–atmosphere system in four frequency ranges that can provide information on various terrestrial objects:

- at a frequency of 19.35 GHz (wavelength of 1.58 cm)—on the visible surface: continents, World Ocean, the densest clouds (with rainfall);
- at a frequency of 22.24 GHz (wavelength of 1.35 cm)—on integral precipitable water (water vapor) in the troposphere;
- at a frequency of 37.00 GHz (wavelength of 0.81 cm)—on storage water in large-droplet clouds and on the visible surface;
- at a frequency of 85.50 GHz (wavelength of 0.35 cm)—on integral storage water of all types of clouds, including liquid (rain) and crystal precipitation (hail and snow).

SSM/I radiometer data can also be used to determine wind velocity, the dynamics of large-scale vortical and frontal structures, (integral) storage water of the atmosphere over World Ocean water basins, precipitable water in separate cloud structures and intensive rainfall precipitation zones, and to estimate heat and momentum fluxes.

Two qualitatively different approaches are applied when using radiometric satellite data: the indirect approach and the direct approach. In the indirect approach, quantitative climatic characteristics are restored from satellite data on the basis of indirect links (or model concepts) between integrated (in altitude) radiothermal measurements in various parts of the spectrum and particular characteristics (temperature, humidity), or on the basis of semi-empirical formulas (e.g., the bulk formulas based on correlations between the temperature and humidity in various layers of the atmosphere). Note that the conditions under which semi-empirical formulas are obtained raises doubt about their application on synoptic scales and in regions where there are large gradients (e.g., fronts or tropical cyclones). Although such an approach is not always suitable for analyzing the variability of dynamical processes on synoptic scales, it can be justified on monthly averaged, seasonal, and larger scales.

The direct, or straightforward, approach considers radiometric satellite data themselves as characteristics of the thermal and dynamic interaction between the ocean and the atmosphere. For example, there exists a close link between the seasonal dynamics of monthly averaged values of the radiobrightness temperature and the difference in temperature between the ocean's surface and the near water atmosphere. In this case, using the satellite data is better than measuring the temperature of the ocean surface and that of the near water atmosphere separately by remote techniques and calculating their difference, which can result in a loss of accuracy. The effectiveness of the direct approach is confirmed not only on intra-annual and larger timescales, but also on synoptic scales in dynamical situations (frontal zones, cyclones, and other regions where there are large gradients), where formalization of the processes themselves becomes inconvenient. However, on the other hand, the direct use of radiothermal observations does not allow estimation of some important power parameters of the ocean–atmosphere system, such as the value of latent heat in the field of integral water vapor or explicit heat in the field of rainfall intensity of dense clouds. Thus, the fundamental question about the energy component in the spatiotemporal evolution of a studied object remains open using such an approach. The only way out of this situation is to search and use indirect links between the required parameter and the intensity of the radiothermal signal. Because the question concerns the radio range of electromagnetic waves and the Rayleigh–Jeans approximation can be used to ascertain thermal radiation, these empirical links turn out to be surprisingly simple and reliable in many cases (Sharkov, 1998, 2003).

Let us now consider in greater detail the possibilities of using the obtained radiothermal fields and the obtained radiobrightness temperature links with the ocean–atmosphere system at three frequencies: 19.35 GHz, 22.24 GHz, and 85.5 GHz (Figure 4.11). Cloudiness, precipitable rainfall in the atmosphere, and SST are among the most important characteristics of the thermodynamic regime

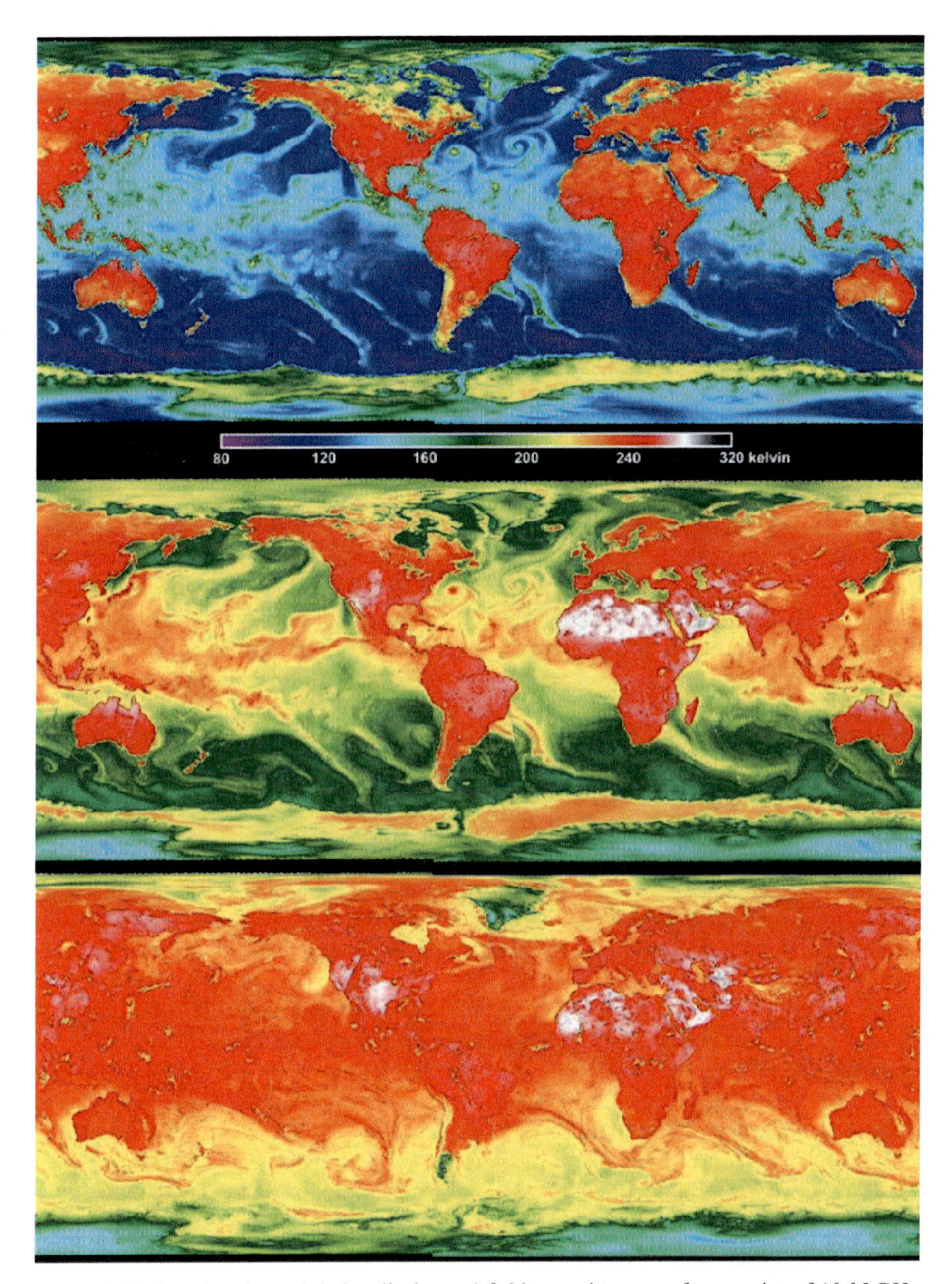

Figure 4.11. Synchronous global radiothermal fields at microwave frequencies of 19.35 GHz (upper diagram), 22.24 GHz (middle diagram), and 85.5 GHz (lower diagram). Patterns were recovered by processing satellite microwave data taken on September 15, 2006. The scale between the upper and middle diagrams is the color radiobrightness scale (Kelvin).

of the climatic system. These parameters are associated with the radiobrightness temperature in the transparency window at a frequency of 19.35 GHz (1.58 cm) and characterize the visible surface (the surface of dense clouds, the World Ocean, and land masses) and the intensity of rainfall. The World Ocean continuously supports power supply to the atmosphere on all scales: local, regional, and planetary. The spatiotemporal structure and variability of this power supply influence the circulation processes in the atmosphere and represent an important component of the climatic system. Change in the thermodynamic conditions on the ocean surface is accompanied by atmospheric disturbances and directly influences the circulation processes in the atmosphere. Detailed analysis of radiothermal fields allows us to advance our understanding of the spatiotemporal dynamics of large-scale processes of heat–mass exchange between the ocean and atmosphere. Figure 4.11 (upper diagram) presents the global radiothermal field constructed from data for September 15, 2006, at a frequency of 19.35 GHz. The lower part of the figure shows the color wedge (the color radio temperature scale) in degrees Kelvin. For convenience, the 360 degrees along which the equator rotates is supplemented by 120 degrees. This enables observation of the structure of atmospheric processes over the water area of each ocean in its entirety. Note that—even on the scale in which the global field is presented in the figure—it is easy to distinguish and identify both the quasi-stationary large-scale structures and dynamical mesoscale structures of the ocean–atmosphere system. Hence, in Figure 4.11 (upper diagram) one can easily identify the frontal zones of various types—climatological, equatorial, Antarctic—as well as a tropical cyclone in the northern Atlantic and a strong midlatitude cyclone in the eastern part of the Northern Atlantic.

An important parameter in the ocean–atmosphere system and climatic system of the planet is atmospheric integral precipitable water—an integrated parameter characterizing the heat–water exchange processes in the system as a whole. A considerable part of the total heat of the ocean–atmosphere system is concentrated in latent heat during transition from vapor to solid. From the remote-sensing point of view it is known that atmospheric integral precipitable water is closely related to the radiobrightness temperature in the resonant line of water vapor radiation (1.35 cm). Radiobrightness temperature variations at a frequency of 22.24 GHz are distinctly recorded in observations from satellites of the DMSP series by the SSM/I multichannel radiometer (Figure 4.11, middle diagram). This figure shows that analysis of atmospheric integral precipitable water is productive on various spatiotemporal scales. As in the case of radiothermal fields at the 19.35 GHz frequency, it is possible to distinguish and identify both the quasi-stationary large-scale structures and dynamical mesoscale structures of the ocean–atmosphere system. The figure also shows that, whereas the pictures are close globally, as a whole (as expected) serious distinctions are observed in the mesoscales. These distinctions are especially clear in the computer animations of radiothermal fields at different frequencies as a result of the high-velocity processes possessing different dynamic properties.

Comparison of these fields with the global radiothermal field at the frequency of 85.5 GHz (Figure 4.11, lower diagram) demonstrates the principal distinction of the spatial structure of this field from previous ones. This is associated with the features of radiothermal radiation formation in the range of 85.5 GHz, which reflects both the water content, and the disperse structure of all types of atmospheric cloud structures, including liquid (rain) and crystal precipitation (hail and snow). Thus, this range allows identification of the general cloud picture in the Earth's atmosphere. This circumstance blocks information on mesoscale processes, however. For example, the images of the tropical cyclone in the Northern Atlantic and the strong midlatitude cyclone in the eastern part of the Northern Atlantic are completely obscured by crystal phase cloud located over the indicated systems (Figure 4.11, upper and lower diagrams).

The global radiothermal fields that are constructed are suitable for the further analysis of thermodynamic processes in the atmosphere both globally and regionally. As an example, the following figures show the radiothermal fields in different regions of the globe. In southern hemisphere early spring the intensive vortical structure has formed and disintegrated over the Indian Ocean's water area in six days (Figure 4.12)—the basic element of this structure was the powerful TC "Elita" (January 27–February 7, 2004) (Pokrovskaya and Sharkov, 2006). The analysis of radiothermal images, presented in the form of a time mosaic (Figure 4.12), evidences that, at the time of highest intensity of TC "Elita" (February 2–6) the vortical system has taken control of the motion of air masses over virtually the whole southern part of the Indian Ocean.

Figure 4.13 shows the detailed radiothermal picture of TC "Podul" in the northwest part of the Pacific Ocean, constructed from data on four frequencies of the SSM/I complex. Note that entrainment by the TC of a large area of water vapor took place in a form of large-scale "ejection" of the water vapor field, rigidly linked with the TC system, from the tropical zone into midlatitudes (see Section 4.4 for more details). Such an ejection is, in essence, one of the components of global poleward latent heat transport from the tropics to high latitudes. Comparative qualitative analysis of these radioimages at various frequencies (i.e., the contribution of water vapor, small and large-drop cloud systems, and precipitation to the energy "ejection" of a tropical cyclone) shows the prevailing contribution of integral water vapor. It is impossible to perform similar estimations from optical and IR data. Figure 4.14 is a successive small-scale mosaic of the evolution of the same tropical cyclone in frequency ranges that demonstrate the visible surface, the integral water vapor content, and total cloudiness. The effect of water vapor field entrainment from the tropical zone, which took place between October 15 and 27, 2001, is well traced not only at the 22.24 GHz frequency (i.e., water vapor) but over all other frequency channels. This implies that transformation also occurs in the field of liquid precipitation and in the field of cloudiness of various phases (large-drop, small-drop, and crystal). We shall consider these effects in more detail in Section 4.4.

4.3.4 Evolution of the radiothermal field of TC "Alberto"

As an example of our study of the radiothermal images of the Northern Atlantic we demonstrate in this section the efficiency of using such dynamic sequences of radiothermal images for studying the TC trajectories and their features in TC motion

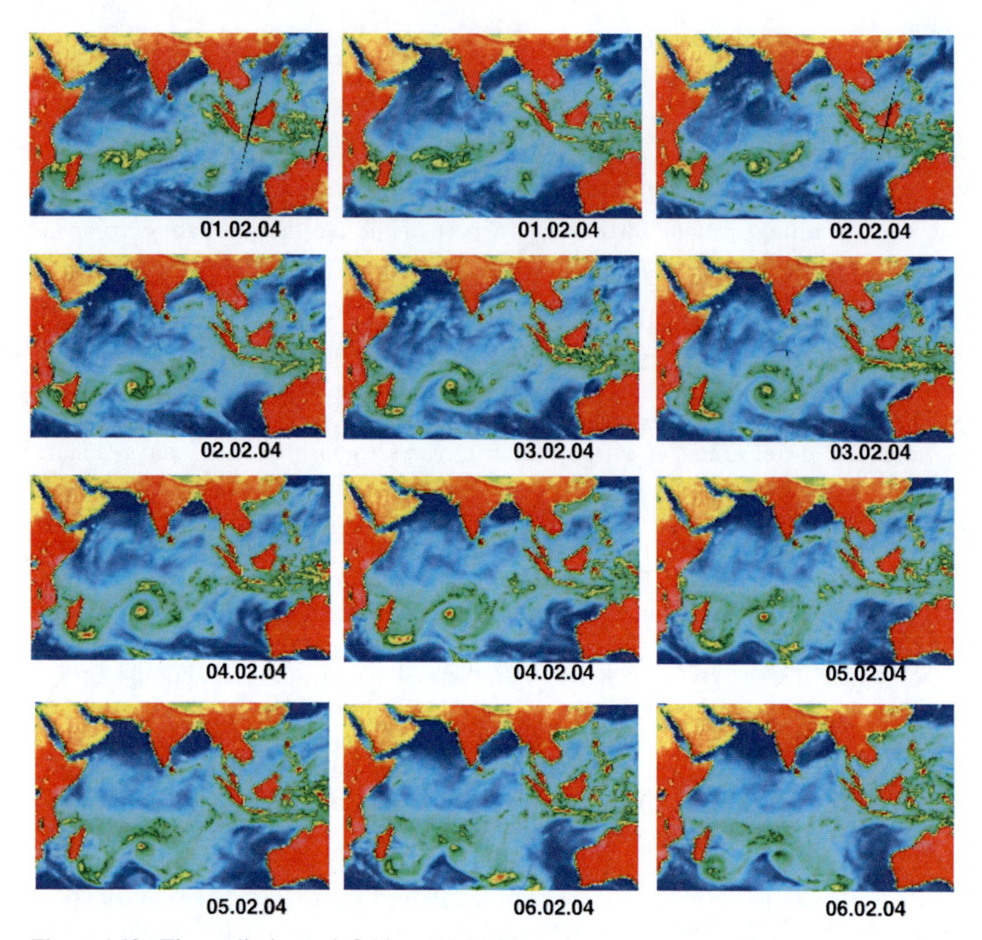

Figure 4.12. The radiothermal field at 19.35 GHz observed over the Indian Ocean basin between February 1 and February 6, 2004.

from the tropical zone to both midlatitudes and middle/high latitudes (Astafyeva and Sharkov, 2008). Basic attention is given to studying the evolution of TC "Alberto" (August 3–23, 2000), which occupies third place in the record of longest lifetime among all Atlantic tropical cyclones and possesses many trajectory features that have not been explained physically until now (Table 4.1). In particular, the trajectory of this cyclone included a large five-day anticyclonic loop, it reached hurricane stage three times, it made an extratropical transition. Comparative analysis of satellite images of the hurricane, obtained at various frequencies in the microwave range reflecting the energy contribution of water vapor, small and large-drop cloud systems, is carried out below. As already mentioned, it is impossible to perform similar estimations based on optical and IR data, because the remote information in these frequency channels "comes" from a rather thin (from tens to

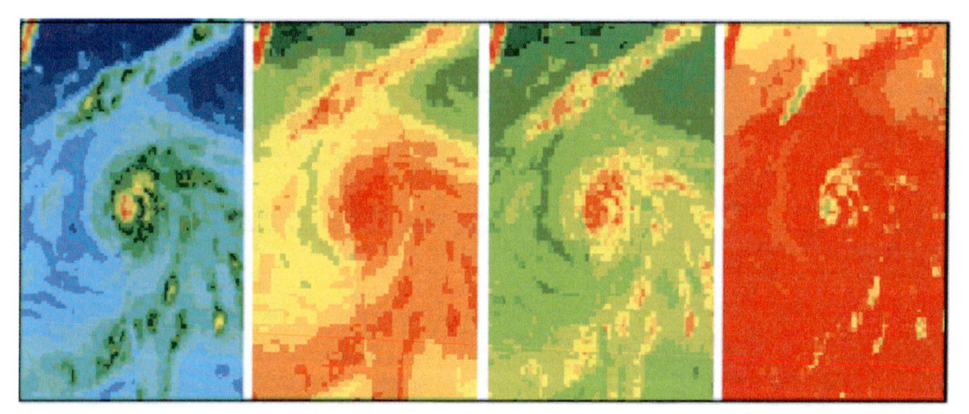

Figure 4.13. Satellite images of TC "Podul" (N0122) in the northwest Pacific taken on October 23, 2001 by SSM/I microwave instruments at 19.35, 22.24, 37.00, and 85.50 GHz.

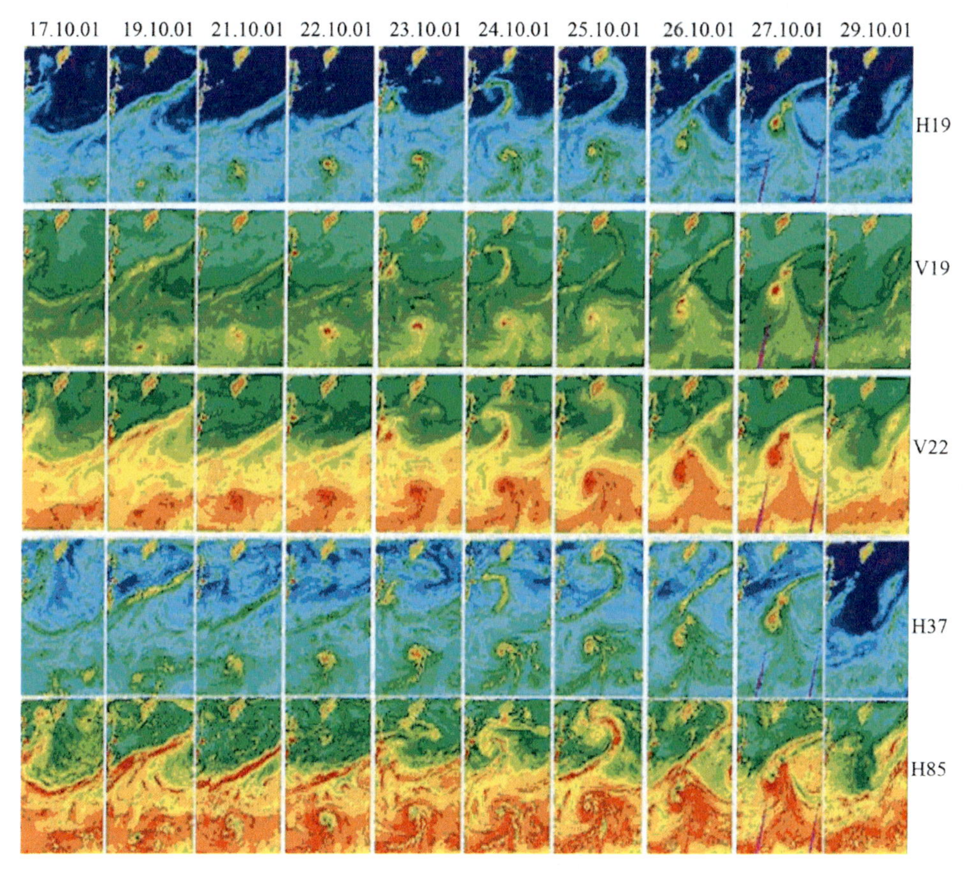

Figure 4.14. Time lapse motion satellite images of TC "Podul" (N 0122) in the northwest Pacific taken over October 17–23, 2001 by SSM/I microwave instruments at 19.35, 22.24, 37.00, and 85.50 GHz at vertical (V) and horizontal (H) polarizations.

Table 4.1. TC "Alberto" evolution.

TC number = ATL 0001				TC name = "Alberto"					All points = 81
No.	*Evolution stage*	*Data* (mm-dd)	*Time* UT	φ (deg)	λ (deg)	*Pressure* (mb)	*Wind velocity* (m/s)	*Shift direction, compass point*	*Shift velocity* (knots)
1	TL	Aug-03	15	10.0	−17.0	1,010	5	W	10
2	TL	Aug-03	21	11.0	−19.0	1,010	5	WNW	15
3	TL	Aug-04	3	12.0	−21.0	1,012	5	WNW	15
4	TD	Aug-04	9	12.2	−22.7	1,007	15	W	16
5	TS	Aug-04	15	12.4	−25.0	1,005	18	W	15
6	STS	Aug-04	21	12.9	−25.9	999	26	W	15
7	STS	Aug-05	3	13.4	−27.5	999	26	WNW	14
8	STS	Aug-05	9	13.7	−28.7	994	28	WNW	13
9	STS	Aug-05	15	14.4	−30.7	1,000	26	WNW	13
10	T	Aug-05	21	14.7	−32.1	990	33	W	13
11	T	Aug-06	3	14.6	−34.1	994	33	W	15
12	T	Aug-06	9	14.4	−35.4	988	36	W	15
13	T	Aug-06	15	14.9	−36.0	987	36	W	11
14	T	Aug-06	21	15.4	−37.2	983	38	WNW	10
15	T	Aug-07	3	16.0	−38.8	983	38	WNW	12
16	T	Aug-07	9	16.2	−40.3	979	41	WNW	13
17	T	Aug-07	15	16.3	−41.7	979	41	W	13
18	T	Aug-07	21	16.6	−42.8	979	41	W	12
19	T	Aug-08	3	16.7	−44.0	981	38	W	11
20	T	Aug-08	9	17.1	−45.6	984	36	WNW	13
21	T	Aug-08	15	17.4	−45.9	984	36	WNW	11
22	T	Aug-08	21	18.8	−47.0	987	33	WNW	11
23	STS	Aug-09	3	20.0	−47.6	990	31	NW	12
24	STS	Aug-09	9	20.8	−48.7	990	31	NW	12
25	STS	Aug-09	15	22.4	−50.5	990	31	NW	15
26	T	Aug-09	21	24.0	−52.0	987	33	NW	9
27	T	Aug-10	3	25.7	−53.5	987	33	NW	19
28	T	Aug-10	9	26.9	−54.6	987	33	NW	18

TC number = ATL 0001				TC name = "Alberto"				All points = 81	
No.	*Evolution stage*	*Data* (mm-dd)	*Time* UT	φ (deg)	λ (deg)	*Pressure* (mb)	*Wind velocity* (m/s)	*Shift direction, compass point*	*Shift velocity* (knots)
29	T	Aug-10	15	28.1	−56.1	987	33	NW	18
30	T	Aug-10	21	29.5	−57.3	987	33	NW	18
31	T	Aug-11	3	30.6	−58.1	981	38	NNW	13
32	T	Aug-11	9	31.6	−58.7	981	38	NNW	16
33	T	Aug-11	15	32.7	−58.7	981	38	N	12
34	T	Aug-11	21	33.8	−58.3	970	46	N	12
35	T	Aug-12	3	35.2	−57.2	965	49	NNE	16
36	T	Aug-12	9	35.7	−56.0	960	51	NE	13
37	T	Aug-12	15	36.3	−54.4	950	57	NE	15
38	T	Aug-12	21	37.1	−53.0	950	65	ENE	15
39	T	Aug-13	3	37.8	−51.0	960	51	ENE	16
40	T	Aug-13	9	38.3	−49.3	970	46	ENE	16
41	T	Aug-13	15	38.7	−47.4	978	46	ENE	16
42	T	Aug-13	21	38.9	−45.8	979	38	E	15
43	T	Aug-14	3	39.1	−43.9	982	36	E	14
44	T	Aug-14	9	39.0	−41.2	988	33	E	20
45	STS	Aug-14	15	39.0	−40.0	990	31	E	18
46	STS	Aug-14	21	39.0	−38.7	997	26	E	14
47	TS	Aug-15	3	38.6	−38.5	998	23	S	6
48	TS	Aug-15	9	37.8	−38.4	998	23	S	7
49	TS	Aug-15	15	36.9	−38.6	998	23	S	9
50	TS	Aug-15	21	36.4	−39.2	1,000	21	SW	8
51	TS	Aug-16	3	35.8	−39.6	1,000	21	SW	7
52	TS	Aug-16	9	35.2	−40.6	1,000	21	SW	8
53	TS	Aug-16	15	34.3	−41.9	1,000	21	SW	11
54	TS	Aug-16	21	33.6	−43.2	1,000	21	WSW	13
55	TS	Aug-17	3	33.2	−44.2	998	23	WSW	11
56	TS	Aug-17	9	32.9	−44.4	998	23	W	6

(*continued*)

Table 4.1 (*cont.*)

TC number = ATL 0001				TC name = "Alberto"				All points = 81	
No.	*Evolution stage*	*Data* (mm-dd)	*Time* UT	φ (deg)	λ (deg)	*Pressure* (mb)	*Wind velocity* (m/s)	*Shift direction, compass point*	*Shift velocity* (knots)
57	TS	Aug-17	15	32.9	−45.3	998	23	W	8
58	STS	Aug-17	21	33.1	−46.4	994	28	W	9
59	STS	Aug-18	3	33.4	−46.8	994	28	WNW	6
60	STS	Aug-18	9	33.7	−47.4	994	28	WNW	6
61	T	Aug-18	15	34.5	−47.8	989	33	NNW	7
62	T	Aug-18	21	35.1	−48.3	986	38	NNW	7
63	T	Aug-19	3	35.2	−48.3	986	36	NNW	5
64	T	Aug-19	9	35.6	−48.4	974	44	NNW	4
65	T	Aug-19	15	36.0	−48.3	970	46	N	6
66	T	Aug-19	21	36.5	−48.1	966	49	N	6
67	T	Aug-20	3	36.7	−48.1	970	46	N	4
68	T	Aug-20	9	37.0	−48.0	970	46	N	5
69	T	Aug-20	15	37.5	−47.9	975	44	N	6
70	T	Aug-20	21	37.9	−47.7	970	44	N	6
71	T	Aug-21	33	8.1	−47.4	975	44	NNE	5
72	T	Aug-21	9	38.8	−47.3	979	38	NNE	5
73	T	Aug-21	15	39.3	−47.1	975	38	NNE	5
74	T	Aug-21	21	40.6	-46.4	975	38	NNE	10
75	T	Aug-22	3	41.8	−45.7	981	38	NNE	12
76	T	Aug-22	9	43.3	−44.9	983	33	NNE	15
77	T	Aug-22	15	45.0	−43.4	985	33	NNE	17
78	L	Aug-22	21	47.2	−41.4	983	33	NNE	24
79	L	Aug-23	3	49.7	−37.8	986	31	NE	32
80	L	Aug-23	9	52.3	−34.7	986	26	NE	40
81	L	Aug-23	15	54.6	−34.1	994	23	NE	32

Note: TC evolution stages: TL is initial tropical disturbance; TD is tropical depression; TS is tropical storm; STS is strong (super) tropical storm; T is typhoon (hurricane); L is extratropical disturbance (in accordance with "best-track" pre-processing published in Pokrovskaya and Sharkov, 2006); φ is latitude; λ is longitude (in degrees).

a few hundred meters) upper layer of cloud systems and does not contain information on water vapor content. In this work the author has used satellite data on Earth monitoring within the DMSP program framework from the Global-RT database (see Section 8.7), on the global radiothermal fields of the Earth from the Global-Field, on global tropical cyclogenesis from the Global-TC database (see Section 8.4), and ground-based data from the NHC-TPC NOAA (National Hurricane Center–Tropical Prediction Center of NOAA). Analysis of the distribution fields of radiobrightness temperature has shown their high information capacity. It was shown that the large-scale atmospheric environment had a noticeable (virtually determining) effect both on the trajectory of TC "Alberto" and on changes in its intensity. Thus, theoretical models should take into account the dynamic and meteorological conditions in distant parts of the large-scale environment of a cyclone to adequately describe the dynamics and power of a TC.

TC "Alberto" turned out to be the most intensive hurricane of 2000, the most long-lived TC of those generated over the Atlantic in August, and took third place in the league of champions—the most long-lived TCs over the Atlantic for all years of observations. Losing and gathering force anew, TC "Alberto" reached the hurricane stage three times (40, 55, and 45 m/s, respectively) and in the intervals dropped down a stage to a tropical storm. Its trajectory included a large 5-day anticyclonic loop, a large part of which was extratropical. There were many times during the lifetime of TC "Alberto" when its trajectory was virtually unpredictable using standard techniques (Astafyeva and Sharkov, 2008). An extraordinary feature of the trajectory of TC "Alberto" was the fact that it remained over Atlantic Ocean basins throughout its life.

Let us now briefly consider the synoptic evolution of TC "Alberto" (Table 4.1). A well-developed tropical wave (the TW stage) was noticed on satellite optical images over central Africa as early as July 30. It moved westward in a stable fashion and on August 3 reached the coastline of the Atlantic Ocean (Figure 4.15). The structure quickly developed over the Atlantic. On that same day the wave became a tropical depression (the TD stage). The cyclone moved west-northwest at a velocity of 8 m/s to 10 m/s, became a tropical storm (the TS stage), and was named "Alberto" at the beginning of the next day (August 4). TC "Alberto" continued to amplify and reached hurricane status (the T stage) early on the morning on August 6. This event took place on a short section of the trajectory when the cyclone changed direction and moved strictly westward. A short while later on the same day (August 6) the hurricane renewed its motion toward the west-northwest and continued moving in this direction until August 7, when it reached its first peak of intensity (at a velocity of 40 m/s). Figure 4.15 presents the trajectory of TC "Alberto" with the fragments of its radiothermal field at a frequency of 19.35 GHz.

The mosaic fragment corresponding to August 7 shows a dry air area north-eastward of TC "Alberto" in the region of the Azores. On August 7 and 8 some of this dry air (the dark blue color on the fragment) penetrated westward and south-westward of TC "Alberto". As a result, a considerable drop in pressure was recorded there at a height of about 10 km. This caused an increase in vertical shift, forcing the

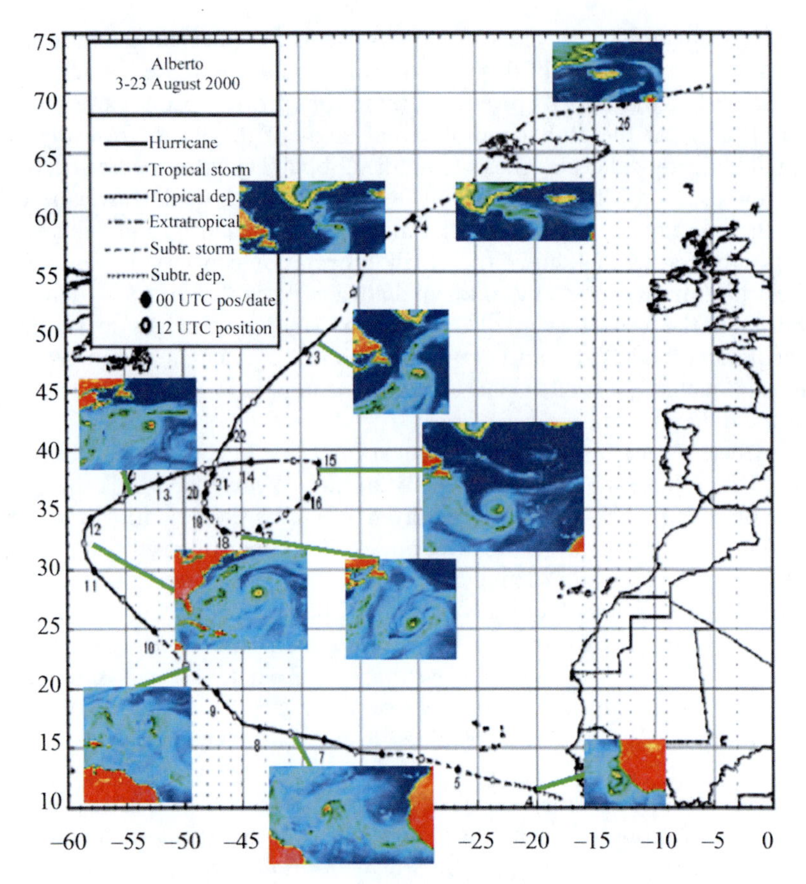

Figure 4.15. Track and intensity evolution of TC "Alberto" (August 3–23, 2000) over the Atlantic Basin with the inclusion of radiothermal field fragments at 19.35 GHz.

hurricane to turn to the northwest on August 8 and losing momentum down to tropical storm stage on August 9.

TC "Alberto" continued moving rapidly northwest on August 10 and restored its hurricane force, bolstered by moist warm air masses coming into this region westward of TC "Alberto" (Figure 4.16). This is clearly seen on the middle diagram of Figure 4.16, which corresponds to the radioimage at the frequency of 22.2 GHz. The light brown color westward of the cyclone center simply demonstrates the presence of warm and moist air masses. Next day, on August 11, "Alberto" passed approximately 550 km east of Bermuda. On August 11–12 the storm gradually turned toward the north and north-northeast through the break in the subtropical crest.

The hurricane reached its second and highest peak of intensity (55 m/s) on August 12. It was on that day that its eye extended more than 90 km in diameter. Having reached the western zonal flux, the cyclone moved very slowly northeastward

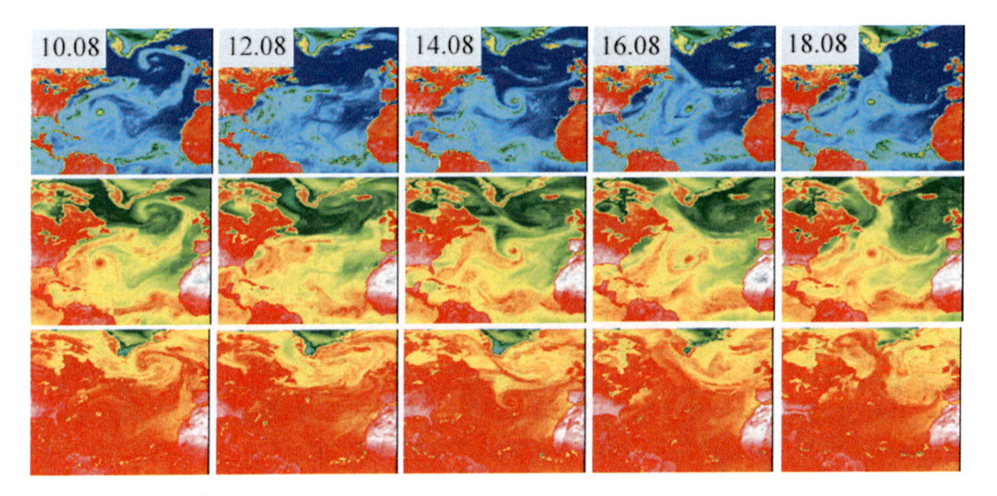

Figure 4.16. Radiothermal images of TC "Alberto" at maximum activity (August 10–18, 2000) at frequencies of 19.35, 22.24, and 85.5 GHz (from top to bottom), which are characteristic of water transfer and water vapor transfer by TCs.

on August 12, 13, and 14, but lost momentum during this motion. The mosaic fragments of August 12 (Figure 4.16) and later clearly depict the extensive area of dry cold air at the north, in the Iceland depression region, which prevented TC "Alberto" from moving northeast (the dark green color of the middle diagram). Finally, on August 15 this area suddenly extended and forced "Alberto" southward. The cyclone then suddenly changed its direction of motion from the northeast to the southeast. In the next few days (August 15 to August 18) the cyclone first moved southward and then southwestward, its trajectory having taken on the form of an anticyclonic loop. Mosaic fragments from August 15 to August 17 show how dry air from the Iceland depression penetrated the structure of the cyclone, weakening it greatly. Over these days the cyclone lost most of its convective energy and reached the second minimum of intensity on August 16.

By moving southwest "Alberto" began to replenish its stock of energy as a result of water vapor flows from both southern and eastern directions (Figure 4.16). The storm began to amplify on August 17 and soon restored its hurricane status—the third time within a few days. On August 20 the hurricane reached its third peak of intensity (45 m/s) and its eye had a diameter of 110 km. By this time "Alberto" had completed its motion along the anticyclonic loop. It moved northwestward over this loop on August 18, northward on August 19, and north-northeastward on August 20 and 21.

The next weakening and then acceleration of TC "Alberto" took place on August 22. The cyclone suddenly weakened to the tropical storm stage prior to becoming an extratropical storm the next day. Extratropical TC "Alberto" continued moving north-northeastward and passed close to Iceland on August 24. On August 25, the storm moved east-northeastward, but its center was scarcely determinable. According to NHC data, "Alberto" dissipated the same day. However, it could be seen on radiothermal fields of August 26 near 75°N.

Meteorological data on TC "Alberto" are very scarce. This was because the hurricane was located over the ocean throughout its lifetime, far from land; moeover, aerial reconnaissance flights were not performed. A few ships witnessed the storm and some drifting beacons sent data relating to "Alberto".

Since TC "Alberto" was believed to miss any continentinental land mass, there was no necessity to issue warnings of its passage. The official forecast of TC "Alberto" development and trajectory was by and large good. However, there were two periods when official forecasts made TC "Alberto" out to be much worse than average hurricanes.

Analysis of the evolution of radiothermal images demonstrates the reason for these forecasting discrepancies. The first period was between 00:00 and 18:00 UTC on August 8, when TC "Alberto" slowed down its westward motion, turned northward and, virtually, began to move toward the northwest, instead of its expected motion toward the west-northwest. At that time the errors of 72-hour forecasts were more than 925 km. Analysis of the radiothermal fields explains such a change in motion by dry air intrusion into the hurricane structure from the Azores Maximum (the atmospheric center of action in the region of the Azores, the North Atlantic Anticyclone) (Figure 4.15).

The second period was between 12:00 UTC on August 11 and 12:00 UTC on August 12. The errors here were even more significant: from 1,110 km to 1,740 km over the 72-hour forecast. At that time, instead of fast motion toward the east-northeast (according to the official forecast), TC "Alberto" actually began to form an anticyclonic loop. Analysis of the radiobrightness temperature fields shows that the reason for the trajectory to take on the form of an anticyclonic loop was an extensive area of dry cold air to the north, in the Iceland depression region, which prevented "Alberto" from moving toward the north-northeast and forced it southward. This area is clearly seen on the mosaic fragments of August 12–14 and later (Figures 4.15 and 4.16).

The two periods in which the official forecasts were inaccurate, TC "Alberto" was under the influence of its large-scale atmospheric environment: the meteorological situation in the Atlantic regions located northward of TC "Alberto". The intrusion of dry air (near the Azores center of action) into the cyclone structure caused sudden turns and many drops in intensity, and the presence of an extensive dry cold air area in the Iceland depression region resulted in the trajectory of TC Alberto forming an anticyclonic loop. The cyclone finally dissipated because of the jet intrusion of cold and dry air masses into its structure from the Iceland depression, where it simply dumped the water vapor mass it brought from the equatorial zone.

Thus, analysis of the radiobrightness temperature fields in the 19.35 GHz and 22.24 GHz ranges demonstrates their capacity to deliver a lot of information. Water vapor (and the radiobrightness temperature field in the corresponding frequency range) can be considered to be a very representative tracer of atmospheric motions, since water vapor is entrained by atmospheric motions, and its propagation in the troposphere is controlled by large-scale and small-scale motions. However, the frequency range of 85.5 GHz turned out to be much less informative (Figure 4.16) by virtue of the fact that in this frequency range almost all cloud systems have a strong effect on thermal radiation within the wide range in which they disperse (small and large drops of water, small ice crystals, snowflakes, hail, etc.) (Sharkov, 2003). Moreover, to adequately describe the dynamics and power of a tropical cyclone, theoretical models need to take into account the dynamic and meteorological conditions throughout the environment of a cyclone.

4.4 GLOBAL WATER VAPOR FIELD AS THE ENERGY SOURCE OF TROPICAL CYCLONES

The optical and IR ranges in the remote study of TCs are known to have serious restrictions. They include the complete absence of data on the internal structure of a tropical cyclone and on the water vapor field (latent heat) contained in the atmosphere around a cyclone. Thus, the general picture of the energy involved in tropical cyclone evolution, both inside the body of the cyclone and in its environment, remains unknown. This mainly relates to energy sources that result in extremely fast (and rather unexpected for observers) intensification of tropical cyclones, as was the case, for example, with TC "Katrina" (2005). Today, it has become clear that these tasks cannot be solved by IR and optical techniques alone (Sharkov, 2006; 2010a, b). As shown in Section 4.3, a principal step toward rectifying this situation is analysis of the data of radiothermal microwave satellite complexes, which can reveal the global picture of the distribution of many meteorological parameters—especially, water vapor fields. This is the gas component of the atmosphere that contains huge stocks of latent heat, which can be transformed into the kinetic and potential energy of tropical cyclones. In the present section we consider the role and significance of tropical cyclogenesis as an effective channel for latent heat energy transfer from the tropics to the midlatitudes in poleward transport mode (Sharkov, 2010; Sharkov *et al.*, 2008a, b, 2010, 2011a, b; Kim *et al.*, 2009, 2010).

4.4.1 Statement of the problem

One of the most important climate-forming factors on the Earth is considered to be the multiscale (in space and time) interaction of the ocean and atmosphere, which entails the diverse processes of energy, momentum, and substance exchange. A fundamental question in this respect is revealing the role and contribution of intensive vortical tropical disturbances—especially, the most powerful of them: tropical cyclones—to this interaction (Henderson-Sellers *et al.*, 1998; Barry *et al.*, 2002; Webster *et al.*, 2005; Vecchi and Soden, 2007; Trenberth and Fasullo, 2007; Sharkov, 1998, 2000, 2006, 2010). In the overwhelming majority of works devoted to this subject, the formation of latent heat fluxes from the ocean's surface layer by means of the convective activity in a hurricane's circulation system is considered to be the basic mechanism by which the TC interacts with the oceanic environment. Expressing the fundamental question in this way is called the "convective hypothesis"—it is traditional and has many adherents—simply deals with the spatiotemporal evolution of the surface temperature field of the World Ocean and its effect on tropical cyclogenesis (Gray, 1979; Palmen and Newton, 1973; Tarakanov, 1980; Webster *et al.*, 2005; Vecchi and Soden, 2007; Trenberth and Fasullo, 2007; Golitsyn, 2008; Hassim and Walsh, 2008; Semmler *et al.*, 2008; Sharkov and Pokrovskaya, 2010). In other words, the convective hypothesis considers the energy stock of the World Ocean's upper layer as the basic energy source for the functioning of tropical cyclogenesis.

However, such a viewpoint has recently shown obvious signs of inadequacy; for example, the speed with which TC "Katrina" was able to rearrange itself and intensify in such a catastrophic way. A fundamental issue in the primary cyclogenesis and intensification of various forms of tropical cyclones is revealing what causes the energy source to be so powerful and (crucially) rearrange itself so quickly, resulting in the extremely rapid intensification and formation of mature forms of tropical cyclones. Sharkov *et al.* (2008a, b, 2010, 2011a, b) and Kim *et al.* (2009, 2010) experimentally show the possibility of the existence of a principally different mechanism of TC intensification and energy transfer from the tropical zone of the ocean–atmosphere system to middle and high latitudes—a mechanism that involves the formation of compact, highly intensive regions of integral water vapor trapped by the TC and transported from the equatorial region (the ITCZ zone) of the global water vapor field. Conventionally, this process can be defined as the effect of entraining part of a region that has a heightened integral value of water vapor and transferring it from the equatorial zone to middle and middle–high latitudes.

This effect was discovered for the first time by Sharkov *et al.* (2008a, b), during study of the multispectral satellite remote observation of the evolution of TC "Gonu" (the Arabian Sea, the Northern Indian Ocean; May 31–June 8, 2007). It was shown that a compact water vapor region was trapped by TC "Gonu" from the equatorial water vapor zone of the ITCZ and was "thrown" into the frontal zone of moderate latitudes stretching from the Rub' al Khali desert up to the Iranian Plateau. A similar result was obtained slightly later by the same team of authors (Kim *et al.*, 2009; Sharkov *et al.*, 2011a, b), who revealed the main energy source behind the functioning of the powerful TC "Hondo" in the Southern Indian ocean basin under hydrometeorological conditions that were essentially different from the conditions under which TC "Gonu" evolved. The mechanism of latent heat transfer of water vapor from the equatorial zone to the midlatitudes by means of multiple tropical cyclogenesis in southern hemisphere oceanic basins (the Southern Indian Ocean and the southwest part of the Pacific Ocean) was studied, based on multispectral satellite observational data, by Sharkov *et al.* (2010, 2011a). This section looks at these works in detail.

Note that in some works (particularly, Liu and Tang, 2005) the zonal and meridian components of the parameter of water vapor transfer parameter were formed by means of 3-monthly time averaging (based on the data of microwave radiothermal satellite missions). This almost certainly excludes from consideration the influence of TC-type formations with a short timescale on polar transfer. Barry *et al.* (2002) take the view that in the baroclinic zone of the atmosphere (the zone of high-temperature gradients and active vortices) the basic elements of polar transfer are midlatitude storms. This point of view is based on numerical simulation results, rather than on experimental natural data. In what follows we shall demonstrate the existence of another essentially different mechanism of energy transfer from the tropical zone of the ocean–atmosphere system to midlatitudes and high latitudes as a formation of compact regions of integral water vapor of heightened intensity, trapped by a TC from the equatorial region of the global water vapor field. One-day global radiothermal fields, constructed and used for analysis in the present section (see Section 8.7), adequately take into account the effect of fast-varying intensive TC-type vortical atmospheric formations on the horizontal transfer of heat and mass (moisture) from the equator toward the poles, which cannot be done without taking account of other spatiotemporal scales.

4.4.2 Initial data of satellite-sounding and information-processing algorithms

The initial data for this study were products in the form of restored microwave images from the website of the NRL Monterey Marine Meteorology Division (*www.nrlmry.navy.mil*), obtained from the SSM/I and AMSR-E microwave complexes and updated in real time, as well as IR images of the northern Indian Ocean basin obtained from the geostationary satellite Meteosat (see Chapter 9). Sharkov *et al.* (2008a, b, 2010, 2011a, b) and Kim *et al.* (2009, 2010) use products in which the integral water vapor content in the atmosphere is restored by merging data from the SSM/I and AMSR-E instruments, since they thoroughly reflect the energy transformation processes occurring in the tropical zone.

The algorithm used to restore data on the full moisture content in the atmosphere is described in detail at *www.ssmi.com/papers/rain.pdf* Data from the SSM/I and AMSR-E instruments were processed in accordance with the traditional techniques of atmospheric parameter restoration based on satellite microwave data.

The same team of authors (mentioned above) analyzed in detail the energy features of TC "Gonu" on the basis of the data-merging method, which is the culmination of the development of the technique of information forming and accumulation based on multiscale satellite remote-sensing data. This method is now actively used for studying virtually stationary objects and fields, such as the field of chlorophyll content in the oceanic surface (e.g., Gregg, 2007). However, when using this method to study such fast processes as TC evolution, the method needs to be essentially updated, as just outlined by the authors.

Since the initial data were raster images (see Figure 4.17a), in order to digitally process them a special program was prepared with the objective of reading the image pixel by pixel and acquiring data in accordance with the color scale (Figure 4.17c) of the image in the form of a five-dimensional data file (from three colors and two coordinates) on each image. Images with TCs were visually chosen as sections over which the calculation would be carried out (Figure 4.17b). To determine the numerical values of the amount of moisture in millimeters, a color-coding scale was used. Each image pixel was compared with this scale by the least squares technique for each of the channels.

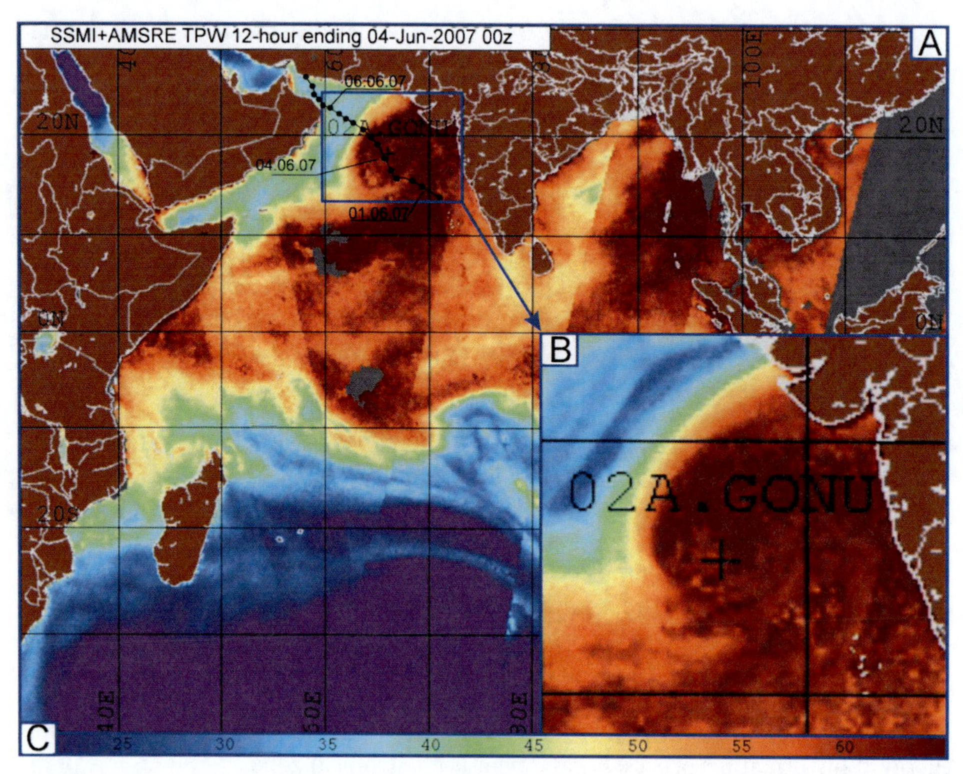

Figure 4.17. TC "Gonu" (02A) on June 4, 2007 over the North Indian Ocean basin in integral water vapor, compiled using the integrated approach from SSM/I observations on NASA's DMSP satellites and the AMSRE on NASA's Aqua satellite: (a) the initial image (NRL Monterey Marine Meteorology Division Total_Precip Water_(TPW)); (b) the image fragment selected for calculation of the energetic properties (latent heat) of the water vapor field entrained by TCs; (c) false-color coding image scale. Black points correspond to the TC's track.

By applying the data-merging method modified by the authors, a time series of combined images was obtained (Figure 4.18), which included superposition of two initially multiscale images—space images from the Meteosat-7 satellite (the black-and-white part of the image) and images of the integral water vapor regions (the color part of the image), from microwave data of the Aqua and DMSP missions—on which further processing was performed. The figure clearly shows that the region of integral water vapor content, which stands out against an atmosphere undisturbed by a cyclone, remains attached to the cloud system (according to the IR data) of the tropical cyclone throughout its lifetime (from June 1 to 6, 2007). This region in the water vapor field is dated to the cyclone's center of action. However, the water vapor region itself has an essentially larger spatial size than that reflected by cloud structures in the IR range. In addition, time analysis of the state of the integral water vapor field (Figure 4.18) shows that the zone, dated directly to the tropical cyclone

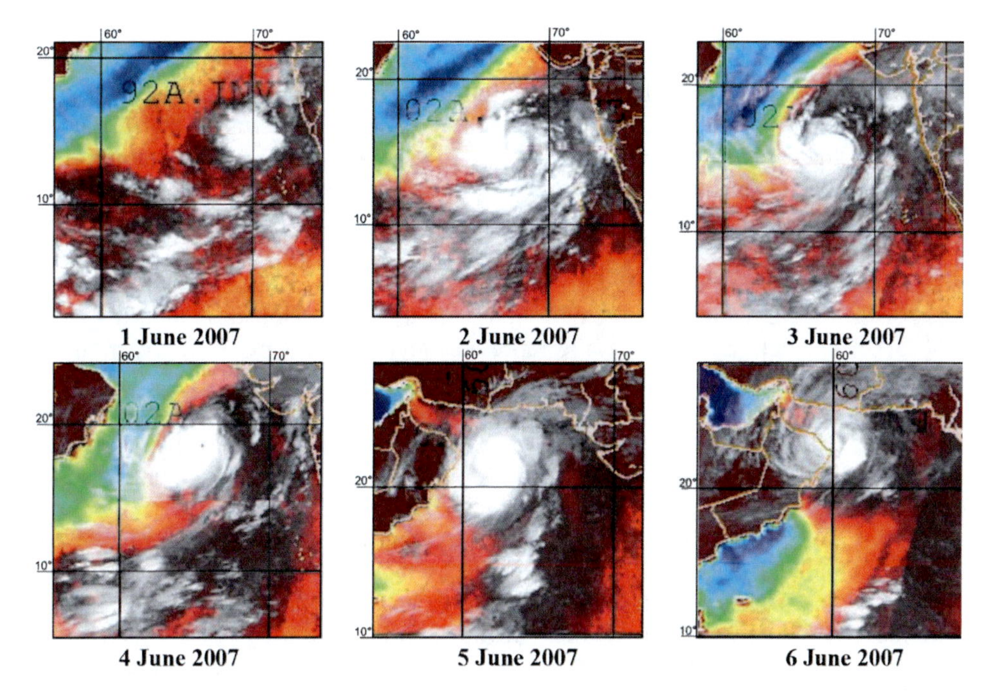

Figure 4.18. TC "Gonu" between June 1 and June 6, 2007 over the North Indian Ocean basin in integral water vapor, compiled using the time lapse motion approach from SSM/I observations by NASA's DMSP satellites and the AMSRE on NASA's Aqua satellite (NRL Monterey Marine Meteorology Division Total_Precip Water_(TPW)) and from IR images from Meteosat (white).

and defined as a system of cloud vortices, has a considerable transition zone (some kind of jet) that joins up with the main water vapor field of the equatorial zone, and the additional "supply" of water vapor to the TC from the basic equatorial zone probably occurs through this transition zone. In order to separate these spheres of influence, a threshold value was empirically estimated and selected, below which the water vapor field was not considered to "belong" to the cyclone. Further, the calculation made an allowance for the size of the chosen section, projection features, and measurement errors.

Translation of the obtained values of moisture quantity into the latent energy value was carried out on the basis of the following assumptions: a precipitated water vapor of thickness 1 mm (according to the initial data) corresponded (with regard to a water density of $1\ \mathrm{g/cm^3}$) to a water mass of 1 kg over an area of $1\ \mathrm{m^2}$, or 10^6 kg over an area of $1\ \mathrm{km^2}$. On the other hand, it is known (Prochorov, 1984) that the specific latent heat of the transition from vapor to solid equals 2.5×10^6 J/kg. Thus, with a precipitated water vapor thickness of 1 mm, the condensation latent heat value will be 2.5×10^{12} J over an area of $1\ \mathrm{km^2}$.

4.4.3 Entrainment effect in the evolution of TC "Gonu"

The purpose of this and subsequent sections is to reveal the main energy source behind TC "Gonu" and TC "Hondo". The choice of the first TC was stipulated by the fact that in the northern basins of the Indian Ocean the integral water vapor content is much lower than in the tropical zone and, consequently, the mode of water vapor intrusion from the basin surrounding a tropical cyclone is completely excluded. Analysis of how the intensity of TC "Gonu" evolved (the Arabian Sea, the Northern Indian Ocean; May 31–June 8, 2007) into the integral water vapor field was performed on the basis of the data-merging method developed by Sharkov *et al.* (2008a, b). The data-merging method represents an advancement of the technique of information forming and accumulation based on the multiscale data of satellite remote sensing in the infrared and microwave ranges.

The category-5 TC "Gonu" developed, functioned, and dissipated in the basin of the Arabian Sea and the Gulf of Oman in immediate proximity to the Arabian desert at Rub' al Khali and the Hindustan Thar desert.

The initial tropical disturbance was recorded on May 31, 2007, westward of the coast of India, at 13.2°N, 72.6°E (see Table 4.2 which was taken from the electronic Global-TC database and compiled using the techniques presented in Pokrovskaya and Sharkov (2006). The wind velocity at the disturbance center was 15 m/s and the pressure was 1,006 mb. On June 2 the tropical disturbance transferred into a tropical storm, the wind velocity was 21 m/s, and the pressure at the center dropped down to 1,000 mb. The tropical cyclone peaked on June 4, when a wind velocity of 140 knots (73 m/s) was recorded; it was situated southeastward of the coast of Oman at that time. The cyclone strengthened as a result of favorable high-altitude outflow and a small shift in high-altitude wind. On June 5, on contact with the land, the cyclone began to weaken and was displaced northward, toward the southern coast of Iran, along the southwest periphery of a strong high-altitude crest located over the southwest of Pakistan. The TC dissipated on June 8 over the Strait of Hormuz and southern Iran, as a result of its internal structure being destroyed on contact with the land and further absorption of the whole structure (including the water vapor region carried by the cyclone) by the large-scale circulation of moderate latitudes, which stretched from the desert at Rub' al Khali up to the Iranian Plateau.

Detailed analysis of the energy features of TC "Gonu"—based on the data-merging method of the multiscale data of satellite remote sensing—has shown that the category-5 TC "Gonu" that formed and developed in the Arabian Sea basin, near the western coast of India, had further functioned in the immediate proximity of the Arabian Desert at Rub' al Khali and the Hindustan Thar desert, under the dry atmospheric conditions of the Arabian Sea. The source of the latent heat energy necessary for it to function and intensify could only be a sizable water vapor region trapped by the TC from a tropical zone experiencing monsoonal circulation and considerably exceeding its size, as routinely determined from optical and IR observation data (i.e., optically and IR-identified cloud systems). This was the zone from which the TC could quickly entrain energy in latent heat form, whereas the mechanism of evaporation from the ocean surface is slow. Such a mechanism of water vapor region entrainment by a tropical cyclone is conventionally called the "camel" model, as it has a similar attribute to the ship of the desert. An effect similar to this was also found in the evolution of tropical cyclones in the North Atlantic as a result of analyzing the dynamics of the microwave natural radiation field in the range of 22.2 GHz (the line of natural radiation of water vapor), without restoring the integral water vapor field and, accordingly, without understanding the energy of the process (see Section 4.3.4). Apparently, all tropical cyclones possess the mentioned property and, owing to this effect, eject a huge quantity of latent heat into the midlatitudes and high latitudes, which just goes to show their undoubtedly cardinal role in forming climatic processes in the Earth's atmosphere.

So, if the entrainment effect exists and provides the energy for a tropical cyclone in accordance with the camel model hypothesis, then it should be reflected, nearly synchronously, in the dynamic properties of a tropical cyclone. We shall show that this is indeed the case. The results of calculating the energy balance are presented in Figure 4.19, from which it follows that—as the intensity of the tropical cyclone increased (in particular, the wind velocity in the eye wall, for the period of June 1–3, 2007; see points 4–11 of Table 4.2)—total latent energy in the water vapor region also increased as a result of the water vapor mass "pulling out" of the equatorial region. By the end of June 3 and the beginning of June 4 the tropical cyclone suddenly intensified (see points 12–16 of Table 4.2 and Figure 4.19) and the energy in the water vapor region dropped down to approximately 0.5×10^{20} J at the maximum development stage. It is reasonable to assume that it was this amount of latent energy that was spent on increasing the cyclone's kinetic energy, especially as in the cyclone vicinity there was no other energy source with the equivalent energy. As the intensity of the cyclone lessened and dissipated the water vapor region gradually broke down, still occupying a large area (Figure 4.18 for June 6, 2007). Final destruction of the whole TC structure, including the water vapor region transported by the cyclone, took place on a large-scale high-altitude crest of pressure, which stretched from the desert at Rub' al Khali up to the Iranian Plateau.

Thus, there exists a strict correlation between the kinematic characteristics of a TC and the total latent heat energy of a TC's integral water vapor zone.

4.4.4 Entrainment effect in the example of TC "Hondo" evolution

The category-5 TC "Hondo" developed, functioned, and dissipated in the Southern Indian Ocean basin from February 2 to 26, 2008. Its evolutionary history was unusual (see Table 4.3, taken from the Global-TC database and compiled using "best-track" pre-processing—a technique developed by Pokrovskaya and Sharkov, 2006—of the initial data presented on the IFA website *http://www.solar.ifa.hawaii.edu*).

The primary tropical disturbance arose on February 2 at 15:00 GMT at the southern periphery of the ITCZ, at about 11°S, 83°E, and represented a smeared poorly organized cloud system. Sea level pressure was 1,004 mb and the wind velocity was about 10 m/s.

Table 4.2. TC "Gonu" evolution.

TC number = NIN 0702			TC name = "Gonu"					All points = 28	
No.	*Evolution stage*	*Data* (mm-dd)	*Time* UT	φ (deg)	λ (deg)	*Pressure* (mb)	*Wind velocity* (m/s)	*Shift direction, compass point*	*Shift velocity* (knots)
1	TL	May-31	12	13.2	72.6	1,006	8	W	10
2	TL	Jun-01	0	13.6	71.4	1,006	11	WNW	10
3	TD	Jun-01	12	14.6	69.7	1,000	13	WNW	10
4	TD	Jun-01	18	15.1	68.8	1,000	13	WNW	8
5	TS	Jun-02	0	15.4	67.7	1,000	21	W	7
6	TS	Jun-02	6	15.4	67.5	1,000	18	W	2
7	STS	Jun-02	12	15.3	67.1	987	28	WNW	3
8	STS	Jun-02	18	15.6	66.9	984	28	WNW	3
9	STS	Jun-03	0	15.8	66.7	987	26	NW	6
10	STS	Jun-03	6	16.8	67.4	980	31	NNW	8
11	T	Jun-03	12	17.2	66.1	976	33	WNW	10
12	T	Jun-03	18	17.5	66.6	954	46	NE	8
13	T	Jun-04	0	18.5	65.5	927	60	NW	7
14	T	Jun-04	6	19.2	64.9	—	68	NW	7
15	T	Jun-04	12	19.9	64.1	—	73	NW	10
16	T	Jun-04	18	20.5	63.2	—	73	WNW	10
17	T	Jun-05	0	20.9	62.5	—	70	WNW	8
18	T	Jun-05	6	21.3	61.9	—	65	NW	7
19	T	Jun-05	12	21.9	61.1	940	54	NW	10
20	T	Jun-05	18	22.1	60.4	954	46	NW	7
21	T	Jun-06	0	22.6	60.0	962	41	NW	8
22	T	Jun-06	6	23.1	59.5	966	38	NW	7
23	T	Jun-06	12	23.9	59.4	971	36	N	8

TC number = NIN 0702				TC name = "Gonu"					All points = 28
No.	*Evolution stage*	*Data* (mm-dd)	*Time* UT	φ (deg)	λ (deg)	*Pressure* (mb)	*Wind velocity* (m/s)	*Shift direction, compass point*	*Shift velocity* (knots)
24	STS	Jun-06	18	24.7	58.8	980	31	NNW	10
25	TS	Jun-07	0	25.1	58.4	991	23	NNW	7
26	TS	Jun-07	6	24.9	58.1	997	18	SW	4
27	TD	Jun-07	12	25.1	58.3	1,000	13	NE	3
28	TL	Jun-08	0	25.9	57.5	1,002	8	NW	6

Note: The TC collapsed over the Iranian Plateau. TC evolution stages: TL is initial tropical disturbance; TD is tropical depression; TS is tropical storm; STS is strong (super) tropical storm; T is typhoon (hurricane) (in accordance with " best-track" pre-processing published in Pokrovskaya and Sharkov, 2006); φ is latitude; λ is longitude (in degrees). In points 7–28, sea level pressure was calculated in accordance with the online processing method presented at *http://www.ssd.noaa.gov/PS/TROP/CI-chart.html*

Within a day, storm activity had strengthened and the pressure at the center of the system dropped by 4 mb, which promoted intensification of the disturbance and its transition into the tropical depression stage. The cloud disk acquired a more distinct outline. On February 5, under favorable environmental conditions, the wind strengthened and the disturbance transferred into a tropical storm. The pressure dropped to 997 mb and the wind velocity grew to 18 m/s. Within another day, the disturbance quickly amplified, and the dense central small core with

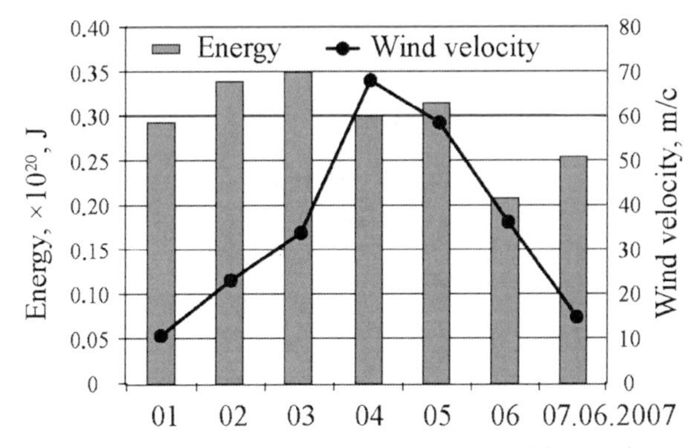

Figure 4.19. Time evolution of latent heat for the entrained integral water vapor field and tropical cyclone intensity during its lifetime (July 1–7, 2007).

Table 4.3. TC "Hondo" evolution.

TC number = SIO 0803				TC name = "Hondo"					All points = 52
No.	*Evolution stage*	*Data* (mm-dd)	*Time* UT	φ (deg)	λ (deg)	*Pressure* (mb)	*Wind velocity* (m/s)	*Shift direction, compass point*	*Shift velocity* (knots)
1	TL	Feb-02	15	−10.9	83.1	1,005	8	W	10
2	TL	Feb-03	6	−11.9	82.1	1,004	11	SW	10
3	TD	Feb-04	0	−12.8	80.1	1,000	13	WSW	8
4	TD	Feb-04	6	−12.3	80.3	999	13	NE	5
5	TD	Feb-04	12	−12.6	80.1	998	13	SW	2
6	TD	Feb-04	18	−13.0	80.5	998	13	SE	4
7	TS	Feb-05	0	−13.4	80.4	997	18	S	3
8	TS	Feb-05	6	−13.4	80.9	990	21	E	3
9	T	Feb-05	12	−13.7	80.9	975	33	S	3
10	T	Feb-05	18	−13.9	81.1	975	33	SE	3
11	T	Feb-06	0	−14.2	81.2	975	33	S	3
12	T	Feb-06	6	−14.1	81.3	965	38	0	0
13	T	Feb-06	12	−14.2	81.7	960	46	ESE	3
14	T	Feb-06	18	−14.4	82.0	934	51	SE	10
15	T	Feb-07	0	−14.7	82.3	925	54	SE	3
16	T	Feb-07	6	−14.7	82.5	925	54	E	3
17	T	Feb-07	12	−15.0	82.7	915	60	SE	3
18	T	Feb-07	18	−15.0	82.9	906	62	E	3
19	T	Feb-08	0	−15.1	83.2	915	60	ESE	3
20	T	Feb-08	6	−15.0	83.6	925	54	E	4
21	T	Feb-08	12	−15.4	84.1	925	54	SE	6
22	T	Feb-08	18	−15.6	84.4	925	54	SE	4
23	T	Feb-09	0	−16.1	85.0	925	54	SE	6
24	T	Feb-09	6	−16.6	85.3	925	54	SE	6
25	T	Feb-09	12	−17.4	85.9	925	54	SE	8
26	T	Feb-09	18	−18.3	86.3	935	49	SSE	9
27	T	Feb-10	0	−19.3	86.5	940	46	SSE	10
28	T	Feb-10	6	−20.3	86.7	945	44	SSE	10

TC number = SIO 0803				TC name = "Hondo"					All points = 52
No.	*Evolution stage*	*Data* (mm-dd)	*Time* UT	φ (deg)	λ (deg)	*Pressure* (mb)	*Wind velocity* (m/s)	*Shift direction, compass point*	*Shift velocity* (knots)
29	T	Feb-10	12	−20.8	86.5	955	38	SSW	8
30	T	Feb-10	18	−21.4	86.7	965	36	SSE	7
31	STS	Feb-11	0	−22.4	85.9	980	28	SW	10
32	STS	Feb-11	6	−23.1	86.0	985	26	S	8
33	TS	Feb-11	12	−23.7	85.7	988	23	SSW	7
34	TS	Feb-11	18	−24.3	85.2	994	18	SW	8
35	TS	Feb-12	0	−24.6	85.1	994	18	S	5
36	TS	Feb-12	6	−24.8	85.2	994	18	S	3
37	TD	Feb-12	12	−24.3	84.8	1,000	15	NW	5
38	TL	Feb-12	21	−23.1	83.9	1,003	8	NW	6
39	TL	Feb-20	12	−17.6	61.1	1,000	8	—	—
40	TL	Feb-21	0	−18.1	59.8	1,005	11	WSW	4
41	TD	Feb-21	6	−18.3	59.2	1,003	13	WSW	6
42	TD	Feb-21	12	−18.0	59.9	1,002	13	ENE	6
43	TD	Feb-21	18	−18.2	59.9	998	15	0	0
44	TD	Feb-22	0	−18.4	59.7	998	15	SW	2
45	TD	Feb-22	12	−18.9	59.0	999	15	SW	3
46	TD	Feb-23	0	−19.3	58.1	999	15	WSW	5
47	TD	Feb-23	12	−20.5	55.8	999	15	WSW	12
48	TD	Feb-24	0	−22.7	54.5	1,002	13	SSW	11
49	TD	Feb-24	12	−24.5	52.5	1,002	13	SW	13
50	TL	Feb-25	0	−26.9	51.6	1,004	11	SSW	10
51	L	Feb-25	12	−29.7	52.3	1,004	8	SSE	10
52	L	Feb-26	6	−31.1	56.2	1,002	8	ESE	15

Note: The TC transformed in the midlatitude system. TC evolution stage: TL is initial tropical disturbance; TD is tropical depression; TS is tropical storm; STS is strong (super) tropical storm; T is typhoon (hurricane); L is extratropical disturbance (a separate area with low subsurface pressure in midlatitudes) (in accordance with "best-track" pre-processing published in Pokrovskaya and Sharkov, 2006); φ is latitude (in degrees); λ is longitude (in degrees).

prominent cloud tails appeared. The small overall size of the disturbance allowed the storm to intensify rapidly. On February 6 at 00:00 GMT the disturbance reached the typhoon stage, the pressure at the center continued to drop quickly, and the cloud eye was formed. A weak shift of wind under a high-altitude crest, located at surface levels of 700 mb and 500 mb, coupled with the high divergence level produced favorable conditions for further strengthening of the disturbance. On February 7, at 18:00 GMT, the disturbance reached its peak intensity. The pressure at the center dropped to 906 mb and the wind velocity reached 62 m/s. A day later, the typhoon continued to be slowly displaced toward the southeast.

During February 8 and 9 the intensity of the typhoon changed slightly, the pressure remained at the level of 925 mb, and the wind velocity varied between 51 m/s and 54 m/s. The overall size of the system remained small. While moving south-southwest, the typhoon displaced into a region with a cold sea surface, where the temperature was about 25°C to 26°C. On February 10 and 11 the disturbance gradually weakened, the pressure at the center quickly grew, the wind velocity decreased to 23 m/s, and the deep convection weakened. On February 12, being displaced west-northwest along the subtropical crest periphery, the whole system weakened dramatically—the deep convection was absent and the disturbance was destroyed.

Note that during February 13–20 the French Meteorological Service on Réunion did not record the disturbance and, accordingly, it was not reflected on IFA websites.

However, on February 21 eastward of Madagascar, at about 18°S, 60°E, an organized cloud system was observed, which was identified by the Réunion Weather Service as a residual form of TC "Hondo". The cloud system was accompanied by active storms. Northwestward displacement along the periphery of the subtropical crest had clearly caused the surface wind to strengthen up to 15 m/s and turn into a tropical depression. During February 22–23 the warm ocean and the favorable atmospheric conditions had stopped the disturbance from weakening. On February 24, while continuing its motion southward, the system displaced to a cold sea surface with a temperature of 26°C and weakened. On February 25, while moving south-southeast around the subtropical crest, the disturbance slowly transformed into a cloud system (typical of moderate latitudes) at about 30°S, 60°E. The unusual evolutionary history of TC "Hondo" was recorded on the IFA website *http://www.solar.ifa.hawaii.edu* and reproduced in artistic form in Kim *et al.* (2009) (Figure 4.20).

The global situation of the integral water vapor field over World Ocean basins for February 8, 2008 is presented in Figure 4.21. It is easy to see that the basic water vapor bulk in the World Ocean was concentrated in the equatorial zones of the southwest Pacific Ocean and the eastern Indian Ocean. In the central Indian Ocean the water vapor field has concentrated (in the form of a vortical system) in the zone affected by TC "Hondo", which acquired maximum intensity at this time (Table 4.3). In the western Indian Ocean the greatest concentration of the water vapor field was observed in the zone in which TC "Ivan" was developing and in the eastern Indian Ocean (near the western coast of Australia) in the zone affected by TC

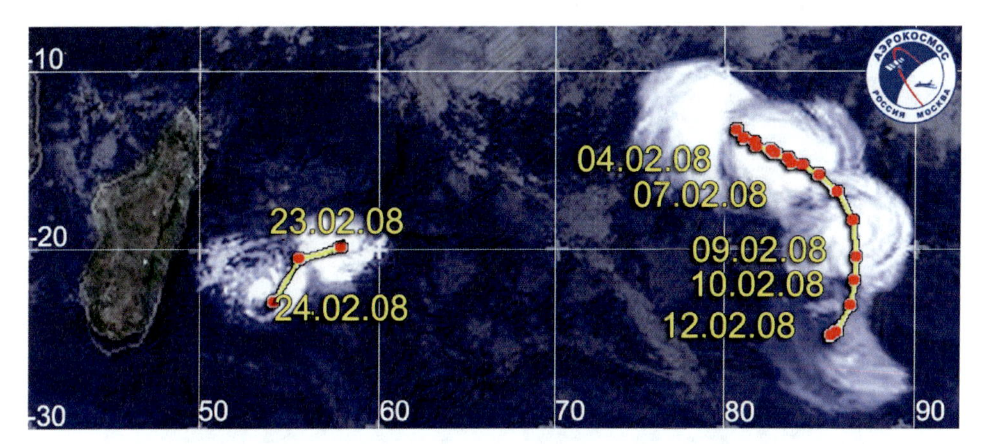

Figure 4.20. Track followed by TC "Hondo" over the periods February 4–12 and February 23–24, 2008 from data reported on the official website (*http://www.solar.ifa.hawaii.edu*).

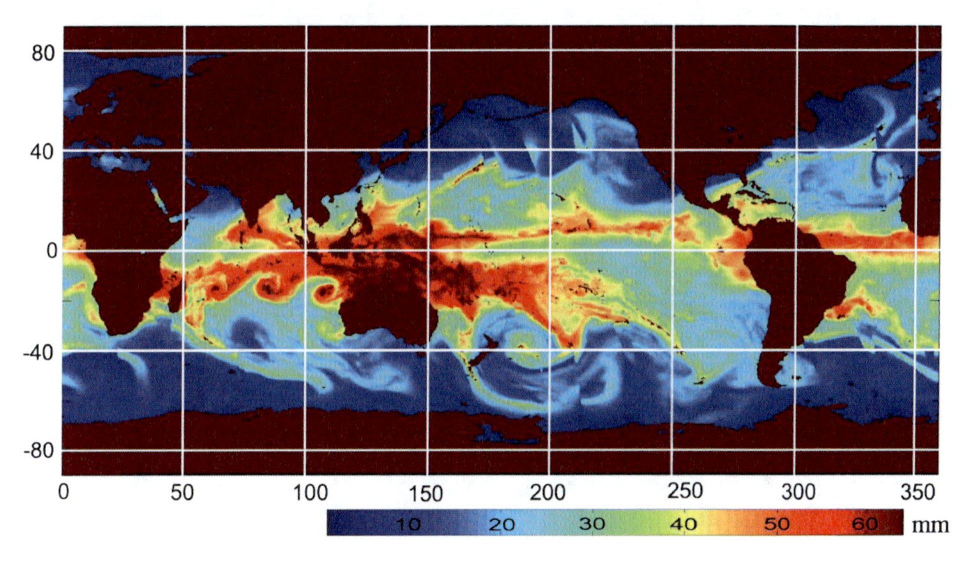

Figure 4.21. Global compiled field of water vapor on February 8, 2008.

"17S-08". These three TCs substantially "absorbed" the bulk of the water vapor field of the Southern Indian ocean, except for a small band of this field near the equator. This was simply the effect of synchronous trapping of water vapor masses by tropical cyclones from the "mother" (i.e., basic) equatorial field.

Let us now consider the entire evolutionary cycle of TC "Hondo" a little more attentively. As a result of applying the modified data-merging method, the time series of double frames for the entire evoluionary cycle of TC "Hondo" was obtained. These frames were combined in a single scale (Figure 4.22), which included initially

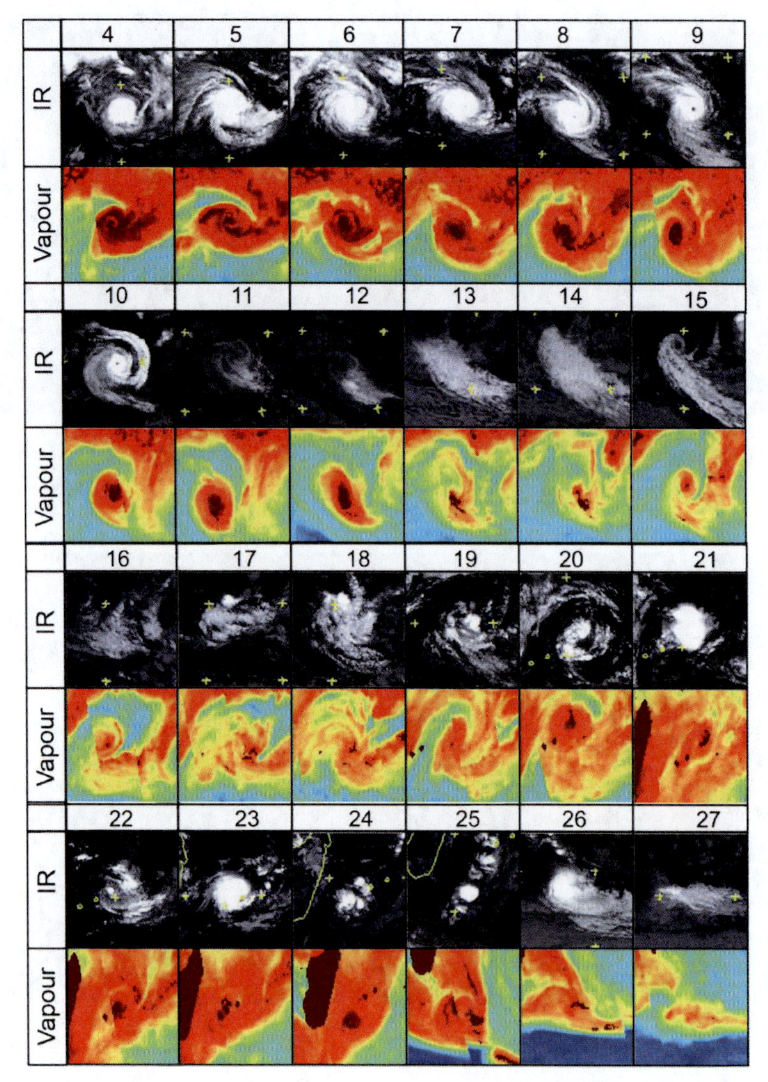

Figure 4.22. Time series of time lapse motion pictures of IR images from Meteosat-7 and water vapor fields adjusted to the specified scale representing the different stages of the evolution of TC "Hondo" from February 4 until February 27, 2008. Dates of observation are shown by numbers above images.

strongly multiscale space IR images obtained from the Meteosat-7 satellite (the black-and-white upper half of a frame) and the images of integral water vapor regions (the color lower half of a frame), based on which more calculations were made (the microwave data product of the Aqua satellite). It is easily seen from the figure that the region of heightened (as against an atmosphere undisturbed by a cyclone) integral water vapor content strictly corresponds to the cloud system (according to IR data) of a tropical cyclone. This region in the water vapor field was dated to the cyclone's center of effect. However, the water vapor region itself had a spatial size essentially larger than that reflected by cloud structures in the IR range. Furthermore, an important element of tropical cyclone evolution is the fact that the zone dated directly to the tropical cyclone (cloud masses) had a considerable transition zone that joined up with the basic water vapor field of the equatorial zone via current structures (jets). Most likely, there is a mechanism that can supply a tropical cyclone with additional water vapor from the main equatorial zone through the transition zone (jet). This can be proved by the break in the supplying jet that occurred between February 9 and 10 (Figure 4.22), which resulted in a fast 2-day (virtually) full dissipation of the cyclone (on February 15). However, at the same time the remnants of TC "Hondo" were trapped by the distant spiral branch of circulation of the strong TC "Ivan" which formed on February 7 near Madagascar and quickly (during 7–8 hours) reached the STS stage. On February 14 a new jet began to be formed (Figure 4.22) from the central equatorial region, which enabled virtually dissipated TC "Hondo" to be born again and form a new tropical disturbance. On February 19 a second supplying jet was formed from the central equatorial region. The next day (February 20) these two jets merged into one. On February 21–22 this region merged with the region of heightened water vapor content, located near the eastern coast of Madagascar, into a single region of heightened water vapor content, which strengthened the tropical formation up to the tropical depression level (on February 21–24) with subsequent dissipation with the break in its supplying jet (on February 25). Later on, the cloud masses of the tropical disturbance and the water vapor field accompanying them were "pulled" into the frontal zone typical of moderate latitudes in the near Antarctic zone (on February 26–27) and their autonomous lifetimes came to an end. It follows from the above analysis that the tropical depression attributed (according to official sources) to TC "Hondo" actually represented an independent tropical disturbance formed at the periphery of TC "Ivan".

Let us now consider the evolution of the daily value of the latent heat energy of the water vapor field accompanying TC "Hondo" and its dynamic characteristics (the maximum wind velocity in the cyclone's eye wall) from February 4 to February 16, 2008 (Figure 4.23).

Analysis of the calculation results presented in Figure 4.23 shows that—as the intensity of the tropical cyclone grew (in particular, the wind velocity at the eye wall) between February 4 and February 6, 2008 (see points 3–13 of Table 4.3)—total latent energy in the water vapor field also gradually increased as a result of water vapor mass being "pulled out" of the equatorial region. During the day of February 6 to February 7, 2008, the tropical cyclone suddenly intensified (see points 11–16 of Table 4.3 and Figure 4.22), and when it reached the stage of maximum development (February 7, 2008) the energy in the water vapor field lowered by approximately 0.5×10^{20} J. It is reasonable to assume that it was this quantity of latent energy that was spent on increasing the cyclone's kinetic energy in light of there being no other energy source with the equivalent power in the cyclone vicinity. During February 9 to February 10, 2008, as already noted, the jet connecting the cyclone region with the

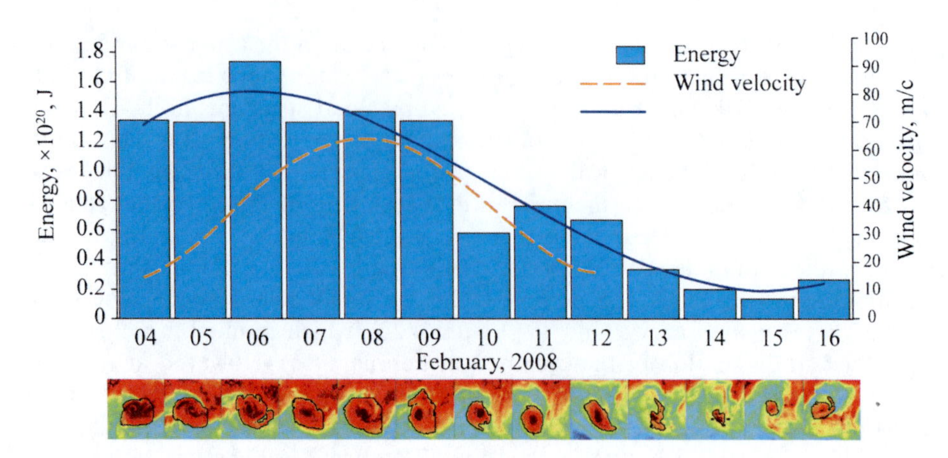

Figure 4.23. Evolution of daily latent heat energy of the water vapor fields that accompanied TC "Hondo" and its dynamics from February 4 to February 16, 2008. Blue bars correspond to water vapor energy contour-integrated over given areas. Domains of integration are contoured by black lines. The solid line is a temporal approximation of the energy of the water vapor fields. The dotted line is an approximation of TC wind velocity.

central equatorial region broke down and as a result the cyclone's intensity decreased. By this time the water vapor region had essentially diffused losing very quickly its accumulated latent heat; so, over 8 days the latent heat stock decreased by 1.3×10^{20} J. Thus, the rate of dissipating energy was 2×10^{14} W. However, as already noted, between February 15 and 16 a new jet was formed that would pump up the remnants of the TC with latent heat, which is clearly seen from the increase in the stock of latent heat on February 16 (Figure 4.23). Later on (February 22–23) the second maximum of the stock of latent heat formed (not shown in Figure 4.23) bringing about the tropical depression stage (Figure 4.22). Final destruction of the new tropical structure, including the water vapor region formed by the tropical depression, occurred as a result of their retraction into the frontal zone of Arctic latitudes.

Thus, detailed analysis of the energetics of TC "Hondo" during its unusual evolution on the basis of the data-merging method according to the multiscale data of satellite remote sensing has shown that the category-5 TC "Hondo" was generated and developed in the Southern Indian Ocean under the complicated conditions of its interaction with the circulation system of the powerful TC "Ivan". The latent heat energy source necessary for its functioning and intensification was likely the considerable water vapor region trapped by the tropical cyclone from a tropical zone experiencing monsoonal circulation and considerably exceeding its size, as routinely determined from optical and IR observation data. This was the zone from which the tropical cyclone was able to quickly entrain energy in the form of latent heat, whereas the mechanism of evaporation from the ocean surface was slow. Thus, the well-known schematic picture of the body of a TC (Figure 4.24a) should

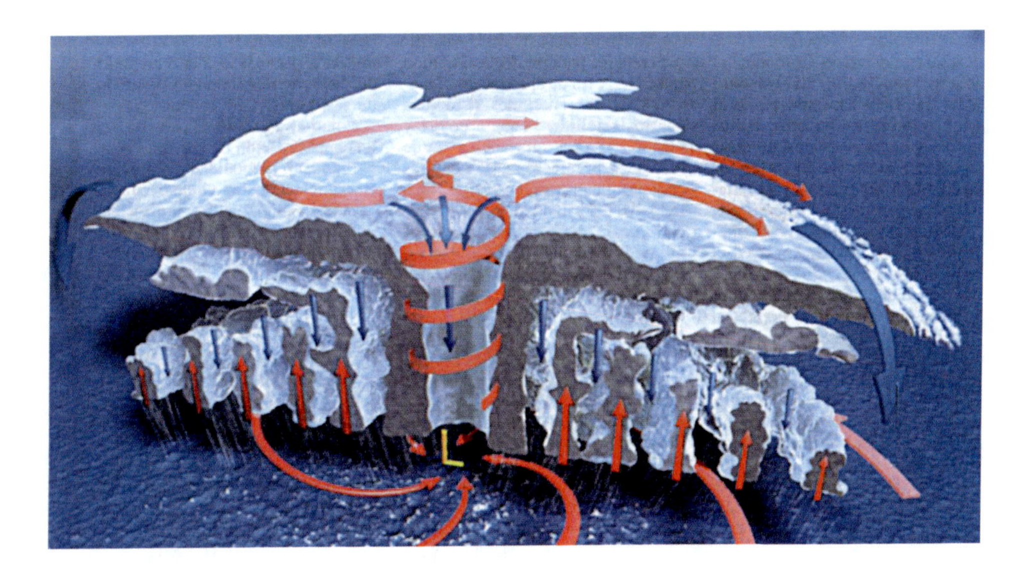

(a)

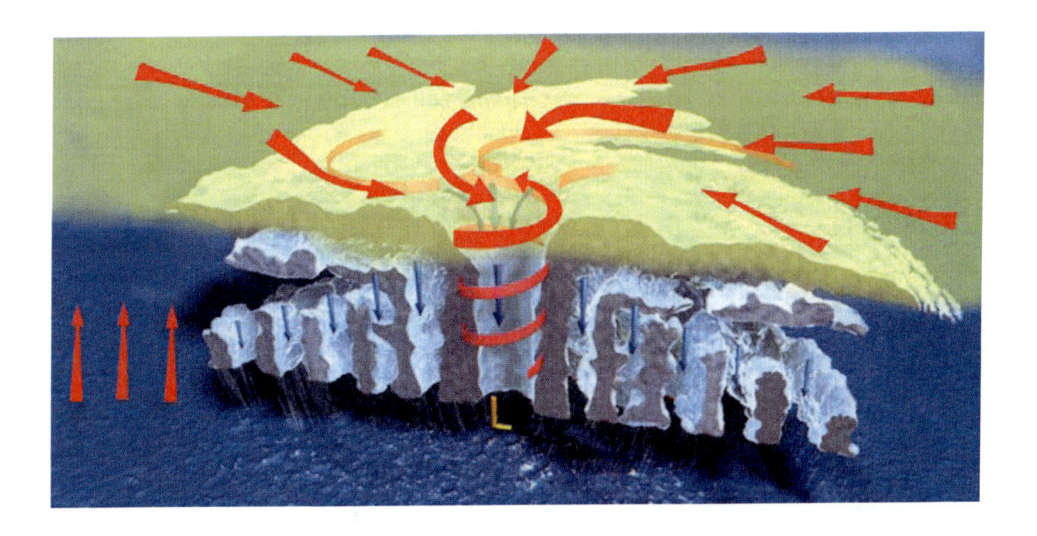

(b)

Figure 4.24. Traditional diagram (a) as exemplified in Hoffman (2004) and adapted diagram (b) of energy transfer in the volume of a TC.

essentially be supplemented by the water vapor cloud enveloping the entire cloud mass of a cyclone (Figure 4.24b).

So, the structure of a TC should be considered as a binary bound medium, a complicated system of cloud masses in which sublimation of latent heat occurs, and as a region of saturated water vapor (the "daughter" field) receiving additional supplies from the "mother" (basic) equatorial water vapor field by means of jet fluxes.

The new result here is the detection of the multijet structure of a water vapor field, which connects the water vapor region dated to a cyclone's cloud body to the central equatorial water vapor zone in the ITCZ. Violation of this jet structure quickly leads to TC dissipation. The formation of such a jet structure results in repeated intensification of a TC and its post-typhoon forms. Note that an effect similar to this was also recorded in the evolution of tropical cyclones in the North Atlantic as a result of analyzing the dynamics of the microwave natural radiation field in the range of 22.2 GHz (the line of natural radiation of water vapor), without restoring the integral water vapor field and, accordingly, without understanding the energy of the process (Astafyeva and Sharkov, 2008). Apparently, all tropical cyclones possess this property and, as a result of this trapping effect, eject a huge quantity of latent heat to the midlatitudes and high latitudes, which stipulates the cardinal role played by cyclones in forming climatic processes in Earth's atmosphere.

Note also that earlier attempts (Liu *et al.*, 1994) were made to use a methodology of estimating the stock of water vapor latent heat energy (i.e., a methodology that closely resembles that offered in this section) in regions dated to a tropical cyclone's cloud body, which, in its turn, was determined as a set of cloud systems of some particular structure. However, this methodology did not lead to positive results because the authors used in their work a scheme in which the field of integration was rigidly fixed spatially (within a radius of 1 or 2° from the cyclone's center), which is traditional in meteorology. As we have shown, the use of such a methodology is inadmissible when studying real cyclones, because the integral vapor region undergoes very strong spatiotemporal variations which depend on the stage of cyclone development.

4.4.5 Time evolution of TC "Hondo" and TC "Ivan" and their interaction in the integral water vapor field

As shown in the previous section, during TC evolution what is important is how one cyclone interacts with another cyclone, located closeby, via spatiotemporal variations in the water vapor field. This interaction can result in serious errors in forecasting the trajectories of TC motion made by official sources. As an example, let us revisit the unusual trajectory of TC "Hondo" (Figure 4.25), taken from official sources, and the physical reasons underlying its strange behavior. What was unusual about this cyclone was that after its dissipation (on February 12, 2008) it was officially "restored" on February 20, 2008, but at a quite different geographical location, some 2,500 km from where it initially dissipated (Figure 4.25). We will

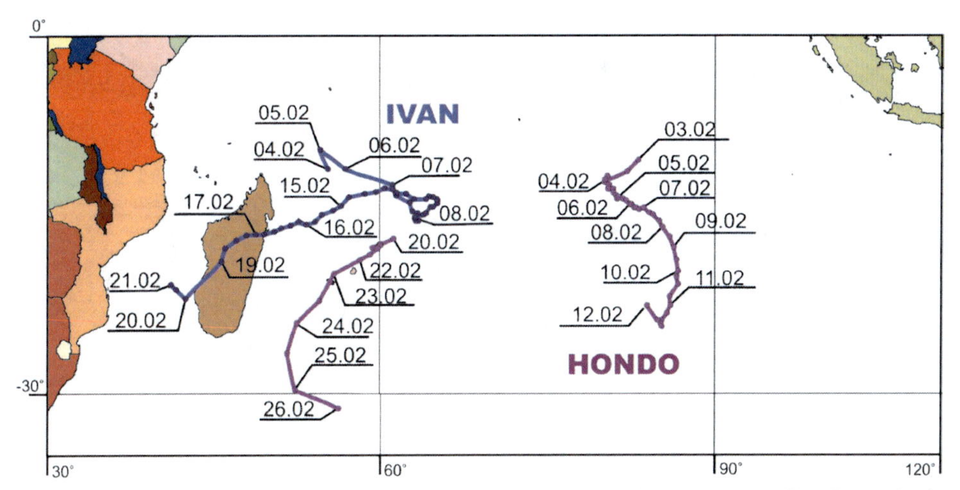

Figure 4.25. Tracks of TC "Ivan" and TC "Hondo" through the southern Indian Ocean basin according to official sources.

now show that this restored TC was in fact a completely new tropical disturbance, but, surprising it may seem, it was associated with the initial TC.

By applying the modified data-merging method, we obtained a time series of double frames for the whole cycle of evolution of TC "Hondo" and TC "Ivan". These frames were combined on the same scale (Figures 4.22 and 4.26) and initially included multiscale space optical images, obtained from the Meteosat-7 satellite (the black-and-white upper half of a frame), and the images of integral water vapor regions (the color lower half of a frame) from the microwave data of the Aqua satellite. Then, further energy calculations were carried out on the integral water vapor fields. The figures readily show that the region of heightened (as compared with an atmosphere undisturbed by a cyclone) integral water vapor content strictly corresponds to the cloud system (from optical data) of a tropical cyclone. This region in the water vapor field is dated to a cyclone's center of effect. However, the water vapor region itself has a spatial size considerably larger than that reflected by cloud structures in the visible range. Note also that the spatiotemporal variability of the optical fields and water vapor fields means that both the cloud field and the water vapor field cannot be completely identified. This is easily seen in the evolution of TC "Ivan" on the joint frames for February 8–9–10, 2008 and on the frames for February 23–24–25, 2008 (Figure 4.26a). Moreover, an important element of tropical cyclone evolution is the fact that the zone dated directly to the tropical cyclone (cloud masses) has a considerable transition zone that connects with the basic water vapor field of the equatorial zone via currents (i.e., jets). It is highly likely that some kind of additional "supply" of water vapor to the TC from the basic equatorial zone probably occurs through this transition zone. This can be proven by the complicated evolution of the water vapor region dated to TC "Hondo". So, after leaving the mother field in tropical depression form (on February 4) and with a

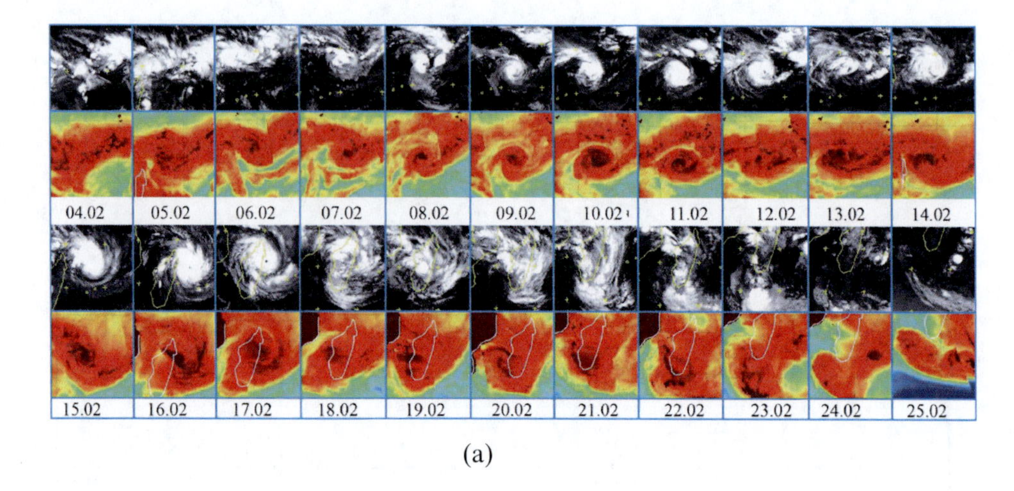

(a)

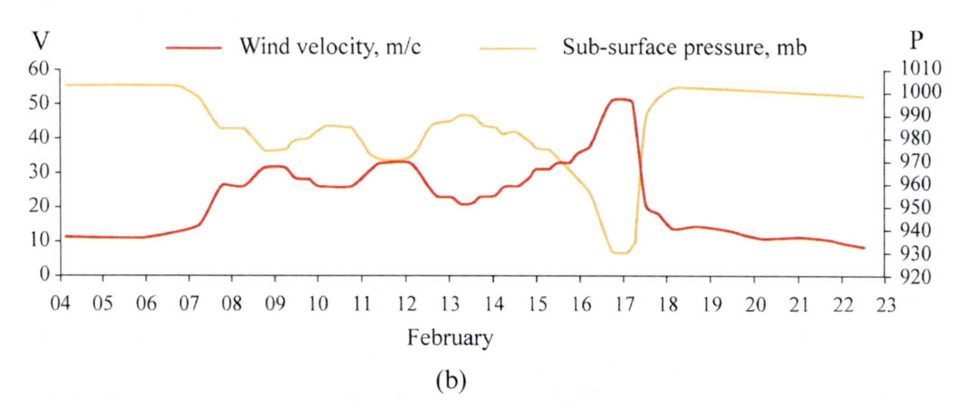

(b)

Figure 4.26. Evolution of TC "Ivan" (February 4–25, 2008). (a) The black-and-white sequence was put together using IR data (Meteosat-7) and the color sequence using microwave satellite data (AMSR-E instrument on Aqua satellite); (b) time series of data on maximum wind velocity V (m/s) and subsurface pressure P (mb).

considerable latent heat stock (about 1.3×10^{20} J) (Figure 4.28) the TC quickly (in 6 hours) intensified up to the hurricane stage and remained at this stage until February 10. On this day the supplying jet broke down (Figure 4.22), which resulted in a fast 2 to 3-day almost complete dissipation of the cyclone (by February 15). During February 10–15 the latent energy of the natural water vapor field of the TC drastically dropped to 2×10^{19} J. But, by the end of February 12 the TC was filled (Figure 4.27) and transferred into the form of a primary tropical disturbance (see Table 4.3), which is clearly recognized on IR images of the evolution of TC "Hondo" (Figure 4.22). The official weather services stopped tracking this TC (Figure 4.27). However, at the same time remnants of the cloud body and of the water vapor field of TC "Hondo" were trapped by a distant spiral branch of circulation of the strong TC

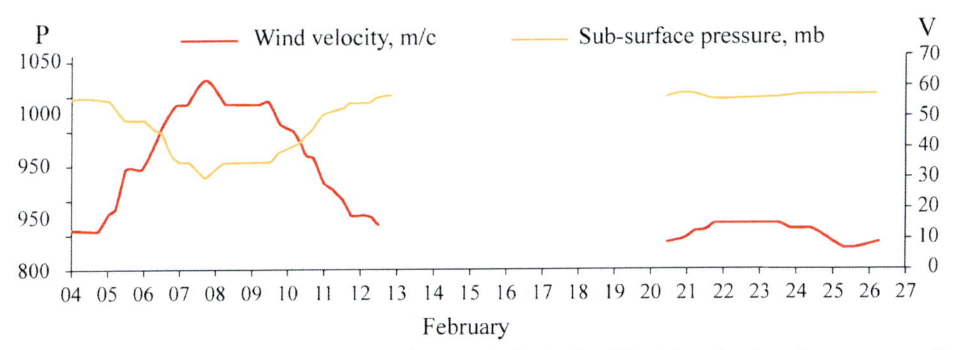

Figure 4.27. Time series of data on maximum wind velocity V (m/s) and subsurface pressure P (mb) during the evolution of TC "Hondo-I" and TD "Hondo-II" (February 4–25, 2008).

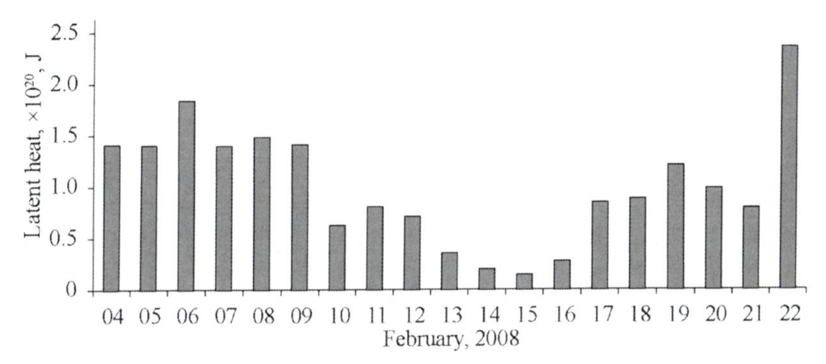

Figure 4.28. Time series of latent heat energy in the water vapor zones of TC "Hondo-I" and TD "Hondo-II" (February 4–22, 2008).

"Ivan". TC "Ivan" formed on February 7 near Madagascar and quickly (7–8 hours) reached hurricane stage, which it kept during February 15–17 (Figure 4.26b). On February 15 a new jet began to be formed (Figure 4.22) from the central equatorial region (the ITCZ), which enabled virtually dissipated TC "Hondo" to be born again and form a new tropical disturbance. On February 19 a second supplying jet was formed from the central equatorial region. The next day (February 20) these two jets merged. On February 21–22 this region merged with a region of heightened water vapor content, off the eastern coast of Madagascar, forming a single region of heightened water vapor content, which elevated the tropical formation up to the tropical depression level (February 21–24) with subsequent dissipation when its supplying jet failed (February 25). Later on, the cloud masses of the tropical disturbance and the water vapor field accompanying them were "pulled" into a frontal zone typical of moderate latitudes in the near Antarctic zone (February 26–27) bringing their autonomous lifetimes to an end. It follows from the above analysis that the tropical depression attributed (according to official sources) to TC "Hondo" actually represented an independent tropical disturbance that formed at the periphery of TC "Ivan" and should be given its own name: TD "Hondo-II".

Let us now consider how the daily value of the latent heat energy of the water vapor field accompanying TC "Hondo" developed along with its dynamic characteristics (such as its maximum wind velocity in the cyclone's eye wall) between February 4 and February 16, 2008 (Figure 4.28).

Analysis of the calculation results presented in Figure 4.23 shows that, as the intensity of the tropical cyclone grew (in particular, wind velocity at the eye wall) over February 4 to 6, 2008 (see points 3–13 of Table 4.3), the total latent energy in the water vapor field also gradually increased due to the water vapor mass "pulling out" from the equatorial region. In a single day (February 6 to February 7, 2008), the tropical cyclone suddenly intensified (see points 11–16 of Table 4.3 and Figure 4.22), and at its height (February 7, 2008) the energy in the water vapor field dropped to about 0.5×10^{20} J. It is reasonable to assume that it was this amount of latent energy that was spent on increasing the cyclone's kinetic energy, especially as in the cyclone vicinity there was no other energy source with the equivalent energy. During February 9 to February 10, 2008, as already noted, the jet connecting the cyclone's region of control with the central equatorial region broke down and, accordingly, the intensity of the cyclone intensity decreased. At this time the water vapor region essentially diffused and quickly lost the latent heat it had built up; so for 8 days the latent heat stock decreased to 1.3×10^{20} J. Thus, the rate of dissipating power was 2×10^{14} W. However, as already noted, over February 15 to 16 a new jet formed that proceeded to pump up the TC's remnants with latent heat, which is clearly seen by the stock of latent heat increasing on February 16 (Figure 4.28). Later on, by February 19–20 a second maximum of the latent heat stock would be reached (Figure 4.28) enabling tropical depression genesis. The merging of the two water vapor regions on February 21 sharply increased the latent heat energy of the integrated field to 2.2×10^{20} J. Final destruction of the new tropical structure, including the water vapor region formed by the tropical depression, occurred as a result of their retraction into a frontal zone typical of the Arctic latitudes.

The new result here is the detection of the multijet structure of a water vapor field, which connects the water vapor regions dated to the cloud body of each cyclone to the central equatorial water vapor zone in the ITCZ. Violation of this jet structure quickly leads to dissipation of one of the interacting TCs. And the additional formation of such a jet structure results in repeated intensification of one of the TCs and its post-typhoon forms. A break in a supplying jet results in fast 1 to 2-day dissipation of one of the interacting cyclones. However, new jets form from the central equatorial region and provide the possibility for secondary genesis—as was the case with the virtually dissipated body of TC "Hondo"—and generation of a new tropical formation at the tropical depression level which would subsequently dissipate as a result of jet failure.

4.5 LATENT HEAT ENERGY TRANSPORT BY PLURAL TROPICAL CYCLOGENESIS

As already noted, one of the most important dynamic elements in poleward latent heat transport is tropical cyclogenesis. In the present section we present the first ever practical quantitative estimates of latent heat energy transport by plural tropical cyclogenesis in the southern Indian Ocean from the zone of equatorial latent heat supply (the ITCZ zone) into the midlatitudes (Sharkov *et al.*, 2010, 2011a). These works present a detailed analysis of the energy features of a sequence of bounded tropical cyclones (plural cyclogenesis) in the oceanic basins of the southern hemisphere (the southern Indian Ocean and the southwest part of the Pacific Ocean) from the end of January–February until the beginning of March, 2008, carried out using the data-merging method with multiscale data from satellite IR and microwave radiothermal remote sensing.

What is new about these studies is the attempt to estimate quantitatively the latent energy of the whole central equatorial zone of water vapor in the ITCZ of the Indian and Pacific Oceans, and to consider the link between the significant time variations of latent heat in the ITCZ and the ejection of bounded water vapor regions into high latitudes by means of plural cyclogenesis.

4.5.1 Statement of the problem

The multiscale (i.e., spatiotemporal) interaction of the ocean and atmosphere, which is composed of various energy, momentum, and substance exchange processes, is considered to be one of the most important climate-forming factors on the Earth. Ascertaining the role and contribution of intensive vortical tropical disturbances is fundamentally important in the ocean–atmosphere interaction—especially the role and contribution of the most powerful of them: tropical cyclones (TCs). The influence of these powerful tropical vortical systems on the state of the ocean–atmosphere system during TC passage is obvious (Riehl, 1979; Gray, 1979; Anthes, 1982; Khain and Sutyrin, 1983; Sharkov, 1997, 1998, 2000, 2010). The basic mechanism in the interaction between TCs and the oceanic environment is considered to be the formation of latent heat fluxes from the ocean's surface layer by means of the direct convective activity in a hurricane's circulating system. Expressing the fundamental question in this way is called the "convective hypothesis", which simply deals with the spatiotemporal evolution of the surface temperature field of the World Ocean and its effect on tropical cyclogenesis (see Section 4.2). Note that the supply of heat to the World Ocean's upper layer is considered to be the key energy source for tropical cyclogenesis. Within the framework of such an approach the question about the long-term impact of tropical cyclogenesis on climate remains moot (*http://news.discovery.com/earth/heavy-hurricanes-and-their-climate-effects.html*).

A principally different mechanism of energy transport from the tropical zone of the ocean–atmosphere system into the midlatitudes and high latitudes taking the form of compact regions of integral water vapor of heightened intensity and entrained by TCs from the equatorial region (the ITCZ) of the global water vapor field was experimentally found for the first time by Sharkov *et al.* (2008a, b, 2011b). Note that these studies were carried out on separate tropical cyclones (i.e., individual cyclogenesis) (see Section 4.4.4). Because of the complicated spatiotemporal redistribution of water vapor fields that takes place (see Section 4.4.5) in the dynamic interaction of tropical systems (i.e., plural cyclogenesis), this can result in the occurrence of "nonlinear" channels that transport latent heat with subsequent considerable time variations of latent heat in the ITCZ. The purpose of the present section is to present the results of an investigation based on the data of multispectral satellite observations of the mechanism of latent heat transport from the equatorial zone of water vapor to the midlatitudes by means of plural tropical cyclogenesis in the oceanic basins of the southern hemisphere (southern Indian Ocean and the southwestern Pacific Ocean).

4.5.2 Plural tropical cyclogenesis evolution

Plural cyclogenesis in the southern Indian Ocean in February 2008 involved three strong tropical cyclones (TC "Hondo-I", TC "Ivan", and TC "Nicholas") and three forms of tropical disturbances (TD 17S, TD "Hondo-II", and TL). The latter three were considerably weaker than the TCs listed above, nevertheless they contributed to the general picture of latent heat transport. The chronology and trajectories of tropical cyclones and disturbances that arose in the southern Indian Ocean and in the southwest Pacific Ocean between January 25 and March 14, 2008 (according to official sources) are presented in Figure 4.29 and—within the background of the global water vapor field (for February 5, 2008)—in Figure 4.30. The basic characteristics of TCs and TDs in the southern Indian Ocean for February 2008 are presented in Table 4.4, which was taken from the Global-TC database and compiled using the pre-processing technique (Pokrovskaya and Sharkov, 2006) for the initial (i.e., raw) data presented on the international website *http://www.solar.ifa.hawaii.edu* Analysis of Figure 4.30 shows that the genesis of all tropical formations in February, 2008 occurred in the key (mother) water vapor field of the equatorial zone and subsequently left this region.

Tropical depression 17S (SIO 08-3) (see Table 4.4 and Figure 4.29a, b) was recorded as a primary tropical disturbance on February 2 at 21:00 GMT in the southern Indian Ocean basin southward of Java island. The near water wind velocity was between 8 m/s and 11 m/s and the pressure was 1,008 mb. After being displaced south-southwest, by February 4 the disturbance strengthened to the tropical depression stage, the wind velocity grew to 15 m/s, and the pressure dropped to 996 mb. During February 4 to 9 the tropical depression remained stationary apart from a slow displacement to the southeast, the pressure varied within the limits of 1,002 mb to 992 mb, and the wind velocity remained constant. On February 9, having reached the latitude of 17°S, the disturbance changed direction and moved north, and on February 10, still at the tropical disturbance stage, it dissipated at about 14°S.

The primary (initial) tropical disturbance (i.e., the TL stage) of TC "Ivan" (SIO 0804) (see Table 4.4) was recorded on February 4 at 15:00 GMT at about 11.7°S, 55.3°E and represented a diffuse poorly organized cloud system. The pressure at the center was 1,003 mb and the surface wind velocity was about 10 m/s. During February 5 and 6 signs of the disturbance strengthening appeared, and at

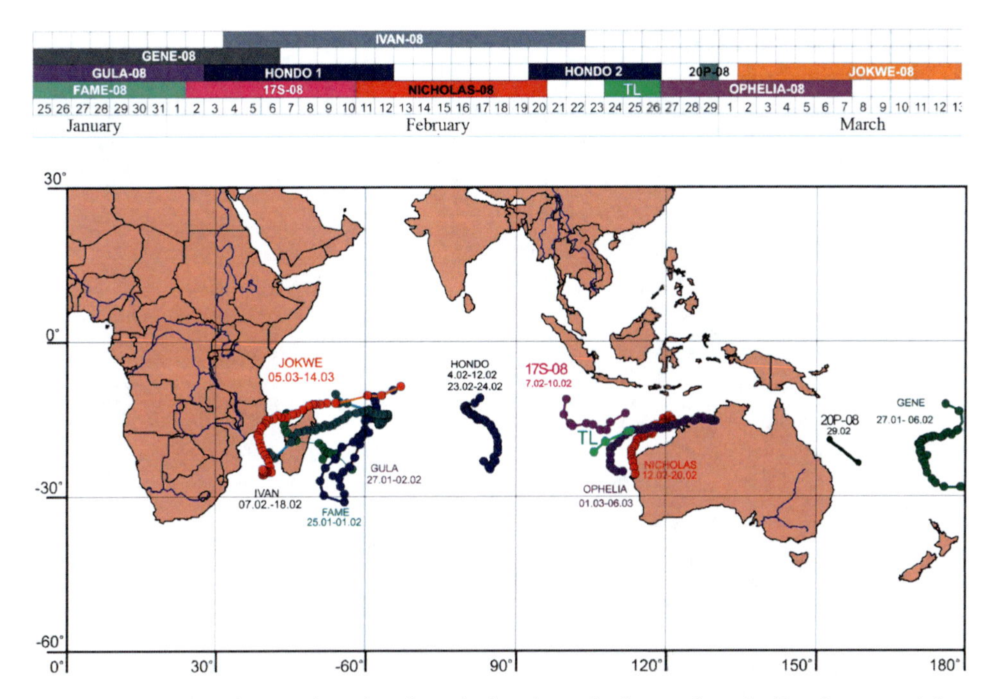

Figure 4.29. Chronology and tracks of tropical cyclones in the southern Indian Ocean and the southwest Pacific Ocean from January 25 to March 14, 2008 according to the official version (*http://www.solar.ifa.hawaii.edu*).

18:00 GMT on February 6 it became a tropical depression. The cloud system became denser and better organized (see the visible range frames in Figure 4.33a). On February 7 at 12:00 GMT the tropical depression became a tropical storm. Intensification was rapid and by 18:00 GMT the disturbance reached the strong tropical storm stage. During February 11 to 14 the storm intensity varied from strong to weak and remained very weak westward of the direction of motion. A cloud eye was recorded on February 12. On February 13 the cloud system became indistinct and the intensity decreased to the tropical storm stage. On February 14 the cloud body assumed distinct outlines with pronounced cloud tails. On February 15, it slowly approached the northeastern coast of Madagascar and, as a result of being displaced over islands in an open warm ocean where there was little wind in the middle atmosphere, the storm reached the typhoon stage and the cloud eye appeared again. During February 16, at the typhoon stage, the wind velocity was about 50 m/s. The TC had a catastrophic effect on the ecology of the northeast coast of Madagascar. On February 17 the position of the cloud eye was recorded in the region of the city of Mananara, on the eastern coast of the island. Being displaced west-southwest over the island along the northwest periphery of a subtropical crest, the system quickly lost its intensity and by February 18 returned to the tropical depression stage with wind velocity dopping to 15 m/s. During February 19 to 20

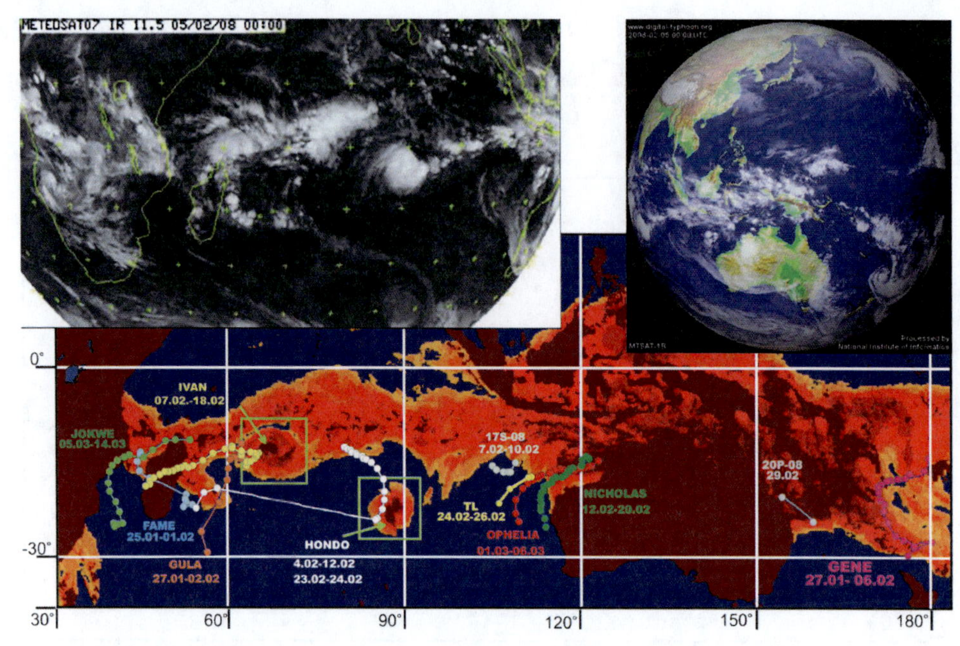

Figure 4.30. Global water vapor field in basins of the Indian and Pacific Oceans on February 11, 2008 (day-averaged) with TC tracks plotted on the map between January 25 and March 14, 2008. The upper-left inset shows an IR image fragment from Meteosat-7 on February 5, 2008; the upper-right inset shows full optical Earth image from MTSAT-1M on February 5, 2008.

the system displaced southwest and then southeast. Within the next three days the system slowly displaced south-southeast and transformed into a system typical of moderate latitudes and finally dissipated at about 30°S.

The primary tropical disturbance of TC "Hondo-I" (SIO 0803) (see Tables 4.3 and 4.4) appeared on February 2 at 15:00 GMT at the southern periphery of the ITCZ at about 11°S, 83°E. It was little more than a diffuse poorly organized cloud system. Sea level pressure was 1,004 mb and wind velocity was about 10 m/s. During the next day storm activity strengthened and the pressure at the system's center dropped by 4 mb, causing the disturbance to intensify and transition to the tropical depression stage. The cloud disk assumed more distinct outlines. On February 5, under favorable environmental conditions, the wind strengthened and the disturbance became a tropical storm. The pressure dropped to 997 mb and the wind velocity grew to 18 m/s.

Over the next day the disturbance quickly intensified and a dense central small core with prominent cloud tails appeared. On February 6 at 00:00 GMT the disturbance reached the typhoon stage, the pressure at the center continued to drop quickly, and the cloud eye formed. On February 7 at 18:00 GMT the disturbance reached its peak. The pressure at the center dropped to 906 mb and the wind velocity reached 62 m/s. Over the next day the typhoon continued displacing at low velocity to the southeast. During February 8 and 9 the typhoon intensity varied only slightly, the

Table 4.4. The main parameters of tropical cyclones in the South Indian Ocean (February 2008).

No.	*Basin, TC number and name*	*Length of TC lifetime*	*Position of TC genesis*		*Date transfer to TS*	*Position of TC transfer to TS*		*Date of TC dissipation*	*Position of TC dissipation*		*Maximal stage of TC evolution and length of lifetime*	V_{max} (m/s)
			φ	λ		φ	λ		φ	λ		
1	SIO 08-3 17S	02.02–10.02	−11.0	100.3	—	—	—	10.02	−13.8	112.2	TD 04.02–10.02	15
2	SIO 0803 "Hondo-I"	02.02–12.02	−10.9	83.1	05.02	−13.4	80.4	12.02	−23.1	83.9	T 05.02–10.02	62
3	SIO 0804 "Ivan"	04.02–22.02	−11.7	55.3	07.02	−13.9	61.3	22.02	−21.8	41.6	T 15.02–17.02	51
4	SIO 0805 "Nicholas"	10.02–20.02	−15.8	122.6	13.02	−14.9	119.8	20.02	−25.6	114.2	T 16.02–18.02	41
5	SIO 0803 "Hondo-II"	20.02–26.02	−17.6	61.1	—	—	—	26.02	−31.1	56.2	TD 21.02–24.02	15
6	SIO 08-6 TL	24.02–26.02	−17.4	113.2	—	—	—	26.02	−21.3	105.9	TL 24.02–26.02	11

Notes: Basins and TCs names are given in accord with international specifications. TC evolution stages: T is typhoon, TD is tropical depression, TL is initial tropical disturbance (in accordance with "best-track" pre-processing published in Pokrovskaya and Sharkov, 2006); V is TC wind velocity (m/s); φ is latitude; λ is longitude.

pressure reached and remained at the level of 925 mb, and the wind velocity was between 51 m/s and 54 m/s. The overall size of the system remained small. While moving south-southwest, the typhoon displaced into a region with a cold sea surface, where the temperature was 25°C to 26°C. On February 10 and 11 the disturbance gradually weakened, the pressure at the center quickly grew, the wind velocity dropped to 23 m/s, and deep convection weakened. On February 12, it displaced to the west-northwest along the subtropical crest periphery, the whole system weakened, deep convection in the system disappeared, and the disturbance dissipated.

The primary tropical disturbance of TC "Nicholas" (SIO 0805) was recorded on February 10 at 21:00 GMT at about 15.8°S, 122.60°E near the northwest coast of Australia, in the Timor Sea. It was a very diffuse cloud system located on the southern side of the ITCZ. On February 12 at 12:00 GMT the tropical disturbance became a tropical depression, and a dense cloud core with diffuse edges was formed. Note that the proximity of the Australian coast (less than 300 km distant) did not prevent the disturbance from strengthening. On February 13 the disturbance reached the tropical storm stage and remained at this stage during February 14. The wind velocity was 21 m/s and the pressure was 980 mb. On February 15 the storm strengthened to the strong tropical storm stage, the wind velocity at the center reached 28 m/s, and the pressure was 966 mb. On February 16 the storm strengthened to the typhoon stage and remained at this stage until February 18. The maximum wind velocity was 44 m/s and the pressure dropped to 44 mb. The cloud disk increased in size with prominent bent cloud tail strips appearing in the northwest sector. The whole vortical system was located just off the coast. On February 19 the vortical TC system continued slowly moving southward reaching landfall and subsequent dissipation on February 20.

As shown in Kim *et al.* (2009) and Sharkov *et al.* (2011b), on February 20 the residual form of TC "Hondo-I" was pulled into the field of influence of TC "Ivan" and, owing to two jet spiral bridges being set up to the basic (mother) water vapor field, this residual form became a tropical depression: "Hondo-II" (SIO 0803) (see Table 4.4). The cloud system was accompanied by active storms. Displacement to the northwest, along the subtropical crest periphery, caused the surface wind to strengthen to 15 m/s and the tropical depression to arise. During February 22–23 the warm ocean and favorable atmospheric conditions prevented the disturbance from weakening. On February 24, while continuing its southward motion, the system displaced to an area with a cold sea surface of 26°C and weakened. On February 25–26, while moving south-southeast around the subtropical crest, the disturbance slowly transformed into a system typical of moderate latitudes at about 30°S, 60°E. The unusual evolutionary history of TC "Hondo" is recorded on the IFA website *http://www.solar.ifa.hawaii.edu*, and is reproduced in artistic form by Kim *et al.* (2009) (see Figure 4.20).

A primary tropical disturbance at the TL stage (SIO 08-6) was recorded on February 24 at 02:00 GMT in the Southern Indian Ocean basin, westward of the northwest coast of Australia. The surface wind velocity was between 11 m/s and 13 m/s and the pressure was 1,004 mb. During February 25–26, as a result of being displaced to the west-southwest, the disturbance gradually weakened, the wind velocity decreased to 5 m/s to 7 m/s, the pressure increased to 1,006 mb, and, having reached 21°S, the disturbance dissipated.

4.5.3 Initial data for satellite sensing

The initial data used in this work came from (1) the Remote Sensing Systems website *http://www.remss.com* where data from the AMSR-E microwave complexes of the Aqua satellite are collated and updated in real time; (2) IR images of basins of the Northern Indian Ocean and of a part of the Pacific Ocean obtained from the geostationary Meteosat-7 and MTSAT-1M satellites; and (3) products in which integral water vapor content is restored based on the data of the AMSR-E microwave complex of the Aqua satellite. In this work were used the spatiotemporal combinations of images of products in which the integral water vapor content in the atmosphere is restored and the IR images of cloud systems reduced to the same scale by the data-merging method, because this is the only technique that most completely reflects the energy transformation processes occurring in the tropical zone.

The full algorithm used to restore data on integrated moisture content in the atmosphere is described in detail at *www.ssmi.com/amsr/amsr_data_description.html#amsre_data* Data from the SSM/I and AMSR-E instruments were processed in accordance with traditional techniques of atmospheric parameter restoration based on satellite microwave data.

4.5.4 Information-processing technique and algorithms

The author has analyzed in detail the energy features of the global field of integral water vapor in the equatorial zone of the Indian and Pacific Oceans under conditions in which tropical disturbances arise. The analysis was carried out on the basis of the data-merging method, which represents development of the information forming and accumulation technique, based on multiscale satellite remote-sensing data, which is applied to study near stationary objects and fields on the Earth surface. However, when this method is used to study such fast processes as the spatiotemporal evolution of the global water vapor field and tropical cyclone evolution, it essentially needs updating. This task was done for the first time by a team of authors (including the author of this book) who analyzed the evolution of TC "Gonu" and TC "Hondo" (Sharkov *et al.*, 2008a, b; Kim *et al.*, 2009). The purpose of the task was essentially to study plural cyclogenesis as presented in this section (Sharkov *et al.*, 2010, 2011a).

As mentioned above, the purpose of this work was to estimate the overall energy of the mother water vapor region of the ITCZ and the regions of integral water vapor accompanying tropical cyclones and tropical disturbances (daughter fields) by using off-the-shelf products of satellite data processing. As noted above, the author used IR channel data of the geostationary Meteosat-7 satellite and a product in which integral water vapor is restored based on data from the microwave AMSR-E complex of the Aqua satellite, which were processed according to the algorithm offered by the author. Detailed analysis of the initial data-processing system, developed by the same team of authors mentioned in the previous paragraph, was presented in Kim *et al.* (2009). Let us now consider some additional features related to the study of plural cyclogenesis, whereas in previous works of the author only individual cyclogenesis has been investigated (Sharkov *et al.*, 2008a, b; Kim *et al.*, 2009).

To restore the complete water vapor picture, the data-merging method was used on data obtained from the descending and ascending orbits of the Aqua satellite which alternate and subsequently eliminate lacunas by the technique of linear data interpolation in coordinates and time. In such a manner, the dataset on total water vapor content for the whole World Ocean was obtained, where one pixel corresponded to an area of 10×10 km to 28×28 km of the Earth's surface (the difference is the result of Mercator projection features). As an example, Figure 4.30 presents the global map of integral water vapor content in the World Ocean, averaged for a day, by separating out the studied TDs and TCs in the Southern Indian Ocean and in the southwest Pacific Ocean.

The next calculation task was separation of the fields of integration—that is, recognition of the image of the global field of water vapor (the mother field) and of integral water vapor regions dated to each stage of the cyclone's evolution (stable daughter zones). The author believes it best to separate the fields of integration in the water vapor region manually for a number of reasons. First of all, it was necessary to separate the zone of heightened water vapor content for each particular image corresponding to a particular stage in the evolution of each disturbance. Since at each TC evolution stage the water vapor field that accompanies the TC evolves in a complicated manner (Figure 4.31), it was still impossible to carry out the recognition procedure using automated identification techniques. Moreover, by separating out the field of integration at the early stages of tropical cyclone genesis, it was difficult to recognize the water vapor region accompanying a cyclone in the water vapor zone of the ITCZ (i.e., it was difficult to distinguish this region from the region it gets its energy supply). To resolve this problem, the authors used IR images from the geostationary Meteosat-7 satellite, because in the IR spectrum region the cyclone can be distinguished by the structural features of cloud masses. The time series of frames (in pairs) of IR images from the Meteosat-7 satellite and from the integral water vapor field, reduced to the same scale and displaying various stages of the evolution and transformation of tropical cyclones and tropical disturbances for February 2008, is presented in Figure 4.31.

The second criterion of cyclone separation was the threshold value of water vapor below which the studied formation did not make a considerable contribution. Thus, on the basis of the criteria listed above, a field of integration was chosen over which the integral water vapor content was calculated with a view to obtaining the total vapor content in the water vapor zone of the ITCZ region of the Indian and Pacific Oceans (Figure 4.30) and the regions accompanying the TCs and TDs (Figure 4.31).

The moisture quantity values obtained were transformed into a value of latent

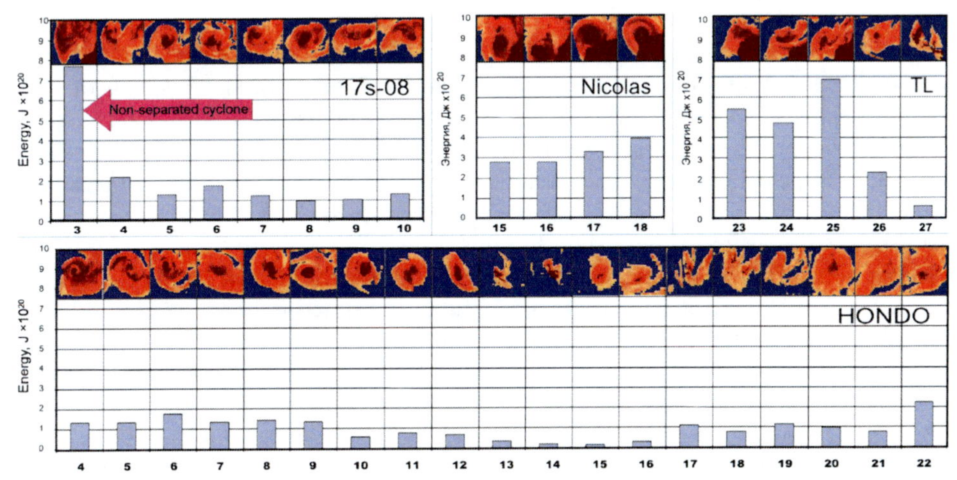

Figure 4.31. Time evolution of latent heat in the daughter water vapor zones joined to TC "Hondo-I" (February 4–16, 2008) and TD "Hondo-II" (February 17–22, 2008); TC "Nicholas" (February 15–18, 2008); TD 17S-08 (February 3–10, 2008); and an initial tropical disturbance (February 24–27, 2008). Color fragments (upper diagram) show the water vapor areas on which integration was carried out.

energy on the basis of the following assumptions: the precipitated water vapor of thickness 1 mm (according to initial data) corresponds (after accounting for water density of 1 g/cm^3) to a water mass of 1 kg over an area of 1 m^2. On the other hand, the specific heat of the transition phase of vapor to water is known to be equal to 2.5×10^6 J/kg. Thus, with a precipitated vapor thickness of 1 mm the condensation latent heat value will be 2.5×10^{12} J over an area of 1 km^2. The integration zone area, expressed in pixels in the initial dataset, was transferred into square kilometers. For this purpose an area transition factor was calculated. As a result of all these steps, we obtained values of the latent heat energy of the mother water vapor field for the ITCZ zone and for the water vapor regions (J) accompanying tropical cyclones at all stages of their evolution. These data are presented in Figures 4.31 and 4.33.

4.5.5 Spatiotemporal evolution of tropical cyclogenesis in the global water vapor field

Let us first make the point that, because of the high spatiotemporal variability of optical fields and water vapor fields, both globally (Figure 4.30) and regionally, identification of the cloud field and the water vapor field is not possible using daily time averaging. This is clearly seen by comparing the water vapor fields of the Indian and Pacific Oceans with images in the IR and visible ranges obtained from geostationary satellites (Figure 4.30).

By applying the modified data-merging method, the time series of double frames was obtained for the whole evolution cycle of five tropical disturbances (see Table 4.4), combined on the same scale and containing initially multiscale space IR images obtained from the Meteosat-7 satellite and images of integral water vapor regions as a microwave data product of the Aqua satellite. Further energy calculations were performed based on integral water vapor fields (Figure 4.31). Analysis of the combined images clearly shows that the region of heightened integral water vapor content corresponds to the main cloud system of the tropical cyclone. This region in the water vapor field is dated to a cyclone's center of influence. However, the water vapor region itself has a spatial size essentially larger than that reflected by cloud structures in the visible range. Note that this statement relates to the situation when a tropical disturbance has left the integral water vapor region of the ITCZ and is associated with this region by a jet spiral bridge. As noted in Sharkov *et al.* (2008) and Kim *et al.* (2009), it is this element of the water vapor field that is most likely responsible for the latent heat supply that the TC transports. It is important to note that analysis of the data in Figures 4.30 and 4.31 has shown that a similar situation takes place in the evolution of primary tropical disturbances and tropical depressions as well. What is remarkable here is the fact that the supply of heat to the daughter water vapor region of the TD is comparable with the supply of heat to a TC and on occasion exceeds it (see the TL and 17S diagrams in Figure 4.31). This also follows from the TC "Hondo" diagram (Figure 4.31), where supplies of heat to TC "Hondo-I" and TD "Hondo-II" are presented simultaneously on the same scale.

Let us now consider changes in the daily value of latent heat energy in the water vapor field accompanying TC and TD structures from February 3 to February 27, 2008. If we assume that the energy interaction between these tropical structures (see Table 4.4) does not occur, then their common daily contribution to the latent heat energy of the southern Indian Ocean, transported to the midlatitudes from the ITCZ, can be presented as the sum of their components (Figure 4.32). Analysis of this figure shows that total daily transport of latent heat by tropical systems is significant and varies widely daily—from 1.5×10^{20} J to 1.4×10^{21} J (i.e., nearly an order of magnitude). So, between February 4 and February 14 latent heat transport hardly changed and was equal to about 4×10^{20} J.

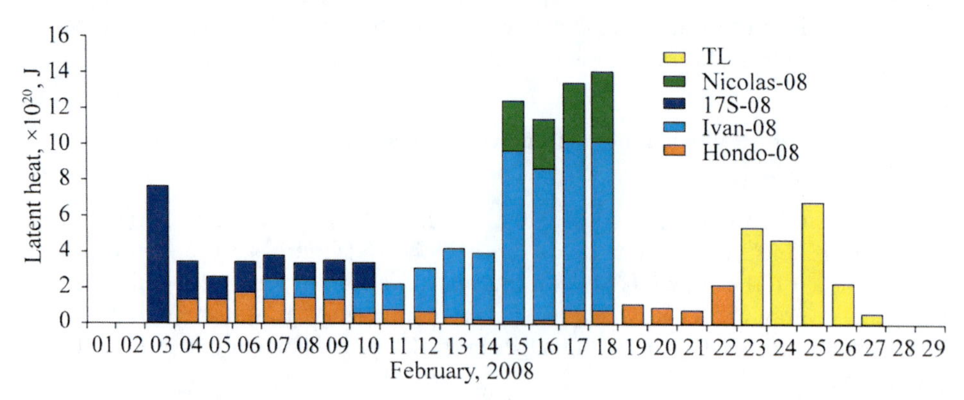

Figure 4.32. Time evolution of latent heat for plural cyclogenesis in the southern Indian Ocean during February 2008 including TD 17S-08, TC "Hondo-I", TD "Hondo-II", TC "Ivan", TC "Nicholas", and TD TL.

Construction of the global water vapor field of the Indian and Pacific Oceans by the technique developed by the team of authors mentioned earlier (see p. 262) allows estimation of the latent heat energy of the central equatorial zone of water vapor in the ITCZ of the Indian and Pacific Oceans (southern hemisphere), as well as detection of variations in latent heat associated with the transfer of bounded water vapor regions to high latitudes by plural cyclogenesis. The energy supply of latent heat of the mother equatorial ITCZ (southern hemisphere, February 2008, daily averaging) was estimated by the same technique as used for individual tropical disturbances. Analysis of Figure 4.33b, given on a linear scale, indicates that the mean value of latent heat of the mother equatorial ITCZ constitutes a stable value of about 10^{23} J with small daily variations of about 1%. Estimate of the transport of latent heat energy by a single TC and a single TD from the mother field (Figure 4.33a, b) equals between 0.1% to 1% of the mother field's total energy. Note that

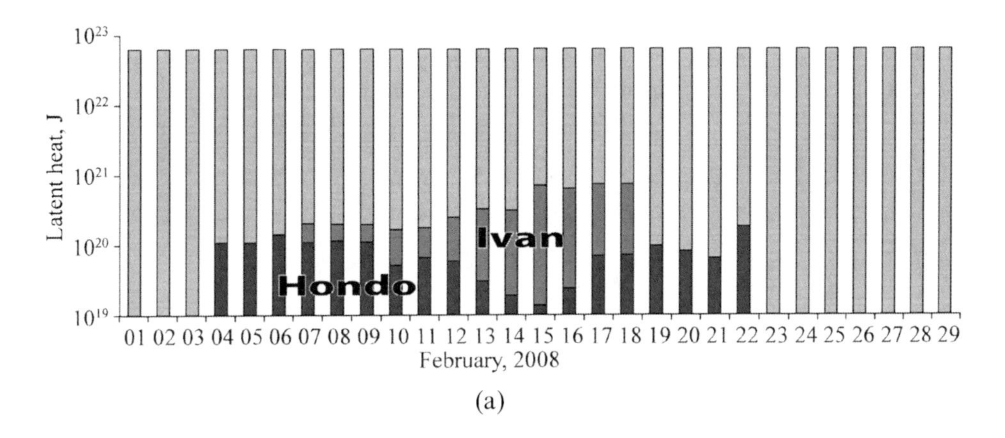

(a)

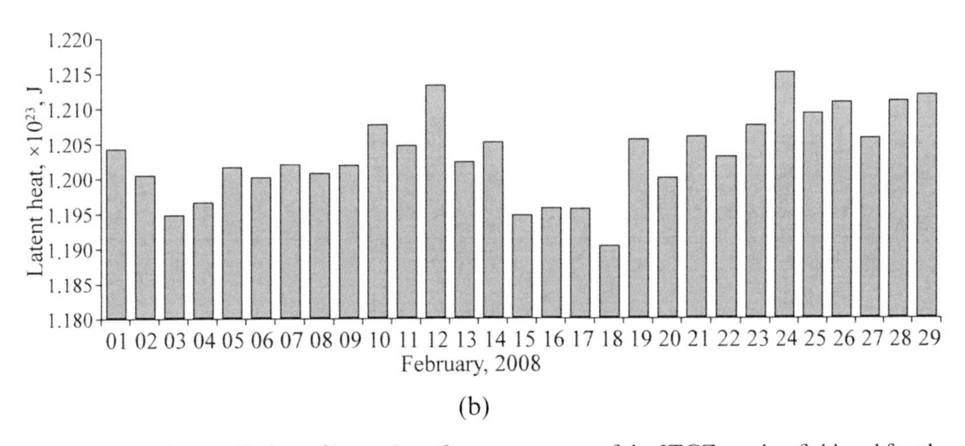

(b)

Figure 4.33. Time evolution of latent heat for water vapor of the ITCZ mother field and for the water vapor field of TC "Hondo" and TC "Ivan": (a) on the semi-logarithmic scale; (b) on the linear scale

the global mother field of integral water vapor is restored by latitudinal transport mechanisms (western transport in the tropics) from Melanesia in about a day.

Let us summarize our findings. Analysis of the results of observation data shows that the effective and quick-response channel of latent heat energy transfer from the tropics to the midlatitudes is a region of water vapor of heightened integral concentration entrained by each cyclone from the tropical zone and sustained by the cyclone throughout its evolution by means of a jet spiral bridge (the "ejection" effect). As a result of the data-merging technique developed by the team of authors and fully discussed in this chapter, new important estimates of the latent heat energy of the central equatorial zone of water vapor in the ITCZ of the Indian and Pacific Oceans (southern hemisphere) and the detection of variations in the latent heat associated with the transfer of bounded water vapor regions to high latitudes by plural cyclogenesis. Estimate of the supply of latent heat energy from the mother equatorial ITCZ (southern hemisphere, January–March) equals about 10^{23} J with daily variations of about 1%. Estimate of the transport of latent heat energy by a single TC and a single TD from the mother field equals between 0.1% to 1% of total mother field energy. The global mother field of integral water vapor is restored by latitudinal transport mechanisms (western transport in the tropics) in about a day.

5

Solar activity and global tropical cyclogenesis

This chapter studies one of the most interesting and intriguing problems in tropical cyclogenesis: ascertaining the influence of solar activity variations on the genesis and functioning of tropical cyclones. This problem has long been studied and researchers have used different approaches to analyze the problem as a whole and various techniques to process the signals. This chapter demonstrates a principally new approach to the solution of such problems: the use of a special procedure of joint (cross) processing (wavelet analysis) of global cyclogenesis data and solar activity data for the 19-year period from 1983 to 2001 (Afonin and Sharkov, 2003). During this period IKI RAN (the Space Research Institute of the Russian Academy of Sciences) has developed and accumulated (based on remote observations) a detailed satellite database containing time series of tropical cyclogenesis intensity (the number of tropical cyclones per day and the length of their lifetimes). Based on this database and on analysis of archive data from the F10.7 solar activity index, it is shown that—in the frequency–time spectra of annual series—the response of global cyclogenesis to external effects from the 27-day solar activity period at frequencies close to this forcing frequency, as well as at its second harmonic and subharmonics. In the spectra, averaged for 19 years in regions that have characteristic cyclogenesis timescales of $S = 10$–80 days, the brightest component is $S = 27$ days, which is almost always observed in northern hemisphere cyclogenesis. In the southern hemisphere the maximum response is most often observed at frequencies close to the harmonics and subharmonics of the basic frequency f27.

5.1 STATEMENT OF THE PROBLEM: SHORT HISTORY OF VIEWPOINTS

It is well known that the tropical zone of the global ocean–atmosphere system plays a key part in the dynamics and evolution of synoptic and climatic meteorological

E. A. Sharkov, *Global Tropical Cyclogenesis* (Second Edition).

processes on the Earth. In this connection, any external (with respect to the troposphere) energy effect (mainly from the Sun) can have critical significance for variations in the dynamic and thermodynamic structural components of the tropical zone of the Earth's atmosphere. This mainly relates to such complicated meteorological tropical objects as catastrophic atmospheric vortices (i.e., tropical cyclones).

Historically, this question was first considered at the end of the 19th century (Meldrum, 1872; Chizhevskii, 1976; Elsner and Kara, 1999). The earliest viewpoints on the possible process involved in the interaction between solar activity and cyclogenesis were to put it mildly naive. For example, one viewpoint stated that each sunspot that arose on the solar disk should strictly correspond to tropical cyclone genesis in some cyclone-generating water basin of the World Ocean. Another viewpoint considered that the number of tropical cyclones in any cyclone-generating water basin for a fixed time interval should correspond to the number of sunspots for the same interval. There also existed a series of combined views on the interaction between solar activity and tropical cyclogenesis. The most long-lived hypothesis was (and still remains the case now) the viewpoint of a possible cross-correlation between the number of sunspots on the solar disk for a certain time interval and the number of tropical cyclones in a certain active water basin for the same time interval. Researchers have normally accepted such a water basin to be primarily the North Atlantic basin (see, e.g., Elsner and Kara, 1999). In the overwhelming number of studies solar activity was (and still is) accepted to be the number of sunspots (Wolf sunspot numbers) in the form of temporal delta-functions for a certain time interval. This was the form in which the quantitative value of cyclogenesis intensity was accepted. However, this overlooked the facts that the number of sunspots is not the best quantitative indicator of solar activity (Elsner and Kara, 1999) and that cyclogenesis intensity, presented as a set of delta-functions, cannot provide satisfactory results in many cases (see Chapters 2 and 3 for more details). Nevertheless, spectral and cross-correlation studies on this basis have been and still are being (Cohen and Sweetser, 1975; Elsner and Kara, 1999) carried out in some research organizations now. However, they have not led to unequivocal and physically evident interpretive results. The reasons for such a situation are associated (as is currently becoming clearer) with the multiscale and nonlinear character of the interaction between solar activity and tropical cyclogenesis, on the one hand, and this is associated with the fact, on the other hand, that the standard cross-correlation approaches, widely used in processing time series of observed phenomena, cannot in principle yield positive results in the analysis of delta-shaped signals, which as a matter of fact follows from known and fundamental results of statistical radiophysics (Rytov, 1966). See Sharkov's (2006) critical analysis of the results of some modern works (Elsner and Kara, 1999; Cohen and Sweetser, 1975).

As already discussed, interest in studying catastrophic geophysical systems can be explained by a number of circumstances. On the one hand, these atmospheric processes represent a direct physical danger to humankind and are accompanied by considerable material damage, in addition to causing administrative and social problems as a result (Pielke and Pielke, 1997). On the other hand, catastrophic atmospheric vortices represent a unique mechanism for the effective rejection of excessive heat in the tropical atmosphere under conditions in which the action of usual mechanisms, the most important being turbulent convection, are clearly insufficient (see Section 7.5). Thus, catastrophic phenomena play an important part in establishing the climatic temperature regime on the Earth (the greenhouse effect), removing excess heat and preventing excessive overheating of the planet in the tropical zone. Problems are aggravated by the difficuty of reliably forecasting the genesis and evolution of these phenomena satisfactorily using modern theoretical models. What is worse, in the theoretical investigation of such geophysical systems a serious crisis state is observed (see Chapter 7) as a result of changing basic theoretical concepts (Rutkevich, 2002; Sharkov, 2010). This is why detailed experimental investigations of the spatiotemporal structure of cyclogenesis including both determinate (e.g., seasonal) and stochastic components of the process are doubtless of interest (Elsner and Kara, 1999; Sharkov, 2000, 2010). It is here that special study should be focused and directed at searching for the components (both stochastic and determinate) of the general complex process determined by external energetic effects, principally caused by solar activity variations. Solar energy flows into the atmosphere both directly, in the form of electromagnetic radiation, and indirectly, in the form of energetic solar particles as a result of numerous solar–magnetosphere–ionosphere–troposphere interactions. Because the Earth's atmosphere represents a complex thermo-aerodynamic system with complicated and nonlinear transfer characteristics, it is reasonable to suppose that global tropical cyclogenesis (GTC) can reflect the links both with direct solar activity indicators and with geomagnetic activity indicators. Researchers have long undertaken—since the end of the 19th century (Meldrum, 1872; Cohen and Sweetser, 1975; Chizhevskii, 1976) right up to today (Elsner and Kara, 1999)—repeated attempts to establish correlation links (using various methods from direct comparison to cross-correlation processing) between the appearance of tropical cyclones (in some World Ocean regions) and solar (the number of sunspots) and magnetospheric activity.

This chapter presents the results of joint analysis of the spectral features of a stochastic time series of global tropical cyclogenesis and two indexes: the most widely used solar activity index F10.7 (radioemission flux at a frequency of 2,800 MHz, $\lambda = 10.7\,\text{cm}$) and the global geomagnetic disturbance index DST. Multiscale wavelet analysis was applied for this purpose.

Focus is given to studying the response of global cyclogenesis (as well as northern and southern hemisphere cyclogenesis) to the external effect of solar activity with a 27-day period (the Sun rotation period) at frequencies close to this forcing frequency, as well as at its second harmonic and subharmonics.

5.2 STUDY TECHNIQUES AND APPROACHES: PROCESSING TECHNIQUE

Study of the genesis and evolution of steady vortical systems against the background of turbulent chaos of the tropical atmosphere is now undertaken by means of two approaches:

- *the "local" approach (individual cyclogenesis)*—which is used to study the genesis of a unitary (individual) vortical structure from turbulent chaos when the ocean–atmosphere system is out of balance locally; and
- *the "global" approach (plural cyclogenesis)*—which considers the formation of vortical systems in the World Ocean basin as a set of relaxation/generation centers in the active medium of the natural ocean–atmosphere system (considered globally).

Normally, theoretical and experimental work is devoted to the study of individual cyclogenesis. Such work is the basis of a number of diverse techniques, from dynamic models to self-organization ideas (e.g., the spiral mechanism). The global approach is also applied here, but studies the problem of atmospheric catastrophe genesis thermodynamically (for more details, see the analytical review of modern theoretical models in Chapter 7). However, tropical cyclogenesis considered globally still remains poorly studied, though some serious and non-trivial results have already been obtained on the basis of plural cyclogenesis (see Chapters 2 and 3). Obviously, the technique used to construct the time series of global tropical cyclogenesis—the physical process considered simultaneously over the whole World Ocean basin (or over regional basins of hemispheres)—should serve as the structural base for plural cyclogenesis investigation. The technique chosen to construct the time series of GTC is undoubtedly important, because its conceptual solution determines the physical significance of the final result.

The complete evolution of a tropical cyclone (TC), with regard to the lifetime of each individual event, will be formed as follows (see Chapter 2). As we are not interested in the detailed structure and dynamics of each individual tropical formation, we shall present each tropical disturbance on the time axis as a pulse of unit amplitude with a random duration (corresponding to the TC functioning time) and with random moments of appearance (individual TC generation). The number of pulses (events) that arrive in the unit time interval chosen (a day in our case) is the natural physical parameter: the intensity of global cyclogenesis determining the energy of the ocean–atmosphere interaction.

Such a procedure of signal formation can be written mathematically as follows:

$$I(t) = \sum_{i=1}^{N} \Xi(t - t_i; \tau_i), \tag{5.1}$$

where $I(t)$ is cyclogenesis intensity; N is the number of TCs appearing during a year; and Ξ is the restricted Heaviside function:

$$\Xi(t - t_i, \tau_i) = \begin{cases} 1, & t_i < t < t_i + \tau_i, \\ 0, & t_i + \tau_i < t < t_i, \end{cases} \tag{5.2}$$

where τ_i is the lifetime (existence) of an individual TC; t_i is the time of its formation (generation); and $t_i + \tau_i$ is the time of its dissipation.

The sequence of pulses generated in such a manner is little more than an integer random time flux of indiscernible events. Thus, we based our representation of a time sequence of the global tropical cyclogenesis intensity as a statistical signal with a complicated structure. Though this approach is only one of many signal formation versions, it nevertheless allowed us to essentially advance our understanding of the stochastic structure of GTC (see Chapters 2–4).

The experimental geophysical data on the spatiotemporal appearance and evolution of TCs over World Ocean basins used in our study were taken from the systematized database Global-TC (see Section 8.4), where the chronological, hydrometeorological, and kinematic characteristics of large-scale tropical disturbances throughout the World Ocean basin are presented in the form of a sequence of events for the period under study: 1983 to 2001. Pokrovskaya and Sharkov (1993a) were the first to show that the random process generated in such a manner represents a typical cable process (on daily scales). Further studies (Pokrovskaya and Sharkov, 1994a; Sharkov, 2000) have shown that the probable structure of fluctuations in the amplitude of a studied flux is very close to the structure of a Poisson-type flux in the presence of certain (and rather symptomatic) deviations from the Poisson model as the timescale of observation increases. It is these deviations from the Poisson model that carry information about the multiscale and nonlinear character of a system.

Solar geophysical parameters (the DST global geomagnetic disturbance index and the F10.7 solar activity index) were taken from two well-known Internet sites: *http://www.ngdc.noaa.gov/stp/GEOMAG/dst.shtml* and *ftp://ftp.ngdc.noaa.gov/STP/GEOMAGNETIC_DATA/INDICES/KP_AP/*

5.3 FEATURES OF WAVELET PROCESSING

As already noted, a principal feature of this work is the use of a specialized (wavelet) spectral analysis technique simultaneously for two time series of signals—cyclogenesis intensity and solar radio emission variations—with subsequent comparison and analysis of the wavelet diagrams obtained. Wavelet transfer is widely applied now for the recognition of images, for the processing and synthesis for all manner of images and signals, for studying the properties of turbulent fields, and for many other purposes.

Wavelet transfer consists in decomposing an initial analyzed signal over a basis constructed from a soliton-like (wavelet) function that has some unique properties by means of scaling transformations and shifts. Each of the functions of this basis characterizes a certain time frequency and its localization in time. Unlike the Fourier transform usually applied to analyze the spectral properties of signals, wavelet transfer (WT) provides two-dimensional (frequency and time) display of the one-dimensional signal studied. As a result, the possibility arises of analyzing signal (process) properties both in the physical (time) and spectral (frequency) spaces simultaneously. Wavelet analysis has also been efficiently applied to the multiscale and multifractal dynamic processes that model the scale transformation and structure formation situations observed in dissipative systems.

As already noted, wavelet transfer consists in expansion of the analyzed function (signal) on the basis of a soliton-like function $\Psi(t)$.

The continuous wavelet transfer (i.e., the two-dimensional wavelet diagram) $W(a,b)$ of the one-dimensional function $f(t)$ is as follows:

$$W(a,b) = a^{-1/2} \int \Psi\left(\frac{t-b}{a}\right) f(t)\, dt. \tag{5.3}$$

Expansion of the analyzed signal over scales is performed by extending or compressing the analyzing wavelet before its convolution with the signal. Wavelet transfer is known to be weakly dependent on the choice of analyzing function; however, using various wavelets it is possible to more clearly reveal features of the analyzed signal.

As usual, the wavelets (sometimes called "caps") are represented by centrally symmetric functions. For example, for isotropic material fields the basis is most often constructed from wavelets that are derivatives of the Gaussian function.

The results presented here are obtained on the basis of the Morlet wavelet which is defined as the product of the complex exponent and the Gaussian function:

$$\Psi(x) = \pi^{-1/4} \exp(i\omega_0 x) \exp\left(-\frac{x^2}{2}\right). \tag{5.4}$$

Wavelet transfer can be used to reveal: (1) the localization and intensity of the features in a signal, (2) the regions of localized periodicities of the signal, (3) the regions of greatest activity of the process, (4) the hierarchical structure of an analyzed system, (5) fracturing and merging of scales, etc.

In what follows we shall show that wavelet transfer can be used to analyze the structure of processes occurring in principally different systems—geophysical and heliophysical—and to ascertain their mutual correlation properties.

Since wavelet transfer spectra ("WT spectra") are usually accepted as mainly qualitative characteristics, it is expedient to perform quantitative "calibration" of our technique or to find out whether it's possible to obtain the quantitative characteristics of the stochastic process under study.

The technique was calibrated by means of a model of the time series:

$$Y = 20 \sin\left(\frac{2\pi T}{5}\right) + 10 \sin\left(\frac{2\pi T}{27}\right) + 20 \sin\left(\frac{2\pi T}{54}\right), \tag{5.5}$$

where $T = 1, \ldots, 365$ (the annual cycle). This series contains terms with prominent timescales (5, 27, and 54 days) and two amplitudes (10 and 20). The results of wavelet transfer (the Morlet wavelet was used here) of this series are shown in Figure 5.1. Figure 5.1 shows the model time series $Y(T)$ for the time period of 365 days (a year). The power spectrum used to carry out wavelet transfer (the WT spectrum) (normalized with respect to the maximal value of spectrum components) is shown in Figure 5.1b. The WT spectrum represents variations in the frequency and time of a signal's components. The relative power of components (i.e., the amplitude) is coded by the false color (the right scale in Figure 5.1b). The time axis (the abscissa

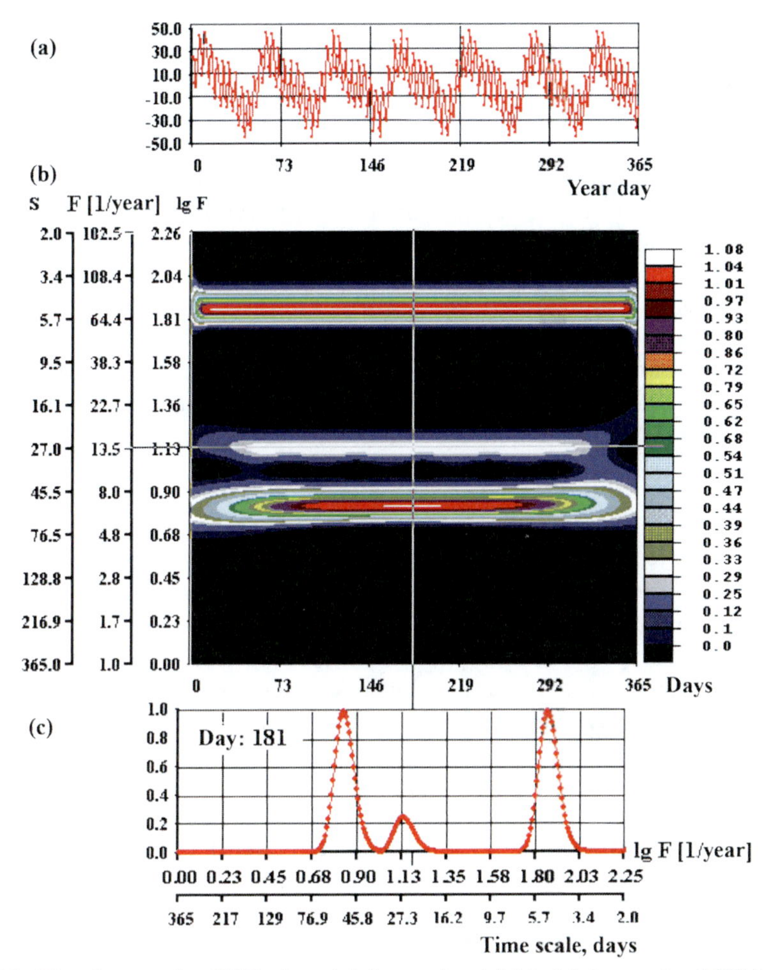

Figure 5.1. Wavelet transfer (WT) of model time series. (a) Model time series $Y(T)$ for a time period of 365 days (year). (b) Three-dimensional WT normalized to maximum component values. Time axis (the abscissa) corresponds to the day of the year (T). Frequency axis (the ordinate lg F) is logarithmic. Minimum frequency lg $F = 0,0$ is at the origin of the coordinate (below). Maximum frequency lg $F = 2,26$ is at the top. For the sake of convenience two other scales (at the left) show corresponding frequencies between $F_{\min} = 1$ (period $T = 1$ year) and $F_{\max} = 182.5$ (period T = 2 days, the Nyquist frequency for the given series) and the corresponding timescale S in days. (c) WT cross-section along the frequency axis for fixed time $T = 181$ days.

axis) corresponds to days of a year (T). The frequency axis (the $\lg f$ ordinate) is logarithmic with the minimum frequency $\lg f$ at the coordinate origin (below) and the maximum frequency $\lg f = 2,26$ (above). For convenience, there are two other scales on the left that can reproduce corresponding frequencies between $f_{\min} = 1$ (period $T = 1$ year) and $f_{\max} = 182.5$ (period $T = 2$ days, the Nyquist frequency of the given series) and the timescales S (in days) corresponding to them. Figure 5.1c presents that part of the WT spectrum that lies along the frequency axis for the fixed time $T = 181$ days. This presentational format is used in all subsequent figures.

As seen from the spectrum itself and the cross-section, wavelet transfer accurately determines the frequency components of an initial signal and, moreover, correctly reproduces the quantitative relationship between the various frequency components of an initial signal. The power of a signal on a 27-day timescale is four times lower than the two other components, in exact conformity with the ratio of amplitudes (the wavelet spectrum normalized with respect to the maximum power).

However, note that, by virtue of those features of the WT technique that apply to a series with a finite number of points, the time resolution at low frequencies (at the bottom of a spectrum) progressively decreases (the boundary effect), which is revealed by frequency lines becoming blurred as the frequency decreases.

5.4 ANNUAL AND INTERANNUAL VARIABILITIES OF SOLAR ACTIVITY AND CYCLOGENESIS

Before investigating the studied time series by means of a special technique (wavelet analysis), it is worth considering the time variability of solar activity in detail (including geomagnetic activity) and cyclogenesis intensity using standard techniques such as trends, accumulating approaches, and the processing of auto-correlations and cross-correlations.

The time dependence of daily values of the global intensity of a tropical cyclogenesis $I(t)$, the F10.7 activity index of the Sun (representing solar radio emission flux at the wavelength of 10.7 cm in units of $10^{-22}\,\mathrm{W\,m^{-2}\,Hz^{-1}}$) and the DST global index of geomagnetic disturbance (representing disturbance of Earth's geomagnetic field in gammas) during 19 years (1983–2001) are shown in Figure 5.2a–c. Figure 5.2d shows the accumulation function $Z(t)$ (see Section 2.3) of the time series of cyclogenesis intensity (the technique was first proposed and described by Sharkov, 2000) and the parameters of linear regressions for the accumulation function of global cyclogenesis and northern and southern hemisphere cyclogenesis. Joint analysis of the time series (Figure 5.2a–c) shows that, despite pronounced 11-year solar activity variation (index F10.7 changes about five times), and quantity $I(t)$ does not have any pronounced large-scale time trends and remains virtually constant (except for some small but pronounced seasonal variations). This tendency can be observed more clearly from the time series of the accumulation function (Figure 5.2d). Below the figure, the correlation factors corr XY and the slope of regression curves ka (1/day) are indicated. The accumulation function of cyclogenesis grows virtually linearly with time (the correlation factor for global cyclogenesis equals 1.00 before 1996 and 0.998 after 1996). This testifies to the constant value of the intensity of the global Poisson process (Section 4.1) (or, putting it simply, this testifies to the constant value of the daily rate of TC formation) and to the complete absence of interannual variations. Small harmonious

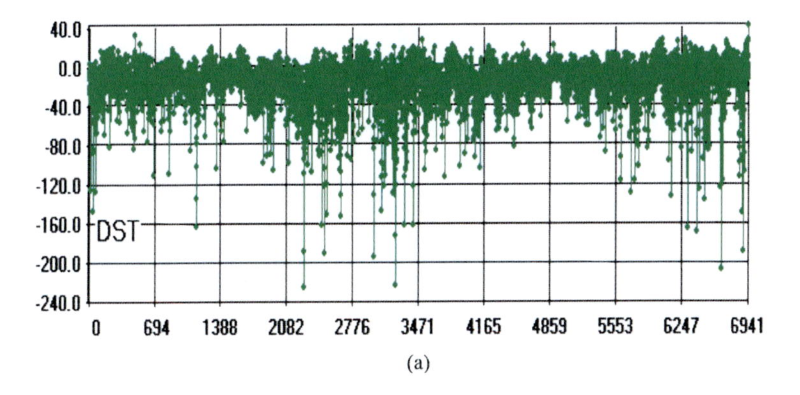

(a)

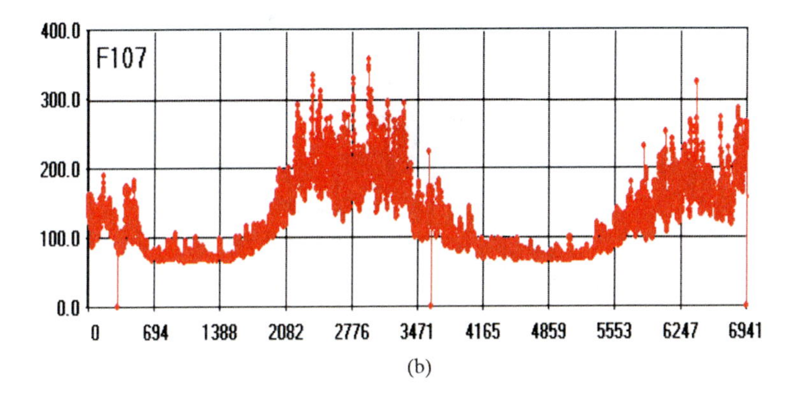

(b)

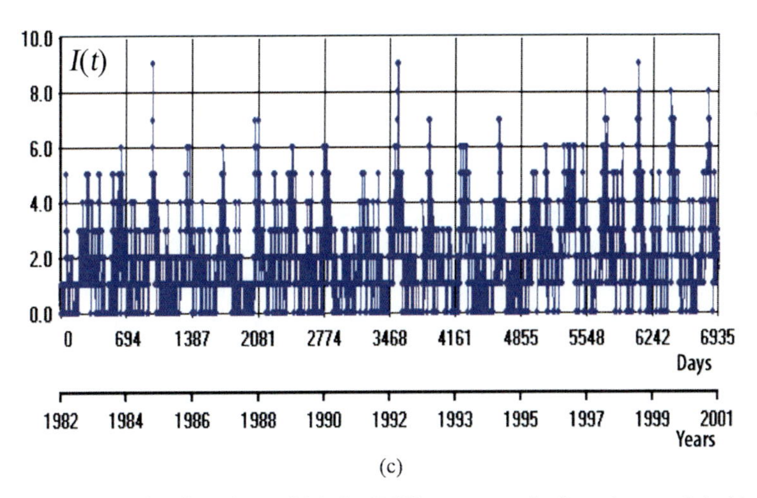

(c)

Figure 5.2. Time series of daily values of (a) the DST geomagnetic disturbance global index (in gammas), (b) the F10.7 solar activity index (in $10^{-22}\ \mathrm{W\,m^{-2}\,Hz^{-1}}$), and (c) the intensity of global tropical cyclogenesis $I(t)$ during 19 years (1983–2001).

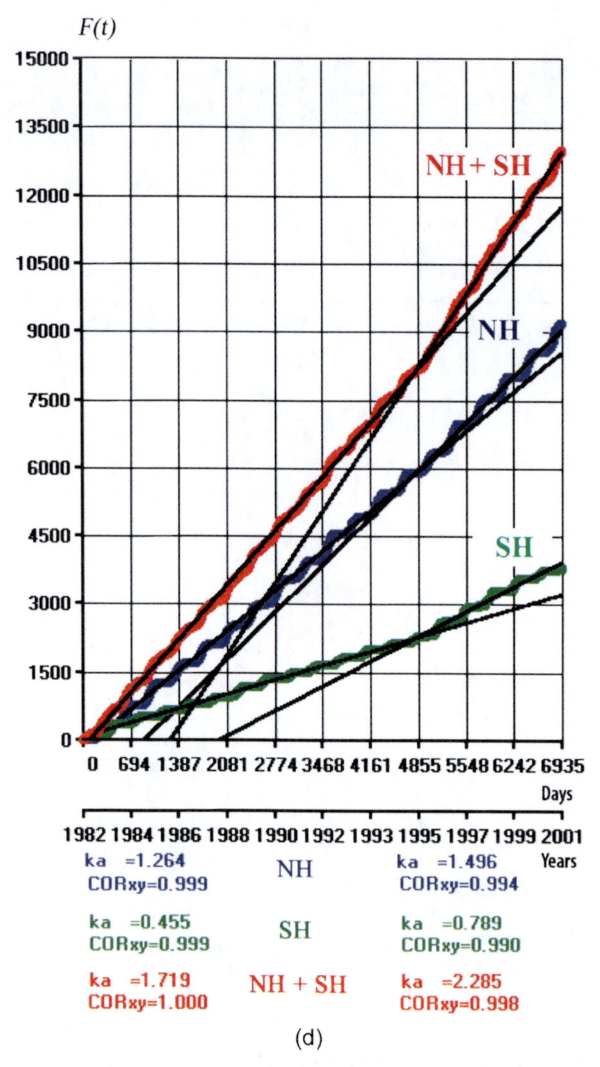

Figure 5.2 (*cont.*). (d) Accumulation function $F(t)$ of the intensity time series for northern hemisphere (NH), southern hemisphere (SH), and global cyclogenesis in the basins of the World Ocean. The slopes of regression lines *ka* and correlation coefficients corr XY until 1995 (at left) and after 1995 (at right) are shown at the bottom.

components on curves correspond to semi-annual variations (they are presented more clearly on the time series of Figure 5.2c). The slope of regression curves—the quantity *ka* (1/day)—physically represents the intensity of the Poisson process (the differential rate of cyclone formation per day). For example, during 1983–1996 the daily differential rate of TC birth on the globe was 1.719 cyclones per day, of which 1.264 cyclones were born in the northern hemisphere and 0.455 in the southern hemisphere. Such an approach using accumulation functions was applied for the first time for time series analysis for a 10-year period (from 1982 until 1992) in Sharkov (2000). The change in cyclone generation rates after 1996 (the global rate was 2.285 and, respectively, 1.496 and 0.789 for the two hemispheres) requires some explanation. Any increase in the TC formation rate could be (at first sight) associated with global processes (climate warming). However, the high values of correlation factors ($\geq$0.999 until 1996 and 0.990 thereafter) testify to the constancy of the TC generation mechanism. Consequently, any change in the slope of curves (the TC formation rate) has to do with using more detailed approaches to systematization and accounting for initial and final forms of TCs and with considerable improvement in meteorological TC tracking services as a result of introducing computer technology in the form of the Internet (this is described in more detail in Pokrovskaya and Sharkov, 2006 and in Section 8.4). Regarding the case in point, what is important is the fact that there were no interannual variations during 19 years. A very similar conclusion can be drawn from the results of Section 4.1 for the 25-year time period as well.

Let us now turn to the question on the correlation between cyclogenesis intensity $I(t)$ and solar activity F10.7 over timescales of more than 1 year. To answer this question, the values of global $I(t)$ for each year (i.e., the given year) of 19, averaged over a year, were compared with F10.7 and DST values calculated in a similar manner (Figure 5.3). To present the data on the same plot, the DST values were multiplied by -15. Figure 5.3b, d shows scatterplots for each of three cyclogenesis parameters (global, north, and south) from F10.7 (Figure 5.3b) and DST (Figure 5.3d). Black lines represent corresponding regressions. Numbers under lines indicate correlation factors—none of which exceed 0.210. It is easily seen that, on timescales of a few years, any correlation between $I(t)$ and solar and geomagnetic activity according to the F10.7 and DST indices is also most likely absent. We emphasize that here the case in point is rather simple analysis techniques (such as scatterplots).

Note that Figure 5.3a points to a seemingly positive trend of $I(t)$ for 19 years. In reality, however, wavelet analysis (and accumulation function analysis) has shown that this trend is undoubtedly an artifact caused by increasing the annual number of cyclones after 1996, as discussed above (Figure 5.2).

Let us now turn to the question of the correlation between cyclogenesis intensity $I(t)$ and solar (F10.7) and geomagnetic (DST) activity on an annual timescale (day-averaged). The time series of annual variations of (a) DST, (b) F10.7, and (c) $I(t)$ in the northern hemisphere and (d) $I(t)$ in the southern hemisphere for minimum (on the left, 1997) and maximum (on the right, 2000) solar activity are shown in Figure 5.4. Figure 5.4f shows scatterplots of the dependence of the global quantity $I(t)$ on F10.7 (Figure 5.4e on the left) and DST (Figure 5.4e on the right) for each day of the year. Analysis of these figures shows that the time series of diurnal values of $I(t)$, F10.7, and DST do not exhibit any essential correlation. The correlation factor for annual series of $I(t)$ and F10.7 (Figure 5.4e on the left) equals -0.113 and that for annual series of $I(t)$ and DST (Figure 5.4e on the right) equals -0.225. Similar results were obtained earlier in the analysis of the cross-correlation properties of the time series of cyclogenesis intensity and magnetospheric activity (Section 2.9). At the same time, the time series of diurnal values of $I(t)$ have a pronounced seasonal

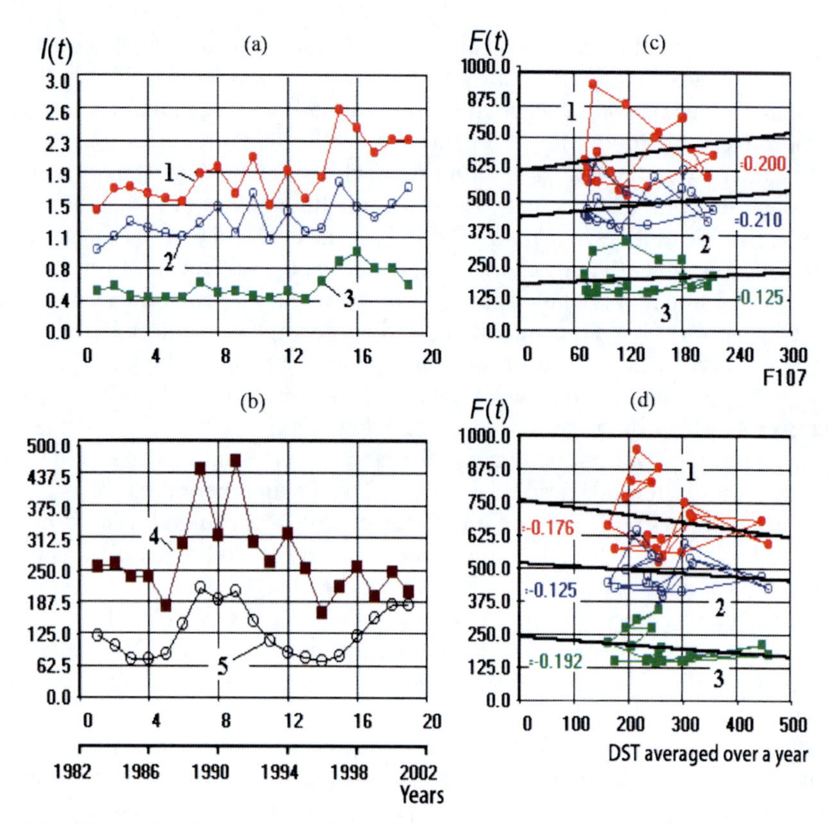

Figure 5.3. Correlation features of daily cyclogenesis intensity (a) averaged over a given year (out of 19 years) and (b) DST and F10.7 values determined similarly: (1) for World Ocean basins; (2) for northern hemisphere basins; (3) for southern hemisphere basins; (4) DST values (multiplied by -15); and (5) F10.7 values. (c), (d) Scatterplots of individual cyclogenesis intensity (WO, NH, SH) with year accumulation $F(t)$ on DST and F10.7 values.

dependence. The annual maxima of $I(t)$ in the northern and southern hemispheres are considerably spaced in time. In both hemispheres $I(t)$ maxima are observed during the summertime. TCs are not observed at all in winter. TC intensity in the southern hemisphere is lower than in the northern hemisphere. Differences between cyclogenesis intensity at the maximum and minimum of solar activity are not revealed, whereas the well-known sudden change of F10.7 radio emission variations and magnetospheric activity are observed (Figure 5.4a, b).

5.5 CORRELATION OF ANNUAL TIME SERIES

The question about the interrelation between stochastic (or "organized") processes in the time series of cyclogenesis has already been considered from the auto-

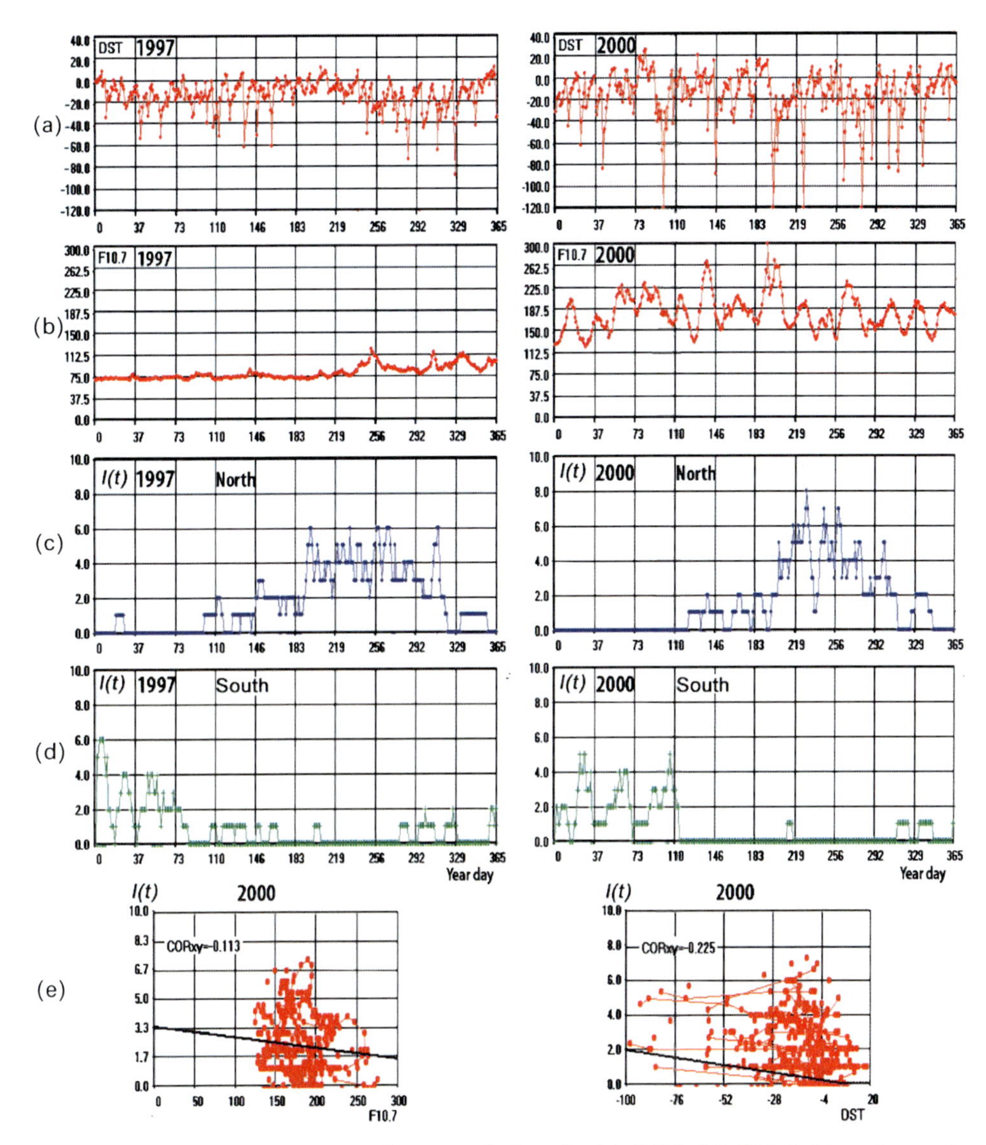

Figure 5.4. Correlation features of year variations in (a) DST, (b) F10.7, (c) cyclogenesis intensity in the NH, and (d) cyclogenesis intensity in the SH for minimum solar activity (at left, 1997) and for maximum solar activity (at right, 2000). (e) Scatterplots of individual cyclogenesis intensity (WO) on DST (at right) and F10.7 (at left) values for each day of the year.

correlation viewpoint in Chapters 2 and 3. In this section we are interested in the interrelation between the autocorrelation characteristics of cyclogenesis and solar activity. Figure 5.5 shows the calculated autocorrelation functions of the time series $I(t)$ and F10.7 for the maximum and minimum of solar activity. In our calculations

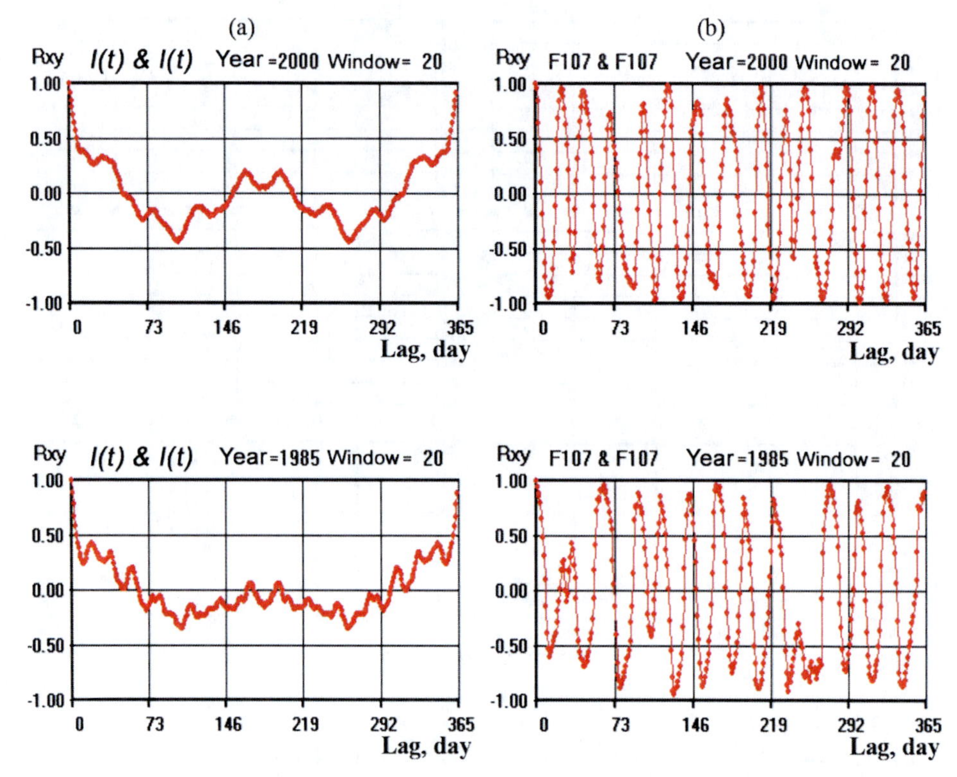

Figure 5.5. Autocorrelation functions of the time series for global cyclogenesis intensity $I(t)$ and solar activity F10.7 for maximum activity (at top) and for minimum activity (at bottom).

we used the cyclic shift of initial series; therefore, the curves are symmetric relative to the shift equal to one half of a year. It would be more correct to extend the initial series from both sides at the expense of series for the previous years and following years; however, this would have little effect on the results obtained. It is seen that the $I(t)$ series only autocorrelate on timescales less than 5 days (i.e., they represent a sequence of short-lived events without a "history"). At greater delays the interrelation between quantities $I(t)$ is absent and each TC can be considered as a random event (Chapter 2). Thus, it is reasonable to preliminarily conclude that possible interactions in the cyclogenesis process exceeding 5 days are not effective. On the other hand, the series of F10.7 quantities autocorrelate well with the period of 27 days. The significant 27-day autocorrelation is kept during the solar activity minimum as well, though it is difficult to obtain this conclusion directly from time series analysis (Figure 5.4a, b).

Thus, both correlation approaches and trend analysis give a negative conclusion about the effect of solar activity on cyclogenesis at first sight. However, more delicate analysis (i.e., wavelet analysis) shows that this is not in fact the reality.

5.6 SEARCHING FOR 27-DAY PERIODICITY IN THE TIME SERIES OF CYCLOGENESIS

Despite the generally accepted viewpoint that the contribution of basic energy to cyclogenesis is the result of the phase transition of water vapor, questions about initial (trigger) mechanisms are far from resolved. A possible trigger mechanism could be variations in solar activity—especially, the known 27-day solar activity variation associated with the Sun's rotation. Wavelet analysis has been applied to the search for 27-day periodicity in the time series of cyclogenesis.

The results of applying wavelet transfer (WT spectra) to the global series of cyclogenesis intensity (on the left) and to the F10.7 parameter (on the right)—covering 1,600 days at solar activity (SA) maximum and 1,200 days at SA minimum—are shown in Figures 5.6 and 5.7, respectively. Horizontal cursors (horizontal lines) on the wavelet diagrams are set either at the f27 frequency (corresponding to the 27-day component) or at bright features, and the vertical cursors are set at the particular day under study.

The general structure of the WT spectrum for cyclogenesis intensity has prominent features for chaotized systems of the class under consideration (see Chapter 2). On small timescales the process behaves like a Poisson random signal; then there comes a nonlinear stage, which is accompanied by nonlinear interactions (including bifurcation types). On great timescales (longer than a year) the system behaves like a linear harmonious oscillator with pronounced seasonal (annual and semi-annual) variations. All these regions show up well on the wavelet diagrams of cyclogenesis intensity (Figures 5.6 and 5.7). As far as solar activity is concerned, only the region of nonlinear interactions (near the f27 frequency) with weak Poisson branches shows up brightly on wavelet diagrams. No seasonal variations in solar activity are naturally observed.

In F10.7 variations the presence of the f27 component is obvious in both cases, and in a solar maximum year it clearly shows up. Spectral maxima do not always lie exactly on the f27 line, but are very close to it. The change in amplitude of this component is significant. However, continuity of this line (i.e., constancy of the f27 amplitude) should hardly be expected because of the chaotized character of solar activity (sunspots) and because of the fact that variation in 27-day solar activity is not the only factor involved in changing the flux of F10.7. These additional factors can displace the f27 frequency seen on the WT spectrum. All that can be said here is that the f27 component in the F10.7 spectrum is present almost permanently and represents its basic most pronounced feature.

There are two basic features that are immediately apparent in cyclogenesis variations: (1) the most intensive components fall in the range of timescales $S = 10, \ldots, 80$ days and (2) the response that has the greatest amplitude is observed at frequencies close to f27. Since cyclogenesis intensity has a pronounced seasonal course, wavelet diagrams can be separated into hemispheres. During solar activity minimum the response at the f27 frequency is greatest in the northern hemisphere, whereas in the southern hemisphere it is absent, and two SA maxima are observed at the first subharmonic ($\sim$54 days) and at the second harmonic ($\sim$15

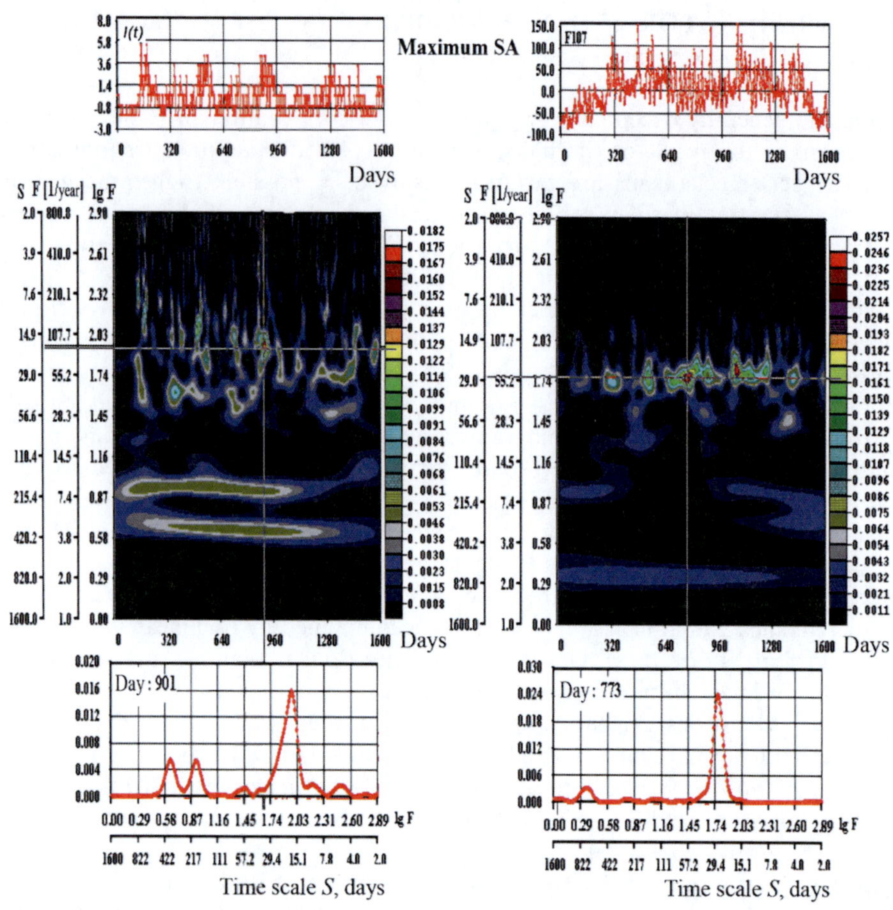

Figure 5.6. Three-dimensional WT spectrum of the time series of cyclogenesis intensity (at left) and solar activity parameter F10.7 (at right) for 1,600 days at solar maximum. Corresponding time series are at the top (with minus year mean value). WT cross-section along the frequency axis for fixed time ($T = 901$ and 773). See Figure 5.1 for details.

days). During solar activity maximum the picture is approximately the same, but richer in details.

5.7 FEATURES OF MULTIYEAR CYCLOGENESIS DATA

WT spectra calculated for each of the 19 years were averaged and the result is shown in Figure 5.8. This figure confirms the conclusions obtained earlier. In the F10.7 spectra the brightest component is the 27-day ($S = 27$) one and, as expected, all other components are virtually absent. The WT spectra of cyclogenesis intensity reveals two regions: the region of $S = 10, \ldots, 80$ days and a semi-annual structure

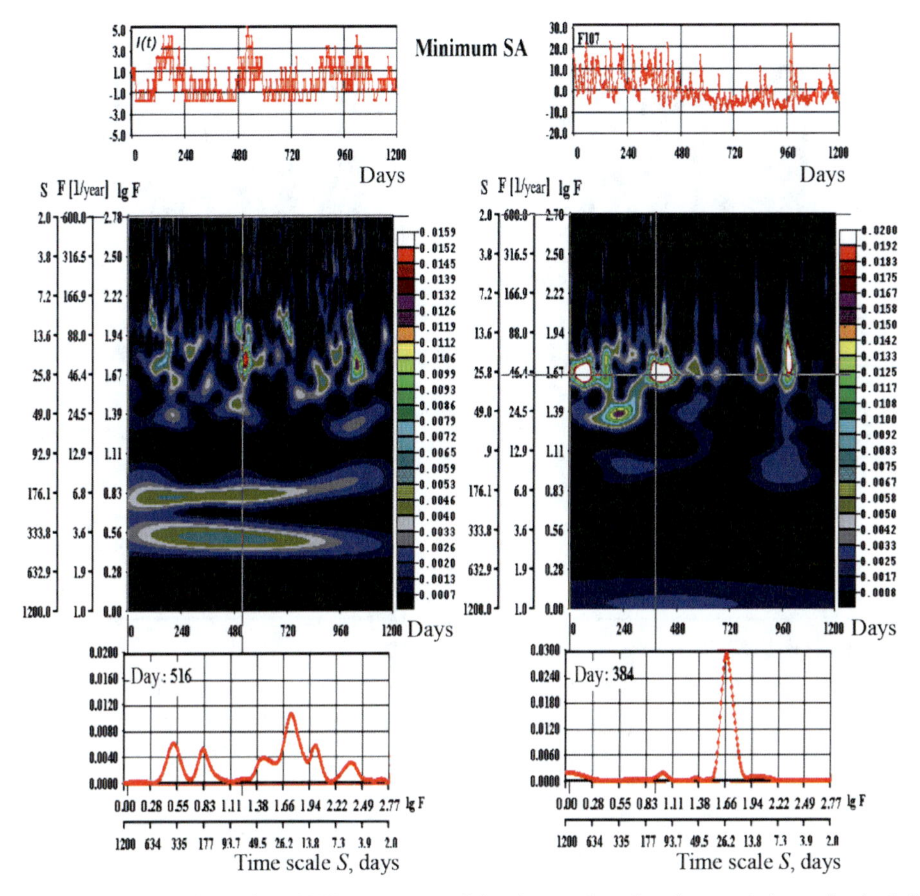

Figure 5.7. Three-dimensional WT spectrum of the time series of cyclogenesis intensity (at left) and solar activity parameter F10.7 (at right) for 1,200 days at solar minimum. See Figure 5.1 for details.

reflecting the presence of a pronounced seasonal effect in tropical cyclogenesis. In the region of $S = 10, \ldots, 80$ days the brightest component is $S = 27$ days, which is almost always observed in the northern hemisphere. In the southern hemisphere the maximum response is most often observed at frequencies close to the harmonics and subharmonics of the basic frequency f27. The response at the f27 frequency is, apparently, forced by the frequency of external solar activity.

The responses of cyclogenesis intensity at the f27 harmonics and subharmonics of external solar activity testify to the nonlinearity of tropical cyclogenesis. The fact that the responses in most cases do not absolutely precisely coincide with f27 and its harmonics and subharmonics leads to the supposition that tropical cyclogenesis is a manifestation of the complicated behavior of the thermohydrodynamic ocean–atmosphere system which has its own natural characteristic (possibly, resonant) frequencies.

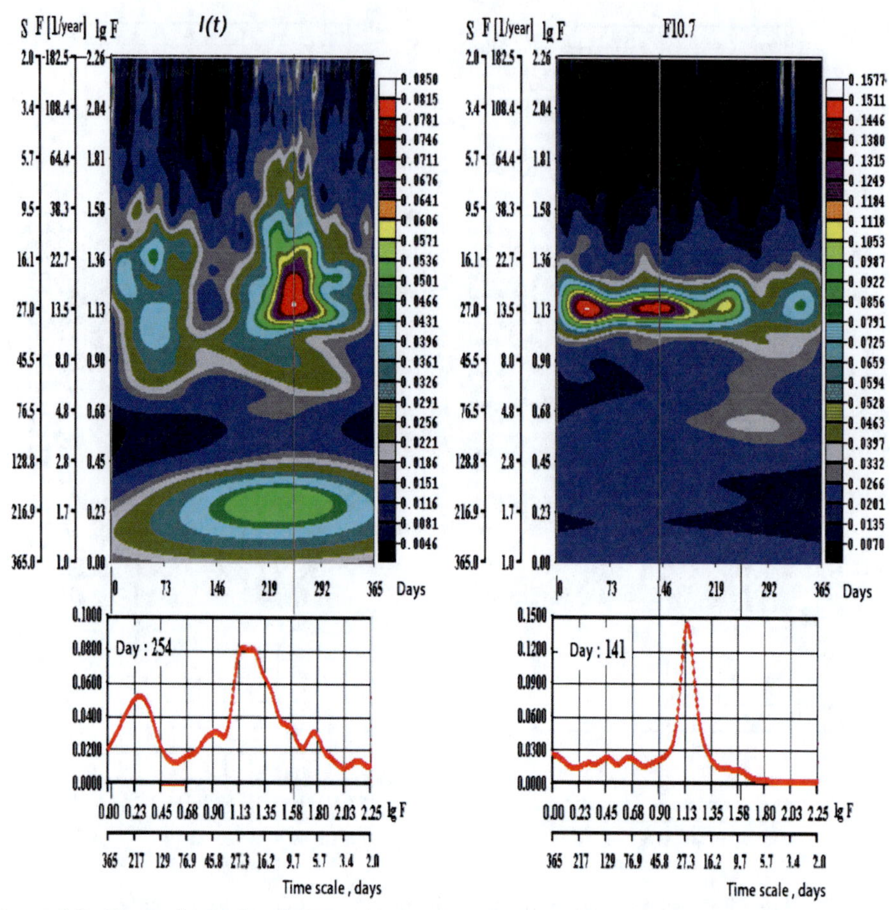

Figure 5.8. Three-dimensional WT spectrum of the time series of cyclogenesis intensity (at left) and solar activity parameter F10.7 (at right), both averaged over 19 years. See Figure 5.1 for details.

Note that the effect of 27-day variations in ultraviolet solar radiation on the periodic warming of the ozone layer was actively investigated in the 1980s. This effect is known to excite atmospheric "resonant" dynamic modes. In a sense, it testifies to the similar physical mechanisms involved in the solar activity effect on upper-troposphere dynamics.

This chapter presents the results of joint special (wavelet) processing of global cyclogenesis data and solar activity data for the period 1983–2001. Global cyclogenesis was found to possess its own timescales. These scales form in the frequency space of the region $S = 10, \ldots, 80$ days, separated both from the Poisson region (with timescales of $S < 10$ days) and from annual and semi-annual variations. The semi-annual structure reflects the presence of pronounced seasonal effects on tropical cyclogenesis. In the spatiotemporal spectra of annual series the response of global cyclogenesis to the external (solar) effect with a 27-day period is observed at frequencies close to this forcing frequency and its second harmonic and subharmonics. In spectra averaged for 19 years in the region $S = 10, \ldots, 80$ days, the brightest component is $S = 27$ days, which is almost always observed in the northern hemisphere. In the southern hemisphere the maximum response is most often observed at frequencies close to the harmonics and subharmonics of the basic frequency f27. Wavelet spectra show tropical cyclogenesis to be a manifestation of the complicated nonlinear behavior of the thermohydrodynamic earth surface–atmosphere system which has its own natural dynamic properties, characteristic timescales, and (possibly) resonant frequencies.

6

Ionosphere and tropical cyclone activity

New results from remote and rocket-based investigations of the state of various ionospheric layers under the effect of tropical cyclones of various intensities, at various horizontal distances both from the observation point and from the trajectories of TCs are considered in this chapter. Experiments have been carried out by Russian researchers in recent years (2006–2011) using fundamentally different ionosphere-sounding technologies: (a) complex processing of data from high-altitude rocket sounding of the equatorial ionospheric region D (in situ mode) from the Tumba rocket range in India in an area subject to strong tropical cyclones; (b) analysis of the signals from oblique high-frequency radar route sounding along the midlatitude Magadan–Irkutsk route (the middle point of which lies south of Yakutsk); (c) study of the features of spatial wave spectra of variations in atmospheric 557.7 nm emissions of atomic oxygen, observed from the Earth surface in the southern part of Eastern Siberia under quiet helio-geomagnetic conditions during tropical cyclone activity.

6.1 STATEMENT OF THE PROBLEM AND MEASUREMENT TECHNIQUES FOR IONOSPHERE SOUNDING

The tropical cyclone model has long been considered a rather closed (in space) dynamic spiral structure isolated from external influence (the "rigid top" model or "rigid washer" model) within strict altitudinal limits inside the troposphere of Earth's atmosphere. However, further investigations have shown that studies of TCs as physical structures should be performed within the framework of the open systems of statistical physics which take account of the effect of the large-scale TC environment (Sharkov, 2000, 2010).

E. A. Sharkov, *Global Tropical Cyclogenesis* (Second Edition).

The idea that studies into how TCs interact with the ocean–atmosphere system should not be limited to the troposphere but should be based on considering the large-scale crisis state as a global phenomenon affecting various geophysical media—beginning with the oceanic surface and the troposphere and finishing with the ozonosphere and ionosphere—was first put forward in detail in 1996 by SRI RAS scientists (Balebanov *et al.*, 1996, 1997). Studying the kinematic, thermodynamic, and electrodynamic relationships between the elements of the ocean–troposphere–upper-atmosphere–ionosphere system under crisis states clearly is a very important component of space research and attempts are currently under way to organize complex investigations using rocket-based radar ionospheric sounding and optical surveying of night-time emissions from the upper atmosphere.

Wave disturbances and ionosphere state variations have long been studied from the purely aeronomic viewpoint. In such studies the search for physical features has usually been carried out by revealing long-period (i.e., 20–50 days or more) wave disturbances and synoptic variations and—as a rule—by discussing the effect of physical space sources on ionospheric disturbances. The possibility of the existence of tropospheric sources—while not excluded—was not given in any great detail in aeronomic works (see, e.g., Shefov *et al.*, 2006). What was surprising about this was that the strongest tropospheric catastrophes—tropical cyclones—were not considered as possible ionosphere excitation sources, especially so because we now know that they can form the different (fast-acting) mechanisms associated with strong ejections of charged particles, neutrals, and acoustic gravitational waves from the zone of stratospheric tropical cyclone "ejection" to considerable altitudes.

Atmospheric gravitational waves (AGWs) are key factors in the dynamics and energy of the atmosphere and ionosphere. They comprise acoustic waves (AWs) whose periods are less than 5 minutes and wavelengths are less than 100 km and inner gravitational waves (IGWs) with characteristic periods from 5 minutes to 3 hours and wavelengths of more than 100 km.

There is now a lot of evidence to support the existence of atmospheric gravitational waves in the Earth's atmosphere. Basic observational facts include the wave structures on tropospheric clouds, near ground pressure variations on microbarograms, orographical waves, moving wave structures on luminescent emissions, oscillations in the energy spectra from airglow observations, wave variations of the content of ozone and other admixtures in the middle atmosphere, wave structures obtained by ionosphere radio-sounding techniques, etc. The modern techniques of upper-atmosphere investigation, such as high-frequency (HF) sounding, radiotomography, and GPS satellites, facilitate simultaneous recording of these wave disturbances in the ionosphere across large spatial regions.

AGWs transfer angular momentum and energy—which arrive during helio-geomagnetic disturbances into the auroral latitudes—from the Earth magnetosphere from high latitudes to low latitudees. They also transfer angular momentum and energy from the lower atmosphere into the upper atmosphere. In the upper atmosphere gravitational waves are observed either directly as neutral gas fluctuations (e.g., by measurements with interferometers, satellite mass-spectrometers, or accelerometers), or indirectly as ionospheric plasma fluctuations, which in principle represent passive indicators (tracers) of neutral gas motion (Hocke and Schlegel, 1996). When penetrating ionospheric altitudes, atmospheric gravitational waves reveal their properties as moving ionospheric disturbances (MIDs).

Two types of MIDs are observed in the ionosphere (Hocke and Schlegel, 1996):

- *mesoscale MIDs*—which propagate at a velocity of between about 100 m/s and 250 m/s (i.e., lower than sound velocity in the lower atmosphere); their periods vary from 15 minutes to 1 hour, their horizontal wavelengths are from ten to a few hundred kilometers, and their vertical wavelengths are no more than 10 km;
- *large-scale MIDs*—which propagate at a velocity of between about 400 m/s and 1,000 m/s (i.e., comparable with sound velocity at these altitudes); their periods vary from 30 minutes to 3 hours and their horizontal wavelengths are more than 1,000 km.

Mesoscale MIDs are observed more often than large-scale ones. Basically, MIDs from auroral sources are observed at night, because in the day ionic deceleration is higher. Large-scale MIDs are usually generated in the auroral zones during magnetic storms. Historically, the first scientific investigations of the ionosphere were associated with the launching of small meteorological rockets to altitudes of 50 km to 70 km with a set of standard meteorological instruments and sensors for measuring the density of neutral and charged components (electrons and ions). Rocket sounding is still going on today but has been completely updated both methodologically and instrumentally. The basic feature of such an approach is the single-cycle (in space) short (in duration) observation of measurements (see Section 6.2).

Principally different approaches to ionospheric investigation are associated with analysis of the propagation of electromagnetic waves, at particular radio wavelengths, through the oblique sounding routes that pass through ionospheric layers. An example is the oblique radiosounding route between midlatitude Magadan and Irkutsk, the middle point of a route at ionospheric altitude located south of Yakutsk (see Section 6.3).

Another method of studying layers of the lower ionosphere is by recording, from the Earth surface, disturbed airglow from the upper atmosphere in the optical range and of spatiotemporal wave disturbances both from identified sources, associated with the helio-magnetic situation, and from sources of unknown nature, including those located in tropospheric layers. See Section 6.4 which outlines the features of spectra of variations of atmospheric 557.7 nm emissions observed in the southern part of Eastern Siberia in 2003 and 2007, under quiet helio-geomagnetic conditions during periods in which tropical cyclones were active in the northwestern and central basins of the Pacific Ocean.

At present, ionospheric tomography techniques are efficiently applied (Kunitsyn *et al.*, 2007) on the basis of signals from low-orbital and high-orbital satellite navigation systems (GLONASS and GPS), which allows revelation of the variations in ionospheric parameters caused by both the helio-magnetic situation and strong cyclone propagation in the Earth troposphere.

6.2 EQUATORIAL LOWER-IONOSPHERE INTERACTIONS WITH TROPICAL CYCLONES STUDIED BY ROCKET SOUNDING

This section presents the results of processing the data of rocket sounding of the lower equatorial ionosphere (layer D) above the Tumba rocket range in India in a region in which tropospheric catastrophes—tropical cyclones—occur. They were observed by satellite sounding in the northern part of the Indian Ocean (Vanina-Dart *et al.*, 2007a, b). The ionospheric measurements considered in this section were carried out by the Central Aerologic Observatory (CAO) of the Federal Hydrometeorology and Environmental Monitoring Service (*Rosgidromet*) located at the rocket-sounding station Tumba (80°N, 77°E) in the equatorial zone of India. The database basis is the bank of electron density measurements taken onboard meteorological M-100B rockets (according to Russian classification) by a technique based on applying an electrostatic Langmuir probe. The probe is installed at the upper end face of the rockets' meteorological spikes, thus protecting it from the effect of the electric and magnetic fields of the rockets themselves. The estimated root-mean-square error of an individual measurement is ~35%. In regular atmosphere sounding, temperature sensors and instruments for measuring wind parameters are also installed on these meteorological rockets along with the electron density probe.

Despite the modeling boom of the last century, the construction of region D models—which are very important for solving practical tasks—still greatly lags behind the results of modeling higher ionospheric regions. The reason for this lies both in the complexity of physical processes (the undoubted influence of tropospheric thermodynamic effects, ionization by various sources, the variety of ionic components, high variations of recombination rate) and in the difficulty of reliable measurements of electron density values $[e]$ in the lower ionosphere.

Figure 6.1 presents the electron density profiles measured in a series of eight rocket launches. Various marks show the changes in $[e]$ depending on altitude h in the corresponding launch. All launches occurred at about 12:00 UT (universal time); accordingly, the values of the Sun's zenith angle χ have a certain but, nevertheless, insignificant scattering ($\chi = 73 \pm 40°$). According to the values of the solar activity index (F10.7 = 82.5 ± 12.5) and the magnetic activity indexes (Ap = 8 ± 5) and (Kp = 1.5 ± 1.5), the helio-geophysical situation during launches was quiet. As a result of the fact that there were two launch groups at $\chi \sim 71°$ and at $\chi \sim 76°$, it is more expedient to construct the average statistical structures for these angles (Figure 6.2). We attribute launch nos. 46, 49, 51 to the first group and launch nos. 47, 48, 53 to the second group. We shall leave unchanged the profiles of launch no. 50, obtained during the active phase of a tropical cyclone at $\chi = 75°$, and the profiles of launch no. 52, measured at $\chi = 74°$, as the closest one to launch no. 50 (according to the zenith angle value). At low helio-geomagnetic disturbance the electron density in the upper region D obeys the zenith angle dependence according to the cosine law and fits the framework of the model considered in the previous section. This testifies to the fact that, in the absence of tropospheric disturbances, the behavior of the equatorial region D of the ionosphere obeys the laws of well-known statistical models.

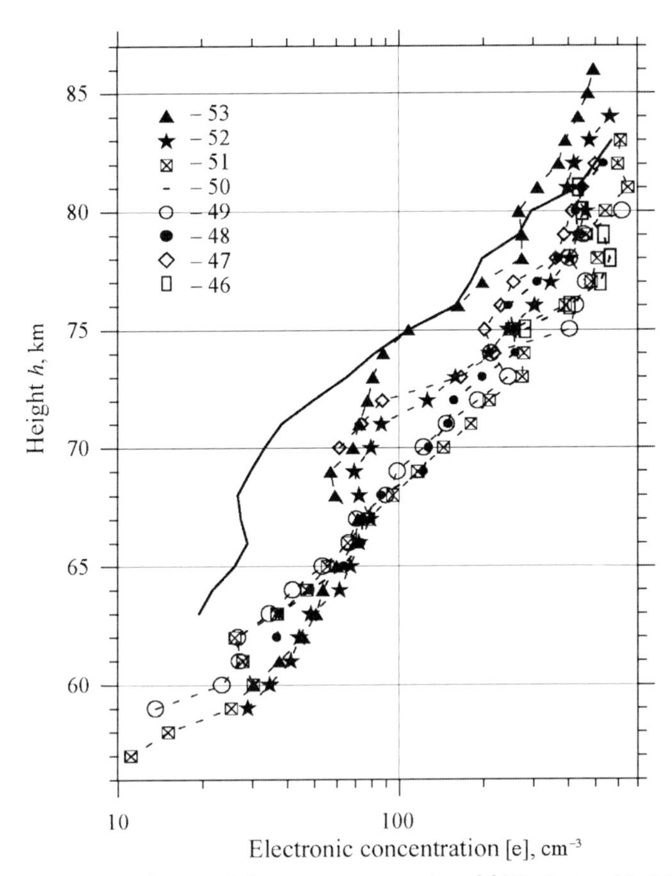

Figure 6.1. Altitude dependences of electron concentrations $[e](h)$ obtained in May–June 1985 by processing high-altitude sounding rocket launch data from the Tumba rocket range (India). Digits specify the number of rocket launches.

Let us now turn to information about existing cyclones for May–June, 1985. The information is taken from the remote-sounding database Global-TC (Pokrovskaya and Sharkov, 2006) with additional reanalysis of particular data. TC no. 8501-01 arose on May 22 at 12:00 GMT (Greenwich Mean Time) in the basin of the Bay of Bengal, at 160°N, 800 km east of the coast of India, as a tropical depression (TD). The wind velocity at the center was 15 m/s and the pressure was 996 mb. After displacing to the northwest, on May 23 the TD became a tropical storm and the pressure dropped to 992 mb. The storm slowly displaced to the northeast strengthening gradually and on May 24, 200km south of the Bangladesh coast, it became a strong tropical storm with a wind velocity of 31 m/s and a pressure of 985 mb. As it approached the land the storm weakened; on May 25 it made landfall and then quickly dissipated over Bangladesh.

The next tropical cyclone was TC no. 8502-02. It arose on May 27 in the Arabian Sea basin, at 16°N, 700 km west of the coast of India. The wind velocity

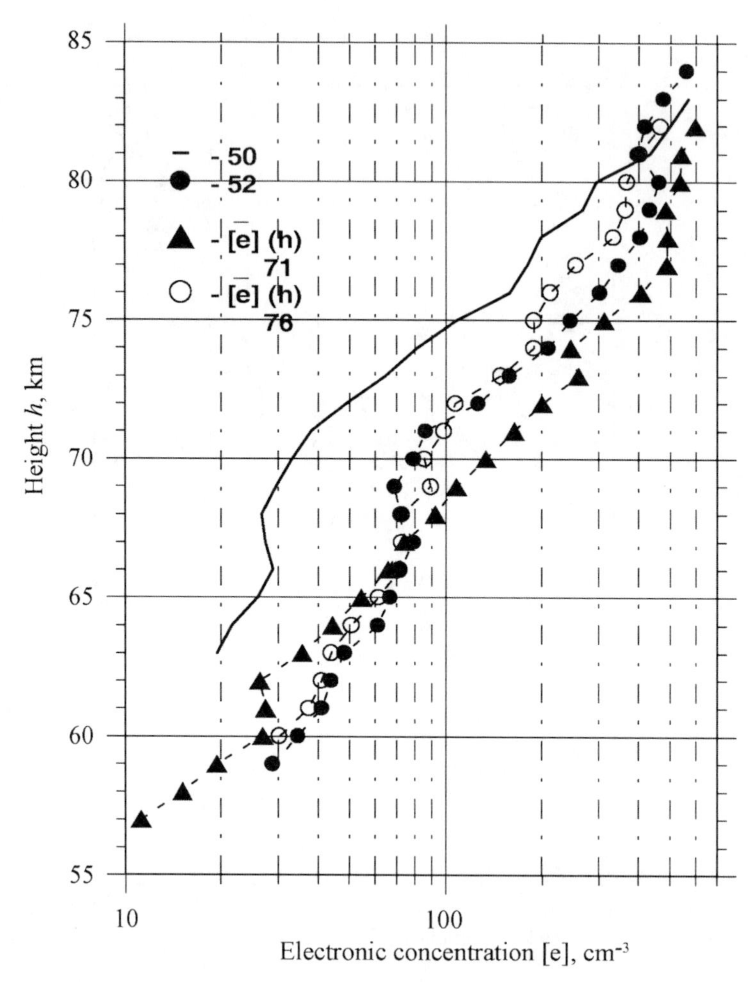

Figure 6.2. Altitude dependences of electron concentrations $[e](h)$ obtained for individual launches (50 and 52) and averaged for specific zenith angles (71 and 76).

was 15 m/s and the pressure was 1,002 mb. After displacing to the northeast, on May 28 the TC became a tropical storm and the pressure dropped to 996 mb. On May 29, 300 km from the coast, it became a strong tropical storm with a wind velocity of 26 m/s and a pressure of 990 mb. As it approached the coast the storm gradually weakened and on May 31 at 18:00 GMT it made landfall as a tropical depression, and on June 1 the storm dissipated over Pakistan. The geographical motion of TC no. 8501-01 and TC no. 8502-02 and the position of the Tumba rocket range are presented in Figure 6.3. This figure also shows the position of TC no. 8502-02 on May 29, 1985. Data on the cyclones were taken at 12:00 UT and rocket measurements were taken at 11:55 UT. Thus, this section presents synchronous data on parameters of the lower ionosphere, troposphere, and stratomesosphere.

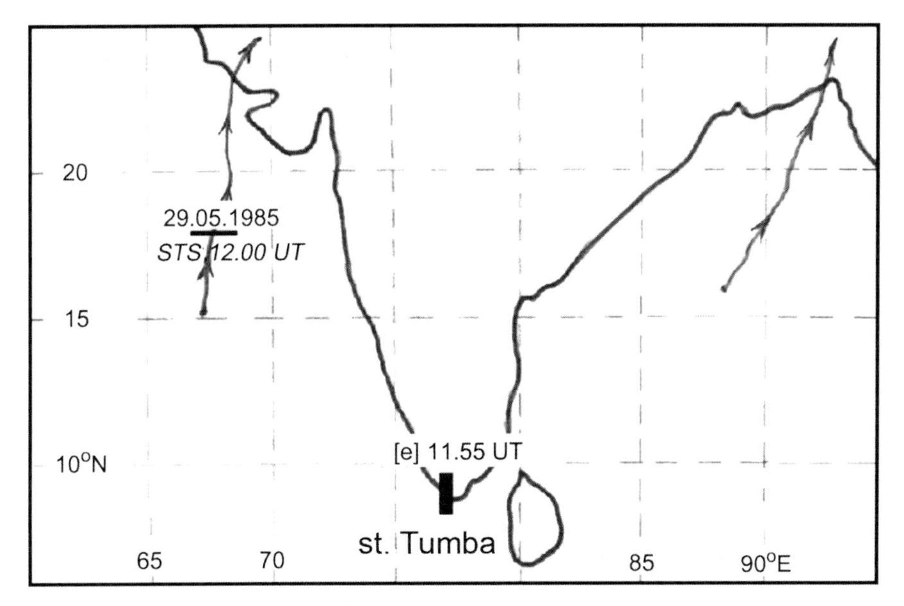

Figure 6.3. Tracks of two tropical cyclones: NIN 8501-01B (unnamed) in the Bay of Bengal and NIN 8502-02A (unnamed) in the Arabian Sea (Pokrovskaya and Sharkov, 2006). TC tracks and Tumba geographical positions are drawn in horizontal projection.

The mechanism behind electron density dropping in the lower ionospheric region D, measured at the periphery of an active tropical cyclone, is possibly the effect of tropospheric disturbance, via a complicated chain of events that takes place through the dynamic channel (by means of both horizontal and vertical transfer), and through the channel in which small constituents (such as O_3) are redistributed. As is well known, electron density drops with increasing ozone concentration. In the final analysis, the decrease of electron density in the lower ionospheric region D may well be shown by an increase in the density of negative ions. However, in the upper ionospheric region D no differences can be seen. Note that the effect of a tropical disturbance, via the whole complicated chain of interactions, is brought about rapidly (in just a few hours, maybe even faster). Proceeding from the results of ground-based, artificial, electromagnetic (weak and virtually instantaneous) excitation of the ionosphere, it is reasonable to suppose that such a rapid excitation really exists under natural conditions. Note that sporadic decreases in the density of plasma components are observed in high F2 layers of the ionosphere as well, though the physical mechanisms of such variations are apparently different in this case and not associated with the troposphere directly. Certainly, detailed analysis is necessary, on the one hand, for more correct determination of the channels through which vortical tropospheric systems affect lower-ionosphere behavior and, on the other hand, for finding a possible ionospheric response to tropical cyclones from the other cyclone-generating basins.

Based on detailed synchronous analysis of a series of measurements of electron density and the thermodynamic parameters of the ionospheric D layer, obtained by rocket sounding in the equatorial region and on remote data on tropical cyclogenesis in the northern Indian Ocean, the experimental fact of electron density dropping in the D region was recorded for the first time at a distance of about 1,000 km horizontally from the tropical cyclone core during its active phase (Vanina-Dart *et al.*, 2007a, b). Electron density drops to its greatest extent (three to four times on average) at altitudes of 71 ± 3 km. Moreover, during tropical cyclone activity, a small rise in temperature (about 3°C) was recorded at the stratopause altitude. Thus, the experimental fact of the direct rapid effect and control of tropospheric intensive vortical structures on overlying lower ionospheric layers were revealed for the first time.

6.3 LARGE-SCALE UPPER-IONOSPHERE VARIABILITY MEASUREMENT BY OBLIQUE PATH RADIOSOUNDING

The wave mechanism involved in the interaction between atmospheric layers and effects taking place at the sides of lower atmospheric layers is highly effective. Atmospheric disturbances excite a wide spatiotemporal spectrum of AWs and IGWs. These waves diverge from the disturbance source in different directions at different velocities as a result of dispersion and are filtered as they propagate through the atmosphere. Studies have shown that AWs are observed over the location of the disturbance and that IGWs are basically observed at long horizontal distances (Kunitsyn *et al.*, 2007).

IGWs transfer energy many hundreds and even thousands of kilometers from the troposphere to the middle atmosphere and ionosphere as they propagate upwards along oblique paths. Since atmospheric density drops with altitude according to the near exponential law, kinetic energy conservation implies that wave amplitude grows with altitude. On reaching the mesosphere, the waves dissipate and this results in local heating and movement of atmospheric gas. However, under certain conditions IGWs can reach the the very top of the ionosphere (~300–350 km). Theoretical calculations and experimental data have shown that IGWs are found at horizontal distances up to thousands of kilometers from the excitation source. They can be captured in the waveguide channel, where they are capable of propagating with very little attenuation. Waveguides are formed as a result of the temperature minimum in the mesopause region. IGWs propagate at various velocities in horizontal directions, which results in the formation of moving packages of waves (Kunitsyn *et al.*, 2007).

The following observational data were used for analysis in Chernigovskaya *et al.* (2008, 2010):

- experimental data of the maximum observed frequencies (MOFs) of single-jump signals of oblique sounding along the Magadan–Irkutsk route that passes through Eastern Siberia and the Far East (measurements were carried out at intervals of ~5 min in the equinoctial September periods of 2005–2007);

- data on tropical cyclones from the Global-TC electronic base of satellite data on global tropical cyclogenesis (Pokrovskaya and Sharkov, 2006).

The length of the midlatitude Magadan–Irkutsk route is about 3,000 km. The middle point of the route lies south of Yakutsk. The route geometry is such that the Magadan emission point and the middle point of the route (the ionosphere region from which radio signals are reflected) in the Yakutsk region have the same geomagnetic latitude of ~51°N. So, radiowave-propagating conditions on this route are mainly determined by the state of the sub-auroral ionosphere and by the dynamics of the boundaries of its large-scale structures as the magnetic disturbance level changes.

Figure 6.4 presents examples of ionograms of oblique sounding along the Magadan–Irkutsk route for November 10, 2005. The figure clearly shows the presence of MIDs associated with IGW propagation in the ionosphere.

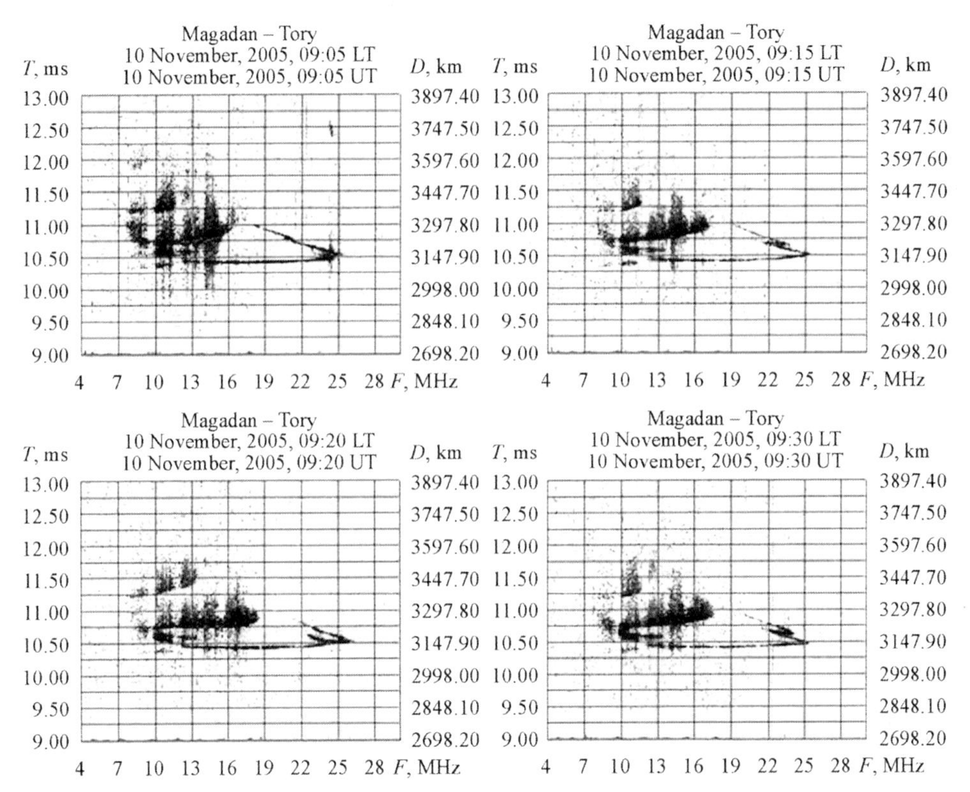

Figure 6.4. Experimental data on maximum observed frequencies (MOFs) of single-jump signals of oblique radio sounding along the Magadan–Irkutsk route passing over Eastern Siberia and the Far East (Russian Federation) (ions per gram). Measurements were carried out with intervals of ~5 min during an equinoctial period (November 10, 2005 from 09:05 until 09:30 UT).

To ascertain whether it is possible to record the effects of strong tropical cyclone activity on variations in upper-atmospheric and ionospheric parameters in a zone far from tropical cyclones, it is important to discover the sources of wave disturbances from below (passage through the solar terminator; seismic activity; the meteorological situation in regions where experimental data are obtained directly such as fronts, cyclones, and jet flows; anthropogenous effects, etc.). It should be kept in mind that the dominating factor in the thermodynamic regime of the ionosphere is helio-geomagnetic activity (electromagnetic emissions of the Sun, corpuscular flows of particles, magnetic fields)—in short, effects from above. So, to efficiently separate the disturbances associated with solar and magnetic activity manifestations and disturbances caused by the effect of lower-atmospheric layers on the upper atmosphere, we have also analyzed the accompanying helio-geomagnetic situation (the F10.7 solar radio emission flux at the wavelength of 10.7 cm characterizing the ionizing ability of the Sun as well as the Kp and Dst geomagnetic indexes).

It is reasonable to suppose that the MIDs revealed in the course of analysis were generated by some meteorological sources located in the troposphere. This is the reason—coupled with our attempts to understand the helio-geomagnetic situation—that we took into account the meteorological situation in the regions under study in the analysis of the effects of wave disturbances on ionospheric parameters because the passages through atmospheric fronts can also be the sources of IGWs. Especially important was taking into account the time it took atmospheric fronts to pass through the Yakutsk region (62°N, 129°E), whose coordinates give the approximate geographical position of the middle point of the oblique sounding route under consideration, where radiowaves are reflected from the ionosphere at various frequencies (6–30 MHz). Variations in atmosphereic and ionosphereic parameters in this region are essentially revealed in the characteristics of radio signals received in Irkutsk. Computer animations were put together using composites from remote monitoring of the cloud cover above Russia. These composites were obtained by NOAA satellites in the infrared range (10.5–11.5 μm) and were used to determine the moments of atmospheric fronts passing through the Yakutsk region. The only other possible meteorological sources located in the troposphere and possessing huge stocks of energy were, according to the Global-TC database (Pokrovskaya and Sharkov, 2006), tropical cyclones acting in the northwest basin of the Pacific ocean. These sources were revealed during radiophysical measurements on the Magadan–Irkutsk oblique sounding route (Septembers of 2005–2007) and the trajectories of cyclone motion were mapped. As an example, Figure 6.5 presents the trajectories of tropical cyclones acting in September of 2006:

- TC “Ioke” (NWP 0613, August 27–September 7, 2006);
- TC “Shanshan” (NWP 0614, September 7–20, 2006);
- TC “Yagi” (NWP 0615, September 14–25, 2006).

Figures 6.6a, b shows calculations of the power of current R_i spectra for the periods $T_i = 1$–5 hours in September, 2006. The lower panels are plots showing how the solar and geomagnetic activity indexes vary. Vertical arrows above the plots of the

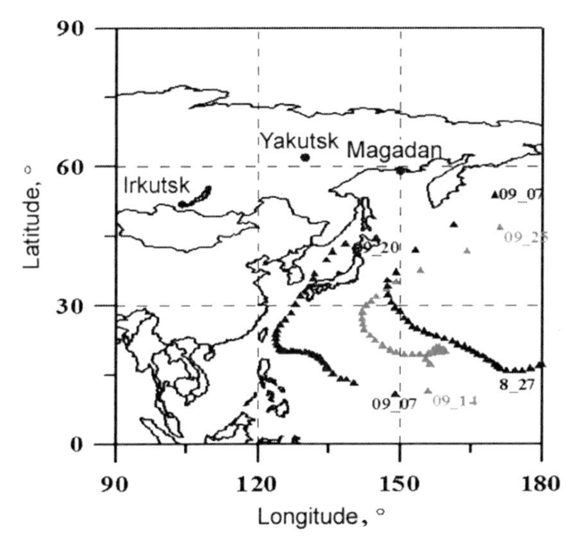

Figure 6.5. Tracks of three tropical cyclones (mature phases) in the northwest Pacific Ocean: NWP 0613 "Ioke" (August 27–September 7, 2006) , NWP 0614 "Shanshan" (September 7–20, 2006); NWP 0615 "Yagi" (September 14–25, 2006) (Pokrovskaya and Sharkov, 2006).

spectrum power of MOF variations indicate the moments of atmospheric fronts passing through the Yakutsk region. Horizontal strips on the time axis of the plot for the Kp index also show the periods of tropical cyclone activity in the northwest basin of the Pacific Ocean.

Analysis of the plots of power R_i of the current spectra of MOF data on various timescales has revealed time intervals in which unique short-period oscillations demonstrate heightened energy over all analyzed periods (wave packages). Such increases in spectrum power can be interpreted as a manifestation of moving ionospheric disturbances the sources of which are IGWs with periods of 1–5 hours.

Comparison of the plots of the spectrum power of MOF variations with the plots of solar and magnetic activity indexes indicates that it is not always the case that the increase in the spectrum power of MOF variations is associated with strengthening of helio-geomagnetic disturbances and with the passage of atmospheric fronts in the area around the route's middle point (e.g., September 5–6 and 8 in Figure 6.6a and September 20 and 23, 2006 in Figure 6.6b). Nevertheless, during these increases in spectrum power, tropical cyclones have always been active in the northwest basin of the Pacific Ocean.

The solar terminator is a constant source of IGWs and the effects of its influence on the atmosphere are also stable and continuous. Every day, at a time associated with passage through the terminator in the morning and at night, the solar terminator effect should show up in time variations in those atmospheric and ionospheric parameters that have approximately identical time duration, intensity, and recorded MID periods. Observed increases in the spectrum power of MOF variations, which can exceed the spectrum power values on neighboring days by as much as twice

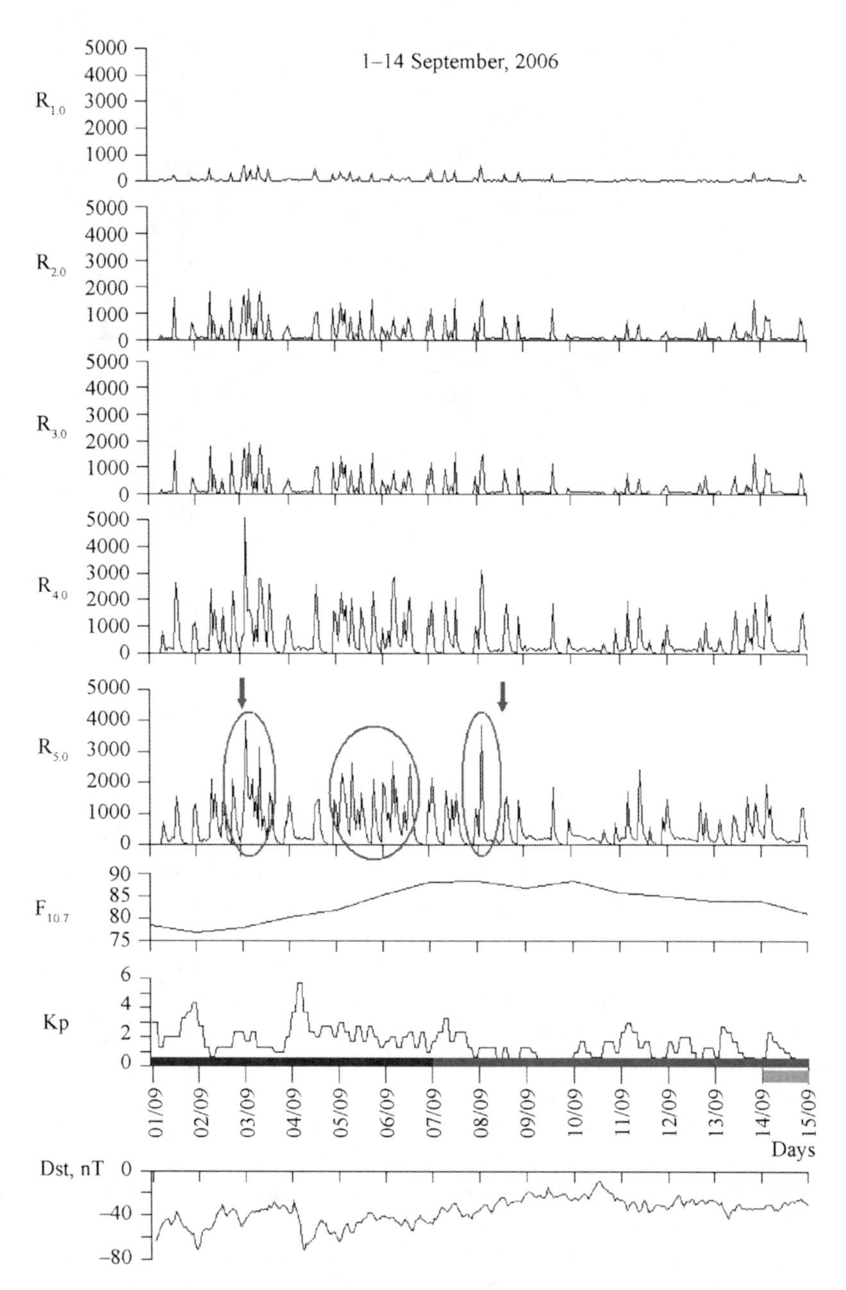

Figure 6.6. Calculations of the power of current R_i spectra for periods of $T_i = 1$–5 hours on September 1–14, 2006. The lower panels show plots of variation of solar and geomagnetic activity indices. Vertical arrows above the plot of the spectrum power of MOF variations indicate the speeds at which atmospheric fronts pass through the Yakutsk region. Horizontal strips on the time axis of the plot for the Kp index reveal periods of tropical cyclone activity in the northwest Pacific Ocean.

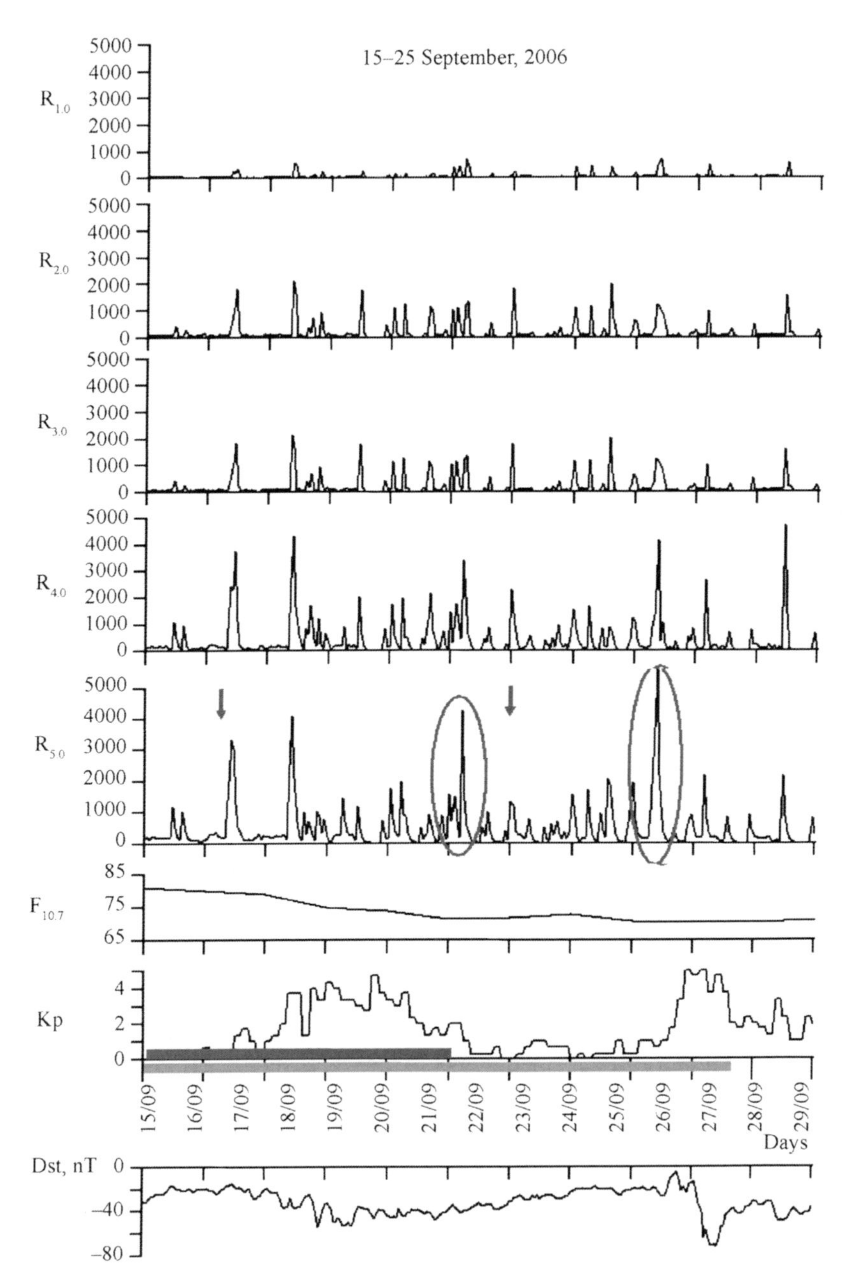

Figure 6.6 (*cont.*). Calculations of the power of current R_i spectra for periods of $T_i = 1$–5 hours on September 15–25, 2006. The lower panels show plots of variation of solar and geomagnetic activity indices. Vertical arrows above the plot of the spectrum power of MOF variations indicate the speeds at which atmospheric fronts pass through the Yakutsk region. Horizontal strips on the time axis of the plot for the Kp index reveal periods of tropical cyclone activity in the northwest Pacific Ocean.

and are steadily revealed during long time periods (more than a day) cannot be associated with the solar terminator effect.

Similar analysis of experimental data was carried out for September 2005 and September 2007, which showed increases in the energy of unique short-period oscillations (wave packages), not associated with increases in solar and magnetic activity or with the meteorological situation in the area around the route's middle point. For September 2005 and September 2007 the increases in the spectrum power of MOF variations were as much as twice as low (in amplitude) as those for September 2007. Similar differences in the energy of short-period oscillations studied for various years may well be associated with possible differences in the conditions under which wave disturbances propagate in the atmosphere, or with the formation, development, and motion of particular tropical cyclones at this time and, consequently, with the various effects TCs have on the overlying atmosphere. Note that many authors studying the effect that lower-atmospheric layers have on the Earth upper atmosphere pay close attention to the fact that the response of the atmosphere and ionosphere to meteorological effects, including massive and very strong atmospheric vortices—tropical cyclones—is always individual and depends on a lot of factors. This testifies to the need for further regular investigations in the topical field of near Earth space physics.

By using a special technique which searches for time series periodicities, we analyzed the short-period (less than a day) time variations of the maximum observed frequencies (MOFs) of oblique sounding signals along the midlatitude Magadan–Irkutsk route (the middle point of the route lies south of Yakutsk) for each September from 2005 to 2007. Analysis revealed time intervals in which unique short-period oscillations have heightened energy (wave packages), which can be interpreted as the manifestation of large-scale moving ionospheric disturbances (MIDs), whose sources are IGWs with periods of 1–5 hours. The possibility of a relationship between these MIDs and any increase in helio-geomagnetic disturbances and meteorological sources in the troposphere was studied. The passage of MIDs was found not always to be associated with growing helio-geomagnetic disturbances or with the passage of local meteorological fronts. Neither are they associated with passage through the solar terminator, because they clearly show up during long time periods (more than a day). Those MIDs that could not be attributed a possible source (out of the IGW sources traditionally discussed) may well be associated with ionospheric responses to tropical cyclones which were active in the northwest basin of the Pacific Ocean during these time periods. Since tropical cyclones have a strong pulse effect on the atmosphere, they do generate IGWs. Under favorable conditions these disturbances, in the form of wave packages of various frequencies, can propagate over oblique paths considerable horizontal distances from their place of generation.

Analysis of the increased energy of short-period MOF oscillations for the months of September between 2005 and 2007 has shown almost twofold differences in the amplitudes of the spectrum power of MOF variations for various years. As stated above, similar differences in the energy of short-period oscillations studied for various years may well be associated with possible differences in the conditions under which wave disturbances propagate in the atmosphere, or with the formation, development, and motion of particular tropical cyclones at this time and, consequently, with the various effects TCs have on the overlying atmosphere.

6.4 TROPICAL CYCLONE ACTIVITY IN MESOSPHERIC AIRGLOW

Important experimental information on the effect wave disturbances have on Earth's upper-atmosphere parameters (at 80–100 km), induced by various sources, can be obtained using optical investigation techniques based on recording atmospheric airglow from the Earth surface (Shefov *et al.*, 2006).

Upper-atmosphere airglow is a characteristic of the atmosphere that reflects numerous variations and disturbances in atmospheric parameters from various helio-geophysical sources within a wide range of timescales. The green 557.7 nm emission of atomic oxygen [O I] is one of the brightest (~300 Rayleighs) discrete lines in the visible region of the spectrum in the airglow of the midlatitude upper atmosphere and one of the most changeable emissions by virtue of its high sensitivity to diverse disturbances at the emission altitudes (85–115 km) of this line. Currently, most characteristics of the basic emissions of Earth's upper atmosphere have been well studied (Shefov *et al.*, 2006). These variations are basically regular: night-time and seasonal dependence; variations caused by solar terminator motion; interannual variations; changes in the solar cycle; lunar diurnal variations and changes during a synodic month (phases of the Moon); and latitudinal and longitudinal variations. These data gave insight into the most probable processes behind the excitation of many emissions in the upper atmosphere and led to the construction of empirical models of variations in atmospheric emissions. Irregular variations of upper-atmosphere airglow are less well studied. "Irregular variations" means those variations whose actual moments of appearance coincide with incidental or random helio-geophysical phenomena. However, immediately after they appear these variations take on a distinctly prominent character (e.g., changes in airglow as a result of solar and geomagnetic disturbances; intrusions of meteoric fluxes; stratospheric warming; and variations associated with IGW propagation from various sources).

Optical techniques facilitate the tasks of identifying the various sources of disturbances including IGWs in the upper atmosphere of Earth. Such investigations are currently being carried out by many research teams (e.g., Taylor, 1997; Brown *et al.*, 2004; Mukherjee, 2003) and are of crucial importance for atmospheric physics at the range of altitudes considered. The red 630 nm emission of atomic oxygen (at altitudes of 180–250 km) is most sensitive to geomagnetic disturbances by virtue of its dependence on ionized components of the upper atmosphere. The green 557.7 nm emission is less subject to the geomagnetic activity effect, since the physical mechanism by which this emission appears is basically determined by neutral atmospheric parameters at the emission altitudes of 85 km to 115 km and is associated with the recombination reactions of atomic oxygen.

By recording IGW manifestations in the Earth atmosphere the problem of identifying the sources of observed wave disturbances is solved to a different degree. Tropospheric meteorological effects (such as cyclonic activity in the Earth's troposphere) in upper-atmosphere emission count among the least studied of effects.

The purpose of this section is to show that recording TC activity based on disturbances in upper-atmosphere airglow in a zone far removed from tropical cyclones under quiet helio-geomagnetic conditions is possible using a special technique to process the signals received (Beletsky *et al.*, 2010a, b).

The following observational data were used for the analysis:

- experimental data of ground-based measurements of airglow from the Earth atmosphere in the 557.7 nm emission line at a time resolution of 25 s (emission altitudes are between 85 km and 115 km—the E region of the ionosphere) in the southern part of Eastern Siberia in 2000–2007;
- TC data from the electron base of satellite data on the Global-TC global tropical cyclogenesis database (SRI RAS) (Pokrovskaya and Sharkov, 2006).

Measurements of upper-atmospheric airglow were carried out in the Geophysical Observatory of the Solar-Terrestrial Physics Institute of the Siberian Branch of the Russian Academy of Sciences (52°N, 103°E, Tory, Buryatiya, 150 km to the southwest of Irkutsk). Measurements were taken by a four-channel zenith Phoenix photometer with interference-shaking optical filters in the 557.7 nm and 630 nm emission lines of atomic oxygen. Emissions in the near-infrared (720–830 nm) and ultraviolet (360–410 nm) regions of the spectrum were also recorded. The angular fields of view of the photometer channels were 4° to 5°. Absolute calibration of the measuring tracks of the instrument was by means of reference stars and was monitored by means of reference light sources during the evening and morning hours of observation. Optical observations of atmospheric airglow were carried out every month on about 7 to 14 nights each month during new moon periods under clear or slightly cloudy weather conditions.

The first step in data processing was obtaining background monthly averaged variations of the 557.7 nm emission spectra from 2000 to 2007. By comparing them with the current spectra of variations in the emission of atomic oxygen's green line, the response of atmospheric airglow to diverse irregular disturbances at upper-mesosphere/lower-thermosphere altitudes can be studied. In the spectrum of variations in the intensity of 557.7 nm emission there exists a wide interval of periods—from a few minutes to hours—which correspond to the periods of atmospheric inner gravitational waves.

To study possible manifestation of cyclonic activity in upper-atmospheric emission we analyzed the measurement data of Earth's atmospheric airglow in the 557.7 nm line for December of 2003 and September of 2007. Figure 6.7 presents the trajectories of tropical cyclone motion in the northwest and central basins of the Pacific Ocean during optical measurement periods.

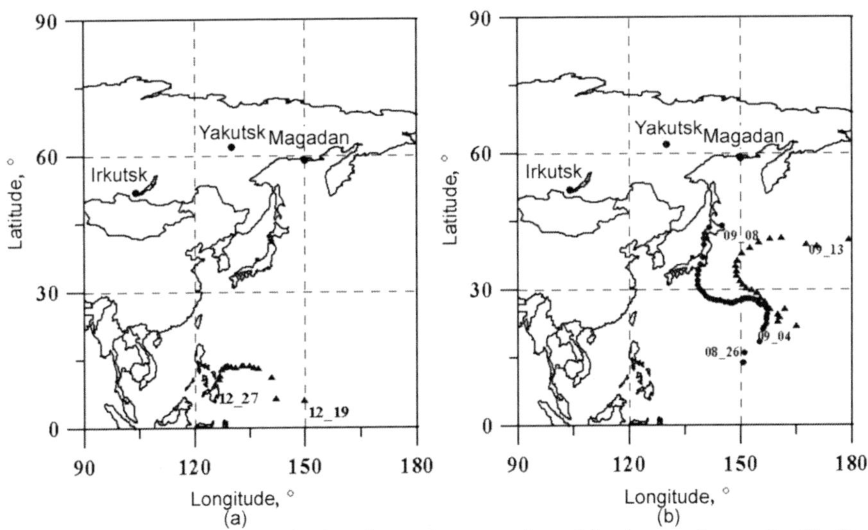

Figure 6.7. Tracks of three tropical cyclones (mature phases) in the northwest Pacific Ocean: (a) NWP 0324 (August 29–September 4, 2003) ; (b) NWP 0710 (August 26–September 8, 2007); NWP 0711 (September 4–13, 2007) (Pokrovskaya and Sharkov, 2006).

Figure 6.8 presents variations in 557.7 nm emission intensity for December 16 and 17, 2003, when no TCs were active in the Pacific Ocean, and for December 20 and 21, 2003, which coincided with the period of activity of TC NWP 0324 in the central basin of the Pacific Ocean (Figure 6.7a).

Despite the fact that powerful helio-geomagnetic events took place in 2003, during optical measurements the situation was by and large quiet: the Dst index was no lower than -25 nT. Over December 16–17 the Kp index was lower than 3, over December 20–21 the Kp index reached a little over 4 but, as noted above, the intensity of 557.7 nm emission was not sensitive to these geomagnetic disturbances. Solar radio emission flux at the wavelength of 10.7 cm (F10.7), characteristic of the ionizing capability of the Sun, varied from 103 to 129 (in units of 10^{-22} W/Hz m^2).

Figure 6.9 presents spectra of the intensity of 557.7 nm emission variations for the same days. Data on the intensity of 557.7 nm emission variations presented in Figures 6.8 and 6.9 reveals an increase in the intensity of 557.7 nm emission variations over 30 to 60-minute periods (and, probably, over 10 to 12-minute periods) on December 20 and 21, 2003 compared with December 16 and 17, when essentially no helio-geomagnetic disturbances occurred, but TC NWP 0324 began to be active. On December 16–17, 2003 there were no tropical cyclones.

Figures 6.10 and 6.11 present similar data on the intensity of 557.7 nm emission variations for September 4, 5, 8, and 9, 2007. The geomagnetic situation was quiet (the Kp and Dst indexes were about 3 and -20 nT, respectively). Solar radio emission flux F10.7 was $\sim$70 (in units of 10^{-22} W/Hz m^2). There were two active TCs in the northwest basin of the Pacific Ocean at this time: TC NWP 0710 (from August 26 to September 8, 2007) and TC NWP 0711 (from September 4–13, 2007)

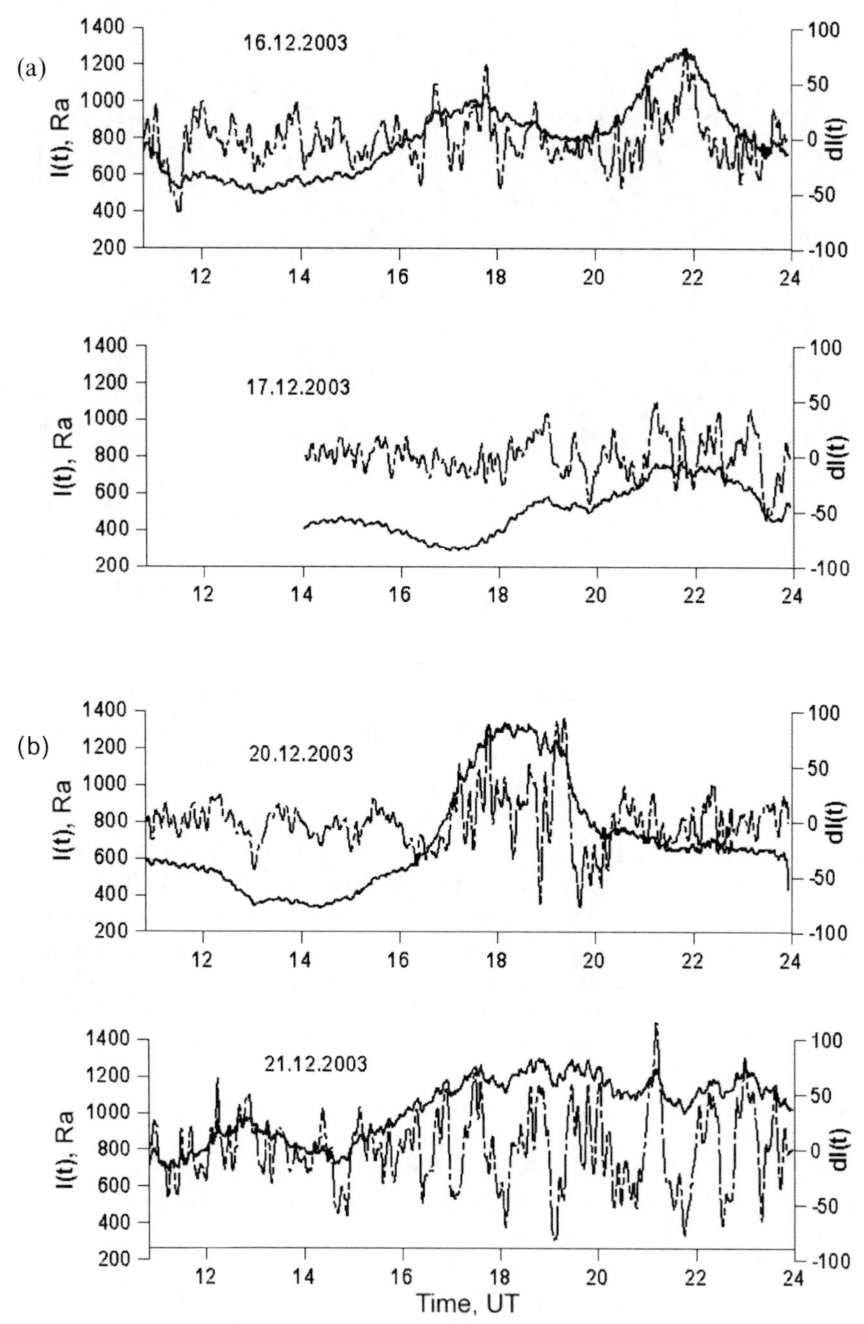

Figure 6.8. Time variations in the intensity of 557.7 nm emission airglow and time derivative intensity from 12:00 to 24:00 (local time): (a) for December 16–17, 2003 when no TCs were present in the Pacific Ocean and (b) for December 20–21, 2003 when TC NWP 0324 was active (Figure 6.7a).

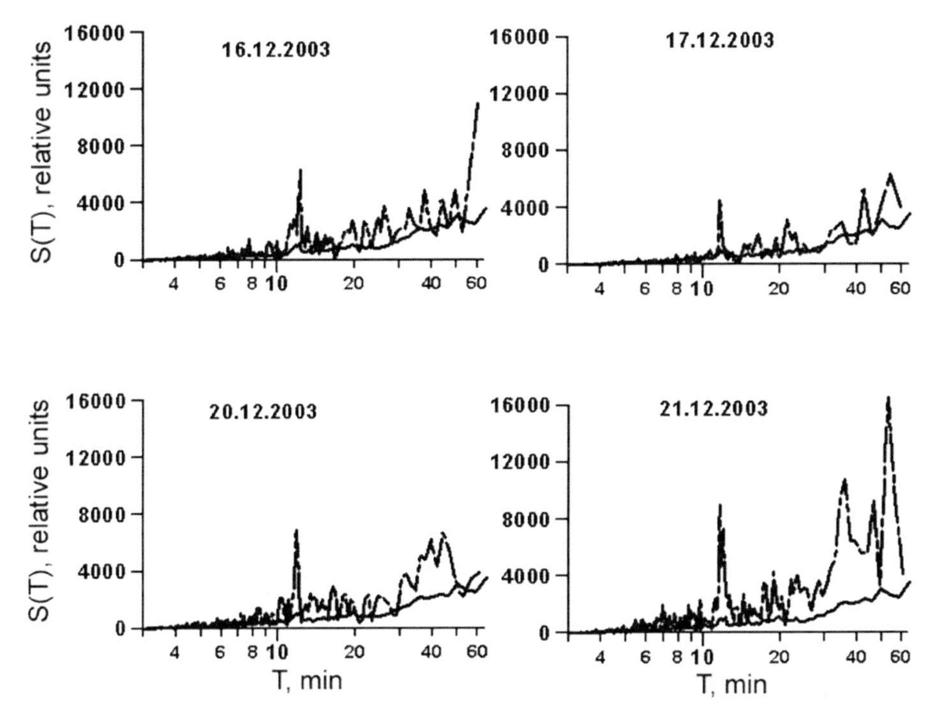

Figure 6.9. Spectra of the intensity of 557.7 nm emission airglow (dotted lines) and of background noise patterns (solid line) during December 16, 17, 20, and 21, 2003.

(Figure 6.7b). It was not possible to distinguish any similar increase in the intensity of 557.7 nm emission variations on these days. Nevertheless, there was a small increase in the intensity of oscillations over 30 to 50-minute periods on September 4 and 5.

If we relate the increase in intensity of 557.7 nm emission variations in December 2003 to tropical cyclone activity, then the absence of any such manifestations of 557.7 nm emission intensity variations in September of 2007 may well have several causes: first, differences in the conditions under which wave disturbances propagate at various seasons of the year; second, the formation, development, and motion of particular tropical cyclones at this time and, consequently, the various effects TCs have on the overlying atmosphere. In particular, it is well known that atmospheric zonal circulation has an essential effect on the upstream propagation of atmospheric inner waves, whose greatest activity is observed in winter. For IGWs obtained from the data of optical observations of the emission of hydroxyl molecules (which emit at altitudes close to those of the 557.7 nm emission), high values of variation intensity in winter are also noted (Shefov *et al.*, 2006).

Figure 6.12 presents seasonal variation in the number of TCs active in the northwest basin of the Pacific Ocean for 2000–2008, according to the data from the Global-TC database.

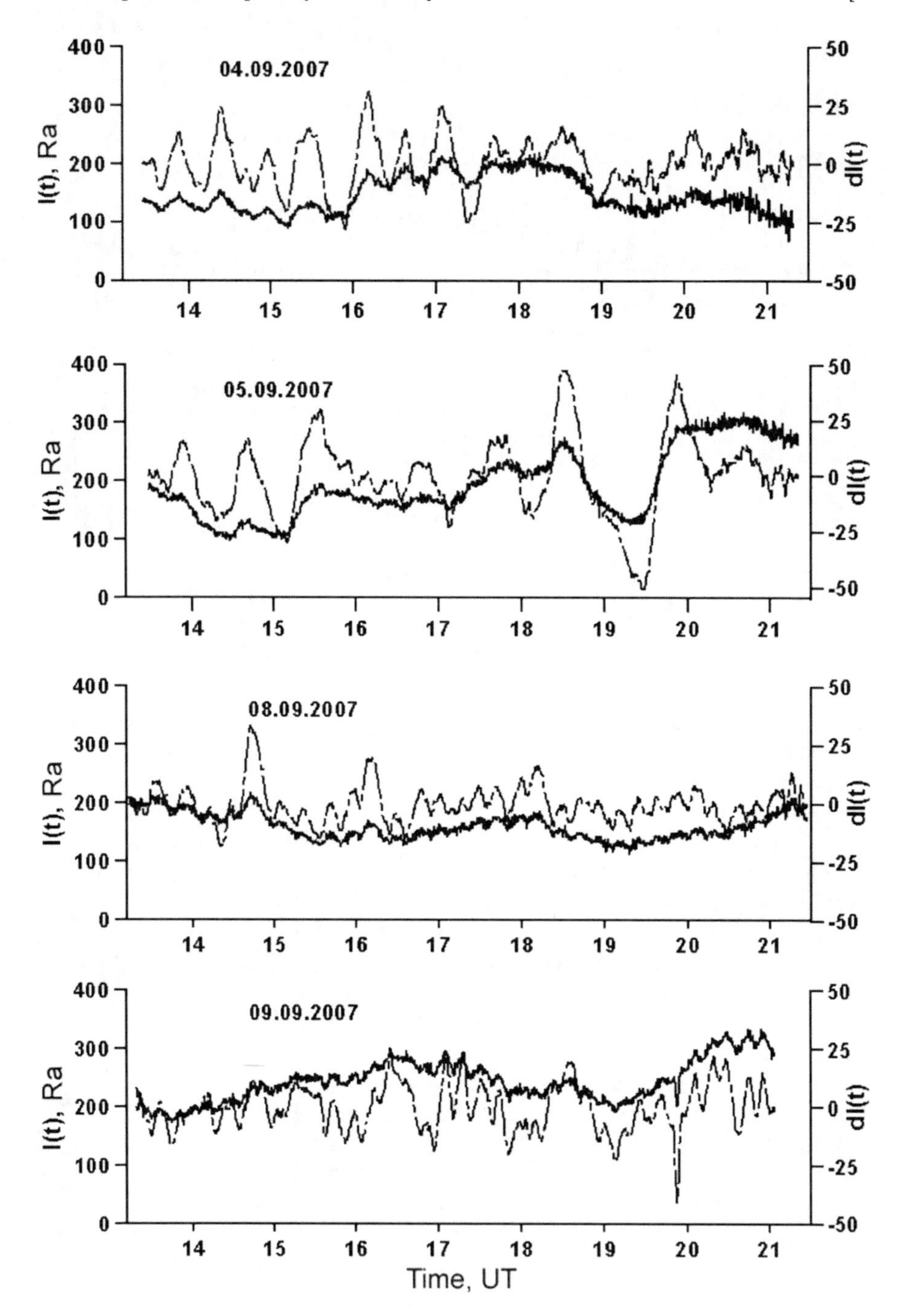

Figure 6.10. Time variations in the intensity of 557.7 nm emission airglow and time derivative intensity from 14:00 to 21:00 (local time) for September 4, 5, 8 and 9, 2007 when TC NWP 0710 and TC NWP 0711 were active (Figure 6.7b).

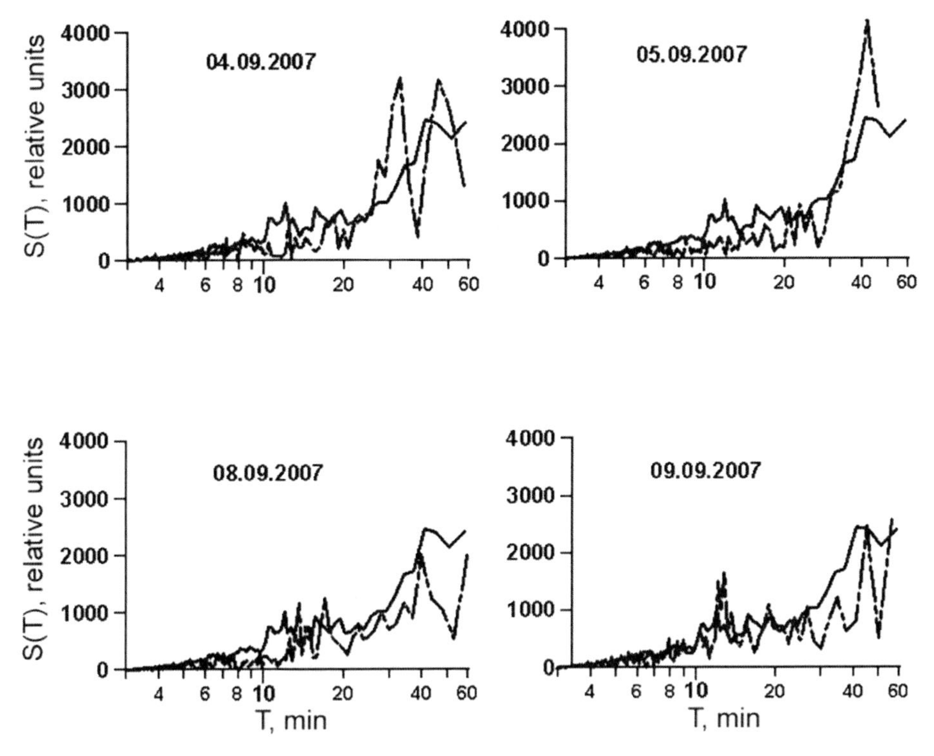

Figure 6.11. Spectra of the intensity of 557.7 nm emission airglow (dotted lines) and of background noise patterns (solid line) during September 4, 5, 8, and 9, 2007.

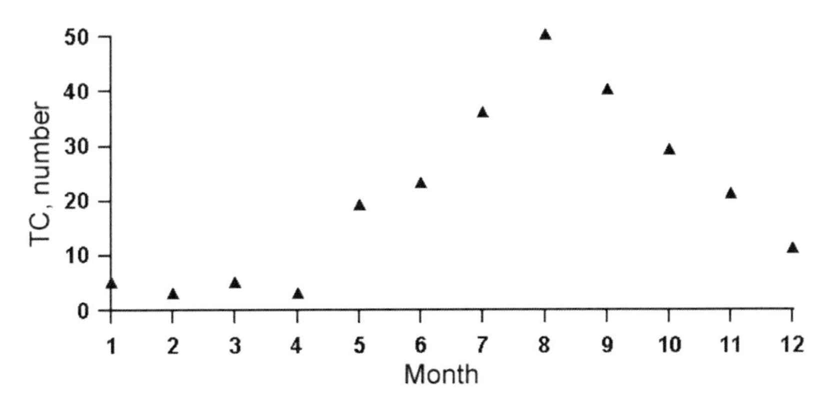

Figure 6.12. Seasonal variation in the number of TCs that were active in the northwest Pacific Ocean for 2000–2008 (Pokrovskaya and Sharkov, 2006).

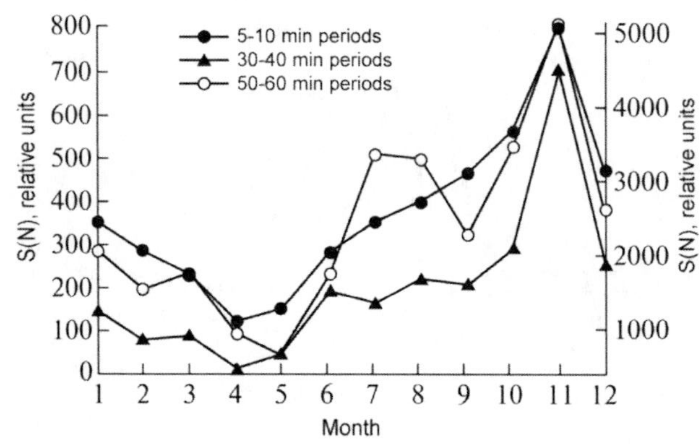

Figure 6.13. Seasonal variation in the intensity of 557.7 nm emission airglow over 5 to 15-minute periods (left axis) and 30 to 40-minute and 50 to 60-minute periods (right axis).

Comparison of the seasonal dependences of the intensity of 557.7 nm emission variations (Figure 6.13) with the repeatability of tropical cyclones (Figure 6.12) leads to the admission that the increase in wave activity of the 557.7 nm emission line in summer, especially for 50 to 60-minute periods (Figure 6.13), could also be associated with the heightened cyclonic activity in this period. The strong maximum in seasonal dependence on the intensity of 557.7 nm emission variations observed in October–November was probably associated with seasonal rearrangements of atmospheric circulation at emission altitudes at this time of year and the concomitant changing conditions for upstream propagation of wave disturbances.

Preliminary analysis of the intensity of 557.7 nm emission variations during TC activity in December 2003 and September 2007 under quiet helio-geomagnetic conditions was also carried out. There was essentially an increase in the intensity (compared with preceding days) of 557.7 nm emission variations in the range of 30 to 60-minute periods on December 20 and 21, 2003, when TC activity began in the central basin of the Pacific Ocean. It was not possible to distinguish a similar noticeable increase in the intensity of 557.7 nm emission variations associated with tropical cyclogenesis manifestations for September 2007.

To reiterate, if we relate the increase in intensity of 557.7 nm emission variations in December 2003 to tropical cyclone activity, then the absence of any such manifestations of 557.7 nm emission intensity variations in September 2007 may well be caused by differences in the conditions under which wave disturbances propagate at various seasons of the year or by the formation, development, and interaction of these cyclones with the overlying atmosphere.

7

Physical models and simulations of global tropical cyclogenesis

Currently, difficulties relating to the construction of physical and synoptical models of tropical cyclogenesis are widely covered in the literature.

The basic lines of investigations are the following:

— synoptical statistical modeling and forecasting based on the relationships that have come to light between tropical cyclone activity and mesoscale and large-scale climatic factors;
— modeling that combines specialized simulation of tropical cyclone–like vortices and highly developed global climate change (GCC) models;
— physical modeling and developing of TC models on the basis of thermodynamic and hydrodynamic principals and on the non-linear theory of chaos (the transition between chaos and order).

Considerable advances have recently been made in the synoptical statistical forecasting model for North Atlantic tropical cyclogenesis (Gray, 1984a, b, 1993; Gray *et al.*, 1997; Elsner and Kara, 1999).

The results of GCC simulations are the subject of a great deal of current interest (Broccoli and Manabe, 1990; Bengtsson *et al.*, 1995, 1996; Knutson *et al.*, 1998; Knutson and Tuleya, 1999; Tsutsui and Kasahara, 1996; Matsuura *et al.*, 1999).

At present the findings of these three lines of investigation in tropical cyclogenesis study are furiously debated in Western literature. Little attention is given to the physical modeling and self-organization approaches that were proposed and developed by Russian scientific groups. The main publications concerning hydrodynamic instabilities in the scientific literature concern turbulence problems. This is the reason more emphasis will be placed on this line of investigation in this chapter.

E. A. Sharkov, *Global Tropical Cyclogenesis* (Second Edition).

7.1 STATISTICAL SYNOPTICAL MODELING AND FORECASTING

Tropical cyclogenesis has long been an important topic in meteorological research and has been intensively studied by many investigators over recent decades (e.g., Riehl, 1954; Gray, 1979; Charney and Eliassen, 1964; Ooyama, 1964; Anthes, 1982; Emanuel, 1986; Zehr, 1992; Elsner and Kara, 1999; Sharkov, 2000).

7.1.1 Seasonal genesis parameters

The physical parameters favorable for cyclogenesis are summarized by Gray (1979). He finds that the climatological frequency of tropical cyclone genesis is related to six environmental factors: (i) large values of low-level relative vorticity, (ii) the Coriolis force (at least a few degrees poleward of the equator), (iii) weak vertical shear of horizontal winds, (iv) high SSTs exceeding 26°C and a deep thermocline, (v) conditional instability through a deep atmospheric layer, and (vi) large values of relative humidity in the lower and middle troposphere.

Although the above six parameters are not sufficient conditions for cyclogenesis, Gray argues that tropical cyclone formation would be most frequent in regions and seasons when the product of the six genesis parameters is at a maximum. Gray defines the product of (i), (ii), and (iii) as the dynamic potential for cyclone development, and the product of (iv), (v), and (vi) as the thermodynamic potential. He derives the seasonal genesis parameter from these six parameters.

However, these conditions are thought (Liu *et al.*, 1995; Henderson-Sellers *et al.*, 1998) to be necessary but not sufficient conditions for tropical cyclogenesis; a particular tropical disturbance may never develop into a hurricane although all the above conditions are met.

On the other hand, using recent observation results Henderson-Sellers *et al.* (1998) emphasize that "the popular belief that the region of cyclogenesis will expand with the 26°C SST isotherm is a fallacy."

A widely accepted list of conditions permitting TC formation and intensification (Gray, 1979) calls for satisfaction of six requirements, one of which places, as stated earlier, a lower limit (about 26°C) on SST. The other five areas follow (Lighthill *et al.*, 1994):

(I) The distance from the equator needs to be at least 5° of latitude to bring the Coriolis force of Earth's rotation into play, which generates cyclonic spiraling (counterclockwise in the northern hemisphere and clockwise in the southern hemisphere).

(II) The gradient of temperature drop with height must be large enough so that air that has become saturated with water vapor near the eyewall is able to continue to rise as it follows the moist air adiabatic (less of a temperature drop than dry air would undergo because of heating associated with precipitation).

(III) Low values of vertical shear (gradient of horizontal wind with height near the cyclone's center) are needed to avoid excessive departure from a vertically coherent axisymmetric vortical structure.

(IV) Relative humidity has to be high enough in the middle troposphere to avoid the drying-out effects of air that is entrained in the updraft of the eyewall mixing with moist air in the deep-convective cloud system.

(V) Finally, there is a requirement for the prior existence at a low altitude of a substantial amount of cyclonic vorticity.

These conditions, derived from extensive observational records, emphasize that much more than just a sufficiently high SST is needed for TC formation and intensification. Condition (III) on low vertical shear is particularly important, with its suggestion that intensification may develop very quickly under conditions when the ambient atmospheric environment is free of features that are likely to disturb vertically coherent axisymmetric spiraling motion.

7.1.2 North Atlantic seasonal forecasts

The North Atlantic experiences the largest interannual variability in tropical cyclone activity of any region. This strong interannual variability suggests that large-scale climate factors acting on seasonal and longer term timescales are involved and that some degree of seasonal predictability may be possible. Research in recent decades (Gray, 1984a, b, 1993; Gray *et al.*, 1997) indicates that there are signals that allow skillful forecasts to be made as early as November the previous year. Gray and collaborators provide forecasts from Colorado State University of seasonal cyclone activity from the previous December, June, and August. Because this is the only region where regular seasonal forecasts are made to the public, the techniques are described in some detail below.

This extended range prediction is made during the late fall of the previous year. The bases of the forecasts are regression relations between several different predictors and six predictands covering the number of named storms (NSs), named storm days (NSDs), hurricanes (Hs), hurricane days (HDs), intense hurricanes (IHs), and intense hurricane days (IHDs). Also predicted is a normalized measure of the square of the maximum windspeed for all hurricanes, called the hurricane destruction potential (HDP):

$$\mathrm{HDP} = \beta_0 + \beta_1(\alpha_1 U_{50} + \alpha_2 U_{30} + \alpha_3 |U_{50} - U_{30}|) + \beta_2(\alpha_4 R_s + \alpha_5 R_G) \tag{7.1}$$

where the β and α parameters are empirically derived coefficients; and U_{50} and U_{30} are winds extrapolated to September of the following year at the 50 hPa and 30 hPa level, to provide an indication of the QBO; R_s is standardized rainfall in the western Sahel (Figure 7.1) during August and September of the prior year; and R_G is standardized rainfall in the Gulf of Guinea (Figure 7.1) during August through November of the prior year.

The relationship is based on observations (Gray, 1993) that the phase of the QBO and African rainfall are strongly related to seasonal hurricane activity in the North Atlantic basin. Using a cross-validated jackknife analysis for hindcast predictions of cyclones between 1950 and 1990, Gray (1993) shows that Eq. (7.1) can explain 45% to 50% of the variance in all seven tropical cyclone predictands. This

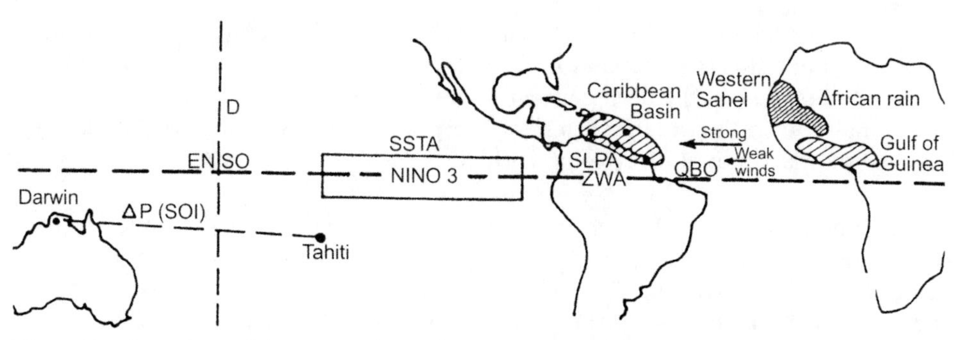

Figure 7.1. Locations of the meteorological parameters used for North Atlantic basin seasonal forecasts of TC activity (from Gray, 1993). See text for notation.

indicates that significant seasonal forecast skill is available for cyclones in the North Atlantic.

Gray (1993) develop a regression relation for prediction of the aforementioned tropical cyclone parameters from early June in the following form:

$$\begin{aligned}\text{HDP} &= \beta_0 + \beta_1(\alpha_1 U_{50} + \alpha_2 U_{30} + \alpha_3 |U_{50} - U_{30}|) \\ &\quad + \beta_2(\alpha_4 R_s + \alpha_5 R_G + \alpha_6 \Delta_x P + \alpha_7 \Delta_x T) \\ &\quad + \beta_3(\alpha_8 \text{SPLA} + \alpha_9 \text{ZWA} + a_{10} \text{SSTA} + \alpha_{11} \Delta_t \text{SSTA} + \alpha_{12} \text{SOI} + \alpha_{13} \Delta_t \text{SOI}) \quad (7.2)\end{aligned}$$

where, in addition to the parameters described for Eq. (7.1) (see Figure 7.1), $\Delta_x P$ and $\Delta_x T$ are the West African zonal pressure and temperature gradient anomalies from February to May; SLPA and ZWA are the sea level pressure anomaly and zonal wind anomaly in the lower Caribbean during April–May; SOI and Δ_tSOI are the SOI in April–May and its change from January-February to April–May; SSTA and Δ_tSSTA are the SST anomaly in NINO3 during April–May and its change from January–February to April–May.

Using a cross-validated jackknife analysis for hindcast predictions of cyclones between 1950 and 1990, Gray (1993) shows that Eq. (7.2) can explain 50% to 70% of the variance in all seven tropical cyclone predictands. All but one (NSs) of the predictands was better than 60% and 71% of hurricane destruction potential was explained.

The prediction equation for August 1 is similar to Eq. (7.2) except that June–July rainfall and other meteorological parameters are used. Only a marginal improvement in skill from the June 1 forecast is obtained from this approach.

Using an algorithm based on the lagged relationships between cyclone numbers and the above indexes of large-scale circulation, Gray issues public forecasts for May of each year from 1984 through 1993 for measures of Atlantic cyclone activity in the following season. Mean values, mean errors, and correlation coefficients between Gray's forecasts and measures of actual cyclone activity are given in Table 7.1. The highest correlation coefficient for this sample of 10 seasons is forecast of the number of tropical cyclones. A time series of the actual versus forecast value of the number of tropical cyclones is shown in Figure 7.2.

Table 7.1. Forecast statistics for four indexes of TC activity for the North Atlantic basin as forecast by Gray (1993) for the month of May from 1984 to 1993. An intense hurricane is calculated according to Saffir–Simpson categories 3, 4, or 5.

Indexes	*Observed mean*	*Mean absolute error*	*Correlation coefficient*
No. of cyclones	9.5	1.8	0.59
No. of intense hurricanes	5.1	1.4	0.42
No. of cyclone days	43.31	1.8	0.26
No. of intense hurricane days	17.1	9.4	0.33

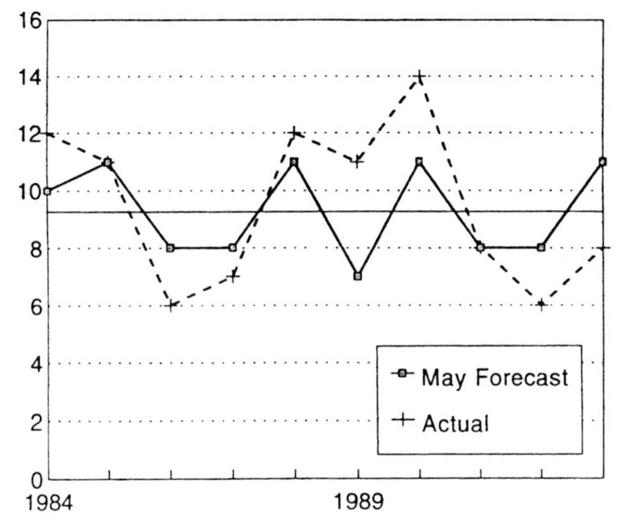

Figure 7.2. Forecast (solid) versus observed (dashed) number of TCs in the North Atlantic basin from 1984 to 1993. The forecasts shown were made public by Gray (1993). In May preceding the cyclone season each year Gray makes these forecasts public; they can be compared with the long-term mean frequency (thin) (from McBride, 1995).

These North Atlantic seasonal forecasts provide useful information on the overall occurrence of tropical cyclones and of hurricane damage potential. The North Atlantic basin best illustrates the potential for seasonal predictions and should be the standard by which seasonal forecast skills of other global basins are measured. A decade ago no one would have imagined that pre-seasonal climate signals would have been so well related to hurricane activity. It is likely that all North Atlantic predictive signals have not yet been fully exploited.

The encouraging degree of skill with seasonal tropical cyclone forecasts for the North Atlantic does not, unfortunately, appear to carry over to the other cyclone basins. The North Atlantic seems to be special because:

1. It has marginal cyclone formation conditions, high interannual variability, and a short season.
2. Seasonal cyclone activity in the North Atlantic is more strongly affected by slowly varying ENSO and QBO conditions than is the activity in other basins.
3. Most cyclones form from easterly wave–type disturbances in the trade winds in such a way that West African conditions play a major role and westerly upper-wind anomalies can act as a strong inhibitor.

A major difficulty with developing seasonal tropical cyclone prediction equations for other basins lies in the lack of good data. Except for the western North Pacific, direct measurements of tropical cyclone intensity (e.g., by aircraft) are not available and the resulting database has to be considered unreliable. The use of satellite estimates since the 1970s has improved quality enormously, but questions remain on the reliability of analyses of intense tropical cyclones.

7.1.3 Development of the statistical synoptical approach

A number of research groups are busy taking active parts in the development of the synoptical statistical approach along a diversity of avenues.

Lu and Chen (1998) use a special case of multiple regression equations called a "multiple linear interdependent model" (MLIM) and apply it to typhoon data from Fujian, China, over the period 1954–1984. MLIMs were established to diagnose losses as a result of typhoons and to forecast the typhoon track, intensity, and windspeed simultaneously after periods of 24 h, 48 h, and 72 h, respectively. MLIMs can include any number of simultaneous multiple regression equations. MLIMs denote systems of linear equations involving several interdependent variables. The results of diagnosing losses as a result of typhoons and forecasting the four main parameters of typhoons are much more accurate than the results of multiple regression (MR) using the same exogenous variables.

The effects of 10 climatological and persistence variables (latitude, maximum windspeed, 12 h change in maximum windspeed, longitude, distance from land, Julian date, sea surface temperature, speed of movement, zonal component of motion, and meridional component of motion) on changes in the intensity of tropical cyclones over the eastern North Pacific Ocean were examined by Hobgood (1998) for 1982–1987 and 1988–1993. Backward multiple regressions were performed to relate these 10 variables to changes in maximum intensity (as determined by windspeed) over periods ranging from 12 h to 72 h. Latitude, maximum windspeed, and 12 h change in maximum windspeed were the most significant variables. Each of the 10 variables was statistically significant at the 95% level during one or more of the time periods. Speed of movement, the component of motion, and the meridional component of motion were the least significant factors. Statistical relationships were tested using independent data from 1994. Mean absolute forecast errors ranged from 3.0 m s^{-1} at 12 h to 13.2 m s^{-1} at 72 h using one of two sets of regression equations developed in Hobgood's study.

A multiple regression scheme with tropical cyclone intensity change as the dependent variable was developed by Fitzpatrick (1997). The new scheme is titled the typhoon intensity prediction scheme (TIPS) and is similar to one used operationally at the National Hurricane Center. However, TIPS contains two major differences: it was developed for the western North Pacific Ocean and utilizes digitized satellite data; the first time such satellite information has been combined with other predictors in a tropical cyclone multiple regression scheme. It is shown that satellite data contain vital information that distinguishes between fast and slow-developing tropical cyclones. The importance of other predictors (such as wind shear, persistence, climatology, and an empirical formula dependent on sea surface temperature) to intensity change are also clarified in the statistical analysis. A normalization technique reveals threshold values useful to forecasters. It is shown that TIPS may be competitive with the Joint Typhoon Warning Center in forecasting tropical cyclone intensity change.

Chan *et al.* (1998) show how operational statistical forecasts of seasonal tropical cyclone (TC) activity have developed over the westem North Pacific (WNP) and the South China Sea (SCS) based on 30 years of data (1965–1994). Predictors include the monthly values of indexes representing (a) El Niño–Southern Oscillation and (b) the environmental conditions over East Asia and the WNP for months beginning with April of the previous year to March of the current year. Trends and short-term oscillations in TC activity are also incorporated.

The prediction equations are derived from the predictors of individual parameters using the projection pursuit regression technique, which is a statistical method that reduces high-dimensional data to a lower dimensional subspace before regression is performed. This technique has been found to explain certain non-linear variations of the predictands. The predictions from individual parameters are then tested using the jackknife technique. Those predictions that have correlations with observations significant at the 95% level or higher are retained. The values of the correlation coefficients are then used as weights in combining the predictions to form a single forecast of each predictand. Forecasts obtained in this way are found to be superior to those from individual parameters.

The combined forecast equations are then used to predict TC activity over the WNP and the SCS for 1997. The prediction is for slightly above-normal activity for the entire WNP but slightly below-normal activity for the SCS. The former is found to be correct and the latter has the right trend although the activity over the SCS was considerably below normal, probably as a result of El Niño of 1997.

An empirical relationship between climatological sea surface temperatures (SSTs) and the maximum intensities of tropical cyclones over the eastern North Pacific Ocean is developed by Whitney and Hobgood (1997) from a 31-year sample (1963–1993). This relationship is compared with an empirical relationship for tropical cyclones over the Atlantic Ocean and with theoretical results. Over the period of study, storms over the eastern North Pacific Ocean reached a lower percentage of their empirical maximum potential intensity (MPI) than tropical cyclones over the Atlantic Ocean. At the time of their maximum intensity, only 11% of eastern North Pacific storms reach 80% of their MPI, while 19% of the

Atlantic tropical cyclones reach that proportion of their MPIs. Poleward recurvature of Atlantic storms over cooler waters appears to be a major factor in the difference between the two regions. The storms were stratified by latitude, longitude, the phase of the quasi-biennial oscillation (QBO), and the status of El Niño. Tropical cyclones that develop west of 110°W tend to reach a higher percentage of their MPI than storms developing farther east. Tropical cyclones also tended to reach a higher percentage of their MPI and to attain higher maximum intensities when the QBO was in its westerly phase.

De Maria and Kaplan (1999) describe updates to the statistical hurricane intensity prediction scheme (SHIPS) for the Atlantic basin. SHIPS combines climatological, persistence, and synoptic predictors to forecast intensity changes using a multiple regression technique. The original version of the model was developed for the Atlantic basin and was run in near real time at the Hurricane Research Division beginning in 1993. In 1996, the model was incorporated into the National Hurricane Center's operational forecast cycle, and a version was developed for the eastern North Pacific basin. Analysis of the forecast errors for the period 1993–1996 shows that SHIPS had little skill relative to forecasts based upon climatology and persistence. However, SHIPS had significant skill in both the Atlantic and East Pacific basins during the 1997 hurricane season.

The regression coefficients for SHIPS were rederived after each hurricane season since 1993 so that the previous season's forecast cases were included in the sample. Modifications to the model itself were also made after each season. Prior to the 1997 season, the synoptic predictors were determined only from an analysis at the beginning of the forecast period. Thus, SHIPS could be considered a "statistical synoptic" model. For the 1997 season, methods were developed to remove tropical cyclone circulation from global model analyses and to include synoptic predictors from forecast fields, so the current version of SHIPS is a "statistical dynamical" model. It was only after modifications for 1997 were carried out that the model showed significant intensity forecast skill.

A simple statistical synoptic technique for tropical cyclone (TC) track forecasting to 72 h in the western North Pacific is derived by Chen *et al.* (1999). This technique applies to the standard (S) pattern/dominant ridge region (S/DR) and poleward/poleward-oriented (P/PO) combinations, which are the two most common and represent about 73% of all situations. Only eight predictors that involve present and past 12 h and 24 h positions, intensity, and date are used. Track predictions are simple to calculate and understand, are available in near real time every 6 h, and apply at all intensities, as compared with the complex global or regional dynamical model predictions that are only available every 12 h at about 3–4 h after the synoptic time, are not calculated for weak TCs, and tend to have accurate predictions only for the tropical storm stage and above. The statistical synoptic technique for S/DR cases has an improvement (skill) relative to the operational climatology and persistence technique of 12% after 12 h and 24% after 72 h if the TC remains in the S/DR pattern/region for the entire 72 h. The statistical synoptic technique for P/PO cases has an improvement relative to WPCLPR of 11% after 12 h and about 13% for 72 h forecasts if the TC remains in the P/PO for the entire 72 h.

Assuming perfect knowledge of the S/DR to P/PO and P/PO to S/DR transitions, a simple blending of a composite post-transition track with the statistical synoptic technique is tested. For the 72 h forecasts initiated 12 h before the S/DR to P/PO transition, the statistical synoptic track error is about 290 kn. (537 km) compared with 410 kn. (759 km) for WPCLPR. For corresponding P/PO to S/DR transition, the statistical synoptic technique 72 h error is 215 kn. (398 km) compared with about 485 kn. (898 km) for WPCLPR.

A statistical method for long-term cyclone risk assessment is proposed. A technique for estimating the accidence probabilities of the maximum windspeed $V_{\max}(T)$ at a given point over a time period T was developed by Golitsyn *et al.* (1999). The technique is based on the well-known relation between windspeed $V(R)$ and the distance R from the typhoon center to a given point: $V(R) = V_{\max}\sqrt{R_m/R}$ where $R \geq R_{\max}$ and $R_{\max}$ is the radius of the typhoon's central part where a maximum windspeed at a given time is observed. The risk assessment method is applied to a catalog of typhoons occurring in the western Pacific during the period 1950–1988. The catalog lists 1,013 events and provides maximum windspeeds every 12 hours. Windspeeds that exceeded the maximum over the time intervals $T = 1$, 2, 5, 10, and 20 years with probabilities of 0.5, 0.9, 0.95, and 0.99 are calculated for Vladivostok and Hong Kong. The method can be useful for designers, builders, insurance companies, and local authorities in regions subject to typhoons.

7.2 GLOBAL CLIMATE CHANGE MODELS AND TROPICAL CYCLONE GENESIS

Understanding how tropical cyclone genesis may change in the greenhouse-warmed climate is certainly a significant challenge to current research.

The problem of predicting how tropical cyclone frequency might respond to greenhouse-induced climate change can be broken (Henderson-Sellers *et al.*, 1998) into two parts: predicting how the environmental capacity to sustain tropical cyclones may change and predicting how the frequency and strength of initiating disturbances may change. The thermodynamic analysis by Holland (1997) indicates that there could be an enhanced environment for tropical cyclone intensification. Global circulation model (GCM) predictions also indicate that the strength of very large-scale tropical circulations such as monsoons and trade winds are expected to be increased, which could provide both an enhanced environment and more initiating disturbances.

GCMs have been used by a number of groups to try to infer changes in tropical cyclone activity by analyzing the resolvable-scale vortices that develop. Three well-known highly developed GCM models, operated by the U.K. Meteorological Organization (UKMO), the Max-Planck-Institut in Germany, and the Geophysical Fluid Dynamics Laboratory (GFDL) in the U.S.A were used in the process of development of early studies (Broccoli and Manabe, 1990; Bengtsson *et al.*, 1995, 1996; Haarsma *et al.*, 1992).

As expected, these studies produce conflicting results: Haarsma *et al.* (1992) find an increase in frequency of tropical cyclones, Bengtsson *et al.* (1996) find large decreases, and Broccoli and Manabe (1990) find that increases or decreases could be obtained by reasonable variations in the model physics. Critical commentaries on these simulations are provided by Evans (1992), Lighthill *et al.* (1994), Landsea (1997), and Henderson-Sellers *et al.* (1998).

It is well known that the development of well-organized tropical vortices is as typical a feature of a GCM as is the development of extra-tropical storms and occurs regularly in the ECHAM model as well as in most other general circulation models.

The most crucial part of the studies is the specification of criteria for automatic determination of hurricane-type vortices in the experiments.

Broccoli and Manabe (1990) use the Geophysical Fluid Dynamics Laboratory GCM to study the response of tropical cyclones to increases in atmospheric CO_2. Two versions of the model, R15 (4.5° lat. × 7.5° long.) and R30 (2.25° lat. × 3.75° long.), were utilized. Fixed cloud and variable cloud amounts were the cloud treatments adopted. In the experiments with fixed cloud, the number and duration of tropical storms increased in a doubled CO_2 climate for the R15 integration. However, a significant reduction in the number and duration of TCs was indicated in the experiments with variable cloud. The response of the simulated number of storms to a doubling of CO_2 is apparently insensitive to the model resolution but crucially dependent on the parameterization of clouds.

Broccoli and Manabe (1990) do not use any criteria on the thermal structure and restrict the search for oceanic gridpoints equatorwards of 30° latitudes. They furthermore limited their search to the 6-month hurricane season, defined as May–October and November–April in the northern and southern hemispheres, respectively. The minimum windspeed was set to $17\,\mathrm{m\,s^{-1}}$. No criterion on vorticity was used.

Haarsma *et al.* (1992) undertake similar experiments for present day and doubled CO_2, concentrations. Haarsma *et al.* (1992) use the 11-layer GCM at the U.K. Met Office coupled with a 50 m mixed layer ocean. Their model resolution was R30 (2.25° lat. × 3.75° long.) with variable cloud amount. Evans (1992) argues that it is important to examine the physical mechanisms involved in generation of the model storm and test the degree to which model vortices have physical similarities with real tropical cyclones. Simulated tropical disturbances for present day climate analyzed by Haarsma *et al.* (1992) have a much larger horizontal extent and weaker intensity than those observed, but some physical features of tropical cyclones, such as low-level convergence, upper-tropospheric outflow, and a warm core, were reproduced by this GCM. Under doubled CO_2 conditions, the number of simulated tropical storms increases by about 50%. They used the same criterion on vorticity as in the study by Bengtsson *et al.* (1995, 1996) as well as criteria on a warm core, but less stringently than the one used here.

Bengtsson *et al.* (1995, 1996) investigate the influence of greenhouse warming on tropical storm climatology, using a high-resolution GCM at T106 resolution (triangular truncation at wavenumber 106, equivalent to 1.1° lat.–long.).

One of the central objectives was to base identification of the vortices on dynamical and physical criteria only, thus avoiding empirical conditions on geographical distribution, sea surface temperature, or specific time of the year. Since the search was limited lo the generation of vortices, the authors limited the search to ocean areas only, as inspection of a large number of maps did not show any land developments. Furthermore, the authors only considered storms with a lifetime of at least 36 hours.

Based on the structure of typical tropical vortices taken from a study at T42 resolution, the following criteria were found suitable:

1. Relative vorticity at 850 hPa $> 3.5 \times 10^{-6}\,s^{-1}$.
2. A maximum velocity of $15\,m\,s^{-1}$ and a minimum surface pressure within a 7×7 gridpoint area around a point that fulfills condition 1.
3. The sum of temperature anomalies (deviation from the mean as defined below) at 700 hPa, 500 hPa, and 300 hPa $> 3°C$.
4. The temperature anomaly at 300 hPa > temperature anomaly at 850 hPa.
5. Mean windspeed at 850 hPa > mean windspeed at 300 hPa.
6. Minimum duration of the event ≥ 1.5 days.

The minimum in surface pressure was determined at the center of the storm. Mean values were calculated within a 7×7 gridpoint area around the point of minimum pressure. It was found that main criteria 3–5, specifying a vertical structure with a warm core and maximum wind at low levels, very efficiently eliminated extra-tropical storms that have a completely different structure. The only differences, which happen a few times a year, were very intense features at high latitudes in both hemispheres in the winter season. Moreover, the authors stress, "that in an experiment like this it is not really possible to specifically separate between hurricanes and intense tropical depressions." These investigations suggest substantial reduction in the number of storms, particularly in the southern hemisphere. They attribute this reduction to a warming in the upper troposphere, enhanced vertical wind shear, and other large-scale changes in tropical circulation such as reduced low-level relative vorticity. Compared with the results of the control experiment, there are no changes in the geographical distribution of GCM-simulated storms. The seasonal variability in storm distribution is said to be in agreement with that of the present climate. However, application of their model results may be limited by their model's sensitivity to its resolution and perhaps also by incompatibilities in the experiment. In the ECHAM3 (T106) doubled CO_2 experiment (Bengtsson *et al.*, 1996), fixed global SSTs were taken from the ECHAM3 (T21) doubled CO_2 experiment in which an enhanced tropical hydrological cycle as a result of a strengthened ITCZ was simulated with a fully coupled ocean model and the SSTs were warmed between 0.5°C and 1.5°C. Surprisingly, with such high global SSTs and noting the results from the underpinning experiment, Bengtsson *et al.* (1996) report a weakened tropical hydrological cycle in their high-resolution experiment. This weakening in tropical circulation appeared to be one of the primary reasons for the decrease in the model's tropical cyclone activity. It appears that this model's tropical climate is very sensitive to its horizontal resolution. Henderson-Sellers *et al.* (1998) suppose that the changes in SSTs in the doubled CO_2 climate, if simulated by the high-resolution

AGCM coupled with the same OGCM, may be different from the ones used in Bengtsson *et al.* (1996) and thus might give a different prediction of the changes in tropical cyclone activity in a greenhouse-warmed climate.

It is notable that, as Lighthill *et al.* (1994) mention in their detailed review, a global model developed at the Japan Meteorological Agency showed no tropical disturbances of any substantial strength. In spite of severe criticism, progress along this line of investigation is gathering force.

Tsutsui and Kasahara (1996) explore the possibility of simulating tropical cyclones (TCs) using the National Center for Atmospheric Research community climate model (CCM2). Daily output from two long-term simulation runs using the standard T42 resolution CCM2 are examined to identify simulated tropical cyclones (STCs) using a search scheme that selects qualified STCs resembling observed TCs. In their study the authors investigated the STC behaviors in two well-documented long-term simulations because these datasets are sufficiently long to get a stable climatology of STCs, and the climatology of global circulations has been well investigated However, the standard version of T42 CCM2 has a coarse horizontal resolution of 250 km to 300 km, so that the scale of circulations smaller than 500 km is not resolved. Hopefully, this survey will serve as a benchmark for similar studies using higher resolution CCMs in the future.

The authors analyze two sets of climate simulations produced by the standard CCM2. The model adopts the generalized terrain-following vertical coordinates and the spherical horizontal coordinates. The dynamical equations are solved using the spectral transform method with a T42 resolution in the horizontal and the finite difference method in the vertical with 18 levels and the topmost at 2.917 hPa. The prediction of specific humidity is performed by a shape-preserving semi-Lagrangian transport scheme.

The two simulation cases are a 20-year run driven by climatological sea surface temperatures (SSTs) and a 10-year run, corresponding to the decade from 1979 to 1988, with the same model configuration except for the use of observed SSTs.

An objective search procedure is set up to detect STCs in the CCM2 outputs and to investigate their structure and climatology. Since not all TC-like vortices that appear in the CCM2 outputs have the characteristics of observed TCs, various criteria are implemented to select qualified STCs for a climatelogical study. The search procedure consists of the following steps: (1) determining the center locations of candidate STCs based on several criteria applied to model variables, (2) tracking the center locations to judge their time continuity, and (3) final testing to select only qualified STCs for a climatological study.

The T42 CCM2 uses a Gaussian grid of 128×64 points in longitude and latitude at approximately 2.8° increments. First, the search grid scans through each field variable, as listed in Table 7.2, at all gridpoints in an equatorial belt between 40°N and 40°S, excluding those over land where the geographical elevation exceeds 400 m. Searching for STCs over low-elevation land areas is necessary to count the number of STCs that move near the coastlines of the continents. The critical elevation of 400 m is determined empirically. Table 7.2 shows various criteria for the selection of candidate STCs. Critical values, such as the drop of

Table 7.2. Criteria for detection of candidate STCs.

Field	*Level*	*Notation*	*Criteria*
Geopotential height	1,000 hPa	$\Phi_{1,000}$	$\Phi_{1,000}$ at the center point is lower than any value of the A and B points. $\Phi_{1,000}$ at the center point is lower than the average over A points by 20 m, and the average over A points is lower than the average over B points by 20 m
Vorticity	900 hPa	ζ_{900}	Average ζ_{900} over the center and A points is cyclonic (positive in the northern hemisphere, negative in the southern hemisphere)
Divergence	900 hPa	δ_{900}	Average δ_{900} over the center and A points is negative
Vertical ρ velocity	500 hPa	ω_{500}	Average ω_{500} over the center and A points is negative (upward motion)
Thickness	Φ_{200}–Φ_{1000}	h	Maximum h between the center and A points is greater than any value at the B points
Longitudinal wind velocity	200 hPa	u_{200}	Average u_{200} over the center and A points is weaker than $10\,\mathrm{m\,s^{-1}}$, or at least one value at the points is negative (easterly)

Note: Center point corresponds to the center of the search grid for detection of STCs. An A point corresponds to the inner region; and a B point corresponds to the periphery of the search grid.

20 m for the 1,000 hPa geopotential height, are empirically determined to filter out weak storms by examining the sequence of daily maps of sea level pressure. Some of the criteria are selected by considering the structure of observed TCs, while the rest are intended to exclude extra-tropical cyclones. The latter type of criteria are used for the thickness between the 200 hPa and 1,000 hPa levels and for the 200 hPa longitudinal wind velocity. The location of a gridpoint where all the search criteria meet is recorded as the location of a candidate STC. The center position of the candidate STC is calculated as the minimum point of the 500 hPa geopotential height field by filling a parabolic surface to the field values at the nine gridpoints surrounding and including the STC location. This search process is applied on each model day for the entire length of the datasets.

Next, a tracking procedure for candidate STCs is implemented to check their temporal continuity. For each candidate STC (STC-A, for example) all the candidate STCs identified on the next model day are examined to see whether or not any of them are located within a critical distance from the position of STC-A. If there are one or more STCs within the critical distance, the closest is considered to be the same

STC as STC-A. The critical distance uses the value of 720 km within the belt equatorward of 22.5°, or 1,320 km elsewhere, by assuming that a STC does not move faster than $30\,\text{km}\,\text{h}^{-1}$ within the equatorial belt or $55\,\text{km}\,\text{h}^{-1}$ elsewhere. After examination of all the candidate STCs, each STC is catalogued as one of the sequential STCs considered to be identical or an incidental STC that appears only once in a particular day. One set of sequential STCs forms an STC trajectory and an STC lifetime is defined as the duration between the endpoints of a set of sequential STCs.

Finally, the following constraints are added to exclude relatively weak and/or incidental vortices: (1) incidental STCs are excluded (i.e., a qualified STC must last for at least 2 days); (2) maximum windspeed over the search grid at the 900 hPa level must exceed a gale force of $17.2\,\text{m}\,\text{s}^{-1}$ throughout the lifetime of an STC; and (3) maximum precipitation rate over the search grid must exceed $100\,\text{mm}\,\text{d}^{-1}$ at least on one day during the lifetime of an STC.

The authors state that the standard T42 CCM2 appears to be potentially useful in finding the relationship between SST and TC genesis, including the overall statistics of seasonal variations in STC frequency in all ocean basins. On the other hand, the model is clearly deficient in studying any aspect of TC intensity and movement.

One significant limitation of the model to simulate realistic TCs comes from the fact that the T42 resolution grid of 2.8° in longitude and latitude is too coarse to resolve the inner structure of TC circulations. Since the composite STC shows no eye-type structures, and the location of maximum windspeed is unrealistically far from the storm center, the validity of STC intensity using the T42 model is uncertain. Concerning the climatology of STC, the impact of a coarse-resolution model on STCs seems to affect their movement and lifetime duration most severely.

Nevertheless, the deficiency of the coarse-resolution model seems not to be detrimental to the genesis of tropical transients over open water. The overall pattern of STC genesis is quite reasonable in terms of geographical distribution and seasonal cycle. In particular, the model's potential usefulness lies in the simulation of TC genesis over such areas as the west North Pacific (WNP), the southwest Pacific (SWP), and the southwest Indian (SWI) Ocean where the seasonal variations of STCs agree very well with those of TCs. The genesis of STCs is clearly influenced by high SST values in the tropical belt outside 5° from the equator. The shape of the histogram of STC formations versus SST values agrees well with that of observed TC formation. Total annual frequencies of STC formation in all basins are stable, meaning that the standard deviations of the annual number of STC formations are small, which is in good agreement with observed values.

However, some local disagreements exist in the spatial distribution of STC genesis. For example, a high rate of TC genesis is not well reproduced over the eastern North Pacific just off the west coast of Central America where the frequency of TC formation per unit area is the highest in the world (see Section 2.6).

Another example of unrealistically sparse STC genesis is found over the North Indian Ocean basin including the Bay of Bengal and the Arabian Sea.

Other disagreements are seen as fewer STCs originate in the eastern South Indian Ocean extending to the Timor Sea and unrealistic STCs form over the northern Australian continent. Since STC searches are carried out over low-lying land as well as seas, STCs are detected there as well. Some of them, such as over the Amazon basin, are obviously unrealistic and will be ameliorated by improvement of model parameterizations. However, STC genesis over land—particularly near coastal regions—is not necessarily considered to be unrealistic. In fact, many observed TCs around Australia evolve from disturbances that form over land and move offshore (Section 2.6). The authors' findings on the relationship between the genesis of STCs and El Niño are not definitive.

Despite obvious shortcomings of the standard CCM2, such as coarse horizontal resolution, the structure and climatology of STCs identified in both climate runs are in reasonably good agreement with those of observed TCs. The annual STC frequency shows better agreement with the observed SST run than the climatological SST run, while many other aspects of STCs in the two climate runs are comparable. The authors suppose that there seems to be little doubt about the capability of the standard CCM2 to reproduce tropical transients that, in many respects, resemble observed TCs.

Another way of looking at the simulation of tropical cyclogenesis is to use the high-resolution coupled ocean–atmospheric general circulation model (CGCM) to investigate the relationship between ENSO and global warning and cyclone-type disturbances regionally—in particular, in the northwestern equatorial Pacific (NWEP) (Matsuura *et al.*, 1999; Knutson *et al.*, 1998; Knutson and Tyleya, 1999).

The relationship between the characteristics of tropical storm (TS) activity in the northwestern equatorial Pacific (NWEP) and El Niño–Southern Oscillation (ENSO) is investigated by Matsuura *et al.* (1999) from the viewpoint of how TS behavior varies interannually. A high-resolution coupled general circulation model (CGCM) capable of simultaneously reproducing both TSs and ENSOs has been developed to study the role played by air–sea interaction in linking TS activity with ENSO.

The atmospheric component of our CGCM is taken from the Japan Meteorological Agency forecast model (GSM 8911). This global spectral model is T106 in resolution (with horizontal grid spacing of approximately 110 km) and has 21 levels in the vertical. This model is classified as a high-resolution atmospheric GCM (AGCM). The ocean model used is the GFDL MOM 2 oceanic general circulation model (OGCM). The MOM 2 grid spacing is 1° in longitude and 0.5° in latitude. In the vertical, there are 37 levels, with 25 of them in the upper 400 m. The authors began the CGCM run using the atmospheric conditions on January 1, 1989, which were obtained from the integration of AGCM from April 1, 1988, using the SST dataset in UKMO as the boundary condition. The ocean model was spun up for 10 years from a static state using Levitus' annual mean temperatures and salinities as initial conditions. The AGCM and OGCM are coupled through daily mean SST and surface fluxes. Flux correction is not used. The CGCM is integrated for 24 years, and output for the 15-year period from year 10 to 24 is analyzed in this study.

The authors define a model tropical storm (TS) using three physical variables: sea surface pressure, windspeed at 850 hPa, and relative vorticity. Although our simulation is conducted using a high-resolution CGCM, the model typhoons are somewhat weaker and have a larger scale than observed typhoons. To investigate TS behavior, they define a model TS as a disturbance meeting the following criteria: is located in the NWEP, has a minimum pressure less than 1,008 hPa, has a windspeed exceeding $17\,\mathrm{m\,s^{-1}}$, and has a relative vorticity in excess of $1.2 \times 10^{-4}\,\mathrm{s}^{-1}$.

The CGCM used in this study is able to reproduce the structure of observed typhoons for both the atmosphere and the ocean, although the simulated typhoons are weaker and larger in scale than observed. Typhoons form and develop while receiving heat energy and humidity from the ocean surface when SSTs exceed 26°C.

Although the model typhoon does not exhibit an eye, sea level pressure drops sharply at the core, and the maximum wind velocity appears at the location where an abrupt gradient in sea level pressure exists. These features are similar to observed typhoons. The maximum tangential wind occurs at around 850 hPa and is located at 2° latitude from the center. The radius of observed maximum winds ranges from about 100 km for typhoons with no eye to tens of kilometers for typhoons with small eyes. It is observed that weak subsidence is present in the center surrounded by strong updrafts near the radius of maxmum winds. Although the model typhoon has the largest ascending vertical motions at the radius of maximum wind, weak subsidence does not occur in the center of the model typhoon. This is because the horizontal resolution of the model is too coarse for its eye to be resolved. Moist air converges in the lower atmosphere, rises in the core of the typhoon, and diverges at 200 hPa—in particular, temperature anomalies exceed 8°C at 200 hPa to 300 hPa in the core of the typhoon. The secondary warm core maximum that appears at 1,000 hPa in the simulated typhoon's center is unrealistic and may occur as a result of using the Kuo scheme as the model's cumulus parameterization scheme.

For the ocean response to the typhoon, strong mixing occurs from the surface to about 50 m and a divergent Ekman drift is induced in the upper ocean by cyclonic wind stresses, so that the seasonal thermocline wells up beneath the core of the typhoon. The characteristic variation of the upper ocean beneath the model typhoon is similar to that beneath observed hurricanes and typhoons. Ocean temperature anomalies below a depth of 50 m (the seasonal thermocline) decrease over 1°C due to upwelling beneath the core of the typhoon.

Analysis of 15 years of simulation and observations shows that TS frequency decreases slightly off the Philippine Islands (120–150°E, 5–20°N) and that the location of their generation shifts toward the east during El Niño years. This results in inactive convection in the NWEP during El Niño years due to an increase in sea level pressure over ocean areas.

During ENSO warm phases the average genesis location of TCs is 146.2°E, 16.4°N and during ENSO cold phases it is 140.6°E, 15.8°N. The simulated location of TC formation corresponds to the observed location during El Niño years: when the SOI is low and the SST in the central and eastern equatorial Pacific is warmer than during La Niña years the genesis location of TC formation in the NWEP shifts eastward. However, in La Niña years, the location of TC formation concentrates east of the maritime continent (140.6 ± 17.2°E, 15.8 ± 5.3°N). The values of standard deviation of TC genesis locations are 19.0°, 5.0° in ENSO warm phases and 17.2°, 5.3° in ENSO cold phases. Therefore, the simulated location of TC formation extends in the longitude during ENSO warm phases. The migration of model TCs decays faster than actual TC migration, and there are too few model TCs moving north over 30°N. Regarding the relationship between TC migration and ENSO in the trade wind zone from 10°N to 25°N, it can be seen that TCs prefer to migrate westward during ENSO warm phases, because there are anticyclonic wind anomalies at 500 hPa in the NWEP and, so, westward migration becomes stronger. During ENSO cold phases, westward migration is decreased and northward migration is increased because there are cyclonic wind anomalies in the NWEP at 500 hPa. Simulated TC migration is somewhat different from that observed during ENSO warm phases because the location of model anticyclone wind anomalies at 500 hPa shifts westward.

Knutson *et al.* (1998) and Knutson and Tyleya (1999) direct their attention to the above-mentioned question as to how the character of these powerful storms could change in response to greenhouse gas–induced global warming.

In contrast to these works, Broccoli and Manabe (1990) and Bengtsson *et al.* (1995, 1996) use a more sophisticated combination of GCM and regional (NWEP) model simulations and another methodology. In this case study approach, the authors selected 51 tropical storm cases from a control climate simulation of a global climate model and 51 cases from a high-CO_2 climate simulation. The global model used was the Geophysical Fluid Dynamics Laboratory (GFDL) R30 coupled ocean–atmosphere climate model that has a resolution of about 2.25° latitude by 3.75° longitude.

The GFDL R30 global coupled ocean–atmosphere climate model has an atmospheric component with a resolution of about 2.25° latitude (250 km) by 3.75° longitude (400 km) and 14 vertical levels. Two 120-year experiments were done with the model: (i) a control integration with CO_2 constant at present day levels and (ii) a transient CO_2 increase experiment in which atmospheric CO_2 levels increased at +1% per year compounded (i.e., by a factor of 2.57 by year 95). Data from years 70 to 120 of these two experiments provided the initial conditions and time-dependent boundary conditions for the regional model case studies. The criteria are similar to those used in other recent GCM studies (Bengtsson *et al.*, 1995).

For the high-CO_2 cases, storms were selected from years 70 to 120 of a +1% per year CO_2 transient experiment, corresponding to CO_2 increases ranging from a factor of 2.0 to 3.3. Tropical storm–like features (weaker and much broader than in real world storms) have previously been analyzed in an R30 global atmospheric model very similar to that used here. The selected storm cases were then rerun as 5-day forecast experiments using the high-resolution GFDL hurricane prediction system, which is currently used at the U.S. National Centers for Environmental Prediction (NCEP).

The hurricane prediction system consists of an 18-level triple-nested movable mesh atmospheric model with a model-generated initial vortex. The outer grid covers a region 75° by 75° at a resolution of 1°, whereas the innermost grid covers a region 5° by 5° at a resolution of 1/6°, or about 18 km. The CO_2 level in the hurricane model was adjusted to the appropriate level for each particular case. As well as differing in spatial resolution, the global and regional models differ in model physics, diurnal variation, and so on.

Before beginning each hurricane model simulation, the crudely resolved global model storm (but not the background environment) was filtered from the global model fields and replaced by a more realistic initial vortex.

In the vortex replacement procedure, the case study storms from the global model were traced back for 2 to 4 days from the time of maximum intensity to an earlier stage of development (at least one closed-surface isobar, using a 4 mb contour interval). The global model storm—but not the background environmental flow fields—was then filtered out and replaced by a more realistic disturbance vortex as an initial condition. The replacement vortex was generated with the GFDL hurricane model's initialization scheme, using an identical target disturbance (maximum wind 17.5 m/s at a radius of 175 km) for each case. This procedure is analogous to that presently used for hurricane prediction at NCEP, except that in the operational case the disturbance target is based on actual storm observations and the global fields are derived from operational analyses.

The storm intensity distributions of the control and high-CO_2 case studies were then compared. Sea surface temperatures (SSTs) were held fixed during hurricane model experiments. The SSTs and initial environmental conditions used for regional hurricane model simulations were derived from the global climate model.

High-CO_2 storms, with SST warmer by about 2.2°C on average and higher environmental convective available potential energy (CAPE), are more intense than the control storms by about 3 m/s to 7 m/s (5–11%) for surface windspeed and 7 hPa to 24 hPa for central surface pressure. Simulated intensity increases are statistically significant according to most of the statistical tests conducted and are robust to changes in storm initialization methods. Near storm precipitation is 28% greater in the high-CO_2 sample. In terms of storm tracks, the high-CO_2 sample is similar to the control. The mean radius of hurricane force winds is 2% to 3% greater for the composite high-CO_2 storm than for the control, and high-CO_2 storms penetrate slightly higher into the upper troposphere. More idealized experiments were also performed in which an initial storm disturbance was embedded in highly simplified flow fields using time mean temperature and moisture conditions from the global climate model. These idealized experiments support the case study results and suggest that, in terms of thermodynamic influences, the results for the NW Pacific basin are qualitatively applicable to other tropical storm basins.

Subsequent study relative to just how sensitive the thermodynamic environment is to GFDL model hurricane activity was performed by Shen *et al.* (2000). In this study, the effect of thermodynamic environmental changes on hurricane intensity is extensively investigated using the National Oceanic and Atmospheric Administration Geophysical Fluid Dynamics Laboratory hurricane model for a suite of experiments with different initial upper-troposphere temperature anomalies up to ±4°C and sea surface temperatures ranging from 26°C to 31°C given the same relative humidity profile.

The results indicate that stabilization in the environmental atmosphere and increase in sea surface temperature (SST) cause opposing effects on hurricane intensity. The offsetting relationship between the effects of atmospheric stability increase (decrease) and SST increase (decrease) is monotonic and systematic in the parameter space. This implies that hurricane intensity increase as a result of possible global warming associated with increased CO_2 is considerably smaller than that expected from warming of the oceanic waters alone. The results also indicate that the intensity of stronger (weaker) hurricanes is more (less) sensitive to atmospheric stability and SST changes. Model-attained hurricane intensity is found to be well correlated with maximum surface evaporation and large-scale environmental convective available potential energy. Model-attained hurricane intensity is highly correlated with the energy available from wet-adiabatic ascent near the eyewall relative to a reference sounding in the undisturbed environment for all the experiments. Coupled hurricane–ocean experiments show that hurricane intensity becomes less sensitive to atmospheric stability and SST changes because ocean coupling causes larger (smaller) intensity reduction for stronger (weaker) hurricanes. This implies that less of the increase in hurricane intensity is related to possible global warming as a result of increased CO_2.

In addition to the study of tropical cyclone genesis as a complex of events (i.e., cyclogenesis), case studies of the simulations of tropical cyclone circulation on the basis of weather forecasting models and of primitive equation models are currently under way.

Rao and Ashok (1999) study a 10-level axi-symmetric primitive equation model with cylindrical coordinates to simulate tropical cyclone evolution from a weak vortex in the Bay of Bengal region. The physics of the model comprises the parameterization schemes of the Arakawa–Schubert cumulus convection planetary boundary layer. Initial conditions were taken from the climate mean data for November of Port Blair (92.4°E, 11.4°N) in the Bay of Bengal, published by the India Meteorological Department. The initial vortex was designed to have a tangential wind maximum of 10 m/s at 120 km radius with a central surface pressure of 1,008 hPa. As a control experiment the model was integrated for 240 h keeping the sea surface temperature (SST) constant at 301 K. The results of the control experiment reveal a slow decrease in central surface pressure (CSP) from the initial value of 1,008 hPa to 970 hPa at 156 h. After 156 h the CSP decreased suddenly until 186 h when it reached 890 hPa. Tangential wind at the 1 km level reached the cyclone threshold intensity (CTI) of 17 m/s around 78 h and a maximum of 87 m/s was found at 210 h. These features indicate a pre-development stage lasting up to 156 h, a deepening stage lasting 30 h (i.e., from 156–186 h) followed by a mature stage. The mature stage is characterized by simulation of the central eye region, warm core, strong cyclonic circulation in the central 300 km with low-level inflow, strong vertical motion at the eye wall, and outflow aloft. The convection features of the different cloud types conform to the circulation features. The control experiment clearly indicates the evolution of a cyclone with hurricane intensity from a weak vortex.

The genesis of TC "Heman" (1996) in the eastern Pacific was investigated by Molinari *et al.* (2000) using gridded analyses from the European Centre for Medium-Range Weather Forecasts and gridded outgoing longwave radiation. TC "Heman" developed in association with an easterly wave that could be tracked back to Africa in longitude–time plots of the filtered component of the wind (2 to 6-day period) at 700 mb. The wave crossed Central America near Lake Nicaragua with little change in its southwest–northeast tilt, but the most intense convection shifted from near the wave axis in the Caribbean to west of the wave axis in the Pacific. The wave intensified as it moved through a barotropically unstable background state (defined by a low-pass filter with a 20-day cut-off) in the western Caribbean and eastern Pacific. A surge in southwesterly monsoons and enhanced convection along 10°N occurred to the west of the 700 mb wave in the Pacific and traveled with the wave. This had the effect of enhancing low-level vorticity over a wide region ahead of the 700 mb wave. Available evidence suggests that additional low-level vorticity was produced by enhanced flow from the north through the Isthmus of Tehuantepec as the 700 mb wave approached. Depression formation did not occur until 6–12 h after the 700 mb wave reached this region of large low-level vorticity in the Gulf of Tehuantepec.

Eastern Pacific SST and vertical wind shear magnitude are typically favorable for tropical cyclone development in northern hemisphere summer and early falls. Because the favorable mountain interaction and the surge in low-level monsoons appear to relate directly to a wave in the easterlies, it is argued that the strength of such waves reaching Central America from the east is the single most important factor in whether subsequent eastern Pacific cyclogenesis occurs.

7.3 KINETIC DIFFUSION APPROACH

In the preceding sections we considered in some detail the approaches associated with vortical disturbances in global circulation models and with formal statistical lines of investigation. Other methods and methodologies are currently being developed. For instance, a number of Russian scientific groups have proposed and developed new physical approaches—namely, the hydrodynamic self-organized approach and the fluctuation kinetic approach. These approaches differ in two crucial respects from known and widely used lines of investigation: GCM simulations and the synoptical statistical method. As these new lines of investigation have only been described in the Russian literature, we will now place greater emphasis on these questions.

Problems relating to the construction of a physical model of the formation of stable eddy systems against a background of turbulent chaos of the atmosphere can be addressed in at least two ways (Sharkov, 1995, 1997, 1998, 1999a):

- a local approach can be used in which a single (individual) eddy structure forms from the turbulent chaos under conditions in which the local ocean–atmosphere system is highly out of equilibrium;
- a global approach can be used to consider the formation of eddy systems in the World Ocean as a set of centers of relaxation oscillations in the active medium of the ocean–atmosphere system (where the latter is viewed globally).

While the local approach involves a self-contained direction for studywork to follow, essentially starting from the concept of self-organization (see, e.g., Moiseev *et al.*, 1983a, b; Moiseev, 1991), tropical cyclogenesis considered globally remains a virtually unstudied physical process. In the first place, this is because there is no unified approach to (or physical model) describing the occurrence and functioning of global tropical cyclogenesis (GTC) in a global ocean–atmosphere system that is out of equilibrium. In this respect, determination of the extent to which the active medium of the ocean–atmosphere system is out of equilibrium in terms of the generation of coherent structures is of great ecological importance, since this is related to possible reconstruction of the regime of mesoscale generation of a sequence of individual TCs into the global synchronous catastrophic regime of the generation of super-typhoons, as happens in the atmosphere of Venus. While this is apparently a hypothetical notion, it is nevertheless fully supported by certain experimental evidence. For example, some time ago the regimes of multiple generations of tropical cyclones (TC) in twos or threes were detected (Minina, 1983; Pokhil, 1990, 1996; Sitnikiv and Pokhil, 1998). In 1995, there was a regime of TC generation that was so active it was given the name "the conveyor TC in the Atlantic" (Carlowicz, 1995). In this instance, as many as four TCs were recorded in a group.

On the other hand, the successes of the kinetic and diffusion approach at describing out-of-equilibrium systems in hydrodynamics and chemical kinetics (see, e.g., Nicolis and Prigogine, 1977; Polak and Michailov, 1983) are well known. The kinetic characteristics of a system are defined in these works as the processes of the birth and death of the elements of the system (reaction mechanism) and diffusion is defined as random walks between an element of the volume and adjacent elements.

Using these analogies, it was expedient to propose and justify (using experimental data) the kinetic and diffusion approach to describe global tropical cyclogenesis as a discrete Markov process. This approach was first proposed and developed by the author of this book (Sharkov, 1995, 1996d, 1997, 1999a).

In this case it is clear that the reaction and diffusion mechanisms may be interpreted as follows: the processes of birth and death of elements of the system correspond to the creation (birth) of a TC and its disappearance (dissipation, death), while diffusion may be interpreted as random walks on the functional lifetime of individual TCs and as a random time series of instants at which they occur.

Based on processing experimental data, the purpose of the following subsections is to present results showing that global tropical cyclogenesis can be described as a kinetic and diffusive stochastic process that develops in the medium of an ocean–atmosphere system that is slightly out of equilibrium.

7.3.1 Principles of the formation of a signal

Recall the principles involved in forming an information signal (Chapter 2). We are not interested in the detailed structure and dynamics of each individual tropical formation and shall therefore represent each tropical disturbance on the time axis as a pulse of unit amplitude with a random duration (corresponding to the functional lifetime of a TC) and with random times of occurrence (generation of an individual TC). The number of pulses (events) occurring in a unit time interval (a day, in our case) is therefore a natural physical parameter, representing the intensity of global cyclogenesis, which defines the energetics of the ocean–atmosphere interaction.

Mathematically, the conjectured procedure for signal formation may be written as follows:

$$I(t) = \sum_i \Theta(t - t_i; \tau_i) \tag{7.3}$$

where $I(t)$ is the intensity of cyclogenesis; and Θ is the restricted Heaviside function:

$$\Theta(t - t_i; \tau_i) = \begin{cases} 1, & t_i < t < t_i + \tau_i, \\ 0, & t_i + \tau_i < t < t_i, \end{cases} \tag{7.4}$$

where τ_i is the lifetime of an individual TC; t_i is the time at which it was formed; and $t_i + \tau_i$ is the time at which it disappeared.

A sequence of pulses formed in this way is no more than an integer-valued time series of indistinguishable events. Thus, our starting point will be representation of the time sequence for the intensity of global tropical cyclogenesis as a statistical signal with a complex structure (Pokrovskaya and Sharkov, 1993a). Furthermore, we propose to carry out a correlation analysis of the signal formed in this way in order to reach conclusions about a possible evolution equation and a corresponding stochastic model of the process.

Experimental geophysical data about the occurrence and temporal and spatial evolution of TCs in the World Ocean were taken from the Global TC database (see Pokrovskaya *et al.*, 1993; Pokrovskaya and Sharkov, 1999d), in which the chronological, hydrometeorological, and kinematic characteristics of large-scale tropical disturbances throughout the World Ocean are represented in the form of a sequence of events during the 5-year period from 1988 to 1992. Pokrovskaya and Sharkov (1993a) show that a random process formed in this way is a typical telegraphic process. In addition, it was also shown there that the probable structure of the amplitude fluctuations of a given series is similar to the structure of a series of Poisson type, but with certain deviations from the Poisson distribution. This fact is known (Nicolis and Prigogine, 1977) to play a fundamental role in analysis of the components of the given stochastic process and affects the relationship (and competition) between the kinetic and diffusive components of the process.

7.3.2 Correlation properties

The results of the correlation analysis carried out for known laws for the stochastic series for the intensity of global tropical cyclogenesis over the period 1988–1992 are shown in Figure 7.3. The main question here is one of finding a suitable approximating rule for the evolution of the correlation properties of the process as a function of the interaction timescale. It follows from analysis of the data presented in Figure 7.3 (linear scale) that for all the yearly intervals studied from 1988 to 1992 the temporal

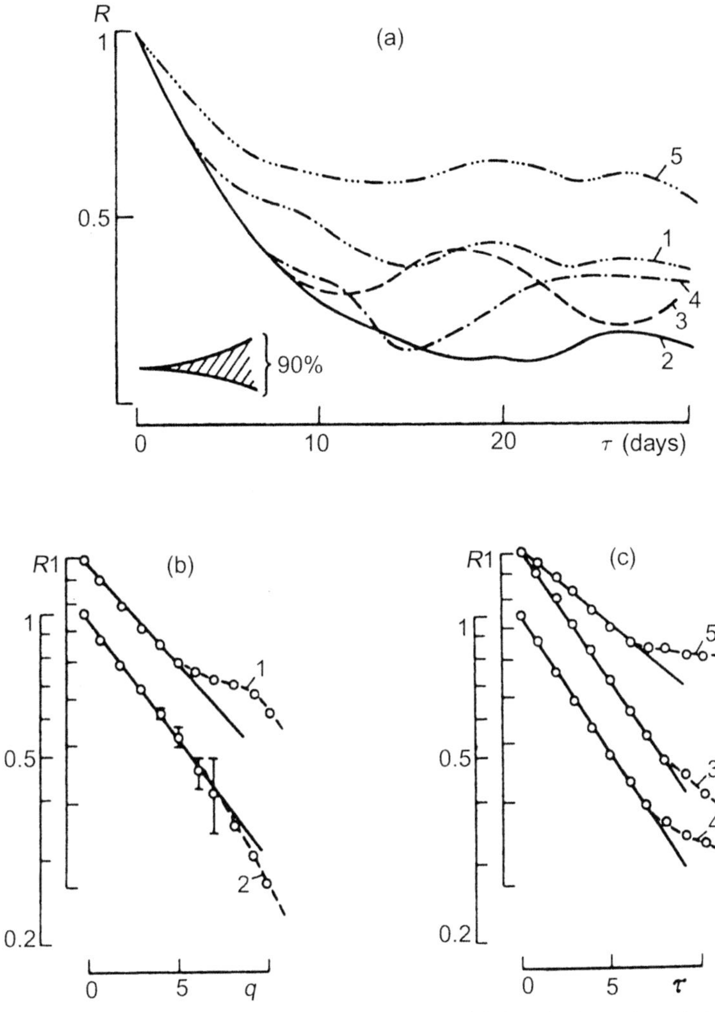

Figure 7.3. Temporal normalized correlation function $R(\tau)$ for the intensity of global tropical cyclogenesis over annual cycles for 1988–1992 on linear (a) and semi-logarithmic scales (b) and (c). The points denote experimental values of the correlation coefficient; the hatched area in (a) and the vertical bars in (b) show the 90% confidence intervals; the continuous lines show the exponential approximation; the dashed lines show deviation from the exponential curve: 1, data for 1988, 2, 1989, 3, 1990, 4, 1991, 5, 1992 (from Sharkov, 1995).

correlation functions for the intensity of cyclogenesis have a characteristic form: up to timescales (time lags) of the order of 6–7 days the correlation coefficient of the process falls strictly exponentially (as shown by the graphs drawn on a semi-logarithmic scale). Later, a "logarithmic plateau"-type dependence with almost harmonic variations and characteristic periods of 12–15 days is observed. Despite the great complexity of the physics involved in the process of global cyclogenesis, on scales from 1 day up to 6–7 days its correlation properties can be approximated very simply: namely, by $R(\tau) = \exp(-\alpha\tau)$, where $R(\tau)$ is the coefficient of correlation and τ is the time lag (scale). It is interesting to note that the values of global cyclogenesis in 1989, 1990, and 1991 are practically the same, $\alpha = 0.125$ day^{-1} (annual difference less than 2%) while for 1988 $\alpha = 0.099$ and for 1992 $\alpha = 0.067$. Despite these differences in the values of α, the timescale for "descent" from the exponential regime and entry into logarithmic dependence is the same for all the yearly cycles studied, $\tau_g \cong 6$–7 days.

One important feature in this analysis is correct estimation of the accuracy of parameter α, the scattering of whose values is associated with statistical error in the construction of the correlation function. To carry out this operation we use the following procedure. First, we estimate confidence intervals for the correlation coefficient (for lags from 0 to 10) using the normalizing Fisher transformation with a corresponding number of degrees of freedom $n = 2N/m$, where N is the size of the complete sample ($N = 365$) and m is the number of steps (lags) of the correlation function (Bendat and Piersol, 1986). The results of estimating confidence intervals for a value of 90% are shown in Figure 7.3 on the linear scale and in Figure 7.3b,c on the semi-logarithmic scale. Estimation of the maximum scattering of values of α (i.e., $\Delta\alpha \cong |\alpha_{\max} - \alpha_{\min}|$) for linear approximation on the semi-logarithmic scale gives the value $\Delta\alpha \cong 0.017$. Assuming that the density of the distribution of the random value α is close to normal, we obtain an estimate of root-mean-square square deviation (r.m.s.) for parameter α. Thus, using the given method of constructing α, the r.m.s. is $\sigma(\alpha) \cong 0.003$.

From the given behavior of time correlation, it follows that the spectral density of the power $G(\omega)$ of the amplitude fluctuations of the intensity of cyclogenesis will exhibit a universal $1/\omega^2$ dependence at high frequencies (by virtue of the exponential correlation regime as $\tau \to 0$) with a transition for $\omega \cong 1/\tau_g$ to a $1/\omega$ dependence (the natural flicker noise in the terminology of Klimantovich, 1982). Thus, we have the following possible approximation for $G(\omega)$:

$$G(\omega) = \frac{D}{\omega^2[L + (D/|\omega|)]} \tag{7.5}$$

where $\tau_g = 1/\omega_g = L^2/D$; τ_g is the time of diffusion; and L and D are parameters characterizing the linear size of the system and the diffusion.

7.3.3 Kinetic diffusion model

The detected exponential nature of the time correlation function (as $\tau \to 0$) and the Poisson properties of fluctuations in the intensity of cyclogenesis clearly indicate that the given system finds itself under conditions that are slightly out of equilibrium and that the non-equilibrium elements that arise during the process of cyclogenesis (dynamically self-organized spiral structures) have practically no effect on the global kinematics and thermodynamics of the ocean–atmosphere system. In the present examination of the formation of individual eddy structures, the infringement of the equilibrium of the ocean–atmosphere system (locally and on ground scales) may be very slight. Ultimately, this also leads to the formation of an individual ordered structure out of the spiral chaos.

As the process of cyclogenesis is slightly out of equilibrium conclusions can be drawn about a possible stochastic model of the phenomenon. For a timescale of less than 6–7 days, we can use a model of the Brownian motion of particles in a viscous medium (Ornstein–Uhlenbeck process) for a finite-dimensional system, which can be described by a known linear equation with a Langevin source (Nicolis and Prigogine, 1977; Polak and Michailov, 1983). It is known (Klimantovich, 1982) that this process includes two stages of relaxation (kinetic and diffusive) with transition to the regime of flicker noise. In the case under study, the kinetic stage is observed for $\tau < \tau_0 = 1/\alpha$ and is characterized by the spectral function:

$$G(\omega) = \frac{\alpha}{\alpha^2 + \omega^2} \tag{7.6}$$

where τ_0 represents a kind of "free running" time (the time between the birth and death of individual structural elements of the system). By "diffusion stage" we mean the chaotic and (temporally) non-uniform filling of the time axis by the time series of births and deaths of structural elements. By "Langevin source" we mean a broad spectrum of effects of various scales on the system (such as internal noise), stimulating the occurrence and controlling the duration of structural elements in the finite-dimensional system (for the effect of "smoothing" Langevin sources). For timescales longer than 6–7 days, non-linear hydrodynamic interactions can be introduced as indicated, on the one hand, by the structural characteristics of the correlation functions for 12–15 days and, on the other hand, by the results of wavelet analysis of the time series for cyclogenesis intensity (Chapter 2).

7.3.4 Radiophysical analogue

These results enable an interesting (and unexpected) radiophysical analogy to be made: the process of global tropical cyclogenesis (on daily timescales) is similar, in terms of its correlation properties, to the passage of white noise through a low-frequency filter (*RC* chain). This process is known to be described by:

$$\dot{x} + \alpha x = \alpha \xi(t) \tag{7.7}$$

where $\alpha = (RC)^{-1}$; $\tau_0 = RC$ is the *RC* constant of the low-frequency filter; and $\xi(t)$ is delta-shaped white noise. In other words, the Earth's atmosphere in the process of the birth of organized eddy structures behaves—with respect to external and internal noise effects of various scales (analogous to white noise)—like a low-frequency filter with a very stable time constant over the five years in question. It is not difficult to determine this value from our data: $\tau_0 \cong 8$–12 days.

The above equation is known to have the following finite-difference analogue—namely, linear transformation to the form of linear expansion with random additive noise:

$$\varphi_i = \lambda \varphi_{i-1} + \varepsilon_i \tag{7.8}$$

where φ_i is the first-order autocorrelation function; and ε_i is uncorrelated white noise.

This finite-difference representation may be used in the mathematical modeling of global tropical cyclogenesis.

7.3.5 Ways of improving the model

A model of relaxation-stimulated autowave media with two variables can provide a more complete description of the evolution of global tropical cyclogenesis. The model is used, for example, to analyze chaotic signals in biological and ecological systems (Romanovskii *et al.*, 1984; Isakov, 1997; Ivanitskii *et al.*, 1998). By "fast variables" we mean hydrodynamic effects on meso and microscales while "slow variations" may be taken to mean slight variations in the global atmospheric circulation on synoptic scales (easterly winds, variations in the indexes of atmospheric circulation, El Niño). In this case, there may be a kind of blocking regime that generates sluggishly (typical for February–April of the year in question) and a triggering regime with a high intensity of relaxation oscillations (June to September). These regimes are clearly identified as a result of processing full-scale data.

Hence, the global ocean–atmosphere system (tropical zone) is very slightly out of equilibrium and, thus, in principle there exists a very slight possibility to reconstruct the mesoscale convective regime of the tropical atmosphere as a global synchronous catastrophic regime of generation of super-typhoons.

The model presented can be used for operational planning of space experiments to study the initial forms of atmospheric catastrophes and for research into variations in global climate.

In addition, the results of this section show that the global ocean–atmosphere system (tropical area) is weakly out of equilibrium. Hence, the possibility of reconstructing (self-organizing) a mesoscale "impulse" regime with operating events (TCs) leading to generation of a form of global "superhurricane" is vanishingly small. In the foreseeable future there will be no such superhurricane to threaten humankind (Sharkov, 1999a, b). The out-of-equilibrium stability of the global atmosphere–ocean system will remain very high in the future.

7.4 CONCEPT OF SELF-ORGANIZATION

The physical mechanism involved in TC onset is a most complicated question. According to the latest views of physicists in the area of chaotic dynamics the atmosphere is a self-organized medium with a great number of appearing and dissipating structures called "structural chaos" (Levich and Tzvetkov, 1985; Hussain, 1986; Lesieur, 1997; Gaspard, 1997). The fundamental task is to discover dominating (and, possibly, competing) mechanisms.

Theoretical studies of the generation of large-scale structures subject to helical turbulence (including in the atmosphere near to the ground surface) were first begun at the Space Research Institute (Russian Academy of Sciences) and have long been actively pursued.

A new physical mechanism was proposed in 1983 (Moiseev *et al.*, 1983a) called the "large-scale vortical dynamo in air–hydrodynamics". Under unstable thermal stratification conditions it provides the possibility of reconstructing the thermo-convection structure in which large-scale ordered vortical systems including the non-trivial topological structure of streamlines appear. The concept gained wide acceptance among hydrodynamic specialists (Moiseev *et al.*, 1983b, Levich and Tzvetkov, 1985; Moiseev *et al.*, 1988, 2000; Moiseev, 1990, 1991; Branover *et al.*, 1999). Following Zimin *et al.* (1992) and Moiseev *et al.* (1996), let us now consider the qualitative foundations of the concept.

7.4.1 Elements of qualitative analysis

Typhoons (tropical cyclones) are intense atmospheric eddies. According to accepted ideas, a typhoon draws energy from comparatively small-scale atmospheric convection. However, we still do not know how this energy from small-scale movements is converted into larger scale movemtnts. In other words, all existing models of the initial stage of development of typhoons mussst be considered phenomenological and cannot pretend to explain the mechanism governing the occurrence of tropical cyclones (TCs).

From our point of view, the failure of all previous models is explained by the fact that they do not take account of the most important physical property of tropical cyclones—namely, the so-called topological non-triviality of the velocity field in TCs. By this we mean the correlation of flow patterns between horizontal and vertical circulation in TCs.

Visually, such a system may be represented as two ring-shaped flow patterns in mutually perpendicular planes. For such a structure, the scalar derivative $H = (\vec{\mathbf{V}}\,\mathrm{rot}\,\vec{\mathbf{V}})$ summed over the whole volume is non-zero ($\mathrm{rot}\,\vec{v} = \vec{\omega}$ is the usual notation for eddy velocity).

The value of H is usually called the helicity, because it is non-zero for movement in a helix, since velocity has a vertical component directed along the axis of the helix and $\mathrm{rot}\,\vec{v}$ also has a vertical component with the same direction as angular velocity. The helicity H characterizes the so-called topological non-triviality of the velocity field. Calculation of this characteristic in theoretic models led to the development of new ideas about the mechanisms governing the formation of tropical cyclones.

The non-zero helicity of the velocity field in TCs leads to a positive inverse correlation between vertical and horizontal circulations, which provides for the growth of initially small eddy disturbances. The hypothesis of Moiseev *et al.* (1983a) and Moiseev (1990) that a powerful typhoon is formed as a result of a set of relatively small weak eddies coalescing in a turbulent atmosphere and taking on the dimensions of a typhoon with the same properties (correlated flow patterns), seems natural.

However, there is a question. Where do eddies of this type in a turbulent atmosphere come from? One notable feature of convection in a rotating non-uniform atmosphere is that there is always a supply of eddies with helicity $H \neq 0$. In fact, let us consider a single element of the air volume rising upwards. This moves into a less dense layer of the atmosphere and begins to spread; furthermore, under the Coriolis effect it twists around the vertical axis. Thus, the movement of the air mass in a convection cell (about 10 km in size) becomes a helix (i.e. for this movement $H \neq 0$).

It is important to stress that if the volume element descends it gets compressed and, as a result, H has the same sign and its total value throughout the whole cell is non-zero. It follows from this that small-scale (compared with typhoon-sized) structures of the required type occur in the intertropical convergence zone where the air is largely rising and in depression outflows where both movements are equally likely.

The following result needs to be understood. The temperature difference in the atmosphere gives rise to small-scale (comparison with typhoon-sized) convection cells, which are converted into helices in a rotating non-uniform atmosphere (i.e., air flows rising and falling into a helix), and these helices form a large-scale eddy (whirlwind). As shown in the previous references, the period of eddy formation is of the order of days (i.e., it corresponds to nature).

These simple physical considerations form the basis of a mathematical theory of the creation of a tropical cyclone, which is described in the references listed below and in other works by the same authors (Moiseev, 1990; Moiseev *et al.*, 1983a, b, 1988, 1996).

7.4.2 Helix mechanism for the formation of large-scale structures

The role of helicity in the generation of large-scale structures has been intensively studied in recent years (see, for example, Moiseev *et al.*, 1983a, b). The mechanism behind the formation of large-scale eddy structures described by Moiseev *et al.* (1983b) reduces to the following. Suppose there exists a small-scale turbulent velocity field $\vec{\mathbf{V}}^t$ with characteristic scale l_t that has a helicity property—that is, $\langle(\vec{\mathbf{V}}^t \operatorname{rot} \vec{\mathbf{V}}^t)\rangle \neq 0$ where $\langle \cdot \rangle$ denotes averaging over the turbulent ensemble—then, by virtue of the helicity, in the averaged velocity field there is an inverse correlation between the degrees of freedom characterizing movement in different planes (i.e., if the averaged velocity field exists, then it is three dimensional). On the other hand, the dissipative inverse correlation provides for a transfer of energy from small to large scales (i.e., an inverse helical cascade).

The equation for mean velocity $\langle \mathbf{V} \rangle$ is obtained as a result of averaging over the small-scale turbulent ensemble. It appears (Moiseev, 1991; Moiseev *et al.*, 1996) that characteristic features of the process can be described by the model equation:

$$\frac{\partial}{\partial t}\langle \mathbf{V} \rangle = \alpha \cdot \operatorname{rot}\langle \mathbf{V} \rangle + \nu \Delta \langle \mathbf{V} \rangle \tag{7.9}$$

where $\alpha = \langle \mathbf{V}^t \cdot \operatorname{rot} \mathbf{V}^t \rangle \tau$; $\mathbf{V}^t$ is the velocity of turbulent pulsations; τ is the correlation time; and ν is viscosity. A new remarkable quality of Eq. (7.9) is the dissipative "antiviscous" force resulting from the term $\alpha \cdot \operatorname{rot} \mathbf{V} \sim L^{-1}$ which provides feedback between motions along different axes. This force decreases more slowly than the viscous force ($\sim L^{-2}$) with increasing characteristic scale L and, hence, Eq. (7.7) describes instability; this is particularly so with perturbations of the form $\langle \mathbf{V} \rangle = V_0 \exp(\gamma t)(\cos qz, \sin qz, 0)$ which increase with $\gamma = \alpha q - \nu q^2$.

Helicity fluctuations lead to negative viscosity (Moiseev, 1991). Indeed, using the representation $\alpha = \langle \alpha \rangle + \tilde{\alpha}$, where $\langle \tilde{\alpha} \rangle = 0$, and letting $\langle \tilde{\alpha}(t)\tilde{\alpha}(t') \rangle = 2D\delta(t - t')$ we average Eq. (7.9) over the helicity fluctuations and obtain the same equation with the substitutions $\alpha \to \langle \alpha \rangle$ and $\nu \to \nu - D$ (i.e., turbulent viscosity decreases). Note also that turbulent viscosity can essentially decrease in the helical turbulence with finite correlation time and non-zero mean helicity.

Let us now qualitatively discuss the causes of helical structure localization, medium inhomogeneity being one of them. If the helicity parameter in Eq. (7.9) is inhomogeneous and, for simplicity, axisymmetric—that is, $\alpha = \alpha(\rho)$—then the solution in the vicinity of $\alpha(\rho)$ maximum α_0 is of the form (Moiseev *et al.*, 1983b):

$$\langle \mathbf{V} \rangle = \langle V_0 \rangle \left\{ 0; \sin\left(\frac{R_\alpha \rho}{2L_\alpha}\right); \cos\left(\frac{R_\alpha \rho}{2L_\alpha}\right) \right\} \times \exp(\gamma_0 t - R_\alpha \rho^2 / 4L_\alpha^2) \tag{7.10}$$

where the parameter $R_\alpha = \alpha_0 L_\alpha / \nu \gg 1$; L_α is the characteristic scale of $\alpha(\rho)$ inhomogeneity; ρ is the cylindrical radius; and $\gamma_0 = (\alpha_0^2/4\nu)(1 - 4/R_\alpha)$ is the increment. Using parameters of the tropical atmosphere and Eq. (7.10), the characteristic scale and the evolution time of a tropical cyclone can be obtained. They are in good agreement with natural ones.

Regarding the Earth's atmosphere, the most interesting case is that of convection in a medium with helical turbulence. This variant occurs in the tropics. Calculation of small-scale helical turbulence may lead to full reconstruction of convective instability in a layer (Moiseev *et al.*, 1988). First, in addition to the poloidal velocity field there is also an associated toroidal component. Second, as helicity increases in a horizontally homogeneous problem the horizontal dimension of a convection cell may become infinite. The horizontal dimension is finite if we assume horizontal heterogeneity.

Looking at the details (Moiseev *et al.*, 1988), we note that the horizontal dimension of a large-scale helical cell is $L_{sb} \sim \sqrt{r_0 h}(S - S_0)^{1/2}$, where h is the height of the layer, r_0 is the characteristic dimension of horizontal heterogeneity, and S is a parameter proportional to helicity. For tropical atmosphere parameters, $L_{sb} \sim 100$–300 km. The helical cell thus formed has properties reminiscent of those of a typhoon. Of course, such features of a typhoon as the structure of the eye can only be studied given an accurate solution of the corresponding non-linear problem.

7.4.3 Effect of thermally insulated boundaries on the formation of large-scale structures

The question of the effect of thermal insulation of boundaries on the horizontal dimension of a convection cell in the case of laminar convection has already been analyzed in a number of papers (see, e.g., Gershuni and Zhukhovitsky, 1972).

Thus, the conditions under which perturbations of the temperature equal heat flow at the upper boundary of a layer may be written (respectively) in the form:

$$\Theta_m = \Theta; \qquad k_1 \frac{d\Theta_m}{dz} - k\frac{d\Theta}{dz} \tag{7.11}$$

where k and k_1 are heat conductivity of the convection layer and of the overlying mass, respectively; and Θ, Θ_m are temperature perturbations. If $(k/k_1) \to \infty$ the horizontal dimension of the convection cell forming in the layer also tends to infinity.

Correspondingly (for a non-compressed fluid), the horizontal velocity of the convection flow increases (i.e., convective transfer occurs mainly across the boundary and not along it).

Consequently, conclusions can be drawn from the above considerations that are quite unexpected. Under conditions of small-scale turbulence (even disregarding helicity), there arise conditions for the growth of the horizontal dimension of convection cells within the system. Indeed, small-scale turbulence may lead to an anomalous increase in heat conductivity, which in turn may lead to an increase in the horizontal dimension of cells. Taking this further, note that if convection causes small-scale turbulence, the latter leads to the convection cells along the layer (auto-thermal insulation of boundaries) getting bigger. The first part of these considerations has already been confirmed by calculations: an anomalous turbulent transfer leads to a growth of the horizontal dimension of convection cells (Zimin *et al.*, 1989). From the previous section and from the considerations given here, it can be seen that—under tropical atmospheric conditions—there are at least two mechanisms for the generation of large-scale structures as a result of the role of small-scale turbulence: helical and thermal turbulence. However, qualitative analysis by Moiseev (1990) usefully confirms the fact that these two mechanisms are not alternatives but rather go side by side.

In the linear stage of development of instability, the helical mechanism is dominant (since, in this case, rotation is a destabilizing effect, while the converse is true for the thermal mechanism). However, in the non-linear stage the direct effect of the Coriolis force on the evolving large-scale structure, the resulting creation of a helical structure, and the growth in the horizontal component of velocity due to the effect of turbulent heat flow are apparently more effective.

7.4.4 Turbulent wave dynamo

In the previous section we showed that large-scale structures might arise as a result of the joint effects of boundary and small-scale turbulence. Here, we briefly mention the intensification of long waves when they interact with small-scale turbulence ($\lambda_t \ll \lambda$). This is the so-called "turbulent wave dynamo", which involves a simultaneous increase in wave amplitudes and in the turbulent energy due to spatiotemporal modulation of the latter by the waves.

Let us now discuss the mechanism governing the turbulent wave dynamo as follows. First, in the presence of a source of non-equilibrium such as convection, intensification of the waves will accompany the process of typhoon development and, thus, may be used as a diagnostic feature for detection of this. Second, we would expect the mechanism governing the turbulent wave dynamo to lead to an acceleration of the development of very large-scale helical cells (typhoon type).

Let us now list some of the features of this process. Suppose that we have a current (i.e., a source of energy in a stratified medium like the ocean or the atmosphere) that is at the threshold of stability (or has relaxed to this state as a result of the development of the initial instability). If long-wave disturbance is propagated by the current from a layer in which the local Richardson number is close to the critical value Ri_{cr} corresponding to the threshold for the generation of turbulence, then a very variable component of turbulent energy density may become apparent. Terms proportional to the gradient of the unperturbed current in turbulent flows acting on the waves and to the variable (in particular, containing harmonics with the frequency and spatial period of the waves) coefficient of turbulent transfer may become critical. Significantly, such parameters of current stratification and waves may be selected so that turbulent flows increase the wave energy as a result of the unperturbed current. As the waves increase in amplitude the corresponding variable component of turbulent energy density also increases. As a result, turbulent wave instability is created whose growth is an increasing function of wave amplitudes.

Thus, stratified currents at the stability threshold may behave anomalously. Furthermore, Moiseev *et al.* (2000) show that the class of wave disturbances intensified by turbulence also includes natural waves generated as a result of the initial instability.

In summary, note that intensified large-scale modes "overgrow" the turbulent "fur" of the earlier stage of evolution as turbulent wave instability develops. Note also that analytical consideration of this problem has been carried out for small-scale turbulent fluctuations ($\lambda_t \ll \lambda$), where λ_t is the typical scale of these fluctuations and λ is the wavelength of the intensified waves.

7.4.5 Diagnostic problems

The results of theoretical studies of the creation of ecologically dangerous whirlwind structures place the problem of the development of methods and facilities for remote monitoring of the inverse cascade of energy in a turbulent atmosphere (as the physical process responsible for the creation and development of large-scale critical ecological situations) firmly on the agenda.

When designing the course that future studies using space facilities should take, we should reflect on recent experimental studies in which the subsatellite part plays almost the leading role. Analysis of the physical grounds for the selected methods of measurement is all that needs to be mentioned here.

It follows from Eq. (7.9) (and analogously in other cases) that in the presence of helical turbulence (see, e.g., Moiseev, 1990) a crucial role in the generation of large-scale structures is played by the averaged measure of helical turbulence $\langle(\vec{\mathbf{V}}' \operatorname{rot} \vec{\mathbf{V}}')\rangle$. Furthermore, using Eq. (7.9), it is easy to show that the helicity of large-scale movement—or, equivalently, its energy—increases during the process of development of the instability. This gives rise to a measurement strategy. Measurements of helicity $(\langle\vec{\mathbf{V}}\rangle \operatorname{rot}\langle\vec{\mathbf{V}}\rangle)$, of turbulent energy $\langle(\mathbf{V}')^2\rangle$, and of the fluxes of energy and the helicity of large-scale movement are required. As far as the helicity of small-scale turbulence is concerned, measurement of this would be more significant—by virtue of Eq. (7.9) and subsequent discussions—since it (together with the energy, the correlation time, and the viscosity of small-scale eddies) determines the growth of the developing instability.

However, the current level of measurement technology is not advanced enough for this to be measured and we may only draw indirect conclusions about the role of small-scale helicity. Thus, at the threshold of stability (critical point), statistical considerations tell us that there should be a sharp increase in the fluctuations of helicity, while, on the other hand, an increase in the fluctuations of helicity decreases turbulent viscosity (see Moiseev, 1990; Moiseev *et al.*, 2000).

A decrease in turbulent viscosity in turn leads to a partial regularization of small-scale motion. In other words, the fractal dimension of the flow pattern (temperature isolines) should fall.

Moreover, the following reasoning favors this. Transfer over time of some of the energy of an eddy to the large-scale region should lead to a more abrupt fall in spectra as the scale of the turbulence decreases than in the Kolmogorov case. On the other hand, the fractal dimension should fall. Thus, study of the dynamics of the fractal dimension of small-scale motion is very important.

Another problem of early remote diagnostics is to fix the initial period of the confluence when energy flow is pumped over small cells into larger ones and when the spirality of a large cell also grows (the so-called reverse cascade in turbulence).

The concept underlying the analysis of pre-crisis states in the atmosphere (of TC type) in summation form entails (e.g., Moiseev *et al.*, 1995; Zimin *et al.*, 1992; Lazarev and Moiseev, 1992):

(1) partial regulation of mesoscale chaotic systems (with spiral turbulence) at the formation and development of large-scale crisis processes such as those that appear on the dynamics and geometrical characteristics of the crisis area;
(2) a reverse spiral cascade, as energy transfers from the small scale to the large scale, of a system's spirality owing to the "attraction" of small vortices providing a generation of larger scale vortices with greater steadiness, a process that can both fluctuate and be coherent;
(3) anomalous intensification of various types of wave in pre-crisis state such as intensification of infra-sound and excitation of electromagnetic bursts;
(4) anomalous behavior of admixtures in pre-crisis areas such as a change in ozone components.

Another problem facing the study of developed crisis structures is the detection and determination of the central spiral structure (against the background of large-scale circulation) and the spatiotemporal hierarchy of spiral structures (spatial position, interaction). There is a need to measure the field of turbulent full energy (on mesoscales) as well as helicity and the stream of helicity (on meso and macroscales).

As far as concepts go, methodical and experimental work is being conducted in many directions:

- detection and measurement of spiral meso-vortices from turbulent chaos (using of results of remote Doppler measurements on research vessels) and building up a certain hierarchy (Klepikov and Sharkov, 1987; Voiskovskii *et al.*, 1987) (see Section 3.4.5);
- study of the spatiotemporal evolution of the meso-turbulent spectrum and exposure of its typical features (spectral and fractal indexes, scales, spectrum "extremes") (Klepikov *et al.*, 1995) (see Section 3.4.5);
- revealing the scale (fractal) of geometric properties of a chaotic system (cloud structures) from space observations (Baryshnikova *et al.*, 1989a, b) (see Section 3.4.4);
- carrying out methods to determine velocity field moments (from radioprobe evidence) and to calculate helicity values and helicity stream values (Veselov *et al.*, 1989; Zimin *et al.*, 1991).

7.5 INSTABILITY GENESIS IN A COMPRESSED AND SATURATED MOIST AIR ATMOSPHERE

In this section a new physical mechanism of vortex disturbance generation in the tropical atmosphere is proposed. The physical essence of the proposed approach consists of the fact that the role played by phase transitions in atmospheric saturated vapor is not only reduced to the energetic factor (condensation energy release), as is accepted in primitive cyclogenesis models, but also results in principal changes in the dynamics of the tropical atmosphere (appearance of the compressibility phase), caused by anomalous behavior of the vertical profile of sound velocity in saturated moist air. On the basis of results of space and surface (ship) radiothermal remote investigations, carried out in zones characterized by intensive generation of tropical cyclones (the northwest Pacific Ocean), as well as on the basis of known model presentations, the vertical profile of water vapor is restored with subsequent calculation of the altitude distribution of sound velocity. Accounting for saturation of the total altitude column of the atmosphere by water vapor, the altitude distribution of sound velocity becomes highly non-monotonous (in contrast to the dry air situation) with the sound velocity value showing a prominent minimum. The latter factor determines the necessary conditions for the generation of vortex structures. The preliminary technical requirements of radiothermal space-based systems, designed to monitor pre-crisis and crisis situations in the tropical atmosphere, are considered.

The ocean–atmosphere system of the Earth's tropical zone possesses a unique ability to generate organized mesoscale vortex profiles—tropical cyclones (TCs)—from the atmospheric turbulent chaos involved in global circulation. TCs belong to one of the most destructive natural phenomena on the globe and represent a serious ecological danger to humankind. Material damage is often accompanied and overshadowed by significant loss of life. TCs occupy second place (after earthquakes) in the league of natural catastrophes that have cost human lives.

It is self-evident that the development of modern techniques of observation, investigation, and forecasting of TCs, as well as formation of the optimum safety strategy under conditions of persistent atmospheric catastrophes that cannot humanly be stopped, represent topical and international issues that need to be addressed.

The role played by aerospace techniques in studying TCs is principal, because the spatiotemporal "randomness" of the process of individual TC genesis and the "uncertainty" of trajectory motion of TCs do not allow full application of reliable ground-based techniques of investigation. The mid-1960s marked the beginning of active TC investigation by space means; at first, there was the incidental detection of a TC from first-generation low-Earth orbit meteorological artificial Earth satellites (AESs) and, then, systematic study of TCs in the visible and IR ranges, both globally and regionally, from geostationary AESs (Sharkov, 2010).

Because the hardware configuration of any space experiment and the technique used to process remote information are determined from the scientific principles underlying the concept of the geophysical object being studied, Section 7.5.1 briefly outlines the descriptions of scientific concepts that relate to the formation and functioning of TCs. In the section a new thermo-hydrodynamic mechanism is proposed (Rutkevich, 2002; Rutkevich and Sharkov, 2004) and the basic requirements necessary to measure parameters are discovered.

7.5.1 Basic mechanisms involved in catastrophe genesis

A tropical cyclone represents a large-scale natural phenomenon and it is natural to consider its genesis to be caused by hydrodynamic large-scale instability. However, the ordinary system of equations of the hydrodynamics of a dry atmosphere presented, for example, in the monograph by Obukhov (1988) does not describe any large-scale instability. Equally, convective instability is not large-scale (compared with the characteristic size of the Earth's atmosphere) and, by itself, cannot be responsible for the appearance and development of such a large-scale structure as a tropical cyclone. Thus, searching for a physical mechanism to adequately describe the appearance of a large-scale TC-type instability is a major physical problem.

Attempts to find such an instability in the hydrodynamics of the atmosphere resulted in constructing the so-called "conventional instability of the second kind" (CISK) model (Ooyama, 1964; Charney and Eliassen, 1964). This theoretical model was considered to be the model of choice in the last 40 years or so and gave rise to a plethora of daughter models. The model was constructed on the basis of the equations of an incompressible liquid and was reduced to a mechanism in which disturbances in the tropical zone interact with disturbances on the cumulus cloud scale via surface friction. The principal element in the CISK model was the technique of small-scale convection parameterization, which stipulated the reverse binding of large-scale instability as a result of geostrophic flow amplification with growing pressure in a tropical depression that has arisen. In the incompressible atmosphere approximation, the CISK model describes large-scale instability. However, by escaping the limits of incompressible atmosphere approximation the sound modes of oscillations in a system are restored. This results in the appearance of the effect of adaptation of the pressure field in the depression in the geostrophic flow (Obukhov, 1988) and, thus, one of the basic conditions of the appearance of large-scale CISK-type instability is violated. In addition, although a lot of experimental work on the structure of TCs from airplane and satellite carriers has been carried out, experimental confirmation of CISK model results is completely absent.

In the 1980s the model of spiral turbulence appeared (Moiseev, 1991; Moiseev *et al.*, 1983a, b, 1988, 1990, 1995). It was found to be capable of causing an essential change in the character of convection in the non-isothermal atmosphere. As the turbulence spirality parameter increases, the transverse dimensions of convective cells become ever greater and, at some spirality parameter value, the convection process turns out to be more profitable energetically with a single cell, whose size is determined by the boundaries of the region heating up in the transverse direction. However, experimental determination of atmospheric turbulence's spirality parameter is difficult. Even today, the results of direct experimental determination of this parameter are unknown.

Recently, the model of tropical cyclone formation appeared. It is based on the potential vortex concept (Montgomery and Farrell, 1993) and is concentrated on the effects of eddy fluxes of angular momentum related to the asymmetry of waves in the upper layers of the atmosphere in the vicinity of the tropical disturbance. The asymmetry of waves in upper-atmospheric layers and the related fluxes of angular momentum can be modeled by means of an invariant called the "potential vortex". The invariance (time independence) of a potential vortex was proved by Obukhov (1988) for the system of equations of the atmosphere linearized on the basic state background.

Numerical calculations on the basis of this model describe large-scale tropical storm–type vortex structures; however, hydrodynamic instability is absent in the model. The generation of vortices in such a system is the result of an external effect, which should act in a stationary manner. In the potential vortex model this effect is supposed to represent a large-scale zonal shear flux.

Note that this model, as well as the question on potential vortex invariance, is closely related to Charney–Obukhov's equation, which is obtained by additionally taking into account the latitudinal dependence of the Coriolis force within the limits of the beta-plane approximation. The extremely beautiful solutions to this equation called "Larichev–Reznik solitons" are well known, and the temptation arises to associate the stable atmospheric tropical storm–type structures with these solutions. However, such a temptation should be resisted, because both sonic and internal waves are filtered out in deriving Charney–Obukhov's equation. The equation describes a non-linear structure moving in a flux that does not appear to be the result of the development of any instability and disappears under inclusion of dissipation. At the same time, it is natural to consider a tropical storm as the result of non-linear saturation of a large-scale instability whose proper existence is caused by a source with considerable power and effluence.

Thus, a basic issue in tropical cyclone genesis is the mechanism and energy source of this phenomenon. All typhoon models agree that the energy source of a cyclone is the release of the latent heat of condensation and the sublimation of atmospheric moisture. However, though the basic models have both dry and moist vortex versions, these versions differ only in their energy characteristics and the role played by phase transformations of moisture in the atmosphere for these models does not escape the auxiliary element framework. In essence, the models of dry atmosphere with phenomenological inclusion of water vapor as a power source are considered here. However, such a consideration misses the possibility that phase transformations of atmospheric moisture can result in principal changes in atmospheric dynamics. It is just this circumstance that is an object of investigation in the present work.

7.5.2 Role played by phase passing

The principal role played by phase transformations of moisture can be studied only on the basis of the first principles of thermodynamics of phase transitions of moisture in saturated air. The physical basis of such an approach consists in using moist hydrodynamics (Rutkevich, 2002) based on the thermodynamics of saturated moist air that takes into account the phase transformations of atmospheric moisture.

The thermodynamic description of a mixture that has dry air and water vapor as its parameters should contain partial pressures, the density of components, and the temperature of a mixture. We shall designate the pressure and density of dry air as P and ρ, the water vapor pressure as E, the ratio of densities of water vapor and dry air as q, and we make the assumption here that $E \ll P$ and $q \ll 1$. The temperature of the mixture will be designated as T. Both components of the system are considered to be ideal gases, the equation of state of dry air is accepted to be $P = \rho RT$, and for the low partial pressure of water vapor we get $q = (R/R_W)(E/P)$, where R and R_W are the specific gas constants of dry air and water vapor. Considering the conditions for water vapor interphase balance with its drop phase to be fulfilled, we shall take the Clapeyron–Clausius equation $dE = EL/(R_W T^2)\, dT$, where L is the latent heat of condensation, into consideration for it.

The equation of the thermal balance of a system will be written in the form (Obukhov, 1988):

$$c_v + P\frac{dV}{dt} + L\frac{dq}{dt} = 0. \tag{7.12}$$

Eq. (7.12) and the aforementioned thermodynamic conditions describe the equilibrium thermodynamics of a mixture that allows determination of all its thermodynamic parameters. Thus, for the thermodynamic sound velocity in a mixture we obtain the formula (Rutkevich, 2002):

$$c^2 = \frac{\partial P}{\partial \rho} = RT \frac{1 + \dfrac{L^2 q}{c_p R_W T^2}}{\dfrac{c_v}{c_p} - \dfrac{L_q}{c_p T} + \dfrac{L_q^2}{c_p R_W T^2}}. \tag{7.13}$$

The tropical atmosphere can be considered in this case as a two-component system consisting of dry air and water vapor. The thermodynamic description of a two-component system contains partial pressures, the densities of components, and the temperature of a mixture as its parameters. Dry air and water vapor are considered to be ideal gases; the water vapor saturation state is described by the Clausius–Clapeyron equation. Successive application of the first principles of thermodynamics to this set of thermodynamic parameters allows determination of sound velocity in saturated moist air, which appears to be distinct from the sound velocity in dry air and dependent on the thermodynamic parameters of water vapor. Unlike the case of dry air, the distribution of sound velocity in the vertical coordinate z in a saturated tropical atmosphere can turn out to be non-monotonous.

In its turn, the non-monotonous vertical dependence of sound velocity results in the problem of genesis (i.e., the development of instability) being formulated by a Schrödinger-type stationary equation, which, in the symmetric "potential hole" approximation, can be presented as:

$$\frac{d^2 U(z)}{dz^2} + \left(\frac{A}{ch^2 \alpha z} - E(\Gamma)\right) U(z) = 0 \tag{7.14}$$

where coefficients A and α depend both on the initial thermodynamic parameters of a problem and on the Coriolis parameter; and quantity $E(\Gamma)$ is a function of the instability increment and determines the spectrum of a problem.

The vertical derivative of sound velocity plays the part of the potential hole in this equation. Eq. (7.14) has solutions corresponding, in terms of quantum mechanics, to the levels of this potential hole. Its form and characteristics, as well as the presence and absence of solutions, are determined by the thermodynamic parameters of moist air. In the limiting transition to the case of dry air the potential hole disappears. Thus, the problem of tropical cyclone generation in the given approach can be reduced to the question of the presence or absence of levels in the potential hole caused by the set of thermodynamic parameters experimentally obtained. The instability increment in this simplest case becomes:

$$\Gamma \approx \left(1 - \frac{c_v}{c_p}\right) \frac{g}{c^2} \Omega L \tag{7.15}$$

where c_v and c_p are thermal capacities at constant volume and pressure; g is free-falling acceleration; c is the sound velocity; Ω is the Coriolis parameter; and L is the characteristic horizontal size of the disturbance.

7.5.3 Physical mechanism involved in rotational instability

The physical essence of this approach consists in the fact that the role played by phase transformations of atmospheric moisture is not only reduced to the energy factor, as is often done in cyclogenesis models, but results in the appearance of principal changes in the dynamics of the tropical atmosphere (the compressibility phase and the dynamic phase of a process) caused by the abnormal behavior of the vertical profile of sound velocity in saturated moist air (Figure 7.4). Sound velocity in dry air is determined, as is well known, by the expression $\sqrt{(c_p/c_v)RT}$, where R is the gas constant of dry air, and T is its temperature. According to Rutkevich (2002), sound velocity in moist air, at sufficiently high values of the mixture ratio, turns out to be lower than $\sqrt{RT}$. With lower moisture, sound velocity assumes intermediate

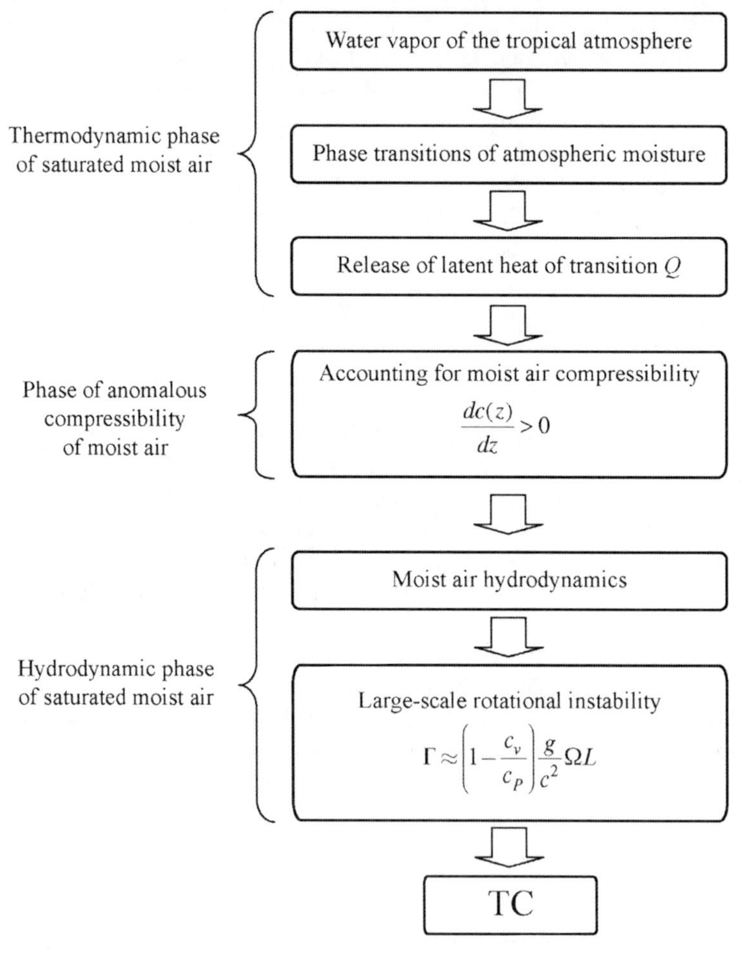

Figure 7.4. Basic phases (elements) of the rotational instability model, caused by anomalous vertical distribution of sound velocity in an atmosphere saturated with water vapor.

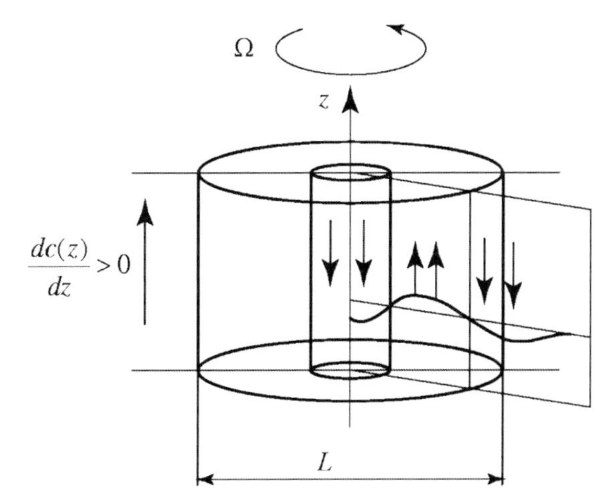

Figure 7.5. Large-scale cell in a layer of the moist tropical atmosphere showing sound velocity growing vertically.

values. Moisture rapidly decreases with altitude in the atmosphere; hence, it appears that sound velocity increases with altitude in the extensive vertical layer of the tropical atmosphere.

Liquid incompressibility approximation is a limiting case of infinite sound velocity where all hydrodynamic processes proceed without any delay. However, when the finiteness of sound velocity is taken into account, all physical processes proceed after a delay and this delay is longer the greater the system size. The large-scale instability mechanism developed in this section can be imagined as follows (Figure 7.5). If sound velocity in some region increases with altitude, then all processes in the upper part of the region will proceed faster with a smaller delay than those in the lower part. Let us imagine the tropical atmosphere region occupied by a tropical depression and let a tropical disturbance arise in this region. In the case of small amplitudes this disturbance can be imagined to be represented by a large-scale cell described by harmonious functions in the horizontal direction. The ascending part of the flow in such a cell will enter less dense layers of the atmosphere and undergo rarefaction, whereas the descending part will undergo compression. The flow in a cell will violate the geostrophic balance in this region, but this will be restored during some characteristic sonic times.

However, in the upper part of the region the geostrophic balance will be restored faster as a result of different sound velocities in the upper and lower layers of the region. In this case rarefaction will proceed faster in the upper part of the disturbance's ascending flux than in the lower part and a drop in pressure (upward-directed thrust) will arise. At the same time, compression in the descending flux of a disturbance cell will proceed faster in the upper part than in the lower part, and once again a drop in pressure (downward-directed thrust) will arise. As a result, the initial disturbance will amplify resulting in an ever greater increase of thrust (Figure 7.5). In this way the positive feedback necessary for instability is established.

If thermodynamic conditions are not fulfilled properly, then instability may not arise and oscillations of a similar nature will take place in the system. These oscillations represent a manifestation of inertial waves in the considered system and, in this sense, point to the formation of the "inertial instability effect" under certain thermodynamic conditions.

The non-linear stage of instability can be described in terms of the amplitude equation for the basic mode $a(t)$:

$$\frac{da(t)}{dt} = \Gamma a(t)\left(1 + N\int_0^t a(t')\,dt'\right) \tag{7.16}$$

with the initial condition $a(0) = a_0$, where N is the non-linearity coefficient that depends on the initial parameters of a problem. Solution of the amplitude Eq. (7.16) is as follows:

$$a(t) = \frac{B^2 e^{Bt}}{N^2\Gamma^2 a_0}\left(\frac{e^{Bt}}{\Gamma + B} - \frac{1}{\Gamma - B}\right)^{-2}, \qquad B = \sqrt{\Gamma^2 - 2a_0 N\Gamma}. \tag{7.17}$$

The explicit expression for function $a(t)$ looks a bit bulky; nevertheless, Eq. (7.17) convincingly describes a function that is equal to a_0 at the initial point and, at the positive value of non-linearity parameter N, monotonously increases up to infinity for finite time. At the negative value of non-linearity the amplitude grows insignificantly, reaches a maximum, and then monotonously decreases down to zero. Positive values of the non-linearity parameter correspond to the descending motion of gas in the central part of the structure under consideration. Non-linear analysis of this problem thus forecasts descending motion of air in the central part of an organized rotating structure. Non-linear analysis further shows that such a structure cannot exist in the case of ascending motion in the central part. This agrees well with the motion of air in a typhoon; such a motion of air in the central part is the very prominent feature called the typhoon's eye.

7.5.4 New requirements for remote-sensing systems

Note that experimental observation of the atmosphere in the presence of these conditions (i.e., inversion of the altitude profile of sound) puts on the agenda requirements for principally new methods of remote sensing of pre-crisis and crisis situations in the Earth atmosphere. This principally concerns remote determination of the spatiotemporal characteristics of the profile of water vapor content and temperature inside cloud systems (both convective and non-convective) on considerable spatial oceanic water areas. Similar investigations can only be performed by means of new-generation passive microwave remote space systems (Sharkov, 2003).

On the basis of the results of surface (ship) radiothermal multifrequency remote investigations of the tropical atmosphere carried out in September–October 1978 in zones in which intensive generation of tropical cyclones was taking place (the northwest Pacific Ocean) (Figure 7.6) and on the basis of known model presentations of the exponential profile, the vertical profile of water vapor is restored along with

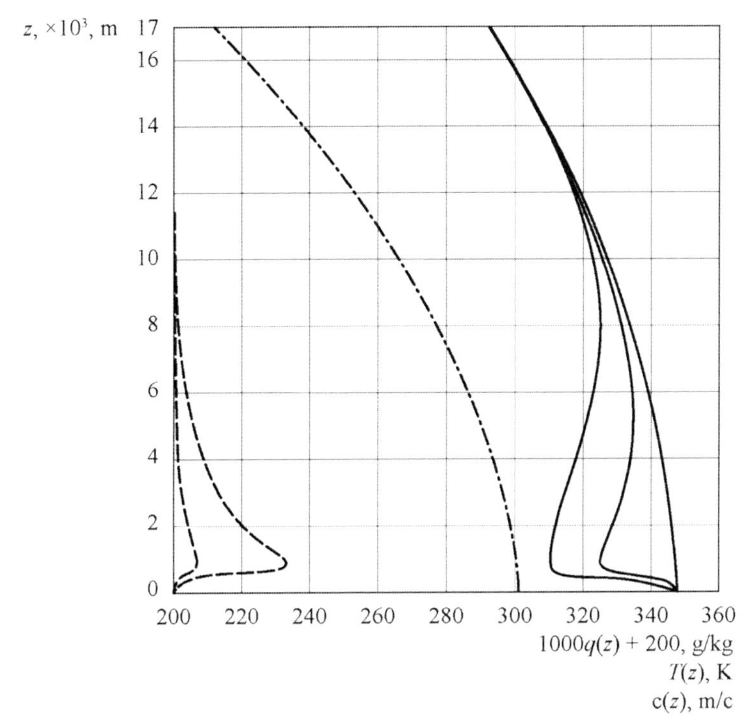

Figure 7.6. Experimental altitude profiles (restored from the data of NOAA series satellites) of the thermodynamic temperature $T(z)$ (K) (dash-dotted line), model profiles of water vapor content (from experimental data) $q(z)$ (g/kg) (dotted line), and calculated profiles of sound velocity in an atmosphere saturated with water vapor $c(z)$ (m/s) (solid line). Digits 1 and 2 mark the profiles related to various values of integral water vapor content (experimental data). Digit 3 marks the sound velocity profile for the dry atmosphere.

subsequent calculation of the altitude distribution of sound velocity by Eq. (7.13). Calculation results are given in Figure 7.6. The figure presents an experimental profile of the thermodynamic temperature of the atmosphere, restored from the data of NOAA series satellites, and two vertical profiles of water vapor concentration, restored by the exponential model for two values of the integral content of water vapor in the atmosphere, obtained from experimental radiothermal observations on the day the tropical depressions formed (Figure 7.6). What is significant here is that the calculated value of the altitude profile of sound velocity, constructed from the experimental data for two studied cases, is non-monotonous (as in the "dry" atmosphere case), and has minima at altitudes of 1 km to 2 km with subsequent fast increase (a positive derivative with respect to altitude) up to values corresponding to a "dry" atmosphere at altitudes of 10 km to 12 km. Thus, these model calculations show that natural conditions surely form under which a considerable part of the atmosphere at altitude is subject to the positive gradient of variations in sound velocity at altitude and, thus, the conditions for the existence of large-scale rotational instability will be met.

On the basis of these model presentations of the altitude profiles of sound velocity, let us now consider the preliminary requirements necessary for measuring space systems to solve these problems. It is known that currently used operative systems of the NOAA mission make it possible to restore 14 altitude levels of the thermodynamic temperature of the atmosphere (though, only under strictly cloudless conditions) and four altitude levels (up to 5 km) of water vapor content (under the same conditions). The accuracy of temperature restoration at each altitude is 1.5% and for vapor content it is between 20% and 30%. Comparing these data with model presentations, it is clear that the restoration of the temperature profile meets (in the first approximation) the requirements for rotational instability problems, but for the measurement of the water vapor content profile the existing operative system in no way meets the necessary requirements. Note that all measurements should be carried out both under cloudless atmospheric conditions and under complex convective cloudiness. The requirements presented here are of course preliminary and require further attentive analysis. However, the microwave space complexes of the CLOUDS and MEGHA-TROPIQUES missions, currently being developed, are likely to satisfy these requirements (though, it has to be said, not fully) (Sharkov, 2003).

This chapter presents a principally new thermohydrodynamic model of large-scale instability in an atmosphere saturated with water vapor, which hopefully will change views about how remote-sensing systems of atmospheric catastrophes should be constructed. On the basis of model presentations, experimental data of microwave complexes, and mindful of the saturation of the total altitude column of the atmosphere with water vapor, the possible existence of a non-monotonous (in contrast to the "dry" atmosphere situation) altitude distribution of sound velocity that has a prominent minimum sound velocity value is shown. This determines the necessary conditions for the generation of vortex structures.

8

Databases of global tropical cyclogenesis

In this chapter we describe up-to-date methods of acquisition, systematization, classification, gathering, and storage of data for global tropical disturbances. Design principals, structure, and performance characteristics of up-to-date datasets and catalogues are also reported.

Note that particular attention has been given to research and development activities directed primarily at the solution of specific problems in the design and use of global tropical cyclogenesis datasets.

8.1 SCIENTIFIC AND APPLIED PHILOSOPHY OF GLOBAL CYCLOGENESIS DATASET DESIGN

When considering the development of modern global cyclogenesis datasets and the application of existing tropical cyclone datasets and archives, both of which play a broad spectrum of roles in radio and TV broadcasts, multimedia applications, and multimedia databases, a number of crucial points should be borne in mind.

One of these crucial points surrounds such defined scientific problems as the interaction between the different degrees of intensity of tropical disturbances and the thermodynamics and kinematics of the tropical atmosphere. This problem is closely associated with global warming and with practical problems of short-term and long-term global forecasts (Kondratyev, 1992; Kondratyev *et al.*, 1995; Hansen *et al.*, 1997; Henderson-Sellers *et al.*, 1998; Lighthill *et al.*, 1994). Systematized remote data on global tropical cyclogenesis can form the experimental base of such studies. What we are talking about here is the physical process considered simultaneously over the whole basin of the World Ocean (rather than on its separate basins). This, of course, does not detract in the very least from the importance of studying the structured features of regional cyclogenesis.

E. A. Sharkov, *Global Tropical Cyclogenesis* (Second Edition).

Another of these crucial points has to do with scientific methodical problems in the classification and identification of different tropical structures, in the exchange of information, and in understanding detailed systematization and acquisition. The difficulties that appear when trying to resolve classification problems are primarily responsible for multiscale processes of dynamic and thermodynamic interaction when simulating the dynamic structure of a tropical cyclone from the helical turbulent chaos. This, in turn, has led to current classification procedures failing to meet fully the requirements for monitoring remote tropical disturbances, giving rise to grave inhomogeneity of initial raw information in databases and archives.

A further crucial point lies on the administrative–economic plane and has to do with broadcasting in the mass media and on the Internet by governmental organizations and commercial companies of reports, fragments of satellite information, and preliminary forecasts of active phases of tropical cyclones with the aim of warning the population of impending danger and advising them to take the necessary precautions. While not wishing to understate the obvious importance of issuing such warnings, it should be noted that fragmentary archiving of this kind is free of any systematic character. This poses major problems when using these information sources for scientific analysis or for building scientific databases.

Despite their fundamental differences, issues like these that crop up when developing scientific databases of tropical disturbances are intimately related. Hence, without detailed analysis of the contribution of each of these crucial points to the general problem of archiving tropical cyclogenesis data it is difficult to understand what is the condition of today's catalogue procedures for global tropical cyclogenesis data.

The whole collection of studies of large-scale tropical disturbances essentially follow two main directions:

— *the local approach*—in which individual tropical disturbances are studied (i.e., the structure, track record, energy, thermo and mass exchange, track features of individual cyclones); and
— *the global approach*—in which the formation and evolution of tropical disturbances as a bound sequence of events are considered on global and regional scales.

The natural physical reason for such a division (which, in a certain sense, is conventional) is the degree of non-equilibrium in the ocean–atmosphere system: it is high for local transformations and weak for global processes. Sharkov (1997, 1998, 2000, 2006; 2010a, b) give a detailed review of the present state of the art of tropical disturbance study.

The fundamental base of the global approach is the study of how structures evolve into tropical disturbances, considered in terms of the model of the stochastic flow of homogeneous events on regional and global scales. The investigation of these problems is an important aspect in the study of the ocean–atmosphere system regarding the contribution of intensive vortex disturbances to the thermodynamics and kinematics of the tropical atmosphere on different timescales. This problem is closely allied to the task of studying the influence of possible climate change on global cyclogenesis (Henderson-Sellers *et al.*, 1998; Lighthill *et al.*, 1994; Bengtsson *et al.*, 1996), to problems of the global ocean–atmosphere system (Hansen *et al.*, 1997), and to the action of large-scale atmospheric circulations and ENSO phenomenon on cyclogenesis (Revell and Goulter, 1986; Elsner and Kara, 1999; Pielke and Landsea, 1999; Sharkov, 2000).

Pioneering purposeful studies of global tropical cyclogenesis as a stochastic flow of events have revealed its combined hierarchical structure ranging from the lifetime scale of a single structure to climatic scales (Astafyeva *et al.*, 1994a, b) and have facilitated the formation of preliminary statistical quantitative models (Pokrovskaya and Sharkov, 1993a, 1994a, b).

As for cyclogenesis considered with respect to the primary stages of a TC (tropical disturbances and tropical depressions), the problem of finding a correlation between primary and mature stages of TC was formulated in Khromov (1948, 1966) and then Minina (1970, 1982). For a variety of reasons (both organizational and methodical) this problem has still not been properly resolved for all timescales.

Close analysis of the geoinformation banks and databases (and their modifications) that are available on large-scale tropical disturbances—see, for instance, the catalogue of climatic databases (WMO, 1992), bulletins of the World Meteorological Organization (WMO, 1995b), broadly quoted reviews (Neumann *et al.*, 1993; Neumann, 1993), and web catalogues (Landsea, 1998)—points to the fact that databanks on TCs are primarily compiled regionally with the aim of resolving the purely administrative–economic problems of a given region or problems of a purely commercial nature. In accordance with international agreements the observational services of nations participating in monitoring TCs are acting consistently with the principle of taking local responsibility for its own observing areas in the World Ocean (Pielke and Pielke, 1997). This is of course fully understandable since the purposes and tasks of regional centers are directed at forecasting the intensity and trajectory of tropical disturbances and passing this information on to the relevant state emergency services. Having said this, recording the initial stages of TCs in existing observational systems is hit and miss (despite initial stages being observed) and, accordingly, have not been acquired in databases. Thus, raw data information is completely lost to researchers studying the different stages of a TC. Moreover, the loss of raw and sometimes very valuable information under transition of the observed object from one area of responsibility to another is something that needs to be addressed.

Consequently, in the experimental study of the time evolution of tropical cyclogenesis there is currently a situation in which the statistics of early stages of TCs and final stages of mature cyclones are highly fragmentary and hinder presenting the full pattern of cyclogenesis of primary and mature stages. A similar situation applies to the study of endpoint (time evolution scale) forms of tropical cyclones (so-called extratropical cyclones), as purposeful remote observations and monitoring of such conditions of tropical forms across the board have not been conducted (excluding individual separate events, of course). Accordingly, information has not been systematized or acquired.

Scientific studies on possible variations of global climate and on the way in which large-scale atmospheric catastrophes evolve require a totally different approach that works along the following lines: strict systematization in pinpointing the moment of formation and in fixing and tracking objects of tropical cyclogenesis at all stages of their evolution (particularly, at their initial stage and when they dissipate or transform). For this to work, observations must be carried out simultaneously on all the basins of the World Ocean and conducted as a united physical process. The results of such a methodological approach will be demonstrated in what follows.

8.2 HISTORICAL PERSPECTIVES

The earliest history of visual and instrumental observations of tropical cyclones dating to the time of Columbus' explorations in the Atlantic can be found in *Early American Hurricanes (1492–1870)* by D. Ludlum (1989). Ludlum's book sets down in chronological order, as far as available historical sources permit, the meteorological situations attending the occurrence of hurricanes prior to 1870 that have either closely approached or actually crossed the Atlantic and Gulf coastlines of the present U.S.A. The publication gives a rough idea of earlier tropical cyclogenesis, but should not be used in scientific investigation.

Elsner and Kara (1999) examine the sources and authenticity of historical hurricane data for the North Atlantic basin.

Between 1871 and 1963, the primary reference for hurricane tracks and associated intensity criteria was the U.S. Weather Bureau (Neumann *et al.*, 1993). Although the main purpose of revisions to Neumann *et al.*'s work was to strictly update track charts, some of the original tracks were modified based on additional environmental data that were available to scientific organizations. Nonetheless, for many years following the establishment of U.S. Government Weather Service in 1870, observational data on precise determination of the location and intensity of individual tropical cyclones were scarce, widely scattered, generally poor quality, and sometimes conflicting.

Technological advances (improved radiosonde equipment, storm-tracking radar, aircraft reconnaissance) since World War II has resulted in more precise tropical cyclone detection, positioning, and intensity determination. However, all of the preceding was specific to just one active basin: the North Atlantic.

The radically new product as a result of the space program was the development of weather satellites, now the standard observational tool for the detection and monitoring of tropical cyclones worldwide. At long last, the possibility of fullscale study of global tropical cyclogenesis study arose.

It is necessary to stress here that those space systems exhibiting the most promise for detecting and monitoring tropical cyclones are geostationary satellites with visible and IR instruments. The reason is that the probability of storm track and sub-orbiting points of a given polar orbiting satellite intersecting is small. Satellites of this type (e.g., NOAA, DMSP, TRMM) are of limited utility for detailed investigation of both individual tropical cyclones and for global tropical cyclogenesis as a whole (see Chapter 6).

Special systematic procedures to estimate the characteristics of storms have been developed and improved, as have the systems for viewing, processing, and analyzing satellite data. At present, a widespread network of land stations, ships, aircraft-based radar, satellites, and data buoys, using complex and sophisticated instrumentation and communication, is available for the detection, tracking, and understanding of tropical cyclones.

Recent years have witnessed an amazing accumulation of raw experimental data on tropical cyclones. The highly speculative nature of the relationship between the execution of high-cost field programs such as BOMEX, GATE, TOGA, etc. (short-term observations) and the use of routinely collected meteorological data (long-term observations) that have been obtained over many years and even now have not been fully processed needs to be pointed out. Opinion differs as to the contribution of each component (short term and long term) to tropical meteorology study (Gray, 1997; King, 1999). The opinion of Prof. W. Gray—deemed by many as the patriarch of tropical cyclone study—is both interesting and instructive:

> "A classic case of a lost research opportunity in past years is seen in the general lack of research on the Guam based military aircraft reconnaissance flights into West Pacific typhoons between 1946–1986. Only a very small fraction of this unique flight data has ever been processed and analyzed. Yet, this flight data contain enormous amounts of information pertaining to tropical cyclone structure, formation, intensity change, etc. that has yet to be sorted out. There has never been a strong enough research commitment to tap this marvelous source of already collected tropical cyclone information which would have cost many orders of magnitude less than other less potentially beneficial research."

Broad historical outlines of the accumulation of knowledge of tropical cyclogenesis as a result of relevant technological advances are summarized pictorially in Figure 8.1. Close inspection of Figure 8.1 shows that solid observational data for global tropical cyclogenesis could only have been available from the early 1980s.

However, there remained a number of problems including setting up a means of reliable global data communication exchange and finding data assimilation techniques. Finally, widespread use of the Internet provided a firm basis for exploratory and advanced development of global tropical cyclogenesis datasets. This led to the creation of a series of prototypes (versions) of global tropical cyclogenesis datasets in the early 1990s (see below).

Note that the pressing trend in current global change studies has been toward historical reconstruction of the location and intensity of hurricanes on the basis of uncovering and interpreting written accounts of tropical cyclones over the North Atlantic from ships' logs, newspapers, chronological lists, and other sources (Ludlum, 1989; Neumann *et al.*, 1993; Fernandez-Partages and Diaz, 1996; Elsner and Kara, 1999). Although such a method is used to reconstruct hurricane large-scale structure (year averaged and monthly averaged) of time series in just one basin

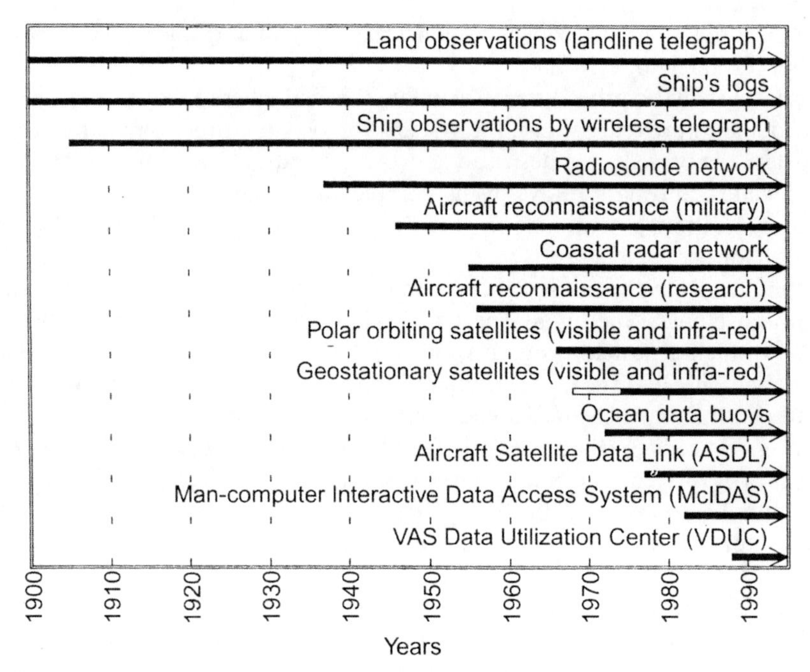

Figure 8.1. Technical advances in systems for observing TCs, 1900 through 1992 (from Neumann *et al.*, 1993).

(the North Atlantic), it is not practical to reconstruct the detailed characteristics (day averaged) of tropical cyclogenesis globally from evidence derived from historical records that are so fragmentary.

8.3 EXISTING ARCHIVING PROCEDURES AND CURRENT ARCHIVES

The variety of problems and tasks facing governmental agencies, the scientific community, and commercial organizations when designing tropical cyclone archives resulted in a multiplicity of lines being pursued in the development of archives and datasets carrying data and information on tropical cyclogenesis. Such a compound hierarchy of archiving procedures was primarily determined by significant variety in the spatiotemporal details for the representation of data products. In this section, existing archiving procedures and current global tropical cyclogenesis archives are examined. The general scheme of existing archiving procedures for tropical cyclogenesis data is shown in Figure 8.2.

8.3.1 Areas of responsibility

The primary sources of initial data on global tropical cyclogenesis are the state centers of the areas of responsibility and the respective warning agencies

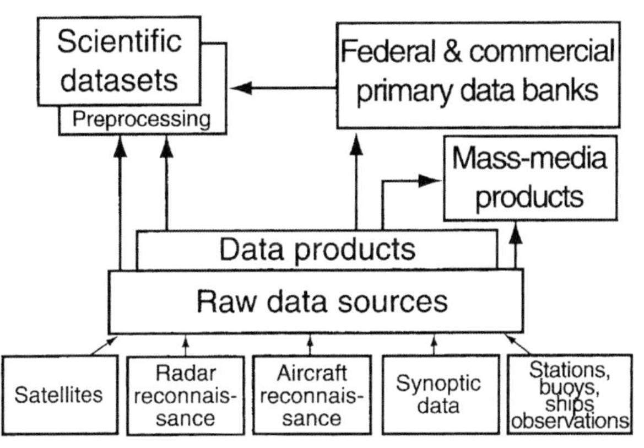

Figure 8.2. Existing archiving procedures for tropical cyclogenesis data.

(Neumann, 1993): Joint Typhoon Warning Center (U.S.A.), National Hurricane Center (U.S.A.), Bureau of Meteorology (Australia), Japanese Meteorological Agency (Japan), Fiji Meteorological Service (Fiji), Philippine Atmospheric, Geophysical, & Astronomical Service Administration (Philippine), Meteorological Service (Réunion), and the India Meteorological Department (India).

In accordance with international agreements and WMO operational plans the areas and centers of responsibility in the World Ocean are distributed in the following manner (Pielke and Pielke, 1997):

- *Regional Association I* (WMO, 1983)—high seas tropical cyclone warning responsibility for the southwest Indian Ocean incorporating the area for Madagascar (Antananarivo) and Mauritius (Vacoas); the area for Réunion (Sainte Clotilde); the area for Mozambique (Maputo); and the area for Kenya (Nairobi).
- *Regional Association IV* (WMO, 1988)—U.S. National Hurricane Center (NHC), Miami, Florida area of high tropical cyclone warning responsibility. This includes (1) the North Atlantic, the Caribbean Sea, and the Gulf of Mexico and (2) the eastern North Pacific east of longitude 140°W. The area of responsibility for the U.S. Central Pacific Hurricane Center (CPHC) located at Honolulu, Hawaii includes the eastern North Pacific west of longitude 140°W to 180°W.
- *U.S. Joint Typhoon Warning Center (JTWC)*,[1] Guam/Pearl Harbor—areas of high seas tropical cyclone warning responsibility west of longitude 180° to the east coast of Africa.
- *Regional Association V* (WMO, 1989)—high seas tropical cyclone warning responsibility for the South Pacific and the southeast Indian Ocean region includes (1) areas for Australia (Perth, Darwin, and Brisbane), Papua New

[1] JTWC initially operated from Guam. It was relocated to Pearl Harbor, Hawaii, in January 1999.

Guinea (Port Moresby), and Indonesia (Jakarta); (2) areas for Fiji (Nadi) and New Zealand (Wellington).

— *WMO/ESCAP Panel on Tropical Cyclones* (WMO, 1986)—high seas tropical cyclone warning responsibility for the Bay of Bengal, the Arabian Sea, and the vicinity includes (1) areas for India (Calcutta, Bay of Bengal) and Bombay (Arabian Sea); (2) areas for Pakistan (Karachi), Bangladesh (Dhaka), Burma (Rangoon), and Sri Lanka (Columbo).
— *ESCAP/Typhoon Committee* (WMO, 1987)—high seas tropical cyclone warning responsibility for the western North Pacific includes (1) areas for China (Dalian, Shanghai, and Guangzhou), for South Korea (Seoul), Vietnam (Hanoi), and Japan (Tokyo) (all relating to the South China Sea and the Yellow Sea); (2) areas for the Philippines (Manila, PAGASA) and Hong Kong (Royal Observatory) include the North Pacific west of longitude 135°E to 115°E and north of latitude 5°N to 25°N.

For a variety of socio-economical and natural reasons, the contribution of each individual association to development of the global tropical cyclogenesis archive varies widely. The JTWC and NHC play key parts in data acquisition processes and in the worldwide transmission of data products. Section 8.3.1 deals with the development and structure of data products provided by the JTWC and NHC in their area of responsibility.

8.3.2 JTWC services and products

The Joint Typhoon Warning Center provides a variety of standard products and services to organizations within its area of responsibility (AOR) as prescribed by special instructions. The JTWC issues the following products:

— *Significant tropical weather advisory*—issued daily, or more frequently as needed, to describe all tropical disturbances and their potential for further development during the advisory period. Separate bulletins are issued for the western Pacific and the Indian Ocean.
— *Tropical cyclone formation alert*—defines the specific area where synoptic, satellite, or other germane data indicate that the development of a significant tropical cyclone (TC) is likely within 24 hours.
— *Tropical cyclone/tropical depression warning*—issued periodically throughout each day to provide forecasts of the position, intensity, and wind distribution of TCs in the JTWC's AOR. The tropical depression warning was dropped in 1998 as a separate product. Post-1997 tropical depressions in the western North Pacific receive regular tropical cyclone warnings.
— *Prognostic reasoning message*—issued in conjunction with warnings for tropical cyclones that have the potential to reach tropical storm or typhoon strength in the western North Pacific. This discusses the rationale for the content of the specific JTWC warning.

The contents and availability of the above JTWC products are set forth in special instructions. Changes to them as well as JTWC products and services are proposed and discussed at the annual U.S. Pacific Command (PACOM) Tropical Cyclone Conference.

The JTWC depends primarily on two reconnaissance platforms, satellite and radar, to provide accurate and timely meteorological information in support of advisories, alerts, and warnings. When available, synoptic and aircraft reconnaissance data are also used to supplement the above. As in past years, optimal use of all available reconnaissance resources to support the JTWC's products remains a primary concern. Weighing the specific capabilities and limitations of each reconnaissance platform against the threat posed by tropical cyclones to life and property both afloat and ashore continues to be an important factor in careful product bpreparation.

Meteorological satellite imagery recorded at USAF/USN group sites and onboard U.S. naval vessels supply day and night coverage in the JTWC's AOR. Interpretation of these satellite data provides TC positions and estimates of current and forecast intensities (Dvorak, 1984). The USAF's tactical satellite sites and the Air Force Weather Agency (AFWA) currently receive and analyze special sensor microwave/imager (SSM/I) data to provide TC locations and estimates of 35 kt (18 m/s) wind radii when the low-level center is obscured by higher clouds.

The Defense Meteorological Satellite Program (DMSP), the National Oceanographic and Atmospheric Administration (NOAA), the Japanese geostationary meteorological satellite (GMS), and the European geostationary meteorological satellite (Meteosat) provide the foundation for reconnaissance.

In addition to imagery, scatterometer data from the European remote-sensing ERS-2 satellite provide valuable insight into the distribution of low-level winds around TCs. When remotely sensed data of this quality became available, the JTWC immediately began using it to supplement other available data. The evolution of algorithms and subsequent display of scatterometer data has occurred rapidly over the past few years and the JTWC has been fortunate to have access to this leading edge technology.

The JTWC retrieves scatterometer data on a routine basis from websites on the NIPRNET/Internet maintained by FNMOC, the Naval Oceanographic Office (NAVOCEANO), and the Ocean Sciences Branch of the NOAA. The scatterometer data available at these sites help to define the TC position and low-level structure. Heavy rain contamination near a TC's center limits the usefulness of intensity estimation to tropical storm strength and below. The JTWC also uses scatterometer data to refine the twice daily manual analyses of surface/gradient-level wind flow and atmospheric structure.

The Pacific Air Force (PACAF) has primary responsibility for providing tropical cyclone reconnaissance for the U.S. Pacific Command (PACOM). The JTWC is in charge of assigning all reconnaissance requirements. Operational control of radar and satellite readout sites engaged in tropical cyclone reconnaissance remains with normal command channels. However, the Guam DMSP site is delegated the authority to manage the Pacific DMSP Tropical

Cyclone Reconnaissance Network (hereafter referred to as the "network") in support of the JTWC. Network control staff and the personnel of satellite operations (SATOPS) are members of the 36 OSS/OSJ, and are co-located with the JTWC at Nimitz Hill, Guam. The PACOM satellite reconnaissance network consists of units based in Hawaii, Japan, Guam, South Korea, Nebraska, and Diego García.

Direct readout network sites provide coverage of the tropical western North Pacific, the South China Sea, and the south central Indian Ocean using DMSP and NOAA TIROS polar-orbiting satellites. PACAF instruction requires each network site to perform a minimum of two fixes per tropical cyclone per day if a tropical cyclone is within the site's coverage. Network direct readout site coverage is augmented by other sources of satellite-based reconnaissance. The Air Force Weather Agency (AFWA) provides AOR-wide coverage to the JTWC using recorder-smooth DMSP and NOAA TIROS imagery. This imagery is recorded and stored on the satellites for later relay to a command readout site, which in turn passes the data via a communications satellite to AFWA. Civilian contract weather support for the U.S. Army at Kwajalein Atoll provides additional polar-orbiting satellite-based tropical cyclone reconnaissance in the Marshall Islands and east of 180°W as needed. The NOAA/NESDIS Satellite Applications Branch at Suitland, Maryland also performs tropical cyclone fix and intensity analysis over the JTWC AOR using Meteosat and GMS geostationary platforms.

The network provides tropical cyclone positions and intensity estimates once the JTWC issues either a tropical cyclone formation alert (TCFA) or a warning using the Dvorak code. Each satellite-derived tropical cyclone position is assigned a position code number (PCN), which is a statistical estimate of fix position accuracy. The PCN is determined by (1) the availability of visible landmarks in the image that can be used as references for precise gridding and (2) the degree of organization of the tropical cyclone's cloud system.

The interpretation of land-based radar, which remotely senses and maps precipitation within tropical cyclones, provides positions in the proximity—usually within 175 nm (325 km)—of radar sites in Kwajalein (Guam), Japan, South Korea, China, Taiwan, the Philippines, Hong Kong, Thailand, and Australia. Where Doppler radars are located—such as the weather surveillance radar-1988 Doppler (WSR-88D) on Guam, in Okinawa, and in South Korea—measurements of radial velocity are also available and observations of the horizontal velocity field and vertical wind structure of TCs are possible.

Land-based radar observations are used to position TCs. Once a well-defined TC moves within range of land-based radar sites, radar reports are invaluable for determination of position, movement, and, in the case of Doppler radar, storm structure and wind information.

The JTWC also determines tropical cyclone positions based on the analysis of conventional surface/gradient-level synoptic data. These positions are an important supplement to fixes derived from remote-sensing platforms and become most valuable in situations where satellite, radar, and aircraft fixes are unavailable or are considered unrepresentative. These datasets are comprised of land and shipboard surface observations, en route meteorological observations from commercial military aircraft (AIREPS) recorded within six hours of synoptic times, and cloud motion winds derived from satellite data. These conventional data are computer plotted and manually analyzed in the tropics for the surface/gradient and 200 mb levels. These analyses are prepared twice daily using 00Z and 12Z synoptic data.

In 1989, the Commander, Naval Meteorology and Oceanography Command (COMNAVMETOCCOM) put the Integrated Drifting Buoy Plan into action to meet USCINCPACFLT requirements that included TC warning support. In 1997, 30 drifting buoys were deployed in the western North Pacific by a NAVOCEANO-contracted C130 aircraft. Of the 30 buoys, 24 were compact meteorological and oceanographic drifters (CMODs) with temperature and pressure sensors and six were windspeed and direction drifters (WSD) with windspeed, wind direction, temperature, and pressure sensors. The buoys were evenly split by type over two deployments: the first in June and the second in September. The purpose of the split deployment was to overlap the expected 3-month lifespan of the CMOD buoys in order to provide continuous coverage during the peak of the western North Pacific TC season.

Through a cooperative effort between COMNAVMETOCCOM, the Department of the Interior, and the NOAA/NWS to increase data availability for tropical analysis and forecasting, a network of 20 automated meteorological observing stations (AMOSs) is being installed in the Micronesian Islands. Since September of 1991, in most of the sites, the capability to transmit data via service ARGOS and NOAA polar-orbiting satellites has been available as a backup to regular data transmission to the geostationary operational environmental satellite (GOES West) and more recently, for sites to the west of Guam, to the GMS. Upgrades to existing sites are being accomplished as opportunities arise. The JTWC receives data from the AMOS sites via the automated weather network (AWN).

Until the summer of 1987, dedicated aircraft reconnaissance was used routinely to locate and determine the wind structure of TCs. Now, aircraft fixes are only rarely available from transiting jet aircraft or from weather reconnaissance aircraft involved in research missions. No aircraft fixes were available in 1997.

Numerical and statistical guidance is available from the USN Fleet Numerical Meteorology and Oceanography Center (FLENUMETOCCEN or FNMOC) at Monterey, California. The FNMOC supplies the JTWC with analyses and prognoses from the Navy Operational Global Atmospheric Prediction System (NOGAPS) via the NIPRNET packet-switched network (Internet gateway). NOGAPS products that are routinely disseminated to the JTWC include surface pressure and winds, upper-air winds, deep-layer mean winds, geopotential height and height change, and sea surface temperature. Moreover, additional various atmospheric components at all standard levels are available. These products are valid for the 00Z and 12Z synoptic times. Selected products from the (U.S.) National Center for Environmental Prediction (NCEP), the European Centre for Medium-Range Weather Forecasts (ECMWF), and the Japanese Meteorological Agency (JMA) are received as electronic files via networked computers and by computer modem

Table 8.1. Regional TC archives.

Area of responsibility	*Products*	*Years*	*Format*	*Website*
Atlantic	Tracks	1886–1996	ASCII	*www.nhc.noaa.gov*
	Tracks	1921–1996	GIF	
	Reports	1995–1996	Text	
	Atlas	1950–1996	HTML	*www.aoml.noaa.gov.hrd*
	Tracks	1886–1999	ASCII, GIF	*www.unisys.com/hurricane*
Northeast Pacific	Tracks	1949–1996	ASCII, GIF	*www.nhc.noaa.gov*
	Reports	1995–1999	Text	
	Tracks	1949–1999	ASCII, GIF	*www.unisys.com/hurricane*
Northwest Pacific	Reports	1959–1998	Text, GIF	*www.npmoc.navy.mil*
	Best tracks	1959–1998	FTP	
	Tracks	1945–1999	ASCII, GIF	*www.unisys.com/hurricane*
South Pacific	Reports Best tracks	1984–1999	Text, GIF FTP	*www.npmoc.navy.mil*
South Indian Ocean	Reports Best tracks	1984–1999	Text, GIF FTP	*www.npmoc.navy.mil*
North Indian Ocean	Reports	1959–1998	Text	*www.npmoc.navy.mil*
	Best tracks	1959–1998	FTP	

connections on government and commercial telephone lines as a backup method for the network.

Detailed description of JTWC products and Annual Tropical Cyclone Reports 1959–1998 (in Adobe Acrobat format) are available at *http://www.npmoc.navy.mil/jtwc.html.* JTWC Annual Tropic Cyclone Reports contain comprehensive information on each storm, including highlights, track and intensity, discussion, damages, and post-analysis best track and verification (Table 8.1).

8.3.3 NHC services and products

The National Hurricane Center (NHC) provides a variety of operational products and services to organizations and U.S. mass media within its area of responsibility. The NHC issues the following products:

— *Tropical Cyclone Public Advisory*
— *Tropical Cyclone Forecast/Advisory*
— *Tropical Cyclone Discussion*
— *Tropical Cyclone Strike Probabilities*
— *Tropical Cyclone Graphics*
— *Tropical Cyclone Update*
— *Tropical Cyclone Position Estimates*

— *Tropical Weather Outlook*
— *Special Tropical Disturbance Statement*
— *Monthly Tropical Weather Summary*
— *Reconnaissance and Dropsonde Observations*
— *Vortex Data Messages.*

The *Tropical Cyclone Public Advisory* contains a list of all current watches and warnings on tropical or subtropical cyclones. It also gives the cyclone position in latitude and longitude coordinates and distance from a selected land point or island, as well as the current motion. The advisory includes the maximum sustained winds in miles per hour and the estimated or measured minimum central pressure in millibars and inches. The advisory may also include information on potential storm tides, rainfall, or tornadoes associated with the cyclone, as well as any pertinent weather observations.

Public advisories are issued for all Atlantic tropical or subtropical cyclones, and for Eastern Pacific tropical or subtropical cyclones that are threatening land. Public advisories are normally issued every six hours. They may be issued every two or three hours when coastal watches or warnings are in effect. Special public advisories may be issued at any time due to significant changes in warnings or in the cyclone.

The *Tropical Cyclone Forecast/Advisory* contains a list of all current watches and warnings on tropical or subtropical cyclones, as well as the current latitude and longitude coordinates, intensity, and system motion. The advisory contains forecasts of the cyclone positions, intensities, and wind fields for 12, 24, 36, 48, and 72 hours from the current synoptic time. The advisory may also include information on any pertinent storm tides associated with the cyclone. All windspeeds in the forecast advisory are given in knots (nautical miles per hour). Forecast/Advisories are issued on all Atlantic and Eastern Pacific tropical and subtropical cyclones every six hours. Special Forecast/Advisories may be issued at any time due to significant changes in warnings or in the cyclone.

The *Tropical Cyclone Discussion* explains the reasoning for the analysis and forecast of tropical or subtropical cyclones. It includes a table of the forecast track and intensity. Tropical Cyclone Discussions are issued on all Atlantic and eastern Pacific tropical and subtropical cyclones every six hours. Special tropical cyclone discussions may be issued at any time due to significant changes in warnings or in the cyclone.

Tropical Cyclone Strike Probabilities give the percentage chance of tropical or subtropical cyclones passing within 75 nm to the right or within 50 nm to the left of a specified point, looking in the direction of cyclone motion. Probabilities are given for time periods of 0–24, 24–36, 36–48, and 48–72 hours, with 0–72 hours given by adding the individual probabilities together. Tropical Cyclone Strike Probabilities are normally issued for Atlantic tropical or subtropical cyclones every six hours when the system is forecast to be within 72 hours of landfall. Special Strike Probabilities may be issued if Special Public/Forecast Advisories are issued. The NHC began creating graphics products based on its Atlantic and East Pacific advisories.

Tropical Cyclone Updates are brief statements issued in lieu of or preceding special advisories to inform of significant changes in a tropical cyclone or to post or cancel watches or warnings.

Tropical Cyclone Position Estimates are issued between 2-hourly intermediate advisories whenever a tropical cyclone with a well-defined radar center is within 200 nautical miles of land-based radar in the United States. These estimates give the center location in map coordinates and distance and direction from a well-known point.

The *Tropical Weather Outlook* is a discussion of significant areas of disturbed weather and their potential for development out to 48 hours. It includes (when possible) a non-technical explanation of the meteorology behind the outlook. Tropical Weather Outlook also includes brief descriptions of any tropical or subtropical cyclones in the region. It also includes the WMO and AFOS headers of where to find more information on an active cyclone during the first 24 hours of existence. Tropical Weather Outlook is issued four times a day during the hurricane season.

The *Special Tropical Disturbance Statement* is used to furnish information on strong formative non-depression systems. These are usually issued for systems strong enough to produce heavy rain and strong winds that do not yet meet the criteria of tropical or subtropical cyclones. These products are transmitted only as needed.

Monthly Tropical Weather Summary is issued on the first of every month during the hurricane season. It describes the previous month's tropical cyclone activity and gives details on cyclones known at that time. The last Tropical Weather Summary of the season gives a brief account of the whole season.

Reconnaissance Observations are coded reports detailing the pressure, wind, temperature, and dewpoint at roughly 30-minute intervals along a reconnaissance aircraft's flight track. Reconnaissance Observations are the primary observations sent during the initial investigation of a potential tropical cyclone and during practice missions. They are also transmitted during tropical cyclone missions: (1) during the trip to and from the cyclone, and (2) at turning points during the flight through the cyclone. Dropsondes are instrument packages ejected from reconnaissance aircraft to take the vertical profiles of pressure, wind, temperature, and dewpoint.

Vortex Data Messages are coded reports issued whenever a reconnaissance aircraft penetrates the center of tropical or subtropical cyclones. They give the position of the center as well as the time of the fix. They include information on winds, temperatures, pressure, and dewpoints encountered during penetration. They also include information on eye size, shape, and status if an eye is present.

Detailed description of NHC products are available at *http://www.nhc.noaa.gov* (Table 8.1). This server maintains a current database of meteorological data, historical data, and written information generated by the NWS or received from other official sources. In addition, this server accesses in real time a selection of current official weather observations, forecasts, and warnings from U.S. government sources for use by the national and international community. It is important to note that in an effort to enhance the science, experimental products may be accessible on this server and care must be taken when using such products as they are intended for research use.

The climatological unit of the server contains links to images (in GIF format) of Atlantic tropical storm and hurricane tracks for 1921–1998 to past hurricane track files (PHTF).

The Atlantic Tracks File is an ASCII file containing the 6-hourly (00:00, 06:00, 12:00, 18:00 UTC) center locations (latitude and longitude in tenths of degrees) and intensities (maximum 1-minute surface windspeed in knots and minimum central pressures in millibars) for all tropical storms and hurricanes from 1886 through 1996.

The East Pacific Tracks File is also an ASCII file containing the 6-hourly (00:00, 06:00, 12:00, 18:00 UTC) center locations (latitude and longitude in tenths of degrees) and intensities (maximum 1-minute surface windspeed in knots and minimum central pressures in millibars) for all tropical storms and hurricanes from 1949 through 1996.

The historical unit of the server contains an important section called "Preliminary Reports" of Atlantic and Eastern Pacific Storms for 1995–1999. The NHC's Preliminary Reports contain comprehensive information on each storm, including synoptic history, meteorological statistics, casualties and damages, and the post-analysis best track (6-hourly positions and intensities).

The NHC publishes three annual summaries of tropical cyclone activity in its area of responsibility. These include:

— *The Atlantic Hurricane Season*
— *The Eastern North Pacific Hurricane Season*, and
— *Atlantic Tropical Systems*.

Hurricane Season summaries include descriptions of named storms along with pertinent meteorological data and satellite imagery. The Atlantic Tropical Systems article gives a description of tropical disturbances in the Atlantic (especially tropical waves) and provides information on non-developing cyclones.

Annual summaries can be found in the following periodicals:

— *Mariner's Weather Log*
— *Monthly Weather Review*
— *Weatherwise* (Clark, 1983; DeAngelis, 1984; Gunther and Cross, 1985; Mayfield and Rappaport, 1992, Mayfield and Avila, 1994; Williams, 1992; Lawrence, 1999; Pasch and Avila, 1999; Lander *et al.* 1999).

However, note that there are many inhomogeneities in these summaries. This is especially true regarding early publications (before 1983).

8.3.4 Global tropical cyclogenesis archives

The main steps in preparing a global tropical cyclogenesis climatology were to develop a global dataset in a common format and to establish reliable and consistent global data communication exchange.

Preparation of a complete global climatology has only been possible since the satellite era began in the mid-1960s. Before then, documentation on tropical cyclones in remote areas of the globe was very fragmented and mostly depended on chance encounters with ships or populated land areas (Neumann, 1993; Elsner and Kara, 1999). Nevertheless, approaches to the unification of fragmented and scarce data obtained from various sources for developing the global pattern are of great value (GTECCA, 1996).

Even with satellite data there were problems including a standardized method for interpretation, establishing suitable documentation procedures and formats, and dissemination of the data—all of which had to be solved. The fact of the matter was that the datasets obtained from the various sources used widely different computerized or hardcopy formats. These format dissimilarities could be circumvented, but it was more difficult to resolve such inhomogeneities as widely different periods of record, missing data, different documentation of the same tropical cyclone in adjacent basins, synoptic observation times, different wind thresholds for stages of tropical cyclones, different practices for naming tropical cyclones, and different wind-averaging times in the various basins (Neumann, 1993).

This lack of a standard for archiving tropical cyclone information was considered by a TCP Technical Co-ordination Meeting (Tokyo, December 1992), which prepared a global tropical cyclone dataset report format, subsequently recommended by the CMM and then approved by the WMO Executive Council in 1993. The standard format is given in Neumann (1993). Nevertheless, a number of problems including standardized methods for interpretation and assimilation of communicated raw data remained to be solved—in particular, global data communication exchange.

Beyond all question, the fundamental step in developing global tropical cyclogenesis datasets was widespread integration of the Internet into communication procedures.

At present the website of Hawaii Solar Astronomy (HSA) (Metcalf, 1996) plays a key part in the global acquisition processes of raw data and in the worldwide transmission of data products including JTWC and NHC data products (Table 8.2).

HSA issues the following products:

- *Current Tropical Advisories*—a list of all current records, watches, and warnings on each tropical cyclone, its pre-hurricane stages and post-hurricane stages (see Section 8.3.2 for details);
- *Storm Track Maps*—in GIF and PS (PostScript) formats;
- *Listings*—a list of quantitatively current records on each tropical cyclone worldwide including year, day, and month of observation, time of observation, name of storm, geographical coordinates, speed and course of storm, estimated central pressure, maximum sustained wind, type of storm, and type of observation.

The historical unit of the server contains Listings for 1994–1999 and Storm Track Archives (in GIF format) for 1996–1999.

Table 8.2. Global tropical cyclogenesis archives and current raw datasets.

Title	*Products*	*Years*	*Format*	*Website*
Tropical Storms Worldwide	Advisories Track maps Listings Storm-track archives	Current 1994–1999 1996–1999	Text, GIF Text GIF	*www.solar.ifa.hawaii.edu/tropical*
Global Tropical and Extratropical Cyclone Climatic Atlas 2.0 (GTECCA)	Storm track cumulative statistics Narratives	NAT 1886–1995 NEP 1965–1995 NWP 1951–1995 SWP 1956–1995 NI 1877–1995 SWI 1933–1995 1980–1995	DOS	CD-ROM *www.ncdc.noaa.gov*
Tropical Cyclones	Advisories Outlooks Winds Images Winds Images	Real time data 1995–1999	Text, GIF	*www.cims.ssec.wisc.edu/tropical*
SUPER Typhoon Worldwide Tropical Cyclone Reconnaissance	Advisories Images	Real time data	Text, GIF	*www.supertyphoon.com*
UKMO Monthly Tropical Bulletin	Monthly summary Catalogue Forecasts	Monthly	Text Booklet form	*www.met-office.gov.uk*

The important base component for development of the advanced global tropical cyclogenesis dataset that can serve as the historical archive is the CD-ROM entitled *Global Tropical and Extratropical Cyclone Climatic Atlas 2.0* produced by the NCDC/NOAA (GTECCA, 1996). This single-volume CD-ROM contains all global historical storm track data available for five tropical storm basins (Table 8.2). The length of time covered by the records vary for each basin from as early as the 1870s and with 1995 as the current year. Northern hemisphere extratropical storm track data are included from 1965 to 1995. Tropical track data include time, position, storm stage (and maximum wind central pressure when available). The user can display tracks and track data for any basin or user-selected geographic area. The user is also able to select storm tracks passing within a user-defined radius of any point. Narratives for all tropical storm for the 1980–1995 period are included along with basin-wide tropical storm climatological statistics (cumulative, monthly averaged, and year averaged).

The monthly averaged global cyclone archive is also available via printed booklets and World Wide Web (WWW) pages of the U.K. Meteorological Office (Table 8.2). U.K. Meteorological Office pages on the WWW contain mainly information on tropical cyclone forecasting at the Meteorological Office. Past issues of this document are held together with observed and forecast track information of recent storms, track prediction error statistics, lists of names, images, movements, and photographs and details of tropical cyclone track prediction development at the Meteorological Office. However, there is no information about initial stages and post-mature stages of the current TC under study. The document only includes the mature (the most active) stages of the TC under consideration.

The "Tropical Cyclones" and "SUPER Typhoon" websites of the University of Wisconsin-Madison, U.S.A. have provided the best collection of satellite imagery, tropical cyclone advisories, and related products including WWW links on the Internet (Table 8.2). The document and webpages mentioned above do not include archiving units in their bodies. Information inserted in these products can be used for preparing global tropical cyclone datasets.

8.4 THE GLOBAL TROPICAL CYCLOGENESIS DATASET GLOBAL-TC

The aim of this section is to illustrate the development of the Global-TC systematic database for chronological and kinematics characterization of large-scale tropical disturbances in the entire World Ocean during a 25-year period (1983–2006) (Pokrovskaya *et al.*, 1993, 1994, Pokrovskaya and Sharkov, 1994c, 1997a, 1999a, b, 2006).

8.4.1 Principles of database design

Pokrovskaya and Sharkov (1993a) put forward the idea of studying and developing the probability characteristics of tropical cyclogenesis intensity as complex structure signals using the theory of random time series. The idea formed the basis of the database under consideration.

The principles of the database are the following:

1. The initial information for Global-TC should contain a chronological (generalized) catalogue of global tropical cyclogenesis in the form of a time series of events (flux of events) from 1983 to 1998.
2. For regional cyclogenesis, analogous catalogues should be formed for the six regions of the World Ocean that are the most active generators of tropical disturbances in global cyclogenesis.
3. The structure of geophysical data on each tropical disturbance should be in the second (independent) information unit of the database.
4. Methods for establishing the database's computational architecture should provide for independent selection and storage of all types of information, rather than a quick search and presentation of information, in a sufficiently concise form of storage for standard PCs.

The central problem with such an information system is thorough systematization of initial material obtained from different regions and observation centers in different information codes and with highly different details.

In accordance with the database principles under consideration, the Global-TC dataset consists of two units: a generalized chronological catalogue and a geophysical data unit. A detailed description of the units, computational architecture, and some examples of the primary processing of data on global cyclogenesis intensity is given below.

8.4.2 Data preparation technique

Primary information for 1983–1992 from Rosgidromettsentr (Russian Hydrometeorological Center) records of tropical cyclones in the World Ocean was thoroughly analyzed, systematized, and brought into one-to-one correspondence. For a more complete characterization of conditions promoting tropical cyclogenesis activization, Pokrovskaya and Sharkov (1993a) made a minimum necessary list of parameters (see below), which allowed tropical processing of material available for the purposes of different approaches in studies of generation, development, and forecasting of tropical disturbances. Unfortunately, the primary raw data obtained by Rosgidromettsentr from regional meteocenters by telex links had serious drawbacks, which created problems in making a universal package of consecutive raw information files. Primary information of the South Pacific Ocean regional meteocenters is scant and inaccurate. The most abundant primary information is obtained from the Tokyo Meteocenter (Japan). It was used to form a set of parameters for an information file in the described database.

Primary information on the tropical disturbances that occurred in the World Ocean over 1993–2010 was derived on a daily basis during 1997–2010 on the Internet (*http://www.solar.ifa.hawaii.edu*) by the Joint Typhoon Warning Center in Guam, where information from regional meteocenters (Tokyo, Miami, Honolulu, Calcutta, Darwin, Perth, Fiji, Réunion) was totalized and entered into the Internet. Obtained information was first subjected to careful selection, critical analysis on the basis of operational meteorological information, systematized as far as possible, complemented, and standardized. Then, with the aim of determining the principal characteristics of tropical cyclone evolution, the minimum necessary list of parameters—on the basis of thematic processing of collected material as a result of different approaches—to study the origin, development, and forecasting of tropical systems was formed. However, note that entering Internet source information on to the database from regional warning centers and generalized information from the JTWC has a number of defects: lack of coordinated definitions on one and the same meteorological occurrence; the lack of hard fixed schemes of presentation information; using non-typical telex English abbreviations; temporary disconnecting information; distortion of information during transmission; subjective evaluations of watchers; using different languages (except English) to describe meteorological occurrences that introduce extraneous issues (with respect to tropical meteorology) in telegrams, and so on. All this creates serious difficulties when standardizing the consequent number of information files and requires serious critical analysis and selection when pre-processing the information entered.

From two to six packages (files) of data parameters were introduced into the database on each tropical cycle in the World Ocean each day of its lifetime. The total number of files on each TC was directly dependent on the TC lifetime and varied from four to forty and more. Altogether 2,086 tropical disturbances of different intensities were registered during the 16-year period, and 38,304 files of primary information were introduced into the database.

8.4.3 Chronological data unit

The aim of this unit of "Global-TC" was to give a chronological picture of successive events of global cyclogenesis during 1983–2010 (without details of each TC state and evolution). The entire World Ocean area was divided into six regions characterized by their own unique circulation of tropical cyclogenesis. Tables of the catalogue contain annual (monthly) data on each TC life period (dates of origin and whirl dissipation) and its maximum according to the international classification. These data were systematized for each region individually. The information stored in the catalogue is shown in generalized form in Appendix A over the 1983–2010 period.

8.4.4 Evolutionary data unit

The unit files contain information on the TC structure as well as the evolution, energy, thermal, and kinetic characteristics of each TC from its origin to dissipation (Table 8.3).

In the top left corner of each file there is the TC name according to the international catalogue, followed by its international number in brackets, which gives information on the region of TC origin, year, and ordinal number in the

Table 8.3. Information structure of a primary information file of the geophysical data unit.

TC "Fabian" (8815)	*Point 33*	*TC "Fabian" (8815)*	*Point 33*
Stage	STS	Route of shift	NE
Date (day/month)	02/09	Rate of shift (knots)	18
Time (h)	18	50 knots	60 miles S
Latitude (degrees)	37.2	30 knots	175 miles SE
Longitude (degrees)	154.8	Wind forecast for 12 h ($m\,s^{-1}$)	—
Pressure (mb)	975	Wind forecast for 24 h ($m\,s^{-1}$)	41
Windspeed in the center ($m\,s^{-1}$)	31		

given year. In the top right corner, there is the ordinal number of the given file in a consecutive line of the entire information on the given TC.

The lines that follow denote (Pokrovskaya and Sharkov, 2006):

1. Whirl development stages: TL, initial tropical disturbance (a separated area with surface low pressure in the tropical latitudes); TD, tropical depression, 15 m/s to 18 m/s; TS, tropical storm, 18 m/s to 23 m/s; STS, strong tropical storm, 26 m/s to 31 m/s; T, HT, TC, typhoon, hurricane, and tropical cyclone (the maximum stage of whirl development) >33 m/s; L, extratropical disturbance (a separated area with low sub-surface pressure in the midlatitudes).
2. Observation data, day, and month.
3. Observation time (hours in Greenwich Mean Time).
4. Whirl center coordinates (degrees).
5. Pressure in the whirl center (millibars).
6. Wind rate in the whirl center (m/s).
7. Trajectory of the cloud and wind systems of whirl (in rhombus of the horizon and movement rate in knots).
8. An area with predominating wind rates of 26 m/s (50 knots) and 15 m/s (30 knots), radius in miles, trajectory in rhombus; within the hemisphere with predominating winds at the rates mentioned, the integers following this information denote the radius of the given wind in the rest space.
9. Wind forecast 12 h and 24 h ahead (m/s).

8.4.5 Computational architecture of dataset

To facilitate rapid search for an information fragment and concise form of storage, the following structure for the database was chosen:

1. All types of information are independently arranged in chronological order according to the timescale.
2. All types of data are written down as binary files with information units of a fixed length and repeating units also of a fixed length, whose numbers are shown in the information unit.
3. A concrete way of subdivision into files is chosen from the expected volume of each type of information.
4. Total files for each type are stored in a hierarchical subdirectory structure in order that directories at each level contain from several dozens to several hundred files.

In addition to this, for information about typhoons files are given names like D:\METEO\gg\15ggNN.dat and have a characteristic file size of 0.5 kB to 4 kB (NN is the international ordinal number of the typhoon for the current year).

Such a database allows easy search for different data corresponding to the spatiotemporal interval of interest and rapid use of data for correcting experimental methods in situ.

The database language is Turbo Pascal (v. 6.0) and the software library is Object Professional.

The data on global cyclogenesis from the Global-TC dataset may be used in studies of global climate, solar–terrestrial interrelations, and crisis situations including forecasting of ecologically dangerous atmospheric events.

8.5 GEOINFORMATION DATASET PACIFIC-TC

In this section we consider the development of a version of the Pacific-TC geoinformation dataset at the Space Research Institute (SRI) of the Russian Academy of Sciences (RAS), designed to study the evolution of large-scale atmospheric disturbances at various stages in time, from the background situation leading to the generation of a tropical disturbance to very late dissipation stages of the TC structure (Klepikov *et al.*, 1994).

There are fundamental differences between this dataset and the Global-TC dataset both in the philosophy of acquisition and the design of archiving, particularly regarding hydrometeorological environmental data in active storm areas and using non-strict time discretization. In other words, meteorological information for various geographical regions (including areas in which tropical disturbances are born, areas in which these structures move, and regions in which they dissipate) is accumulated and stored.

8.5.1 Principles used to develop the database

The theoretical basis for the development of the database is the recent concept of the self-organization of turbulent movements, in which some of the kinetic energy of a helical turbulence is converted by an inverse spiral cascade mechanism into the energy of a large-scale eddy structure (eddy dynamo) (Moiseev *et al.*, 1983a, b, 1988). Such a self-organization process involves scales from the convective (2–20 km) to the mesoscale (20–200 km). The "launch" of instability requires a large-scale but weak kinematic interaction on the synoptic scale (200–2,000 km). Further theoretical estimates (Moiseev *et al.*, 1990) show that taking into account the cascade processes of a humid convective medium (convective scale and mesoscale) considerably decreases the threshold for the "response" mechanism for generation of a large-scale eddy.

Based on the requirements of the scientific problem to be solved, the following principles were applied to the construction of the database:

1. The raw information should contain the fields of thermal, humidity-related, and kinematic meteorological parameters on convective, meso, and synoptic scales, as required to form and study the informative geophysical parameters of the problem—namely, the spatial and spectral distribution of turbulent kinetic energy, the spatial distribution of humidity-related static energy, and the spatial evolution of the large-scale (relative to the initial form of the tropical disturbance) velocity field.
2. The structure of the database should permit attachment of the required meteorological information to the time dynamics (evolution) of the specific atmospheric formation of interest in the study. In other words, it should be possible to track a dangerous situation throughout all stages of its development, including the very earliest stage of its formation, background circumstances at the place and time of its birth, its exit from this region, and the trajectory of motion of the structure.
3. The design of the database should allow for these facilities:
   (1) independent collection and storage of all forms of information entered in a single information system, including the possibility of unlimited independent accumulation of experimental information;
   (2) sufficiently fast retrieval and display of an interesting fragment of information of a specific form (independently of other forms of information stored) in a user-friendly manner;
   (3) sufficiently compact facility for storage on standard PCs.

The main difficulty in forming such a database is that information with various scales of averaging is acquired from diverse initial carriers (satellites, research vessels, island and coastal radiosonde stations) and in different codes.

8.5.2 Structure of geophysical data

In connection with the above problem, the SRI RAS team developed a geoinformation base of satellite and hydrometeorological information on an IBM PC.

The region chosen for study was the northwestern area of the Pacific Ocean (one of the most energy-active zones of the Pacific Ocean and a zone of intense tropical cyclogenesis) bounded by the coordinates 5–30°N, 100–180°E. This region is comprehensively covered by information from the Japanese Meteorological Satellite Center, has a ramified network of ground (coastal, insular) aerological stations, and is quite frequently visited by scientific research vessels of the Far Eastern Regional Meteorological Center; thus, it presents a rare possibility for joint collection of various types of information in a single database.

The choice of time intervals (September–December 1986, July–October 1988, May–August 1989) for the database was determined by the time of intensification of the cyclogenesis processes in this region and the availability of the information listed below to the developers of the system.

The database includes four information modules:

1. The chronological unit contains series of data relating to the dynamics of the development of tropical eddy disturbances that may or may not turn into tropical cyclones. The information about each TC begins with the international index for the year and extends from the time of its onset (stage of tropical disturbance) to the time of its destruction or conversion (dissipation) into a midlatitude cyclone. Each day in the life of a TC is characterized by four observation times: 00:00, 06:00, 12:00 and 18:00 GMT. The following data are entered at each observation time: the stage of development of the eddy (TL, TD, TS, STS, T), the coordinates of its center, the pressure and the maximum windspeed at the center, the direction and rate of drift of the eddy, the size of the zone experiencing windspeeds from 30 to 50 knots, forecasts of the intensity of its development over the next 12 or 24 hours, and the presence of photographic information for the region from NOAA and GMS satellites. The chronological unit contains data on about 36 TCs and 30 disturbances that did not reach the TC stage. Data about tropical eddies born here which did not become TCs are not usually published and in this sense the database provides a unique possibility for the data it contains to be used to analyze the state of the tropical atmosphere that leads up to the dissipation of eddies in their initial stage.
2. The geophysical data module contains data recovered from space information provided by NOAA satellites. Each point (location) of a recovered data profile is marked by the coordinates of the point and the time of the session of the satellite link. Each day includes four observation times, which cover 35–50 points in the whole test area. The database includes the vertical profiles of the air temperature for 14 height levels from the ocean surface to 30 mb (sea level 1,000, 850, 700, 500, 400, 300, 250, 200, 150, 100, 70, 50, 30 mb), the vertical profiles of the overall content of water vapor in the atmosphere in four layers (sea level 400, 850–400, 700–400, 500–400 mb). The database includes data relating to a total of 18,700 profiles.
3. The module for data from radar sensing from ground stations (coastal, insular). Each station is entered in the database with a standard international index together with the coordinates of its position. For each station, observations are taken twice per day at 00:00 and 12:00 GMT; there are on average 70 stations in the region. The following are entered for each station in the database: vertical

air temperature profiles, vertical profiles of relative humidity (dewpoint deficit), vertical profiles of wind direction and speed. All the profiles include 11 height levels (land surface, 1,000, 850, 700, 500, 400, 300, 250, 200, 100 mb surface). If the level of the tropopause lies below the 100 mb surface then it is also included in the vertical profile. The total number of systematic entries in the database is 33,000.

4. The module for detailed radar sensing from scientific research vessels located within the test area during the given period. Data from each sensing station on an RV is marked with the date, the time of sensing, and the coordinates of the RV, and includes a variable number of height levels (on average from 25 to 40), depending on the state of the atmosphere at the time of observation. There are two to four times of observation daily at 00:00, 06:00, 12:00, and 18:00 GMT from which data are entered in the database. The database includes vertical pressure profiles, vertical temperature profiles, vertical profiles of relative humidity (dewpoint deficit), and vertical profiles of windspeed and direction. The total number of systematic measurements is 586.

8.5.3 Computational architecture

To facilitate fast retrieval of interesting fragments of information and compact storage, the following structure was chosen for the database:

1. All forms of information are formed independently and in chronological order.
2. All types of data are stored in the form of binary files with a fixed length information block followed by repeated fixed length blocks, the number of which is specified in the information block.
3. A specific facility for splitting into files is selected from the expected amount of information of each type.
4. All files of each type are stored in a hierarchical subdirectory structure so that each level of directories contains tens to hundreds of files.

Based on the above, we selected the following structure for satellite information. The information for each day is entered in a single file that takes the form Si*yymmdd*.dat, where *yy* denotes the year, *mm* denotes the month, and *dd* denotes the day. At the same time, all the satellite information is stored in a directory SI, which contains two levels of directories *yy* and *yymm*. All the files corresponding to a given year and month are stored in the latter. The filename of satellite information in the computer environment used by the authors takes the form D:\METEO\SI*yy**yymm*\Si*yymmdd*.dat. Typically, a file of this type is between 1 kB and 3 kB.

Schemes for constructing two other forms of information were chosen analogously. For information about typhoons, the filename takes the form D:\METEO\TS*yy*\Nc*yy*NN.dat and has a typical size of 0.5–4 kB (NN denotes the international serial number of the typhoon in the year). Information from ground meteorological stations is stored in files with filenames of the form D:\METEO\ZOND*yymm*\Zn*yymmdd*.dat and with typical file sizes of 15–25 kB.

For radar-sensing data from RVs, files with filenames of the form D:\METEO\PROF*yy**yymm**yymmddhh*.znd are used, where the typical file size is 200–500 kB. This notation also shows the time of observation in hours GMT. This scheme for constructing the database enables the use of simple programs to retrieve data of various types corresponding to interesting spatiotemporal intervals and enables these data to be used to implement corrections to the methods of large-scale experiments in operational mode.

The database software is written in Turbo Pascal (v. 6.0) and makes broad use of the facilities of the Object Professional software library. The software package consists of the main program bank and a number of high-level libraries: Menulib provides for dialog in the form of a multilevel menu, Dirlib supports real time interaction with the directories of disk devices, Editlib is used for data input and editing, Rwreclib is used to read and write files; there are a number of other service libraries. A description of the data formats in CD-ROM form is contained in the Rwreclib library.

8.5.4 Visualization of information

To represent the facilities of the database and the software package graphically a computer animation was created, a fragment of which is given below.

Figure 8.3 shows the trajectories of typical tropical cyclones in 1988 for the given region, information about which is held in the database. Each cyclone is tracked from the moment of its onset (TD stage denoted by "O") to the time when, on landfall, it is destroyed or converted into a midlatitude cyclone. Information about the characteristics of the trajectory of a TC is denoted by "o", observations of which are taken every 6 hours. Figure 8.3 also provides useful information about the fact that in 1988 in the region of the Pacific studied, the processes of cyclogenesis were intense and varied. The characteristics of cyclogenesis include the northern (18–20°N) position of the region of birth of many eddies and the anomalous nature of trajectories (the presence of a southern component) in the initial stages of development. TS 8823 had an altogether unusual trajectory in the South China Sea. The storm proceeded on its own (previous) trajectory with increasing intensity.

Figure 8.4 shows a fragment of a file reflecting an episode in the study of the track of TC "Warren" (8806) in the first half of July from the RV *Akademik Shirshov* and the provision of space data for the ocean in this time interval. The course of the RV is denoted by a continuous line with dots, which relate to 6 hourly intervals and correspond to the points at which radiosondes were launched. Figure 8.4 demonstrates the following important spatial characteristics of the space data recovered. If stationary points of aerial radiosonde sensing are spread over the open ocean at distances greater than 1,500 km to 2,000 km, then the points at which hydrometeorological parameters are recovered from space data cover the studied waters more or less uniformly with an average network step of 50 km to 100 km, despite the presence

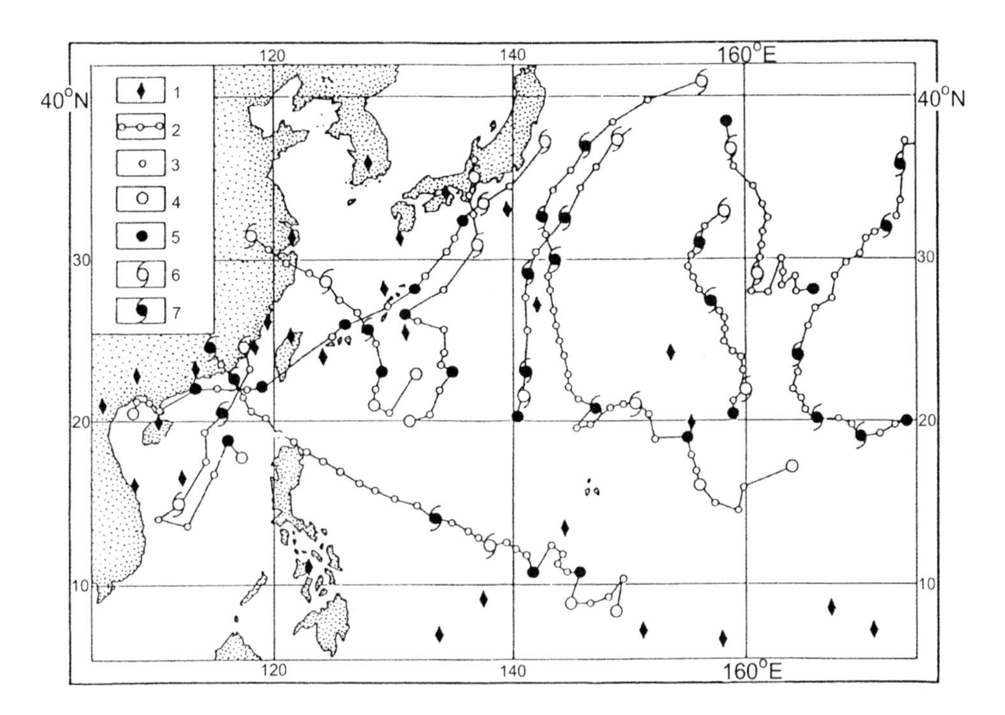

Figure 8.3. Trajectories (tracks) of TCs in July 1988 in the NWP: 1, coastal meteostations; 2, trajectories (tracks) of TCs; 3, position of the center of TC every 6 hours (data on every intensity is stored in the database; 4, stage of tropical disturbance; 5, stage of tropical depression; 6, stage of tropical storm; 7, stage of typhoon (from Klepikov *et al.*, 1994).

of dense fields of cloud cover in the region of the ITCZ and the passage of a tropical cyclone.

Figure 8.5 reflects an episode in the study of TS "Mamie" (8823) which formed and dissipated in the South China Sea in September 1988. One important characteristic of the experiment was the strictly synchronous study of the thermal structure of the atmosphere using aerial radiosondes from island meteorological stations directly after the passage of a tropical storm there and using aerial radiosondes from an RV located more than 500 km from the tropical storm and from the island station. These measurements may be assumed to have been made in an undisturbed atmosphere (calibration mode). Analysis of the thermal structures (in Figure 8.5b they are shown on a bi-logarithmic scale) graphically shows that the effect of the tropical storm cannot be traced in the thermal structure either in the troposphere as a whole or in the lower stratosphere. In other words, no significant traces of the tropical disturbance remain in the thermal structure of the atmosphere.

As these fragments show, the architectural structure of the database and the variety of forms of information held in it provide users with a wide range of facilities to make any necessary calculations using the corresponding software.

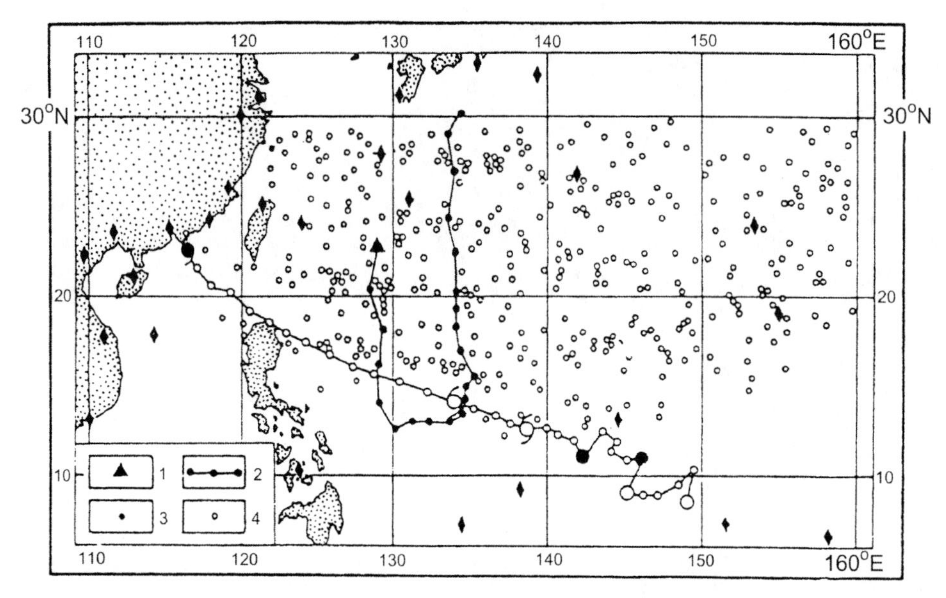

Figure 8.4. Study of the track of TC "Warren" (8806) from the RV *Akademik Shirshov* and the position of space data (NOAA satellite data) in the basin from July 10 to 20, 1988: 1, RV; 2, course of RV; 3, position of points at which radiosondes were launched; 4, position of points where satellite data (meteo-profiles) were recovered (cf. Figure 8.3 for some of the notation) (from Klepikov *et al.*, 1994).

8.6 COMBINED SATELLITE AND IN SITU SCENARIO-TC DATABASE

The present section outlines non-traditional approaches to the formation and development of research-thematic databases of satellite and ground-based information designed for the solution of problems by studying the physical conditions under which the genesis and evolution of atmospheric catastrophic phenomena (tropical cyclones) take place. The first versions of these experimental databases are presented. They were produced in the Space Research Institute of the Russian Academy of Sciences (SRI RAS) on the basis of scenario principles proposed by Russian researchers (Pokrovskaya *et al.*, 2002, 2004). These principles include the use of geophysical and thermohydrometeorological information restored both from space observation data and from the data of complex expedition work in the tropical zone of the Pacific Ocean.

8.6.1 Methodological problems of observation of atmospheric catastrophes

The term "catastrophic atmospheric vortex" is usually applied to such natural phenomena as tropical cyclones (typhoons). This is because of the life-threatening danger posed by these atmospheric processes, the concomitant significant material damage caused, and the administrative–social problems that arise. Problems are

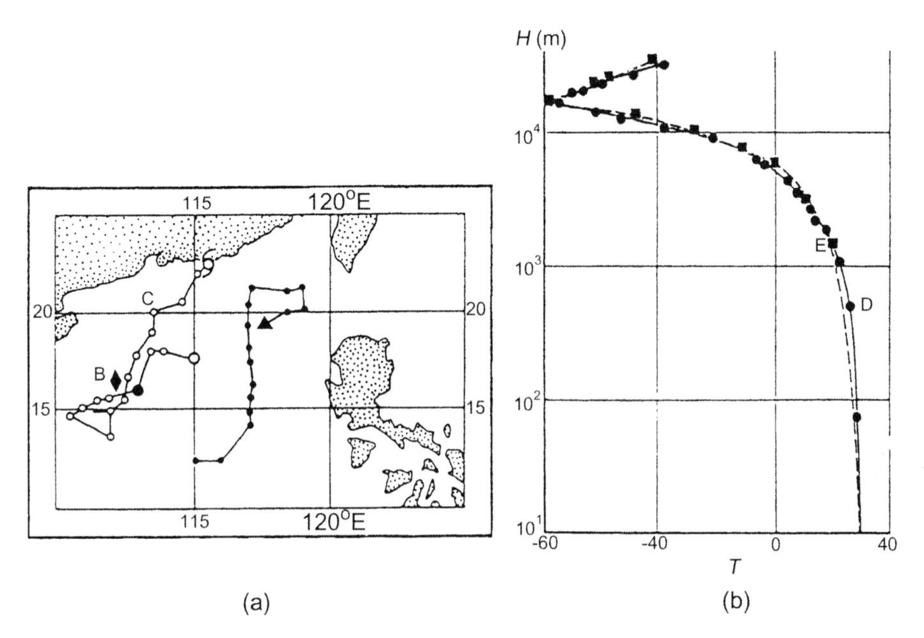

Figure 8.5. Study of TC "Mamie" (8823) in the South China Sea. (a) Trajectories of the TC and the RV (notation as in Figures 8.3 and 8.4): A, B, points of synchronous sensing on the RV and on the island station; C, position of the center of the storm at the time of sensing. (b) Profile of thermal stratification of the atmosphere based on radiosonde data from the RV (D) and from the island station (E) (from Klepikov *et al.*, 1994).

aggravated by the impossibility of reliable forecasting of either the genesis or evolution of these phenomena. Today there exist a considerable number of theoretical models of these phenomena with various degrees of complexity, but none of them is considered to be finally consistent in the theoretical respect and, moreover, none of them has been confirmed experimentally at any satisfactory degree of reliability. As is known, the problem of comparing theoretical models with observational data is both complicated and ambiguous, as a result not only of the imperfection of theoretical models, but also of the structure and quality of primary observational data which, as a rule, are gathered and formed under completely different principles that are in no way associated with the theoretical models in the overwhelming majority of cases. In an ideal world, the accumulation and exploration of experimental data when studying complicated geophysical phenomena must—from the viewpoint of the spatiotemporal scales of the measured parameters—adequately correspond to the considered theoretical model (as is done, for instance, in experimental physics). When studying atmospheric catastrophes the situation is diametrically opposite: the gathering and accumulation of experimental material proceeds according to strict rules prescribed by purely meteorological practice, whereas the authors of theoretical concepts do not prescribe sufficiently clear and definite criteria for gathering and accumulating the necessary experimental information (Sharkov, 2000). Moreover, in many cases, even the necessary physical parameters are not determined. In Pokrovskaya *et al.* (2002, 2004) an attempt at overcoming this "break" is undertaken by means of the proposed scenario principle of accumulation and exploration of existing archival satellite and ground-based information when studying experimental problems involved in the investigation of physical conditions under which the genesis and evolution of catastrophic atmospheric phenomena (tropical cyclones) take place. To reiterate, this section presents the first version of an experimental database produced on the basis of the proposed scenario principle, which includes geophysical and thermohydrometeorological information restored both from the space observational data and from data garnered during complex expedition work in the tropical zone of the Pacific Ocean.

The phrase "large-scale catastrophic vortices in the atmosphere" refers to typhoons (tropical cyclones) that represent mesoscale formations supported by cyclonic-scale systems. They arise and develop at tropical latitudes, take their energy from heat in the tropical part of the ocean–atmosphere system and rapidly transfer it to midlatitudes and high latitudes, thus performing so-called "polar transfer" in the global greenhouse effect. In essence, catastrophic vortices represent a unique mechanism that effectively dumps excessive heat in the atmosphere under conditions in which the action of usual mechanisms—the basic one being turbulent convection—becomes insufficient. Thus, catastrophic phenomena play an important part in establishing the climatic temperature regime of the Earth, removing excessive heat and preventing excessive overheating of the planet in the tropical zone. These are the reasons that the basic challenge for tropical cyclone theory is now considered to consist in describing the processes of two (and, probably, more) interacting structures as a single entity. Full description requires using almost all meteorological objects and concepts—from the physics of clouds in turbulent convection up to the general circulating mechanism in the tropics, and from the atmosphere–ocean interaction up to radiation transfer into space. Comprehension of the role played by moist convection processes, naturally present in other atmospheric systems, is considered to be one of the most difficult problems. Despite small-scale details changing continuously, the tropical cyclone as a whole is a stable system that lasts as a whole for many days over the warm ocean. Compared with other atmospheric systems, the relative stability of the dynamic structure of tropical cyclones is an important feature (and, we add, a scientific riddle). Detailed analysis of modern theoretical concepts and hypotheses related to studying the genesis of atmospheric catastrophes is presented in Chapter 7.

Let us now consider in more detail one of the approaches to studying the genesis of large-scale atmospheric instabilities associated with the so-called "spiral reverse cascade" of energy in atmospheric turbulence. According to classical notions long held in turbulence theory, turbulent motions destroy structures of all types resulting in energy transfer from large-scale organized motions to chaotic small-scale vortices, a process known as the "direct cascade of energy". However, in recent decades situations have been found in which energy is transferred in the opposite direction, from the small scale to the large scale, known as the "reverse cascade of energy". Under real atmospheric conditions the most important mechanism is the spiral reverse cascade in three-dimensional turbulence. Such a mechanism of energy transfer to large-scale motion exists in turbulence that—being homogeneous and isotropic—is characterized by symmetry violation that reflects the coordinate system. In such a case, turbulence is described by a non-zero pseudo-scalar (spirality) and is called "spiral turbulence" (see Section 8.7). Theoretical study of the generation properties of a small-scale hydrodynamic spiral turbulence, as a consequence of exciting and maintaining large-scale long-lived vortical disturbances (the vortical dynamo), is basically associated with the work of a team of authors from the SRI RAS. The possibility of implementing the vortical dynamo for an incompressible liquid was substantiated for the first time by Moiseev *et al.* (1983a, b), where some additional factors of symmetry violation were shown to be necessary, along with spirality (e.g., the force of gravity, temperature gradient, stationary flow). The appearance of models that take into account the spiral properties of turbulence opened up prospects for solving a wide class of problems on the generation of large-scale structures; one possible application might be description of the evolution of cyclonic formations in the tropical atmosphere of the Earth. The key point in describing tropical cyclone onset is explanation of the conditions and factors under which small-scale convective atmospheric cells, having the size and lifetime of a normal cumulus cloud (a few kilometers and a few hours), quickly reorganize themselves into fast-rotating spiral vortices of synoptic scale, 1,000 km and more, which last for some days. The fundamental bases for a completely new direction in theoretical investigations can be attibuted to Moiseev *et al.* (1983a, b), who show that the effect of small-scale spiral turbulence on the usual convection process results in the need for completely rearranging how motion is perceived—namely, in the horizontal layer of a liquid warmed from below a single large vortex forms instead of many small convective cells. Moreover, in this vortex the toroidal field of velocity gets entangled (hooking) with a weaker poloidal field (horizontal and vertical circulation). In such a case the large-scale instability arising against a background of small-scale spiral turbulence turns out to be caused by positive feedback between toroidal and poloidal components of the velocity field as a result of the turbulence spirality factor. Rutkevich (1993, 1994) shows that—applied to atmospheric conditions—the spiral turbulence concept parameterizes (reformulates) such important cyclogenesis factors as convection and the Coriolis force. At the present time, a large series of theoretical works (Moiseev, 1990, 1991) has already appeared in which attempts have been made to construct models of large-scale vortical structures and the scenario necessary for their development. A principal feature of these models consisted in the presence of hooking of vortical current lines of the large-scale structure velocity. Discovery of this feature represents the basic achievement of spiral models and something that had never been contained in previous models of tropical cyclone genesis, because these models did not take into account the feedback between the solenoidal components of the velocity field.

The first attempts at experimental estimation of spirality on macroscales at the initial stage of tropical cyclone onset, carried out within the framework of natural experiments during the Typhoon-89 and Typhoon-90 expeditions (Zimin *et al.*, 1991) in the tropical zone of the Pacific Ocean, undoubtedly stimulated the experimental part of the considered problem. On the other hand, however, these experiments also revealed serious methodological and experimental complexities in determining this parameter under natural conditions. Hence, the spatiotemporal scales on which to perform experimental estimations of spirality were not clear; the phases of development and state of tropical meso-structures, in which the measurements were carried out, were not distinctly determined; what is more, errors in performing separate estimations of spirality were not indicated and comparisons with theoretical estimates of spirality were not performed. All this testifies to the fact that, despite of optimism of Zimin *et al.* (1991), one should treat the quantitative results of these works with a great deal of care. Moreover, the question about the validity of the spiral mechanism to describe the primary forms of atmospheric catastrophes remains moot. All these problems are mainly associated with the initial data being obtained meteorologically, with no account of the specific features of the physical hypothesis under study (i.e., the spiral mechanism).

Data from Doppler radar observations, carried out earlier under similar pre-typhoon conditions in the tropical atmosphere (Klepikov *et al.*, 1995; see also Section 3.4.5), testify to fast evolution of spectral components in the energy spectrum of the velocity field (measured inside the cloud masses) toward the large scale, which could be interpreted as the possible existence of a reverse cascade of energy in the presence of meso-spirality. However, more detailed analysis has shown that these results (because of a fast unknown reconstructive mode that works on timescales of 2–10 min) relate to the simultaneous coexistence and antagonism of several mechanisms (Kolmogorov's spiral mechanisms, flicker noise mode, and Brownian motion) rather than to large-scale structure formation in turbulent chaos (Sharkov, 1999b).

While recognizing the theoretical achievements in this field of research, it has to be admitted that many questions of undoubtedly great scientific interest still remain open: reliable experimental confirmation of the role played by spirality in the cyclogenesis process; revealing and studying the reverse spiral cascade under real natural conditions; the development of experimental techniques to process the data of hydrometeorological observations with the purpose of measuring meso-spirality values on various spatiotemporal scales. Experimental study of these problems should principally be directed at adequate (for the accepted hypothesis) formation of a geophysical database of dynamic fields in the tropical convective atmosphere both in pre-typhoon situations and during the evolution of developed cyclones.

However, there now exists another thermodynamic direction taken by studies in the atmospheric catastrophe genesis problem, which is substantially different from the dynamic approach (i.e., from the spiral mechanism). The majority of authors of these theoretical models of catastrophic phenomena (as well as many scientists) converge in the opinion that the release of latent heat of atmospheric moisture condensation in the cloud of tropical systems represents their basic energy source. Note that, in contrast to these long-held beliefs concerning tropical cyclogenesis, the successive approach to phase transformations of moisture in the atmosphere has only developed comparatively recently; the first works in this direction appeared in the early 1990s (Rutkevich, 2002; Rutkevich and Sharkov, 2004).

Water vapor in tropical depressions reaches a saturated state. Hydrodynamic motions in such a medium are accompanied by the release and absorption of heat as a result of phase transitions, which is considered to be the basic energy source of a tropical cyclone. However, the heat of phase transitions results in a change in the basic stationary state of the atmosphere. Under these conditions, the hydrodynamic motions of saturated moist air can acquire properties that are unusual in dry-air hydrodynamics. It is reasonable to expect that the new hydrodynamic instability, associated with phase transformations of moisture (Rutkevich, 2002; Rutkevich and Sharkov, 2004), will turn out to be responsible for the appearance and development of natural large-scale atmospheric vortices. The existence of tornadoes and typhoons suggests that one of the most important concomitant factors in this case is the rotation of a vortex, which is inseparably related to process development. This property of instability, causing the appearance of tornadoes and typhoons, sharply contradicts the properties of usual convection in dry air. However, because instability energy (in initially motionless air) should nevertheless be associated with the instability of temperature stratification in the atmosphere, it is also reasonable to expect that the new instability in the moist air will turn out to be a new and non-trivial channel of convective instability development. As is the case with the dynamic approach, adequate measurement of the necessary spatiotemporal parameters also demands the formation and accumulation of a database of space and ground-based data with regard to specific features of the physical hypothesis.

8.6.2 Scenario principles of geophysical data formation

The complexity of revealing the geophysical situations in which the usual turbulent state of the atmosphere transitions into the clearly defined self-organized structure of an atmospheric catastrophe (a tropical cyclone) consists in extremely rapid spatiotemporal rearrangement of dynamic modes of atmospheric convection. Nobody has managed to record experimentally the moment of jump-wise transition in basic characteristics sufficiently clearly. The most efficient way of doing this was long considered to be local techniques (radio sensing and Doppler radar techniques from research vessels situated sufficiently closely to the active cyclogenesis zone), because remote optical techniques already recorded the re-organization and rearrangement of cloud systems (which, as a matter of fact, underlie the modern remote techniques of TC forecasting; Dvorak, 1975, 1984) with the resulting great aftereffect. On the other hand, ship-borne radio sensing possesses its own peculiar features related to the practice and technology of radio sensor production according to a strictly specified time grid with time constants exceeding the characteristic times of the qualitative rearrangement of dynamic modes. In this sense, the optimum observation means would be the use of powerful Doppler radar ship-borne systems with heightened sensitivity for measuring the velocities of wind fluxes in combination with ship-borne radio sensing according to the "free" scheme determined by the physics of processes and their spatiotemporal features. The appropriate possibilities were demonstrated experimentally in 1987–1988 by SRI RAS specialists during the Typhoon-87 and Typhoon-88 tropical ship expeditions (Zimin *et al.*, 1991; Klepikov *et al.*, 1990, 1995; see also Section 3.4.5). However, this work has not been further developed for well-known financial–organizational reasons.

In connection with the methodological features of radio sensing, as noted above, the following experimental local-sensing methodology was developed. It consists in constructing a database in which scenario-type experimental data on active geophysical events are collected as well as the necessary information from spacecraft (NOAA-type and GMS-type satellites), from meteorological centers (maps of the baric situation transmitted to RVs), and from other carriers (ships of all types, stationary meteorological stations). Detailed data on tropical cyclone evolution were obtained via the Internet and processed using a special technique (Pokrovskaya and Sharkov, 2006). We believe this methodological scenario-type approach will be useful in studying other fast-acting geophysical large-scale processes.

The spatiotemporal structure of geophysical data obtained on board ships in the Pacific Ocean's tropical zone is formed under the effect of many conflicting circumstances (scientific tasks, navigation rules, the rigidly fixed techniques and rules in which meteorological observations must be carried out, the random place and time of occurrence of a studied geophysical system, the uncertainty in forecasting the trajectory of motion of a tropical system); hence, the obtained information has principally been systematized with respect to operation ranges, for each of which temporal and spatial (coordinate) attribution was carried out, and the correctness and completeness of messages checked. On the basis of verified information, five scenarios were formed, which reflected particular features of the geophysical (natural) evolution of crisis anomalies and features of the hardware–technical situation in performing the experiments:

- *the "retreat" scenario*, in which the RV took measurements from the area of activity of TC "Brenda" (8903) to its outermost edge;
- *the "pursuit" scenario*, in which the RV carried out geophysical observations while pursuing TC "Dot" (8905);
- *the "socket" scenario*, in which the RV carried out geophysical sounding while criss-crossing the area of activity of TC "Elsie" (8906) with the purpose of obtaining multi-scale information;
- *the "conglomerate" scenario*, in which the RV was located in the area of activity of two successive tropical disturbances (TDs) that arose and faded in this area and three strong tropical cyclones—TC "Gordon" (8908), TC "Irving" (8910), and TC "Judy" (8911);
- *the "periphery" scenario*, in which the RV was located on the outermost edge of TC "Lola" (8912) and TC "Mac" (8913).

8.6.3 Two experiments: the conglomerate scenario and the pursuit scenario

The meteorological conditions under which two experiments (the pursuit scenario and the conglomerate scenario) were carried out are considered below in detail.

Detailed data on the evolution of these tropical cyclones can be found in the Global-TC database prepared and issued by Pokrovskaya and Sharkov (2006).

The pursuit scenario (June 2–12, 1989). In June 1989, cyclonic activity strengthened further over the Asian continent and weakened over the moderate latitudes of northwest Pacific basins. In the tropical zone the processes became more active than they were in May, as revealed by the onset of two TCs over the Philippine Sea. The mean depth of the equatorial hollow changed to 1,008 mb to 1,010 mb. The axis of the intratropical convergence zone (ITCZ) was situated mainly between latitudes 5°N and 10°N, and was activated by upper-tropospheric hollow intrusion into the equatorial area (as in May). The development of a monsoonal hollow (MH) over the Philippine Sea was only observed between June 18 and 24. In the frontal part of the hollow there was weak cyclonic activity, which promoted MH development over the East China Sea and Japan, including the northern part of the Philippine Sea. Stabilization of the upper-tropospheric hollow promoted destruction of a subtropical crest over the western part of the Pacific Ocean and development of a lower pressure region with western and southwestern fluxes in the lower layers of the troposphere, all of which is typical of MHs.

This process coincided with the development of TC No. 8906. The development of tropical disturbances in the ITCZ is also known to occur under the effect of high-altitude meso-vortices. The mean diameter of cloud clusters associated with a developing tropical disturbance varies within wide limits (from 2° to 10° of latitude), but most frequently (in more than 60% of cases) within the limits of 4° to 6° of latitude. At the instant of TC onset the diameters of cloud clusters are not less than 6° of latitude.

For a month the structure of tropospheric fluxes over the Philippine Sea did not undergo any essential changes compared with May. Eastern (passage) fluxes mainly prevailed and western ones were noticed during upper-tropospheric hollow intrusion from moderate latitudes, and during the development of cyclonic activity in the tropical zone the latter fluxes were periodic.

Neither were there essential changes in stratospheric circulation, where eastern fluxes over the Philippine Sea reached velocities of 40 m/s as a result of the summer stratospheric anticyclone in the northern hemisphere.

The onset of TC No. 8905 was preceded by upper-tropospheric hollow intrusion into the northwest part of the ocean and by ITCZ aggravation. The tropical disturbance was not traced at sea level and represented a high-altitude cold tropospheric meso-vortex with a temperature of −30°C. The cloud cluster diameter at this moment exceeded 10° of latitude in size. High relative humidity was noticed throughout the whole layer of the troposphere. At this time tropical cyclogenesis was accompanied by upper-tropospheric hollow strengthening over the Philippine Sea. On June 5 at 00:00 GMT a warm core with a temperature of −28°C formed at the altitude of 300 mb. Sea level pressure dropped to 1,006 mb and wind velocity increased to betweeen 13 m/s and 15 m/s (see the map of near ground analysis for June 5, 1989 at 00:00 GMT). TC onset occurred under the effect of horizontal divergence of upper-tropospheric fluxes on the southwest periphery of the upper-tropospheric hollow, where vertical shifts of wind in the layer between 200 mb and

900 mb did not exceed 5–10 m/s and the maximum velocity of eastern fluxes at an altitude between 12 km and 14 km reached no more than 15 m/s. On June 5 at 12:00 GMT in the region of 10°30′N, 128°00′E the TD reached the TS stage, pressure having dropped to 1,000 mb. Later, the TS entered the water area of the South China Sea, where it reached the T stage over central areas of the sea and dissipated on reaching landfall in northern Vietnam.

Radio sensing on the RV in the first half of July was carried out southward of the arising tropical disturbance and, later, at tropical disturbance displacement to the west-northwest on its southeast periphery.

In the course of the experiment, radio sensing of the atmosphere was carried out four times a day, whereas meteorological observations, hydrological sensing of the ocean, and VHF measurements were carried out every hour. Synoptic information was gathered from the accompanying satellite and consisted of NOAA photographs, facsimile images of GMS photographs (three to four times a day), baric topography maps of 850, 700, 500, 250 mb (twice a day), and restored profiles of the temperature and humidity of the air mass. The scale of the temperature profile restoration zone (the grid scale) was equal to 100 × 100 km at 14 levels from the ocean surface up to 30 mb from the surface. Restoration accuracy was equal to ±1–1.5°C at each altitude level. The humidity profile was restored at four levels: 1,000–500, 850–500, 700–500, 500–400 mb.

Figure 8.6 schematically shows the trajectory of RV motion and the trajectory along which the TC was moving at the time the pursuit scenario was carried out. The numbers of points (representing radio probe release instants on the RV trajectory and geographic positions of the TC along its trajectory of motion) correspond to those presented in Table 8.4, which indicates their date, time, and coordinates, respectively.

The conglomerate scenario (July 7–22, 1989). In July 1989, cyclonic activity

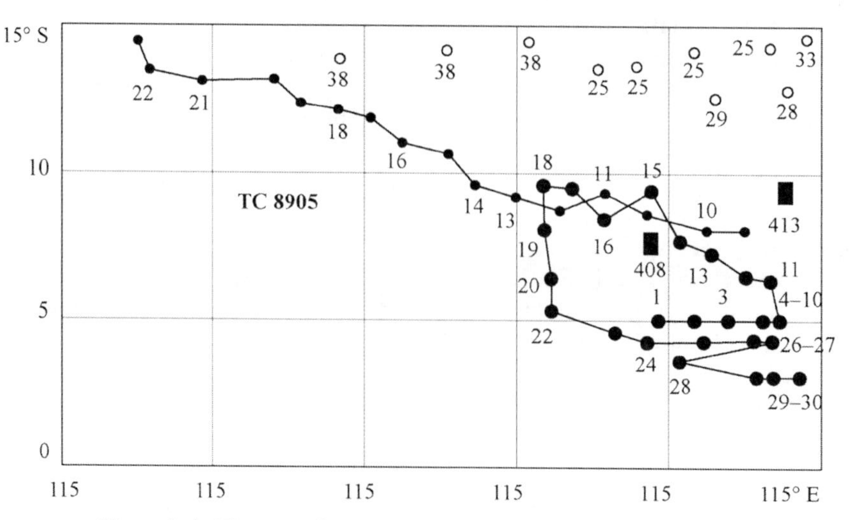

Figure 8.6. The pursuit scenario. See Table 8.4 for notation.

over Asia reached its maximum, while anticyclonic activity over the ocean developed further as well. The power of the North Pacific Ocean anticyclone was between 1,036 mb and 1,038 mb, and a subtropical crest extended from it westward. Its power depended on thermal features of the ocean surface and global processes developing over the northwest part of the ocean. Under meridional processes, which became more active in the second half of July, anticyclone power increased to between 1,028 mb and 1,032 mb, and its center shifted from 30°N to 40°N. The North Pacific Ocean upper-tropospheric hollow was also fully developed, as a result of which the upper-tropospheric hollow could frequently be seen to be cut off with cold high-altitude meso-vortices arising on its southern periphery. As a result of these processes, cyclonic activity in the tropical zone including the Philippine Sea grew dramatically. The upper-tropospheric hollow formed most frequently over the central part of the ocean, near 30° latitude, and then it developed toward the south and southwest, propagating to the tropical zone and often reaching equatorial latitudes. The monsoonal hollow fully developed. All these processes reflect the cyclogenesis of the region.

The development of tropical disturbances mainly occurred when cold meso-vortices at the ITCZ and a monsoonal hollow spread over a huge area from equatorial latitudes up to 20°N, delimited at 160°E. The mean diameter of TD cloud clusters varied widely, but at the time of TC onset cloud cluster diameters up to 2° of latitude in size were observed, which is abnormal for the northwest Pacific Ocean.

In July the ITCZ position moved between 10°N and 20°N, its northward displacement was observed as monsoonal circulation strengthened, which just goes to show the wide scatter of the TC onset region. No essential changes in stratospheric fluxes circulation were observed in July compared with June. Seven TCs arose in July 1989, corresponding to the long-term average value.

TC No. 8908 arose at the time of a cold high-altitude meso-vortex. This process was preceded by upper-tropospheric hollow intrusion and a meso-vortex being cut off it. The latter existed as a high-altitude vortex between July 8 and July 10 and was not traced in the lower layer of the troposphere. On July 11 at about 18.0°N, 149.5°E the pressure at the ground suddenly dropped to 1,008 mb, since at this time a high-altitude cyclonic vortex propagated all the way to the ocean surface. At the same time, a convergence of fluxes was observed in the upper layers of the troposphere, and the velocity of eastern and northeastern fluxes along the trajectory of a developing vortex at AT200 reached between 20 m/s and 25 m/s. The vertical shift of wind in the layer from 200 mb to 900 mb exceeded 15 m/s. On July 9 at 18:00 GMT the diameter of clouds in the developing vortex reached 2° of latitude. By 18:00 GMT on July 11, the TD had developed such that its mean diameter increased to 4°; the outflow of clouds from the upper circle was traced southward of the cloud vortex, which indicated upper-tropospheric hollow divergence and the existence of high velocity fluxes in the upper layers of the troposphere. On July 14, the vortex reached its maximum, the diameter of clouds was 8° of latitude, an eye appeared at its center, the pressure dropped to 935 mb, and maximum wind velocity reached 45 m/s.

Table 8.4. The pursuit scenario.

	RV radiosondes					*FSR*		*TC No. 8905*			*Attendant data*			
Number	*Date*	*GMT*	*No.*	φ	λ	*No. station*		φ	λ	*Evolution stage*	*NOAA*	*GMS*	*Baric maps*	*Data recovery profiles*
						408	*413*							
1	2	3	4	5	6	7	8	9	10	11	12	13	14	15
1	2.06	00	66	5.0	135.0							+	+	
2		06	67	5.0	136.0						+	+	+	
3		12	68	5.0	137.0							+	+	
4		18	69	5.0	138.0							+	+	
5	3.06	00	70	5.0	138.1	+						+	+	
6		06	71	5.0	138.2						+	+	+	
7		12	72	5.0	138.3							+	+	
8		18	73	5.0	138.4							+	+	
9	4.06	00	74	5.0	138.4			8.0	137.5	TD		+	+	
10		06	75	5.2	138.5			8.0	136.0	TD	+	+	+	
11		12	76	5.7	138.1			9.0	133.0	TD		+	+	
12		18	77	6.7	137.4			9.0	131.0	TD		+	+	
13	5.0	60	07	87.5	136.5	+	+	9.1	130.5	TD		+		
14		06	79	8.41	35.7			9.5	129.0	TD	+	+	+	
15		12	80	8.8	134.4			10.5	128.0	TS		+	+	
16		18	81	9.1	133.2			11.3	126.8	TS		+	+	
17	6.0	60	08	29.4	132.2	+	+	12.0	125.5	TS		+	+	

18		06	83	9.61	31.1			12.3	124.1	TS		+	+	
19		12	84	8.7	131.2			12.6	123.0	TS			+	
20	7.0	60	08	56.3	131.3	+	+	13.4	120.0	TS			+	
21		06	86	6.3	131.4			13.8	118.4	STS	+	+	+	
22		12	87	5.31	31.5			14.1	118.0	STS			+	
23	8.0	60	08	84.7	133.7			15.0	116.7	T		+	+	
24		06	89	4.5	134.5			15.8	116.0	T	+	+	+	
25		12	90	4.5	136.1			16.2	115.0	T			+	+
26		00	91	4.5	138.1	+	+	16.6	113.5	T	+		+	+
27	9.0	60	69	24.4	138.2			17.0	112.7	T		+	+	
28		12	93	4.3	135.2			17.3	111.9	T			+	+
29		00	94	4.3	138.4			18.0	110.3	T	+	+	+	+
30	10.0	60	69	54.4	138.5			18.4	109.6	T			+	
31		12	96	4.7	137.8			19.0	108.4	T		+	+	+
32		18	97	4.71	37.9			19.4	107.6	STS			+	
33		00	98	4.7	138.1			19.7	106.8	STS	+	+	+	+
34	11.0	60	69	94.7	138.2			20.0	106.3	STS			+	+
35		12	100	4.7	138.3			20.3	105.7	STS			+	+
36		00	101	4.6	138.5	+	+	20.9	104.2	TD	+	+	+	
37	12.06	06	102	4.6	138.5			21.0	103.0	TL		+	+	
38		12	103	4.6	138.5			23.0	104.0	TL		+	+	+

Note: TC evolution stages: TL, initial tropical disturbance; TD, tropical depression; TS, tropical storm; STS, strong tropical storm; T, typhoon (Pokrovskaya and Sharkov, 2006); φ, latitude; λ, longitude (in degrees); FSR, radiosondes of fixed station.

Note that the development of this TC exemplifies well the ocean's contribution to the increase in TC energy. According to synoptic analysis, the structure of fluxes in the troposphere and its stratification (the weakness of moist penetrating convection during the onset) clearly did not favor TC development.

Another TC (No. 8911) arose when a high-altitude cold meso-vortex started to interact with the ITCZ. Its development at the early stage was observed when there was upper-tropospheric hollow divergence caused by the southern periphery of the upper-tropospheric hollow, which generated in the north Philippine Sea. On July 21, a tropical disturbance with a depth of 1,006 mb arose near the ground. Between July 25 and 26, the TC reached its maximum, the pressure at the center dropped to 940 mb, and the wind velocity was 43 m/s. Radio sensing on the RV was mainly carried out at this time near the ITCZ and on the southern periphery of the developing TC.

Figure 8.7 schematically shows the trajectories along which the RV and the TC were moving, and Table 8.5 gives their dates, times, and coordinates. Meteorological observations and the accompanying information that was gathered are similar to those listed in the pursuit scenario.

A detailed description of the other scenarios is contained in Pokrovskaya *et al.* (2002).

In short, this section presents the results of systematization and classification of the large-scale crisis anomalies of the tropical atmosphere in the water area of the northwest Pacific Ocean that took place in 1989. The classification is based on

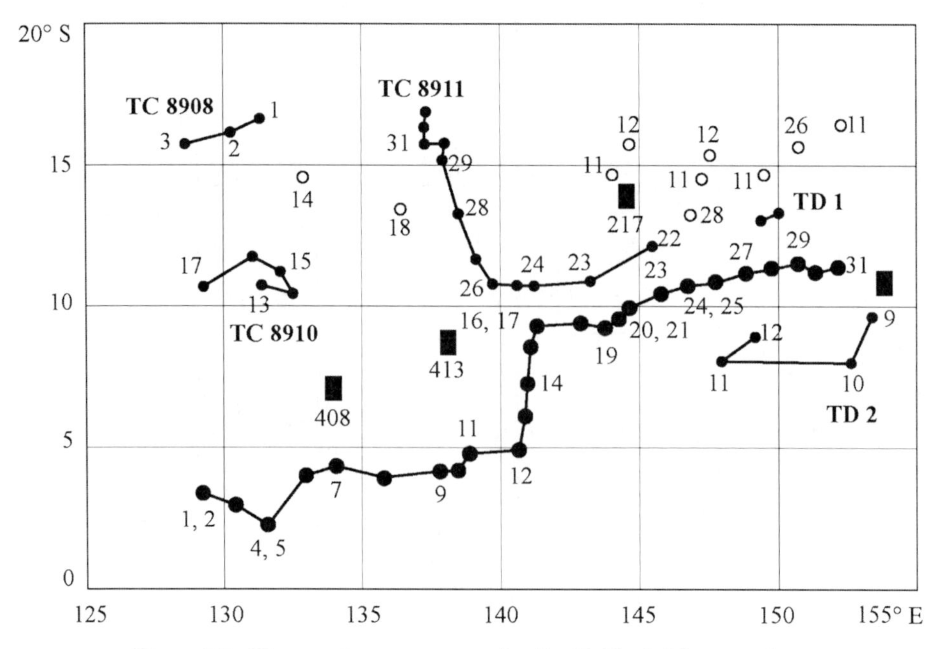

Figure 8.7. The conglomerate scenario. See Table 8.5 for notation.

the scenario principle proposed by the authors. This principle reflects features of the geophysical (natural) evolution of crisis anomalies and features of the hardware–technological situation in carrying out the experiments. On the basis of the critical analysis of primary (raw) data that was carried out, the unique thematic geophysical Scenario-TC database of dynamic and thermodynamic fields of the tropical atmosphere was formed, which has been successfully used in studying the detailed features of turbulence in the crisis zones of the tropical atmosphere (Sharkov, 1999a, b; Klepikov *et al.*, 1990, 1995) (see also Section 3.4.5) and in identifying the experimental tasks that need to be carried out in order to study the large-scale fields of spirality (Zimin *et al.*, 1991; Pokrovskaya *et al.*, 2002).

8.7 THE LONG-STANDING GLOBAL-RT DATABASE FOR THE GLOBAL RADIOTHERMAL FIELDS OF THE EARTH

The section describes the Global-RT electronic database which collects long-standing measurements carried out by satellite microwave instruments (e.g., SSM/I). The database is founded on a principally new methodology, which differs from standard meteorological databases. The database was developed at the Space Research Institute of the Russian Academy of Sciences. The development of this database was triggered by the need to use data on the global radiothermal field of the ocean–atmosphere system to solve the problems that face multiscale investigation of the Earth climate—particularly in studying the stochastic modes of atmospheric catastrophes (tropical cyclones) and polar transfer in the atmosphere. In this section the original principle on which satellite data are to be interpreted is formulated (i.e., the ideological basis of the database), the special software developed and introduced to date is described, and the future prospects for database development are presented.

8.7.1 The need for global databases to provide computer animations

One of most important climate-forming factors on the Earth is considered to be the multiscale (in space and time) interaction of the ocean and atmosphere, which is composed of the diverse processes of energy, momentum, and substance exchange. The basic means of obtaining instantaneous information about this interaction (temperature of the atmosphere and ocean surface, near surface wind velocity, integrated moisture content and water content of the atmosphere, intensity of precipitation, etc.) globally is satellite microwave radiometry. As recent investigations have shown, radiothermal space observations are pretty effective at studying the ocean–atmosphere interaction on seasonal and synoptic scales, and tropospheric catastrophes on diurnal scales. Among presently functioning satellite microwave radiothermal complexes the SSM/I instruments stand out; they are popular with the science community as a result of the unprecedented reliability of their measurements and the long time of stable operation on orbit within the framework of the U.S. Defense Meteorological Satellite Program (DMSP) (*http://dmsp.ngdc.noaa.gov/*

Table 8.5. The conglomerate scenario.

Part 1														
	RV radiosondes NIS					*FSR*			*TL No. 1*			*TC No. 8908*		
Number	*Date*	*GMT*	*No.*	φ	λ	*No. station*			φ	λ	*Evolution stage*	φ	λ	*Evolution stage*
						408	*413*	*217*						
1	2	3	4	5	6	7	8	9	10	11	12	13	14	15
1	14.07	00	166	4.0	129.5	+		+	13.5	153.0	TL	16.6	131.5	T
2		06	167	4.0	129.5				13.0	149.5	TL	16.4	130.3	T
3		12	16	83.7	130.5			+				16.3	128.7	T
4	15.07	00	169	3.0	131.7		+	+						
5		12	170	2.9	131.7									
6	16.0	70	017	14.9	132.7	+		+						
7		06	172	4.7	133.9									
8		13	173	4.61	35.3			+						
9	17.0	70	017	44.5	137.5	+		+						
10		06	175	4.51	38.2									
11		12	176	5.0	138.5									
12	18.0	70	017	75.0	140.9	+	+	+						
13		06	178	6.2	140.8									
14		12	179	7.2	140.9			+						
15		19	180	8.7	141.0									

16	19.0	70	018	19.3	141.2	+	+	+						
17		06	182	9.3	141.2									
18		12	183	9.3	142.4			+						
19		18	184	9.3	143.5									
20	20.0	70	018	59.6	144.0		+	+						
21		08	186	9.7	144.2									
22		12	187	10.0	144.8			+						
23		18	188	10.2	145.9									
24	21.07	00	189	10.7	146.8	+		+						
25		06	190	10.8	146.8									
26		12	191	10.9	147.8			+						
27		18	192	11.2	149.0									
28	22.07	00	193	11.5	150.0	+	+	+						
29		06	194	11.5	150.0									
30		12	195	11.6	151.0			+						
31		18	196	12.0	152.0									

(continued)

Table 8.5 *(cont.)*

Part 2

Number	*TL No. 2*			*TC No. 8910*			*TC No. 8911*			*Attendant data*			
	φ	λ	*Evolution stage*	φ	λ	*Evolution stage*	φ	λ	*Evolution stage*	*NOAA*	*GMS*	*Baric*	*Data recovery profiles*
1	16	17	18	19	20	21	22	23	24	25	26	27	28
1										+	+	+	
2											+	+	
3											+	+	
4										+	+	+	
5											+	+	
6										+	+	+	
7											+	+	+
8											+	+	
9	10.0	153.0	TL							+	+	+	+
10	8.5	152.5	TL								+	+	
11	9.0	148.0	TL							+	+	+	
12	9.0	149.0	TL							+	+	+	+
13				11.0	132.0	TL					+	+	
14				10.5	133.0	TL					+	+	+
15				11.0	132.5	TL					+	+	

16				11.5	131.0	TL					+	+	+
17				11.0	129.5	TL					+	+	
18											+	+	+
19											+	+	
20										+	+	+	+
21											+	+	
22							12.0	145.5	TL		+	+	+
23							11.0	143.0	TL		+	+	
24							11.0	141.0	TD	+	+	+	
25							11.0	140.5	TD		+	+	
26							11.0	140.0	TD		+	+	+
27							12.0	139.5	TD		+	+	
28							13.5	139.0	TD	+	+	+	
29							15.5	138.0	TD		+	+	
30							15.8	138.0	TD		+	+	
31							15.9	137.5	TD		+ +		

Note: See Table 8.4 for meanings of symbols.

dmsp.html). The information saturation of the global radiothermal field of the ocean–atmosphere system required for climatological and atmosphere–ocean investigations, carried out at various institutes of the Russian Academy of Sciences including the SRI, puts the formation of a specialized SSM/I remote database firmly on the agenda. Construction of the database should be based on the principle of considering remote data as long datasets: both in spatial (global coverage of the Earth with the possibility of sensing important regions) and temporal (long-standing daily observations of separate zones and the globe as a whole) concepts. As the global situation in the ocean–atmosphere system varies rapidly, the most representative and visually revealing of multiscale interactions in the ocean–atmosphere system could be accomplished by studying computer-animated representations of spatiotemporal images of space data. However, the use of raw data, taken directly from the open databases of the DMSP system by combining data mechanically, does not allow (for a number of reasons) generation of representative datasets of global radiothermal fields which, in turn, would allow computer-animated representations (video clips) to be implemented.

The purpose of the present section is detailed consideration of the original principle of satellite data interpretation, which became the ideological basis of the database of interest, and description of the special software developed and introduced to date (Ermakov *et al.*, 2007).

8.7.2 Principles of the design of the Global-RT database

The SSM/I complex represents a seven-channel radiometer receiving linearly polarized radiation at frequencies of 19.35, 22.235, 37.0, and 85.5 GHz. Both horizontally and vertically polarized radiation is measured at all these frequencies, except 22.235 GHz, which is exclusively the domain of vertically polarized radiation. The spatial step of measurements on the Earth surface equals 12.5 km for the 85.5 GHz channel and 25 km for other channels (for various sizes of the resolution spot). The observation band equals about 1,400 km in width, with conic scanning geometry. The SSM/I is carried by U.S. satellites of the DMSP series whose measurements virtually completely cover the Earth's surface (see, e.g., *http://podaac.jpl.nasa.gov:2031/SENSOR_DOCS/ss mi.html*).

A series of electronic archives (mainly in the U.S.A.) specializes in storing SSM/I data. The data can be obtained via the information portals of these archives on the Internet; members of the science community may order the necessary information online. The authors of this work have ongoing contact with the U.S. Global Hydrological Research Center (GHRC) (*http://ghrc.msfc.nasa.gov*)—one of the basic keepers of satellite data from DMSP series vehicles. This center kindly provided raw data on the SSM/I measurements carried out for the 1995–2005 period by F10/F15 DMSP satellites. The total volume of data was about 150 GB. The days for which data are available in the database are marked for each complex and for each year.

The significance of information on the global radiothermal field of the ocean–atmosphere system for climatological and atmosphere–ocean investigations carried out at the SRI was the trigger for construction of a specialized base of global radiothermal SSM/I data (Global-RT). Construction of this database was based on the principle of considering global remote data as a long series of spatiotemporal observations. The long sequence of radiothermal measurements is not considered here as a mechanical association of the data from several files corresponding to successive instants of survey or neighboring points at the Earth surface, but represents a fundamental database record generated by user queries which allows subsequent processing operations to be applied to this unit. Particular characteristics of the datasets (data sources, spatiotemporal extent, digitization, averaging, etc.) are determined by the parameters of user queries. Output data can be written in a single or several files. The most natural method of visualizing the obtained data seems to be the formation of a series of images or a video clip (i.e., computer animation).

Such an approach to database construction is most effective when analyzing global changes in the ocean–atmosphere system on seasonal, annual, and multiyear timescales. At the same time, it does not have full analogues among the open electronic archives of remote observation data, which generally select and configure data files according to user selection criteria without essentially updating the data structure and techniques of data representation in files. Consistent implementation of this approach allows determination of the best database structure and the format in which internal data are represented, and classification of the basic types of data generated as a result of user inquiries. This resulted in original software that combines the required flexibility and efficiency of the user's work with stored data. On the basis of the global radiothermal fields obtained, the SRI put together—for the first time in radiothermal global observation practice—a computer-animated representation. Two short clips can be viewed on the SRI's website. They were produced in the Earth Studies from Space Department of the Climatic Investigations Laboratory (*http://www.iki.rssi.ru/asp/lab_555.htm*).

8.7.3 Structure and data presentation in the Global-RT database

The structure of the database was determined by what topical climatology scientists needed to resolve problems relating to processes in the ocean–atmosphere system on various temporal and spatial scales. The principle of satellite radiometric information interpretation as long datasets formulated above suggests fast access to database elements (i.e., the storage of the greater part of the database on hard disks at the server used to process user inquiries, or on computers in a local network with a server). Data storage on independent carriers (i.e., CDs, magnetic tapes, etc.) was initially considered here basically as a means of backup, whose automation was dependent on future developments. Thus, the database structure can be represented as a "tree" of catalogues on hard disks with local and network catalogues and logic disks as "branches" of the tree. Inside the tree of catalogues the data are grouped according to the sources (numbers of satellites), time of surveying, and/or other parameters.

The satellite information delivered to SRI from electronic archives that are accessible via the Internet is written in HDF format, which is the world standard of remote data storage. The advanced typology of HDF format makes it a natural choice as the meta-language (i.e., a data description language) for a database. HDF format was developed at the National Center for Supercomputing Applications (NCSA) (*http://www.ncsa.uiuc.edu/*), at Illinois University, as the standard by which the center could organize data for wide free exchange in the science community. The basic purpose of introducing such a standard is to relieve users of the necessity to accompany data by explanations and comments describing the context in which particular data files should be interpreted and used.

With this purpose in mind a limited set of basic types of data were introduced into HDF: number and string attributes, the dataset (a multidimensional array), a bitmap, a palette (an indexed set of colors), and a V-group. For data of arbitrary origin, adequate representation in terms of one or several types of data determined in HDF can be selected. Hence, universal constants can be stored as number attributes, experimental results as a multidimensional array of integers or fractional numbers, plots as bitmaps, etc. V-groups are used for logic association of several datasets by analogy with the file system's directories. Users can equip written datasets and files as a whole with textual comments by adding string attributes.

Internal data representation in HDF suggests that the most convenient way of organizing data storage in files is not linear (successively recording information as it arrives), but hierarchical. Each HDF file starts with a heading that describes the general contents of the file and gives subheadings of the separate types of data. The latter contain information on the quantity of datasets of their particular type and give their subheadings. Each new subheading details the parameters of a particular dataset according to the hierarchical principle down to the lowermost level—the data themselves. As a whole, the structure resembles a book's table of contents, and the data themselves play the part of the basic text.

The recommended way of reading and recording HDF file structure information is the use of standard libraries. Such libraries are supplied by the developer as initial codes adapted for various platforms and operating systems. The authors have used libraries compiled on IBM-compatible personal computers for the OC Windows family in the software development medium MS Visual Studio 6.0.

Thus, the database's internal data representation standards are files in HDF format. Standard zip compression is applied to stored information for economy of disk space. Note that, along with files containing measurement data on the initial coordinate grid of an instrument, it is necessary to store in the database information on the geographical attribution of these data as well, which is supplied as special files—two for each data file—giving the coordinates of measurements in the low-frequency channel and high-frequency channel separately).

8.7.4 Basic data types generated as a result of user queries

The data stored in the database contain calculated (restored from initial satellite measurements) values of radiobrightness temperatures in all SSM/I channels: on the initial grid of measurements (swath-type data) and on the regular grid, transfer into which is accomplished by averaging initial data in cells of size 0.5° latitude by 0.5°

longitude (grid-type data). Swath-type data bear more complete and adequate information, as these are the data used to generate output data in response to user queries. Output data are in the form of compressed HDF files, but they have two principal distinctions from initial data:

(1) when generating output data only those initial data that satisfy the selection criteria specified by the user are used (filtration is performed not only on separate files, but also on parts of a file);
(2) output data are represented much more simply in files than initial data, which considerably facilitates the user's work (especially as knowledge of HDF is not required here).

In accordance with the accepted principle of stored information interpretation, output data are represented as long datasets of measurements performed under fixed (specified by the user) conditions. At present, such datasets are formed as a sequence of several (probably, many) files. There is also the possibility of synthesizing a single file of more complicated type from the obtained sequence in the long term.

Since there is a long series of measurements made by satellite instruments during long time intervals, to combine them into a sequence of identical datasets it is reasonable to transfer from measurements made on the instrument's own coordinate grid (which has an irregular structure because of scanning geometry features and Earth's surface curvature) to the regular grid (which has a constant step in latitude and longitude). Data smoothing in cells is used for this purpose, which is similar to the standard technique applied at transition from swath-type data to grid-type data. The important feature is the fact that it is the user who chooses the averaging scales. If necessary, having specified the corresponding small scale, the user obtains real data of one-time satellite measurements.

So, the output data generated represent the values of radiobrightness temperatures of the ocean–atmosphere system restored from SSM/I measurements, stacked on a regular grid, and written as a sequence of files of simple format. Detailed classification of the output data is then carried out for two categories:

— conditions of selection (filtration) of the initial data;
— the method of data representation in data files.

The selection conditions, determined by user queries, allow sampling of the required data to be carried out according to the following (independent) criteria:

(1) the data source (selection of information from one of the DMSP series satellites: F8, F10–F15, ...);
(2) the channel of observations (from one to seven SSM/I measuring channels);
(3) the interval of observations (beginning and end of observations to an accuracy of a year, month, and day);
(4) the zone of observations (global coverage or a rectangle with specified latitude and longitude boundaries);
(5) the spatial scale of averaging (step of the regular grid in degrees or fractions of a degree adjusted independently in latitude and longitude);

(6) timescale of averaging in units of a day (used to accumulate and average the data of successive flights of a satellite over the chosen area of the Earth).

Any data, selected according to these criteria, can now be presented in one of two types: PCK and RAW.

Data in PCK format are written as a sequence of PCK files containing the values of radiobrightness temperatures, calculated by averaging, at the nodes of a regular grid in floating point double precision–type format of 8 bytes per number. Actual accuracy is determined by the maximum accuracy of initial data (to hundredths of a degree). The nodes of a RAW grid are supposed to be ordered in strings in the direction of decreasing western and/or increasing eastern longitude (i.e., from west to east) at the fixed latitude, and successive strings in decreasing northern and/or increasing southern latitude (i.e., from north to south). Each PCK file contains the data for a single time-averaging interval (or, without time averaging, the data for one day). These data are filtered, averaged, and combined with an allowance made for other selection conditions: the data source (a satellite), the channel chosen (if several channels are chosen the corresponding number of separate files is formed), the given area of the Earth surface. The sequence of such files covers the whole time interval of observations chosen by the user (with limitations imposed by possible absence of the necessary initial data). The semantics of the PCK filename reflects the frameworks described above (which restrict the selected initial data) and the processing applied to them. In addition, each PCK file is supplemented by one index NDX file. The NDX file is arranged similarly to the PCK file and has the same name (apart from the extension: NDX instead of PCK). Each value of the mean temperature in the PCK file has a corresponding integer (4 bytes per value) that shows the number of separate measurements included when calculating the mean value for the given node of a grid. The NDX file is a service file used to generate PCK files. Nevertheless, it can be used by a user, for example, to estimate the reliability of calculated mean values at separate nodes. PCK-type data are optimum for use in further calculations, since they are presented as floating point numbers and, if necessary, can be supplemented by index files characterizing the quality of averaging.

Data in RAW format are arranged similarly to data in PCK format. The difference consists in the fact that calculated values of radiobrightness temperatures are multiplied by a scaling factor rounded up to integers and written down in integer format (one or two bytes per value). These data are optimum for direct visualization of averaged radiothermal fields (RAW files can be looked at using popular software like Adobe Photoshop). They are suitable for fast analysis of selected information, but do not contain the absolute values of temperatures and, consequently, are not very suitable for further calculations.

8.7.5 Special software to cope with accumulated data

The database formulated above required development of special software (the database driver) allowing problems relating to the representation and processing of stored data as a long series of measurements to be efficiently solved. The huge volume of stored information (several hundred thousand files) makes addressing separate data files too time consuming and unproductive for the user or database operator. It is necessary to have some universal means of addressing immediately a distinct class (set) of files or all the database's files needed within the framework of one processing session. Simultaneously, a certain flexibility or "contextual orientation capability" (in geek speak) in the way this is done is required that would allow various processing procedures to be chosen and adjusted and to apply them only to those files that contain the data of interest to the user. Such a means must also be easily expanded to satisfy the new requirements of output data that arise when working with a database.

Such a database driver consists of two parts: a library of specialized modules, each of which contains a description of (1) a particular operation applied to database files or to the files of derivative types of data, and (2) a universal module that organizes "conveyor" processing of a distinct class of data files as a result of applying a library operation.

The universal module (the conveyor) is called a Stamper. It represents a Windows application that has a simple graphical user interface in dialogue window form, which allows the operator to load one of the library modules, to indicate a root directory of processing, to determine the class of processed files (such as *.gz or *.dat, etc.), and to actuate the conveyor adjusted in such a way. The task of Stamper is to construct a processing tree (by locating and including in the processing list all files satisfying the given pattern and placing them in the root directory of processing and in its subdirectories), to extract specific processing instructions from the loaded library module, and to apply these instructions to the whole list of files found. Thus, Stamper effectively automates the processing of a long series of files without limiting its capabilities; the arbitrary processing algorithm, described according to specific standards, will be loaded and applied to a particular file or to an arbitrary set of files on choosing the database operator.

The library modules used for processing represent special dynamic link Windows libraries. To distinguish them from other dynamic link libraries (system libraries and those supplied as part of the software), they have a file extension of "stp" (from the word "stamp"). A specific feature of these modules consists in the obligatory presence of an input function of processing with an established name and syntax. The duty of the module, along with any processing as such, includes also checking the file contents to see it they correspond to established processing criteria. So, the module, which searches remote data for the given region within the given time interval, must check the data actually contained in a file to see if these restrictions correspond. Module settings (the data selection criteria specific for the given operation) are stored in automatically generated sections of a system registry. If necessary, the operator is capable of editing the records of the registry directly, by means of the reedit application situated in the Windows system directory. However, special applications are produced for existing library modules, which allow these records to be edited in dialogue mode by means of a simple graphical user interface.

A set of processing modules constitutes the library of modules, which can be infinitely expanded by producing new dynamic link libraries satisfying fixed standards. Among modules developed and introduced up to now at the SRI, the Picker.stp and Mapper.stp modules should be mentioned because they generate the output data in PCK and RAW formats, respectively.

8.7.6 Future developments

The practice of working with this database in the SRI expands and updates the set of requirements placed on generated output data and on the techniques to process them. The following requirements are the most topical:

- introduction of a data type similar to PCK or RAW, but equipped with time labels characterizing the surveying instant;
- expansion of Picker's capability to synthesize output data based on information from several satellites with various types of association/supplementation logic;
- expansion of Stamper by getting a program written so that it automatically loads and successively produces several library modules of processing;
- developing the library of modules for thematic processing of radiometric data and for preparing and backing up information.

So, issues relating to the study of Earth climate variations requires satellite remote-sensing data if they are to be resolved—in particular, regular and long-term global microwave radiothermal observations of the ocean–atmosphere system. The database of satellite SSM/I instrument measurements, produced at the SRI, is called on to provide this; it is organized according to the principle of interpreting data as a series of long-term global measurements. Such long datasets are constructed from initial data according to the selection criteria specified by the user and represent from the user's viewpoint a fundamental database record. To date, the stored information basically includes the data of continuous measurements during 1995–2005, performed on the F10/F15 spacecraft of the DMSP series. The total volume of information equals about 150 GB.

To generate datasets of long-term global measurements (i.e., the principle underlying the database), special software has been developed and partially introduced. This software combines the efficiency of a universal approach to searching and gathering initial data with the flexibility of applied processing operations and the possibility of unlimited expansion of a library of such operations.

The database has already been given a number of tasks to resolve within the SRI, which have facilitated update and expansion of the list of requirements of both the format of output data and the format of developed software.

8.8 INTEGRATED OBJECT-RELATED DATABASE OF TROPICAL CYCLONES IN THE GLOBAL WATER VAPOR FIELD

The advantages of object-related technology for checking the hypothesis on the interrelation between integrated water vapor concentration and multiple tropical cyclogenesis are considered in this section. By means of the ENVI 4.3 and Microsoft Visual Studio 2008 software the first version of the object-related EVA-00 database is considered. This database includes remote satellite information on two stochastic processes possessing essentially different spatiotemporal scaling and structural characteristics. The first process—tropical cyclogenesis—is considered as a stochastic set of random events (objects)—tropical cyclones—and the second process as the spatial global field of integrated water vapor that had considerable spatiotemporal variability in 2001. The section experimentally presents the interrelation between the water vapor areas of heightened concentration and tropical cyclogenesis, which became evident only after applying object-related technologies. A computer-animated representation based on data from the database clearly demonstrates the interrelation between water vapor areas of heightened concentration and tropical cyclogenesis.

8.8.1 Considerations for the design of an integrated object-related database

The remote study of primary forms of tropical cyclones (TCs) and the geophysical medium surrounding them occupies a special place in programs that remotely monitor tropical disturbances. Mention should first be made of the problems of forecasting the appearance of primary forms of disturbance and subsequent transition of one of them into a TC, and of the problems involved in the detailed remote study of structural, dynamic, and thermodynamic features of a tropical disturbance directly at the instant of drastic (and often unexpected for observers) intensification of a primary form. The drastic intensification of TC "Katrina" in 2005, which entailed a loss of lives and great material damage, can serve as an impressive example of such a situation (Sharkov, 1997, 1998, 2000, 2010).

However, attempts at the remote study of the primary forms of tropical disturbances face a series of difficulties, not least of which is the absence of a generally accepted physical model of this complicated geophysical phenomenon and, accordingly, of the necessary geophysical parameters for measurement. Despite considerable efforts by researchers to observe and record separate (and fragmentary) optical and IR images of tropical vortical disturbances at various phases (see, e.g., Sharkov, 1997, 1998, 2000, 2006), the generally accepted remote criteria of geophysical medium proximity to individual tropical disturbance generation and to the critical moment of transition into the developed form are still absent. So, a principally new way of studying the remote criteria of TC genesis has been developed. It considers the results of complex multifrequency optics–IR–microwave satellite investigations of the way in which optical images of TCs evolve in the integrated water vapor field. Analysis of these results revealed the fundamental contribution of a quick-response energy source, as a result of which mature forms of typhoons are formed, and horizontal transfer of water vapor by global circulation and jet fluxes which sustains mature TC forms (Sharkov, 2010; Sharkov *et al.*, 2008a, b; Kim *et al.*, 2009). This explanation for TC genesis is clearly of major importance. Moreover, it was obtained after three separate samplings of TCs were studied. Full experimental proof of this statement requires studying the evolution of multiple tropical cyclogenesis when integrated water vapor is intensively migrating in the tropical zone of Earth's atmosphere.

The complexity of the problem lies in the need for synchronous analysis of the remote satellite information of two stochastic processes possessing essentially different spatiotemporal scaling and structural characteristics. As already mentioned, the first process—tropical cyclogenesis—is considered as a stochastic set of random events (objects)—namely, the stochastic genesis of tropical cyclones (Sharkov, 2000)—and the second process as the spatial global field of integrated water vapor that has considerable spatiotemporal variability (Sharkov *et al.*, 2008a, b). Combining these two processes should be performed over a minimum time interval (a daytime pixel in this case). If, however, the time interval increases, the efficiency of the offered technique drastically drops because of the finite lifetime of tropical disturbances and the high spatiotemporal variability of the water vapor field. Attempts at generating such complex databases have recently been undertaken (Pokrovskaya *et al.*, 2004); however, they have not been completely computerized. Analysis of the methodology used to construct modern databases has shown that the necessary temporal combination of the data of two stochastic processes is possible only by using object-related technology when constructing complex integrated databases, something that is currently being actively developed (*http://www.s-networks.ru/index-686.html*).

The purpose of this section is to describe the first version of the integrated EVA-00 database set up using elements from object-related technology, which includes remote satellite information on the two stochastic processes (Shramkov *et al.*, 2010). Synchronous analysis of these stochastic fields shows that the formation of multiple tropical cyclogenesis during a year interval (both in the northern and southern hemispheres) occurs in the field of heightened concentration of integrated water vapor. This interrelation only becomes clear after applying the object-related technologies of satellite information processing.

8.8.2 EVA-00 database structure

Multimedia types of data—such as satellite images, different-scale maps, and videoclips—are usually processed by specialist software. However, many web applications have appeared now that demand server databases that are interactive so that such data can be controlled (despite the complex character of the data). Moreover, new software systems to process stochastic fast-varying processes also are needed to store the data. We have already discussed two such types of data and the necessity of controlling them. Object-related technology has been developed to satisfy these requirements; it provides simple techniques for the development, deployment, and control of applications operating with complex data.

Servers that support object-related technology can be adjusted by developers such that their own types of data specific to the field of application can be produced. These servers have been expanded to support the full capabilities of object modeling, including inheritance and multilevel collections, as well as for the evolution of data types. All these elements of object modeling are fully revealed on statement of the problem as a synchronous combination of databases of stochastic processes of essentially different spatiotemporal form, such as global tropical cyclogenesis and the global field of integrated water vapor. The complexity of the problem lies in the fact that applications of these newly produced types of data should operate with abstractions peculiar to the domain of the given subject (tropical cyclone evolution as a stochastic set of random events and the evolution of a spatiotemporal field of water vapor). It is highlly desirable to integrate these new types with server databases as closely as possible, so that they could be processed along with built-in standard types of data.

A block diagram of the first version of the EVA-00 database showing object-related technology elements, including remote satellite information about the two stochastic processes referred to, is presented in Figure 8.8.

The EVA-00 database uses information from its precursors: the Global-TC and Global-Fields databases. The first represents a storehouse of systematized remote data on global tropical cyclogenesis (i.e., it contains information on the physical process considered over all water areas of the World Ocean; Pokrovskaya and Sharkov, 2006). The information was first systematized over separate regions, in

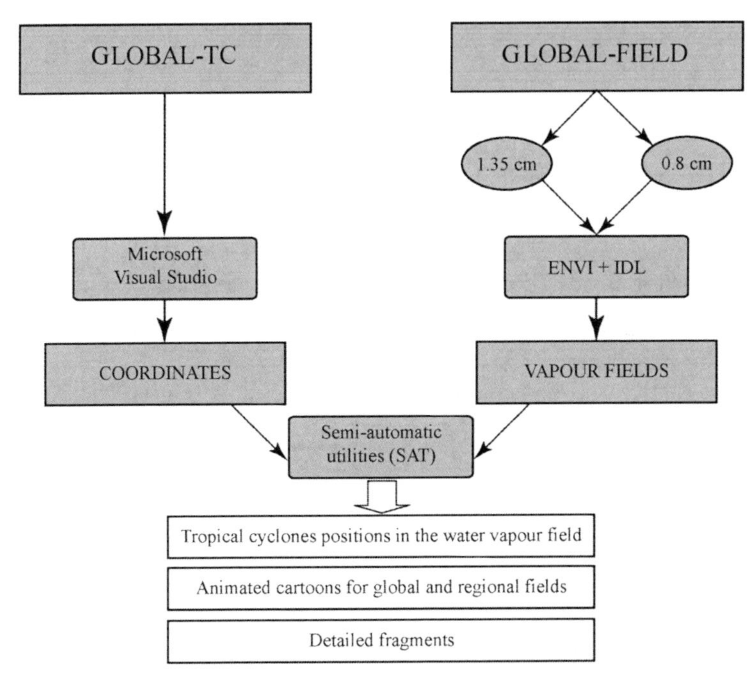

Figure 8.8. Block diagram of the first version of the EVA-00 database incorporating object-related technology. This database includes remote satellite information on two stochastic processes: tropical cyclogenesis and the spatial global fields of integrated water vapor. Global-TC collects initial data on global tropical cyclogenesis. Global-Fields collects initial data on global radiothermal fields derived from satellite data (from the Global-RT database). Types of EVA-00 database products are illustrated at the bottom of the diagram.

each of which temporal and spatial attributions were carried out, the correctness and completeness of messages relating to the characteristic climatic features of each region were checked, and pre-processing of raw information was performed. Each newly formed tropical cyclone, or tropical disturbance that did not transfer later into the developed TC form, constitutes a separate file of information in the database.

The Global-Fields database contains information on the global fields of radiobrightness temperatures (Ermakov *et al.*, 2007) obtained by means of the SSM-I complex—a seven-channel radiothermal instrument receiving linearly polarized radiation at frequencies of 19.35, 22.235, 37.0, and 85.5 GHz. At all frequencies (except 22.235 GHz which is reserved for vertical radiation), both vertical and horizontal polarized radiation are measured. The spatial view field of measurements on the Earth surface equals 12.5 km for the 85.5 GHz channel and 25 km for other channels. For information on water vapor content, data on radiobrightness temperatures in two channels—22.235 GHz and 37.0 GHz—are required.

The EVA-00 database represents a software complex that performs all the necessary functions at a given instant for processing global water vapor fields and information on tropical cyclones. Successive processing of data from the Global-TC base is first performed using the Microsoft Visual Studio programming language. As a result a text file is generated that contains data on the geographical position, time of origin, and existence of the disturbance and some meteorological information for 2001 (the year of interest).

The fields of radiobrightness temperatures, obtained at two frequencies—22.235 and 37.0 GHz—and taken from the Global-Fields database, are processed by means of the IDL programming language using the empirical relation (which represents a kind of linear algorithm of the given inverse problem) presented in Ruprecht (1996), according to the formula:

$$W = 131.95 - 39.50 \ln(280 - T_{22V}) + 12.49 \ln(280 - T_{37V}) \tag{8.1}$$

where W is the integrated water vapor value in kg/m^2 (or in millimeters) in the spatial resolution pixel of the SSM/I instrument; and T_{22V} and T_{37V} are the radiobrightness temperature values for the 22.235 and 37.0 GHz channels (vertical polarization) in the spatial resolution pixel of the SSM/I instrument. Special validation—carried out in this work between restored water vapor values and those measured from radio-sensing data (250 profiles) in the Atlantic Ocean water area and with difference in the times of measurements between the satellites and radio probes better than two hours—has shown that the root-mean-square deviation of results equaled about 2.58 kg/m^2. As a result, the global water vapor fields were obtained, which were then used for further processing.

The processing and structure of water vapor fields are based on the principle of considering global remote data as a long series of spatiotemporal observations. The long sequence of water vapor fields is not considered here as a mechanical association of the data from several files corresponding to survey instants, but represents from the user's point of view the basic structural unit of the database generated as a result of user inquiries and allowing for application of further processing operations. Particular characteristics of the data series (data sources, spatiotemporal extent, digitization and averaging, etc.) are determined by the user's inquiry parameters (the elements of object-related technologies). The output data can be written in one or several files. The best way of visualizing the data obtained seems to be by a series of images or a computer-animated representation.

The computer-animated representation that was produced on the basis of this database is available on the SRI website of the Earth Studying from Space Department of the Russian Academy of Sciences (*http://www.iki.rssi.ru/asp/dep_coll.htm*). This film visually demonstrates the relation between the areas of heightened concentration of water vapor and the genesis of tropical cyclones.

8.8.3 EVA-00 database products

To illustrate the products available from this database, we present frames from the computer animation of the global water vapor field, tropical cyclogenesis, and the time sequence of local fields and detailed frames. Figure 8.9 presents the global field of integrated water vapor on World Ocean water areas on September 5, 2001. The color scheme of water vapor intensities is presented at the bottom of the figure, where crimson areas correspond to the maximum values of water vapor concentration. These pictures were obtained by digitizing once a day. The current date of the picture is presented in the lower left corner of the figure in month–day–year format. The dark blue squares in the figure display mature tropical cyclones. On September 5, 2001, four tropical cyclones—"Danas", "Erin", "Gil", and "Henriette"—were observed simultaneously in the cyclone-generating water areas of the World Ocean at various stages of development (Table 8.6). Note that all tropical cyclones have

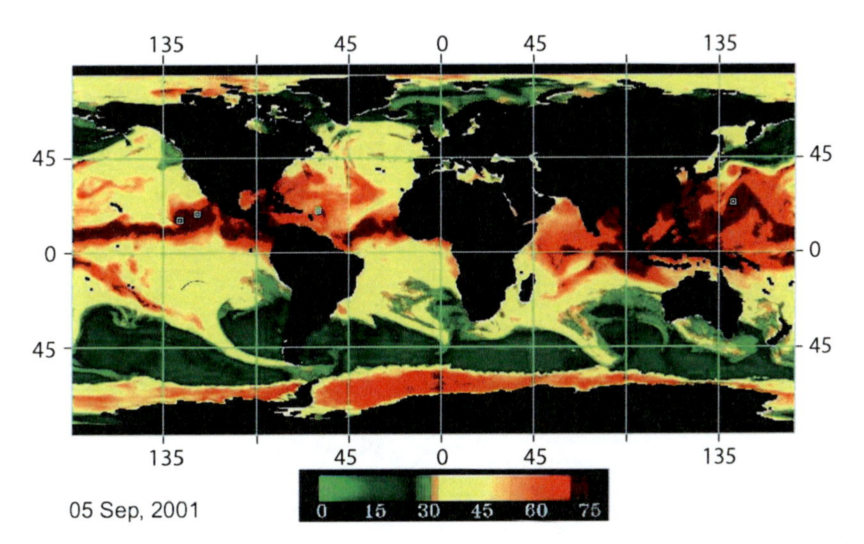

Figure 8.9. The water vapor global field above World Ocean basins on September 5, 2001. Centers of cloud structures for the four tropical cyclones that were active on September 5, 2001 are shown by white squares.

Table 8.6. The main parameters of tropical cyclones in the World Ocean on September 5, 2001.

Number	*Basin, TC number, and name*	*TC lifetime*	*Date of TC transfer to TS stage*	*Position of TC transfer to stage TS*		*Maximal stage of TC evolution and lifetime*	*Evolution stage on September 5, 2001*	*TC position on September 5, 2001*		*V*
				φ	λ			φ	λ	(m/s)
1	NWP 0116 "Danas"	02.09–12.09	04.09	18.5	152.8	T 04.09–10.09	T	21.5	150.8	41
2	NEP 0107 "Henriette"	01.09–09.09	05.09	17.7	−116.5	STS 06.09–07.09	TS	17.7	−116.5	21
3	NEP 0108 "Gil"	03.09–12.09	04.09	15.7	−123.7	T 06.09–08.09	STS	15.4	−124.7	28
4	ATL 0105 "Erin"	27.08–15.09	02.09	13.7	−39.7	T 08.09–14.09	TS	18.3	−57.5	21

Note: Basins and TC names are given in accord with international specifications. NWP, northwest Pacific Ocean; NEP, northeast Pacific Ocean; ATL, North Atlantic Ocean. TC evolution stages: T, typhoon; STS, strong tropical storm; TS, tropical storm; V, TC wind velocity (m/s); φ, latitude (degrees); λ, longitude (degrees).

entered the crimson areas of water vapor content (i.e., water vapor regions with intensities higher than 60 kg/m^2). Water vapor content is not presented on continents because of the complexity of solving the inverse problem over continents. Analysis of the computer-animated representation of the global field of integrated water vapor shows it possesses very high spatiotemporal variability. Westward transfer of the water vapor field can be from 300 km to 400 km per day causing the shape of isolines of various intensity to change considerably. It is this circumstance that likely stipulates the high value of root-mean-square deviation between water vapor field values restored from satellite data and those measured by radio sensing. It needs to be emphasized that only timescales of the order of a day are considered here, because they are necessary for studying how the energy content of the water vapor field evolves during cyclogenesis. If investigations are carried out on monthly and seasonal scales (Ruprecht, 1996), such a circumstance does not arise.

Other computer-animated products of local fields and detailed fragments are demonstrated in Figure 8.10 showing evolution of the water vapor field in the northwest Pacific Ocean in the presence of TC "Francisco" at various stages of its development and dissipation. The basic characteristics of TC "Francisco", which was active in the northwest Pacific Ocean during September 2001, are presented in Table 8.7, which was taken from the Global-TC database and was compiled using pre-processing techniques (Pokrovskaya and Sharkov, 2006) for the initial (raw) data presented on the IFA site *http://www.solar.ifa.hawaii.edu*

Table 8.7. The main parameters of TC "Francisco" (NWP 0117) in the northwest Pacific over the period September 15–25, 2001.

Date of evolution stage	*Evolution stage*	*Position of evolution stage*		*V*
		φ	λ	(m/s)
15.09	TL	10.0	171.0	10
19.09	TS	14.6	161.5	18
20.09	TS	16.7	157.2	23
21.09	T	19.2	151.6	33
22.09	T	22.6	148.9	46
23.09	T	26.3	147.5	51
24.09	T	31.4	148.7	46
25.09	L	39.6	152.7	28

Note: TC evolution stages: TL, initial tropical disturbance; TS, tropical storm; T, typhoon; L, extratropical disturbance in the midlatitudes (Pokrovskaya and Sharkov, 2006); V, wind velocity of the given stage of TC evolution; φ, latitude (degrees); λ, longitude (degrees).

The primary tropical disturbance of this TC was recorded on September 15 in the Marshall Islands region at 10.0°N, 171.0°E and represented an indistinct poorly organized cloudy system. The pressure at the center was 1,007 mb and near ground wind velocity varied from about 5 m/s to 7 m/s. During four days, from September 15 to 18, the disturbance remained at the initial stage as a result of being displaced in the west–northwest direction. In the course of time the cloud massif gradually isolated, but its edges remained blurred.

On September 19, having reached the 15° latitude, the disturbance quickly intensified and passed into the tropical depression stage, wind velocity increased to 15 m/s, and by 12:00 GMT the disturbance has become a tropical storm, pressure at the center continued to drop and reached 996 mb, and wind velocity increased to 23 m/s (Figure 8.9a). Note that transition from the TL stage to the TS stage was accomplished in a water vapor field whose intensity was more than 70 kg/m^2. On September 20 at 18:00 GMT the storm became a strong tropical storm, the pressure at the center dropped to 992 mb, and wind velocity increased to 31 m/s. Deep convection at the storm center acquired the distinct configuration with characteristic bending of cloud tail bands in the southern sector and less prominent tail bands in the northern sector. On September 21, at 12:00 GMT the strong tropical storm became a typhoon. This transition to TC stage occurred in a water vapor field whose intensity exceeded 70 kg/m^2 (Figure 8.10b–c).

During September 21–24 the typhoon intensified, wind velocity reached 51 m/s, and pressure dropped to 945 mb. According to optical satellite observation data, the cloud massif of the TC had adopted the characteristic TC formationof a compact, small central core and well-developed bent cloudy tail bands in the southern and southeast sectors. Note that it was during this particular time interval that the TC formed its own (daughter) field of integrated water vapor in the form of a compact central core (corresponding to the optical image) with a jet spiral bridge between the daughter core and mother field of water vapor (Figure 8.10e–f). This is especially clearly seen in Figure 8.10f. This was the moment (September 23) the storm circulation system, having reached 27.0°N at the typhoon stage, came under the influence of a stationary front located east of Japan. Storm displacement velocity reached 35 knots in the north–northeast direction. The southern tail bands have almost completely disappeared, the central core remaining compact. This was the moment when water vapor intensity in the jet spiral structure suddenly decreased and virtually broke down (Figure 8.10g) but the central core of the daughter field was kept. Continuing to be displaced northward, on September 24 at 18:00 GMT, the tropical storm transformed into the system of moderate latitudes eastward of the Kurile Islands (stage L; Table 8.7). The way in which the basic stages of TC "Francisco" evolved fully corresponds to the qualitative "camel" model of TC evolution offered in Sharkov *et al.* (2008a, b) and Kim *et al.* (2009).

Thus, the first version of the EVA-00 database incorporating elements of object-related technologies has been developed, using ENVI 4.3 and Microsoft Visual Studio 2008 software. This database includes remote satellite information on two stochastic processes possessing essentially different spatiotemporal scales and structural characteristics. The first process—tropical cyclogenesis—is considered as a

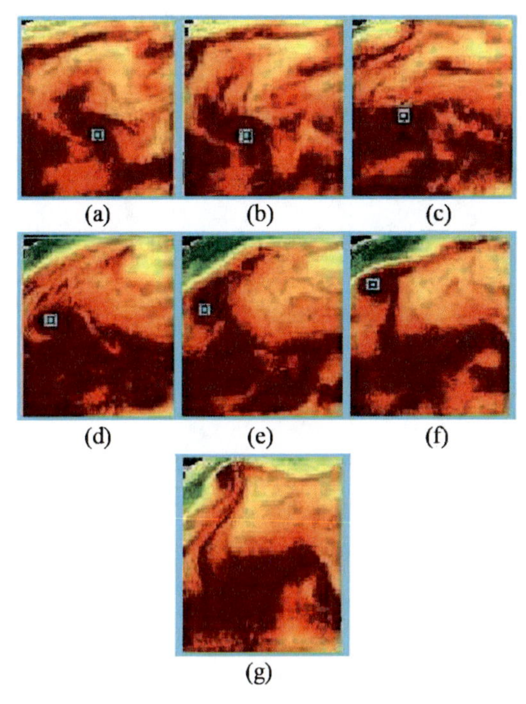

Figure 8.10. The time evolution of the integral water vapor field in the northwest Pacific Ocean in the presence of TC "Francisco" during September 19–25, 2001. The center of the TC cloud system is shown by the white square: (a) on September 19, 2001; (b) September 20, 2001; (c) September 21, 2001; (d) September 22, 2001; (e) September 23, 2001; (f) September 24, 2001; (g) September 25, 2001.

stochastic set of random events (objects)—tropical cyclones—and the second process as a spatial global field of integrated water vapor with considerable spatiotemporal variability in 2001. This work has experimentally presented the interrelation between the water vapor areas of heightened concentration and tropical cyclogenesis, which became evident only after applying object-related technologies. Despite all the advantages of the Global-TC and Global-RT databases, their use in analytical investigations and in gathering data automatically is not effective enough. As shown in this section, object-related technology databases will doubtless take the lead in processing and storage of such complex information.

9

Remote sensing activity for cyclogenesis studies

The wide variety of problems related to remote sensing of tropical regions and the difficulties that emerge when attempting to resolve them has led to the need to use essentially the whole spectrum of up-to-date spacecraft and potential space missions. Nonetheless, the whole spectrum of space missions is divisible into two major marked groups: the first serves substantially ecological and climatic tasks, the second is directed of mesoscale short-term atmospheric processes.

So, the existing satellite operations do not allow reliable estimations of radiative budget and atmospheric water parameters on time and space scales corresponding to the extent and lifetimes of typical tropical systems (TSs). They are able to provide monthly means over areas of some square degrees for clouds in the International Satellite Cloud Climatology Project (ISCCP) or for precipitation in the Global Precipitation Climatology Project (GPCP), while the systems extend only a few tens of kilometers and last only a few hours. Geostationary satellites provide good temporal resolutions, but suffer from inappropriate instrumentation, while polar orbiters, carrying more adequate instruments, provide insufficient time sampling. Small satellite systems with low-inclination orbits and a special space mission to study tropical structures are proposed to fill this gap. At present a number of interesting projects on tropical cyclone (TC) study using complex rocket–satellite systems have been proposed and developed by Russian researchers.

This chapter covers the current status of operational satellite systems and gives a synopsis of projected developments for special missions for tropical study.

The Western reader, as a rule, has a weak grasp of Russian scientific literature, where particular emphasis has been placed on programs relevant to climatic and ecological studies and of Russian missions for TS observations. This chapter will add to the substantial information on the Russian space program published by Harvey (1995).

E. A. Sharkov, *Global Tropical Cyclogenesis* (Second Edition).

9.1 POSITION OF TROPICAL STUDIES IN EXISTING SPACE PROGRAMS

9.1.1 Ongoing missions and development trends

The general structure of the international Earth Observing System's (EOS) (with the exception of the Russian section) ongoing missions and technical instrumention parameters are included in the *EOS Reference Handbook* (Asrar and Dokken, 1993) and the *EOS Science Plan: The State of Science in the EOS Program* (King, 1999). In this section we look at Western and Japanese missions in space that have a direct bearing on remote-sensing problems in the tropics. Operational satellite missions, their sensors, and short instrument descriptions are presented in Tables 9.1 and 9.2.

The major contribution made by geostationary systems to investigation and monitoring of large-scale phenomena (large-scale flows, cloud structures, TC interactions with environmental fields) is well known (Harries, 1995). However, there are distinct restrictions in the use of geostationary systems: active and passive radiophysical microwave instruments cannot be applied (they need spatial resolutions); poor resolution in visible and infrared (IR) bands for imaging and sounding modes; difficulty of solving ecological tasks in the Earth sciences. In spite of these facts geostationary systems are and will remain an integral part of future remote-sensing systems.

The Meteosat[1] Operational Program (MOP) was intended to provide an operational service to the European meteorological community until 1995, but has extended well into the 21st century. However, jointly with Eumetsat,[2] a so-called Meteosat Second Generation (MSG) was launched at the end of the 20th century (Goldsmith and Readings, 1994; Battrick, 1999).

MSG will also be spin-stabilized but will have an enhanced capability over the MOP, including both additional channels and a high resolution (0.50 km) option. The channel selection is as follows:

A. *Imaging*: 0.56–0.71, 0.71–0.95, 1.44–1.79, 3.40–4.20, 9.80–11.80, 11.00–13.00 μm.
B. *Liquid water*: 5.35–7.15, 6.65–7.85 μm.
C. *Pseudo-sounding*: 4.42–4.62, 9.46–9.94, 13.04–13.76, 13.60–14.40 μm.
D. *Liquid water/ice distinction*: 8.30–9.10 μm.

The MOP has only three channels.

With the European Space Agency's (ESA) MSG-1, launched in mid-2001, and MSG-2, launched in 2002 as its in-orbit standby, the spacecraft development and manufacturing program now extends well beyond 2003, the year in which MSG-3 was placed in ground storage nominally for 5 years. This scenario will allow Eumetsat to guarantee an uninterrupted operational geostationary imaging and

[1] Meteosat, European geostationary meteorological satellite.
[2] Eumetsat, European Organization for the Exploitation of Meteorological Satellites.

Table 9.1. Operational satellite missions relevant to tropical study.

Satellite	*Country/ Organization*	*Longitude*	*Sensor useful to tropical study*
Geostationary			
GOES-E	U.S.A.	75°W	Visible/infrared imaging radiometer (VISSR), atmospheric sounder (VAS)
GOES-W	U.S.A.	135°W	Visible/infrared imaging radiometer (VISSR), atmospheric sounder (VAS)
GOES-4/5	Japan	140°E	Visible/infrared imaging radiometer (VISSR)
Meteosat-3/6	ESA	75°W	Visible/infrared imaging radiometer
Insat-2B	India	74°E	Visible/infrared imaging radiometer (VISSR)
GOMS	Russia	76°E	Visible/infrared imaging radiometer (VIS-IR)
Polar orbiting			
NOAA series	U.S.A.	—	Visible/infrared imaging radiometer (AVHRR), infrared/microwave atmospheric sounders (HIRS, MSV)
Meteor series	Russia	—	Visible/infrared imaging radiometer
SPOT series	France	—	High-resolution visible sensor (HRV)
Landsat series	U.S.A.	—	Thematic mapper (TM), multispectral scanner
ERS-1/2	European	—	Active microwave instrument/synthetic aperture radar, scatterometer, radar altimeter (RA), along-track scanning radiometer (ATSR)
Topex/Poseidon	U.S.A./France	—	Altimeter (ALT)
DMSP series	U.S.A.	—	Operational linescan system (OLS), microwave imager (SSM/I)
ADEOS	Japan/U.S.A./ France	—	Advanced visible near infrared Radiometer (AVNIR), ocean color and temperature sensor (OCTS), NASAscatterometer (NSCAT), total ozone mapping spectrometer (TOMS), polarization and directionality of Earth's reflectance (POLDER), interferometric monitor for greenhouse gases (IMG)
JERS-1	Japan	—	Synthetic aperture radar (SAR), optical sensor (OPS)

Table 9.2. Operational/Experimental satellite sensors, key to abbreviations, instrument descriptions, capabilities.

ALT	TOPEX altimeter (2 bands: Ku 13.5 GHz, C-band 5.3 GHz, 2 cm precision, atmospheric correction provided by on-board microwave radiometer)
AMI/SAR	Active microwave instrument/synthetic aperture radar (C-band 5.3 GHz, 30×30 m resolution); scatterometer (wind mode); three-beam C-band, VV polarization, 50 km resolution, range 4–24 m s^{-1}, accuracy 2 m s^{-1} or 10%; scatterometer (wave mode); 5×5 km area every 100 km
ATSR	Along-track scanning radiometer (4 visible bands: 3, 7, 11, and 12 μm bands, 1×1 km resolution; accuracy +0.5 K; and microwave sounder (23.8 and 36.5 GHz) for water vapor content correction
AVHRR	Advanced very high resolution radiometer (5 bands: 0.7, 0.5, 3.7, 11, and 12 μm; 11 km resolution; ±1 K precision)
HIRS	High resolution infrared sounder (20 bands: 4.3–15 μm; field of view: 17 km)
HRV	High resolution visible sensor (3 bands: 0.5–0.89 μm, 20 m resolution; panchromatic band, 0.51–0.73 μm, 10 m resolution)
MSS	Multispectral scanner (4 bands: 0.5–1.1 μm, 80 m resolution)
MSU	Microwave sounding unit (4 bands: 50.3–57.9 GHz; 7.5° field of view; 0.3 K precision)
NSCAT	NASA scatterometer (14.6 GHz, 6 antennas, dual/single polarization, 50 km resolution; accuracy 2 m s^{-1}; ±20° direction; 3–30 m s^{-1} range)
OLS	Operational linescan system (2 bands: 0.40–1.01, 10.0–13.4 μm; 0.55/2.7 km resolution)
OPS	Optical sensor (3 visible near IR bands; 4 shortwave IR bands; 1 stereoscopic band; 18×4 m resolution; 75 km swath width
RA	ERS-1 radar altimeter (13.8 GHz, 10 cm precision)
SAR	Synthetic aperture radar (1.2 GHz frequency; 18×18 m resolution; 75 km swath width)
SSM/I	Spectral sensor microwave/imager (4 bands: 19, 22, 37, and 85 GHz; spatial resolution: 12.5 km at 85 GHz and 25 km other channels)
SSU	Stratospheric sounding unit (3 bands about 15 m)
TM	Thematic mapper (6 bands: 0.45–2.35 μm, 30 m resolution; 10.40–12.50 μm, 120 m resolution)
VAS	VISSR atmospheric sounder (0.6 μm, 1 km resolution; 11.5 μm, 7 km resolution; 12 sounding channels; 3.6–14.7 μm, 14 km resolution)
VHRR	Very high resolution radiometer (2 bands: 0.6–0.9 μm, 2.8 km resolution; 10.5 μm, 11 km resolution)
VISSR	Visible and infrared spin scan radiometer (2 bands: 0.55–0.7 μm, 1 km resolution; 10.5–12.6 μm, 7 km resolution)

data dissemination service from 2000 until 2012, each spacecraft having a design lifetime of 7 years.

The new generation of GOES I-M[3] satellites were launched in the mid-1990s and is scheduled to continue well into the second decade of the 21st century. Each element of the mission has been designed to meet all in-orbit performance requirements for at least 5 years. The GOES I-M system performs the following basic functions:

- aquisition, processing, and dissemination of imaging and sounding data;
- acquisition and dissemination of Space Environment Monitor (SEM) data;
- reception and relay of data from ground-based data collection platforms (DCPs) that are situated in carefully selected urban and remote areas by the NOAA Command and Data Acquisition (CDA) station;
- continuous relay of weather facsimile (WEFAX) and other data to users, independently of all other functions;
- relay of distress signals from people, aircraft, or marine vessels to the search-and-rescue ground stations of the Search and Rescue Satellite Aided Tracking (SARSAT) system.

Each satellite in the series carries two major instruments: an imager and a sounder. These instruments resolve visible and infrared data, as well as the temperature and moisture profiles of the atmosphere. They continuously transmit these data to ground terminals where the data are processed for rebroadcast to primary weather services both in the United States and around the world, including the global research community.

The GOES system produces a large number of primary data products. They include:

- basic day/night cloud imagery and low-level cloud and fog imagery;
- upper and lower-troposphere water vapor imagery;
- sea surface temperature (SST) data;
- winds from cloud motions at several levels and hourly cloud-top heights and amounts;
- albedo and infrared radiation flux to space, important for climate monitoring and climate model validation;
- detection and monitoring of forest fires resulting from natural causes and/or human-made causes and monitoring of smoke plumes;
- precipitation estimates;
- total column ozone concentration;
- relatively accurate estimates of total outgoing longwave radiation flux.

These data products enable users to accurately monitor severe storms, determine winds from cloud motion and, when combined with data from conventional

[3] GOES, Geostationary Operational Environmental Satellites.

meteorological sensors, produce improved short-term weather forecasts. The major operational use of 1 km resolution visible and 4 km resolution IR multispectral imagery from the 1 to 15-minute interval GOES imager is to provide early warnings of threatening weather. Forecasting the location of probable severe convective storms and the landfall position of TCs and hurricanes is heavily dependent upon the GOES IR and visible picture. Quantitative temperature, moisture, and wind measurements are useful for isolating areas of potential storm development (Purdom, 1996).

Major changes to the surface of the planet can be detected, measured, and analyzed using Landsat[4] and SPOT[5] data. The effects of desertification, deforestation, pollution, cataclysmic volcanic activity, and other natural and anthropogenic events can be examined using data acquired from the Landsat and SPOT series of Earth-observing satellites. The information obtainable from historical and current Landsat and SPOT data play a key role in studying surface changes through time.

The Landsat platforms (Landsat-5 is operational) operate from a Sun-synchronous near polar orbit imaging the same 185 km (115 miles) ground swath every 16 days. Since 1972 these satellites have provided repetitive synoptic global coverage of high-resolution multispectral imagery. The characteristics of the MSS[6] and TM[7] bands were selected to maximize their capabilities for detecting and monitoring different types of Earth resources. For example, MSS band 1 can be used to detect green reflectance from healty vegetation, and band 2 of MSS is designed to detect chlorophyll absorption in vegetation. MSS bands 3 and 4 are ideal for recording near IR reflectance peaks in healthy green vegetation and for detecting water–land interfaces.

In 1992 the U.S. Congress authorized procurement, launch, and operation of a new Landsat satellite, resulting in Landsat-7 being launched in May 1998. At the time it was the latest in a series of Earth-observation satellites dating back to 1972. The 22-year record of data acquired by Landsat satellites constitutes the longest continuous record of the Earth's continental surfaces. Preservation of the existing record and continuation of the Landsat capability were identified by law as critical to land surface monitoring and global change research.

Landsat-7 played an essential role in the realm of Earth-observing satellites in orbit for the remainder of the two-decade period (Asrar and Dokken, 1993). No other system will match Landsat's combination of synoptic coverage, high spatial resolution, spectral range, and radiometric calibration. In addition, the Landsat program is committed to provide Landsat digital data to the user community in greater quantities, more quickly, and at lower cost than at any previous time in the history of the program.

The Earth-observing instrument on Landsat-7, the enhanced thematic mapper plus (ETM+), replicates the capabilities of the highly successful thematic mapper instruments on Landsat-4 and Landsat-5. The ETM+ also includes new features that

[4] Landsat, land satellite (U.S.A.).

[5] SPOT, Système pour l'Observation de la Terre.

[6] MSS, multispectral scanner.

[7] TM, thematic mapper.

make it a more versatile and efficient instrument for global change studies, land cover monitoring and assessment, and large-area mapping than its design forebears. The primary features on Landsat-7 are:

- a panchromatic band with 15 m spatial resolution;
- onboard full-aperture 5% absolute radiometric calibration;
- a thermal IR channel with 60 m spatial resolution.

The SPOT program supports commercial remote sensing on an international scale, establishing a global network of control centers, receiving stations, processing centers, and data distributors (Revah, 1994). The French space agency, *Centre National d'Etudes Spatiales* (CNES), operates the SPOT satellite system while worldwide commercial operations are anchored by private companies (i.e., SPOT IMAGE Corporation in the United States, SPOT IMAGE in France, SATIMAGE in Sweden, and distributors in over 40 countries).

The SPOT satellites carry high-resolution visible (HRV) sensors, two high-density tape recorders and a telemetry transmitter. HRVs, constructed with multi-linear array detectors, operate in an across-track direction. Operating independently of each other, the two HRVs acquire imagery in either multispectral and/or panchromatic modes at any viewing angle within $\pm 27^\circ$. This off-nadir viewing enables acquisition of stereoscopic imagery.

The SPOT system provides global coverage between 87°N and 87°S. The Earth nominal scene covers a 60×60 km area.

Carrying HRV sensors that operate in either single or dual mode, SPOT satellites maintain a near polar, near circular, Sun-synchronous orbit with a mean altitude of 832 km (at 45°N which corresponds to continental France), an inclination of 98.7°, and a mean revolution period equaling 101.4 minutes. SPOT satellites orbit the same ground track every 26 days with a nominal cycle of 386 revolutions. Crossing the equator from north to south at 10:30 AM mean local solar time, the satellites' reference tracks are 108.6 km apart. The reference tracks draw closer at higher altitudes.

SPOT data are transmitted in direct mode or recording mode to ground receiving stations (SPOT's worldwide ground station network comprises 16 direct receiving stations). Data transmitted in direct mode are received in real time during daytime passes. Data transmitted in recording mode (i.e., via onboard tape recorders) are received during night-time passes by SPOT's main receiving stations in Toulouse (France) and Kiruna (Sweden).

The European remote-sensing satellites (ERS-1/2) were launched in 1991 and 1995. With their unique set of all weather microwave instruments, ERS-1/2 are basically part of a meteorological ocean climate mission with particular emphasis on ocean and ice processes. It provides data intended to address a wide range of primarily environmental problems contributing directly to studies of the ocean–atmosphere interaction, ocean circulation, both Arctic and Antarctic ice sheets, coastal processes, and land use. The *SAR Ocean Feature Catalogue* (Johannessen *et al.*, 1994) contains a selection of ERS-1 synthetic aperture radar (SAR) images, which illustrate impressive and mysterious radar-backscattering signatures over the Atlantic Ocean and coastal waters under a variety of environmental conditions.

To achieve this, ERS-1/2 make all weather day and night measurements on a global scale of many parameters not covered by optical satellite systems. Significantly, a lot of data are being collected from remote areas such as the Southern Ocean, for which there is little comparable information.

ERS-2 was conceived not only as a copy of ERS-1 but it added an important new capability—namely, the Global Ozone Monitoring Experiment (GOME)—to address atmospheric chemistry (an area of growing concern). This instrument is a nadir-viewing spectrometer covering the spectral range of 250 nm to 790 nm of atmospheric chemistry. It observes ozone and some related species in both the troposphere and the stratosphere.

An ERS-1/2 mission of very great scientific interest was started during 1995 by operating the two ERS satellites in tandem. The orbits of the two satellites were carefully phased to have the same ground track, but with a 1-day interval. This allowed generation of (interferometric) SAR image pairs, which are used for production of a global digital terrain map, among other novel applications.

While ERS-1/2 was seen as a major step forward, it is clear that its potential for both research and operational applications can only be fully realized if long-term continuity of its important and unique data can be assured. Many of the processes relevant to climate studies have timescales which exceed the duration of a single mission.

The next important step in radiophysical sensing of the Earth was carried out by the launch of the Canadian satellite Radarsat-1 (November 1995) with SAR which, as distinct from its predecessors ERS-1/2 and JERS-1, provided many new modes in terms of: spatial resolutions (from 10 m to 100 m); incident angles (less than 20° to more than 50°); swath widths (from 45 km to 500 km). Certainly, such multimodel SAR regimes will be in operation on other microwave SAR missions (particularly for tropical studies) (Tack, 1994).

The Defense Meteorological Satellite Program (DMSP) is a Department of Defense (DoD) program (Asrar and Dokken, 1993). The DMSP program designs, builds, and launches satellites to observe oceanographic and solar–terrestrial physics environments. DMSP satellites are in a near polar, Sun-synchronous orbit at an altitude of approximately 830 km above the Earth. Each satellite crosses every point on the Earth twice a day and has an orbital period of about 101 minutes thus providing complete global coverage of clouds every 6 hours. DMSP satellites monitor the atmospheric, oceanographic, and solar–geophysical environment of the Earth. The visible and IR sensors collect images of global cloud distribution across a 3,000 km swath during both daytime and night-time conditions. The coverage of microwave imagery and sounders are one half the visible and infrared sensor coverage. Thus, they cover the polar regions above 60° on a twice daily basis and the equatorial region on a daily basis. The spaceborne environmental sensors record along-track plasma densities, velocities, compositions, and drifts.

Visible and IR imagery from DMSP Operational Linescan System (OLS) instruments are used to monitor the global distribution of clouds at low resolution and high resolution, regional coverage, imagery recorded along a 3,000 km scan, satellite ephemerides, and solar and lunar data. IR pixel values vary from 190 K to 310 K in 256 equally spaced steps. Onboard calibration is performed during each scan. Visible pixels are currently relative values ranging from 0 to 63 rather than absolute values in watts per m^2. Instrumental gain levels are adjusted to maintain constant cloud reference values under varying conditions of solar and lunar illumination. Telescope pixel values are replaced by photomultiplier tube (PMT) values at night. A telescope pixel is 0.55 km at high resolution and 2.7 km at low resolution. Low-resolution values are the mean of the appropriate 25 high-resolution values. A PMT pixel is 2.7 km at nadir. In addition to cloud images, ground-based sources (such as fires) and upper-atmospheric sources (such as the northern lights) are recorded. Night-time imagery records the aurora, city lights, human-made and natural fires, and natural gas flaring. TC images are the most prominent examples of the efficiency of the visible and IR OLS system.

The special sensor microwave imager (SSM/I) is a seven-channel, four-frequency, linearly polarized, passive microwave radiometric system which measures atmospheric, ocean, and terrain microwave brightness temperatures at 19.35, 22.235, 37.0, and 85.5 GHz. The data are used to obtain synoptic maps of critical atmospheric, oceanographic, and selected land parameters on a global scale. SSM/I data are used to derive geophysical parameters: notably, ocean surface windspeed, area covered by ice, age of ice, ice edge, precipitation over land, cloud liquid water, integrated water vapor, precipitation over water, soil moisture, land surface temperature, snow cover, and SST.

The special sensor microwave/temperature (SSMT-2) sensor is a five-channel, total power microwave radiometer with three channels situated symmetrically about the 183.31 GHz water vapor resonance line and window channels. This instrument was flown on all DMSP Block 5D-2 satellites starting with F11 launched in 1991. SSM/T-2 is designed to provide global monitoring of the concentration of water vapor in the atmosphere under all sky conditions by taking advantage of the reduced sensitivity of the microwave region to cloud attenuation.

The objectives of the ADEOS (Advanced Earth Observation Satellite) mission are to acquire data on worldwide environmental changes such as the greenhouse effect, ozone layer depletion, tropical rainforest deforestation, and abnormal climatic conditions, in order to contribute to international monitoring and to develop platform bus technology, interorbital data relay technology, etc. which are necessary for the development of future Earth observation systems (Harujama, 1994; Kondratyev and Tanaka, 1997).

To make continuous thorough observations of the Earth surface and its atmosphere, the spacecraft carries two core sensors developed by the National Space Development Agency of Japan (NASDA): the advanced visible and near infrared radiometer (AVNIR) and the ocean color and temperature sensor (OCTS). In addition, the spacecraft carries six announcement of opportunity (AO) sensors: the NASA scatterometer (NSCAT) and total ozone-mapping spectrometer (TOMS) provided by NASA, polarization and directionality of the Earth's reflectance (POLDER) radiometer/polarimeter provided by CNES, the interferometric monitor for greenhouse gases (IMG) provided by the Japanese Ministry of International Trade and Industry, and the improved limb atmospheric spectrometer (ILAS) and retroreflector in space (RIS) provided by the Environment Agency of Japan.

ADEOS is a large satellite with a mass of approximately 3,500 kg. On August 17, 1996 ADEOS was launched into a Sun-synchronous subrecurrent orbit at an altitude of approximately 830 km by the H-II launch vehicle from the Tanegashima Space Center.

AVNIR is a high spatial resolution optical sensor for observing land and coastal zones in visible and near IR regions. AVNIR has four spectral bands with 16 m spatial resolution and one panchromatic band with 8 m spatial resolution. AVNIR data are useful for environmental awareness and monitoring of such phenomena as desertification, destruction of tropical forests, and pollution of coastal zones as well as for resource exploration, land use, etc. AVNIR's field of view (FOV) is about 80 km, picked up by lines of small pixels. Its FOV scans the entire Earth's surface as the satellite moves. There are only a few other similar sensors, such as TM on Landsat. The many features of AVNIR include the sensor's high spatial resolution, a pointing function to change the observation field by $\pm 40^\circ$ along-track, a 0.4 mm band useful for coastal zones and lakes, and an optical calibration function using solar light and lamps.

OCTS is an optical radiometer devoted to the frequent global measurement of ocean color and SST. OCTS shows the amount of chlorophyll and dissolved substances in water, as well as temperature distribution. OCTS data are used to determine ocean primary production and the carbon cycle, as well as for getting information about ocean conditions for fishery and environment monitoring, etc. OCTS is a successor to CZCS (coast zone color scanner), a U.S. project which was the first real optical sensor for ocean observation onboard Nimbus-7, launched in 1978. OCTS has eight bands in the visible and near IR region and four bands in the thermal region, and achieves highly sensitive spectral measurements with them. The observation bands are determined by the characteristics of spectral reflectance of the object's substance, atmospheric windows, and atmospheric correction. Its spatial resolution is about 700 m. This is applicable to observation of the coastal zone and land, as the features of this boundary can be quickly compared with the open ocean. As its swath width is about 1,400 km on the ground, OCTS can observe the same area every 3 days and can monitor rapidly changing phenomena. OCTS has an optical calibration function using solar light and a halogen lamp as the calibration source.

NSCAT measures windspeeds and directions over at least 90% of the ice-free global oceans every 2 days, under all weather and cloud conditions. Winds are a critical factor in determining regional weather patterns and global climate. At present, weather data can be acquired over land, but our only knowledge of surface winds over oceans comes from infrequent and sometimes inaccurate reports from ships. Since oceans cover approximately 70% of the Earth's surface, NSCAT data play a key role in the researcher's efforts to understand and predict complex global weather patterns and climate systems. NSCAT is a specialized microwave radar designed to measure winds over the oceans and uses an array of antennas that radiate microwave pulses at a frequency of 14 GHz across broad regions of the Earth's surface. This array of six 3 m long antennas scan two 600 km bands of ocean, one band on each side of the instrument's orbital path, separated by a gap of approximately 330 km. Obtaining backscatter and wind vector information almost continuously, NSCAT can make 190,000 wind measurements per day, more than 100 times the amount of ocean wind information presently available from ships. Studies using NSCAT data have led to improved methods of global weather forecasting and modeling and the possibility of better understanding of environmental phenomena (such as El Niño) that greatly affect world economies.

TOMS measures the albedo of the Earth's atmosphere at six narrow spectral bands in the near ultraviolet. Albedo is measured by comparing the reflectivity of the Earth with the reflectivity of a calibrated onboard diffuser plate. Total ozone is derived from the differential albedo in three pairs of the spectral bands, which are selected to function at all latitudes and under solar illumination conditions, as well as measurement of surface reflectivity at a longer non-absorbed wavelength band. The five shortest TOMS wavelengths show the absorption spectra of ozone and sulfur dioxide in the near-ultraviolet. Sulfur dioxide, which normally is below the TOMS detection threshold, is measured using the same six wavelengths. Ozone and sulfur dioxide absorption spectra are sufficiently different that these two absorbing constituents can be separated by analysis.

The POLDER (polarization and directionality of the Earth's reflectances) instrument will observe the polarization, directional and spectral characteristics of the solar light reflected by aerosols, clouds, oceans and land surfaces. POLDER is a push broom, wide field of view, multiband imaging radiometer/polarimeter developed by CNES. Multi-angle viewing is achieved by along-track migration at spacecraft velocity of a quasi-square footprint intercepted by the total instantaneous 114° wide field of view. This footprint is partitioned into 242×274 elements of quasi-constant 7×6 km resolution, imaged by a CCD matrix in the focal plane. Simultaneously, a filter/polarized wheel rotates and scans eight narrow spectral bands in the visible and near infrared (443, 490, 565, 665, 763, 765, 865, and 910 nm), and three polarization angles at 443, 665, and 865 nm.

The interferometric monitor for greenhouse gases (IMG) is a sensor to monitor the Earth's radiation balance, the temperature profile of the atmosphere, the temperature of the Earth's surface, and the physical properties of clouds. The Japan Resources Observation System Organization (JAROS) developed it for the Ministry of International Trade and Industry (MITI). IMG obtains detailed spectra of thermal IR radiation from the Earth's surface and atmosphere. Detailed spectra measured by the IMG will be used to infer atmospheric concentrations of water vapor and greenhouse gases.

A global increase in tropospheric concentrations of trace gases, such as carbon dioxide, methane, nitrous oxide, and chlorofluorocarbons (CFCs), has been noted. These increases have been brought about by human activities. However, we have limited knowledge of the magnitude or distribution of the anthropogenic sources of these gases. Two sources whose magnitude must be investigated are deforestation and biomass burning. IMG maps the global and regional distribution of emission sources by measuring variation in the concentrations of trace gases. Moreover, natural sources and sink strengths of trace gases can vary widely with different terrestrial and oceanic ecosystems.

9.1.2 Missions relevant to tropical studies

As already pointed out (see Chapters 2 and 4), global tropical cyclogenesis serves as an example of a physico-geographical system that possesses a very wide range of spatiotemporal scales of interaction with the environment. However, it should be noted that the relationships of global tropical cyclogenesis with the global atmosphere–ocean system are not clearly understood. Because of this, the role of the coming generation of integrated concepts and programs for observing Earth as a whole system is unquestionably great.

We should first note the Earth Science Enterprise at NASA and its Earth Observation System (EOS) program (King, 1999). The main goal of this concept is "to bring congruency to multiple disciplines within Earth sciences through integrated observations, interdisciplinary scientific research and analysis, and modelling" (Greenstone and King, 1999).

The development of EOS can be traced back to 1979 and the establishment of the World Climate Research Program (WCRP), which is an international effort to understand the physical basis of climate in response to droughts and floods that highlighted societal vulnerability to climate variability. In the following years it began to be recognized that an integrated programmatic framework would have to be the central paradigm of both national and international programs for global change to be understood.

EOS is an integral element of the U.S. Global Change Research Program (USGCRP). Four key USGCRP global change issues are:

(1) seasonal to interannual climate variability;
(2) climate change over decades to centuries;
(3) changes in ozone, ultraviolet radiation, and atmospheric chemistry; and
(4) changes in land cover and in terrestrial and aquatic systems.

In addition to these areas of particular scientific and practical importance, the USGCRP has defined an overarching objective (often referred to as the "human dimension") to identify, undertand, and analyze how human activities contribute to changes in natural systems, how the consequences of natural and human-induced change affect the health and well-being of humans and their institutions, and how humans respond to problems associated with environmental change (Greenstone and King, 1999).

EOS science in support of the USGCRP is well described by seven topical scientific lines:

(1) radiation, clouds, water vapor, precipitation, and atmospheric circulation;
(2) ocean circulation, productivity, and exchange with the atmosphere;

(3) atmospheric chemistry and greenhouse gases;
(4) land ecosystems and hydrology;
(5) cryospheric systems;
(6) ozone and stratospheric chemistry; and
(7) volcanoes and climate effects of aerosols.

As pointed out in the text of *EOS Science Plan: Executive Summary* (Greenstone and King, 1999), EOS observations of atmospheric phenomena will provide for validation of the size of cloud systems from regional to global scales and will support modeling efforts at these spatial scales as well as temporal scales ranging, correspondingly, from less than a day to 10 days and to 100 years. Several EOS interdisciplinary investigations are pursuing improved understanding of cloud–climate feedback where great uncertainties exist for predictions of climate change.

EOS measurements will shed new light on both physical and biological aspects of the World's oceans, including their circulation, their productivity, and their role in gas exchange (such as carbon dioxide) with the atmosphere. EOS observations related to greenhouse gases and the chemistry of the lower atmosphere will lead to new understanding of the events that control the production of tropospheric ozone including the occurrence of biomass burning and changes in land cover that are taking place worldwide.

EOS measurements of land ecosystems and hydrology will provide unprecedented worldwide coverage of vegetation extent and characteristics, thus improving our knowledge of terrestrial biospheric dynamics and vegetation process activity. Long-term consistent datasets related to the cryosphere will be able to document changes and validate transient model simulations using global climate models. EOS laser altimetry through repeated measurements will reveal changes in the surface elevation of ice sheets and, thereby, improve our knowledge of changes in their mass balance with implications for sea level rise.

EOS will contribute greatly, through measurements of trace gas chemistry and temperatures in the stratosphere, to our understanding of changes in the ozone layer and related climate forcing. EOS will enable the first truly global inventory of volcanic eruptions with their implications for climate change through the introduction of stratospheric aerosols. In addition, timely EOS observations of volcanic phenomena can lead to significant warnings of hazards for people on the ground or for aircraft that may encounter volcanic plumes or clouds, possibly causing the aircraft to crash.

EOS observations and modeling studies clearly will support numerous surface observation efforts and many major field campaigns: the Global Energy and Water Cycle Experiment (GEWEX), the Climate Variability Project (CLIVAR) of the WCRP and the Joint Global Ocean Flux Study (JGOFS) of the International Geosphere–Biosphere Program (IGBP).

EOS has also developed a measurement strategy that includes:

- simultaneity of observations taken by a group of sensors preferably on the same satellite platform or else closely coordinated in space and time;

- overlap between measurements by successive sensors for intercomparison and intercalibration, in order to construct useful long-term climate records;
- diurnal sampling to detect changes in rapidly varying systems such as clouds;
- high-quality calibration to permit intra and intersystem data comparisons; and
- a data continuity strategy that ensures successive sensors will maintain a data stream for long-term analyses of changing phenomena.

Global tropical cyclogenesis as a global Earth system can make itself evident in various lines of investigation by EOS. Climate study should primarily note the role played by convection and clouds developing in TSs.

Clouds and especially tropical cloud systems not only affect radiative energy fluxes in the atmosphere through scattering, absorption, and reradiation but also vertical motions associated with them produce important convective transport of energy and moisture. Interactions of clouds with the clear environment around them play a critical role in determining both the amount of water vapor that is retained in the clear atmosphere and the amount of precipitation reaching the surface. Convection associated with clouds also affects the exchange of heat between the surface and the atmosphere.

Many atmospheric circulation systems are organized on the regional scale or mesoscale (defined as encompassing horizontal length scales of 20 km to 500 km). Mesoscale convective systems dominate weather over most of the tropics and the summertime Midwestern United States. Mesoscale models (MMs) have become powerful tools for understanding and forecasting regional atmospheric circulations, taking into account such features as severe midlatitude cyclones, hurricanes, orographically forced flows, fronts, and thermally forced flows such as land–sea breezes.

The increased use of MMs has particular promise in the areas of tropical convection, cirrus cloud evolution, midlatitude cyclonic cloud systems, prediction of boundary layer fog, and cloud prediction. Several EOS interdisciplinary investigations are pursuing improved understanding of cloud–climate feedback. Higher spatial resolution, the availability of new cloud variables, and the greater accuracy of EOS cloud observations will enable better validation of regional cloud simulations.

Modeling of the atmosphere including clouds is currently conducted on three scales—global, regional, and cloud resolving—and the corresponding horizontal resolutions range from 100 km (global) to about 10 km (regional) and to about 1 km (cloud resolving). EOS observations will provide data for validation on all these horizontal scales, and EOS interdisciplinary investigations will be conducted at all these scales. Model runs can range from 100-year integrations (global) to 10-day integrations (regional) to 1-day (cloud resolving).

However, there are many questions regarding the proper methodology for the study of global tropical cyclogenesis on the basis on EOS remote-sensing data.

The Envisat-1 polar mission is the most challenging ESA has ever undertaken in the field of Earth observation (Asrar and Dokken, 1993; Bruzzi, 1995; Louet and Levrini, 1998). The second-generation polar-orbiting Earth observation satellite

Envisat-1[8] was launched in 2002. The most challenging Earth observation satellite ever undertaken by ESA, it not only provides continuity of ERS observations, but adds important new capabilities for gathering valuable information to contribute to the understanding and monitoring of the Earth's environment, particularly in the areas of atmospheric chemistry and ocean biological processes.

The instrument payload is a combination of the following ESA-developed instruments (EDIs):

- advanced synthetic aperture radar (ASAR);
- radar altimeter (RA);
- microwave radiometer (MWR);
- medium-resolution imaging spectrometer (MERIS);
- Michelson interferometer for passive atmospheric sounding (MIPAS);
- global ozone monitoring by occultation of stars (GOMOS);

and other instruments:

- advanced along-track acanning radiometer (AATSR);
- scanning imaging absorption spectrometer for atmospheric cartography (SCIAMACHY); and
- Doppler orbitography and radio-positioning integrated by satellite (DORIS).

The second series of missions (Metop-1,[9] Metop-2, etc.), which are of prime concern to Eumetsat, have been making a major contribution to operational meteorology by providing operational meteorological data since 2003 (Goldsmith and Readings, 1994; Readings and Reynolds, 1996). It has a climate-monitoring role as well. Metop-1 was the first of this series of satellites, which will basically be identical, replacing the current series of "morning" NOAA satellites.

Operational instruments are improved versions of the instruments currently flying on the NOAA Tiros-N[10] polar-orbiting operational meteorological satellites:

VIRSR	Visible and infrared scanning radiometer
IRTS	Infrared temperature sounder
MTS/MNS	Advanced microwave sounding unit (temperature and humidity)
MCP	Meteorological communications package
ARGOS	Data collection and location system
S&R	Search and rescue package

[8] Envisat, environmental satellite (ESA).

[9] Metop, meteorological operational satellite (ESA).

[10] Tiros, television and infrared observation satellite.

These instruments are provided through Eumetsat (jointly with NOAA[11]) as a means of continuing observations in the morning orbit after 2000 when NOAA only operated spacecraft in the afternoon orbit.

ARISTOTELES (which stands for "applications and research involving space techniques observing Earth fields from low-Earth spacecraft") is a program that involved cooperation with NASA. It focuses on possibly the most essential prerequisite for our understanding of the structure, dynamics, and evolution of Earth: precise determination of the Earth's gravity and magnetic fields. Both fields have been measured and mapped in a general way but not in sufficient detail and more importantly not homogeneously on a global scale (Goldsmith and Readings, 1994).

ARISTOTELES was therefore conceived as a mission dedicated to achieve these objectives. The measuring instruments are a gradiometer (a set of micro-accelerometers that determine the tensor of the gravity gradient) and a magnetometer package. The primary needs for such data lie in four areas:

- solid Earth geophysics;
- geodesy;
- physical oceanography;
- climatology and global change.

It is appropriate and important to highlight the contributions that ARISTOTELES has made to the latter two domains. In physical oceanography a precise geoid, as derived from high-quality gravity data, is an indispensable reference for the monitoring of large-scale ocean currents, which are themselves the vehicle responsible for the very large poleward transport of tropical heat input from the Sun. In turn this transport of heat has a major influence on our climate.

ARISTOTELES contributed directly to research in climatology and global change by using the resulting high-accuracy height reference level for determination of global sea level rise or for monitoring the El Niño–Southern Oscillation (ENSO) phenomenon (particularly in the tropics).

As far as Earth observation research in the post-2000 era is concerned, ESA (European Space Agency) intends to depart from the multi-objective program approach leading to large satellites, and to pursue a "dual-mission strategy" more suited to the needs of both scientific and operational users. This strategy comprises (Readings and Reynolds, 1996; ESA, 1996a–c):

- *Earth Explorer Mission (EEM)*—these are research demonstration missions with the emphasis on advancing understanding of different Earth system processes. Each mission focused on a particular research field or regrouped a limited number of research fields. The demonstration of specific new observing techniques also fell under this category.
- *Earth Watch Mission (EWM)*—these are preoperational missions addressing the requirements of specific Earth observation application areas. After the devel-

[11] NOAA, National Oceanic and Atmospheric Administration.

opment/preoperational phase (funded by ESA), the responsibility for this type of mission was eventually transferred to (European) entities providing operational services. To ensure such a scenario, data continuity over a period of at least 10 years was required.

The two categories of mission cross-fertilized in as much as the Earth Explorer mission demonstrated new technologies for use later in routine operational applications, while the Earth Watch mission provided data that also supported scientific activities (e.g., long-term monitoring of specific processes or parameters).

All nine potential candidates for the first Earth Explorer mission (Readings and Reynolds, 1996) were of immediate interest to studies of topical regions. Candidates for the first Earth Explorer mission were the following:

- *Earth Radiation Mission*—to advance understanding of the Earth's radiation balance, which is of fundamental importance to the Earth's climate.
- *Precipitation Mission*—to observe precipitation especially in tropical regions.
- *Atmospheric Dynamics Mission*—to observe three-dimensional wind fields in clear air in both the troposphere and stratosphere.
- *Atmospheric Profiling Mission*—to observe temperature profiles in the troposphere and stratosphere for climate research.
- *Atmospheric Chemistry Mission*—to advance understanding of the chemistry of the atmosphere, including the study of active chlorine species and hydrogen oxides.
- *Gravity Field and Ocean Circulation Mission*—to derive a highly accurate global and regional model of the Earth's gravity field and its geoid.
- *Magnetometry Mission*—to observe the Earth's magnetic field.
- *Surface Processes and Interactions Mission*—to advance understanding of biospheric processes and their interactions with the other processes that occur in the Earth–atmosphere system.

Concerning applications, two Earth Watch–type missions are currently in preparation by ESA: MSG, the successor to the current Meteosat geostationary satellites, and Metop, a series of polar-orbiting operational meteorological satellites.

As for candidate fields for the first Earth Watch mission, the following are top contenders:

- *coastal zones*—this covers a wide field of applications, including bathymetry, oil spill monitoring, sea state monitoring and forecasting, mineral and hydrocarbon exploration, flood surveillance and prevention, inland waters, fisheries, coastal erosion and surveillance of river discharge, coastal land use, etc.;
- *land surface*—this spans a variety of uses on both the international and European level, including crop forecasting, crop damage assessment, forestry, land use, cartography, etc. (some of these relate directly to enforcement of regulatory measures);

- *atmospheric chemistry*—the long-term need for atmospheric chemistry monitoring is only partially addressed by the ozone instrument on Metop (it is also necessary to consider the monitoring of halogens in the stratosphere).

The main purpose of the EOS, ADEOS, and Envisat projects is focused on understanding long-term changes in the Earth's environment.

Basically different from these missions, the dramatic project of developing the World Environment and Disaster Observation System (WEDOS) and the Global Disaster Observation System (GDOS) were carried out by Japanese researchers (Kuroda and Koizumi, 1996). The purpose of these systems is to construct a global tool for continuous operational monitoring of the Earth's environment in order to detect and mitigate natural disasters (including TCs, eruptions, and earthquakes) and human-induced accidents by launching 26 Earth observation satellites in low-altitude, Sun-synchronous, and Earth polar orbits together with six data relay satellites in geostationary orbits. All locations in the World are observed by WEDOS at least once a day at 20 m resolution, and so irregularities and environmental changes are detected immediately, and more precise observations of damaged areas are also possible several times a day at 2 m resolution (GDOS system). These systems provide immediate availability of information to any country in the world.

Previously, similar designs were constructed by Russian researchers (Avduevskii *et al.*, 1983), who showed that—for continuous observations of spontaneous natural phenomena (particularly in tropical regions)—it is desirable to use satellites on geosynchronous orbit with 24-hour period and 65° inclination equipped with optical telescope instruments (1–5° viewing angles, 20–50 m spatial resolutions, 0.5-hour periodicity of imaging, observing area of 1,000 × 1,000 km). The concept of making use of satellite constellations for optimal continuous coverage was first mentioned by Beste (1978).

9.2 MISSIONS FOR TROPICAL CONVECTIVE SYSTEM STUDY

9.2.1 TRMM mission

The Tropical Rainfall Measuring Mission (TRMM) (Simpson, 1988; Simpson *et al.*, 1988) is a joint space mission between Japan and the United States, with the cooperation of several other nations. As originally planned during the mid-1980s, TRMM has on board a synergistic complement of three instruments (see Table 9.3). A key feature is the first rain radar to be flown in space. The other instruments are a multichannel dual-polarized passive microwave radiometer and a high-resolution visible–IR radiometer. The TRRM satellite was launched in November 1997.

TRMM can be regarded as a "flying rain gauge" because the improved measurement capability of the microwave instruments can be used to calibrate techniques based on IR brightness temperatures. Calibrations are likely to be

Table 9.3. TRMM sensor summary.

Sensor	*System parameters*	
Precipitation radar (PR)	Radar type	Active phased array radar
	Frequency	13.796 and 13.802 GHz (two-channel frequency agility)
	Swath width	215 km
	Observable range	From surface up to height 15 km
	Range resolution	250 m
	Horizontal resolution	4.3 km (nadir)
	Sensitive	S/N per pulse ≥ 0 dB for rain of 0.7 mm h^{-1}
	Data rate	93.5 kbit s^{-1}
	Beam width	0.71°
	Scan angle	±17°
	Peak power	500 W
	Pulse width	1.6 μs
	Pulse repetition frequency	2,776 Hz
TRMM microwave imager (TMI)	Observation frequency	10.7, 19.4, 21.3, 37, and 85.5 GHz
	Polarization	V/H
	Resolution	6–50 km
	Swath width	760 km
	Scan mode	Conical scan (49°)
	Data rate	8.8 kbit s^{-1}
Visible infrared scanner (VIRS)	Observation bands	0.63, 1.6, 3.75, 10.8, and 12 μm
	Resolution	2 km (nadir)
	Swath width	720 km
	Scan mode	Across-track scan
	Data rate	50 kbit s^{-1}
Cloud and the Earth's Radiant Energy System (CERES)	Observation bands	0.3–5 μm (shortwave channel) 8–12 μm (longwave channel) 0.3–50 μm (total channel)
	Resolution	10 km (nadir)
	Swath width	Scan angle ±80°
	Scan mode	Across-track scan or biaxial scan
	Data rate	9 kbit s^{-1}
Lighting imaging sensor (LIS)	Observational band	0.77765 μm
	Resolution	10 km (nadir)
	Swath width	600 km
	Data rate	6 kbit s^{-1}

different in different climate regimes. After calibration, the improved IR techniques can be applied to fill in between TRMM swaths using geosynchronous data, and to upgrade many of the past rain estimates which used the proxy-variable approach.

One basic TRMM data product is mean monthly rainfall over areas ($5^\circ \times 5^\circ$ in size) for climate studies. The orbit was selected to optimize the effectiveness of the instruments. The low 350 km altitude obtains good resolution for the instruments (the 19 GHz channel of the passive microwave instrument has a resolution of about 10 km). The orbit inclination of 35° ensures overflights at different local times every day, covering the entire 24 hours in a month, permitting documentation of the diurnal variability of tropical rain. More extensive discussion of the TRMM motivation, design, and scientific basis are provided in the *Report of the Science Steering Group for a Tropical Rainfall Measuring Mission* (Simpson, 1988) and in annual publications of the TRMM Office.

In late 1990, the urgent need to improve climate models by means of pre-EOS radiative measurements from an inclined orbit was recognized. This recognition led to addition to TRMM of the Earth's Radiation Budget Experiment's (ERBE) instrumentation. Redesign of the spacecraft and data system was carried out. The present TRMM sensor summary is shown in Table 9.3.

The precipitation radar (PR) on board TRMM is the first spaceborne rain radar in the world. Major objectives of the PR are:

(1) to provide three-dimensional rainfall structure;
(2) to achieve quantitative rainfall measurement over land as well as the ocean; and
(3) to improve the accuracy of TMI (TRMM microwave images) measurements by providing rain structure information.

The TMI is a multichannel/dual-polarized microwave radiometer, which provides data related to rainfall rates over the ocean. TMI data together with PR data make up the primary dataset of precipitation measurements.

The VIRS (visible infrared scanner) is a passive across-track scanning radiometer, which measures scene radiance in five spectral bands operating in the visible through the infrared spectral regions. VIRS data provide information about convective cloud fields (cloud type, convective conditions).

CERES (Cloud and Earth's Radiant Energy System) is a passive broadband scanning radiometer, which has three spectral bands in the visible through the infrared spectral regions and measures the Earth's radiation budget and atmospheric radiation from the top of the atmosphere to the surface of the Earth.

LIS (lightning imaging sensor) is an optical staring telescope and filter imaging system that acquires and investigates the distribution and variability of both intra-cloud and cloud-to-ground lightning over the Earth. LIS data are used with PR, TMI, and VIRS data to investigate the correlation of global incidence of electrical activity with rainfall and other storm properties (including TCs).

9.2.2 TROPIQUES mission

The overall characteristics of the TROPIQUES mission follow up the above needs (Desbois, 1995):

(1) the 3-hour average sampling time for tropical regions can only be obtained from low orbit by at least four polar satellites with low inclination to the equator;
(2) this constraint also implies that instruments must be able to observe sufficiently wide swaths to cover the area several times a day;
(3) radiative budget components can be obtained, as their measurement is relatively direct, with best accuracy (instruments like ScaRaB[12] or CERES, with their associated data-processing systems, can fulfill the specifications);
(4) integrated water vapor–microwave radiometers have proved to be efficient over oceans and can fulfill the specifications;
(5) the temperature and height of cloud tops can be obtained (every half-hour and in the future every quarter of an hour) from geosynchronous satellite data, at a spatial resolution better than the TROPIQUES requirements; however, for the sake of systematic synergetic use, the possibility of getting AVHRR-type visible–IR cloud imagery from TROPIQUES is being studied;
(6) condensed water in the atmosphere (cloud liquid water and ice, precipitation, their vertical distribution); measurements in the microwave domain are fundamental for these quantities, as has been demonstrated, for example, by SSM/I instruments (measurement in the optical domain is used as a complement);
(7) convective activity—a useful complementary measurement could be the electric activity of systems, which could also improve separation and localization of the convective/stratiform parts of clouds.

The two fundamental instruments are the radiative budget instrument (ScaRab type) and the microwave imager (SSM/I or derived type). Two useful complementary instruments are the visible–IR imager and the lightning detector–locator. Provisional mission specifications are given below:

Geographic area to cover	20°N–20°S
Mission duration	3 years
Time-sampling constraints	3-hour sampling or seven passes per day
Space resolution average	40 km; from 15 to 100 km
Swaths	2,000 km
Orbit inclination	15°
Orbit height	1,000–1,250 km
Required pointing accuracy	0.3°
Total data flow rate	Some tens of kbit s^{-1}

[12] ScaRaB, scanning radiation budget.

TROPIQUES instruments are mainly derived from already existing systems because it is possible to assess their main characteristics, although detailed studies of modifications relevant for TROPIQUES are just starting.

A version of the radiative budget instrument ScaRaB has already been launched on a Russian Meteor satellite, giving satisfactory results. An improved model was later flown.

Briefly, in order to derive fluxes in the shortwave domain (0.2–4 μm) and in the longwave domain (4–50 μm), ScaRaB measures radiances in a solar channel (0.2–4 μm) and in a "total" channel (0.2–200 μm). Two auxiliary channels are also available for scene identification: an IR window channel (10.5–12.5 μm) and a visible channel (0.6–0.7 μm). The procedures for calibration and processing of the data in order to transform directional radiances in fluxes were carefully defined and were made available for following missions. The accuracy reached met the specifications of TROPIQUES, although some progress could still be made in shortwave fluxes with better scene-dependent angular corrections. The gross characteristics of a ScaRaB-type instrument are: mass 50 kg, electric power 90 W, data flow rate 3 kbit s^{-1}.

The microwave instrument which is presently the most used for atmospheric water studies is the SSM/I, flying on the U.S. DMSP. A derivative of this instrument, called TMI, has been adapted to TRMM needs and specifications. These two instruments are used here as models for the TROPIQUES microwave radiometer, but a specific instrument is being evaluated to fly on the TROPIQUES orbit.

The channels that can be used to derive the required water parameters of the tropical atmosphere are already well known: 19 GHz (two polarizations), 21 GHz (one polarization), 37 and 85 GHz (two polarizations). The 150 GHz channel is being evaluated to study the ice phase of clouds and to provide better space resolution.

Taking account of the present state of development of algorithmic research for deriving geophysical parameters, the accuracies that can be obtained for instantaneous quantities are:

- *surface wind (sea)*—2 m s^{-1} to 10 m s^{-1} (depending on precipitation);
- *integrated water vapor content*—accuracy 3 kg m^{-2} for contents of 50 kg m^2;
- *cloud liquid water (theoretical)*—30% for contents of 0.5 kg m^{-2};
- *cloud ice*—0.15 kg m^{-2} for contents from 0 kg m^{-2} to 2 kg m^{-2};
- *precipitation*—100% over the range 0.1 mm h^{-1} to 50 mm h^{-1};
- *vertical structure*—qualitative information over five levels.

The gross characteristics of TMI-type instruments are: mass 60 kg, electric power 70 W, data flow rate 8 kbit s^{-1}. Taking account of an additional channel at 150 GHz and the specific characteristics of the TROPIQUES mission, the mass of the radiometer as the data flow rate will probably increase relative to TMI. The microwave radiometer is crucial for the TROPIQUES mission and much effort has been devoted to the concept of this lightweight instrument.

We stated above that the adjunct of an optical (visible–IR) instrument with better resolution than the core instruments could be useful for interpretation of both radiative budget and microwave instruments. The optimum characteristics needed for this AVHRR-type instrument need to be evaluated.

The electrical activity of a storm has been shown to be a good indicator of its convective activity and even of the precipitations released; therefore, a lightning-monitoring system would constitute a good complement to other measurements. A lightning detector–locator working in the VHF domain was preliminarily defined. It was able to discriminate intra-cloud flashes from cloud-to-ground flashes, and to perform localization of flashes within 20 km. The gross characteristics of this instrument are: mass 15 kg, electric power 35 W, data flow rate 1 kbit s^{-1}.

As for complementarity with other satellite programs, we reiterate that efficient utilization of TROPIQUES requires combining its instruments with the new generation of instruments onboard geostationary satellites. TROPIQUES' instruments should, of course, also be used complementarily and simultaneously with equivalent instruments existing on polar satellites and platforms currently in operation simultaneously in order to further increase measurement sampling. This benefits both the study of tropical systems and global coverage of the Earth.

The only other scheduled mission for the study of the tropical atmosphere is the U.S.–Japanese TRMM, which carried the first precipitation radar in space plus different passive instruments: a microwave imager (TMI, a derivative from SSM/I), a visible–IR radiometer derived from AVHRR, and a lightning detector–locator. The presence of the radar constrained the orbit to a low altitude (350 km), thus limiting the swath width of all the instruments. Moreover, the chosen orbit had an inclination of 35° over the equator. These two characteristics were the cause of limited sampling of areas close to the equator, far removed from TROPIQUES sampling requirements. TRMM was launched in 1997; its exploitation will undoubtedly be profitable for TROPIQUES, particularly for the improvement of microwave determination of precipitation. But the philosophy of the two missions is different with TRMM favoring sophisticated means for improving determination of instantaneous precipitation rates and vertical profiles, but neglecting spatiotemporal coverage, while TROPIQUES favors sampling only by passive instruments.

The complementarity of TROPIQUES with oceanographic satellites is mainly through the study of atmosphere–ocean exchanges; moreover, the repetitive rate of TROPIQUES allows it to find numerous spacetime measurements coincident with those made by other low-orbiting satellites; thus, we can study at different timescales the interaction between tropical cloud systems, atmosphere–ocean exchange, and oceanic phenomena.

9.3 RUSSIAN SATELLITE SYSTEMS RELEVANT TO TROPICAL STUDIES

During the 1990s a number of Russian meteorological and environmental satellite missions were flown on the Meteor, Goms, Resurs, and Okean series of satellites and the *Priroda* ("Nature") mission provided observations and soundings relevant to radiation budget and cloud studies, ocean–atmosphere interaction, and middle-atmosphere monitoring. Current and planned sensor capabilities for carrying out some of these observational programs are presented in this section. An outline of the Meteor-3/TOMS (total ozone mapping spectrometer, NASA) project is given as an example of mutually fruitful international cooperation to provide valuable input about acquisition of data related to global change and climate studies (Karpov, 1994; Kondratyev, 1992; Armand, 1993; Harvey, 1995).

9.3.1 Meteor-2 and Meteor-3 series

Remote-sensing instruments on board Meteor-2 and Meteor-3 were intended for investigation and monitoring of water energy cycles in the atmosphere–ocean system including cloud formation, dissipation, and radiative properties, which influence the response of atmosphere to greenhouse forcing, large-scale hydrology, and moist processes including precipitation and evaporation.

Meteor-2 No. 19 (launched June 28, 1990) and No. 20 (launched September 28, 1990) were operational satellites that experienced, unfortunately, anomalies with sounding missions and IR imagery (noisy downlinked data for both spacecraft). They were still capable of performing routine radiometric measurements and visible imaging including APT transmissions worldwide. Originally designed for an operational lifetime of 1 year in orbit, these spacecraft, however, remained operational for several more months and, depending on their performance, were on standby mode or deorbited. The Meteor-2 system, which came into operational use in 1975, was fully replaced by the Meteor-3 system by the mid-1990s.

Meteor-3 No. 6 (launched January 24, 1992) was the principal operational satellite providing the main and regional centers at least twice daily with global data on the distribution of clouds, snow, and ice conditions in visible and IR bands, atmospheric temperature sounding data, cloud-top heights, SST, and radiation flux data from outer space.

Table 9.4 lists the basic technical characteristics of instrument packages flown on board Meteor-2 and Meteor-3 satellites. As seen from the table, the higher altitude of Meteor-3 extends the instrument swath width, therefore providing complete coverage of the Earth surface. As compared with Meteor-2, an advanced scanning radiometer with better spectral and spatial resolution and spectrometer for total ozone content measurements are installed on board Meteor-3.

Meteor-3 No. 5 was the first spacecraft to carry NASA's TOMS since it initially flew aboard NASA's Nimbus-7 satellite in 1978. The Meteor-3/TOMS mission continues bilateral cooperation begun in the early 1960s with an agreement on meteorology, communications, magnetic surveys, and life sciences. Since its successful launch in August 1991, the Meteor-3/TOMS 2-year mission made it possible to provide monitoring of global ozone levels by measuring backscattered radiation. TOMS data are downlinked to a receiving station at NASA's Wallops Flight Facility and Roskomgidromet's ground station in Obninsk. From there information is sent over data links to NASA's Goddard Space Flight Center and Roskomgidromet's Central Aerological Observatory for processing and analysis. The end

Table 9.4. Characteristics of sensors flown on Meteor-2 and Meteor-3 and output data (Meteor-3 data are given in parens).

Instruments	*Spectral band*	*Swath width at 900 (1,200) km altitude (km)*	*Resolution (km)*	*Output products*
Scanning TV system for direct relay of cloud and underlying surface imagery	0.5–0.7 μm	2,100 (2,600)	2(1 × 2)	Individual images, photomosaics of images from two to three passes over receiving station within 200 km in radius
Scanning TV system with onboard data recorder to provide global coverage	0.5–0.7 μm	2,400 (3,00)	1(0.7 × 0.4)	Individual images, global photomosaics of images of various regions of the globe (two to three times daily), cloud-free photomosaics of the Arctic and Antarctic Oceans once every 5 days
Scanning IR radiometer for global coverage with direct broadcast capabilities	8–12 (10.5–12.5) μm	2,600 (3,100)	8(3 × 3)	Global photomosaics: individual images; digital SST and cloud-top height charts, TC coordinates, cloud amount data
Scanning 8-channel IR radiometer for atmospheric thermal sounding	11.1–18.7 (9.65–18.7) μm	1,000 (2,000)	32 × 32 (angular 2 × 2°)	SATEM messages with atmospheric thermal sounding data (total ozone content)
NASA's 6-channel total ozone mapping spectrometer (TOMS)	312.5–380.0 nm	(2,900)	63 × 63	Global ozone charts
Microwave scanner	1.5, 0.86, 0.32 cm	(1,500)	80, 50, 20	Integral humidity, rain rate, snow ice slope, sea roughness
Radiative metric complex	0.15–90 MeV	—	—	Data on radiation fluxes

products of Meteor-3/TOMS in the form of a TOMS grid (i.e., grid array form) are available to the domestic and international user community (grid array of total ozone content data with 1 × 1.25 mesh over the Earth's surface). These data are available on CDs. The volume of data produced daily is 300 kB.

9.3.2 Resurs-01 and Resurs-F series

Resurs-01 No. 3 (launched November 4, 1994 on a Sun-synchronous orbit at an altitude of 678 km and an inclination of 98.0°) was an operational satellite. The remote-sensing systems of Resurs-01 were used to investigate and monitor land surface hydrology and ecosystem processes including improved estimates of runoff from the surface into oceans and change in land cover. Table 9.5 summarizes Resurs-01's baseline specifications. Its successor Resurs-01 No. 4 was launched in 1997.

The Resurs-01 mission was the first time that digital Russian satellite imagery had been received at a station outside Russia (i.e., the Esrange ground station near

Table 9.5. Resurs-01 current and projected (*) baseline characteristics.

Instrument	*Spectral band* (μm)	*Resolution* (m)	*Swath width* (km)
Optico-electronic high-resolution scanner (MSU-E)	0.5–0.6 0.6–0.7 0.8–0.9	45 × 30	45 within 600–700-km coverage (80 in double instrument mode)
Five-channel medium-resolution scanner (MSU-SK)	0.5–0.6 0.6–0.7 0.7–0.8 0.8–1.1 10.4–12.4	170 170 170 170 600	600
(*) High-resolution scanner (MSU-V2)	0.45–0.49 0.53–0.59 0.65–0.69 0.70–0.74 0.83–0.87 1.55–1.75 2.10–2.35 10.5–12.5	15 15 15 15 15 15 15 45	200
(*) Medium-resolution scanner (MSU-SK-M)	0.5–0.6 0.6–0.7 0.7–0.8 0.8–1.1 1.4–1.7 3.1–4.2 10.2–12.6	500	1,500

Kiruna, Sweden). It provides worldwide, medium-resolution data with 600 km swath and up to 15 m pixels. Resurs-01 is able to provide frequently repeated imagery at medium spatial resolution. This provides an ideal source of data for mapping at scales in the range 1:500,000 to 1:1,000,000 and for operational environment monitoring. Such characteristics bridge the gap between the capabilities of Landsat and AVHRR NOAA. All data are received by SSC Satellitbild straight from satellite at Esrange (Bjerkesjo *et al.*, 1996). Imagery of Europe is received in direct mode, while scenes from elsewhere in the World (including tropical regions) are from the onboard tape recorder. Quick looks from Resurs-01 can be browsed on the WWW at Eurimage's EiNet (http://www.eurimage.it/einet/einet-home.html) and on the Russian Space Science Internet (RSSI) (http:/smis.iki.rssi.ru).

The Russian space natural resource system Resurs-F was (Lukashevich, 1994) developed for periodical photographic survey of the Earth from space to resolve Earth science tasks. The system includes two series of spacecraft—Resurs-F1 and Resurs-F2—equipped with different sets of photo instrumentation. After completion of the survey program the exposed films are returned to Earth.

Spacecraft of the Resurs-F system are launched from the Plesetsk Cosmodrome by Soyuz-type launchers into near circular and near polar orbits with an altitude of about 250 km and an inclination of 82.3°. These orbits make survey of almost all of the Earth's surface possible (with the exception of the polar regions) within the belt restricted by the northern and southern latitudes equal to orbit inclination.

The spacecraft's angular orientation is carried out by its orbital coordinate system with an accuracy not less than $\pm 1.5^{\circ}$ for roll, pitch, and yaw angles and $\pm 0.01\,\mathrm{deg\,s^{-1}}$ for angular velocities. Both types of spacecraft are equipped with an orbit correction engine. Resurs-F1 can operate in active flight mode (when surveying) for 14 days and Resurs-F2 can operate in active flight mode for 30 days. Three KFA-200 space cameras (focal length $f = 200\,\mathrm{mm}$), two KFA-1000 cameras ($f = 1{,}000\,\mathrm{mm}$), and a stellar camera are installed on Resurs-F1. Star images are used for determination of the angular elements of the exterior orientation of the Earth's surface photographs. KFA-200 cameras are used for multiband survey and KFA-1000 for spectrozonal survey. The optical axes of the former are directed at the nadir, those of the latter are declined from the nadir in the plane that is perpendicular to the flight direction so that the swath width can be increased.

Resurs-F2 is equipped with a specially developed four-band MK-4 camera with a built-in stellar camera. All four lenses have a focal length of 300 mm and are equipped with replaceable light filters. The MK-4 includes a set of six light filters four of which are selected and installed depending on the objects under survey.

The photos obtained by the MK-4 are characterized by the high accuracy of their geometric and photometric parameters. The photometric scale is printed in every frame for all four bands for photometric measurements (Tables 9.6 and 9.7).

Launches of the Resurs-F system spacecraft are carried out mainly in spring and summer in the northern hemisphere. Annually between two and three Resurs-Fl spacecraft and one and two Resurs-F2 spacecraft are launched. The specific operating orbit and survey program are defined in accordance with the requests

Table 9.6. Geometric characteristics of photo information (Resurs-F system).

	Photo cameras			
Characteristics	*KFA-200*	*KFA-100*	*MK-4*	*KFA-3000*
Flight altitude (km)	275/235	275/235	240	275
Focal length (mm)	200	1,000	300	3,000
Picture format (cm)	30 × 30	18 × 18	30 × 30	—
Film length (m)	250	580	500	600
Swath width in flight altitude fractions	0.9	0.3 × 2/0.3 × 3	0.6	0.1 × 2
Spatial resolution (m) for survey in one spectral band for spectrozonal survey	 25–30/23–25 —	 —/4–6 8–10/6–8	 10–12/7–10 12–14/8–11	 3 —

Note: For double data the nominator contains the system characteristics after modernization.

Table 9.7. Spectral characteristics of onboard photo instruments (Resurs-F system).

Satellite	*Photo camera*	*Light filters*	*Films*	*Spectral bands* (nm)
Resurs-F1	KFA-200	ZhS-18/SZS-23	T-42	510–600
		KS-19	1-840K	700–840
		KS-10	T-38	600–700
	KFA-100	OS-14	SN-10	570–800
Resurs-F2	MK-4	KS-13	T-30M	640–690
		KS-17		
		Interferometer	T-30M	515–565
		Interferometer	T-30M	460–510
		OS-14	SN-10	610–750
		ZH-11	TsN-4	435–680
Resurs-F1M	KFA-200	KS-10	T-38	600–700
	KFA-1000	OS-14	SN-10	570–800
Resurs-F2M	MK-4M	KS-13	T-92	640–690
		Interferometer	T-92	520–560
		KS-17	1-860K	800–870
		OS-14	SN-18	610–760

Note: Different spectral bands are stipulated by the different optical characteristics of lenses and portholes.

of space photo-information users. The date and time of spacecraft launch are chosen taking into account user requests for photo-information and statistical meteorological data for the regions under survey.

Together with the Resurs spacecraft, "spinoff" satellites are used for Earth photo survey. Survey of the Earth's surface is carried out according to the orders of scientific and economic institutions by two KFA-3000 space cameras (focal length 3,000 mm) at an altitude of 275 km. The spatial resolution of the images obtained by these cameras is about 3 m.

Modernization of the Resurs-F space system is being carried out. For modernized Resurs-F1M spacecraft the active flight mode has been increased by 5 days and the flight altitude has been decreased by 40 km. The onboard set of photo-instruments has been considerably changed and is oriented towards obtaining photos with high spatial resolution and widening swath width. One KFA-200 and three KFA-1000 cameras are installed on *Resurs-F1M*. The optical axes of the KFA-200 and of one of the KFA-1000 are directed at the nadir. Unlike Resurs-F system spacecraft, spinoff satellites can be used both for nadir and perspective survey by means of a programmed turn with a roll up to 42°.

Updating of the MK-4 camera and usage of new types of films provide for a 1.4 times increase in spatial resolution of multiband photo-information received from Resurs-F2M spacecraft.

9.3.3 Geostationary Operational Meteorological Satellite Electro (GOMS)

The launch (October 31, 1994) of the three-axis stabilized spacecraft GOMS, stationed over 76°E, undoubtedly represented a major change and advance from the then current national environmental satellite observing system. The onboard instruments of GOMS provided continuous observations of the Earth in visible and IR spectral bands. The mission also included collection and relay of hydrometeorological and other environmental data from DCPs, as well as information exchange between main regional centers including data from polar orbits and end products. The following services are available to various users:

- *imaging*—GOMS acquires images of the full Earth in two spectral channels (this became three with GOMS-2) up to 48 times per day (images are pre-processed in the main center before distribution to users);
- *analogue image dissemination*—pre-processed data is retransmitted from the main center via spacecraft to national and foreign user stations;
- *data collection and relay*—environmental data from national and foreign DCPs are collected and retransmitted to regional centers;
- *space environment monitoring*—various parameters of magnetic and radiation fields in space at the geostationary orbital altitude are measured and relayed to regional centers;
- *meteorological data dissemination*—image fragments, charts, and other meteorological products in alphanumerical form are retransmitted from regional

centers via spacecraft to national and foreign users. Table 9.8 presents the basic characteristics of GOMS.

9.3.4 *Priroda* mission

Priroda is a Russian multisensor remote-sensing mission in which active and passive sensors in the microwave and optical range are used on board a *Priroda* module. This module was launched and docked to the Russian space station *Mir* on April 23, 1996. The goal of the mission was development and verification of multisensor remote-sensing methods for the investigation of land, oceans, atmosphere, and ecological problems of the environment (Armand, 1993).

The *Mir* station orbits at an altitude of 400 km at an inclination of 51.6°. *Priroda* is a multisensor mission with a broad variety of remote-sensing sensors:

- SAR TRAVERS;
- microwave radiometric system IKAR;
- visible and near IR spectrometric system MOS OBZOR;
- IR spectrometric system ISTOK-1;
- precision radiometer (PR) GREBEN;
- visible and near IR spectrometric systems MSU-SK and MSU-E;
- UV and IR spectrometric system OZON-M.

The IKAR system consists of 15 passive microwave radiometers:

- three panoramic radiometers of subsystem IKAR-P operate at the 13.33 GHz frequency (2.25 cm wavelength), each with horizontal and vertical polarization, at an incidence angle of 40°;
- five panoramic radiometers of subsystem IKAR-P operate at the 5.0 GHz frequency (6.0 cm wavelength), each with horizontal and vertical polarization, at an incidence angle of 40°;
- five nadir-looking radiometers of subsystem IKAR-N operate at the 5.0, 13.33, 22.235, 37.5, and 100 GHz frequencies (6.0, 2.25, 1.35, 0.8, and 0.3 cm wavelengths, respectively), each with two orthogonal polarizations (in the 2.25 cm band both module and direction of emission polarization vector are registered);
- the Delta scanning radiometer operates at three frequencies, each with horizontal and vertical polarization, 22.235, 37.5, and 100 GHz (1.35, 0.8, and 0.3 cm wavelengths, respectively) at an incidence angle of 40°;
- the R-400 scanning radiometer operates at the 7.5 GHz frequency (4.0 cm wavelength) with horizontal and vertical polarization at an incidence angle of 40°.

Over the oceans, the IKAR system is used to measure SST, windspeed, precipitable water content (water vapor content), cloud liquid water content, and rain rate. Over

Table 9.8. Characteristics of Electro's instruments and information radio complex.

TV radiometer	Spectral bands	
	VIS	0.46–0.7 μm
	IR I	10–12.5 μm
	IR II	6–7 μm
	IFOV/ground resolution	
	VIS	31.5 μrad/1.25 km
	IR II	160 μrad/6.5 km
	Data transmission sessions	
	Session frequency	30 min
	Frame length	12 min
	Band	1.7 and 7.5 GHz (S-band and X-band)
	Informativity	2.56 Mbit s^{-1}
Radiation measurement system	Energy bands for measuring the density of the electron, proton, α-particle fluxes and galactic radiation	0.04–600 MeV
	Four bands for measuring solar radiation	0.4–1.3 μm
	Internal induction measurement of the Earth's magnetic field along three axes	±180 nT
	Frequency of data transmission sessions	60 min
Radio relay complex	Data collection platforms	
	Sampling band	401–403 MHz
	Number simultaneously sampled	133
	Sampled platform data rate	100 bit/s
	Band for data transmission from data collection platforms to weather service centers	1.7 GHz (S-band)
	Relay bands for processing data (WEFAX mode) between weather service centers	1.7 and 7.5 GHz (S-band and X-band)
	Data rate	1.200 bit/s
	Band for transmitting the same data to small stations	1.7 GHz (S-band)
	Band for relaying high-rate digital data	7.3 GHz (X-band)
	Data rate	Up to 0.96 Mbit/s
	Band for platform calling	469 MHz

the land, IKAR will complement visible and IR (MOS[13]-Obzor, MSU-SK, MSU-E) and active microwave (SAR TRAVERS) observations of vegetation status, biomass, and soil moisture. Over snow and ice-covered areas IKAR data in conjunction with data of SAR TRAVERS can be used for ice-type determination and for estimations of ice concentration.

TRAVERS is a SAR operating at two wavelengths—9 cm and 23 cm—with horizontal and vertical polarization at an incident angle of 35°. The swath width is 50 km at spatial resolution of 100 m and at minimum registered effective cross-section reflectivity of −25 dB to −30 dB. TRAVERS includes the study of vegetation canopy type and condition and soil variation over land areas. It is also used in mapping the topography of the land roughness of ice and the sea surface.

GREBEN is a nadir-looking radar altimeter operating at 13.76 GHz at a spatial resolution on the surface of 9.7×2.5 km. The circular polarization used in GREBEN permits transmission and reception circuits up to 20 dB o be decoupled, at the linear frequency modulation mode of the sounding signal. It provides measurements of ocean wave height with standard fluctuation error of sea wave height determination of 0.5 m. In addition, GREBEN gives information about spacecraft altitude and sea level variations when the trajectory of the satellite is known. Simultaneous determination of these parameters is also possible.

The modular optoelectronic scanner (MOS) is a spaceborne imaging spectrometer in the visible and near IR range. It was especially designed for remote sensing of the ocean–atmosphere system. MOS within the *Priroda* payload provides spectral high-resolution measurements in 17 channels with high radiometric accuracy (of 0.3–1.0%). The *Priroda* instrument has a 12-bit digital resolution and is a hermetically sealed version (for CCD focal plane cooling) for use inside the manned *Mir* space station. With the advent of MOS a specialist instrument concept was realized for remote sensing of the atmosphere–ocean system, which had been tested using non-imaging spectrometers on several preceding missions. MOS consists of two separate spectrometer blocks, MOS-A and MOS-B. MOS-A was designed for measurement of atmospheric parameters in four narrow channels of $\Delta\lambda = 1.4$ nm in the O_2-A-absorption band at 760 mm. From the data in one window channel outside the O_2-A-band and three channels of different absorption the optical thickness of aerosols can be estimated, as well as the amount of strato-spheric aerosols and the height of the upper border of clouds. MOS-B has 13 channels of $\Delta\lambda = 10$ nm in the range from 408 nm to 1,010 nm and is designed (through channel selection and spatial resolution) for remote sensing of ocean and coastal zones. The data are used for quantitative retrieval of water constituents. The common use of MOS-A aerosol parameters and the data from the two MOS-B water vapor channels (at 815 and 945 nm) gives an estimation of atmospheric influence on the other MOS-B channels and enables atmospheric correction of "water data". MOS-B channels also allow vegetation signature determination.

The IR ISTOK-1 system provides spectral measurements of thermal emission by the atmosphere and by the underlying Earth surface on a variety of incident angles as

[13] MOS, modular optoelectronic scanner.

well as measurements of transmission spectra of the atmosphere while tracking the Sun. In addition, ISTOK-1 includes a TV camera to survey cloud cover and the Earth surface. ISTOK-1 has its own computer unit, which controls system functionality and provides signal processing and transmission. Its IR spectroradiometric parameters are:

Operational wavelength range (μm)	4–16
Number of spectral channels	64
Spectral resolution (μm) in 4–8 μm wavelength range in 8–16 μm wavelength range	 0.125 0.25
Threshold sensitivity	5×10^{-6} W/(cm$^2 \times$ step $\times$ μm)
Integration times 0.5 s 1 s	 8–16 μm (0.3 K at 10 μm while object temperature 300 K) 4–8 μm
Field of view	$12' \times 48'$ (angular minutes)
Linear resolution ($H = 300$ km) (km) Along-track, nadir-looking mode In altitudes, horizon-looking mode	 1–4 < 7
Time of scene detection (ms)	20

MSU-SK and MSU-E instruments are used to obtain visible and IR images of land and ocean surfaces with spatial resolutions of 25 m (for MSU-E), 120 m (for MSU-SK), and 300 m (for IR channel). Detailed characteristics of these instruments are in Section 9.3.2 on Resurs systems.

OZON-M is a four-channel scanning diffraction spectrometer with an autonomous tracking system that permits Sun-occultation measurements to be conducted. OZON-M is used to evaluate measurements of trace gas concentrations. Its UV and IR spectrometric parameters are:

Spectral band	0.26–1.05 (four channels 0.26–0.3, 0.36–0.42, 0.60–0.60, 0.91–1.05 μm)
Spectral resolution (A)	2–7
Beamwidth	$2 \times 25'$ (angular minutes)
Vertical resolution (km)	Approximately 1 (tangent to a track)

Interesting results were obtained from visual and instrumental observations of TCs performed on board the Russian space station *Mir* (Bondur *et al.*, 1997).

9.3.5 Meteor-3M system

Roskomgidromet has contracted with industry to develop a new generation of meteorological satellites called the Meteor-3M series. To upgrade the capabilities and compatibility of operational international observing systems, Meteor-3M carries a multispectral scanning complex, comprising visible, IR, and microwave scanners to determine more accurately cloud-top heights and temperatures, integral water vapor, and condensed moisture content, and to evaluate rain rate and measure sea surface roughness. Meteor-3M has on board a spectrometric complex including a scanning Fourier spectrometer and a microwave module to determine the temperature and humidity profiles of the atmosphere (Cherney and Raizer, 1998).

In addition to this, a side-looking radar (SLR) and a radiation budget instrument complex are on board. A scanning radiometer (ScaRaB) is also in the payload of Meteor-3M. The major characteristics of the spacecraft and its instruments are summarized in Table 9.9. The essential feature of the subsequent generation of Meteor was its ability to make satellite data directly accessible for the user community. Data transmissions are made at the S-band (1,670–1,770 MHz) in three modes with rates of 15 Mbit s^{-1}, 960 Kbit s^{-1}, and 96 Kbit s^{-1}. The last two modes are worldwide direct transmissions in standard format.

9.3.6 Some aspects of the Russian remote-sensing program

Russian researchers like Antonov *et al.* (1995) and Ziman and Krasnopevtseva (1996) believe that image data should be used to solve the five main groups of environment-monitoring tasks:

(1) forecasting of weather and climate change;
(2) pollution control of the atmosphere, water, and soils;
(3) emergency situation control during natural and man-made disasters;
(4) information support in agriculture, forestry, fishery, mining, building, and land management;
(5) information support in the Earth sciences, and in the creation and development of a dynamic model of the Earth as an ecological system.

Experience in remote sensing shows that different users request different image data with a spatial resolution that varies from several meters up to several kilometers. However, image data of 20 m to 30 m spatial resolution is in greatest demand when studying the Earth's surface.

As for spectrometry, it is desirable to image in approximately 30 bands of the visible and IR ranges with a spectral resolution of about 0.025 μm. If TV imaging and spectrometry are simultaneously conducted within the same swath then the

Table 9.9. Summary of instruments flown on Meteor-3M.

Instruments	*Application*	*Spectral range*	*Ground resolution*	*Swath width* (km)
Multichannel visible and IR radiometer	Cloud, ice, snow imaging; SST, NDVI	0.5–0.7 μm 1.5–1.75 μm 3.55–3.93 μm 8.25–14.5 μm 10.3–11.3 μm 11.5–12.5 μm	1.5 × 2.0 km	3,100
Microwave scanner (LR)	Integral humidity, water content, rain rate, snow and ice boundaries, sea surface roughness	1.5 cm (V, H) 0.86 cm (V, H) 0.32 cm (H)	80 km 55 km 20 km	1,500
Scanning spectrometer–interferometer	Vertical temperature and humidity soundings	15-μm absorption band, spectral resolution 0.5 cm^{-1}	20 × 20 km (31 elements per line)	2,500
Scanning microwave sounder	Vertical temperature soundings (cloud-contaminated FOV)	10–12 channels in 52–56 GHz band (H)	50 km	1,500
ScaRaB	Radiation budget measurements	0.2–4.0 μm 0.2–50 μm 0.5–0.7 μm 10.5–12.5 μm	60 km	3,000
Active cavity irradiance monitor	Total solar irradiance, solar constant	0.001–1.00 μm	Solar printing	—
Side-looking radar	Operational sea ice and snow cover mapping	3 cm	200 × 400 m	Two symmetric 400–500 km swath widths

spatial resolution of spectrometric data can be an order lower than the resolution of TV data.

The periodicity of imaging the same site is an important characteristic of the Earth remote-sensing system. For the majority of tasks where image data of 25 m to 30 m spatial resolution is used it is desirable to observe the studied region at least once every 2–3 days.

It is only possible to achieve the required periodicity of imaging of the same surface site from space at a high spatial resolution by creating a cluster of satellites strictly spaced in orbit. When a task is solved using not just one satellite but a constellation of satellites it becomes impossible to simplify and reduce the cost of the satellites, minimize their mass and power consumption, and, in particular, simplify the housekeeping systems and optimize the parameters of remote-sensing instrumentation. The miniaturization of onboard instrumentation for observing the Earth's surface in the visible and IR spectral bands makes it possible to create space information remote-sensing complexes with parameters satisfying the requirements of the majority of users and with a mass of less than 100 kg. Small spacecraft with a total mass of about 300 kg can serve as a platform for such complexes.

It is possible to solve the problem of imaging any terrestrial site every 2–3 days by creating a cluster of four to seven satellites equally spaced in a common Sun-synchronous orbit at an altitude of 400 km to 1,000 km.

Centralized image data reception in remote-sensing systems of the Earth and decentralization of secondary (thematic) data processing by multiple users amount to a contradiction that cannot be compensated by using ground communication facilities. Personal spaceborne image data that are received and processed by users themselves may be the solution to this problem. Additionally, receiving stations that are available to users are not able to receive bulky data streams. It is possible to resolve this contradiction by implementation of the Local Space Service (LOCSS) (Ziman and Krasnopevtseva, 1996). The main aspects of this concept (whose aims are to further develop Earth remote sensing from space) are:

- optimization of parameters for the onboard remote-sensing information complex such that image data can be acquired by combining both high spatial and the required spectral resolution;
- in-orbit creation of a cluster of small spacecraft with image spectrometric complexes that can provide imaging of the specified regions with the required periodicity;
- onboard compression of image spectrometric data for data downlinking via a low-rate radio channel;
- direct reception of image spectrometric data by users at small personal receiving stations and data storage in their computers; and
- decompression, processing, and thematic interpretation of obtained image spectrometric data using personal computers equipped with special processors.

The LOCSS concept solves the contradiction between the amount of data of interest to users and the limited data-downlinking rate. First, image data are compressed on board and then restored on the ground. Second, the set of chosen data parameters is optimized (number of spectral bands, swath, spatial resolution, etc.). In addition, the data parameters are chosen to satisfy the majority of users.

The creation of a universal technique to meet the requirements of the huge diversity of tasks in the field of Earth remote sensing from space is a very complex problem. One way of solving this consists in simultaneous imaging by a multispectral TV camera and an imaging spectrometer. High-resolution data acquisition within a small number of comparatively wide spectral bands is proposed for the main TV imaging. In contrast, spectrometry is performed with low spatial and high spectral resolution. It is advantageous to use an imaging spectrometer with the same swath as the TV camera. In this way, these imaging spectrometric complexes can be considered as a base for future systems now being designed both for remote sensing from small spacecraft and nominal environmental monitoring from spacecraft of the Resurs-O, Landsat, and SPOT series.

Russian space corporations have created a new generation of spacecraft for Earth remote sensing (Dolgopolov *et al.*, 1996). These spacecraft are equipped with optoelectronic and radar observation systems, providing for imagery at a spatial resolution down to 5 m in the optical and radio bands, in which it is possible to change the observation spectral band to satisfy the demands of data users. These spacecraft have significantly smaller dimensions and mass compared with existing vehicles.

In order to reduce the time needed for development and to make the spacecraft cheaper at the first stage many parts of the Yamal and Signal communications satellites were used in these designs; moreover, many key components were tested at the *Mir* orbital station under spaceflight conditions. The next stage involved creating a second generation of these spacecraft, making wider use of key components and systems recently developed. Table 9.10 gives the main characteristics of these spacecraft.

The spacecraft were developed for two fields of study: optoelectronic observation and radar observation. The remote-sensing spacecraft were equipped with a new generation of instrumentation—imaging spectrometers—making it possible to acquire imagery at a high spatial and spectral resolution and with a bandwidth of 100 Å at the first stage and 10 Å at the second stage. These spacecraft were launched into Sun-synchronous orbits from the Plesetsk, Baikonur, and Svobodny Cosmodromes by both Soyuz and Zenit launchers and spinoff missiles.

On the other hand, there are Russian researchers who hold an alternative viewpoint toward environment-monitoring tasks; this is to use heavy spacecraft, like *Mir* and *Almaz* (Antonov *et al.*, 1995) and like the geostationary satellite GOMS with its complicated optical telescopes (Avduevskii *et al.*, 1983). The advantage of this line of attack consists in implementation of the concept of complex instrumentation in single-position multiwave observations in different electromagnetic bands. Small spacecraft cannot make complex observations, and such a type of instrumentation method has often been the decisive factor for a number of remote-sensing experiments in the study of complicated meteoprocesses (like TCs). So, in general, the problem is sophisticated and ambiguous and will need detailed design for its solution.

Table 9.10. Characteristics of Earth remote-sensing spacecraft.

	First-generation spacecraft		*Second-generation spacecraft*	
	With optoelectronic systems	*With radar systems*	*With optoelectronic systems*	*With radar systems*
Operation orbit parameters Altitude (km) Inclination (deg)	300 96.3	575 97.5	575 97.5	575 97.5
Spacecraft mass in operational orbit (kg)	1,100	850	400	700
Orbit average power supply system capacity (W)	400	320	200	300
Remote-sensing instrumentation mass (kg)	150	135	170	135
Remote-sensing instrumentation power consumption (W) during observation session	700	1,135	80	700
Orbit average	200	25	30	50
Lifetime (year)	1	3	7	7

9.4 SPECIAL RUSSIAN MISSIONS FOR TC STUDY

In recent decades, the problem of studying large-scale atmospheric hazards has been one of the main trends in a number of potential space programs (Anfimov *et al.*, 1995, 1996; Avanesov *et al.*, 1992; Balebanov *et al.*, 1996, 1997). The means of obtaining useful information is totally dependent on remote techniques. Nevertheless, theoretical and experimental studies show that certain specific features in the structure of wind turbulence cannot be analyzed with adequate accuracy by using remote-sensing data in isolation.

For example, the wind velocity field reconstructed from traversing cloud systems—technically called "cloud vector winds" (CVWs)—possesses a spatiotemporal grid scale of 100 km and 30 min and a root-mean-square error (r.m.s.e.) of reconstruction ranging within $4\,\mathrm{m\,s^{-1}}$ and $7\,\mathrm{m\,s^{-1}}$ (Velden *et al.*, 1984). Planned space systems for lidar sensing of the situation from the grid scale will not change, though they will substantially improve the r.m.s.e. value of reconstructed wind velocity by $1\,\mathrm{m\,s^{-1}}$ (Betout *et al.*, 1989).

At the same time, special field experiments using Doppler sensing (Klepikov *et al.*, 1995; Sharkov, 1996a, b) showed that to form the structural characteristics of turbulence on mesoscales (1–100 km) a reasonable accuracy (r.m.s.e.) for wind velocity determination should be of the order of 0.3 m/s to 0.5 m/s at a spatio-temporal grid scale of 1 km to 2 km and 1 min. So, projects set up to study large-scale hazards must comprehensively combine purely remote techniques with contact sounding of the object under investigation.

On the other hand, solid experimental results proving that TCs as global phenomena interact with geophysical media of various physical natures have been obtained (see Chapters 2 and 3). Prospective projects for the study of large-scale events should entail execution of combined satellite missions intended for measurement of the characteristics of these geophysical media.

In this section we outline a project to investigate large-scale critical structures with the help of contact sounding by meteorological probes and balloons brought to the region under investigation by ballistic rockets—the Zodiak mission (Anfimov *et al.*, 1995, 1996)—and another project to study the kinematic, thermodynamic, and electrodynamic links between the elements of the ocean–troposphere–upper-atmosphere system in a crisis situation by a special experimental satellite—the Helix mission (Balebanov *et al.*, 1996, 1997).

9.4.1 Preliminary background to the projects

The new physical concept of an inverse helical cascade ("vortex dynamo"), proposed and developed at the Space Research Institute (SRI) of the Russian Academy of Sciences, and field experiments carried out to date provide the necessary basis for running special detailed large-scale experimental investigations into the kinematic and thermodynamic characteristics of metastable states of the ocean–atmosphere system.

The essence of this physical concept is the existence of a large-scale vortex dynamo in aerohydrodynamics under conditions of unstable thermal stratification, which provides a possibility for fundamental reconstruction of the thermal convection conditions giving rise to origination of an ordered large-scale vortex system whose stream lines have a non-trivial topological structure. In other words, individual convective cells of the velocity field moving helically tend to merge and form one large helical cell, which is actually a TC. The main experimental task of early diagnostics is to detect the initial period of this merging (before formation of the TC), when the flow of energy is transferred from small cells to larger ones as the helicity of the large cells increases (i.e., the so-called inverse cascade in turbulence).

Other specific features accompanying the formation of a critical situation are intensification of waves in the unstable zone of cyclogenesis (as a result of pent-up non-equilibrium energy) and abnormal behavior of admixtures in this zone (a change in diffusion coefficient and rearrangement of its isotropic characteristics).

An important feature in the process of evolution of a TD into a three-dimensional TC structure is the formation of a high-altitude (stratospheric) anticyclonic "outflow", which shows up in optical space images in the form of a weak "veil" of spindrift clouds. The role played by anticyclonic circulation (AC) is very important because it provides for the kinematic effect of TCs on stratospheric air flows, on the formation of specific features in the trajectory of TC motion, and on the interaction of these features with the ozonosphere and, possibly, with ionosphere layers.

However, for a number of reasons there is practically no reliable experimental information on the structure and characteristics of anticyclonic circulation (in contrast with cyclonic circulation in the "tropospheric" body of the TC). Such reasons include the absence of tracers in the form of cloud systems at these altitudes, impossibility of radar sounding, difficulties in using meteorological probes at stratospheric heights, etc. Therefore, the urgency of solving the problem of experimental sounding of the AC zone using meteorological systems, parachuted from space vehicles, into the region to be investigated is self-evident.

To generalize, this physical mechanism recommends that data obtained in detailed measurements of the kinematic and turbulent characteristics of velocity fields should be used to reconstruct the readiness of a non-equilibrium atmosphere (in its active zones) to generate large-scale structures. These characteristics comprise the three-dimensional structure of the velocity field (revealing its poloidal and toroidal components), the latitudinal and spatial distribution of the turbulent energy stock and of the field helicity, scale and time estimates of the energy flows and helicity, and mesoscale measurements of the diffusion coefficient and its fluctuations. Furthermore, it is necessary to investigate the spatiostatistical thermal and humidity characteristics of thermal convection, including high-altitude (at 30 km) temperature and humidity profiles.

The technical facilities currently used to obtain experimental data on TDs and TCs fail to provide an operational detailed three-dimensional pattern characterizing the atmospheric parameters across the total volume occupied by the object in space with an acceptable horizontal resolution (20–50 km) and vertical resolution (hundreds of meters). Instrumental measurements in the anticyclonic part of TCs at altitudes exceeding 10 km to 13 km are practically absent. A similar situation takes place with experimental investigations into the leading flow in a radius of 100 km from the TC center.

Experimental information on the central part of TCs is most important, as it is vital to carrying out fundamental investigations of them. Information on the leading flow and AC is of primary importance for operative prediction of direction and velocity of TC travel: therefore, a databank on the leading flow and AC would be invaluable to the creation of new efficient prediction schemes.

Objects of study

In accordance with the main tasks of the project, atmospheric critical states can be classified into the following types:

- *Type A*—developed active TC with a definitely pronounced AC (the remote attributes for recognition and identification of the TC evolutionary stage, seasonal features of cyclogenessis activity and geographic zones of activity, and the trajectories of TC travel in the northern hemisphere are well-known).

- *Type B*—tropical disturbance (TD) at the stage of transition from tropical depression TDE to developed TC: namely, a TS (i.e., essentially the moment of TC generation). They have a sharply inhomogeneous structure up to the formation of active centers in generation zones (northern hemisphere).
- *Type C*—TD at the initial stage. TDs can either develop into TCs or remain undeveloped. The detailed quantitative characteristic of TDs as well as the zones and trajectories along which they travel are inadequately known today, though we can infer from preliminary data that TDs originate and exist every day in active seasons.

9.4.2 Mission Zodiak

Requirements for measuring procedure and composition of measurements

To solve the scientific tasks of the project, it is necessary to perform regular monitoring of the turbulent atmosphere in the zones of active cyclogenesis both under quiescent conditions and when TDEs exist and especially when they change into mature forms (TCs and TSs).

The experiment should be based on two mutually correlated procedures: investigation of the kinematic and thermodynamic structure of the free atmosphere and of the TD body by means of "sections" (vertical samples) and "drifts" (balloons). The first technique determines the "instantaneous" thermodynamic characteristics and the small-scale part played by the kinematics of turbulent flows of the TC body. The second technique must provide information on the kinematic features of mesoscale and synoptic-scale turbulent flows in the zone of TD action.

When developing a methodology for the experiment, it is necessary to carry out preliminary mathematical modeling of the process of direct measurement of thermodynamic and kinematic parameters of the tropical perturbation body by using vertical samples (sections) whose spatial distribution is random, and freely drifting tracers (balloons). Mathematical modeling should be carried out both in direct problem mode and inverse problem mode. It is necessary here to estimate the possibilities of reconstructing and varying the characteristics that are crucial to the concept of the vortex dynamo: small-scale helicity and turbulent stock of energy (from individual samples); estimation of the inverse energy cascade and of helicity flow (mesoscales); measuring dimensionally the correlation between spatial spectra of temperature fluctuation and the data of thermal sections (mesoscales); measurements of the diffusion coefficient from trajectory data of neutral buoyancy balloons (mesoscales).

Preliminary estimates, based on field investigation data (Anfimov *et al.*, 1995, 1996), show that thc accuracy of estimating atmospheric parameters, measured by the present-day generation of meteorological probes and balloons, is satisfactory for studying mesoscale kinematic and thermodynamic parameters (at least at the first stages of the project). Absolute geographic control of the spatial attitude for meteorological probes and balloons should be 1 km. Acceptable horizontal spatial resolution can be 20 km to 40 km when the diameter of the probed region is 200 km. The total number of measuring cycles for different TD stages must be about 100 for 4–5 years of observations.

Scenario of the project

Operational transportation of diagnostic sensors using rocket and space facilities to the region under investigation is the best way of solving the project's scientific problems.

Implementation of the method illustrated schematically in Figure 9.1 comprises the following stages: transportation of diagnostic module 1 along a ballistic or orbital trajectory to the region to be investigated; separation of parachute capsules (PCs) 2, provided with radioprobes (meteorological probes) 3 and repeater unit 4, from the diagnostic module 1 beyond atmospheric boundaries; carrying away the diagnostic module and delivery of the PCs to the region under investigation; measuring the parameters of the atmosphere, the Earth surface, and the ocean using radioprobes during their descent and after their landing (sea or land); transmission of information from the radioprobes to reception stations 5 and communication satellites 6. The set of parachute capsules is separated from the diagnostic module to ensure their prescribed spatial distribution and delivery to the upper boundary 7 of the region to be investigated.

The use of rocket and space facilities for simultaneous delivery of a considerable number of measuring probes directly to the zone of TC action is very promising aspect of the Zodiak mission. Using these facilities, measuring probes can be distributed along the upper boundary of the investigated region of the atmosphere, so that later up to several tens of vertical profiles—such atmospheric parameters as wind velocity, temperature, humidity, pressure, etc.—can be measured simultaneously during descent at the required in-height resolution.

The proposed method of delivering the diagnostic equipment can be implemented with the help of a rocket and space complex (RSC) that delivers several tens of small-size (with a mass of 10–15 kg) automatic radioprobes to the required region and distributes them over the prescribed area. Critical analyses showed (Lukiashchenko *et al.*, 1994; Utkin *et al.*, 1997) that this complex could be built around the rocket complex SS-19, which is no longer used for defense purposes.

The RSC comprises the Rokot launch vehicle (LV) complex; a space nosecone (SNC) containing a Briz booster unit (BU) accommodating the PCs and meteorological probes (MPs); a repeater unit (RU); devices for separating these capsules and probes; the technical complex; and the launching complex.

Adequate functioning of the RSC is provided by satellite facilities for detection of TCs and their coordinates; facilities for transmission of information to the RSC control center; Glonass and Navstar space navigation systems; aircraft, marine, and satellite means for receiving information from the repeater unit.

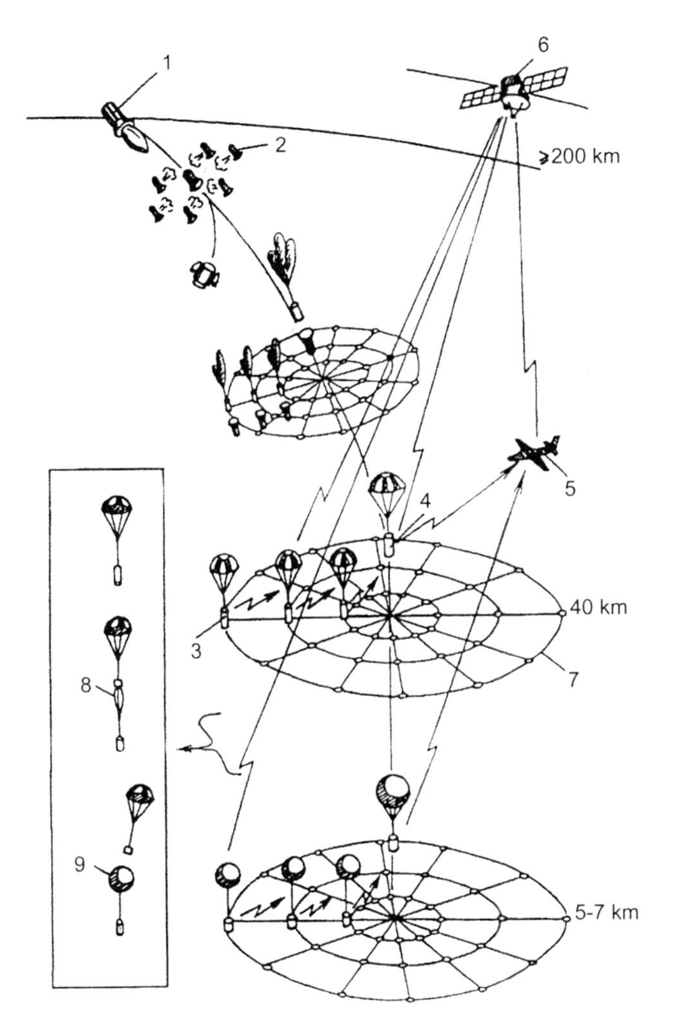

Figure 9.1. Diagram illustrating implementation of an operational investigation into atmospheric parameters: 1, diagnostic module; 2, parachute capsules; 3, radioprobes (meteorological probes); 4, repeater unit; 5 (aircraft) and 6 (communications satellite), information-receiving stations; 7, upper boundary of investigated region; 8, extraction and gas filling of balloon envelope; 9, balloon probe (from Anfimov *et al.*, 1995).

The main specifications of the measuring part of the RSC are presented in Table 9.11.

The separation system is intended to disperse the PCs and their radioprobes over the area of the region under study. This area is viewed from the plane of the local horizon at altitudes of 40 km down to 30 km—the altitudes at which the measuring instruments start to work—and has the shape of a circle of diameter 500 km or of a 500×500 km square (see Figure 9.1).

Table 9.11. Main characteristics of RSC for operational contact sounding.

Characteristics	*Value*
Quantity of delivered meteorological probes	Up to 50
Mass of meteorological probe (kg)	Up to 2
Mass of parachute capsule (PC) (kg)	12–15
Diameter of probed region (km)	Up to 500
Probed altitudes (km)	35 ... 0
Measurement atmospheric parameters:	
Pressure (mbar)	1,050 ... 100 (±0.5)
Relative humidity (%)	40 ... 95 (5–20)
Temperature (°C)	−50 ... +40 (±0.5)
Wind velocity components ($m\,s^{-1}$)	2 ... 100 (±0.5–0.75)

Note: Parameters measuring r.m.s.'s are enclosed in parens in the right-hand column.

Analysis of different ways of loosing the PCs—swiveling the space vehicle (SV), cartridge pressure accumulators, and solid propellant rocket motors (SPRMs)—suggested that SPRMs were the best option.

The RSC functions according to a strict program. The preparation of the LV through launch to a circular orbit including the way in which the space vehicle (SV) functions with the Briz BU, are carried out in accordance with the routine cyclorama of Rokot LV operation. On reaching final orbit altitude, a circular orbit is effected carrying the SV to the right. On approaching the viewing point, the SV is oriented into position by means of a braking impulse required to align the descent trajectory with the target point in the TC epicenter. The power plant of the Briz BU operates the braking impulse. After that the SV is oriented into position for separation of the PCs by means of their radioprobes and repeater unit.

When the PCs are released from the SV, their end switches operate triggering the time devices of the PC automatic systems. The PC power plants used to separate the PCs from the radioprobes are switched on with a calculated time delay after separation from the SV (pyre delay); as a result of power plant operation the PCs acquire the radial velocity required for their distribution over the prescribed area.

A typical PC descent trajectory is illustrated in Figure 9.2. The points of impact of the PCs at $H = 25$ km in the region with $R = 250$ km are also presented in Figure 9.2. Separation of the capsules is ensured by their individual power plants which operate by means of correcting pulses $\Delta V = 210, 682, 435, 952\,m\,s^{-1}$ perpendicular to the SV's velocity vector under corresponding angles of inclination to the local horizon $\alpha_{PP_p} = -8 \div -26^{\circ}$, where PP_p denotes power plant pulses.

After the PCs enter the dense layers of the atmosphere by aerodynamic braking and "cascade" braking by parachute at a height of $H = 40$ km, the thermal protective

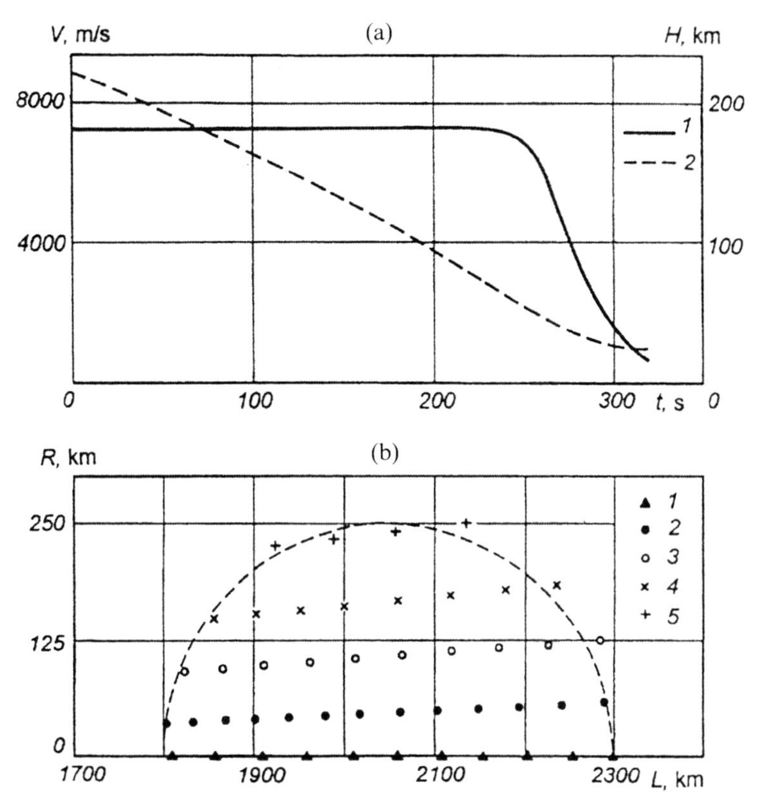

Figure 9.2. (a) Typical trajectory of PC descent: 1, velocity; 2, height from orbital altitude $H_0 = 220$ km at the velocity and angle of entry into the atmosphere $V_{ent} = 7{,}300\ \text{m s}^{-1}$ and $\theta_{ent} = -4^{\circ}$, inclination of the thrust vector of the PC PP to the local horizon $\alpha_{pp} = 0^{\circ}$ and correcting impulse $\Delta V = 952\ \text{m s}^{-1}$. (b) Results of estimating the distribution of PC landing points at the altitude $H = 25$ km in the region $R = 250$ km: 1, $\Delta V = 0\ \text{m s}^{-1}$; 2, $\Delta V = 210\ \text{m s}^{-1}$; 3, $\Delta V = 435\ \text{m s}^{-1}$; 4, $\Delta V = 682\ \text{m s}^{-1}$; 5, $\Delta V = 952\ \text{m s}^{-1}$ (from Anfimov *et al.*, 1995).

shells are released. Further descent and functioning of the radioprobes take place when the main parachute reaches a vertical velocity of $V = 10\ \text{m s}^{-1}$. After the main parachute becomes operational and the thermal protective shells are released, air temperature and humidity sensor rods disengage from the housing, and the airborne radiotechnical system of the radioprobes starts up. On reaching a height of 5 km to 7 km, the main parachute separates (Figure 9.1), and the envelope of the neutral buoyancy balloon (8 in diagram) dissolves. After that, the pyrotechnic valve that fills the balloon with a lift gas from the pressurized gas system is opened on command from the time device. After filling the balloon envelope with the lift gas, the gas-filling line is made tight near the upper pole of the balloon, and the parachute with the gas-filling system is separated from the balloon probe (9 in diagram). Having reached the height of equilibrium with the environment, estimated to be between 5 km and 7 km, the balloon with the meteorological probe drifts, slows with respect to the Earth as a result of atmospheric streams, and performs the required measurements.

The radioprobes measure meteorological parameters whose values are transmitted either to the repeater unit (in real time) or to the onboard repeater unit of the SV (from the radioprobe's memory device). In this case the BU must be at least 5 km above the radioprobe, a requirement that ensures direct visibility when receiving information from radioprobes farthest from the BU.

The information collected from the BU is transferred either to aircraft laboratories or to balloonborne information-gathering facilities and is then relayed to observation stations on Earth. The current location of the radioprobes can be determined by using Navstar (Glonas) space navigation systems. The accuracy of determining plan coordinates is about 50 m to 70 m, the accuracy of determining the height is about 100 m to 120 m (in the case of differential treatment the accuracy is 20–30 m and 30–40 m, respectively). Wind velocity components within the entire specified range are determined by radioprobes with an error not exceeding $0.5\,\mathrm{m\,s^{-1}}$ in plan and $0.75\,\mathrm{m\,s^{-1}}$ in height.

Operational delivery of technical facilities to active areas is an essential factor in carrying out the tasks of the present project. The time of delivery can be reduced by appropriate choice of the takeoff point of the LV's launch azimuth; the takeoff point is also critical to carrying out three-dimensional maneuvers of the vehicle when launching the SV into orbit, during its orbital flight, and during descent from the orbit.

The limitations imposed on launching routes because of the small number of regions allocated for dropping boosters of the first and second stages of LVs rule out an arbitrary choice of launch azimuth. At present Rokot LVs can be launched from the Baikonur Cosmodrome with a single azimuth $A_0 = 35°$, corresponding to an inclination of $i = 64°$; from the Plesetsk Cosmodrome these LVs can be launched with two azimuths $A_0 = 42.6°$ ($i = 72°$) and $A_0 = 16.6°$ ($i = 82.5°$).

The operational characteristics of the RSC and the accessibility of TC origination regions from the Baikonur and Plesetsk Cosmodromes were calculated from the following main initial data: orbital altitude $H_0 = 200$ km; mass of useful load of the LV (PCs, repeater units, and facilities for their separation) 1,000 kg; conditions for SNC (space nosecone) entry into the atmosphere: $H_{ent} = 100$ km, $V_{ent} = 7{,}304\,\mathrm{m\,s^{-1}}$, $\theta_{ent} = -4°$.

To illustrate the results of calculations, Figure 9.3 shows the zones of TC activity regions (1–6 in diagram) that were accessible to PCs 6 hours after the LV was launched from Plesetsk at azimuth $A_0 = 42.6°$ and the boundaries of the lateral maneuver of the SV to reach a distance $R_z = 2{,}300$ km (continuous lines), drawn with a constant interval perpendicular to the trajectories of SNC motion (the numerals in parens characterize the relative frequency of TC origination).

The minimum time of TC radio probing (40–140 min) was for regions 3 and 4 of the Pacific Ocean, where about 50% of all typhoons are observed; the maximum time of TC radio probing (480–570 min) was in the western region of the Indian Ocean, where the annual number of typhoons is about 14% of the ocean's total number. A

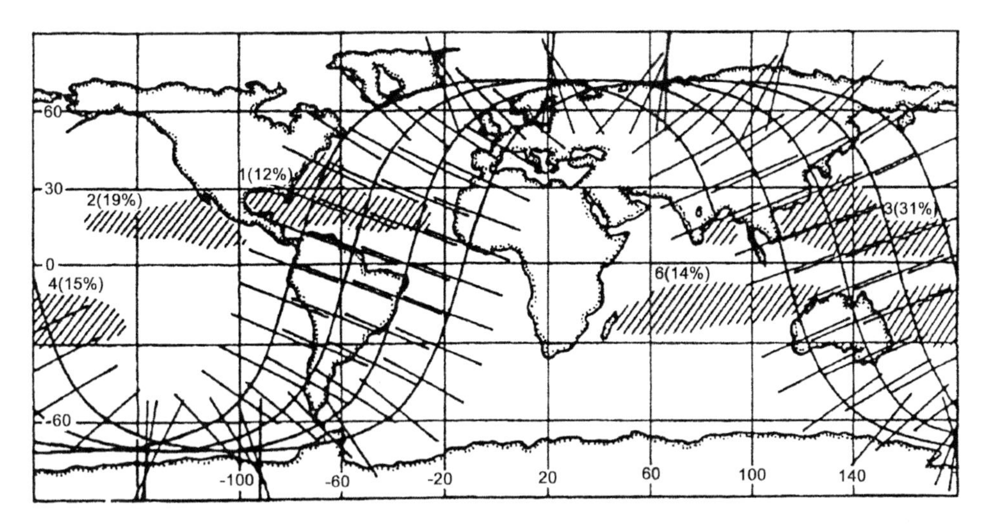

Figure 9.3. Zones of TC origination regions (1–6) for the deployment of parachute capsules (shown by lines perpendicular to the orbit traces, which indicate the lateral maneuver $R_z = 2{,}300$ km) about 6 h after launch from the Plesetsk ground station with azimuth $A_0 = 42.6°$ into orbit at altitude $H_0 = 200$ km. Numerals in parens characterize the relative frequency of TC origination in regions 1–6 (from Anfimov *et al.*, 1995).

further increase in the operativity of radio probing of TCs could be attained by launching ballistic missiles from territories located near to the investigated region.

The next step taken in this line of TC investigation was development of the Volna-TC Russian project or, as Lukiashchenko *et al.* (1999) call it, "Rocket-space system based on N-18 spacecraft missiles for sounding of tropical cyclones." The radically new element of the project consists in using a movable platform (submarines) for launching research rockets equipped with intercontinental N-18 spacecraft missiles (Table 9.12). Such an approach has made it possible to adopt different tactics in the pursuance of research.

Lukiashchenko *et al.* (1999) believe the financial prospects to be favorable as long as international cooperation can be developed.

In the 1980s a group of Russian researchers (Aleksashkin *et al.*, 1988) proposed another version of this approach by sounding TCs with probes and balloons delivered from satellites.

9.4.3 Mission Helix

In short, the "vortex dynamo" theory recommends using kinematic measurements of the turbulent field to calculate the possibility of large-scale vortex generation. Such kinematic features include the altitude and space distribution of turbulent energy as well as the helicity, structure, and fractal properties of IR images of cloud fields. It is also necessary to scan the state of the ocean surface and the spatiostatistical properties of deep convection, including the temperature and humidity profiles,

Table 9.12. Basic characteristics of the Volna-TC system.

Launch mass of the rocket (ton)	35.2
Range of payload delivery (km)	2,500–7,600
Operational areas of submarine	Without limitation
Duration of an operation of TC sounding from the moment of decision making about rocket launch until transmission of information product to the customer: length of duty of submarine in ocean (min)	115 ... 135
Number of descent capsules on a rocket	12
Number of rockets for one TC sounding	3
Number of rockets on board the submarine	Not more than 16
DC mass (kg)	10–15
Size of area of DC separation (km × km)	1,800 × 400
Range of operation altitudes of radiosondes (km)	20–0
Parameters of the atmosphere measured by radiosondes inside TCs: Temperature of air (°C) Relative humidity (%) Pressure (mb) Components of windspeed ($m\,s^{-1}$)	 (−90 ... + 40) ±0.2 (0 ... 100) ± 2.5 (1,060 ... 20) ± 0.5 (0 ... 150) ± 0.5
Periodicity of radioprobe interrogation (s)	0.5–3

both as structural features of wavefronts in the atmosphere and their reflection in the structure of the ocean surface. To monitor the upper atmosphere it is necessary to carry out simultaneous measurements of the spatial fields of airglow emissions of neutral and ionized gas components in the upper atmosphere.

A schematic presentation of the interaction between TCs and the various geophysical media investigated in this project is shown in Figure 9.4.

The main purpose of the Helix mission is the investigation of the pre-crisis thermodynamical and kinematical states of the ocean–atmosphere system in order to develop a satellite-based system capable of discovering large-scale atmospheric vortex processes. This purpose determines the following specific goals:

- investigation of spatiotemporal changes and the geometrical structure of convective cloud field borders in the tropical atmosphere;
- investigation of spatiotemporal changes in the altitude distribution of turbulent energy on meso and macroscales in convective cloud fields;
- investigation of processes in the ocean–atmosphere system in metastable zones

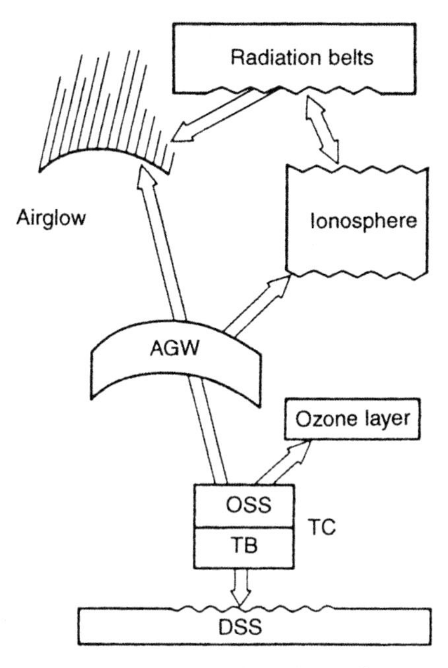

Figure 9.4. Schematic illustration of TC interaction with various geophysical media (within the Helix project): TC, tropical cyclone; TB, tropospheric body; OSS, outward-spiraling (stratospheric) surge; AGW, acousto-gravitational waves; DSS, disturbed sea surface (from Balebanov *et al.*, 1997).

using their display in the ocean surface structure, including impurity fields on the ocean surface;

- investigation of spatiotemporal variation in the humidity and temperature stratification fields of the tropical atmosphere, as well as precipitation fields over the ocean;
- investigation of spatiotemporal variation in airglow emissions of neutral and ionized gas components in the upper atmosphere based on analysis of panoramic monochromatic images in the visible and IR ranges.

Experiments carried out by satellite monitoring include the following measurements:

- observations of IR radiation in the tropical zone of the ocean and analysis of the structural and fractal properties of cloud fields (the TIR complex);
- measurement of altitude distributions of the turbulent energy and phase composition of convective cloud systems using the Doppler atmospheric 8 mm scatterometer (the DAS complex);
- thermal radiation mapping and monitoring of the ocean–atmosphere system, including determination of the thermodynamical and humidity features of the atmosphere and the ocean surface using the scanning multispectral heat

radiation complex of millimeter wavelength range (the Panorama complex) and side-looking radar (SLR);

- panoramic spectrophotometric observations of emissions of the upper atmosphere and ionosphere.

Thermal IR cameras (the TIR complex) are used for observations of the thermal IR radiation of cloud systems in oceanic tropical zones and have these characteristics: energetic sensitivity $< 0.5°C$; scanning bandwidth 1,600 km to 2,000 km; spatial resolution < 1 km.

The DAS complex is used to register the power, Doppler shift, and spectral width of the signal reflected from cloud and the background surface along the scanning route. By using the inverse scattering method the altitude profiles of the turbulent energy distribution and phase composition of cloud fields can be found. Analysis of the turbulent energy distribution allows localization of anomalies connected to crisis phenomena in the atmosphere. The characteristics of DAS are:

Wavelength	8 mm
Distance resolution	1 km
Frequency of addresses	15 kHz
The directional pattern width	0.2°

The DAS complex includes an emitter–receiver module, a power supply module, an aerial, and a data-processing control module.

The Panorama complex is used for thermal radiation mapping of the ocean–atmosphere system. The radiometer/spectrometer receives thermal emissions at wavelengths 13.5 (water vapor band), 8.6, 8.8, 6.2, 4.0, and 3.1 mm. Both spectral measurement of oxygen and polarization measurement are obtained using multi-frequency aerial emitters. The fluctuational sensitivity of the radiometric channels is 0.03 K to 0.05 K. Spatial resolution is ~ 10 km; total mass is no more than 100 kg; diameter of the aerial is 2 m to 3 m; power supply is 150 W to 200 W; and information flow is no more than 0.4 Mbit s^{-1}.

The radiometer/spectrometer is used for radiolocation mapping of the ocean–atmosphere system. The working wavelength is 3 cm; polarization is vertical; spatial resolution is 1 km to 2 km; the survey range is 500 km to 550 km; total mass is 160 kg; power supply is 1.6 kW; and size is 1 m^3. The radiometer/spectrometer monitors upper-atmosphere emission activity using a special diagnostic airglow imaging system (DAIS), which was developed at SRI. It is intended for observations of instantaneous spatial fields of emission intensities of specific neutral and ionized gas components excited by both accelerated electron fallout from radiation belts and acoustic gravity waves generated by TCs.

9.4.4 International and practical aspects of the projects

Owing to the importance of the projects in solving weather-related problems facing humankind, it is reasonable to finance their development from the funds of the international organization GEOWARN, established as a specialist agency of the U.N. to develop a global emergency observation and warning system.

These projects can become a component of several international research programs—such as the Tropical Ocean and Global Atmosphere (TOGA) program and the Global Atmosphere Warning (GAW) program—providing them with information on the mesoscale turbulent characteristics of active metastable zones in the tropical atmosphere. Other projects do not stipulate provision of such information. Taking into account the importance of these investigations, the potential ecological effect, and the use of procedures for early prediction of a critical situation in the atmosphere, based on information furnished by the projects, we believe that it is reasonable to establish an international regional association of the countries located in the tropical zone (the U.S.A., China, Japan, Malaysia, Philippines, Vietnam, India, Bangladesh, and others) for implementation of the projects.

Carrying out these investigations into critical states of the atmosphere with the help of rocket and space facilities and using their results will make it possible:

- to obtain unique science-intensive information on the dynamics of vortex processes developing in any region of the globe, primarily on the structure and characteristics of ACs, and to transmit this information to interested consumers;
- to establish a databank on the three-dimensional structure of TCs and develop accurate mathematical models of the physical processes occurring in TCs;
- to predict the origination, intensity, and paths of travel of TCs.

Optimization of the procedure for early prediction of origination and development of TCs will make it possible to drastically reduce the damage caused by TCs; within the framework of the projects, experimental optimization of the space and rocket complex for operative radio probing of TCs and warning about their approach is envisaged.

Technical and methodological developments carried out within the framework of these projects can be used to construct a global system for detection, control, and mitigation of the consequences of natural hazards (e.g., within the scope of such scientific and technological international programs as EOS and WEDOS).

Appendix A

Tables: Quantitative data on the spatiotemporal features of global and regional tropical cyclogenesis for 1983–2010

This appendix reviews quantitative results of the spatiotemporal features of global and regional tropical cyclogenesis for 1983–2010. The initial raw data derived from operational meteorological information on the tropics (*http://www.solar.ifa.hawaii.edu*) were subjected to the special objective analysis and procedure developed by Dr. I. V. Pokrovskaya and Prof. E. A. Sharkov (Space Research Institute, Russian Academy of Sciences) (Pokrovskaya and Sharkov, 2006). The procedure is described in Section 8.4. Appendix A was prepared by Dr. I. V. Pokrovskaya and involves unique information that is nowhere to be found in standard TC databases.

Table A.1 constitutes information on the annual distribution of the number of TCs in areas of incipient tropical cyclones (active basins of the World Ocean) over the 27-year period (1983–2010). The annual distribution of the number of TCs in active basins (Pacific, Atlantic, Indian Oceans) is tabulated in Tables A.2–A.4. Tables A.5–A.12 carry information on the monthly distribution of the number of TCs in more detail—averaged through the given month in a given year for a specific basin.

The detailed annual distribution of supertyphoons generated in the World Ocean basin and in regional active basins during 1990–2010 are listed in Tables A.13–A.16.

Details on the overall number of TC transformations in midlatitude circulation and how they developed in World Ocean basins (1990–2010) for each specific active area are tabulated in Tables A.17–A.22.

Tables A.23–A.33 carry unique information on the annual distribution of TC dissipation conditions in more detail—averaged through a given year for specific active basins.

Important information on the number of initial tropical disturbances (TLs) and the number of TCs formed from TL in World Ocean basins (1999–2010) are listed in Tables A.34–A.35.

E. A. Sharkov, *Global Tropical Cyclogenesis* (Second Edition).

The length of time that tropical systems last in a given geographical region is fundamentally important for research into remote sensing and global climate change. The global tropical cyclogenesis data tabulated in the following tables have been used to great advantage in many applications: for research into global climate change and into meso and macroscale interactions in the atmosphere–ocean system to facilitate atmospheric emergency warning.

Table A.1. The annual distribution of the number of TCs in the World Ocean (WO), in the northern hemisphere (NH), and in the southern hemisphere (SH) (1983–2010).

Year	*NH*		*SH*		*WO*
	$\sum TC$	% *WO*	$\sum TC$	% *WO*	$\sum TC$
1983	51	68	24	32	75
1984	59	67	29	31	88
1985	65	71	26	29	91
1986	53	69	24	31	77
1987	56	71	23	29	79
1988	58	77	17	23	75
1989	64	68	30	32	94
In all	296		120		416
MY	59	71	24	29	83
1990	68	73	25	27	93
1991	54	69	24	31	78
1992	74	74	25	26	99
1993	52	69	23	31	75
1994	64	74	22	26	86
In all	312		119		431
MY	62	72	24	28	86
1995	58	74	20	26	78
1996	59	64	33	36	92
1997	60	62	36	38	96
1998	52	59	36	41	88
1999	49	65	26	35	75
In all	278		151		429
MY	56	65	30	35	86

Year	*NH*		*SH*		*WO*
	$\sum TC$	% *WO*	$\sum TC$	% *WO*	$\sum TC$
2000	62	68	29	32	91
2001	62	70	26	30	88
2002	59	71	24	29	83
2003	59	66	30	34	89
2004	64	75	21	25	85
In all	306		130		436
MY	61	70	26	30	87
2005	72	74	25	26	97
2006	57	70	26	30	83
2007	57	67	28	33	85
2008	65	76	21	24	86
2009	58	67	28	33	86
In all	309		128		437
MY	62	71	26	29	87
$\sum TC$	1,611		701		2,312
MY	60	70	26	30	86
2010	46	68	22	32	68

Notes: $\sum TC$ is the total annual number of TCs in a given year; % *WO* is the annual number of TCs as a percentage of the number of TCs in World Ocean basins. All data are divided into five annual intervals. MY is the mean value of the annual number of TCs in the given 5-year period and in the given 27-year period (second line from the bottom of the table).

Table A.2. The annual distribution of the number of TCs in World Ocean basins (1983–2010).

Year	*Oceans*						
	Pacific		*Atlantic*		*Indian*		*World*
	$\sum TC$	% WO	$\sum TC$	% WO	$\sum TC$	% WO	$\sum TC$
1983	55	73	4	5	16	22	75
1984	55	63	12	14	21	23	88
1985	57	63	11	12	23	25	91
1986	54	70	6	8	17	22	77
1987	51	64	7	9	21	27	79
1988	49	65	12	16	14	19	75
1989	65	69	11	12	18	19	94
In all	276		47		93		416
MY	55	66	9	11	19	23	83
1990	59	63	14	15	20	22	93
1991	49	63	7	10	21	27	78
1992	70	71	7	7	22	22	99
1993	52	69	8	11	15	20	75
1994	60	70	7	8	19	22	86
In all	290		43		95		431
MY	58	68	9	10	19	22	86
1995	38	49	19	24	21	27	78
1996	49	53	13	14	30	33	92
1997	70	73	7	7	19	20	96
1998	46	52	14	16	28	32	88
1999	42	56	12	16	21	28	75
In all	245		65		119		429
MY	49	57	13	15	24	28	86
2000	54	59	14	15	23	26	91

Year	*Oceans*						
	Pacific		*Atlantic*		*Indian*		*World*
	$\sum TC$	% *WO*	$\sum TC$	% *WO*	$\sum TC$	% *WO*	$\sum TC$
2001	53	60	15	17	20	23	88
2002	50	60	12	14	21	26	83
2003	50	56	16	20	23	24	89
2004	49	58	15	18	21	24	85
In all	256		72		108		436
MY	51	59	14	16	22	25	87
2005	49	50	27	28	21	22	97
2006	53	64	9	11	21	25	83
2007	46	54	15	18	24	28	85
2008	46	53	16	19	24	28	86
2009	54	63	9	10	23	27	86
In all	248		76		113		437
MY	50	57	15	17	23	26	86
TC in all	1,425		320		565		2,312
MY	53	62	12	14	21	24	86
2010	33	48	19	28	16	24	68

Notes: See Table A.1.

Table A.3. The annual distribution of the number of TCs in the Pacific and Pacific basins (1983–2010).

Year	*Pacific basins*							
	NWP		*NEP*		*SWP*		*Pacific*	
	$\sum TC$	*% PO*	$\sum TC$	*% PO*	$\sum TC$	*% PO*	$\sum TC$	*% WO*
1983	23	42	21	38	11	20	55	73
1984	27	49	16	29	12	22	55	63
1985	27	47	21	37	9	16	57	63
1986	28	42	17	31	9	17	54	70
1987	23	45	20	39	8	16	51	65
1988	31	63	12	24	6	13	49	65
1989	32	49	18	28	15	23	65	69
In all	141		88		47		276	
MY	28	51	18	33	9	16	55	66
1990	29	49	21	36	9	15	59	63
1991	29	59	14	29	6	12	49	63
1992	31	44	27	39	12	17	70	71
1993	28	54	14	27	10	19	52	69
1994	35	58	19	32	6	10	60	70
In all	152		95		43		290	
MY	30	52	19	32	9	16	58	68
1995	25	66	10	26	3	8	38	49
1996	30	61	8	16	11	23	49	53
1997	30	43	19	27	21	30	70	73
1998	17	37	13	28	16	35	46	52
1999	23	55	9	21	10	24	42	56
In all	125		59		61		245	
MY	25	52	12	24	12	24	49	57

Year	*Pacific basins*							
	NWP		*NEP*		*SWP*		*Pacific*	
	$\sum TC$	*% PO*	$\sum TC$	*% PO*	$\sum TC$	*% PO*	$\sum TC$	*% WO*
2000	26	48	18	33	10	19	54	59
2001	28	53	15	28	10	19	53	60
2002	27	54	15	30	8	16	50	60
2003	24	48	16	32	10	20	50	56
2004	32	65	12	25	5	10	49	57
In all	137		76		43		256	
MY	27	53	15	29	9	18	51	59
2005	24	49	15	31	10	20	49	50
2006	24	45	18	34	11	21	53	64
2007	25	54	11	24	10	22	46	54
2008	25	54	17	37	4	9	46	54
2009	24	44	20	37	10	19	54	63
In all	122		81		45		248	
MY	24	49	16	33	9	18	50	57
In all	727		436		262		1,425	
MY	27	51	16	30	10	19	53	62
2010	14	43	8	24	11	33	33	48

Notes: See Table A.1. NWP, North-West Pacific; NEP, North-East Pacific; and SWP, South-West Pacific.

Table A.4. The annual distribution of the number of TCs in the Indian Ocean and the Indian Ocean basins (1983–2010).

Year	*Basins*				*Indian Ocean*	
	NIO		*SIO*			
	$\sum TC$	*% IO*	$\sum TC$	*% IO*	$\sum TC$	*% IO*
1983	3	19	13	81	16	21
1984	4	19	17	81	21	24
1985	6	26	17	74	23	25
1986	2	12	15	88	17	22
1987	6	29	15	71	21	27
1988	3	21	11	79	14	19
1989	3	17	15	83	18	19
In all	20		73		93	
MY	4	21	15	79	19	23
1990	4	20	16	80	20	22
1991	3	14	18	86	21	27
1992	9	41	13	59	22	22
1993	2	11	13	87	15	20
1994	3	16	16	84	19	24
In all	21		76		97	
MY	4	21	15	79	19	22
1995	4	19	17	81	21	27
1996	8	27	22	73	30	33
1997	4	21	15	79	19	20
1998	8	29	20	71	28	32
1999	5	24	16	76	21	28
In all	29		90		119	
MY	6	25	18	75	24	28

Year	*Basins*				*Indian Ocean*	
	NIO		*SIO*			
	$\sum TC$	*% IO*	$\sum TC$	*% IO*	$\sum TC$	*% IO*
2000	4	17	19	83	23	26
2001	4	20	16	80	20	23
2002	5	24	16	76	21	26
2003	3	13	20	87	23	24
2004	5	24	16	76	21	24
In all	21		87		108	
MY	4	18	18	82	22	25
2005	6	29	15	71	21	22
2006	6	29	15	71	21	25
2007	6	25	18	75	24	28
2008	7	29	17	71	24	28
2009	5	22	18	78	23	26
In all	30		83		113	
MY	6	26	17	74	23	26
In all	128		437		565	
MY	5	21	16	79	21	24
2010	5	31	11	69	16	24

Notes: See Table A.1. NIO, North Indian Ocean; and SIO, South Indian Ocean.

Table A.5. The monthly distribution of the number of TCs in the Pacific basins (1983–2010).

Year	*Month*												*In all TCs*
	1	*2*	*3*	*4*	*5*	*6*	*7*	*8*	*9*	*10*	*11*	*12*	
1983	1	4	3	2	1	2	9	8	7	9	6	3	55
1984	1	4	5	1	2	6	7	9	8	7	3	2	55
1985	6	1	4	—	1	7	8	11	9	7	1	2	57
1986	1	3	—	3	4	4	7	9	8	5	4	6	54
1987	2	2	2	2	1	3	10	10	10	4	2	2	51
1988	3	3	—	1	1	6	6	11	11	4	2	1	49
1989	4	5	1	3	6	4	11	11	10	5	3	1	65
In all	16	14	7	9	13	24	42	52	48	24	12	13	276
MY	3.2	2.8	1.4	1.8	2.6	4.8	8.4	10.4	9.6	4.8	2.4	2.4	55
1990	4	1	3	1	2	7	8	9	10	6	5	3	59
1991	—	1	3	1	4	4	6	7	9	5	8	2	49
1992	4	3	4	1	—	6	11	11	11	12	3	3	70
1993	1	6	3	—	—	3	8	11	8	6	2	4	52
1994	1	1	3	—	1	5	12	11	14	8	1	3	60
In all	10	12	16	3	7	25	43	49	52	37	19	15	290
MY	2.0	2.4	3.2	0.6	1.4	5.0	8.6	9.8	10.4	7.4	3.8	3.0	58
1995	—	—	2	2	—	3	5	9	8	7	1	1	38
1996	4	1	3	1	2	2	9	7	10		33	4	49
1997	4	4	3	4	5	6	8	10	11	7	5	3	70
1998	7	1	3	2	—	2	5	6	6	6	3	3	44
1999	5	5	1	3	2	1	4	10	6	3	2	—	47
In all	20	11	12	12	9	14	31	42	41	26	14	11	245
MY	4.0	2.2	2.4	2.4	1.8	2.8	6.2	8.4	8.2	5.2	2.8	2.2	49

Year	*Month*												*In all TCs*
	1	*2*	*3*	*4*	*5*	*6*	*7*	*8*	*9*	*10*	*11*	*12*	
2000	3	2	4	1	3	3	8	13	8	7	1	—	53
2001	—	6	—	1	2	4	8	7	10	6	4	5	49
2002	2	3	2	—	4	4	9	11	6	5	3	1	53
2003	4	1	2	2	5	5	8	6	8	5	2	1	48
2004	—	3	3	1	4	5	5	12	6	4	3	3	49
In all	9	15	11	5	18	21	38	49	38	27	12	10	256
MY	1.8	3.0	2.2	1.0	3.6	4.2	7.6	9.8	7.6	5.4	2.4	2.0	51
2005	3	5	3	2	2	2	7	9	11	2	2	1	49
2006	3	2	3	1	2	2	9	10	6	8	5	2	53
2007	4	1	2	1	4	—	7	7	7	7	5	1	46
2008	3	—	—	2	5	4	5	9	10	3	4	1	46
2009	3	1	4	2	1	3	7	12	12	6	2	1	54
In all	16	9	12	8	14	11	35	47	46	26	18	6	248
MY	3.2	1.8	2.4	1.6	2.8	2.2	7.0	9.4	9.2	5.2	3.6	1.2	50
In all	73	69	67	40	65	104	208	257	241	157	85	59	1,425
MY	2.7	2.6	2.4	1.5	2.4	3.8	7.7	9.5	8.9	5.8	3.1	2.1	53
2010	4	3	4	—	1	3	2	8	4	2	—	2	33

Notes: See Table A.1.

Table A.6. The monthly distribution of the number of TCs in the North-West Pacific (1983–2010).

Year	*Month*												*In all TCs*
	1	*2*	*3*	*4*	*5*	*6*	*7*	*8*	*9*	*10*	*11*	*12*	
1983	—	—	—	—	—	1	3	5	2	6	5	1	23
1984	—	—	—	—	—	2	5	5	4	7	3	1	27
1985	2	—	—	–	1	3	2	7	5	4	1	2	27
1986	—	1	—	1	2	2	4	4	3	4	4	3	28
1987	1	—	–	1	—	2	4	4	5	2	2	1	23
1988	1	—	—	—	1	3	3	7	9	4	2	1	31
1989	1	—	—	1	1	2	7	5	6	4	3	1	32
In all	5	1	—	3	5	12	20	27	28	18	12	8	141
MY	1.0	0.2	—	0.6	1.0	2.4	4.0	5.4	5.6	3.6	2.4	1.6	28
1990	1	—	—	1	1	3	4	5	5	3	4	2	29
1991	—	—	2	1	1	1	4	5	6	3	6	—	29
1992	1	—	—	—	—	3	4	7	5	7	3	—	31
1993	—	—	1	—	—	1	5	6	6	4	2	3	28
1994	—	—	1	—	1	2	8	7	8	6	—	2	35
In all	2	—	4	2	3	10	25	31	30	23	15	7	152
MY	0.4	—	0.8	0.4	0.6	2.0	5.0	6.2	6.0	4.6	3.0	1.4	30
1995	—	—	—	1	—	2	2	6	5	7	1	1	25
1996	—	—	—	1	2	—	7	7	6	3	2	2	30
1997	1	—	—	2	3	3	4	6	5	4	1	1	30
1998	—	—	—	—	—	—	2	2	5	3	3	2	17
1999	1	1	—	3	1	—	2	6	5	2	2	—	23
In all	2	1	—	7	6	5	17	27	26	19	9	6	125
MY	0.4	0.2	—	1.4	1.2	1.0	3.4	5.4	5.2	3.8	1.8	1.2	25

Year	*Month*												*In all TCs*
	1	*2*	*3*	*4*	*5*	*6*	*7*	*8*	*9*	*10*	*11*	*12*	
2000	—	—	—	—	2	1	5	7	5	4	1	1	26
2001	—	—	—	—	1	3	5	6	4	3	3	3	28
2002	1	1	—	—	2	3	6	6	4	2	2	—	27
2003	1	—	—	1	4	1	4	3	3	4	2	1	24
2004	—	1	2	—	3	5	2	8	4	2	3	2	32
In all	2	2	2	1	12	13	22	30	20	15	11	7	137
MY	0.6	0.4	—	0.8	2.0	1.6	4.4	5.6	4.2	3.0	2.0	1.0	27
2005	1	—	1	1	1	—	5	5	5	2	2	1	24
2006	—	—	1	—	1	2	3	6	4	4	1	2	24
2007	—	—	1	—	1	—	4	4	5	6	4	—	25
2008	—	—	—	1	4	1	2	5	6	2	3	1	25
2009	—	—	—	1	1	2	3	4	8	3	2	—	24
In all	1	—	3	3	8	5	17	24	28	17	12	4	122
MY	0.6	0.4	1.0	0.4	2.0	3.0	3.2	5.6	4.8	3.2	2.6	1.2	24
In all	12	4	9	17	34	48	109	149	138	105	67	34	727
MY	0.4	0.1	0.3	0.6	1.3	1.8	4.0	5.5	5.1	3.9	2.5	1.3	27
2010	—	—	1	—	—	—	2	6	3	2	—	—	14

Notes: See Table A.1.

Table A.7. The monthly distribution of the number of TCs in the North-East Pacific (1983–2010).

Year	*Month*												*In all TCs*
	1	*2*	*3*	*4*	*5*	*6*	*7*	*8*	*9*	*10*	*11*	*12*	
1983	—	—	—	—	1	1	6	3	5	3	1	1	21
1984	—	—	—	—	2	4	2	4	4	—	—	—	16
1985	—	—	—	—	—	4	6	4	4	3	—	—	21
1986	—	—	—	—	1	2	3	5	5	1	—	—	17
1987	—	—	—	—	—	1	6	6	5	2	—	—	20
1988	—	—	—	—	—	3	3	4	2	—	—	—	12
1989	—	—	—	—	1	2	4	6	4	1	—	—	18
In all	—	—	—	—	2	12	22	25	20	7	—	—	86
MY	—	—	—	—	0.4	2.4	4.4	5.0	4.0	1.4	—	—	18
1990	—	—	—	—	1	4	4	4	5	3	—	—	21
1991	—	—	—	—	1	3	2	2	3	2	1	—	14
1992	1	—	1	—	—	3	7	4	6	5	—	—	27
1993	—	—	—	—	—	2	3	5	2	2	—	—	14
1994	—	—	—	—	—	3	4	4	6	2	—	—	19
In all	1	—	1	—	2	15	20	19	24	14	1	—	95
MY	0.2	—	0.2	—	0.4	3.0	4.0	3.8	4.8	2.8	0.2	—	19
1995	—	—	—	—	—	1	3	3	3	—	—	—	10
1996	—	—	—	—	—	2	2	—	4	—	—	—	8
1997	—	—	—	—	1	2	4	4	6	—	2	—	19
1998	—	—	—	—	—	2	3	4	1	3	—	—	13
1999	—	—	—	—	—	1	2	4	1	1	—	—	9
In all	—	—	—	—	1	8	14	15	15	4	2	—	59
MY	—	—	—	—	0.2	1.6	2.8	3.0	3.0	0.8	0.4	—	12

Year	*Month*												*In all TCs*
	1	*2*	*3*	*4*	*5*	*6*	*7*	*8*	*9*	*10*	*11*	*12*	
2000	—	—	—	—	1	2	3	6	3	3	—	—	18
2001	—	—	—	—	1	1	3	1	6	3	—	—	15
2002	—	—	—	—	1	1	3	5	2	3	—	—	15
2003	—	—	—	—	1	2	4	3	5	1	—	—	16
2004	—	—	—	—	1	—	3	4	2	2	—	—	12
In all	—	—	—	—	5	6	16	19	18	12	—	—	76
MY	—	—	—	—	1.0	1.2	3.2	3.8	3.6	2.4	—	—	15
2005	—	—	—	—	1	2	2	4	6	—	—	—	15
2006	—	—	—	—	1	—	6	4	2	3	2	—	18
2007	—	—	—	—	2	—	3	3	2	1	—	—	11
2008	—	—	—	—	1	3	3	4	4	1	1	—	17
2009	—	—	—	—	—	1	4	8	4	3	—	—	20
In all	—	—	—	—	5	6	18	23	18	8	3	—	81
MY	—	—	—	—	1.0	1.2	3.6	4.6	3.6	1.6	0.6	—	16
In all	1	—	1	—	18	52	98	108	104	48	7	1	436
MY	—	—	—	—	0.7	1.9	3.6	4.0	3.9	1.8	0.3	—	16
2010	—	—	—	—	1	3	—	2	1	—	—	1	8

Notes: See Table A.1.

Table A.8. The monthly distribution of the number of TCs in the South-West Pacific (1983–2010).

Year	*Month*												*In all TCs*
	1	*2*	*3*	*4*	*5*	*6*	*7*	*8*	*9*	*10*	*11*	*12*	
1983	1	4	3	2	—	—	—	—	—	—	—	1	11
1984	1	4	5	1	—	—	—	—	—	—	—	1	12
1985	4	1	4	—	—	—	—	—	—	—	—	—	9
1986	1	2	—	2	1	—	—	—	—	—	—	3	9
1987	1	2	2	1	1	—	—	—	—	—	—	1	8
1988	2	3	—	1	—	—	—	—	—	—	—	—	6
1989	3	5	1	2	4	—	—	—	—	—	—	—	15
In all	11	13	7	6	6	—	—	—	—	—	—	4	47
MY	1.2	2.6	1.4	1.2	1.2	—	—	—	—	—	—	0.8	9
1990	3	1	3	—	—	—	—	—	—	—	1	1	9
1991	—	1	1	—	1	—	—	—	—	—	1	2	6
1992	2	3	3	1	—	—	—	—	—	—	—	3	12
1993	1	6	2	—	—	—	—	—	—	—	—	1	10
1994	1	1	2	—	—	—	—	—	—	—	1	1	6
In all	7	12	11	1	1	—	—	—	—	—	3	8	43
MY	1.6	2.4	2.2	0.2	0.2	—	—	—	—	—	0.6	1.6	9
1995	—	—	2	1	—	—	—	—	—	—	—	—	3
1996	4	1	3	—	—	—	—	—	—	—	1	2	11
1997	3	4	3	2	1	1	—	—	—	3	2	2	21
1998	7	1	3	2	—	—	—	—	—	—	—	3	16
1999	4	4	1	—	1	—	—	—	—	—	—	—	10
In all	18	10	12	5	2	1	—	—	—	3	3	7	61
MY	3.6	2.0	2.4	1.0	0.4	0.2	—	—	—	0.6	0.6	1.4	12

Year	*Month*												*In all TCs*
	1	*2*	*3*	*4*	*5*	*6*	*7*	*8*	*9*	*10*	*11*	*12*	
2000	3	2	4	1	—	—	—	—	—	—	—	—	10
2001	—	6	—	1	—	—	—	—	—	—	1	2	10
2002	1	2	2	—	1	—	—	—	—	—	1	1	8
2003	3	1	2	1	—	2	—	—	—	—	—	1	10
2004	—	2	1	1	—	—	—	—	—	—	—	1	5
In all	7	13	9	4	1	2	—	—	—	—	2	5	43
MY	1.4	2.6	1.8	0.8	0.2	0.4	—	—	—	—	0.4	1.0	9
2005	2	5	2	1	—	—	—	—	—	—	—	—	10
2006	3	2	2	1	—	—	—	—	—	1	2	—	11
2007	4	1	1	1	1	—	—	—	—	—	1	1	10
2008	3	—	—	1	—	—	—	—	—	—	—	—	4
2009	3	1	4	1	—	—	—	—	—	—	—	1	10
In all	15	9	9	5	1	—	—	—	—	1	3	2	45
MY	3.0	1.8	1.8	1.0	0.2	—	—	—	—	0.2	0.6	0.4	9
In all	62	65	56	24	11	3	—	—	—	4	11	28	262
MY	2.3	2.4	2.1	0.9	0.4	0.1	—	—	—	0.2	0.4	1.0	9.7
2010	4	3	3	—	—	—	—	—	—	—	—	1	11

Notes: See Table A.1.

Table A.9. The monthly distribution of the number of TCs in the North Atlantic (1983–2010).

Year	*Month*												*In all TCs*
	1	*2*	*3*	*4*	*5*	*6*	*7*	*8*	*9*	*10*	*11*	*12*	
1983	—	—	—	—	—	—	—	2	2	—	—	—	4
1984	—	—	—	—	—	—	—	3	6	1	1	1	12
1985	—	—	—	—	—	—	2	3	3	2	1	—	11
1986	—	—	—	—	—	2	—	1	2	—	1	—	6
1987	—	—	—	—	—	—	—	3	3	1	—	—	7
1988	—	—	—	—	—	—	—	4	6	1	1	—	12
1989	—	—	—	—	—	1	3	3	2	1	1	—	11
In all	—	—	—	—	—	3	5	14	16	5	4	—	47
MY	—	—	—	—	—	0.6	1.0	2.8	3.2	1.0	0.8	—	9
1990	—	—	—	—	—	—	2	6	2	4	—	—	14
1991	—	—	—	—	1	—	1	1	3	2	—	—	8
1992	—	—	—	1	—	—	—	1	4	1	—	—	7
1993	—	—	—	—	—	1	—	4	3	—	—	—	8
1994	—	—	—	—	—	1	—	2	2	—	2	—	7
In all	—	—	—	1	1	2	3	14	14	7	2	—	44
MY	—	—	—	0.2	0.2	0.4	0.6	2.8	2.8	1.4	0.4	—	9
1995	—	—	—	—	—	1	4	7	3	4	—	—	19
1996	—	—	—	—	—	1	2	4	2	3	1	—	13
1997	—	—	—	—	—	1	3	0	0	1	2	—	7
1998	—	—	—	—	—	—	1	4	7	1	1	—	14
1999	—	—	—	—	—	1	—	4	3	3	1	—	12
In all	—	—	—	—	—	4	10	19	15	12	5	—	65
MY	—	—	—	—	—	0.8	2.0	3.8	3.0	2.4	1.0	—	13

Year	*Month*												*In all TCs*
	1	*2*	*3*	*4*	*5*	*6*	*7*	*8*	*9*	*10*	*11*	*12*	
2000	—	—	—	—	—	—	—	5	6	3	—	—	14
2001	—	—	—	—	—	1	—	4	3	5	2	—	15
2002	—	—	—	—	—	—	1	3	8	—	—	—	12
2003	—	—	—	1	—	1	—	3	5	2	—	2	14
2004	—	—	—	—	—	—	2	6	4	2	1	—	15
In all	—	—	—	1	—	2	5	21	26	12	3	2	72
MY	—	—	—	0.2	—	0.4	1.0	4.2	5.2	2.4	0.6	0.4	14
2005	—	—	—	—	—	3	4	6	5	5	3	1	27
2006	—	—	—	1	—	1	2	2	4	—	—	—	10
2007	—	—	—	—	2	—	1	3	7	1	—	1	15
2008	—	—	—	—	1	—	3	5	3	3	1	—	16
2009	—	—	—	—	—	—	—	5	1	3	—	—	9
In all	—	—	—	1	3	3	10	21	20	12	4	2	76
MY	—	—	—	0.2	0.6	0.6	2.0	4.2	4.0	2.4	0.8	0.4	15
In all	—	—	—	3	4	14	33	94	99	49	19	5	320
MY	—	—	—	0.1	0.2	0.6	1.2	3.5	3.7	1.8	0.7	0.2	12
2010	—	—	—	—	—	1	11	5	7	5	—	—	19

Notes: See Table A.1.

Table A.10. The monthly distribution of the number of TCs in the Indian Ocean (1983–2010).

Year	*Month*												*In all TCs*
	1	*2*	*3*	*4*	*5*	*6*	*7*	*8*	*9*	*10*	*11*	*12*	
1983	2	2	—	3	—	—	—	—	—	3	3	3	16
1984	5	7	1	1	1	—	—	—	—	1	3	2	21
1985	3	7	2	3	1	1	—	—	1	2	1	2	23
1986	5	4	4	1	1	—	—	—	—	—	2	—	17
1987	4	4	1	3	—	2	—	—	—	2	3	2	21
1988	3	2	2	—	1	1	—	—	—	—	3	2	14
1989	5	4	3	2	1	1	—	—	—	—	1	1	18
In all	20	21	12	9	4	5	—	—	1	4	10	7	93
MY	4.0	4.2	2.4	1.8	0.8	1.0	—	—	0.2	0.8	2.0	1.4	19
1990	5	2	6	3	2	—	—	—	—	1	—	1	20
1991	2	6	3	3	1	1	—	—	—	1	2	3	21
1992	—	7	—	2	1	1	—	—	2	1	4	3	22
1993	4	—	1	2	1	—	—	—	—	—	3	2	13
1994	6	1	6	2	—	1	—	—	—	—	2	1	19
In all	17	16	16	12	5	3	—	—	2	3	11	10	95
MY	3.4	3.0	2.2	2.4	1.0	0.6	—	—	0.4	0.6	2.2	2.0	19
1995	4	3	4	2	—	—	—	—	1	1	3	3	21
1996	3	5	2	3	1	3	—	1	—	4	3	4	29
1997	7	4	—	—	2	—	1	—	1	—	2	2	19
1998	3	4	3	4	3	—	—	—	2	2	3	4	28
1999	5	3	5	1	1	1	—	—	—	2	—	3	21
In all	22	20	14	10	7	4	1	1	4	8	11	16	119
MY	4.4	3.8	2.8	2.0	1.4	0.8	0.2	0.2	0.8	1.6	2.2	3.2	24

Year	Month												In all TCs
	1	*2*	*3*	*4*	*5*	*6*	*7*	*8*	*9*	*10*	*11*	*12*	
2000	5	4	2	4	0	—	1	—	—	2	3	2	23
2001	3	1	2	3	1	1	—	—	1	3	4	1	20
2002	3	2	2	3	3	—	—	—	1	—	4	3	21
2003	2	6	4	2	—	—	—	—	1	—	2	5	22
2004	3	3	3	—	3	—	—	1	1	1	3	3	21
In all	16	16	13	12	7	1	1	1	4	6	16	14	108
MY	3.2	3.2	2.6	2.4	1.4	0.2	0.2	0.2	0.8	1.2	3.2	2.8	21
2005	8	1	3	1	—	—	—	—	—	2	3	3	21
2006	4	3	3	3	—	1	—	—	2	1	1	3	21
2007	1	5	4	—	2	2	1	—	—	1	3	5	24
2008	2	4	4	3	—	—	—	—	1	3	4	3	24
2009	5	3	4	2	1	—	—	—	3	—	1	4	23
In all	20	16	18	9	3	3	1	—	6	7	12	18	113
MY	4.0	3.2	3.6	1.8	0.6	0.6	0.2	—	1.2	1.4	2.4	3.6	22
In all	102	98	74	56	27	16	3	2	17	32	66	70	565
MY	3.8	3.6	2.7	2.1	1.0	0.6	0.1	—	0.6	1.2	2.4	2.6	21
2010	2	2	3	2	4	—	—	—	—	3	—	—	16

Notes: See Table A.1.

Table A.11. The monthly distribution of the number of TCs in the North Indian Ocean (1983–2010).

Year	*Month*												*In all TCs*
	1	*2*	*3*	*4*	*5*	*6*	*7*	*8*	*9*	*10*	*11*	*12*	
1983	—	—	—	—	—	—	—	—	—	2	1	—	3
1984	—	—	—	—	1	—	—	—	—	1	2	—	4
1985	—	—	—	—	1	1	—	—	—	2	1	1	6
1986	—	—	—	—	—	—	—	—	—	—	2	—	2
1987	1	—	—	—	—	2	—	—	—	2	1	—	6
1988	—	—	—	—	—	1	—	—	—	—	1	1	3
1989	—	—	—	—	1	1	—	—	—	—	1	—	3
In all	1	—	—	—	2	5	—	—	—	4	6	2	20
MY	0.2	—	—	—	0.4	1.0	—	—	—	0.8	1.2	0.4	4
1990	—	—	—	1	1	—	—	—	—	1	—	1	4
1991	—	—	—	1	1	—	—	—	—	—	1	—	3
1992	—	—	—	—	1	1	—	—	1	1	4	1	9
1993	—	—	—	—	—	—	—	—	—	—	1	1	2
1994	—	—	—	1	—	1	—	—	—	—	1	—	3
In all	—	—	—	3	3	2	—	—	1	2	7	3	21
MY	—	—	—	0.6	0.6	0.4	—	—	0.2	0.4	1.4	0.6	4
1995	—	—	—	—	—	—	—	—	1	1	2	—	4
1996	—	—	—	—	1	3	—	—	—	2	2	—	8
1997	—	—	—	—	1	—	—	—	1	—	2	—	4
1998	—	—	—	—	3	—	—	—	1	1	2	1	8
1999	1	—	—	—	1	1	—	—	—	2	—	—	5
In all	1	—	—	—	6	4	—	—	3	6	8	1	29
MY	0.2	—	—	—	1.2	0.8	—	—	0.6	1.2	1.6	0.2	6

Year	*Month*												*In all TCs*
	1	*2*	*3*	*4*	*5*	*6*	*7*	*8*	*9*	*10*	*11*	*12*	
2000	—	—	—	—	—	—	—	—	—	2	1	1	4
2001	—	—	—	—	1	—	—	—	1	1	1	—	4
2002	—	—	—	—	2	—	—	—	—	—	2	1	5
2003	—	—	—	—	1	—	—	—	—	—	1	1	3
2004	—	—	—	—	2	—	—	—	1	1	1	—	5
In all	—	—	—	—	6	—	—	—	2	4	6	3	21
MY	—	—	—	—	1.2	—	—	—	0.4	0.8	1.2	0.6	4
2005	1	—	—	—	—	—	—	—	—	2	1	2	6
2006	1	—	—	1	—	1	—	—	2	1	—	—	6
2007	—	—	—	—	2	2	—	—	—	1	1	—	6
2008	—	—	—	1	—	—	—	—	1	2	2	1	7
2009	—	—	—	1	1	—	—	—	1	—	1	1	5
In all	2	—	—	3	3	3	—	—	4	6	5	4	30
MY	0.4	—	—	0.6	0.6	0.6	—	—	0.8	1.2	1.0	0.8	6
In all	4	—	—	6	21	14	—	—	10	25	35	13	128
MY	0.1	—	—	0.2	0.8	0.5	—	—	0.4	0.9	1.3	0.9	5
2010	—	—	—	—	3	—	—	—	—	2	—	—	5

Notes: See Table A.1.

Table A.12. The monthly distribution of the number of TCs in the Southern Indian Ocean (1983–2010).

Year	*Month*												*In all TCs*
	1	*2*	*3*	*4*	*5*	*6*	*7*	*8*	*9*	*10*	*11*	*12*	
1983	2	2	—	3	—	—	—	—	—	1	2	3	13
1984	4	7	1	1	—	—	—	—	—	—	1	2	17
1985	3	7	2	3	—	—	—	—	1	—	—	1	17
1986	5	4	4	1	1	—	—	—	—	—	—	—	15
1987	3	4	1	3	—	—	—	—	—	—	2	2	15
1988	3	2	2	—	1	—	—	—	—	—	2	1	11
1989	5	4	3	2	—	—	—	—	—	—	—	1	15
In all	19	21	12	9	2	—	—	—	1	—	4	5	73
MY	3.8	4.2	2.4	1.8	0.4	—	—	—	0.2	—	0.8	1.0	15
1990	5	2	6	2	1	—	—	—	—	—	—	—	16
1991	2	5	3	2	—	1	—	—	1	1	1	3	19
1992	—	7	—	2	—	—	—	—	—	—	—	2	13
1993	4	—	1	2	1	—	—	—	—	—	2	1	11
1994	6	1	6	1	—	—	—	—	—	—	1	1	16
In all	17	15	16	9	2	1	—	—	1	1	4	7	74
MY	3.4	3.0	3.2	1.8	0.4	0.2	—	—	0.2	0.2	0.8	1.4	15
1995	4	3	4	2	—	—	—	—	—	—	1	3	17
1996	3	5	2	3	—	—	—	1	2	2	1	4	22
1997	7	4	—	—	1	—	1	—	—	—	—	2	15
1998	3	4	3	4	—	—	—	—	1	1	1	3	20
1999	4	3	5	1	—	—	—	—	—	—	—	3	16
In all	21	19	14	10	1	—	1	1	3	3	3	15	90
MY	4.2	3.8	2.8	2.0	0.2	—	0.2	0.2	0.6	0.6	0.6	3.0	18

Year	*Month*												*In all TCs*
	1	*2*	*3*	*4*	*5*	*6*	*7*	*8*	*9*	*10*	*11*	*12*	
2000	5	4	2	4	—	—	1	—	—	—	2	1	19
2001	3	1	2	3	—	1	—	—	2	2	3	1	16
2002	3	2	2	3	1	—	—	—	—	—	2	2	16
2003	2	6	4	2	—	—	—	—	—	—	1	4	20
2004	3	3	3	—	1	—	—	1	—	—	2	3	16
In all	16	16	13	12	2	1	1	1	2	2	10	11	87
MY	3.2	3.2	2.6	2.4	0.4	0.2	0.2	—	0.4	0.4	2.0	2.2	17
2005	7	1	3	1	—	—	—	—	—	—	2	1	15
2006	3	3	3	2	—	—	—	—	—	—	1	3	15
2007	1	5	4	—	—	—	1	—	—	—	2	5	18
2008	2	4	4	2	—	—	—	—	—	1	2	2	17
2009	5	3	4	1	—	—	—	—	—	—	2	3	18
In all	18	16	18	6	—	—	1	—	—	1	9	14	83
MY	3.6	3.2	3.6	1.2	—	—	0.2	—	—	0.2	1.8	2.8	16
In all	97	96	74	50	7	2	3	2	7	8	33	57	437
MY	3.6	3.6	2.7	1.8	0.3	0.1	0.1	0.1	0.3	0.3	1.2	2.1	16
2010	2	2	3	2	1	—	—	—	—	1	—	—	11

Notes: See Table A.1.

Table A.13. The annual distribution of the number of supertyphoons in the World Ocean (WO), in the northern hemisphere (NH), and in the southern hemisphere (SH) (1990–2010).

Year	*NH*	*SH*	*WO*
1990	2	—	2
1991	—	—	—
1992	3	—	3
1993	1	1	2
1994	9	—	9
In all	15	1	16
MY	3.0	0.2	3.2
1995	9	2	11
1996	8	2	10
1997	13	1	14
1998	5	2	7
1999	5	3	8
In all	40	10	50
MY	8.0	2.0	10.0
2000	7	—	7
2001	3	—	3
2002	11	1	12
2003	6	1	7
2004	8	—	8
In all	35	2	37
MY	7.0	0.8	7.6

Year	*NH*	*SH*	*WO*
2005	9	—	9
2006	7	1	8
2007	6	—	6
2008	5	—	5
2009	7	—	7
In all	34	1	35
MY	6.8	0.2	7.0
In all	124	14	138
MY	6.2	0.7	6.9
2010	4	—	4

Notes: See Table A.1. Wind velocity in the wake of supertyphoons is $65\,m\,s^{-1}$.

Table A.14. The annual distribution of the number of supertyphoons in World Ocean basins (1990–2010).

Year	*World Ocean*			*In all World Ocean*
	Pacific	*Atlantic*	*Indian*	
1990	2	—	—	2
1991	—	—	—	—
1992	2	1	—	3
1993	1	—	1	2
1994	9	—	—	9
In all	14	1	1	16
MY	2.8	0.2	0.2	3.2
1995	6	3	2	11
1996	6	2	2	10
1997	14	—	—	14
1998	4	2	1	7
1999	1	3	4	8
In all	31	10	9	50
MY	6.2	2.0	1.8	10.0
2000	5	2	—	7
2001	3	—	—	3
2002	12	—	—	12
2003	5	1	1	7
2004	7	1	—	8
In all	32	5	1	38
MY	6.4	1.0	0.2	7.6

Year	*World Ocean*			*In all World Ocean*
	Pacific	*Atlantic*	*Indian*	
2005	4	5	—	9
2006	8	—	—	8
2007	4	—	2	6
2008	2	3	—	5
2009	7	—	—	7
In all	25	8	2	35
MY	5.0	1.6	0.4	7.0
In all	102	24	13	139
MY	5.1	1.2	0.65	7.0
2010	2	1	1	4

Notes: See Table A.1.

Table A.15. The annual distribution of the number of supertyphoons in the Pacific basins (1990–2010).

Year	*Pacific basins*			*Pacific Ocean*
	NWP	*NEP*	*SWP*	
1990	1	1	—	2
1991	—	—	—	—
1992	—	2	—	2
1993	—	1	—	1
1994	5	4	—	9
In all	6	8	—	14
MY	1.2	1.6	—	2.8
1995	5	1	—	6
1996	6	—	—	6
1997	9	4	1	14
1998	2	1	1	4
1999	1	—	—	1
In all	23	6	2	31
MY	4.6	1.2	0.4	6.2
2000	4	1	—	5
2001	3	—	—	3
2002	8	3	1	12
2003	5	—	—	5
2004	6	1	—	7
In all	26	5	2	32
MY	5.2	1.0	0.4	6.4

Year	*Pacific basins*			*Pacific Ocean*
	NWP	*NEP*	*SWP*	
2005	4	—	—	4
2006	6	1	1	8
2007	4	—	—	4
2008	2	—	—	2
2009	5	2	—	7
In all	21	3	1	25
MY	4.2	0.6	0.2	5.0
In all	76	22	5	102
MY	3.8	1.1	0.2	5.1
2010	1	1	—	2

Notes: See Tables A.1 and A.3.

Table A.16. The annual distribution of the number of supertyphoons in the Indian Ocean (1990–2010).

	Indian Ocean basins		*Indian Ocean*
Year	*NIP*	*SIP*	
1990	—	—	—
1991	—	—	—
1992	—	—	—
1993	—	1	1
1994	—	—	—
In all	—	1	1
MY	—	0.2	0.2
1995	—	2	2
1996	—	2	2
1997	—	—	—
1998	—	1	1
1999	1	3	4
In all	1	8	9
MY	—	1.6	1.8
2000	—	—	—
2001	—	—	—
2002	—	—	—
2003	—	1	1
2004	—	—	—
In all	—	1	1
MY	—	0.2	0.2

Year	*Indian Ocean basins*		*Indian Ocean*
	NIP	*SIP*	
2005	—	—	—
2006	—	—	—
2007	2	—	2
2008	—	—	—
2009	—	—	—
In all	2	—	2
MY	0.4	—	0.4
In all	3	10	13
MY	0.15	0.5	0.65
2010	1	—	1

Notes: See Tables A.1 and A.4.

Table A.17. TC transformations into midlatitude circulation and their development in World Ocean basins (1990–2010).

Year	*World Ocean basins*		
	$\sum TC$	*TC transformations into midlatitude circulation*	*% WO*
1990	93	17	18
1991	78	18	23
1992	99	24	24
1993	75	19	25
1994	86	20	23
In all	431	98	
MY	86	20	23
1995	78	18	23
1996	92	21	23
1997	96	34	35
1998	88	25	28
1999	75	17	23
In all	429	115	
MY	86	23	27
2000	91	26	29
2001	88	29	33
2002	83	28	34
2003	89	32	36
2004	85	35	41
In all	43	61	50
MY	87	30	35

Year	*World Ocean basins*		
	$\sum TC$	*TC transformations into midlatitude circulation*	*% WO*
2005	97	34	35
2006	83	21	25
2007	85	26	31
2008	86	25	29
2009	86	21	24
In all	437	127	
MY	87	25	29
In all	1,733	490	
MY	87	25	29
2010	68	16	23

Notes: See Table A.1.

Table A.18. TC transformations into midlatitude circulation and their development in the northern hemisphere (NH) and in the southern hemisphere (SH) (1990–2010).

Year	*NH*			*SH*		
	$\sum TC$	*TC number transformed into midlatitude circulation*	*% NH*	$\sum TC$	*TC number transformed into midlatitude circulation*	*% SH*
1990	68	16	24	25	1	4
1991	54	17	31	24	1	4
1992	74	21	28	25	3	13
1993	52	15	29	23	4	17
1994	64	17	27	22	3	14
In all	31	28	61	19	12	
MY	62	17	27	24	2.4	10
1995	58	17	29	20	1	5
1996	59	19	32	33	2	6
1997	60	21	35	36	13	36
1998	52	13	25	36	12	33
1999	49	12	25	26	5	19
In all	278	82		151	33	
MY	56	16	29	30	6.6	22
2000	62	20	32	29	6	21
2001	62	23	37	26	6	23
2002	59	20	34	24	8	33
2003	59	19	32	30	13	43
2004	64	29	45	21	6	29
In all	306	111		130	39	
MY	61	22	36	26	7.8	30

Year	*NH*			*SH*		
	$\sum TC$	*TC number transformed into midlatitude circulation*	*% NH*	$\sum TC$	*TC number transformed into midlatitude circulation*	*% SH*
2005	72	22	31	25	12	48
2006	57	15	26	26	6	23
2007	57	18	32	28	8	29
2008	65	19	29	21	6	29
2009	58	13	22	28	8	29
In all	309	87		128	40	
MY	62	17	27	26	8	32
In all	1,205	366		528	124	
MY	60	18	30	26	6	23
2010	46	12	26	22	4	18

Notes: See Table A.1.

Table A.19. TC transformations into midlatitude circulation and their development in the Pacific (1990–2010).

Year	*Pacific Ocean*		
	$\sum TC$	*TC number transformed into midlatitude circulation*	*% P*
1990	59	11	19
1991	49	12	24
1992	70	20	28
1993	52	13	25
1994	60	15	25
In all	290	71	
MY	58	14	24
1995	38	10	26
1996	49	15	31
1997	70	28	40
1998	46	11	24
1999	42	11	26
In all	245	75	
MY	49	15	31
2000	54	17	31
2001	53	19	36
2002	50	18	36
2003	50	19	39
2004	49	20	41
In all	256	93	
MY	51	19	37

Year	*Pacific Ocean*		
	$\sum TC$	*TC number transformed into midlatitude circulation*	*% P*
2005	49	15	31
2006	53	12	23
2007	46	15	33
2008	46	14	30
2009	54	13	24
In all	248	69	
MY	50	14	28
In all	1,037	308	
MY	52	15.5	30
2010	33	8	24

Note: See Table A.1.

Table A.20. TC transformations into midlatitude circulation and their development in the Pacific basins (1990–2010).

Year	*NWP*			*NEP*			*SWP*		
	$\sum TC$	*TC number transformed into midlatitude circulation*	*% P*	$\sum TC$	*TC number transformed into midlatitude circulation*	*% P*	$\sum TC$	*TC number transformed into midlatitude circulation*	*% P*
1990	29	11	19	21	—	—	9	—	—
1991	29	12	24	14	—	—	6	—	—
1992	31	17	24	27	1	1	12	2	3
1993	28	11	21	14	2	4	10	—	—
1994	35	14	23	19	1	2	6	—	—
In all	152	65		95	4		43	2	
MY	30	13	22	19	1	1	9	0.4	0.6
1995	25	10	26	10	—	—	3	—	—
1996	30	13	27	8	—	—	11	2	4
1997	30	17	24	19	2	3	21	9	13
1998	17	5	11	13	—	—	16	6	13
1999	23	7	17	9	—	—	10	4	10
In all	125	52		59		2	61	21	
MY	25	10	20	12	—	0.6	12	4	8

2000	26	14	26	18	—	—	10	3	6
2001	28	15	28	15	—	—	10	4	8
2002	27	12	24	15	3	6	8	3	6
2003	24	12	24	16	—	—	10	7	14
2004	32	18	37	12	—	—	5	2	4
In all	137	71		76	3		43	19	
MY	27	14	27	15	1	1.2	9	4	8
2005	24	11	22	15	—	—	10	5	10
2006	24	7	13	18	—	—	11	5	9
2007	25	12	26	11	—	—	10	3	7
2008	25	12	26	17	—	—	4	2	4
2009	24	9	17	20	1	1	10	3	6
In all	122	51		81	1		45	18	
MY	24	10	20	16	—	0.2	9	47	
In all	536	239	23	311	10	1	192	60	6
MY	27	12	80	16	0.5	1	10	3	19
2010	14	5	15	8	1	3	11	2	6

Notes: NWP, North-West Pacific; NEP, North-East Pacific; SWP, South-West Pacific; % P is the number of TCs transformed into midlatitude circulation as a percentage of the number of TCs formed in the Pacific basins.

Table A.21. TC transformations into midlatitude circulation and their development in the Indian Ocean and its basins (1990–2010).

Year	*NIO*			*SIO*			*IO*		
	$\sum TC$	*TC number transformed into the midlatitude circulation*	*% IO*	$\sum TC$	*TC number transformed into the midlatitude circulation*	*% IO*	$\sum TC$	*TC number transformed into the midlatitude circulation*	*% WO*
1990	4	—	—	16	1	5	20	1	1.0
1991	3	—	—	18	1	5	21	1	1.2
1992	9	—	—	13	1	5	22	1	1.0
1993		—	—	11	4	31	13	4	5.5
1994	3	—	—	16	3	16	19	3	3.5
In all	21	—	—	74	10		95	10	
MY	4	—	—	15	2	12	19	2	2.3
1995	4	—	—	17	1	5	21	1	1.3
1996	8	—	—	22	—	—	30	—	—
1997	4	—	—	15	4	21	19	4	4.2
1998	8	—	—	20	6	21	28	6	6.8
1999	5	—	—	16	1	5	21	1	1,3
In all	29	—	—	90	12		119	12	
MY	6	—	—	18	2	10	24	2	2.8

2000	4	—	—	19	3	13	23	3	3.3
2001	4	—	—	16	2	10	20	2	2.3
2002	5	—	—	16	5	24	21	5	6.0
2003	3	—	—	20	6	26	23	6	6.7
2004	5	—	—	16	4	19	21	4	4.7
In all	21	—	—	87	20		108	20	
MY	4	—	—	18	4	18	22	4	5.7
2005	6	—	—	15	7	33	21	7	7.2
2006	6	—	—	15	1	5	21	1	1.2
2007	6	—	—	18	5	21	24	5	5.9
2008	7	—	—	17	4	17	24	4	4.6
2009	5	—	—	18	5	22	23	5	5.8
In all	30	—	—	83	22		113	22	
MY	6	—	—	17	4	19	23	4	5.1
In all	101	—	—	334	64		435	64	
MY	5	—	—	17	3.2	15	22	3.2	3.7
2010	5	—	—	11	2	12	16	2	3

Notes: IO, Indian Ocean; NIO, North Indian Ocean; and SIO, South Indian Ocean; *% WO* and *% IO* are the number of TCs transformed into the midlatitude circulation as a percentage of the number of TCs formed in the World Ocean (WO) and in the Indian Ocean (IO).

Table A.22. TC transformations into midlatitude circulation and their development in the Atlantic Ocean (AO) (1990–2010).

Year	*Atlantic Ocean*		
	$\sum TC$	*TC number transformed into midlatitude circulation*	*% AO*
1990	14	5	36
1991	7	5	71
1992	7	3	42
1993	8	2	24
1994	7	2	29
In all	43	17	
MY	8	3	38
1995	19	7	36
1996	13	6	46
1997	7	2	29
1998	14	8	57
1999	12	5	42
In all	65	28	
MY	13	6	43
2000	14	6	43
2001	15	8	53
2002	12	6	50
2003	16	8	50
2004	15	11	74
In all	72	39	
MY	14	856	

Year	*Atlantic Ocean*		
	$\sum TC$	*TC number transformed into midlatitude circulation*	*% AO*
2005	27	11	33
2006	9	8	67
2007	15	6	40
2008	16	7	44
2009	9	3	33
In all	76	35	
MY	15	7	48
In all	257	115	
MY	13	5.8	45
2010	19	6	32

Notes: *% AO* is the number of TCs transformed into midlatitude circulation as a percentage of the number of TCs formed in the Atlantic Ocean (AO).

Table A.23. TC dissipation conditions in the World Ocean (1990–2010).

Year	$\sum TC$	*TC dissipation conditions*					
		Land		*High seas*		*Transformation*	
		$\sum TC$	*% WO*	$\sum TC$	*% WO*	$\sum TC$	*% WO*
1990	93	25	27	50	54	17	19
1991	78	14	18	45	58	18	24
1992	99	20	20	53	54	24	26
1993	75	21	28	35	47	19	25
1994	86	24	28	42	49	20	23
In all	431	104		225		98	
MY	86	21	24	45	52	20	24
1996	92	35	38	36	39	21	23
1997	96	21	22	40	43	34	35
1998	88	33	38	30	34	25	28
1999	75	28	37	30	40	17	23
In all	429	144		169		115	
MY	86	29	34	34	40	23	26
2000	91	26	29	39	43	26	28
2001	88	25	28	34	39	29	33
2002	83	21	25	33	40	29	35
2003	89	29	33	27	30	33	37
2004	85	13	15	37	43	35	42
In all	436	114		170		152	
MY	87	23	26	34	39	30	35

Year	$\sum TC$	*TC dissipation conditions*					
		Land		*High seas*		*Transformation*	
		$\sum TC$	*% WO*	$\sum TC$	*% WO*	$\sum TC$	*% WO*
2005	97	32	33	33	34	32	33
2006	83	29	35	35	42	19	23
2007	85	25	29	34	40	26	31
2008	86	26	30	35	41	25	29
2009	86	22	26	43	50	21	24
In all	437	134		180		123	
MY	86	27	31	36	41	25	28
In all	1,733	497		746		489	
MY	86	25	29	37	44	24	27
2010	68	24	35	28	41	16	24

Notes: Land implies that TC dissipation has occurred over land. High seas means that TC dissipation has occurred under high seas conditions. Transformation means that TC dissipation was due to TC transformation into midlatitude circulation.

Table A.24. TC dissipation conditions in the northern hemisphere (1990–2010).

Year	$\sum TC$	*TC dissipation conditions*					
		Land		*High seas*		*Transformation*	
		$\sum TC$	*% NH*	$\sum TC$	*% NH*	$\sum TC$	*% NH*
1990	68	20	29	31	45	16	24
1991	54	12	22	24	44	17	31
1992	74	15	20	37	50	21	28
1993	52	19	36	18	35	15	29
1994	64	20	31	27	42	17	26
In all	312	87		138		87	
MY	62	17	28	28	44	17	28
1995	58	20	34	21	36	17	29
1996	59	27	47	13	22	19	32
1997	60	15	25	23	38	21	35
1998	52	25	48	14	27	13	25
1999	49	19	39	18	37	12	24
In all	278	106		89		83	
MY	56	21	38	18	32	16	29
2000	62	18	29	24	39	20	32
2001	62	12	19	27	43	23	37
2002	59	16	27	23	37	20	35
2003	59	24	41	16	25	19	34
2004	64	10	16	25	39	29	45
In all	306	80		114		112	
MY	61	16	26	23	38	22	36

Year	$\sum TC$	*TC dissipation conditions*					
		Land		*High seas*		*Transformation*	
		$\sum TC$	*% NH*	$\sum TC$	*% NH*	$\sum TC$	*% NH*
2005	72	30	42	20	30	22	31
2006	57	20	35	22	40	15	26
2007	57	19	33	20	35	18	32
2008	65	25	38	21	32	19	29
2009	58	18	31	27	47	13	22
In all	309	112		110		87	
MY	62	22	35	22	37	17	28
In all	1,205	384		452		369	
MY	60	19	32	23	38	18	30
2010	46	18	39	16	35	12	26

Notes: Land implies that TC dissipation has occurred over land. High seas means that TC dissipation has occurred under high seas conditions. Transformation means that TC dissipation was due to TC transformation into midlatitude circulation.

Table A.25. TC dissipation conditions in the southern hemisphere (1990–2010).

Year	$\sum TC$	*TC dissipation conditions*					
		Land		*High seas*		*Transformation*	
		$\sum TC$	*% SH*	$\sum TC$	*% SH*	$\sum TC$	*% SH*
1990	25	5	20	19	76	1	4
1991	24	2	8	21	88	1	4
1992	25	5	20	16	67	3	13
1993	23	2	9	17	74	4	17
1994	22	4	18	15	68	3	14
In all	119	18		88		12	
MY	24	4	16	18	74	3	10
1995	20	7	35	12	60	1	5
1996	33	8	24	23	70	2	6
1997	36	6	17	17	47	13	36
1998	36	8	22	16	45	12	33
1999	26	9	35	12	46	5	19
In all	151	38		80		33	
MY	30	8	27	16	53	7	20
2000	29	8	27	15	52	6	21
2001	26	13	50	7	27	6	23
2002	24	5	21	11	47	8	32
2003	30	5	17	12	40	13	43
2004	21	3	14	12	57	6	29
In all	130	34		57		39	
MY	26	7	26	11	44	8	30

Year	$\sum TC$	*TC dissipation conditions*					
		Land		*High seas*		*Transformation*	
		$\sum TC$	*% SH*	$\sum TC$	*% SH*	$\sum TC$	*% SH*
2005	25	2	8	11	44	12	48
2006	26	9	35	11	42	6	23
2007	28	6	21	14	50	8	29
2008	21	1	5	14	66	6	29
2009	28	4	14	16	57	8	29
In all	128	22		66		40	
MY	26	4	16	13	52	8	32
In all	537	112		301		124	
MY	27	6	22	15	55	6	23
2010	22	6	27	12	54	4	19

Note: Land implies that TC dissipation has occurred over land. High seas means that TC dissipation has occurred under high seas conditions. Transformation means that TC dissipation was due to TC transformation into midlatitude circulation.

Table A.26. TC dissipation conditions in the Pacific Ocean (1990–2010).

Year	$\sum TC$	*TC dissipation conditions*					
		Land		*High seas*		*Transformation*	
		$\sum TC$	*% P*	$\sum TC$	*% P*	$\sum TC$	*% P*
1990	59	18	31	29	50	11	19
1991	49	9	19	28	57	12	24
1992	70	13	19	36	53	20	28
1993	52	15	29	24	46	13	25
1994	60	15	25	30	50	15	25
In all	290	70		149		71	
MY	58	14	24	30	52	14	24
1995	38	13	34	15	40	10	26
1996	49	17	35	17	34	15	31
1997	70	14	21	27	39	28	40
1998	46	18	39	17	37	11	24
1999	42	14	33	17	41	11	26
In all	245	76		93		75	
MY	49	15	30	19	39	15	31
2000	54	14	26	23	43	17	31
2001	53	11	21	23	43	19	36
2002	50	10	20	22	44	18	36
2003	50	16	31	15	30	19	39
2004	49	6	12	23	47	20	41
In all	256	57		106		93	
MY	51	11	22	21	41	19	37

Year	$\sum TC$	*TC dissipation conditions*					
		Land		*High seas*		*Transformation*	
		$\sum TC$	*% P*	$\sum TC$	*% P*	$\sum TC$	*% P*
2005	49	15	31	18	38	15	31
2006	53	16	30	25	47	12	23
2007	46	10	21	21	46	15	33
2008	46	15	33	17	37	14	33
2009	54	13	24	28	52	13	24
In all	248	69		109		69	
MY	50	14	28	22	44	14	28
In all	1,039	272		458		309	
MY	52	14	26	23	44	15	30
2010	33	13	39	12	36	8	25

Note: Land implies that TC dissipation has occurred over land. High seas means that TC dissipation has occurred under high seas conditions. Transformation means that TC dissipation was due to TC transformation into midlatitude circulation.

Table A.27. TC dissipation conditions in the North-West Pacific (1990–2010).

Year	$\sum TC$	*TC dissipation conditions*					
		Land		*High seas*		*Transformation*	
		$\sum TC$	*% NWP*	$\sum TC$	*% NWP*	$\sum TC$	*% NWP*
1990	29	13	46	4	15	11	39
1991	29	8	28	9	31	12	41
1992	31	8	27	5	17	17	56
1993	28	10	36	7	25	11	39
1994	35	13	37	8	23	14	40
In all	152	52		33		65	
MY	30	10	33	7	24	13	43
1995	25	10	40	5	20	10	40
1996	30	10	33	7	23	13	44
1997	30	7	24	5	17	17	59
1998	17	11	65	1	6	5	29
1999	23	10	44	6	26	7	30
In all	125	48		24		52	
MY	25	10	40	5	19	10	41
2000	26	7	27	5	19	14	54
2001	28	9	32	4	14	15	54
2002	27	7	26	8	30	12	44
2003	24	10	42	2	8	12	50
2004	32	5	16	9	28	18	56
In all	137	38		28		71	
MY	27	8	30	6	20	14	52

Year	$\sum TC$	*TC dissipation conditions*					
		Land		*High seas*		*Transformation*	
		$\sum TC$	*% NWP*	$\sum TC$	*% NWP*	$\sum TC$	*% NWP*
2005	24	11	46	2	8	11	46
2006	24	10	42	7	29	7	29
2007	25	8	32	5	20	12	48
2008	25	12	48	14	12	48	
2009	24	10	42	5	21	9	38
In all	122	51		20		51	
MY	24	10	42	4	17	10	41
In all	536	190		106		240	
MY	27	10	36	5	19	12	45
2010	14	7	50	2	14	5	36

Note: Land implies that TC dissipation has occurred over land. High seas means that TC dissipation has occurred under high seas conditions. Transformation means that TC dissipation was due to TC transformation into midlatitude circulation.

Table A.28. TC dissipation conditions in the North-East Pacific (1990–2010).

Year	$\sum$ *TC*	*TC dissipation conditions*					
		Land		*High seas*		*Transformation*	
		$\sum$ *TC*	*% NEP*	$\sum$ *TC*	*% NEP*	$\sum$ *TC*	*% NEP*
1990	21	2	10	19	90	—	—
1991	14	1	7	13	93	—	—
1992	27	2	7	24	89	1	4
1993	14	4	29	8	57	2	14
1994	19	1	5	17	90	1	5
In all	95	10		91		4	
MY	19	2	11	16	84	1	5
1995	10	1	10	9	90	—	—
1996	8	5	62	3	38	—	—
1997	19	5	26	12	63	2	11
1998	13	2	15	11	85	—	—
1999	9	2	22	7	78	—	—
In all	59	15		42		2	
MY	12	3	25	8	70	—	5
2000	18	4	22	14	78	—	—
2001	15	—	—	15	100	—	—
2002	15	2	13	10	67	3	20
2003	16	6	37	10	63	—	—
2004	12	1	8	11	92	—	—
In all	76	13		60		3	
MY	15	3	17	12	80	1	3

Year	$\sum TC$	*TC dissipation conditions*					
		Land		*High seas*		*Transformation*	
		$\sum TC$	*% NEP*	$\sum TC$	*% NEP*	$\sum TC$	*% NEP*
2005	15	2	13	13	87	—	—
2006	18	4	22	14	78	—	—
2007	11	2	18	9	82	—	—
2008	17	3	18	14	82	—	—
2009	20	1	5	18	90	1	5
In all	81	12		68		1	
MY	16	2	15	14	85	—	—
In all	311	50		251		10	
MY	16	2	17	14	80	0.5	3
2010	8	2	25	5	63	1	12

Note: Land implies that TC dissipation has occurred over land. High seas means that TC dissipation has occurred under high seas conditions. Transformation means that TC dissipation was due to TC transformation into midlatitude circulation.

Table A.29. TC dissipation conditions in the South-West Pacific (1990–2010).

Year	$\sum TC$	*TC dissipation conditions*					
		Land		*High seas*		*Transformation*	
		$\sum TC$	*% SWP*	$\sum TC$	*% SWP*	$\sum TC$	*% SWP*
1990	9	3	33	6	67	—	—
1991	6	—	—	6	100	—	—
1992	12	3	25	7	58	2	17
1993	10	1	10	9	90	—	—
1994	6	1	17	5	83	—	—
In all	43	8		33		2	
MY	9	2	19	7	76	0.4	5
1995	3	2	67	1	33	—	—
1996	11	2	18	7	64	2	18
1997	21	2	10	1	48	9	42
1998	16	5	31	5	31	6	38
1999	10	2	20	4	40	4	40
In all	61	13		27		21	
MY	12	3	21	5	45	4	34
2000	10	3	30	4	40	3	30
2001	10	2	20	4	40	4	40
2002	8	1	12	4	50	3	38
2003	10	—	—	3	22	7	78
2004	5	—	—	3	60	2	40
In all	43	6		18		19	
MY	9	1	14	4	41	4	45

Year	$\sum TC$	*TC dissipation conditions*					
		Land		*High seas*		*Transformation*	
		$\sum TC$	*% SWP*	$\sum TC$	*% SWP*	$\sum TC$	*% SWP*
2005	10	2	20	3	30	5	50
2006	11	2	18	4	36	5	46
2007	10	—	—	7	70	3	30
2008	4	—	—	2	50	2	50
2009	10	2	20	5	50	3	30
In all	45	6		21		18	
MY	9	1	13	4	47	4	40
In all	192	33		99		60	
MY	10	2	17	5	51	3	32
2010	11	4	36	5	45	2	19

Note: Land implies that TC dissipation has occurred over land. High seas means that TC dissipation has occurred under high seas conditions. Transformation means that TC dissipation was due to TC transformation into midlatitude circulation.

Table A.30. TC dissipation conditions in the Atlantic Ocean (1990–2010).

Year	*∑ TC*	*TC dissipation conditions*					
		Land		*High seas*		*Transformation*	
		∑ TC	*% AO*	*∑ TC*	*% AO*	*∑ TC*	*% AO*
1990	14	3	21	6	43	536	
1991	7	—	—	2	30	5	70
1992	7	2	29	2	29	3	42
1993	8	3	37	3	38	2	25
1994	7	3	42	2	29	2	29
In all	43	11		15		17	
MY	8	2	24	3	38	3	38
1995	19	6	32	6	32	7	36
1996	13	5	38	2	16	6	46
1997	7	—	—	5	71	2	29
1998	14	5	36	1	7	8	57
1999	12	3	25	4	33	542	
In all	65	19		18		28	
MY	13	4	29	3	28	6	43
2000	14	5	36	3	21	6	43
2001	15	3	20	4	27	8	53
2002	12	4	33	2	17	6	50
2003	16	5	32	3	18	8	50
2004	15	2	13	2	13	11	74
In all	72	19		14	39		
MY	14	4	27	3	20	8	53

Year	$\sum TC$	*TC dissipation conditions*					
		Land		*High seas*		*Transformation*	
		$\sum TC$	*% AO*	$\sum TC$	*% AO*	$\sum TC$	*% AO*
2005	27	13	48	5	19	11	33
2006	9	1	11	2	22	8	67
2007	15	4	27	5	33	6	40
2008	16	5	32	4	24	7	44
2009	9	2	23	4	44	3	33
In all	76	25		20		35	
MY	15	5	33	4	26	7	41
In all	256	73		66		117	
MY	13	4	29	3	26	6	45
2010	19	6	32	7	36	6	32

Note: Land implies that TC dissipation has occurred over land. High seas means that TC dissipation has occurred under high seas conditions. Transformation means that TC dissipation was due to TC transformation into midlatitude circulation.

Table A.31. TC dissipation conditions in the Indian Ocean (1990–2010).

Year	$\sum TC$	*TC dissipation conditions*					
		Land		*High seas*		*Transformation*	
		$\sum TC$	*% IO*	$\sum TC$	*% IO*	$\sum TC$	*% IO*
1990	20	4	20	15	75	1	5
1991	21	5	24	15	71	1	5
1992	22	5	23	15	68	1	9
1993	15	3	20	8	53	4	27
1994	19	6	32	10	53	3	15
In all	95	23		63		10	
MY	19	5	24	13	67	2	9
1995	21	8	38	12	57	1	5
1996	30	13	43	17	57	—	—
1997	19	7	36	8	42	4	22
1998	28	10	36	12	43	6	21
1999	21	11	52	9	43	1	5
In all	119	49		58		12	
MY	24	10	42	12	50	2	8
2000	23	7	30	13	56	3	14
2001	20	11	55	7	35	2	10
2002	21	7	33	9	43	5	24
2003	23	8	35	9	39	6	26
2004	21	5	24	12	57	4	19
In all	108	38		50		20	
MY	22	8	36	10	46	4	18

Year	$\sum TC$	*TC dissipation conditions*					
		Land		*High seas*		*Transformation*	
		$\sum TC$	*% IO*	$\sum TC$	*% IO*	$\sum TC$	*% IO*
2005	21	4	19	10	48	7	33
2006	21	12	57	8	38	1	5
2007	24	11	46	8	33	5	21
2008	24	6	25	14	58	4	17
2009	23	7	30	11	48	5	22
In all	113	40		51		22	
MY	23	8	36	10	45	4	19
In all	435	150		221		64	
MY	22	8	36	11	50	3	14
2010	16	5	31	9	56	2	13

Note: Land implies that TC dissipation has occurred over land. High seas means that TC dissipation has occurred under high seas conditions. Transformation means that TC dissipation was due to TC transformation into midlatitude circulation.

Table A.32. TC dissipation conditions in the Northern Indian Ocean (1990–2010).

Year	$\sum TC$	*TC dissipation conditions*					
		Land		*High seas*		*Transformation*	
		$\sum TC$	*% NIO*	$\sum TC$	*% NIO*	$\sum TC$	*% NIO*
1990	4	2	50	2	50	—	—
1991	3	3	100	—	—	—	—
1992	9	3	33	6	67	—	—
1993	2	2	100	—	—	—	—
1994	3	3	100	—	—	—	—
In all	21	13		8			
MY	4	3	65	1	35	—	—
1995	4	3	75	1	25	—	—
1996	8	7	88	1	12	—	—
1997	4	3	75	1	25	—	—
1998	8	7	88	1	12	—	—
1999	5	4	80	1	20	—	—
In all	29	24		5			
MY	6	5	80	1	20	—	—
2000	4	2	50	2	50	—	—
2001	4	—	—	4	100	—	—
2002	5	3	60	2	40	—	—
2003	3	3	100	—	—	—	—
2004	5	2	40	3	60	—	—
In all	21	10		11			
MY	4	2	50	2	50	—	—

Year	$\sum TC$	*TC dissipation conditions*					
		Land		*High seas*		*Transformation*	
		$\sum TC$	*% NIO*	$\sum TC$	*% NIO*	$\sum TC$	*% NIO*
2005	6	4	67	2	33	—	—
2006	6	5	83	1	17	—	—
2007	6	5	83	1	17	—	—
2008	7	5	71	2	29	—	—
2009	5	5	100	—	—	—	—
In all	30	24		6			
MY	6	5	90	1	10	—	—
In all	101	71		30			
MY	5	3	60	2	40	—	—
2010	5	3	60	2	40	—	—

Note: Land implies that TC dissipation has occurred over land. High seas means that TC dissipation has occurred under high seas conditions. Transformation means that TC dissipation was due to TC transformation into midlatitude circulation.

Table A.33. TC dissipation conditions in the Southern Indian Ocean (1990–2010).

Year	∑ *TC*	*TC dissipation conditions*					
		Land		*High seas*		*Transformation*	
		∑ *TC*	*% SIO*	∑ *TC*	*% SIO*	∑ *TC*	*% SIO*
1990	16	2	12	13	81	1	7
1991	18	2	11	15	83	1	6
1992	13	2	17	9	75	1	8
1993	13	1	8	8	62	4	30
1994	16	3	19	10	62	3	19
In all	76	10		55		10	
MY	15	2	13	11	74	2	13
1995	17	5	29	11	65	1	6
1996	22	6	27	16	73	—	—
1997	15	4	27	7	47	4	26
1998	20	3	15	11	55	6	30
1999	16	7	44	8	50	1	6
In all	90	25		53		12	
MY	18	5	28	10	59	2	13
2000	19	5	26	11	58	3	16
2001	16	11	69	3	19	2	12
2002	16	4	25	7	44	5	31
2003	20	5	25	9	45	6	30
2004	16	3	19	9	56	4	25
In all	87	28		39		20	
MY	18	6	32	8	45	4	23

Year	$\sum TC$	*TC dissipation conditions*					
		Land		*High seas*		*Transformation*	
		$\sum TC$	*% SIO*	$\sum TC$	*% SIO*	$\sum TC$	*% SIO*
2005	15	—	—	8	53	7	47
2006	15	7	47	7	47	1	6
2007	18	6	33	7	39	5	28
2008	17	1	6	12	70	4	24
2009	18	2	11	11	61	5	28
In all	83	16		45		22	
MY	17	3	19	9	54	4	27
In all	336	80		192		64	
MY	17	4	24	9	58	3	18
2010	11	2	18	7	64	2	18

Note: Land implies that TC dissipation has occurred over land. High seas means that TC dissipation has occurred under high seas conditions. Transformation means that TC dissipation was due to TC transformation into midlatitude circulation.

Table A.34. Number of initial tropical disturbances (TLs) and number of TCs formed from TLs in the World Ocean (WO), in the northern hemisphere (NH), and in the southern hemisphere (SH) (1999–2010).

Year	*Number*					
	NH		*SH*		*WO*	
	TL	*TC*	*TL*	*TC*	*TL*	*TC*
1999	313	49	126	26	439	75
2000	203	62	93	29	296	91
2001	202	62	68	26	270	88
2002	175	59	66	24	241	83
2003	177	59	66	30	243	89
2004	139	64	60	21	199	85
In all	896	306	353	130	1,249	436
MY	179	61	71	26	250	87
2005	134	72	60	25	194	97
2006	143	57	65	26	208	83
2007	122	57	64	28	186	85
2008	137	65	50	21	187	86
2009	101	58	54	28	155	86
In all	637	309	293	128	930	437
MY	127	62	59	26	186	87
In all for 10 year	1,533	615	646	257	2,179	873
MY	153	62	65	26	218	87
2010	99	46	58	22	157	68

Table A.35. Number of initial tropical disturbances (TLs) and number of TCs formed from TLs in World Ocean basins (1999–2010).

Year	*Ocean*							
	Pacific		*Atlantic*		*Indian*		*World*	
TL	*TC*	*TL*	*TC*	*TL*	*TC*	*TL*	*TC*	
1999	280	42	36	12	123	21	439	75
2000	173	54	43	14	80	23	296	91
2001	163	53	34	15	73	20	270	88
2002	153	50	29	12	59	21	241	83
2003	142	50	42	16	59	23	243	89
2004	114	49	29	15	56	21	199	85
In all	745	256	177	72	327	108	1,249	436
MY	149	51	35	14	65	22	250	87
2005	149	49	48	27	46	21	194	97
2006	136	53	30	9	42	21	208	83
2007	105	46	30	15	51	24	186	85
2008	101	46	26	16	49	24	187	86
2009	96	54	13	9	46	23	155	86
In all	587	248	147	76	234	113	930	437
MY	117	50	29	15	47	23	186	86
In all for 10 years	1,332	503	324	148	561	221	2,179	873
MY	133	50	32	15	56	22	218	87
2010	83	33	26	19	58	16	167	68

Appendix B

Saffir–Simpson Hurricane Scale

The Saffir–Simpson Hurricane Scale is a 1–5 rating based on the hurricane's intensity. This is used to give an estimate of the potential property damage and flooding expected along the coast from a hurricane landfall. Windspeed is the determining factor in the scale, as storm surge values are highly dependent on the slope of the continental shelf in the landfall region.

CATEGORY 1 HURRICANE

Windspeed	64–82 kt (33–42 m s^{-1}); central pressure: $\geq$980 mb; storm surge: 4–5 ft above normal.
Types of damage	Damage to unanchored mobile homes, shrubs, and trees; no real damage to buildings; some coastal road flooding and minor pier damage.

CATEGORY 2 HURRICANE

Windspeed	83–95 kt (43–49 m s^{-1}); central pressure: 965–979 mb; storm surge: 6–8 ft above normal.
Types of damage	Roof, door, and window damage to buildings; considerable damage to vegetation, mobile homes, and piers; coastal and low-lying escape routes flood 2–4 hours before the arrival of the hurricane center.

CATEGORY 3 HURRICANE

Windspeed	96–113 kt (50–58 m s^{-1}); central pressure: 945–964 mb; storm surge: 9–12 ft above normal.
Types of damage	Structural damage to small houses and utility buildings, some curtain wall failures; mobile homes destroyed; flooding destroys smaller structures near the coast, larger structures damaged by floating debris; terrain lower than 5 feet above sea level may flood up to 6 miles inland.

CATEGORY 4 HURRICANE

Windspeed	114–135 kt (59–69 m s^{-1}); central pressure: 920–944 mb; storm surge: 13–18 ft above normal.
Types of damage	Extensive curtain wall failures, some complete roof structure failure on small residences; major erosion of beach areas; major damage to lower floors of structures near shore; terrain lower than 10 feet above sea level floods requiring massive evacuation of residental areas as far as 6 miles inland.

CATEGORY 5 HURRICANE

Windspeed	$>$135 kt ($>$ 69 m s^{-1}); central pressure: $<$ 902 mb; storm surge: $>$18 ft above normal.
Types of damage	Complete roof failure on many houses and industrial buildings; some buildings completely destroyed, small utility buildings blown over; major damage to lower floors of all structures located less than 15 feet above sea level and within 500 yards of the shoreline; massive evacuation of residential areas on low ground up to 10 miles inland.

ABBREVIATIONS USED

n.m. = nautical mile

kt = knot

mb = millibar

ft = foot

Pa = pascal

CONVERSION FACTORS

1 statute mile = 1.609 kilometers

1 nautical mile = 1.15 statute miles = 1.85 kilometers

1 knot = 1.15 statute miles per hour

= 1.85 kilometers per hour

= 0.1514 meters per second

1 foot = 0.3048 meters

1 yard = 3 feet

= 0.9144 meters

1 mb = 10^2 Pa

= 1 hectopascal

References and bibliography

Afonin V. V. and Sharkov E. A. (2003) Helioactivity and properties of global tropical cyclogenesis. *Proc. of the Conference in Memory of Yuri Galperin "Auroral Phenomena and Solar-terrestrial Relations", February 3–7, 2003, Moscow*, pp. 421–429 (*http://www.iki.rssi.ru/conf/galperin.pdf*).

Alcamo J., Leemans R., and Kleileman E. (Eds.) (1998). *Global Change Scenarios of the 21st century: Results from the IMAGE 2.1 Model.* Elsevier Science, Amsterdam, p. 250.

Aldinger W. T. and Stapler W. (1999). *1998 Annual Tropical Cyclone Report.* U.S. Naval Pacific Meteorology and Oceanography Center West/Joint Typhoon Warning Center, Pearl Harbor, Hawaii (*http://metoc.npmoc.navy.mil/jtwc*)

Aleksashkin S. N., Baibakov S. N., Karyagin V. P., Kremnev R. S., Linkin V. M., Lopatkin A. I., Martynov A. I., Pinchkphadze K. M., Rogovskii G. N., Skuridin G. A. *et al.* (1988). In-situ sounding of tropical cyclones from AES-delivered balloons. *Earth Research from Space* (*Earth Obs. Rem. Sens.—English transl.*), N4, 3–11.

Allison L. J., Rodgers E. B., Wilheit T. T., and Fett R. W. (1974). Tropical cyclone rainfall as measured by the Nimbus-5 electrically scanning microwave radiometer. *Bull. Amer. Meteorol. Soc.*, Vol. 55, N9, 1074–1089.

Alves P. G. (1992). *Access to Outer Space Technologies: Implications for International Security* (Research Paper No. 15). United Nations, New York, 156 pp.

Anfimov N. A., Gordeev S. P., Senkevich V. P., Moiseev S. S., Tsyboulskii G. A., and Sharkov E. A. (1995). Project "Zodiak": Contact sounding of critical states in the atmosphere with the help of rocket and space facilities. *Earth Obs. Rem. Sens.*, Vol. 13, N2, 177–190.

Anfimov N. A., Gordeev S. P., Tsyboulskii G. A., Moiseev S. S., and Sharkov E. A. (1996). Technique of balloon investigations of tropical disturbances on the ballistic missiles transportation base. *Adv. Space Res.*, Vol. 17, N9, 95–98.

Anon. (1993) The global climate in 1992. *World Climate News*, N3, 9–10.

Anthes R. A. (1982). *Tropical Cyclones: Their Evolution, Structure and Effects* (Meteorological Monograph No. 41). American Meteorological Society, Boston, 208 pp.

Antonov V. V., Armand N. A., Bobylev L. P., Bunkin F. V., Viter V. V., Volkov A. M., Volyak K. I., Vorobiev P. M., Voyakin S. N., Galeev A. A. *et al.* (1995). *Russian Space-*

craft for the Study of Earth as a Common Ecological System (Preprint No: Pr-1916). Space Research Institute, Moscow, 73 pp. [in Russian].

Apanasovich V. V., Kolyada A. A., and Chernyavskii A. F. (1988). *Statistical Analysis of Random Flows in Physical Experiments*. Universitetskoe, Minsk, 254 pp. [in Russian].

Arkin P. A. and Xie P. (1994). The Global Precipitation Climatology Project: First algorithm intercomparison project. *Bull. Amer. Meteorol. Soc.*, Vol. 75, 401–419.

Armand N. A. (Ed.) (1993). *International Project PRIRODA: Instrument Reference Handbook*. Institute of Radioengineering and Electronics, Russian Academy of Sciences, Moscow, 9 pp.

Arnett W. A. (1999). *The Nine Planets: A Multimedia Tour of the Solar System* (*http://www.seds.org/billa/tnp/*)

Asrar Gh. and Dokken D. J. (Eds.) (1993). *EOS Reference Handbook*. NASA, Washington, D.C., 145 pp.

Astafyeva N. M. (1996). Wavelet analysis: Basic theory and some applications. *Physics-Uspekhi*, Vol. 39, N11, 1085–1108.

Astafyeva N. M. and Sharkov E. A. (1998) Hierarchical structure of a public railway service. *Matematicheskoe modelirovanie (Mathematical Modeling)*, Vol. 10, N7, 37–47 [in Russian].

Astafyeva N. M. and Sharkov E. A. (2008) Track lines and evolution of hurricane Alberto from tropical latitudes to middle and high latitudes: Satellite microwave remote sensing. *Earth Research from Space*, N6, pp. 60–66 [in Russian].

Astafyeva N. M., Pokrovskaya I. V., and Sharkov E. A. (1994a). Scaling features of global tropical cyclogenesis. *Doklady Akad. Nauk (Trans. of Russian Academy of Sciences/Earth Science Section—English transl.)*, Vol. 337, N4, 517–520.

Astafyeva N. M., Pokrovskaya I. V., and Sharkov E. A. (1994b). Hierarchical structure of global tropical cyclogenesis. *Earth Research from Space*, N2, 14–23 [in Russian].

Astafyeva N. M., Pokrovskaya I. V., and Sharkov E. A. (1994c). Global tropical cyclogenesis scaling structure with wavelets. *Annales Geophysicae*, Suppl. II to Vol. 12, Part II, C494.

Astafyeva N. M., Raev M. D., and Sharkov E. A. (2006a) Global radiothermal fields for the atmosphere–ocean system: The method of construction using microwave satellite instruments. *Earth Research from Space*, N4, pp. 64–69 [in Russian].

Astafyeva N. M., Raev M. D., and Sharkov E. A. (2006b) The Earth's portrait from space: Global radiothermal field. *Nature*, N9, pp. 17–27 [in Russian].

Astafyeva N. M., Raev M. D., and Sharkov E. A. (2008) Interannual changes of a radiothermal field of the Earth according to microwave satellite monitoring. *Earth Research from Space*, N5, pp. 9–15 [in Russian].

Atlas (1977) *Atlas of Oceans*. M. GUNO USSR, 340 pp. [in Russian].

Avanesov G. A., Galeev A. A., Zhukov B. S., Ziman Ya. L., and Mitrofanov I. G. (1992). Ecos-A Project: Space research and modeling of global environmental and climatic processes and calamities. *Earth Research frm Space*, N2, 3–14 [in Russian].

Avduevskii V. S., Grishin S. D., Uspenskii V. K., Astashkin A. A., and Saulskii V. K. (1983). On the effectiveness and principal parameters of satellites for continuous observation of spontaneous natural phenomena. *Earth Research from Spaee*, N2, 117–124 [in Russian].

Bak P., Tang C., and Wiesenfeld K. (1987). Self-organised criticality: An explanation of $1/f$ noise. *Phys. Rev. Lett.*, Vol. 59, N5, 381–384.

Balebanov V. M., Moiseev S. S., Sharkov E. A., Lupian E. A., Kuzmin A. K., Chikov K. N., Smirnov N. K., Zabychnyi A. I., Kalmykov A. I., and Zymbal V. N. (1996). "Helix" Project: Space monitoring of the ocean–troposphere–upper atmosphere system under large-scale hazard situations. *31st Scientific Assembly of COSPAR* (Abstract). University of Birmingham, U.K., p. 42.

Balebanov V. M., Moiseev S. S., Sharkov E. A., Lupian E. A., Kalmykov A. I., Zabyshnyi A. I., Kuzmin A. K., Smirnov N. K., Tsymbal V. N., and Chikov K. N. (1997). The "Helix" Project: Space monitoring of the ocean–troposphere–upper atmosphere system under large-scale crisis conditions. *Earth Obs. Rem. Sens.*, Vol. 14, N5, 823–835 [in Russian].

Barry L., Craig G. C.,, and Thuburn J. (2002) Poleward heat transport by the atmospheric heat engine. *Nature*, Vol. 415, N6873, 774–777.

Baryshnikova Yu. S., Lupian E. A., and Sharkov E. A. (1988). *Determination of Dynamical Characteristics of Wind Flows from Temperature Variations at Upper Cloud Levels* (Preprint No. 1350). Space Research Institute, Russian Academy of Sciences, Moscow, 25 pp. [in Russian].

Baryshnikova Yu. S., Zaslavskii G. M., Lupian E. A., Moiseev S. S., and Sharkov E. A. (1989a). Fractal dimensionality of cloudiness IR images and turbulent atmosphere properties. *Earth Research from Space*, N1, 17–26 [in Russian].

Baryshnikova Yu. S., Zaslavsky G. M., Lupian E. A., Moiseev S. S., and Sharkov E. A. (1989b). Fractal analysis of the pre-hurricane atmosphere from satellite data. *Adv. Space Res.*, Vol. 9, N7, 405–408.

Battrick B. (Ed.) (1999). *Meteosat Second Generation: Satellite Development* (ESA BRM153). European Space Agency, Noordwijk, The Netherlands, 55 pp.

Beletsky A. B., Mikhalev A. V., Chernigovskaya M. A., Sharkov E. A., and Pokrovskaya I. V. (2010a). The possibility of tropical cyclone activity manifesting itself in atmospheric airglow. *Earth Research from Space*, N4, 41–49 [in Russian].

Beletsky A. B., Mikhalev A. V., Tatarnikov A. V., Tashchilin M. A., Chernigovskaya M. A., Sharkov E. A., Pokrovskaya I. V., and Xu Jiyao (2010b). Investigation of nightglow upper-atmosphere parameter variations connected with disturbances in the troposphere and stratosphere of the Earth. *Modern Problems of Earth Remote Sensing from Space*, Vol. 7. N1, 75–82 [in Russian].

Bendat J. S. and Piersol A. G. (1986). *Random Data: Analysis and Measurement Procedures*. John Wiley & Sons, New York, p. 350.

Bengtsson L. (1997). A numerical simulation of anthropological climate change. *Ambio*, Vol. 26, N1, 58–65.

Bengtsson L., Botzet M., and Esch M. (1995). Hurricane-type vortices in a general circulation model. *Tellus*, Vol. 47A, N2, 175–196.

Bengtsson L., Botzet M., and Esch M. (1996). Will greenhouse gas-induced warming over the next 50 years lead to higher frequency and greater intensity of hurricanes? *Tellus*, Vol. 48A, N1, 57–73.

Bengtsson L., Botzet M., and Esch M. (1997). Reply to comments on: "Will greenhouse gas-induced warming over the next 50 years lead to higher frequency and greater intensity of hurricanes?" by C. W. Landsea. *Tellus*, Vol. 49A, N5, 624–625.

Beste D. C. (1978). Design of satellite constellations for optimal continuous coverage. *IEEE Trans. Aerospace Electronic Systems*, Vol. 14, N3, 466–473.

Betout P., Burridge D., and Werner Ch. (1989). *ALADIN: Atmospheric Laser Doppler Instrument* (Doppler Lidar Working Group Report ESA SP-1112). European Space Agency, Noordwijk, The Netherlands, 45 pp.

Beven J. (1994). *Tropical Cyclone Name Lists*. National Hurricane Center (*http://www.nws.fsu.edu/tropname.html*).

Bevilacqua R.M., Kriebel D. L., Pails T. A., Aelling C. P., Siskind D. E., Daehler M., Olivero J. J., Puliafito S. E., Hartmann G. K., Kampfer N. *et al.* (1996). MAS measurements of the latitudinal distribution of water vapor and ozone in the mesosphere and lower thermosphere. *Geophys. Res. Lett.*, Vol. 23, N17, 2317–2320.

Bharucha-Reid A. T. (1960). *Elements of the Theory of Markov Processes and Their Applications.* McGraw-Hill, New York, 360 pp.

Bielli S. and Roux F. (1999). Initialization of a cloud-resolving model with airborne Doppler radar observations of an oceanic tropical convective system. *Monthly Weather Rev.*, Vol. 127, N6, 1038–1055.

Bjerkesjo L., Mikhailov V. V., Selivanov A. S., and Stern M. (1996). Reception and processing of the Russian spacecraft "Resurs-01" data in Sweden. *Russian Space Bulletin*, Vol. 3, N4, 18–19.

Blier W. and Ma Q. (1997). A Mediterranean sea hurricane? *22nd Conf. on Hurricane and Tropical Meteorology, 19–23 May 1997, Ft. Collins, Colorado.* American Meteorological Society, Boston, pp. 592–595.

Boccaletti S., Grabogi C., Lai Y. C., Mancini H., and Maza D. (2000). The control of chaos: Theory and applications. *Physics Reports*, Vol. 329, N3, 103–197.

Bochnicek J., Hejda P., Buch V., and Pycha J. (1999). Possible geomagnetic activity effects on weather. *Annales Geophysicae*, Vol. 17, 925–932.

Bondur V. G., Kaleri A. Yu., and Lazarev A. I. (1997). *Observation of the Earth from Space: The Orbiting Station "Mir" in March–August 1992.* Hydrometeoizdat, St Petersburg, 92 pp.

Bouman C. and Newell A. C. (1998). Natural patterns and wavelets. *Review of Modern Physics*, Vol. 70, N1, 289–301.

Branover H., Eidelman A., Golbraich E., and Moiseev S. (1999). Turbulence and structures. *Chaos, Fluctuations and Helical Self-organization in Nature and the Laboratory.* Academic Press, Orlando, FL, 296 pp.

Brindly J., Kapitaniak T., and Barcilon A. (1992). Chaos and noisy periodicity in forced ocean–atmosphere models. *Physics Lett. A.*, Vol. 167, N2, 179–184.

Broccoli A. J. and Manabe S. (1990). Can existing climate models be used to study anthropogenic changes in tropical cyclone climate? *Geophys. Res. Lett.*, Vol. 17, N11, 1917–1920.

Brown L. B., Gerrard A. J., Meriwether J. W., and Makela J. J. (2004) All-sky imaging observations of mesospheric fronts in OI 557.7 nm and broadband OH airglow emissions: Analysis of frontal structure, atmospheric background conditions, and potential sourcing mechanisms. *J. Geophys. Res.*, Vol. 109, No. D19, D19104/1–D19104/19.

Bruzzi S. (1995). The Envisat mission in the context of present and future Earth observation programs of the European Space Agency. *EARSeL Newsletter*, N24, 2–12.

Bryant E. (1993). *Natural Hazards.* Cambridge University Press, 294 pp.

Burgel van J. L., Le Marshall J., and Lynch M. J. (1994). Upper temperature anomalies near Australian tropical cyclones using satellite microwave data. *Proceedings of PORSEC '94, Melbourne, Australia, 1–4 March, 1994* (edited by J. Le Marschall and J. D. Lasper). Bureau of Meteorology, Melbourne, pp. 219–225.

Burluzkiy R. F. (1987). On the processes involved in tropical cyclone development. *Tropical Meteorology, Trudy 3 Meshdun. Simp., Jalta, 1985.* Gidrometeoizdat, Leningrad, pp. 58–63 [in Russian].

Cahalan R. F. and Joseph J. H. (1989). Fractal statistic of cloud fields. *Monthly Weather Rev.*, Vol. 117, N3, 261–272.

Carlowicz M. (1995). Winds of change stir up near-record hurricane season. *EOS, Trans. Amer. Geophys. Union*. Vol. 76, N50, 513–514.

Cerveny R. S. and Balling R. C. (1998). Weekly cycles of air pollutants, precipitation and tropical cyclones in the coastal NW Atlantic region. *Nature*, Vol. 394, N6693, 561–563.

Chahine M. T., Haskins R., and Fetzer E. (1997). Observation of the recycling rate of moisture in the atmosphere: 1988–1994. *GEWEX News*, Vol. 7, N3, 1, 3–4.

Chan J. C. L. (1985). Tropical cyclone activity in the Northwest Pacific in relation to the El Niño/Southern Oscillation Phenomenon. *Monthly Weather Rev.*, Vol. 113, N4, 599–606.

Chan J. C. L., Shi J., and Lam C. (1998). Seasonal forecasting of tropical cyclone activity over the Western North Pacific and the South China Sea. *Weather and Forecasting*, Vol. 13, N4, 997–1004.

Chang A. T. C., Chiu L. S., Liu G. R., and Wang K. H. (1995). Analyses of 1994 typhoons in the Taiwan region using satellite data. *COSPAR Colloquium Space Remote Sensing of Subtropical Oceans, Taipei, Taiwan*, Taipu, pp. 13B2-26–13B2-18.

Chang P., Ji L., and Flugel M. (1996). Chaotic dynamics versus stochastic processes in El Niño–Southern Oscillation in coupled ocean–atmospheric models. *Physica D*, Vol. 98, N6, 301–320.

Charney J. I. and Eliassen A. (1964). On the growth of the hurricane depression. *J. Atmos. Sci.*, Vol. 21, N2, 68–75.

Chen G. and Dong X. (1996). From chaos to order. *Perspectives and Methodologies*. World Scientific, Singapore, 301 pp.

Chen J. M., Elsberry R. L., Boothe M. A., and Carr L. E. (1999). A simple statistical–synoptic track prediction technique for Western North Pacific tropical cyclones. *Monthly Weather Rev.*, Vol. 127, N1, 89–102.

Chen S. S. and Houze R. A. (1997a). Diurnal variation and life-cycle of deep convective systems over the tropical Pacific warm pool. *Quart. J. Roy. Meteorol. Soc.*, Vol. 123(B), 357–388.

Chen S. S. and Houze R. A. (1997b). Interannual variability of deep convection over the tropical warm pool. *J. Geophys. Res.*, Vol. 102, ND22, 25783–25795.

Chen S. S., Houze R. A., and Mapes B. E. (1996). Multiscale variability of deep convection in relation to large-scale circulation in TOGA COARE. *J. Atmos. Sci.*, Vol. 53, N10, 1380–1409.

Cheng M. D. (1989a). Effects of downdrafts and mesoscale convective organization on the heat and moisture budgets of tropical cloud clusters, Part I: A diagnostic cumulus ensemble model. *J. Atmos. Sci.*, Vol. 46, N11, 1517–1538.

Cheng M. D. (1989b). Effects of downdrafts and mesoscale convective organization on the heat and moisture budgets of tropical cloud clusters, Part II: Effects of convective-scale downdrafts. *J. Atmos. Sci.*, Vol. 46, N11, 1540–1561.

Cherney I. V. and Raizer V. Yu. (1998). *Passive Microwave Remote Sensing of Oceans*. John Wiley & Sons/Praxis, Chichester, U.K., 195 pp.

Chernigovskaya M. A., Sharkov E. A., Kurkin V. I., Orlov I. I., and Pokrovskaya I. V. (2008) Short-term temporal variations of ionospheric parameters in Siberia and the Far East. *Earth Research from Space*, Vol. 6, 17–24 [in Russian].

Chernigovskaya M. A., Kurkin V. I., Orlov I. I., Sharkov E. A., and Pokrovskaya I. V. (2010) Study of coupling ionospheric parameter short-period variations in the northeastern region of Russia with the manifestation of tropical cyclones. *Earth Research from Space*, N5, 32–41 [in Russian].

Cheung N. K. W. and Kyle W. J. (2000). Trends in seasonal forecasting of tropical cyclone activity. *Australian Met. Mag.*, Vol. 49, N3, 201–221.

Chigirinskaya Y., Scherzer D., Lovejoy S., Lazarev A., and Ordanovich A. (1994). Unified multifractal atmospheric dynamics tested in the tropics, Part I: Horizontal scaling and self criticality. *Nonlinear Processes in Geophysics*, Vol. 1, N2/3, 105–114.

Chizhevskii A. L. (1976). *Terrestrial Echo of Solar Storms*. Mysl, Moscow, 366 pp.

Chong M. and Bousquet O. (1999). A mesovortex within a near-equatorial mesoscale convective system during TOGA COARE. *Monthly Weather Rev.*, Vol. 127, N6, 1145–1156.

Choudhury J. R. (1994). *The Impact of Natural Disasters on Urban Infrastructure* (Report of Technical Conference on Tropical Urban Climates, WMO/TD N647). World Meteorological Organization, Geneva, Switzerland, pp. 139–148.

Chu P. S. and Wang J. (1998). Tropical cyclone occurrence in the vicinity of Hawaii: Are the differences between El Niño and non-El Niño years significant? *J. Climate*, Vol. 10, N10, 2683–2689.

Chui Ch. K. (1992). *An Introduction to Wavelets*. Academic Press, New York, 266 pp.

Chylek P. and Lesins G. (2008). Multidecadal variability of Atlantic hurricane activity: 1851–2007. *J. Geophys. Res.*, Vol. 113, ND22106, doi: 10.1029/2008JD010036.

Clark G. B. (1983). Atlantic hurricane season of 1982. *Monthly Weather Rev.*, Vol. 111, N5, 1071–1079.

Cohen T. J. and Sweetser E. I. (1975). The "spectra" of the solar cycle and of data for Atlantic tropical cyclones. *Nature*, Vol. 256, N5515, 295–296.

Coniglio A. (1987). Scaling approach to multifractality. *Philosophical Magazine B*, Vol. 56, N6, 785–790.

Cox D. R. and Lewis P. A. W. (1966). *The Statistical Analysis of Series of Events*. John Wiley & Sons, New York, 310 pp.

Cox D. R. and Oakes D. (1984). *Analysis of Survival Data*. Chapman & Hall, London, 250 pp.

Cressie N. A. C. (1993). *Statistics for Spatial Data*. John Wiley & Sons, New York, 900 pp.

Danilov A. D., Kazimirovskii E. S., Vergasov G. V., and Khachikyan G. Y. (1987). *Meteorological Effects in the Ionosphere*. Hydrometeoizdat, Leningrad, 268 pp [in Russian].

Danson F. M. and Plummer S. E. (Eds.) (1995). *Advances in Environmental Remote Sensing*. John Wiley & Sons, Chichester, U.K., 184 pp.

DeAngelis D. (1984). Hurricane Alley. *Mar. Weather Log.*, Vol. 28, N4, 247–255.

DeMaria M. and Kaplan J. (1999). An updated statistical hurricane intensity prediction scheme (SHIPS) for the Atlantic and Eastern North Pacific basins. *Weather and Forecasting*, Vol. 14, N3, 326–337.

Desbois M. (1995). TROPIQUES: A small satellite for studying water and energy cycles in the intertropical band, and atmosphere–ocean interactions. *COSPAR Colloquium "Space Remote Sensing of Subtropical Oceans", Taipei, Taiwan*, pp. 1-15A4-1–1-15A4-5.

Dessler A. E., Hintsa E. J., Weinstock E. M., Anderson J. G., and Chan K. R. (1995). Mechanisms controlling water vapour in the lower stratosphere: "A tale of two stratospheres". *J. Geophys. Res.*, Vol. 100, ND11, 23167–23172.

Diaz H. F. and Margraf V. (Eds.) (1993). *El Niño: Historical and Paleoclimatic Aspects of the Southern Oscillation*. Cambridge University Press, Cambridge, U.K., 490 pp.

Diaz H. F. and Pulwarty R. S. (Eds.) (1997). *Hurricanes: Climate and Socio-economic Impacts*. Springer-Verlag, Berlin, 292 pp.

Dillon C. P. and Andrews M. J. (1998). *1997 Annual Tropical Cyclone Report*. U.S. Naval Pacific Meteorology and Oceanography Center West/Joint Typhoon Warning Center, Guam (*http://metoc.npmoc.navy.mil/jtwc*).

Dobryshman E. M. (1994). Some statistical parameters and characteristics of typhoons. *Russian Meteorology and Hydrology*, N11, 83–98.

Dobryshman E. M. (1995). Non-stationary model of the eye of a typhoon. *Russian Meteorology and Hydrology*, N12, 5–19.

Dolgopolov G. A., Danilkin A. P., Diomin A. V., Dolgikh V. N., Ivanov K. V., Baskov S. M., and Zemskov E. F. (1996). Earth remote sensing spacecraft future developments. *Russian Space Bulletin*, Vol. 3, N3, 8–9.

Donoso C. I., LeMehaute B., and Long R. B. (1987). Data base of maximum sea states during hurricanes. *J. Waterway Port. Coast. and Ocean Eng.*, Vol. 113, N4, 311–326.

Doviak R. J. and Zrnič D. S. (1984). *Doppler Radar and Weather Observation*. Academic Press, Orlando, FL, 458 pp.

Duane G. S. (1997). Synchronized chaos in extended systems and meteorological teleconnections. *Physical Review E*, Vol. 56, N6, 6475–6493.

Dubois J. (1998). *Non-Linear Dynamics in Geophysics*. John Wiley & Sons/Praxis, Chichester, U.K. 261 pp.

Dvorak V. F. (1975). Tropical cyclone intensity analysis and forecasting from satellite imagery. *Monthly Weather Rev.*, Vol. 103, N5, 420–430.

Dvorak V. F. (1984). *Tropical Cyclone Intensity Analysis Using Satellite Data* (NOAA Tech. Rep. NESDIS 11). U.S. Department of Commerce, Washington, DC, 47 pp.

Elsberry R. (1987). Observation and analysis of tropical cyclones. *A Global View of Tropical Cyclones* (edited by R. L. Elsberry). Office of Naval Research, Monterey, CA, 1–12.

Elsner J. B. (1997). A multi-season prediction algorithm for Atlantic hurricanes. *22nd Conf. on Hurricane and Tropical Meteorology, 19–23 May, 1997, Ft. Collins, Colorado*. American Meteorological Society, Boston, pp. 590–591.

Elsner J. B. and Kara A. B. (1999). *Hurricanes of the North Atlantic: Climate and Society*. Oxford University Press, New York, 488 pp.

Elsner J. B., Kara A. B., and Owens M. A. (1999). Fluctuations in North Atlantic hurricane frequency. *J. Climate*, Vol. 12, N2, 427–437.

Emanuel K. A. (1986). An air–sea interaction theory for tropical cyclones, Part I: Steady-state maintenance. *J. Atmos. Sci.*, Vol. 43, N6, 585–604.

Emanuel K. A. (1987). The dependence of hurricane intensity on climate. *Nature*, Vol. 326, N6112, 483–485.

Emanuel K. A. (1991). The theory of hurricanes. *Annual Rev. Fluid. Mech.*, Vol. 23, 179–196.

Emanuel K. A. (1996). *Atmospheric Convection*. Oxford University Press, 580 pp.

Emanuel K. A. (1999). Thermodynamic control of hurricane intensity. *Nature*, Vol. 401, N6754, 665–669.

Emiliani C. (1992). *Planet Earth: Cosmology, Geology and the Evolution of Life and Environment*. Cambridge University Press, Cambridge, U.K., 717 pp.

Engle M. (1987). Space station hurricane alert. *Spaceflight*, Vol. 29, N3, 115–117.

Ermakov D. M., Raev M. D., Suslov A. I., and Sharkov E. A. (2007). Electronic long-standing database for the global radiothermal field of the Earth in context of multiscale investigation of the atmosphere–ocean system. *Earth Research from Space*, N1, 7–13 [in Russian].

ESA (1996a). *Atmospheric Dynamic Mission* (ESA SP–1196(4)). European Space Agency, Noordwijk, The Netherlands, 70 pp.

ESA (1996b) *Atmospheric Profiling Mission* (ESA SP–1196(7)). European Space Agency, Noordwijk, The Netherlands, 58 pp.

ESA (1996c). *Precipitation Mission* (ESA SP–1196(8)). European Space Agency, Noordwijk, The Netherlands, 82 pp.

ESA (1997). *Meteosat Collection: The Weather Machine, N4* (ESA SR–1213). European Space Agency, Noordwijk, The Netherlands (CD-ROM).

Erokhin N. S., Moiseev S. S., Lupian E. A., Lazarev A. A., Pankov V. M., Sharkov E. A., Pokrovskaya I. V., Shkurkin Yu. G., Vinogradov B. V., Shpuntov M. A. *et al.* (1995). Aerospace block of monitoring system for terrestrial natural and technogenic disasters. *Turkish Journal of Physics*, Vol. 19, N8, 1087–1092.

Evans J. L. (1992). Comment on "Can existing climate models be used to study anthropogenic changes in tropical cyclone climate?" *Geophys. Res. Lett.*, Vol. 19, 1523–1524.

Fagundes P. R., Takahashi H., Sahai Y., and Gobbi D. (1995) Observations of gravity waves from multispectral mesospheric nightglow emissions observed at 23°S. *J. Atmos. Terr. Phys.*, Vol. 57, N4, 395–405.

Falkovich A. I. (1979). *Dynamics and Energetics of the Intertropical Convergence Zone*. Gidrometeoizdat, Leningrad, 274 pp. [in Russian].

Fedorov A. V., Brierley C. M., and Emanuel K. (2010) Tropical cyclones and permanent El Niño in the early Pliocene epoch. *Nature*, Vol. 463, N7284, 1066–1070.

Feller W. (1968). *An Introduction to Probability Theory and Its Applications*, Vol. I, II, Third Edition. John Wiley & Sons, New York, 506 pp.

Fernandez-Partages J. and Diaz H. F. (1996). Atlantic hurricanes in the second half of the nineteenth century. *Bull. Amer. Meteorol Soc.*, Vol. 77, N11, 2899–2906.

Fischer S., Dornbusch R., and Schmalensee R. (1988). *Economics*. McGraw-Hill, New York, 720 pp.

Fitzpatrick P. J. (1997). Understanding and forecasting tropical cyclone intensity change with the typhoon intensity prediction scheme (TIPS). *Weather and Forecasting*, Vol. 12, N4, 826–846.

Frank W. (1987). Tropical cyclone formation. *A Global View of Tropical Cyclones* (edited by R. L. Elsberry). Office of Naval Research, Monterey, CA, pp. 53–90.

Frankignoul C. and Hasselmann K. (1977). Stochastic climate models, Part 2: Application to sea-surface temperature anomalies and thermocline variability. *Tellus*, Vol. 29, N4, 289–305.

Frisch V., Sulem P. I., and Nellin M. (1978). A simple dynamical model of intermittent fully developed turbulence. *J. Fluid Mech.*, Vol. 87, N4, 719–736.

Gabis I. P. and Troshichev O. A. (2000). Influence of short-term changes in solar activity on baric field perturbations in the stratosphere and troposphere. *J. Atmos. Solar-Terr. Physics*, Vol. 62, N9, 725–735.

Gamache J. F. and Houze R. A. (1983). Water budget of a mesoscale convective system in the tropics. *J. Atmos. Sciences*, Vol. 40, N7, 1835–1850.

Gaspard P. (1997). Chaos and hydrodynamics. *Physica A*, Vol. 240, N1–2, 54–67.

GCOS (Global Climate Observing System) (1995). *GCOS Plan for Space-Based Observations* (Version 1.0, GCOS-15, WMO/TD-N684). World Meteorological Organization, Geneva, Switzerland., 50 pp.

Gdalevich G. L., Pokrovskaya I. V., and Sharkov E. A. (1994). Magnetospheric processes as stimulators of global tropical cyclogenesis. *Kosmicheskie issledovania* (*Space Research—English transl.*), Vol. 32, N2, 108–111.

Gentry R. C., Rodgers E. B., Steranka J., and Shenk W. E. (1980). Predicting tropical cyclone intensity using satellite measured equivalent blackbody temperature of cloud tops. *Monthly Weather Rev.*, Vol. 108, N4, 445–455.

Gershuni G. Z. and Zhukhovitsky E. M. (1972). *Convective Stability of Incompressible Fluid.* Nauka, Moscow, 366 pp. [in Russian]. English translation by Keterpress, Jerusalem, 1976.

Ghil M., Bensi R., and Parisi G. (Eds.) (1985). *Turbulence and Predictability in Geophysical Fluids Dynamics and Climate Dynamics.* North–Holland, Amsterdam, 205 pp.

Ghil M., Kimoto M., and Neelin J. D. (1991). Nonlinear dynamics and predictability in the atmospheric sciences. *Rev. Geophysics*, Suppl., April, 46–55.

Gilmore R. (1998). Topological analysis of chaotic dynamical systems. *Reviews of Modern Physics*, Vol. 70, N4, 1455–1529.

Glantz M., Katz R., and Nicholls N. (Edss) (1991). *Teleconnections Linking Worldwide Climate Anomalies.* Cambridge University Press, Cambridge, U.K., 545 pp.

Glynn P. W. (Ed.) (1990). *Global Ecological Consequences of the 1982–83 El Niño–Southern Oscillation.* Elsevier, Amsterdam, 350 pp.

Gnedenko B. W. and Kovalenko I. N. (1987). *An Introduction to Mass Service Theory.* Nauka, Moscow, 334 pp. [in Russian].

Gohara K. and Okuyama A. (1999). Fractal transition: Hierarchical structure and noise effect. *Fractals*, Vol. 7, N3, 313–326.

Goldsmith P. and Readings C. J. (1994). The plans of the European Space Agency for Earth observation. *Adv. Space Res.*, Vol. 14, N1, 11–16.

Golitsyn G. S. (1997) Statistics and energy of tropical cyclones. *Doklady Akad. Nauk (Trans. of Russian Academy of Sciences/Earth Science Section—English transl.)*, Vol. 354, N4, 535–538.

Golitsyn G. S. (2008) Polar lows and tropical hurricanes: Their energy and sizes and a quantitative criterion for their generation. *Fizika atmosphery i okeana (Izvestya, Atmospheric and Oceanic Physics—English transl.)*, Vol. 44, N5, 579–590.

Golitsyn G. S., Pisarenko V. F., Rodkin M. V., and Yaroshevich M. I. (1999). Statistical characteristics of tropical cyclone parameters and the risk assessment problem. *Izvestiya, Atmospheric and Oceanic Physics*, Vol. 35, N6, 663–670.

Grassia P. S. (2000). Delay, feedback and quenching in financial markets. *Europ. Phys. J. B*, Vol. 17, N2, 347–362.

Gray W. M. (1979) Hurricanes: Their formation, structure and likely role in the tropical circulation. In: D. B. Shaw (Ed.), *Meteorology over the Tropical Oceans.* Royal Meteorological Society, Bracknell, U.K., pp. 155–218.

Gray W. M. (1984a). Atlantic seasonal hurricane frequency, Part I: El Niño and 30 mb quasibiennial oscillation influences. *Monthly Weather Rev.*, Vol. 112, N9, 1649–1668.

Gray W. M. (1984b). Atlantic seasonal hurricane frequency, Part II: Forecasting its variability. *Monthly Weather Rev.*, Vol. 112, N9, 1669–1683.

Gray W. M. (1993). Seasonal forecasting. In *Global Guide to Tropical Cyclone Forecasting* (WMO Technical Document N560, Tropical Cyclone Program Report N31). World Meteorological Organization, Geneva, Switzerland, 5.1–5.21.

Gray W. M. (1997) A personal (and perhaps unpopular) view of tropical meteorology over the last 40 years and future outlook. *22nd Conf. on Hurricane and Tropical Meteorology, May 19–23, 1997, Ft. Collins, Colorado.* American Meteorological Society, Boston, pp. 19–24.

Gray W. M., Sheaffer J. D., and Landsea Ch. W. (1997). Climate trends associated with multidecadal variability of Atlantic hurricane activity. *Hurricanes. Climate and Socioeconomic Impacts* (edited by H. F. Diaz and R. S. Pulwarty). Springer-Verlag, Berlin, pp. 15–53.

Greenstone R. and King M. D. (Eds.) (1999). *EOS Science Plan: Executive Summary* (NP-1998-12-070-GSFC). NASA Goddard Space Flight Center, Greenbelt, MD, 64 pp.

Gregg W. (Ed.) (2007). *Ocean-Color Data Merging* (IOCCG Report N6). International Ocean-Color Coordinating Group, Dartmouth, Canada, 68 pp.

GTECCA (1996). *Global Tropical and Extratropical Cyclone Climatic Atlas 2.0*, NCDC/NOAA, Asheville, NC (CD-ROM).

Gunther E. B. and Cross R. L. (1985). Eastern North Pacific tropical cyclones, 1984. *Mar. Weather Log*, Vol. 29, N2, 63–71.

Gurney R. J., Foster J. L., and Parkinson C. L. (1993). *Atlas of Satellite Observations Related to Global Change*. Cambridge University Press, Cambridge, U.K., 461 pp.

Haarsma R. J., Mitchell J. F. B., and Senior C. A. (1992). Tropical disturbances in a GCM. *Climate Dyn.*, Vol. 8, N3, 247–257.

Haken H. (1978). *Synergetics: An Introduction*, Second Edition, Vol. 1. Springer-Verlag, Berlin, 420 pp.

Halverson J. B., Ferrier B. S., Rickenbach T. M., Simpson J., and Tao Wei-kuo (1999). An ensemble of convective systems on 11 February 1993 during TOGA COARE: Morphology, rainfall characteristics, and anvil cloud interactions. *Monthly Weather Rev.*, Vol. 127, N6, 1208–1228.

Hansen J., Sato M., and Ruedy R. (1997). Radiative forcing and climate response. *J. Geophys. Res.*, Vol. 102, ND6, 6831–6864.

Hanstrum B. N., Reader G., and Bate P. W. (1999). The South Pacific and southern Indian Ocean tropical cyclone season 1996–97. *Australian Met. Mag.*, Vol. 48, N3, 197–210.

Harr P. A. and Elsberry R. L. (2000). Extratropical transition of tropical cyclones over the Western North Pacific, Part I: Evolution of structural characteristics during the transition process. *Monthly Weather Rev.*, Vol. 128, N8, 2613–2633.

Harr P. A., Elsberry R. L., and Hogan T. F. (2000). Extratropical transition of tropical cyclones over the Western North Pacific, Part II: The impact of midlatitude circulation characteristics. *Monthly Weather Rev.*, Vol. 128, N8, 2634–2653.

Harries J. E. (1995). *Earthwatch: The Climate from Space*. Wiley/Praxis, Chichester, U.K., 216 pp.

Harujama Y. (1994). Progress of Japan's earth observation satellites. *Adv. Space Research*, Vol. 14, N1, 21–24.

Harvey B. (1995). *The New Russian Space Programme*. Wiley/Praxis, Chichester, U.K.

Hasselmann K. (1976). Stochastic climate models, Part I: Theory. *Tellus*, Vol. 28, N6, 473–485.

Hasselmann K. (1979). On the signal-to-noise problem in atmospheric response studies. *Meteorology of Tropical Oceans* (edited by D. B. Shaw). Royal Meteorological Society, Bracknell, U.K., 251–259.

Hassim M. E. E. and Walsh K. J. E. (2008). Tropical cyclone trends in the Australian region. *Geochem. Geophys. Geosyst.* G^3, Vol. 8, N7, 1–17, Q07V07, doi: 10129/2007GC001804.

Hastings P. A. (1990). Southern oscillation influences on tropical cyclone attivity in the Australian/south-west Pacific region. *Int. J. Climatol.*, Vol. 3, N5, 291–298.

Hawkins H. F. and Rubsam D. T. (1967). Hurricane Inez: A classic "micro-hurricane". *Mar. Weather Log*, Vol. 11, 157–160.

Henderson-Sellers A., Zhang H., Berz G., Emanuel K., Gray W., Landsea C., Holland G., Lighthill J., Shieh S. L., Webster P., and McGuffie K. (1998). Tropical cyclones and global climate change: A post-IPCC assessment. *Bull. Amer. Meteorol. Soc.*, Vol. 79, 19–38.

Hobgood J. S. (1986). The effects of climatological and persistence variables on the intensities of tropical cyclones over the Eastern North Pacific Ocean. *Weather and Forecasting*, Vol. 13, N3, 632–639.

Hocke K. and Schlegel K. (1996). A review of atmospheric gravity waves and travelling ionospheric disturbances: 1982–1995. *Ann. Geophys*, Vol. 14, 917–940.

Hoffman R. N. (2004). Controlling hurricanes. *Scientific American*, October, N10, 68–75.

Holdom B. (1998). From turbulence to financial time series. *Physica A*, Vol. 254, N3MM4, 569–576.

Holland G. J. (1997). The maximum potential intensity of tropical cyclones. *J. Atmos. Sci.*, Vol. 54, N10, 2519–2541.

Holliday C. R. and Thompson A. H. (1986). An unusual near-equatorial typhoon. *Monthly Weather Rev.*, Vol. 114, N12, 2674–2677.

Holton J. R. and Alexander M. J. (1999) Gravity waves in the mesosphere generated by tropospheric convection. *Tellus*, Ser. A and B, Vol. 51, N1, 45–58.

Holton J. R., Haynes P. H., McIntyre M. E., Douglass A. R., Rood R. B., and Pfister L. (1995). Stratosphere–troposphere exchange. *Rev. Geophys.*, Vol. 33, 403–439.

Hoskins B. and Pearce R. (Eds.) (1983). *Large-scale Dynamical Processes in the Atmosphere*. Academic Press, London, 421 pp.

Houze R. A. (1993). *Cloud Dynamics*. Academic Press, London, 573 pp.

Huffman G. J., Adler R. F., Arkin P., Chang A., Ferraro R., Gruber A., Janowiak J., McNab A., Rudolf B., and Schneider U. (1997). The global precipitation climatology project (GPCP) combined precipitation data set. *Bull. Amer. Meteorol. Soc.*, Vol. 78, N1, 5–20.

Hunt J. C. R. (1995). Searching for certainty. *Nature*, Vol. 374, N6517, 23.

Hussain A. K. M. F. (1986). Coherent structures and turbulence. *J. Fluid Mech.*, Vol. 173, 303–356.

Inoue T. (1997). Contrast of 87/88 Indian summer monsoon observed by split window measurements. *Adv. Space Res.*, Vol. 19, N3, 447–455.

IPCC (Intergovernmental Panel of Climate Change) (1996) *Climate Change 1995: The Science of Climate Change* (edited by J. T. Houghton, F. G. Meira Filho, B. A. Callander, N. Harris, A. Kattenberg, and K. Maskell). Cambridge University Press, Cambridge, U..K., 572 pp.

Irisova T. A. (1974). Cloud features in cyclogenesis over the Eastern Mediterranean. *Trans. of Hydrometeocentre*, Vol. 148, 42–52. [in Russian].

Irwin R. P. and Davis R. E. (1999). The relationship between the Southern Oscillation index and tropical cyclone tracks in the eastern North Pacific. *Geophys. Res. Lett.*, Vol. 26, N15, 2251–2254.

Ivanitskii G. R., Medvinskii A. B., Deev A. A., and Tsyganov M. A. (1998). From Maxwell's demon to the self-organization of mass transfer processes in living systems. *Physics-Uspekhi*, Vol. 41, N11, 115–1126.

Isakov M. N. (1992). Satellite measurement of negentropy influx for ecological studies. *Sov. J. Remote Sensing*, Vol. 9, N4, 541–561.

Isakov M. N. (1997). Self-organization and information for planets and ecosystems. *Physics-Uspekhi*, Vol. 40, N10, 1035–1042.

Isakov M. N. and Zhukov B. S. (1993). Global satellite monitoring of space-time changes in biota. *Sov. J. Remote Sensing*, Vol. 10, N6, 1016–1036.

Jacobs G. A., Hulburt H. E., Kindle J. C., Metzger E. J., Mitchell J. L., Teague W. J., and Wallcraft A. J. (1994). Decade-scale trans-Pacific propagation and warning effects of an El Niño anomaly. *Nature*, Vol. 370, N6488, 360–363.

Jensen H. J. (1998). *Self-Organized Criticality: Emergent Complex Behavior in Physical and Biological Systems*. Cambridge University Press, New York, 153 pp.

Jin F. F., Neelin J. D., and Ghil M. (1994). El Niño on the devil's staircase: Annual subharmonic steps to chaos. *Science*, Vol. 264, N4, 70–72.

Jin F., Neelin J. D., and Ghil M. (1996). El Niño–Southern Oscillation and the annual cycle: Subharmonic frequency-locking and aperiodicity. *Physica D*, Vol. 98, N6, 442–465.

Johannessen J. A., Digranes G., Espedal H., Johannessen O. M., Samuel P., Browne D., and Vachon P. (1994). *SAR Ocean Feature Catalogue* (ESA-SP-1174). European Space Agency, Noordwijk, The Netherlands, 106 pp.

Jorgensen D. P. (1984a). Mesoscale and convective-scale characteristics of mature hurricanes, Part I: General observations by research aircraft. *J. Atmos. Sci.*, Vol. 41, N8, 1268–1285.

Jorgensen D. P. (1984b). Mesoscale and convective-scale characteristics of mature hurricanes, Part II: Inner core structure of Hurricane Allen (1980). *J. Atmos. Sci.*, Vol. 41, N8, 1287–1311.

JSTC (Joint Scientific and Technical Committee) (1995). *Plan for the Global Climate Observing System (GCOS)* (Version 1.0, May 1995, GCOS-14, WMO/TD-N681). World Meteorological Organization, Geneva, Switzerland, 60 pp.

Kadanov L. P. (1993). From order to chaos. *Essays: Critical, Chaotic and Otherwise*. World Scientific, Singapore, 576 pp.

Kalmykov A. I., Pichugin A. P., Zimbal V. N., and Shestopalov V. P. (1984). Radiophysical space observations of mesoscale structures on ocean surface. *Doklady USSR Akad. Sci.* (*Trans. of USSR Acad. of Sciences—English transl.*), Vol. 279, N4, 860–862.

Karpov A. V. (1994). National satellite systems capabilities relevant to climate studies. *Adv. Space Res.*, Vol. 14, N1, 37–46.

Kazimirovsky E. S., Herraiz M., and De la Morena B. A. (2003). Effects on the ionosphere due to phenomena occurring below it. *Survey in Geophysics*, Vol. 24, N1, 139–184.

Khain A. P. (1984). *Mathematical Modeling of Tropical Cyclones*. Gidrometeoizdat, Leningrad, 247 pp. [in Russian].

Khain A. P. and Sutyrin G. G. (1983). *Tropical Cyclones and Their Interactions with the Ocean*. Gidrometeoizdat, Leningrad, 271 pp. [in Russian].

Khinchin A. J. (1963). *Mathematical Methods in Mass Service Theory*. Phismatgiz, Moscow, 120 pp.

Khromov S. P. (1948). *Foundations of Synoptic Meteorology*. Gidrometeoizdat, Leningrad, 696 pp. [in Russian].

Khromov S. P. (1966). *Atmospheric Circulation in the Tropics*. Gidrometeoizdat, Leningrad, 280 pp. [in Russian].

Kidder S. Q. and Shyu K. (1984). On the potential use of satellite sounder data in forecasting tropical cyclone motion. *Monthly Weather Rev.*, Vol. 12, N10, 1977–1984.

Kidwell K. (Ed.) (1988). *NOAA Polar Orbiter User's Guide*, NOAA/NESDIS/NCDC/SDSD, Washington, DC, 50 pp.

Kim G. A., Sharkov E. A., and Pokrovskaya I. V. (2009). Evolution and energy structure of TC "Hondo" using optical and microwave satellite data. *Modern Problems of Earth Remote Sensing from Space: Physical Foundations, Methods and Technology of the Monitoring of Environment, Potential Dangerous Occurrences and Objects* (Collected Book of Scientific Papers, Issue 6. Vol. II). Azbuka-2000, Moscow, pp. 126–136 [in Russian].

Kim G. A., Sharkov E. A., and Pokrovskaya I. V. (2010) Features of the interaction between TC "Hondo" and TC "Ivan" in water vapor fields. *Modern Problems of Earth Remote Sensing from Space: Physical Foundations, Methods and Technology of Monitoring the Environment, Potential Dangerous Occurrences and Objects* (Collected Book of Scientific Papers, Vol. 7, N4). DoMira, Moscow, pp. 287–295 [in Russian].

King M. D. (Ed.) (1999). *EOS Science Plan: The State of Science in the EOS Program* (NP-1998-12-069-GSFC). NASA Goddard Space Flight Center, Greenbelt, MD, 397 pp.

Klein P. M., Harr P. A., and Elsberry R. L. (2000). Extratropical transition of Western North Pacific tropical cyclones: An overview and conceptual model of the transformation stage. *Weather and Forecasting*, Vol. 15, N4, 373–396.

Klepikov I. N. (1990). Satellite method of measurements for spectral characteristics of Earth atmosphere turbulence. *Earth Research from Space*, N5, 3–11.

Klepikov I. N. and Sharkov E. A. (1987). *On Possibilities of Experimental Investigation for Spiral Features of Turbulent Atmosphere* (Preprint Pr-1318). Space Research Institute, Russian Academy of Sciences, Moscow, 80 pp. [in Russian].

Klepikov I. N., Moiseev S. S., and Sharkov E. A. (1990). Experimental evidence of large intermittency in convective turbulence with high Reynolds numbers. *Pisma v Zurnal Tech. Fiziki (Technical Physics Letters—English Transl.)*, Vol. 16, N16, 81–87.

Klepikov I. N., Pokrovskaya I. V., and Sharkov E. A. (1994). Structure of a data base of space and hydrometeorological observations of mesoscale topical disturbances. *Sov. J. Remote Sensing*, Vol. 11, N3, 468–479.

Klepikov I. N., Pokrovskaya I. V., and Sharkov E. A. (1995). Satellite and radio remote sensing of the mesoscale atmospheric turbulence during pre-typhoon conditions. *Earth Obs. Rem. Sens.*, Vol. 13, 349–364.

Klimantovich Yu. L. (1982). *Statistical Physics*. Nauka, Moscow, 606 pp.

Knutson T. R. and Tuleya R. E. (1999). Increased hurricane intensities with CO_2-induced warming as simulated using the GFDL hurricane prediction system. *Climate Dynamics*. Vol. 15, 503–519.

Knutson T. R., Tuleya R. E., and Kurihara Y. (1998). Simulated increase of hurricane intensities in a CO_2-warmed climate. *Science*. Vol. 279, 1018–1020.

Knutson T. R., Sirutis J. J., Garner S. T., Vecchi G. A., and Held I. M. (2008). Simulated reduction in Atlantic hurricane frequency under twenty-first-century warming conditions. *Nature Geoscience*, Vol. 1, N6, 359–364.

Kodama Y. and Asai T. (1988). Large–scale distributions and their seasonal variations as derived from GMS–IR observations. *J. Meteorol. Soc. Japan*, Vol. 66, N1, 87–100.

Kondragunta Ch. and Gruber A. (1996). Seasonal and annual variability of the diurnal cycle of clouds. *J. Geophys*, Vol. 101, ND16, 21377–21390.

Kondratyev K. Ya. (1980). *Radiation Factors in Recent Changes of the Global Climate*. Gidrometeoizdat, Leningrad, 280 pp.

Kondratyev K. Ya. (1992). *Global Climate*. Nauka, St. Petersburg, 359 pp [in Russian].

Kondratyev K. Ya. (1994). Global Climate Observation System. *Sov. J. Remote Sensing*, Vol. 11, N6, 1038–1056.

Kondratyev K. Ya. and Tanaka T. (1997). Advanced Earth observing satellite-II (ADEOS-II): New perspectives of global environmental monitoring. *Earth Research from Space* (*Earth Observ. Rem. Sens.—English transl.*), N1, 105–121.

Kondratyev K. Ya., Kozoderov V. V., and Smokty O. I. (1992). *Remote Sensing of the Earth from Space: Atmospheric Correction*. Springer-Verlag, Heidelberg, Germany, 478 pp.

Kondratyev K. Ya., Buznikov A., and Pokrovsky O. (1995). *Global Change and Remote Sensing*. Wiley/Praxis, Chichester, U.K., 300 pp.

Kononovich E. V. and Shefov N. N. (1999) On solar activity influence on long-term climate variations. *Doklady Akad. Nauk*. (*Trans. of Russian Academy of Sciences/Earth Science Section—English transl.*), Vol. 367, N1, 108–111.

Korn G. A. and Korn T. M. (1961). *Mathematical Handbook for Scientists and Engineers: Definitions, Theorems and Formulas for References and Review*. McGraw-Hill, New York, 720 pp.

Kunitsyn V. E. and Tereshchenko E. D. (2003) *Ionospheric Tomography*. Springer-Verlag. Berlin, 335 pp.

Kunitsyn V. E., Tereshchenko E. D., and Andreeva E. S. (2007). *Radiotomography of the Ionosphere*. Fizmatgiz, Moscow, 305 pp.

Kuo H. L. (1965). On formation and intensification of tropical cyclones through heat release by cumulus convection. *J. Atmos. Sci.*, Vol. 22, pp. 40–63.

Kuroda T. and Koizumi Sh. (1996). The plan of the World Environment and Disaster Observation System (WEDOS) and the Global Disaster Observation System (GDOS). *Earth Research from Space*, N4, 56–66.

Lajoie F. A. and Butterworth I. J. (1984). Oscillation of high-level cirrus and heavy precipitation around Australian region tropical cyclones. *Monthly Weather Rev.*, Vol. 112, N3, 535–544.

Landa P. S. (1996). *Nonlinear Oscillations and Waves in Dynamical Systems*. Kluwer Academic, Dordrecht, The Netherlands, 556 pp.

Lander M. A. (1994). An exploratory analysis of the relationship between tropical storm formation in the western North Pacific and ENSO. *Monthly Weather Rev.*, Vol. 122, N5, 636–651.

Lander M. A., Trehubenko E. J., and Guard Ch. P. (1999). Eastern hemisphere tropical cyclones of 1996. *Monthly Weather Rev.*, Vol. 127, N6, 1274–1300.

Landsea Ch. W. (1993) A climatology of intense (or major) Atlantic hurricanes. *Monthly Weather Rev.*, Vol. 121, N6, 1703–1713.

Landsea Ch. W. (1997). Comments on: "Will greenhouse gas-induced warming over the next 50 years lead to higher frequency and greater intensity of hurricanes". *Tellus*, Vol. 49A, N5, 622–632.

Landsea Ch. W. (1998). *FAQ: Hurricanes, Typhoons and Tropical Cyclones* (Parts 1 and 2, Version 2.7). NOAA/AOML/Hurricane Research Division, Miami, FL (*http://www.aoml.noaa.gov/hrd/tcfaq/tcfaqI.html*)

Landsea Ch. W., Gray W. M., Mielke P. W., and Berry K. J. (1994). Seasonal forecasting of Atlantic hurricane activity. *Weather*, Vol. 49, N8, 273–284.

Landsea Ch. W., Nicholls N., Gray W. M., and Avila L. A. (1996). Downward trends in the frequency of intense Atlantic hurricanes during the past five decades. *Geophys. Res. Lett.*, Vol. 23, N23, 1697–1700.

Landsea Ch. W., Pielke R. A., Mestas-Nunez A. M., and Knaff J. A. (1999). Atlantic basin hurricanes: Indices of climatic changes. *Climatic Change*, Vol. 42, N2, 89–129.

Lastovicka J. (2006). Forcing of the ionosphere by waves from below. *J. Atmos. and Solar–Terr. Phys.*, Vol. 68, 479–497.

Lawrence M. B. (1999). Eastern North Pacific hurricane season of 1997. *Monthly Weather Rev.*, Vol. 127, N10, 2440–2454.

Lazarev A. A. and Moiseev S. S. (1992). *Geophysical Precursors of Early States of Cyclogenesis* (Preprint No. Pr–1844). Space Research Institute, Russian Academy of Sciences, Moscow, 42 pp.

Lazarev A. A., Scherzer D., Lovejoy S., and Chigirinskaya Y. (1994). Unified multifractal atmospheric dynamics tested in the tropics, Part II: Vertical scaling and generalized scale invariance. *Nonlinear Processes in Geophysics*, Vol. 1, N2/3, 115–123.

Lesieur M. (1997). *Turbulence in Fluids*. Kluwer Academic, Dordrecht, The Netherlands, 548 pp.

Levich E. and Tzvetkov E. (1985). Helical inverse cascade in three-dimensional turbulence as a fundamental dominant mechanism in meso-scale atmospheric phenomena. *Phys. Reports*, Vol. 128, 1–37.

Levina G. V., Moiseev S. S., and Rutkevich P. B. (2000). Hydrodynamic alpha-effect in a convective system. *Nonlinear Instability, Chaos and Turbulence* (edited by L. Debnath and D. N. Riahi). VIT Press, Southamption, U.K., Vol. II., pp. 111–161.

Lighthill J. (1998) Tropical cyclone disasters. *Science International*, N67, April, 1.

Lighthill J., Holland G., Gray W., Landsea Ch., Crain G., Evans J., Kurihara Y., and Guard Ch. (1994) Global climate change and tropical cyclones. *Bull. Amer. Meteorol. Soc.*, Vol. 75, N11, 2147–2157.

Lilly D. K. and Gal-Chen T. (Eds.) (1983). *Meso-Scale Meteorology: Theories, Observations and Models*. D. Reidel, Hingham, MA, 305 pp.

Liu G., Curry J. A., and Weadon M. (1994). Atmospheric water balance in typhoon Niña as determined from SSM/I satellite data. *Meteorol. Atmos. Phys.*, Vol. 54, N2, 141–156.

Liu G., Curry J. A., and Clayson C. A. (1995). Study of tropical cyclogenesis using satellite data. *Meteorol. Atmos. Phys.*, Vol. 56, N2, 111–123.

Liu W. T., Tang W., and Dunbar R. S. (1997). Scatterometer observed extratropical transition of Pacific typhoons. *EOS. Trans. Amer. Geophys. Union*, Vol. 78, N23, 237–240.

List R. (1988). A linear radar reflectivity–rainrate relationship for steady tropical rain. *J. Atmos. Sci.*, Vol. 45, N23, 3564–3572.

Loginov V. F. (1973). *Features of Solar–Terrestrial Links*. Hydrometeoizdat, Leningrad, 86 pp [in Russian].

Louet J. and Levrini G. (1998). Envisat: Europe's Earth-observation mission for the new millennium. *Earth Observation Quarterly*, N60, 1–39.

Lovejoy S. (1982). Area-perimeter relation for rain and cloud areas. *Science*, Vol. 216, N9, 185–187.

Lovejoy S. and Schertzer D. (1985). Generalized scale invariance in the atmosphere and fractal models of rain. *Water Resources Research*, Vol. 21, N8, 1233–1250.

Lovejoy S., Schertzer D., and Ladoy P. (1986). Fractal characterization of inhomogeneous geophysical measuring networks. *Nature*, Vol. 319, N6048, 43–44.

Lu C.-L. and Chen S.-H. (1998). Multiple linear interdependent models (MLIM) applied to typhoon data from China. *Theor. Appl. Climatology*, Vol. 61, N3/4, 143–149.

Ludlum D. M. (1989). *Early American Hurricanes (1492–1870)* (Historical Monograph, AMS Code EAH). American Meteorology Society, 198 pp.

Lukashevich E. L. (1994). The space system "Resurs-F" for photographic survey of the Earth. *Russian Space Bulletin*, Vol. 1, N4, 21.

Lukiashchenko V. I., Tsibulsky G. A., Karrask V. K., Medvedev A. A., Melyankov N. A., Philosopov S. N., and Karmasin V. P. (1994). Experimental investigation of large-scale natural hazards using rocket and space facilities. *COSPAR '94: 30th COSPAR Scientific Assembly, Hamburg, Germany, July 11–21, 1994*. V.II, p. 42 (Abstract).

Lukiashchenko V. I., Tsyboulsky G., and Sytyi G. (1999). "Volna-TC" rocket space system based on SSN-18 missile for sounding of tropical cyclones (*http://www.rka.ru/cp1251/english/e_volna_tc.html*).

Lupian E. A., Pokrovskaya I. V., Shavva I. I., and Sharkov E. A. (1987). *Influence of Large-Scale Atmosphere Flows on Onsets of Tropical Cyclones* (Preprint No. Pr-1313). Space Research Institute, Russian Academy of Sciences, Moscow, 29 pp. [in Russian].

Mandelbrot B. B. (1977). *Fractals: Forms, Chance and Dimension*. W. H. Freeman, pp. 1755–1758.

Matveev Yu. L. (1998). Dynamical and statistical analysis of global cloud fields from satellite data. *Earth Obs. Rem. Sens.*, Vol. 15, N1, 115–127.

Matveev Yu. L., Matveev L. T., and Soldatenko S. A. (1986) *Global Cloudiness*. Gidrometeoizdat, Leningrad, 288 pp. [in Russian].

May P. T. and Rajopadhyay D. K. (1999). Vertical velocity characteristics of deep convection over Darwin, Australia. *Monthly Weather Rev.*, Vol. 127, N6, 1056–1071.

Mayfield M. and Avila L. (1994). North Atlantic hurricanes: 1993. *Mar. Weather Log*, Vol. 38, N2, 10–16.

Mayfield M. and Rappaport E. N. (1992). Eastern Pacific hurricanes: Long season, but normal numbers. *Weatherwise*, Vol. 45, N1, 42–46.

McBride J. L. (1981a). Observational analysis of tropical cyclone formation, Part I: Basic description of data sets. *J. Atmos. Sci.*, Vol. 38, N6, 1117–1131.

McBride J. L. (1981b). Observational analysis of tropical cyclone formation, Part III: Budget analysis. *J. Atmos. Sci.*, Vol. 328, N6, 1152–1166.

McBride J. L. (1995). Tropical cyclone formation. In *Global Perspectives on Tropical Cyclones*, (WMO Technical Document N693, Tropical Cyclone Program Report Njmber 38). World Meteorological Organization, Geneva, Switzerland, pp. 63–105.

McBride J. L. and Zehr R. (1981). Observational analysis of tropical cyclone formation, Part II: Comparison of non-developing versus developing systems. *J. Atmos. Sci.*, Vol. 38, N6, 1132–1151.

McCown S., Graumann A., and Ross T. (1998). Mitch: The deadliest Atlantic hurricane since 1780 (*http://www.ncdc.noaa.gov/ol/reports/mitch/mitch.html*).

McGuirk J. P., Thompson A. H., and Smith N. R. (1987). Moisture bursts over the tropical Pacific Ocean. *Monthly Weather Rev.*, Vol. 115, N4, 787–798.

McPhaden M. J., Zebiak S. E., and Glantz M. H. (2006). ENSO as an integrating concept in Earth science. *Science*, Vol. 314, N5806, 1740–1745.

Meldrum C. (1872). On aperiodicity in the frequency of cyclones in the Indian Ocean south of the Equator. *Nature*, Vol. 6, 357–358.

Menzel W. P. and Purdom J. F. W. (1996). Introduction to GOES-1: The first of a new generation of geostationary operational environmental satellites. 1. The satellite sensor systems, quality control procedures for converting data into products. *Earth Obs. Rem. Sens.*, Vol. 14, N1, 81–99.

Merrill R. T. (1988a). Characteristics of the upper-tropospheric environmental flow around hurricanes. *J. Atmos. Sci.*, Vol. 45, N6, 1665–1677.

Merrill R. T. (1988b). Environmental influence on hurricane intensification. *J. Atmos. Sci.*, Vol. 45, N6, 1678–1687.

Merrill R. T. and Velden C. S. (1996) A three-dimensional analysis of the outflow layer of supertyphoon Flo (1990). *Monthly Weather Rev.*, Vol. 124, N1, 47–63.

Metcalf T. R. (1996). *Tropical Storms* (*http://www.solar.ifa.hawaii.edu/Tropical/*).

Minina L. S. (1970). *Practice of Nephanalysis.* Gidrometeoizdat, Leningrad, 335 pp [in Russian].

Minina L. S. (1982). Tropical cyclones. *Guide on Using Satellite Data for Weather Analysis and Forecasting* (edited by I. P. Vetlov and N. F. Veltishev). Gidrometeoizdat, Leningrad, pp. 253–285 [in Russian].

Minina L. S. (1983). On generating and developing typhoons. *Russian Meteorology and Hydrology*, N11, 5–13.

Minina L. S. (1987). Features of tropical cyclogenesis. In *"Tropical Meteorology", Trans. III Int. Symposium on Tropical Meteorology, Yalta, March 1985.* Gidrometeoizdat, Leningrad, 1987, pp. 36–51.

Mock C. J. (2008). Tropical cyclone variations in Louisiana, USA, since the late eighteenth century. *Geochem. Geophys. Geosyst. G^3*, Vol. 9, N5, Q05V02, doi: 10.1029/2007GC001846.

Moiseev S. S. (1990). The helical mechanism of generation of large-scale structures in continuous media. *Nonlinear World* (edited by V. G. Bar′ykhtar *et al.*), World Scientific, Singapore, Vol. I, pp. 541–560.

Moiseev S. S. (1991). Structure and intermittent chaos in random media. In *Non-Linear Dynamics of Structures* (edited by R. Z. Sagdeev *et al.*). World Scientific, Singapore, pp. 81–93.

Moiseev S. S., Sagdeev R. Z., Tur A. V., Khomenko G. A., and Shukurov A. M. (1983a). Physical mechanism of developing vortex disturbances in atmosphere. *Doklady USSR Akad. Sci. (Doklady-Physics—English transl.).* Vol. 273, N3, 549–553.

Moiseev S. S., Sagdeev R. Z., Tur A. V., Khomenko G. A., and Yanovskii V. V. (1983b). Theory of onset of large-scale structure in hydrodynamic turbulence. *J. Exp. Tech. Physics* (English transl.), Vol. 85, N6, 1979–1987.

Moiseev S. S., Rutkevich P. B., Tur A. V., and Yanovskii V. V. (1988). Vortex dynamo in convective media with spiral turbulence. *J. Exp. Tech. Physics* (English transl.), Vol. 94, N2, 144–153.

Moiseev S. S., Oganyan K. R., Rutkevich P. B., Tur A. V., and Yanovskii V. V. (1990). An eddy dynamo and spiral turbulence. *Integrability and Kinetic Equations for Solitons* (edited by V. G. Bar′yakhtar *et al.*). Naukova Dumka, Kiev, pp. 280–332.

Moiseev S. S., Pungin V. G., and Sharkov E. A. (1995). Aerospace monitoring of natural hazards and turbulence. *Proc. 5th Int. Symp. Recent Advances in Microwave Technology, Kiev, Ukraine, September 11–16, 1995*, Vol. I, pp. 307–310.

Moiseev S. S., Onishchenko O. G., Sharkov E. A., Branover H. H., and Eidelman A. E. (1996). Influence of variations of dissipation and pressure on formation of coherent structures in geophysical media. *Phys. Chem. Earth*, Vol. 21, N5–6, 545–547.

Moiseev S. S., Pungin V. G., and Oraevskii V. N. (2000). *Nonlinear Instabilities in Plasmas and Hydrodynamics*. IOP, Bristol, U.K., 162 pp.

Molinari J., Vollaro D., Skubis S., and Dickinson M. (2000). Origins and mechanisms of Eastern Pacific tropical cyclogenesis: A case study. *Monthly Weather Rev.*, Vol. 128, N1, 125–139.

Montgomery M. and Farrell B. (1993). Tropical cyclone formation. *J. Atmos. Sci.*, Vol. 50, 285–310.

MSC (Meteorological Satellite Center) (1986). *Monthly Report of the Meteorological Satellite Center*, June/September, 1985, Tokyo, 250 pp.

Mukherjee G. K. (2003). The signature of short-period gravity waves imaged in the OI 557.7 nm and near infrared OH nightglow emissions over Panhala. *J. Atmos. and Solar–Terr. Phys.*, Vol. 65, N14–15, 1329–1335.

NCDC (1991). *Global Tropical and Extratropical Cyclone Climatic Atlas*. National Climatic Data Center, Asheville, NC, 46 pp. (diskette documentation).

Neumann C. J. (1993). *Global overview: Global Guide to Tropical Cyclone Forecasting* (WMO Technical Document N560, Tropical Cyclone Program Report N31). World Meteorological Organization, Geneva, Switzerland, pp. 1.1–1.43.

Neumann C. J., Jarvinen B. R., McAdie C. J., and Elms J. D. (1993). *Tropical Cyclones of the North Atlantic Ocean, 1871–1992* (Historical Climatology Series 6-2). National Climatic Data Center, Asheville, NC, 193 pp.

Nicholls J. M. (1996). *Economic and Social Benefits of Climatological Information and Services: A Review of Existing Assessments* (WCASP-38, WMO/TD-N780). World Meteorological Organization, Geneva, Switzerland, 33 pp.

Nicholls N. (1979). A possible method for predicting seasonal tropical cyclone activity in the Australian region. *Monthly Weather Rev.*, Vol. 107, N5, 1221–1224.

Nicholls N. (1984). The Southern Oscillation, sea-surface temperature and interannual fluctuations in Australian tropical cyclone activity. *J. Climatol.*, Vol. 4, 661–670.

Nicholls N. (1985). Predictability of interannual variations of Australian seasonal tropical cyclone activity. *Monthly Weather Rev.*, Vol. 1113, N4, 1144–1149.

Nicholls N. (1992). Recent performance of a method for forecasting Australian region tropical cyclone activity. *Australian Met. Mag.*, Vol. 40, N2, 105–110.

Nicolis G. and Prigogine I. (1977). *Self-Organization in Nonequilibrium Systems*. Wiley-Interscience, New York, 300 pp.

Nihoul J. C. J. and Jamart B. M. (Eds.) (1989). *Mesoscale/Synoptic Coherent Structure in Geophysical Turbulence*. Elsevier Science, Amsterdam, 842 pp.

Noin D. (1997). *People on Earth: World Population Map*. UNESCO, Paris, 50 pp.

Oates S. (2000). The South Pacific and southeast Indian Ocean tropical cyclone season 1998–99. *Australian Met. Mag.*, Vol. 49, N3, 223–244.

Ohring G. (1994). The current status of operational satellite products for climate studies. *Adv. Space Res.*, Vol. 14, N1, 55–59.

Ooyama K. (1964). A dynamical model for the study of tropical cyclone development. *Geophysica Intern.*, Vol. 4, N4, 187–198.

Palmen E. and Newton C. W. (1969). *Atmospheric Circulation Systems: Their Structure and Physical Interpretation* (International Geophysics Series 13). Academic Press, New York, 603 pp.

Pasch R. J. and Avila L. A. (1999). Atlantic hurricane season of 1996. *Monthly Weather Rev.*, Vol. 127, N5, 581–610.

Penland C. (1996). A stochastic model of Indo-Pacific sea surface temperature anomalies. *Physica D*, Vol. 98, N6, 534–558.

Petrova L. I. (1987). The moisture content of the disturbances, developed and non-developed in tropical cyclones. In "*Tropical Meteorology": Trans. III Int. Symposium on Tropical Meteorology, Yalta, March 1985*. Gidrometeoizdat, Leningrad, pp. 220–225.

Philander S. G. H. (1990). *El Niño, La Niña and the Southern Oscillation*. Academic Press, New York, 289 pp.

Philander S. G. (1999). A review of tropical ocean–atmosphere interactions. *Tellus*, Ser. A and B, N1, 71–90.

Pielke R. A. Jr. and Landsea C. N. (1998). Normalized hurricane damage in the United States: 1925–95. *Weather and Forecasting*, Vol. 13, N3, 621–631.

Pielke R. A. Jr. and Landsea C. N. (1999). La Niña, El Niño, and Atlantic hurricane damages in the United States. *Bull. Amer. Meteorol. Soc.*, Vol. 80, N10, 2027–2033.

Pielke R. A. Jr. and Pielke R. A. Sr. (1997). *Hurricanes. Their Nature and Impacts on Society*. John Wiley & Sons, Chichester, U.K., 279 pp.

Pietronero L. and Tosatti E. (Eds.). (1986). Fractals in physics. *Proceedings of the Sixth Trieste Int. Symposium on Fractals in Physics, ICTP, Trieste, Italy, July 9–12, 1985*. North-Holland, Amsterdam, 490 pp.

Pokhil A. E. (1990). On any cases of interactions of tropical cyclones in Pacific for 1988. *Russian Meteorology and Hydrology*, N6, 25–30.

Pokhil A. E. (1996). On unusual season of typhoons in Pacific. *Russian Meteorology and Hydrology*, N3, 32–39.

Pokrovskaya I. V. and Sharkov E. A. (1988). *Properties of the Thermal Stratification of the Tropical Atmosphere (Based on Satellite Data)*, (Preprint No. Pr-1426), Space Research Institute, Russian Academy of Sciences, Moscow, 54 pp. [in Russian].

Pokrovskaya I. V. and Sharkov E. A. (1990). *Thermal Stratification Conditions of the Tropical Atmosphere in the Cyclogenesis Process* (Preprint No. Pr-1701). Space Research Institute, Russian Academy of Sciences, Moscow, 44 pp. [in Russian].

Pokrovskaya I. V. and Sharkov E. A. (1991). Effect of tropical cyclones on the thermal stratification of a tropical atmosphere. *Sov. J. Remote Sensing*, Vol. 9, N2, 296–314.

Pokrovskaya I. V. and Sharkov E. A. (1993a). Global tropical cyclogenesis as an accidental Poisson process. *Doklady Akad. Nauk USSR (Trans. of USSR Acad. of Sciences/Earth Science Section—English Transl.)*, Vol. 331, N5, 625–627.

Pokrovskaya I. V. and Sharkov E. A. (1993b) Remote studies of the thermal stratification of the tropical atmosphere in the process of cyclogenesis. *Sov. J. Remote Sensing*, Vol. 11, N1, 93–105.

Pokrovskaya I. V. and Sharkov E. A. (1994a). Poisson features of global tropical cyclogenesis by satellite data. *Earth Research from Space (Earth Obs. Rem. Sens.—English transl.)*, N2, 24–39.

Pokrovskaya I. V. and Sharkov E. A. (1994b). Interannual variabiity of global tropical cyclogenesis. *Russian Meteorology and Hydrology*, N4, 20–28.

Pokrovskaya I. V. and Sharkov E. A. (1994c). *Catalogue of Global Tropical Cyclogenesis Data for 1983–1987 and 1993* (Preprint No. Pr-1890). Space Research Institute, Russian Academy of Sciences, Moscow, 30 pp.

Pokrovskaya I. V. and Sharkov E. A. (1995a). Study of the center-spatial structure of tropical cyclogenesis as applied to satellite monitoring problems. *Earth Obs. Rem. Sens.*, Vol. 12, N6, 793–806.

Pokrovskaya I. V. and Sharkov E. A. (1995b). Scaling structure of tropical cyclogenesis areas. *Annales Geophysicae*, Suppl. II to Vol. 13, Part II, C569.

Pokrovskaya I. V. and Sharkov E. A. (1996a). Study of temporal intermittancy of the Pacific Ocean basin tropical cyclogenesis as applied to satellite monitoring problems. *Earth Obs. Rem. Sens.*, Vol. 14, N1, 17–30.

Pokrovskaya I. V. and Sharkov E. A. (1996b). Interannual variability of tropical cyclogenesis in the Pacific. *Russian Meteorology and Hydrology*, N3, 40–49.

Pokrovskaya I. V. and Sharkov E. A. (1997a). *Data Set of Global Tropical Cyclogenesis: Design Principles, Structure, Catalogues* (Preprint No. Pr-1969). Space Research Institute, Russian Academy of Sciences, Moscow, 40 pp.

Pokrovskaya I. V. and Sharkov E. A. (1997b). Study of the original forms of tropical disturbances of the Pacific Ocean basin by remote sensing. *Earth Obs. Rem. Sens.*, Vol. 14, N5, 777–792.

Pokrovskaya I. V. and Sharkov E. A. (1997c). Remote study of spatial fields of moisture content in the tropical atmosphere in the process of cyclogenesis. *Earth Obs. Rem. Sens.*, Vol. 14, N6, 875–889.

Pokrovskaya I. V. and Sharkov E. A. (1999a). *Data set of Global Tropical Disturbances for 1997–1998: Design Formation, Structure, Catalogue* (Preprint No. Pr-2004). Space Research Institute, Russian Academy of Sciences, Moscow, 41 pp. [in Russian].

Pokrovskaya I. V. and Sharkov E. A. (1999b). Structural features of tropical cyclogenesis for northern and southern hemispheres in application to satellite monitoring problems. *Earth Research from Space (Earth Obs. Rem. Sens.—English transl.)*, N1, 18–27.

Pokrovskaya I. V. and Sharkov E. A. (1999c). Structural features of global tropical cyclogenesis for initial forms in application to satellite monitoring problems. *Earth Research from Space* (*Earth Obs. Rem. Sens.—English transl.*), N3, 1–13.

Pokrovskaya I. V. and Sharkov E. A. (1999d). *Catalogue of Tropical Cyclones and Tropical Disturbances of the World Ocean for 1983–1998* (Version 1.1). Poligraph Service, Moscow, 160 pp. [in Russian].

Pokrovskaya I. V. and Sharkov E. A. (2000). Global features of the generation rate of tropical cyclones. *Doklady Akad. Nauk.* (*Trans. of Russian Academy of Sciences/Earth Science Section—English transl.*), Vol. 373, N5. 851–854.

Pokrovskaya I. V. and Sharkov E. A. (2001). *Tropical Cyclones and Tropical Disturbances of the World Ocean: Chronology and Evolution* (Version 2.1, 1983–2000). Poligraph Service, Moscow, 548 pp. [in Russian/English].

Pokrovskaya I. V. and Sharkov E. A. (2006). *Tropical Cyclones and Tropical Disturbances of the World Ocean: Chronology and Evolution* (Version 2.1, 1983–2005). Poligraph Service, Moscow, 728 pp. [in Russian/English].

Pokrovskaya I. V., Sharkov E. A., Klepikov I. N., and Karaseva I. A. (1993). *Catalogue and Data Base of Global Tropical Cyclogenesis for 1988–1992* (Preprint No. Pr-1869). Space Research Institute, Russian Academy of Sciences, Moscow, 29 pp. [in Russian].

Pokrovskaya I. V., Sharkov E. A., Klepikov I. N., and Karaseva I. A. (1994). Structure of database for global tropical cyclogenesis remote sensing. *Sov. J. Remote Sensing*, Vol. 11, N6, 981–989.

Pokrovskaya I. V., Rutkevich P. B., and Sharkov E. A. (2002). *Dynamical Fields Database and Algorithmic Approaches to the Problem of Turbulence Study in the Tropical Atmosphere* (Preprint Pr-2048). Institute Space Research, Russian Academy of Sciences, 43 pp.

Pokrovskaya I. V., Rutkevich P. B., and Sharkov E. A. (2004). Scenario principle behind the assimilation of satellite and *in situ* data in the context of problems of atmospheric catastrophe investigation. *Earth Research from Space*, N3, 32–42 [in Russian].

Polak L. S. and Michailov A. S. (1983). *Self-Organization in Non-Equilibrium Physico-Chemical Systems.* Nauka, Moscow, 248 pp.

Ponzi A. and Aizawa Y. (2000). Criticality and punctuated equilibrium in a spin system model of a financial market. *Chaos, Solitons and Fractals*, Vol. 11, N11, 1739–1746.

Powell M. D. (1982). The transition of the Hurricane Frederic boundary-layer wind field from the open Gulf of Mexico to landfall. *Monthly Weather Rev.*, Vol. 110, N12, 1912–1932.

Powell M. D., Houston S. H., Amat L. R., and Morisseau-Leroy N. (1998). The HRD real-time hurricane wind analysis systems. *J. Wind Engineering and Industrial Aerodynamics*, Vol. 77/78, N1, 53–64.

Prochorov A. M. (Ed.) (1984). *Physical Encyclopedia.* Soviet Encyclopedia, Moscow, 944 pp.

Pudovkin M. I. and Raspopov O. M. (1993). Physical mechanism of solar activity influence on atmosphere state, meteocharacteristics and climate. *Uspekhi Fizicheskikh Nauk* (*Physics-Uspekhi—English transl.*), Vol. 163, N7, 113–116.

Purdom J. F. W. (1996). Nowcasting with a new generation of GOES satellites. *31st Scientific Assembly of COSPAR Abstracts.* University of Birmingham, U.K., pp. 6.

Raevskii A. N. and Vetroumov V. A. (1987). Spatial distribution of characteristics of North Atlantic tropical cyclogenesis. *Meteorologia i Hidrologia* (*Russian Meteorology and Hydrology—English transl.*), N3, 47–52.

R. A. K. (1982). A few lessons learned. *Science*, Vol. 217, N4559, 520.

Raizonville P., Cugny B., Zanife O. Z., Jaulhac Y., and Richard J. (1991). *Poseidon Radar Altimeter Flight Model Design and Tests Results* (ESA SP-328). European Space Agency, Noordwijk, The Netherlands, pp. 93–98.

Ramiras M. R. (1987). On source of energy onset for tropical cyclones. *"Tropical Meteorology", Trudy 3 Meshdun. Simp. Jalta, 1985.* Gidrometeoizdat, Leningrad, pp. 170–184 [in Russian].

Rao D. V. B. and Ashok K. (1999). Simulation of tropical cyclone circulation over Bay of Bengal using the Arakawa–Schubert cumulus parametrization, Part I: Description of the model, initial data and results of the control experiments. *Pure Appl. Geophys.*, Vol. 156, N3, 525–542.

Rappaport E. N. and Fernandez-Partagas J. (1995). *The Deadliest Atlantic Tropical Cyclones, 1492–1994* (NOAA Technical Memorandum NWS NHC-47). National Hurricane Center, Miami, FL, 41 pp.

Raschke E. and Jacob D. (Eds.) (1993). *Energy and Water Cycles in the Climate System.* Springer-Verlag, Berlin, 467 pp.

Rassadovsky V. A. and Troitsky A. V. (1981). Remote radiometric atmospheric studies in the zone of tropical cyclone origin. *Fizika atmosphery i okeana* (*Izvestya, Atmospheric and Oceanic Physics—English transl.*), Vol. 17, N7, 698–705.

Readings Ch. J. and Reynolds M. (1996). ESA's future "Earth Explorer" missions. *Earth Obs. Quarterly*, N53, 1–6.

Revah I. (1994). The French program in Earth Observations from Space. *Adv. Space Res.*, Vol. 14, N1, 29–35.

Revell C. and Goulter S. (1986). South Pacific tropical cyclones and the Southern Oscillation. *Monthly Weather Rev.*, Vol. 114, N6, 1138–1145.

Riehl H. (1954). *Tropical Meteorology.* McGraw-Hill, 392 pp.

Riehl H. (1979). *Climate and Weather in the Tropics.* Academic Press, London, 400 pp.

Ritchie E. A. and Holland G. J. (1997). Scale interactions during the formation of Typhoon Irving. *Monthly Weather Rev.*, Vol. 125, N7, 1377–1396.

Roberts D. C. and Turcotte D. L. (1998). Fractality and self-organized criticality of wars. *Fractals*, Vol. 6, N4, 351–357.

Rodgers E. B. and Adler R. F. (1981). Tropical cyclone rainfall characteristics as determined from a satellite passive microwave radiometer. *Monthly Weather Rev.*, Vol. 109, 506–521.

Rodgers E. B., Chang S. W., and Pierce H. F. (1994). A satellite observational and numerical study of precipitation characteristics in western North Atlantic tropical cyclones. *J. Appl. Meteorol.*, Vol. 33, pp. 573–593.

Romanovskii Yu. M., Stepanova N. V., and Chernavskii D. S. (1984). *Mathematical Biophysics.* Nauka, Moscow, 315 pp. [in Russian].

Rosenlof K. H., Tuck A. F., Kelly K. K., Russell J. M., and McCormick M. P. (1997). Hemispheric asymmetries in water vapor and inferences about transport in the lower stratosphere. *J. Geophys. Res.*, Vol. 102, ND11, 13213–13234.

Rossow W. B. and Schiffer R. A. (1991). ISCCP cloud data product. *Bull. Amer. Meteorol. Soc.*, Vol. 72, N1, 2–20.

Ruprecht E. (1996). Atmospheric water vapour and cloud water: An overview. *Adv. Space Res.*, Vol. 18, N7, 5–16.

Rutkevich P. B. (1994). Generation properties of convective turbulence in a Coriolis force field. *Doklady–Physics*, Vol. 334, N1, 44–46.

Rutkevich P. P. and Sharkov E. A. (2004). *Physical Mechanism of Rotational Instability Genesis in a Compressed and Saturated Moist Air Atmosphere* (Preprint NPr-2102). Space Research Institute, Moscow, 12 pp. [in Russian].

Rytov S. M. (1966). *Introduction to Statistical Radiophysics.* Nauka, Moscow, 404 pp. [in Russian].

Samuelson P. A. (1961). *Economics: An Introductory Analysis.* McGraw-Hill, New York, 790 pp.

Sanada T. (1993). Helicity production in the transition to chaotic flow simulated by Navier–Stokes equation. *Phys. Rev. Lett.*, Vol. 20, N20, 3035–3038.

Schneider T., Gorman P. A. O., and Levine, X. J. (2010), Water vapor and the dynamics of climate changes. *Rev. Geophys.*, Vol. 48, RG3001, 22 pp., doi: 10.1029/2009RG000302.

Schmetz J., Tjemkes S. A., Gube M., and van de Berg L. (1997). Monitoring deep convection and convective overshooting with Meteosat. *Adv. Space Res.*, Vol. 19, N3, 433–441.

Schuster H. G. (1984). *Deterministic Chaos: An Introduction.* Physik-Verlag, Weinheim, Germany, 240 pp.

Schweitzer F. (Ed.) (1997). *Self-Organization of Complex Structures: From Individual to Collective Dynamics.* Gordon & Breach, London, 622 pp.

Seidov D. G. (1989). *Synergetics of the Ocean Processes.* Hydrometeoizdat, Leningrad, 287 pp. (in Russian).

Semmler T., Varghese S., McGrath R., Nolan P., Wang S., Lynch P., and O'Dowd C. (2008). Regional model simulation of North Atlantic cyclones: Present climate and idealized response to increased sea surface temperature. *J. Geophys. Res.*, Vol. 113, ND02107, doi: 10/1029/2006JD008213.

Shaik H. A. and Bate P. W. (1999). The tropical circulation in the Australian/Asian region: November 1998 to April 1999. *Australian Met. Mag.*, Vol. 48, N3, 211–222.

Shapiro L. J. (1982a). Hurricane climatic fluctuations, Part I: Pattern and cycles. *Monthly Weather Rev.*, Vol. 110, N8, 1007–1013.

Shapiro L. J. (1982b). Hurricane climatic fluctuations, Part II: Relation to large-scale circulation. *Monthly Weather Rev.*, Vol. 110, N8, 1014–1023.

Shapiro L. J. (1987). Month-to-month variability of the Atlantic tropical circulation and its relationship to tropical storm formation. *Monthly Weather Rev.*, Vol. 115, N10, 2598–2614.

Sharkov E. A. (1995). *Kinetic–Diffusion Approach to Description of Global Tropical Cyclogenesis Processes* (Preprint No. Pr-1910). Space Research Institute, Russian Academy of Sciences, Moscow, 14 pp.

Sharkov E. A. (1996a). Temporal intermittence as main feature of tropical cyclone activity. *Annales Geophysicae*, Suppl. II to Vol. 14, Part II, C654.

Sharkov E. A. (1996b). On the spatio–temporal convective turbulent model. *Annales Geophysicae*, Supplement II to Vol. 14, Part II, C460.

Sharkov E. A. (1996c). Global tropical cyclogenesis as relaxation oscillator of the kinetic–diffusional type. *Annales Geophysicae*, Suppl. II to Vol. 14, Part II, C54.

Sharkov E. A. (1996d). Global tropical cyclogenesis model for climate applications. *31st Sci. Assembly of COSPAR, July 14–21, 1996.* University of Birmingham, U.K., 8 pp. (Abstract).

Sharkov E. A. (1997). Global tropical cyclogenesis as a geophysical system slightly out of equilibrium. *Earth Obs. Rem. Sens.*, Vol. 14, N6, 865–874.

Sharkov E. A. (1998). *Remote Sensing of Tropical Regions.* John Wiley & Sons/Praxis, Chichester, U.K., 310 pp.

Sharkov E. A. (1999a). Evolution of global tropical cyclogenesis: Humankind in danger? *PACON '99 Symposium "Humanity and the World Ocean: Interdependence at the Dawn of the New Millennium".* Russian Academy of Sciences, Moscow, 244 pp.

Sharkov E. A. (1999b) Spatiotemporal evolution of convective tropical atmospheric turbulence with Doppler radar technique. *International Workshop on Ratio Method for Studying Turbulence.* University of Illinois, Urbana, pp. 36–37.

Sharkov E. A. (2000). *Global Tropical Cyclogenesis.* Springer/Praxis, Heidelberg, Germany/Chichester, U.K., 361 pp.

Sharkov E. A. (2003). *Passive Microwave Remote Sensing of the Earth: Physical Foundations.* Springer/Praxis, Heidelberg, Germany/Chichester, U.K., 613 pp.

Sharkov E. A. (2005) Radiothermal air–space remote sensing of the Earth: Past, present and plans in the future. *Earth Research from Space*, N5, 75–92 [in Russian].

Sharkov E. A. (2006). Global tropical cyclogenesis: The evolution of scientific concepts and the role of remote sensing. *Earth Research from Space*, N1, 68–76 [in Russian].

Sharkov E. A. (2010). Remote sensing of atmospheric catastrophes. *Earth Research from Space*, N1, 52–68 [in Russian].

Sharkov E. A. and Pokrovskaya I. V. (2006). Tropical disturbance genesis in surface sea temperature fields using remote sensing and *in situ* data. *Earth Research from Space*, N6, 3–9 [in Russian].

Sharkov E. A. and Pokrovskaya I. V. (2010). Regional tropical cyclogenesis in the surface sea temperature fields of the World Ocean. *Earth Research from Space*, N2, 54–62 [in Russian].

Sharkov E. A., Kim G. A., and Pokrovskaya I. V. (2008a). Evolution of TC "Gonu" and its interaction with the precipitable water field in the equatorial zone. *Earth Research from Space*, N6, 25–30 [in Russian].

Sharkov E. A., Kim G. A., and Pokrovskaya I. V. (2008b). Evolution and energy features of TC "Gonu" using the Data Merging Method for multiscale remote-sensing data. *Modern Problems of Earth Remote Sensing from Space*, Issue 5. Vol. I, 530–538 [in Russian].

Sharkov E. A., Kim G. A., and Pokrovskaya I.V. (2010). The plural generation of tropical cyclogenesis in the South Indian Ocean. *Modern Problems of Earth Remote Sensing from Space*, Vol. 7, N3, 75–86 [in Russian].

Sharkov E. A., Kim G. A., and Pokrovskaya I. V. (2011a). TC "Hondo" evolution in equatorial water vapor fields using the multispectral approach. *Earth Research from Space*, N1, 22–29 [in Russian].

Sharkov E. A., Kim G. A., and Pokrovskaya I. V. (2011b). Energy properties of the plural tropical cyclogenesis in global water vapor field using multispectral satellite observations. *Earth Research from Space*, N2, 1–8 [in Russian].

Shefov N. N., Semenov A. I., and Khomich V. Yu. (2006). *Airglow as an Indicator of Upper-Atmospheric Structure and Dynamics*. GEOS, Moscow, 740 pp.

Shen W., Tuleya R. E., and Ginis I. (2000). A sensitivity study of the thermodynamic environment on GFDL model hurricane intensity: Implications for global warning. *J. Climate*, Vol. 134, N1, 109–121.

Sherwood S. C. and Wahrlich R. (1999). Observed evolution of tropical deep convective events and their environment. *Monthly Weather Rev.*, Vol. 127, N8, 1777–1795.

Shramkov Y. N., Sharkov E. A., Pokrovskaya I. V., and Raev M. D. (2010) Satellite data base of tropical cyclogenesis in the global field of water vapor using object-related technology. *Earth Research from Space*, N6, 52–58 [in Russian].

Simpson J. (Ed.) (1988). *Report of the Science Steering Group for a Tropical Rainfall Measuring Mission (TRMM)*. NASA Goddard Space Flight Center, Greenbelt, MD, 94 pp.

Simpson J., Adler R. F., and North G. R. (1988). A proposed Tropical Rainfall Measuring Mission (TRMM) satellite. *Bull. Amer. Meteorol. Soc.*, Vol. 69, N3, 278–295.

Simpson R. H. (1998). Stepping stones in the evolution of a nation's hurricane policy. *Weather and Forecasting*, Vol. 13, N3, 617–620.

Smith E. (1999). Atlantic and East coast hurricanes 1900–98: A frequency and intensity study for the twenty-first century. *Bull. Amer. Meteorol. Soc.*, Vol. 80, N12, 2717–2720.

Sitnikov I. G. and Pokhil A. E. (1998). Interaction of tropical cyclones between each other and with other weather systems (Part I). *Meteorologia i Gidrologia* (*Russian Meteorology and Hydrology—English transl.*), N5, 36–44.

Stendal M., Christy J. R., and Bengtsson L. (2000). Assessing levels of uncertainty in recent temperature times series. *Climate Dynamics*, Vol. 16, 587–601.

Steranka L., Rodgers E. B., and Gentry R. (1986). The relationship between satellite measured convective bursts and tropical cyclone intensification. *Monthly Weather Rev.*, Vol. 114, N8, 1539–1546.

Svensmark H. and Friis-Christensen E. (1997). Variation of cosmic ray flux and global cloud coverage: A missing link in solar–climate relationships. *J. Atmos. Solar-Terr. Physics*, Vol. 59, N11, 1225–1232.

Tack R. E. (1994). Commercial distribution of data from RADARSAT, Canada's Earth observation satellite. *Earth Research from Space*, N2, 118–124.

Tarakanov G. G. (1980). *Tropical Meteorology*. Gidrometeoizdat, Leningrad, 220 pp. [in Russian].

Taylor M. J. (1997). A review of advances in imaging techniques for measuring short period gravity waves in the mesosphere and lower thermosphere. *Adv. Space Res.*, Vol. 19. N4, 667–676.

Taylor M. J. and Edwards R. (1991). Observations of short period mesospheric wave patterns: In situ or tropospheric wave generation? *Geophys. Res. Lett.*, Vol. 18, N7, 1337–1340.

Tchijevsky A. L. (1976). *Terrestrial Echo of Solar Storms* (Second Edition). Mysl′, Moscow, 366 pp.

Theon J. S. and Fuguno N. (Eds.) (1988). *Tropical Rainfall Measurements*. A. Deepak, Hampton, VA, 528 pp.

Theon J. S., Matsuno T., Sakato T., and Fuguno N. (Eds.) (1992). *The Global Role of Tropical Rainfall*. A. Deepak, Hampton, VA, 350 pp.

Tilford S. G., Asrar G., and Backlund P. W. (1994). Mission to planet Earth. *Adv. Space Res.*, Vol. 14, N1, 5–9.

Timmermann A. (2003). Decadal ENSO amplitude modulation: A nonlinear paradigm. *Global and Planetary Change*, Vol. 37, 135–156.

Travis W. P. (1964). *The Theory of Trade and Protection*. Harvard University Press, Cambridge, MA, 269 pp.

Trenberth K. E. (1997). Atmospheric moisture residence times and cycling: Implications for how precipitation may change as climate changes. *GEWEX News*, Vol. 7, N3, 1, 4–6.

Trenberth K. (2005). Uncertainty in hurricanes and global warming. *Science*, Vol. 308, N5729, 1753–1757.

Trenberth K. E. and Fasullo J. (2007). Water and energy budgets of hurricanes and implications for climate change. *J. Geophys. Res.*, Vol. 112, D23107, doi: 10.1029/2006JD008304.

Tsinober A. (1994). Anomalous diffusion in geophysical and laboratory turbulence. *Nonlinear Processes in Geophysics*, Vol. 1, N2/3, 80–94.

Tsutsui J.-I. and Kasahara A. (1996). Simulated tropical cyclones using the National Center for Atmospheric Research community climate model. *J. Geophys. Res.*, Vol. 101, ND10, 15013–15032.

U.K. Met. Office (1999). *Monthly Summary of Tropical Cyclone Activity and Forecasts* (*http://www.met-office.gov.uk*).

U.N. (1992). *Space Activity of the United Nations and International Organizations* (Document A/AC.105/521). United Nations, New York, 318 pp.

Utkin V. F., Senkevich V. P., Lukiashchenko V. I., Semenenko E. G., and Tsyboulsky G. A. (1997). Rocket/space system for fast direct sounding of large ecological and natural disaster areas. *48th International Astronautical Congress, October 6–10, 1997, Turin, Italy* (IAF–97C.2.03). International Astronautical Federation, Paris, pp. 2–7.

Vanina-Dart L. B., Pokrovskaya I. V., and Sharkov E. A. (2007a). Investigations of the interactions of the equatorial lower ionosphere with tropical cyclones using remote sensing and rocket soundings. *Earth Research from Space*, N1, 19–27 [in Russian].

Vanina-Dart L. B., Pokrovskaya I. V., and Sharkov E. A. (2007b). Solar activity influence on the equatorial lower ionosphere reply during active phase of tropical cyclones. *Earth Research from Space*, N6, 3–10 [in Russian].

Vecchi G. A. and Soden B. J. (2007). Effect of remote sea surface temperature change on tropical cyclone potential intensity. *Nature*, Vol. 450, N7172, 1066–1070.

Velden C. S. and Smith W. L. (1983). Monitoring tropical cyclone evolution with NOAA satellite microwave observations. *J. Climate and Applied Meteorol.*, Vol. 22, N5, 714–724.

Velden C. S., Smith W. L., and Mayfield M. (1984). Application of VAS and TOVS to tropical cyclones. *Bull. Amer. Meteorol. Soc.*, Vol. 65, N10, 1059–1067.

Velden C. S., Hayden C. M., and Menzel W. P. (1992). The impact of satellite-derived winds on numerical hurricane track forecasting. *Weather and Forecasting*, Vol. 7, 107–118.

Velden C. S., Olander T. L., and Zehr R. M. (1998). Development of an objective to estimate tropical cyclone intensity from digital geostationary satellite infrared imagery. *Weather and Forecasting*, Vol. 13, N1, 172–186.

Veselov V. M., Gerbek E. E., Zabyshnii A. I., Zimin V. D., Lazarev A. A., Levina G. V., Lupian E. A., Mazurov A. A., Moiseev S. S., Rutkevich P. B. *et al.* (1989). *On Verification of Physical Model for Onset of Large-Scale Vortices and Nonzero Helicity* (Preprint No. Pr-1604). Space Research Institute, Russian Academy of Sciences, Moscow, 10 pp [in Russian].

Vetlov I. P. and Veltishev N. F. (Eds.) (1982). *Guide on Using Satellite Data for Weather Analysis and Forecasting*. Gidrometeoizdat, Leningrad, 292 pp [in Russian].

Voiskovskii M. I., Gurin L. S., Klepikov I. N., Rezanova E. E., and Sharkov E. A. (1987). *Detection and Parameter Evaluation of Vortex Structures by Remote Measurements of the Radial Component of Turbulent Fields* (Preprint No. Pr-1308), Space Research Institute, Russian Academy of Sciences, Moscow, 35 pp. [in Russian].

Voiskovskii M. I., Gurin L. S., Klepikov I. N., Pokrovskaya I. V., and Sharkov E. A. (1989). *Vortex Structure in the Wind Field of Tropical Atmosphere* (Preprint No. Pr-1516). Space Research Institute, Russian Academy of Sciences, Moscow, 37 pp [in Russian].

Voloshuk V. M., Ingel L. Ch., Lebedev S. L., Karmasin V. P., Milechin L. I., Nerushev A. F., Pavlov N. I., Petrova L. I., Pudov V. D., Sitnikov I. G. *et al.* (1989). *Tropical Cyclones: Results of Investigations of U.S.S.R. Scientists*. Gidrometeoizdat, Leningrad, 53 pp. [in Russian].

Voss R. (1989). Random fractals: Self-affinity in noise, music, mountains and clouds. *Physica D*, Vol. 38, 362–371.

Walsh K. J. E. and Kleeman R. (1997) Predicting decadal variations in Atlantic tropical cyclone numbers and Australian rainfall. *Geophys. Res. Lett.*, Vol. 24, N24, 3249–3252.

Weatherford C. L. and Gray W. M. (1988a). Typhoon structure as revealed by aircraft reconnaissance, Part I: Data analysis and climatology. *Monthly Weather Rev.*, Vol. 116, N5, 1032–1043.

Weatherford C. L. and Gray W. M. (1988b). Typhoon structure as revealed by aircraft reconnaissance, Part II: Structural variability. *Monthly Weather Rev.*, Vol. 116, N5, 1044–1056.

Webster P. J. (1994). The role of hydrological processes in ocean–atmosphere interactions. *Reviews of Geophysics*, Vol. 32, N4, 427–476.

Webster P. J. and Lukas R. (1992). TOGA COARE: The coupled ocean–atmosphere response experiment. *Bull. Amer. Meteorol. Soc.*, Vol. 73, N10, 1377–1416.

Webster P. J. and Palmer T. N. (1997). The past and the future of El Niño. *Nature*, Vol. 390, N6660, 562–564.

Webster P. J., Holland G. J., Curry J. A., and Chang H. R. (2005). Changes in tropical cyclone number, duration, and intensity in a warming environment. *Science*, Vol. 309, N5712, 1844–1816.

Weinstock E. M., Hintsa E. J., Dessler A. E., and Anderson J. G. (1995). Measurement of water vapor in the tropical lower stratosphere during the CEPEX campaign: Results and interpretation. *Geophys. Res. Lett.*, Vol. 22, N23, 3231–3234.

Whitney L. D. and Hobgood J. S. (1997). The relationship between sea surface temperatures and maximum intensities of tropical cyclones in the Eastern North Pacific Ocean. *J. Climate*, Vol. 10, N11, 2921–2930.

Williams J. (1992). Tracking Andrew from ground zero. *Weatherwise.*, Vol. 45, N6, 8–13.

Williams M. and Houze R. A. (1987). Satellite-observed characteristics of winter monsoon cloud cluster. *Monthly Weather Rev.*, Vol. 115, N5, 505–519.

Willoughby H. E. (1999). Hurricane heat engines. *Nature*, Vol. 401, N6754, 649–650.

Willoughby H. E. and Black P. G. (1996). Hurricane Andrew in Florida: Dynamics of a Disaster. *Bull. Amer. Meteorol. Soc.*, Vol. 77, N3, 543–549.

Willoughby H. E. and Chelmow M. B. (1982). Objective determination of hurricane tracks from aircraft observations. *Monthly Weather Rev.*, Vol. 110, N9, 1298–1305.

Willoughby H. E., Jorgensen D. P., Black R. A., and Rosenthal S. L. (1985). Project STORMFURY: A scientific chronicle 1962–1983. *Bull. Amer. Meteorol. Soc.*, Vol. 66, N5, 505–514.

Wilson R. M. (1997). Comment on "Downward trends in the frequency of intense Atlantic hurricanes during the past 5 decades" by C. W. Landsea *et al. Geophys. Res. Lett.*, Vol. 24, N17, 2203–2204.

WMO (1979). *Operational Techniques for Forecasting Tropical Cyclone Intensity and Movement* (Tropical cyclone sub-project N6/N528). World Meteorological Organization, Geneva, Switzerland, 60 pp.

WMO (1983). *Tropical Cyclone Operational Plan for the Southwest Indian Ocean* (WMO-N618/TCP-12). World Meteorological Organization, Geneva, Switzerland, 95 pp.

WMO (1986). *Tropical Cyclone Operational Plan for the Bay of Bengal and the Arabian Sea* (WMO/TD-N84/TCP-21). World Meteorological Organization, Geneva, Switzerland, 80 pp.

WMO (1987) *Typhoon Committee Operational Manual: Meteorological Component* (WMO/TD-N196/TCP-23). World Meteorological Organization, Geneva, Switzerland, 57 pp.

WMO (1988). *Regional Association IV (North and Central America) Hurricane Operational Plan* (WMO/TD-N494/TCP-30). World Meteorological Organization, Geneva, Switzerland, 68 pp.

WMO (1989). *Tropical Cyclone Operational Plan for the South Pacific and Southeast Indian Ocean* (WMO/TD-N292/TCP-24). World Meteorological Organization, Geneva, Switzerland, 90 pp.

WMO (1990a). *WCRP Global Precipitation Climatology Project: Implementation and Data Management Plan* (WMO/TD-N367). World Meteorological Organization, Geneva, Switzerland, 47 pp.

WMO (1990b). *Scientific Plan for the TOGA Coupled Ocean–Atmosphere Response Experiment* (WCRP Publ. Series N3 Addendum, WMO/TD-N64). World Meteorological Organization, Geneva, Switzerland, p. 75.

WMO (1992). *Infoclima Catalogue of Climate System Data Sets* (WCDP-5 WMO/TD N293). World Meteorological Organization, Geneva, Switzerland, 31 pp.

WMO (1994). *Report of the Technical Conference on Tropical Urban Climates* (WCASP-30, WMO/TD-N647). World Meteorological Organization, Geneva, Switzerland, 558 pp.

WMO (1995a). *Stormy Times: Tropical Cyclones, Typhoons, Hurricanes, Tornadoes* (Global Climate System Review N819). World Meteorological Organization, Geneva, Switzerland, pp. 69–81.

WMO (1995b). Climate system monitoring. *Monthly Bulletin*, Issue N1/12.

WMO (1999). *Annual Review of the World Climate Research Program and Report of the Twentieth Session of the Joint Scientific Committee, Kiel, Germany, March 15–19, 1999* (WMO/TD-No. 976). World Meteorological Organization, Geneva, Switzerland, 120 pp.

Yu Cheng-Ku, Jou Jong-Dao, and Smull B. F. (1999). Formative stage of a long-lived mesoscale vortex observed by airborne Doppler radar. *Monthly Weather Rev.*, Vol. 127, N5, 838–857.

Zehr R. M. (1992). *Tropical Cyclogenesis in the Western North Pacific* (NOAA Tech. Report NESDIS-61). U.S. Department of Commerce, Washington, D.C., 181 pp.

Zeldovich Ya. B. and Sokolov D. D. (1985). Fractal, scaling and intermediate asymptotics. *Uspekhi Fiz. Nauk* (*Physics-Uspekhi—English transl.*), Vol. 146, N3, 493–506.

Zhang G. Z., Drosdowsky W., and Nicholls N. (1990). Environmental influence on northwest Pacific tropical cyclone numbers. *Acta Meteorologica Sinica*, Vol. 4, 180–188.

Ziman I. L. and Krasnopevtseva E. B. (1996). Earth remote sensing using small spacecraft. *Russian Space Bulletin*, Vol. 3, N2, 28–30.

Zimin V. D., Levina G. V., Moiseev S. S., and Tur A. V. (1989). Creation of large-scale structures with turbulent convection in a rotating layer heated from below. *Doklady Akad. Nauk* (*Trans. of Russian Academy of Sciences/Doklady Physics—English transl.*), Vol. 309, N1, 88–92.

Zimin V. D., Levina G. V., Veiber E. E., Veselov V. M., Gerbic E. E., Zabishniy A. I., Lazarev A. A., Mazurov A. N., Moiseev S. S., Pokrovskaya I. V. *et al.* (1991). Experimental studies of large-scale structure origination in tropical atmosphere (Expedition '89). *Nonlinear Dynamics of Structures* (edited by R. Z. Sagdeev *et al.*), World Scientific, Singapore, pp. 327–336.

Zimin V. D., Klepikov I. N., Lazarev A. A., Moiseev S. S., Cherny I. V., and Sharkov E. A. (1992). Study of large-scale, ecologically dangerous whirlwind flows in the Earth's atmosphere. *Sov. J. Remote Sensing*, Vol. 10, N1, 1–14.

Zolotarev V. M. (1983). *One-Dimensional Tolerance Distributions*. Nauka, Moscow, 210 pp. [in Russian].

Index